D0321774

Third Edition
FRACTURE MECHANICS

Fundamentals and Applications

Third Edition
FRACTURE MECHANICS

Fundamentals and Applications

T.L. Anderson, Ph.D.

Taylor & Francis
Taylor & Francis Group

Boca Raton   London   New York   Singapore

A CRC title, part of the Taylor & Francis imprint, a member of the
Taylor & Francis Group, the academic division of T&F Informa plc.

Published in 2005 by
CRC Press
Taylor & Francis Group
6000 Broken Sound Parkway NW, Suite 300
Boca Raton, FL 33487-2742

© 2005 by Taylor & Francis Group, LLC
CRC Press is an imprint of Taylor & Francis Group

No claim to original U.S. Government works
Printed in the United States of America on acid-free paper
12

International Standard Book Number-10: 0-8493-1656-1 (Hardcover)
International Standard Book Number-13: 978-0-8493-1656-2 (Hardcover)

This book contains information obtained from authentic and highly regarded sources. Reprinted material is quoted with permission, and sources are indicated. A wide variety of references are listed. Reasonable efforts have been made to publish reliable data and information, but the author and the publisher cannot assume responsibility for the validity of all materials or for the consequences of their use.

No part of this book may be reprinted, reproduced, transmitted, or utilized in any form by any electronic, mechanical, or other means, now known or hereafter invented, including photocopying, microfilming, and recording, or in any information storage or retrieval system, without written permission from the publishers.

For permission to photocopy or use material electronically from this work, please access www.copyright.com (http://www.copyright.com/) or contact the Copyright Clearance Center, Inc. (CCC) 222 Rosewood Drive, Danvers, MA 01923, 978-750-8400. CCC is a not-for-profit organization that provides licenses and registration for a variety of users. For organizations that have been granted a photocopy license by the CCC, a separate system of payment has been arranged.

Trademark Notice: Product or corporate names may be trademarks or registered trademarks, and are used only for identification and explanation without intent to infringe.

Library of Congress Cataloging-in-Publication Data

Catalog record is available from the Library of Congress

Taylor & Francis Group is the Academic Division of T&F Informa plc.

Visit the Taylor & Francis Web site at
http://www.taylorandfrancis.com

and the CRC Press Web site at
http://www.crcpress.com

UNIVERSITY
OF SHEFFIELD
LIBRARY

Dedication

To
Vanessa, Molly, Aleah, and Tom

Preface

The field of fracture mechanics was virtually nonexistent prior to World War II, but has since matured into an established discipline. Most universities with an engineering program offer at least one fracture mechanics course on the graduate level, and an increasing number of undergraduates have been exposed to this subject. Applications of fracture mechanics in industry are relatively common, as knowledge that was once confined to a few specialists is becoming more widespread.

While there are a number of books on fracture mechanics, most are geared to a specific audience. Some treatments of this subject emphasize material testing, while others concentrate on detailed mathematical derivations. A few books address the microscopic aspects of fracture, but most consider only continuum models. Many books are restricted to a particular material system, such as metals or polymers. Current offerings include advanced, highly specialized books, as well as introductory texts. While the former are valuable to researchers in this field, they are unsuitable for students with no prior background. On the other hand, introductory treatments of the subject are sometimes simplistic and misleading.

This book provides a comprehensive treatment of fracture mechanics that should appeal to a relatively wide audience. Theoretical background and practical applications are both covered in detail. This book is suitable as a graduate text, as well as a reference for engineers and researchers. Selected portions of this book would also be appropriate for an undergraduate course in fracture mechanics.

This is the third edition of this text. The first two editions were published in 1991 and 1995. Although the overwhelming response to the earlier editions was positive, I have received a few constructive criticisms from several colleagues whose opinions I respect. I have tried to incorporate their comments in this revision, and I hope the final product meets with the approval of readers who are acquainted with the first or second edition, as well as those who are seeing this text for the first time.

Many sections have been revised and expanded in this latest edition. In a few cases, material from the second edition was dropped because it had become obsolete or did not fit within the context of the revised material. Chapter 2, which covers linear elastic fracture, includes a new section on crack interaction. In addition, a new section on so-called plane strain fracture has been added to Chapter 2 in an attempt to debunk certain myths that have arisen over the years. Chapter 7 and Chapter 8 have been updated to account for recent developments in fracture toughness testing standards. Chapter 9 on application to structures has been completely reorganized and updated. In Chapter 10, the coverage of fatigue crack closure, the fatigue threshold, and variable amplitude effects has been expanded and updated. Perhaps the most noticeable change in the third edition is a completely new chapter on environmental cracking (Chapter 11). The chapter on computational fracture mechanics, which was formerly Chapter 11, is now Chapter 12. A number of problems have been added to Chapter 13, and several problems from the second edition have been modified or deleted.

The basic organization and underlying philosophy are unchanged in the third edition. The book is intended to be readable without being superficial. The fundamental concepts are first described qualitatively, with a minimum of higher level mathematics. This enables a student with a reasonable grasp of undergraduate calculus to gain physical insight into the subject. For the more advanced reader, appendices at the end of certain chapters give the detailed mathematical background.

In outlining the basic principles and applications of fracture mechanics, I have attempted to integrate materials science and solid mechanics to a much greater extent compared to those in other fracture mechanics texts. Although continuum theory has proved to be a very powerful tool in fracture mechanics, one cannot ignore microstructural aspects. Continuum theory can predict the stresses and strains near a crack tip, but it is the material's microstructure that determines the critical conditions for fracture.

The first chapter introduces the subject of fracture mechanics and provides an overview; this chapter includes a review of dimensional analysis, which proves to be a useful tool in later chapters. Chapter 2 and Chapter 3 describe the fundamental concepts of linear elastic and elastic-plastic fracture mechanics, respectively. One of the most important and most often misunderstood concepts in fracture mechanics is the single-parameter assumption, which enables the prediction of structural behavior from small-scale laboratory tests. When a single parameter uniquely describes the crack tip conditions, fracture toughness—a critical value of this parameter—is independent of specimen size. When the single-parameter assumption breaks down, fracture toughness becomes size depen- dent, and a small-scale fracture toughness test may not be indicative of structural behavior. Chapter 2 and Chapter 3 describe the basis of the single-parameter assumption in detail, and outline the requirements for its validity. Chapter 3 includes the results of recent research that extends fracture mechanics beyond the limits of single-parameter theory. The main bodies of Chapter 2 and Chapter 3 are written in such a way as to be accessible to the beginning student. Appendix 2 and Appendix 3, which follow Chapter 2 and Chapter 3, respectively, give mathematical derivations of several important relationships in linear elastic and elastic-plastic fracture mechanics. Most of the material in these appendices requires a graduate-level background in solid mechanics.

Chapter 4 introduces dynamic and time-dependent fracture mechanics. The section on dynamic fracture includes a brief discussion of rapid loading of a stationary crack, as well as rapid crack propagation and arrest. The C^*, $C(t)$, and C_t parameters for characterizing creep crack growth are introduced, together with analogous quantities that characterize fracture in viscoelastic materials.

Chapter 5 outlines micromechanisms of fracture in metals and alloys, while Chapter 6 describes fracture mechanisms in polymers, ceramics, composites, and concrete. These chapters emphasize the importance of microstructure and material properties on the fracture behavior.

The application portion of this book begins with Chapter 7, which gives practical advice on fracture toughness testing in metals. Chapter 8 describes fracture testing of nonmetallic materials. Chapter 9 outlines the available methods for applying fracture mechanics to structures, including both linear elastic and elastic-plastic approaches. Chapter 10 describes the fracture mechanics approach to fatigue crack propagation, and discusses some of the critical issues in this area, including crack closure and the behavior of short cracks. Chapter 11 is a completely new chapter on environmental cracking. Chapter 12 outlines some of the most recent developments in compu- tational fracture mechanics. Procedures for determining stress intensity and the J integral in structures are described, with particular emphasis on the domain integral approach. Chapter 13 contains a series of practice problems that correspond to material in Chapter 1 to Chapter 12.

If this book is used as a college text, it is unlikely that all of the material can be covered in a single semester. Thus the instructor should select the portions of the book that suit the needs and background of the students. The first three chapters, excluding appendices, should form the foundation of any course. In addition, I strongly recommend the inclusion of at least one of the material chapters (5 or 6), regardless of whether or not materials science is the students' major field of study. A course that is oriented toward applications could include Chapter 7 to Chapter 11, in addition to the earlier chapters. A graduate level course in a solid mechanics curriculum might include Appendix 2 and Appendix 3, Chapter 4, Appendix 4, and Chapter 12.

I am pleased to acknowledge all those individuals who helped make all three editions of this book possible. A number of colleagues and friends reviewed portions of the draft manuscript, provided photographs and homework problems, or both for the first and second editions, including W.L. Bradley, M. Cayard, R. Chona, M.G. Dawes, R.H. Dodds Jr., A.G. Evans, S.J. Garwood, J.P. Gudas, E.G. Guynn, A.L. Highsmith, R.E. Jones Jr., Y.W. Kwon, J.D. Landes, E.J. Lavernia, A. Letton, R.C. McClung, D.L. McDowell, J.G. Merkle, M.T. Miglin, D.M. Parks, P.T. Purtscher, R.A. Schapery, and C.F. Shih. Mr. Sun Yongqi produced a number of SEM fractographs especially for this book. I am grateful to the following individuals for providing useful comments and literature references to aid in my preparation of the third edition: R.A. Ainsworth, D.M. Boyanjian, S.C. Daniewicz, R.H. Dodds Jr., R.P. Gangloff, R. Latanision, J.C. Newman, A.K. Vasudevan, K. Wallin, and J.G. Williams. I apologize to anyone whose name I have inadvertently omitted from this list. When preparing the third edition, I received valuable assistance from number of colleagues at my current company, Structural Reliability Technology. These individuals include Devon Brendecke, Donna Snyman, and Greg Thorwald. Last but certainly not least, Russ Hall, formerly with CRC Press and now a successful novelist, deserves special mention for convincing me to write this book back in 1989.

Ted L. Anderson

Table of Contents

Part I
Introduction 1

Chapter 1
History and Overview 3
1.1   Why Structures Fail...3
1.2   Historical Perspective 6
        1.2.1   Early Fracture Research 8
        1.2.2   The Liberty Ships...9
        1.2.3   Post-War Fracture Mechanics Research 10
        1.2.4   Fracture Mechanics from 1960 to 1980...10
        1.2.5   Fracture Mechanics from 1980 to the Present..12
1.3   The Fracture Mechanics Approach to Design 12
        1.3.1   The Energy Criterion..12
        1.3.2   The Stress-Intensity Approach 14
        1.3.3   Time-Dependent Crack Growth and Damage Tolerance........................15
1.4   Effect of Material Properties on Fracture..16
1.5   A Brief Review of Dimensional Analysis 18
        1.5.1   The Buckingham Π-Theorem...18
        1.5.2   Dimensional Analysis in Fracture Mechanics 19
References 21

Part II
Fundamental Concepts..23

Chapter 2
Linear Elastic Fracture Mechanics 25
2.1   An Atomic View of Fracture..25
2.2   Stress Concentration Effect of Flaws...27
2.3   The Griffith Energy Balance 29
        2.3.1   Comparison with the Critical Stress Criterion..31
        2.3.2   Modified Griffith Equation 32
2.4   The Energy Release Rate...34
2.5   Instability and the R Curve 38
        2.5.1   Reasons for the R Curve Shape 39
        2.5.2   Load Control vs. Displacement Control 40
        2.5.3   Structures with Finite Compliance..41
2.6   Stress Analysis of Cracks 42
        2.6.1   The Stress Intensity Factor..43
        2.6.2   Relationship between K and Global Behavior......................................45
        2.6.3   Effect of Finite Size 48
        2.6.4   Principle of Superposition...54
        2.6.5   Weight Functions..56

2.7 Relationship between K and $\mathcal{G}$ 58
2.8 Crack-Tip Plasticity...61
    2.8.1 The Irwin Approach 61
    2.8.2 The Strip-Yield Model 64
    2.8.3 Comparison of Plastic Zone Corrections...66
    2.8.4 Plastic Zone Shape 66
2.9 K-Controlled Fracture..69
2.10 Plane Strain Fracture: Fact vs. Fiction 72
    2.10.1 Crack-Tip Triaxiality 73
    2.10.2 Effect of Thickness on Apparent Fracture Toughness 75
    2.10.3 Plastic Zone Effects 78
    2.10.4 Implications for Cracks in Structures..79
2.11 Mixed-Mode Fracture...80
    2.11.1 Propagation of an Angled Crack 81
    2.11.2 Equivalent Mode I Crack...83
    2.11.3 Biaxial Loading..84
2.12 Interaction of Multiple Cracks 86
    2.12.1 Coplanar Cracks...86
    2.12.2 Parallel Cracks 86
Appendix 2: Mathematical Foundations of Linear Elastic
            Fracture Mechanics..88
    A2.1 Plane Elasticity 88
         A2.1.1 Cartesian Coordinates 89
         A2.1.2 Polar Coordinates..90
    A2.2 Crack Growth Instability Analysis 91
    A2.3 Crack-Tip Stress Analysis 92
         A2.3.1 Generalized In-Plane Loading 92
         A2.3.2 The Westergaard Stress Function 95
    A2.4 Elliptical Integral of the Second Kind...100
References 101

Chapter 3
Elastic-Plastic Fracture Mechanics 103
3.1 Crack-Tip-Opening Displacement..103
3.2 The J Contour Integral 107
    3.2.1 Nonlinear Energy Release Rate 108
    3.2.2 J as a Path-Independent Line Integral 110
    3.2.3 J as a Stress Intensity Parameter 111
    3.2.4 The Large Strain Zone 113
    3.2.5 Laboratory Measurement of J...114
3.3 Relationships Between J and CTOD 120
3.4 Crack-Growth Resistance Curves 123
    3.4.1 Stable and Unstable Crack Growth...124
    3.4.2 Computing J for a Growing Crack 126
3.5 J-Controlled Fracture...128
    3.5.1 Stationary Cracks...128
    3.5.2 J-Controlled Crack Growth 131
3.6 Crack-Tip Constraint Under Large-Scale Yielding...133
    3.6.1 The Elastic T Stress 137
    3.6.2 J-Q Theory..140

               3.6.2.1 The *J-Q* Toughness Locus..142

               3.6.2.2 Effect of Failure Mechanism

                           on the *J-Q* Locus...144

      3.6.3   Scaling Model for Cleavage Fracture 145

               3.6.3.1 Failure Criterion 145

               3.6.3.2 Three-Dimensional Effects..147

               3.6.3.3 Application of the Model...148

      3.6.4   Limitations of Two-Parameter Fracture Mechanics..................................149

Appendix 3:   Mathematical Foundations

              of Elastic-Plastic Fracture Mechanics 153

      A3.1   Determining CTOD from the Strip-Yield Model 153

      A3.2   The *J* Contour Integral 156

      A3.3   *J* as a Nonlinear Elastic Energy Release Rate...158

      A3.4   The HRR Singularity..159

      A3.5   Analysis of Stable Crack Growth

              in Small-Scale Yielding..162

               A3.5.1 The Rice-Drugan-Sham Analysis 162

               A3.5.2 Steady State Crack Growth...166

      A3.6   Notes on the Applicability of Deformation Plasticity

              to Crack Problems 168

References 171

Chapter 4

Dynamic and Time-Dependent Fracture...173

4.1   Dynamic Fracture and Crack Arrest 173

     4.1.1   Rapid Loading of a Stationary Crack 174

     4.1.2   Rapid Crack Propagation and Arrest 178

             4.1.2.1 Crack Speed...180

             4.1.2.2 Elastodynamic Crack-Tip Parameters.............................182

             4.1.2.3 Dynamic Toughness 184

             4.1.2.4 Crack Arrest 186

     4.1.3   Dynamic Contour Integrals 188

4.2   Creep Crack Growth 189

     4.2.1   The *C** Integral 191

     4.2.2   Short-Time vs. Long-Time Behavior 193

             4.2.2.1 The *C_t* Parameter 195

             4.2.2.2 Primary Creep 196

4.3   Viscoelastic Fracture Mechanics..196

     4.3.1   Linear Viscoelasticity 197

     4.3.2   The Viscoelastic *J* Integral 200

             4.3.2.1 Constitutive Equations...200

             4.3.2.2 Correspondence Principle..200

             4.3.2.3 Generalized *J* Integral 201

             4.3.2.4 Crack Initiation and Growth 202

     4.3.3   Transition from Linear to Nonlinear Behavior..204

Appendix 4:   Dynamic Fracture Analysis...206

      A4.1   Elastodynamic Crack Tip Fields 206

      A4.2   Derivation of the Generalized Energy

              Release Rate 209

References 213

Part III
Material Behavior **217**

Chapter 5
Fracture Mechanisms in Metals 219
5.1   Ductile Fracture 219
         5.1.1   Void Nucleation 219
         5.1.2   Void Growth and Coalescence 222
         5.1.3   Ductile Crack Growth 231
5.2   Cleavage 234
         5.2.1   Fractography 234
         5.2.2   Mechanisms of Cleavage Initiation 235
         5.2.3   Mathematical Models of Cleavage Fracture
                     Toughness 238
5.3   The Ductile-Brittle Transition 247
5.4   Intergranular Fracture 249
Appendix 5:   Statistical Modeling of Cleavage Fracture 250
                     A5.1   Weakest Link Fracture 250
                     A5.2   Incorporating a Conditional Probability
                                 of Propagation 252
References 254

Chapter 6
Fracture Mechanisms in Nonmetals 257
6.1   Engineering Plastics 257
         6.1.1   Structure and Properties of Polymers 258
                     6.1.1.1   Molecular Weight 258
                     6.1.1.2   Molecular Structure 259
                     6.1.1.3   Crystalline and Amorphous Polymers 259
                     6.1.1.4   Viscoelastic Behavior 260
                     6.1.1.5   Mechanical Analogs 263
         6.1.2   Yielding and Fracture in Polymers 265
                     6.1.2.1   Chain Scission and Disentanglement 265
                     6.1.2.2   Shear Yielding and Crazing 265
                     6.1.2.3   Crack-Tip Behavior 267
                     6.1.2.4   Rubber Toughening 268
                     6.1.2.5   Fatigue 270
         6.1.3   Fiber-Reinforced Plastics 270
                     6.1.3.1   Overview of Failure Mechanisms 271
                     6.1.3.2   Delamination 272
                     6.1.3.3   Compressive Failure 275
                     6.1.3.4   Notch Strength 278
                     6.1.3.5   Fatigue Damage 280
6.2   Ceramics and Ceramic Composites 282
         6.2.1   Microcrack Toughening 285
         6.2.2   Transformation Toughening 286
         6.2.3   Ductile Phase Toughening 287
         6.2.4   Fiber and Whisker Toughening 288
6.3   Concrete and Rock 291
References 293

Part IV
Applications **297**

Chapter 7
Fracture Toughness Testing of Metals 299
7.1   General Considerations 299
        7.1.1   Specimen Configurations...299
        7.1.2   Specimen Orientation 301
        7.1.3   Fatigue Precracking...303
        7.1.4   Instrumentation 305
        7.1.5   Side Grooving..307
7.2 K_{Ic} Testing 308
        7.2.1   ASTM E 399 309
        7.2.2   Shortcomings of E 399 and Similar Standards 312
7.3 K-R Curve Testing...316
        7.3.1   Specimen Design...317
        7.3.2   Experimental Measurement of K-R Curves...318
7.4 J Testing of Metals...320
        7.4.1   The Basic Test Procedure and J_{Ic} Measurements 320
        7.4.2 J-R Curve Testing...322
        7.4.3   Critical J Values for Unstable Fracture...324
7.5   CTOD Testing...326
7.6   Dynamic and Crack-Arrest Toughness 329
        7.6.1   Rapid Loading in Fracture Testing 329
        7.6.2 K_{Ia} Measurements 330
7.7   Fracture Testing of Weldments 334
        7.7.1   Specimen Design and Fabrication...334
        7.7.2   Notch Location and Orientation..335
        7.7.3   Fatigue Precracking...337
        7.7.4   Posttest Analysis...337
7.8   Testing and Analysis of Steels in the Ductile-Brittle Transition Region.....................338
7.9   Qualitative Toughness Tests 340
        7.9.1   Charpy and Izod Impact Test..341
        7.9.2   Drop Weight Test..342
        7.9.3   Drop Weight Tear and Dynamic Tear Tests..344
Appendix 7:   Stress Intensity, Compliance, and Limit Load Solutions
                        for Laboratory Specimens..344
References 350

Chapter 8
Fracture Testing of Nonmetals...353
8.1   Fracture Toughness Measurements in Engineering Plastics...353
        8.1.1   The Suitability of K and J for Polymers 353
                    8.1.1.1 K-Controlled Fracture..354
                    8.1.1.2 J-Controlled Fracture..357
        8.1.2   Precracking and Other Practical Matters 360
        8.1.3 K_{Ic} Testing...362
        8.1.4 J Testing...365
        8.1.5   Experimental Estimates of Time-Dependent Fracture Parameters....................369
        8.1.6   Qualitative Fracture Tests on Plastics 371
8.2   Interlaminar Toughness of Composites...373

8.3   Ceramics 378
     8.3.1   Chevron-Notched Specimens 378
     8.3.2   Bend Specimens Precracked by Bridge Indentation 380
References 382

Chapter 9
Application to Structures 385
9.1   Linear Elastic Fracture Mechanics..385
     9.1.1 K_I for Part-Through Cracks...387
     9.1.2   Influence Coefficients for Polynomial Stress Distributions 388
     9.1.3   Weight Functions for Arbitrary Loading 392
     9.1.4   Primary, Secondary, and Residual Stresses 394
     9.1.5   A Warning about LEFM...395
9.2   The CTOD Design Curve 395
9.3   Elastic-Plastic *J*-Integral Analysis...397
     9.3.1   The EPRI *J*-Estimation Procedure 398
          9.3.1.1   Theoretical Background 398
          9.3.1.2   Estimation Equations...399
          9.3.1.3   Comparison with Experimental *J* Estimates....................401
     9.3.2   The Reference Stress Approach 403
     9.3.3   Ductile Instability Analysis 405
     9.3.4   Some Practical Considerations..408
9.4   Failure Assessment Diagrams 410
     9.4.1   Original Concept 410
     9.4.2   *J*-Based FAD 412
     9.4.3   Approximations of the FAD Curve..415
     9.4.4   Estimating the Reference Stress..416
     9.4.5   Application to Welded Structures 423
          9.4.5.1   Incorporating Weld Residual Stresses....................423
          9.4.5.2   Weld Misalignment..426
          9.4.5.3   Weld Strength Mismatch 427
     9.4.6   Primary vs. Secondary Stresses in the FAD Method 428
     9.4.7   Ductile-Tearing Analysis with the FAD....................................430
     9.4.8   Standardized FAD-Based Procedures 430
9.5   Probabilistic Fracture Mechanics..432
Appendix 9:   Stress Intensity and Fully Plastic *J* Solutions
               for Selected Configurations 434
References 449

Chapter 10
Fatigue Crack Propagation...451
10.1   Similitude in Fatigue 451
10.2   Empirical Fatigue Crack Growth Equations 453
10.3   Crack Closure 457
     10.3.1   A Closer Look at Crack-Wedging Mechanisms.......................460
     10.3.2   Effects of Loading Variables on Closure.................................463
10.4   The Fatigue Threshold..464
     10.4.1   The Closure Model for the Threshold......................................465
     10.4.2   A Two-Criterion Model...466
     10.4.3   Threshold Behavior in Inert Environments 470
10.5   Variable Amplitude Loading and Retardation...473

        10.5.1   Linear Damage Model for Variable Amplitude Fatigue 474
        10.5.2   Reverse Plasticity at the Crack Tip 475
        10.5.3   The Effect of Overloads and Underloads 478
        10.5.4   Models for Retardation and Variable Amplitude Fatigue 484
10.6   Growth of Short Cracks 488
        10.6.1   Microstructurally Short Cracks 491
        10.6.2   Mechanically Short Cracks 491
10.7   Micromechanisms of Fatigue 491
        10.7.1   Fatigue in Region II 491
        10.7.2   Micromechanisms Near the Threshold 494
        10.7.3   Fatigue at High ΔK Values 495
10.8   Fatigue Crack Growth Experiments 495
        10.8.1   Crack Growth Rate and Threshold Measurement 496
        10.8.2   Closure Measurements 498
        10.8.3   A Proposed Experimental Definition of ΔK_{eff} 500
10.9   Damage Tolerance Methodology 501
Appendix 10:   Application of The J Contour Integral to Cyclic Loading 504
        A10.1   Definition of ΔJ 504
        A10.2   Path Independence of ΔJ 506
        A10.3   Small-Scale Yielding Limit 507
References 507

Chapter 11
Environmentally Assisted Cracking in Metals 511
11.1   Corrosion Principles 511
        11.1.1   Electrochemical Reactions 511
        11.1.2   Corrosion Current and Polarization 514
        11.1.3   Electrode Potential and Passivity 514
        11.1.4   Cathodic Protection 515
        11.1.5   Types of Corrosion 516
11.2   Environmental Cracking Overview 516
        11.2.1   Terminology and Classification of Cracking Mechanisms 516
        11.2.2   Occluded Chemistry of Cracks, Pits, and Crevices 517
        11.2.3   Crack Growth Rate vs. Applied Stress Intensity 518
        11.2.4   The Threshold for EAC 520
        11.2.5   Small Crack Effects 521
        11.2.6   Static, Cyclic, and Fluctuating Loads 523
        11.2.7   Cracking Morphology 523
        11.2.8   Life Prediction 523
11.3   Stress Corrosion Cracking 525
        11.3.1   The Film Rupture Model 527
        11.3.2   Crack Growth Rate in Stage II 528
        11.3.3   Metallurgical Variables that Influence SCC 528
        11.3.4   Corrosion Product Wedging 529
11.4   Hydrogen Embrittlement 529
        11.4.1   Cracking Mechanisms 530
        11.4.2   Variables that Affect Cracking Behavior 531
                11.4.2.1   Loading Rate and Load History 531
                11.4.2.2   Strength 533
                11.4.2.3   Amount of Available Hydrogen 535
                11.4.2.4   Temperature 535

11.5   Corrosion Fatigue...538
        11.5.1 Time-Dependent and Cycle-Dependent Behavior 538
        11.5.2 Typical Data 541
        11.5.3 Mechanisms..543
                11.5.3.1 Film Rupture Models 544
                11.5.3.2 Hydrogen Environment Embrittlement 544
                11.5.3.3 Surface Films 544
        11.5.4 The Effect of Corrosion Product Wedging on Fatigue................................544
11.6   Experimental Methods 545
        11.6.1 Tests on Smooth Specimens 546
        11.6.2 Fracture Mechanics Test Methods 547
References 552

Chapter 12
Computational Fracture Mechanics..553
12.1   Overview of Numerical Methods 553
        12.1.1 The Finite Element Method 554
        12.1.2 The Boundary Integral Equation Method..556
12.2   Traditional Methods in Computational Fracture Mechanics 558
        12.2.1 Stress and Displacement Matching..558
        12.2.2 Elemental Crack Advance 559
        12.2.3 Contour Integration 560
        12.2.4 Virtual Crack Extension: Stiffness Derivative Formulation 560
        12.2.5 Virtual Crack Extension: Continuum Approach 561
12.3   The Energy Domain Integral...563
        12.3.1 Theoretical Background 563
        12.3.2 Generalization to Three Dimensions 566
        12.3.3 Finite Element Implementation..568
12.4   Mesh Design 570
12.5   Linear Elastic Convergence Study 577
12.6   Analysis of Growing Cracks 585
Appendix 12:   Properties of Singularity Elements...587
                A12.1   Quadrilateral Element 587
                A12.2   Triangular Element 589
References 590

Chapter 13
Practice Problems..593
13.1   Chapter 1..593
13.2   Chapter 2..593
13.3   Chapter 3..596
13.4   Chapter 4..598
13.5   Chapter 5..599
13.6   Chapter 6..600
13.7   Chapter 7..600
13.8   Chapter 8..603
13.9   Chapter 9..605
13.10   Chapter 10..607
13.11   Chapter 11 608
13.12   Chapter 12 609

Index 611

Part I

Introduction

1 History and Overview

Fracture is a problem that society has faced for as long as there have been man-made structures. The problem may actually be worse today than in previous centuries, because more can go wrong in our complex technological society. Major airline crashes, for instance, would not be possible without modern aerospace technology.

Fortunately, advances in the field of fracture mechanics have helped to offset some of the potential dangers posed by increasing technological complexity. Our understanding of how materials fail and our ability to prevent such failures have increased considerably since World War II. Much remains to be learned, however, and existing knowledge of fracture mechanics is not always applied when appropriate.

While catastrophic failures provide income for attorneys and consulting engineers, such events are detrimental to the economy as a whole. An economic study [1] estimated the annual cost of fracture in the U.S. in 1978 at $119 billion (in 1982 dollars), about 4% of the gross national product. Furthermore, this study estimated that the annual cost could be reduced by $35 billion if current technology were applied, and that further fracture mechanics research could reduce this figure by an additional $28 billion.

1.1 WHY STRUCTURES FAIL

The cause of most structural failures generally falls into one of the following categories:

1. Negligence during design, construction, or operation of the structure.
2. Application of a new design or material, which produces an unexpected (and undesirable) result.

In the first instance, existing procedures are sufficient to avoid failure, but are not followed by one or more of the parties involved, due to human error, ignorance, or willful misconduct. Poor workmanship, inappropriate or substandard materials, errors in stress analysis, and operator error are examples of where the appropriate technology and experience are available, but not applied.

The second type of failure is much more difficult to prevent. When an "improved" design is introduced, invariably, there are factors that the designer does not anticipate. New materials can offer tremendous advantages, but also potential problems. Consequently, a new design or material should be placed into service only after extensive testing and analysis. Such an approach will reduce the frequency of failures, but not eliminate them entirely; there may be important factors that are overlooked during testing and analysis.

One of the most famous Type 2 failures is the brittle fracture of World War II Liberty ships (see Section 1.2.2). These ships, which were the first to have all-welded hulls, could be fabricated much faster and cheaper than earlier riveted designs, but a significant number of these vessels sustained serious fractures as a result of the design change. Today, virtually all steel ships are welded, but sufficient knowledge was gained from the Liberty ship failures to avoid similar problems in present structures.

Knowledge must be applied in order to be useful, however. Figure 1.1 shows an example of a Type 1 failure, where poor workmanship in a seemingly inconsequential structural detail caused a more recent fracture in a welded ship. In 1979, the Kurdistan oil tanker broke completely in two

3

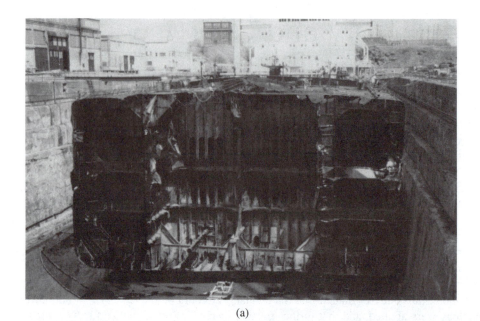

(a)

(b)

FIGURE 1.1 The MSV Kurdistan oil tanker, which sustained a brittle fracture while sailing in the North Atlantic in 1979: (a) fractured vessel in dry dock and (b) bilge keel from which the fracture initiated. (Photographs provided by S.J. Garwood.)

while sailing in the North Atlantic [2]. The combination of warm oil in the tanker with cold water in contact with the outer hull produced substantial thermal stresses. The fracture initiated from a bilge keel that was improperly welded. The weld failed to penetrate the structural detail, resulting in a severe stress concentration. Although the hull steel had adequate toughness to prevent fracture initiation, it failed to stop the propagating crack.

Polymers, which are becoming more common in structural applications, provide a number of advantages over metals, but also have the potential for causing Type 2 failures. For example, polyethylene (PE) is currently the material of choice in natural gas transportation systems in the U.S. One advantage of PE piping is that maintenance can be performed on a small branch of the line without shutting down the entire system; a local area is shut down by applying a clamping tool to the PE pipe and stopping the flow of gas. The practice of pinch clamping has undoubtedly saved vast sums of money, but has also led to an unexpected problem.

In 1983, a section of a 4-in. diameter PE pipe developed a major leak. The gas collected beneath a residence where it ignited, resulting in severe damage to the house. Maintenance records and a visual inspection of the pipe indicated that it had been pinch clamped 6 years earlier in the region where the leak developed. A failure investigation [3] concluded that the pinch clamping operation was responsible for the failure. Microscopic examination of the pipe revealed that a small flaw apparently initiated on the inner surface of the pipe and grew through the wall. Figure 1.2 shows a low-magnification photograph of the fracture surface. Laboratory tests simulated the pinch clamping operation on sections of PE pipes; small thumbnail-shaped flaws (Figure 1.3) formed on the inner wall of the pipes, as a result of the severe strains that were applied. Fracture mechanics tests and analyses [3, 4] indicated that stresses in the pressurized pipe were sufficient to cause the observed time-dependent crack growth, i.e., growth from a small thumbnail flaw to a through-thickness crack over a period of 6 years.

The introduction of flaws in PE pipe by pinch clamping represents a Type 2 failure. The pinch clamping process was presumably tested thoroughly before it was applied in service, but no one anticipated that the procedure would introduce damage in the material that could lead to failure after several years in service. Although specific data are not available, pinch clamping has undoubtedly led to a significant number of gas leaks. The practice of pinch clamping is still widespread in the natural gas industry, but many companies and some states now require that a sleeve be fitted to the affected region in order to relieve the stresses locally. In addition, newer grades of PE pipe material have lower density and are less susceptible to damage by pinch clamping.

Some catastrophic events include elements both of Type 1 and Type 2 failures. On January 28, 1986, the Challenger Space Shuttle exploded because an O-ring seal in one of the main boosters did not respond well to cold weather. The shuttle represents relatively new technology, where

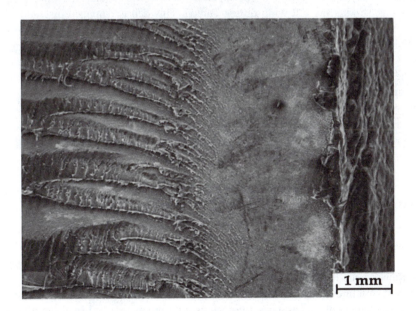

FIGURE 1.2 Fracture surface of a PE pipe that sustained time-dependent crack growth as a result of pinch clamping. (Taken from Jones, R.E. and Bradley, W.L., *Forensic Engineering,* Vol. I, 1987.) (Photograph provided by R.E. Jones, Jr.)

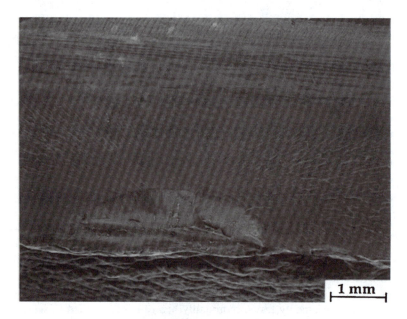

FIGURE 1.3 Thumbnail crack produced in a PE pipe after pinch clamping for 72 h. (Photograph provided by R.E. Jones, Jr.)

service experience is limited (Type 2), but engineers from the booster manufacturer suspected a potential problem with the O-ring seals and recommended that the launch be delayed (Type 1). Unfortunately, these engineers had little or no data to support their position and were unable to convince their managers or NASA officials. The tragic results of the decision to launch are well known.

On February 1, 2003, almost exactly 17 years after the Challenger accident, the Space Shuttle Columbia was destroyed during reentry. The apparent cause of the incident was foam insulation from the external tank striking the left wing during launch. This debris damaged insulation tiles on the underside of the wing, making the orbiter vulnerable to reentry temperatures that can reach 3000°F. The Columbia Accident Investigation Board (CAIB) was highly critical of NASA management for cultural traits and organizational practices that, according to the board, were detrimental to safety.

Over the past few decades, the field of fracture mechanics has undoubtedly prevented a substantial number of structural failures. We will never know how many lives have been saved or how much property damage has been avoided by applying this technology, because it is impossible to quantify disasters that *don't* happen. When applied correctly, fracture mechanics not only helps to prevent Type 1 failures but also reduces the frequency of Type 2 failures, because designers can rely on rational analysis rather than trial and error.

1.2 HISTORICAL PERSPECTIVE

Designing structures to avoid fracture is not a new idea. The fact that many structures commissioned by the Pharaohs of ancient Egypt and the Caesars of Rome are still standing is a testimony to the ability of early architects and engineers. In Europe, numerous buildings and bridges constructed during the Renaissance Period are still used for their intended purpose.

The ancient structures that are still standing today obviously represent successful designs. There were undoubtedly many more unsuccessful designs with much shorter life spans. Because knowledge of mechanics was limited prior to the time of Isaac Newton, workable designs were probably achieved largely by trial and error. The Romans supposedly tested each new bridge by requiring the design engineer to stand underneath while chariots drove over it. Such a practice would not only provide an incentive for developing good designs, but would also result in the social equivalent of Darwinian natural selection, where the worst engineers were removed from the profession.

The durability of ancient structures is particularly amazing when one considers that the choice of building materials prior to the Industrial Revolution was rather limited. Metals could not be produced in sufficient quantity to be formed into load-bearing members for buildings and bridges. The primary construction materials prior to the 19th century were timber, brick, and mortar; only the latter two materials were usually practical for large structures such as cathedrals, because trees of sufficient size for support beams were rare.

Brick and mortar are relatively brittle and are unreliable for carrying tensile loads. Consequently, pre-Industrial Revolution structures were usually designed to be loaded in compression. Figure 1.4 schematically illustrates a Roman bridge design. The arch shape causes compressive rather than tensile stresses to be transmitted through the structure.

The arch is the predominant shape in pre-Industrial Revolution architecture. Windows and roof spans were arched in order to maintain compressive loading. For example, Figure 1.5 shows two windows and a portion of the ceiling in King's College Chapel in Cambridge, England. Although these shapes are aesthetically pleasing, their primary purpose is more pragmatic.

Compressively loaded structures are obviously stable, since some have lasted for many centuries; the pyramids in Egypt are the epitome of a stable design.

With the Industrial Revolution came mass production of iron and steel. (Or, conversely, one might argue that mass production of iron and steel fueled the Industrial Revolution.) The availability of relatively ductile construction materials removed the earlier restrictions on design. It was finally feasible to build structures that carried tensile stresses. Note the difference between the design of the Tower Bridge in London (Figure 1.6) and the earlier bridge design (Figure 1.4).

The change from brick and mortar structures loaded in compression to steel structures in tension brought problems, however. Occasionally, a steel structure would fail unexpectedly at stresses well below the anticipated tensile strength. One of the most famous of these failures was the rupture of a molasses tank in Boston in January 1919 [5]. Over 2 million gallons of molasses were spilled, resulting in 12 deaths, 40 injuries, massive property damage, and several drowned horses.

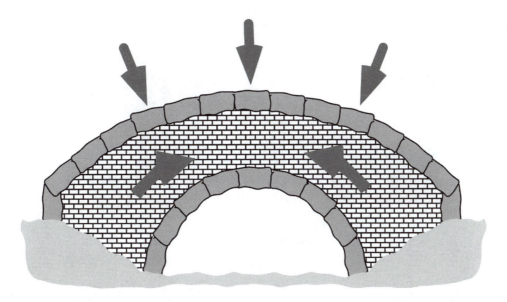

FIGURE 1.4 Schematic Roman bridge design. The arch shape of the bridge causes loads to be transmitted through the structure as compressive stresses.

FIGURE 1.5 Kings College Chapel in Cambridge, England. This structure was completed in 1515.

The cause of the failure of the molasses tank was largely a mystery at the time. In the first edition of his elasticity text published in 1892, Love [6] remarked that "the conditions of rupture are but vaguely understood." Designers typically applied safety factors of 10 or more (based on the tensile strength) in an effort to avoid these seemingly random failures.

1.2.1 EARLY FRACTURE RESEARCH

Experiments performed by Leonardo da Vinci several centuries earlier provided some clues as to the root cause of fracture. He measured the strength of iron wires and found that the strength varied inversely with wire length. These results implied that flaws in the material controlled the strength;

FIGURE 1.6 The Tower Bridge in London, completed in 1894. Note the modern beam design, made possible by the availability of steel support girders.

a longer wire corresponded to a larger sample volume, and a higher probability of sampling a region containing a flaw. These results were only qualitative, however.

A quantitative connection between fracture stress and flaw size came from the work of Griffith, which was published in 1920 [7]. He applied a stress analysis of an elliptical hole (performed by Inglis [8] seven years earlier) to the unstable propagation of a crack. Griffith invoked the first law of thermodynamics to formulate a fracture theory based on a simple energy balance. According to this theory, a flaw becomes unstable, and thus fracture occurs, when the strain-energy change that results from an increment of crack growth is sufficient to overcome the surface energy of the material (see Section 2.3). Griffith's model correctly predicted the relationship between strength and flaw size in glass specimens. Subsequent efforts to apply the Griffith model to metals were unsuccessful. Since this model assumes that the work of fracture comes exclusively from the surface energy of the material, the Griffith approach applies only to ideally brittle solids. A modification to Griffith's model, that made it applicable to metals, did not come until 1948.

1.2.2 THE LIBERTY SHIPS

The mechanics of fracture progressed from being a scientific curiosity to an engineering discipline, primarily because of what happened to the Liberty ships during World War II [9].

In the early days of World War II, the U.S. was supplying ships and planes to Great Britain under the Lend-Lease Act. Britain's greatest need at the time was for cargo ships to carry supplies. The German navy was sinking cargo ships at three times the rate at which they could be replaced with existing ship-building procedures.

Under the guidance of Henry Kaiser, a famous construction engineer whose previous projects included the Hoover Dam, the U.S. developed a revolutionary procedure for fabricating ships quickly. These new vessels, which became known as the Liberty ships, had an all-welded hull, as opposed to the riveted construction of traditional ship designs.

The Liberty ship program was a resounding success, until one day in 1943, when one of the vessels broke completely in two while sailing between Siberia and Alaska. Subsequent fractures occurred in other Liberty ships. Of the roughly 2700 Liberty ships built during World War II, approximately 400 sustained fractures, of which 90 were considered serious. In 20 ships the failure was essentially total, and about half of these broke completely in two.

Investigations revealed that the Liberty ship failures were caused by a combination of three factors:

- The welds, which were produced by a semi-skilled work force, contained crack-like flaws.
- Most of the fractures initiated on the deck at square hatch corners, where there was a local stress concentration.
- The steel from which the Liberty ships were made had poor toughness, as measured by Charpy impact tests.

The steel in question had always been adequate for riveted ships because fracture could not propagate across panels that were joined by rivets. A welded structure, however, is essentially a single piece of metal; propagating cracks in the Liberty ships encountered no significant barriers, and were sometimes able to traverse the entire hull.

Once the causes of failure were identified, the remaining Liberty ships were retrofitted with rounded reinforcements at the hatch corners. In addition, high toughness steel crack-arrester plates were riveted to the deck at strategic locations. These corrections prevented further serious fractures.

In the longer term, structural steels were developed with vastly improved toughness, and weld quality control standards were developed. Also, a group of researchers at the Naval Research Laboratory in Washington, DC. studied the fracture problem in detail. The field we now know as fracture mechanics was born in this lab during the decade following the war.

1.2.3 POST-WAR FRACTURE MECHANICS RESEARCH[1]

The fracture mechanics research group at the Naval Research Laboratory was led by Dr. G.R. Irwin. After studying the early work of Inglis, Griffith, and others, Irwin concluded that the basic tools needed to analyze fracture were already available. Irwin's first major contribution was to extend the Griffith approach to metals by including the energy dissipated by local plastic flow [10]. Orowan independently proposed a similar modification to the Griffith theory [11]. During this same period, Mott [12] extended the Griffith theory to a rapidly propagating crack.

In 1956, Irwin [13] developed the energy release rate concept, which was derived from the Griffith theory but in a form that was more useful for solving engineering problems. Shortly afterward, several of Irwin's colleagues brought to his attention a paper by Westergaard [14] that was published in 1938. Westergaard had developed a semi-inverse technique for analyzing stresses and displacements ahead of a sharp crack. Irwin [15] used the Westergaard approach to show that the stresses and displacements near the crack-tip could be described by a single constant that was related to the energy release rate. This crack-tip characterizing parameter later became known as the "stress-intensity factor." During this same period of time, Williams [16] applied a somewhat different technique to derive crack tip solutions that were essentially identical to Irwin's results.

A number of successful early applications of fracture mechanics bolstered the standing of this new field in the engineering community. In 1956, Wells [17] used fracture mechanics to show that the fuselage failures in several Comet jet aircraft resulted from fatigue cracks reaching a critical size. These cracks initiated at windows and were caused by insufficient reinforcement locally, combined with square corners that produced a severe stress concentration. (Recall the unfortunate hatch design in the Liberty ships.) A second early application of fracture mechanics occurred at General Electric in 1957. Winne and Wundt [18] applied Irwin's energy release rate approach to the failure of large rotors from steam turbines. They were able to predict the bursting behavior of large disks extracted from rotor forgings, and applied this knowledge to the prevention of fracture in actual rotors.

It seems that all great ideas encounter stiff opposition initially, and fracture mechanics is no exception. Although the U.S. military and the electric power generating industry were very supportive of the early work in this field, such was not the case in all provinces of government and industry. Several government agencies openly discouraged research in this area.

In 1960, Paris and his coworkers [19] failed to find a receptive audience for their ideas on applying fracture mechanics principles to fatigue crack growth. Although Paris et al. provided convincing experimental and theoretical arguments for their approach, it seems that design engineers were not yet ready to abandon their S-N curves in favor of a more rigorous approach to fatigue design. The resistance to this work was so intense that Paris and his colleagues were unable to find a peer-reviewed technical journal that was willing to publish their manuscript. They finally opted to publish their work in a University of Washington periodical entitled *The Trend in Engineering*.

1.2.4 FRACTURE MECHANICS FROM 1960 TO 1980

The Second World War obviously separates two distinct eras in the history of fracture mechanics. There is, however, some ambiguity as to how the period between the end of the war and the present should be divided. One possible historical boundary occurred around 1960, when the fundamentals of linear elastic fracture mechanics were fairly well established, and researchers turned their attention to crack-tip plasticity.

---

[1] For an excellent summary of early fracture mechanics research, refer to *Fracture Mechanics Retrospective: Early Classic Papers (1913–1965)*, John M. Barsom, ed., American Society of Testing and Materials (RPS 1), Philadelphia, PA, 1987. This volume contains reprints of 17 classic papers, as well as a complete bibliography of fracture mechanics papers published up to 1965.

Linear elastic fracture mechanics (LEFM) ceases to be valid when significant plastic deformation precedes failure. During a relatively short time period (1960–1961) several researchers developed analyses to correct for yielding at the crack tip, including Irwin [20], Dugdale [21], Barenblatt [22], and Wells [23]. The Irwin plastic zone correction [20] was a relatively simple extension of LEFM, while Dugdale [21] and Barenblatt [22] each developed somewhat more elaborate models based on a narrow strip of yielded material at the crack tip.

Wells [23] proposed the displacement of the crack faces as an alternative fracture criterion when significant plasticity precedes failure. Previously, Wells had worked with Irwin while on a sabbatical at the Naval Research Laboratory. When Wells returned to his post at the British Welding Research Association, he attempted to apply LEFM to low- and medium-strength structural steels. These materials were too ductile for LEFM to apply, but Wells noticed that the crack faces moved apart with plastic deformation. This observation led to the development of the parameter now known as the crack-tip-opening displacement (CTOD).

In 1968, Rice [24] developed another parameter to characterize nonlinear material behavior ahead of a crack. By idealizing plastic deformation as nonlinear elastic, Rice was able to generalize the energy release rate to nonlinear materials. He showed that this nonlinear energy release rate can be expressed as a line integral, which he called the J integral, evaluated along an arbitrary contour around the crack. At the time his work was being published, Rice discovered that Eshelby [25] had previously published several so-called conservation integrals, one of which was equivalent to Rice's J integral. Eshelby, however, did not apply his integrals to crack problems.

That same year, Hutchinson [26] and Rice and Rosengren [27] related the J integral to crack-tip stress fields in nonlinear materials. These analyses showed that J can be viewed as a nonlinear, stress-intensity parameter as well as an energy release rate.

Rice's work might have been relegated to obscurity had it not been for the active research effort by the nuclear power industry in the U.S. in the early 1970s. Because of legitimate concerns for safety, as well as political and public relations considerations, the nuclear power industry endeavored to apply state-of-the-art technology, including fracture mechanics, to the design and construction of nuclear power plants. The difficulty with applying fracture mechanics in this instance was that most nuclear pressure vessel steels were too tough to be characterized with LEFM without resorting to enormous laboratory specimens. In 1971, Begley and Landes [28], who were research engineers at Westinghouse, came across Rice's article and decided, despite skepticism from their co-workers, to characterize the fracture toughness of these steels with the J integral. Their experiments were very successful and led to the publication of a standard procedure for J testing of metals 10 years later [29].

Material toughness characterization is only one aspect of fracture mechanics. In order to apply fracture mechanics concepts to design, one must have a mathematical relationship between toughness, stress, and flaw size. Although these relationships were well established for linear elastic problems, a fracture design analysis based on the J integral was not available until Shih and Hutchinson [30] provided the theoretical framework for such an approach in 1976. A few years later, the Electric Power Research Institute (EPRI) published a fracture design handbook [31] based on the Shih and Hutchinson methodology.

In the United Kingdom, Well's CTOD parameter was applied extensively to fracture analysis of welded structures, beginning in the late 1960s. While fracture research in the U.S. was driven primarily by the nuclear power industry during the 1970s, fracture research in the U.K. was motivated largely by the development of oil resources in the North Sea. In 1971, Burdekin and Dawes [32] applied several ideas proposed by Wells [33] several years earlier and developed the CTOD design curve, a semiempirical fracture mechanics methodology for welded steel structures. The nuclear power industry in the UK developed their own fracture design analysis [34], based on the strip yield model of Dugdale [21] and Barenblatt [22].

Shih [35] demonstrated a relationship between the J integral and CTOD, implying that both parameters are equally valid for characterizing fracture. The J-based material testing and structural design approaches developed in the U.S. and the British CTOD methodology have begun to merge in recent years, with positive aspects of each approach combined to yield improved analyses. Both parameters are currently applied throughout the world to a range of materials.

Much of the theoretical foundation of dynamic fracture mechanics was developed in the period between 1960 and 1980. Significant contributions were made by a number of researchers, as discussed in Chapter 4.

1.2.5 FRACTURE MECHANICS FROM 1980 TO THE PRESENT

The field of fracture mechanics matured in the last two decades of the 20th century. Current research tends to result in incremental advances rather than major gains. The application of this technology to practical problems is so pervasive that fracture mechanics is now considered an established engineering discipline.

More sophisticated models for material behavior are being incorporated into fracture mechanics analyses. While plasticity was the important concern in 1960, more recent work has gone a step further, incorporating time-dependent nonlinear material behavior such as viscoplasticity and viscoelasticity. The former is motivated by the need for tough, creep-resistant high temperature materials, while the latter reflects the increasing proportion of plastics in structural applications. Fracture mechanics has also been used (and sometimes abused) in the characterization of composite materials.

Another trend in recent research is the development of microstuctural models for fracture and models to relate local and global fracture behavior of materials. A related topic is the efforts to characterize and predict geometry dependence of fracture toughness. Such approaches are necessary when traditional, so-called single-parameter fracture mechanics break down.

The continuing explosion in computer technology has aided both the development and application of fracture mechanics technology. For example, an ordinary desktop computer is capable of performing complex three-dimensional finite element analyses of structural components that contain cracks.

Computer technology has also spawned entirely new areas of fracture mechanics research. Problems encountered in the microelectronics industry have led to active research in interface fracture and nanoscale fracture.

1.3 THE FRACTURE MECHANICS APPROACH TO DESIGN

Figure 1.7 contrasts the fracture mechanics approach with the traditional approach to structural design and material selection. In the latter case, the anticipated design stress is compared to the flow properties of candidate materials; a material is assumed to be adequate if its strength is greater than the expected applied stress. Such an approach may attempt to guard against brittle fracture by imposing a safety factor on stress, combined with minimum tensile elongation requirements on the material. The fracture mechanics approach (Figure 1.7(b)) has three important variables, rather than two as in Figure 1.7(a). The additional structural variable is flaw size, and fracture toughness replaces strength as the relevant material property. Fracture mechanics quantifies the critical combinations of these three variables.

There are two alternative approaches to fracture analysis: the energy criterion and the stress-intensity approach. These two approaches are equivalent in certain circumstances. Both are discussed briefly below.

1.3.1 THE ENERGY CRITERION

The energy approach states that crack extension (i.e., fracture) occurs when the energy available for crack growth is sufficient to overcome the resistance of the material. The material resistance may include the surface energy, plastic work, or other types of energy dissipation associated with a propagating crack.

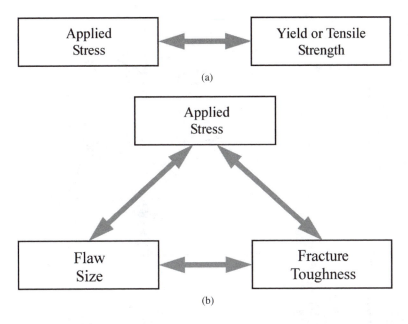

FIGURE 1.7 Comparison of the fracture mechanics approach to design with the traditional strength of materials approach: (a) the strength of materials approach and (b) the fracture mechanics approach.

Griffith [7] was the first to propose the energy criterion for fracture, but Irwin [13] is primarily responsible for developing the present version of this approach: the energy release rate $\mathcal{G}$ which is defined as the rate of change in potential energy with the crack area for a linear elastic material. At the moment of fracture $\mathcal{G} = \mathcal{G}_c$, the critical energy release rate, which is a measure of fracture toughness.

For a crack of length $2a$ in an infinite plate subject to a remote tensile stress (Figure 1.8), the energy release rate is given by

$$\mathcal{G} = \frac{\pi \sigma^2 a}{E} \tag{1.1}$$

where E is Young's modulus, σ is the remotely applied stress, and a is the half-crack length. At fracture $\mathcal{G} = \mathcal{G}_c$, and Equation (1.1) describes the critical combinations of stress and crack size for failure:

$$\mathcal{G}_c = \frac{\pi \sigma_f^2 a_c}{E} \tag{1.2}$$

Note that for a constant $\mathcal{G}_c$ value, failure stress σ_f varies with $1/\sqrt{a}$. The energy release rate $\mathcal{G}$ is the driving force for fracture, while $\mathcal{G}_c$ is the material's resistance to fracture. To draw an analogy to the strength of materials approach of Figure 1.7(a), the applied stress can be viewed as the driving force for plastic deformation, while the yield strength is a measure of the material's resistance to deformation.

The tensile stress analogy is also useful for illustrating the concept of similitude. A yield strength value measured with a laboratory specimen should be applicable to a large structure; yield strength does not depend on specimen size, provided the material is reasonably homogeneous. One of the fundamental assumptions of fracture mechanics is that fracture toughness ($\mathcal{G}_c$ in this case) is

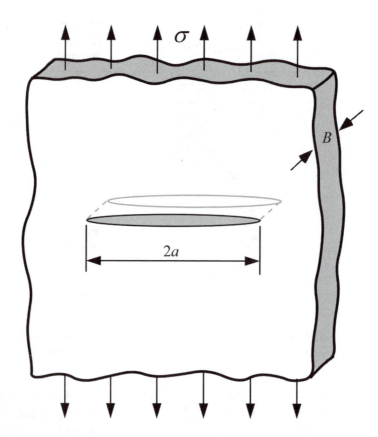

FIGURE 1.8 Through-thickness crack in an infinite plate subject to a remote tensile stress. In practical terms, "infinite" means that the width of the plate is $\gg 2a$.

independent of the size and geometry of the cracked body; a fracture toughness measurement on a laboratory specimen should be applicable to a structure. As long as this assumption is valid, all configuration effects are taken into account by the driving force $\mathcal{G}$. The similitude assumption is valid as long as the material behavior is predominantly linear elastic.

1.3.2 THE STRESS-INTENSITY APPROACH

Figure 1.9 schematically shows an element near the tip of a crack in an elastic material, together with the in-plane stresses on this element. Note that each stress component is proportional to a single constant K_I. If this constant is known, the entire stress distribution at the crack tip can be computed with the equations in Figure 1.9. This constant, which is called the stress-intensity factor, completely characterizes the crack-tip conditions in a linear elastic material. (The meaning of the subscript on K is explained in Chapter 2.) If one assumes that the material fails locally at some critical combination of stress and strain, then it follows that fracture must occur at a critical stress intensity K_{Ic}. Thus, K_{Ic} is an alternate measure of fracture toughness.

For the plate illustrated in Figure 1.8, the stress-intensity factor is given by

$$K_I = \sigma \sqrt{\pi a} \tag{1.3}$$

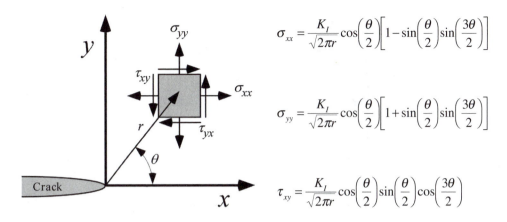

FIGURE 1.9 Stresses near the tip of a crack in an elastic material.

Failure occurs when $K_I = K_{Ic}$. In this case, K_I is the driving force for fracture and K_{Ic} is a measure of material resistance. As with $\mathcal{G}_c$, the property of similitude should apply to K_{Ic}. That is, K_{Ic} is assumed to be a size-independent material property.

Comparing Equation (1.1) and Equation (1.3) results in a relationship between K_I and $\mathcal{G}$:

$$\mathcal{G} = \frac{K_I^2}{E} \qquad (1.4)$$

This same relationship obviously holds for $\mathcal{G}_c$ and K_{Ic}. Thus, the energy and stress-intensity approaches to fracture mechanics are essentially equivalent for linear elastic materials.

1.3.3 TIME-DEPENDENT CRACK GROWTH AND DAMAGE TOLERANCE

Fracture mechanics often plays a role in life prediction of components that are subject to time-dependent crack growth mechanisms such as fatigue or stress corrosion cracking. The *rate* of cracking can be correlated with fracture mechanics parameters such as the stress-intensity factor, and the critical crack size for failure can be computed if the fracture toughness is known. For example, the fatigue crack growth rate in metals can usually be described by the following empirical relationship:

$$\frac{da}{dN} = C(\Delta K)^m \qquad (1.5)$$

where da/dN is the crack growth per cycle, ΔK is the stress-intensity range, and C and m are material constants.

Damage tolerance, as its name suggests, entails allowing subcritical flaws to remain in a structure. Repairing flawed material or scrapping a flawed structure is expensive and is often unnecessary. Fracture mechanics provides a rational basis for establishing flaw tolerance limits.

Consider a flaw in a structure that grows with time (e.g., a fatigue crack or a stress corrosion crack) as illustrated schematically in Figure 1.10. The *initial* crack size is inferred from nondestructive examination (NDE), and the *critical* crack size is computed from the applied stress and fracture toughness. Normally, an *allowable* flaw size would be defined by dividing the critical size by a safety factor. The predicted service life of the structure can then be inferred by calculating the time required for the flaw to grow from its initial size to the maximum allowable size.

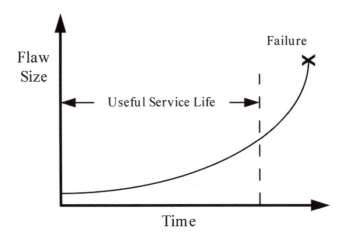

FIGURE 1.10 The damage tolerance approach to design.

1.4 EFFECT OF MATERIAL PROPERTIES ON FRACTURE

Figure 1.11 shows a simplified family tree for the field of fracture mechanics. Most of the early work was applicable only to linear elastic materials under quasistatic conditions, while subsequent advances in fracture research incorporated other types of material behavior. Elastic-plastic fracture mechanics considers plastic deformation under quasistatic conditions, while dynamic, viscoelastic, and viscoplastic fracture mechanics include time as a variable. A dashed line is drawn between linear elastic and dynamic fracture mechanics because some early research considered dynamic linear elastic behavior. The chapters that describe the various types of fracture behavior are shown in Figure 1.11. Elastic-plastic, viscoelastic, and viscoplastic fracture behavior are sometimes included in the more general heading of *nonlinear fracture mechanics*. The branch of fracture mechanics one should apply to a particular problem obviously depends on material behavior.

Consider a cracked plate (Figure 1.8) that is loaded to failure. Figure 1.12 is a schematic plot of failure stress vs. fracture toughness K_{Ic}. For low toughness materials, brittle fracture is the governing

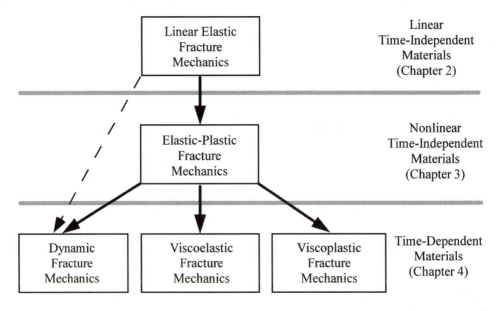

FIGURE 1.11 Simplified family tree of fracture mechanics.

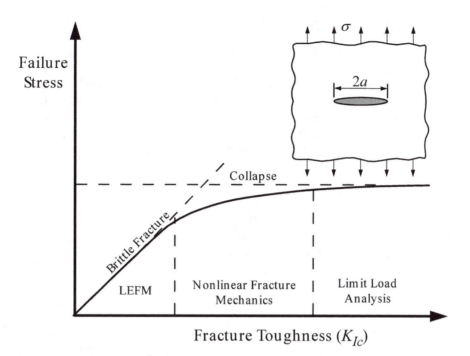

FIGURE 1.12 Effect of fracture toughness on the governing failure mechanism.

failure mechanism, and critical stress varies linearly with K_{Ic}, as predicted by Equation (1.3). At very high toughness values, LEFM is no longer valid, and failure is governed by the flow properties of the material. At intermediate toughness levels, there is a transition between brittle fracture under linear elastic conditions and ductile overload. Nonlinear fracture mechanics bridges the gap between LEFM and collapse. If toughness is low, LEFM is applicable to the problem, but if toughness is sufficiently high, fracture mechanics ceases to be relevant to the problem because failure stress is insensitive to toughness; a simple limit load analysis is all that is required to predict failure stress in a material with very high fracture toughness.

Table 1.1 lists various materials, together with the typical fracture regime for each material.

TABLE 1.1
Typical Fracture Behavior of Selected Materials[a]

Material	Typical Fracture Behavior
High strength steel	Linear elastic
Low- and medium-strength steel	Elastic-plastic/fully plastic
Austenitic stainless steel	Fully plastic
Precipitation-hardened aluminum	Linear elastic
Metals at high temperatures	Viscoplastic
Metals at high strain rates	Dynamic/viscoplastic
Polymers (below T_g)[b]	Linear elastic/viscoelastic
Polymers (above T_g)[b]	Viscoelastic
Monolithic ceramics	Linear elastic
Ceramic composites	Linear elastic
Ceramics at high temperatures	Viscoplastic

[a] Temperature is ambient unless otherwise specified.
[b] T_g — Glass transition temperature.

1.5   A BRIEF REVIEW OF DIMENSIONAL ANALYSIS

At first glance, a section on dimensional analysis may seem out of place in the introductory chapter of a book on fracture mechanics. However, dimensional analysis is an important tool for developing mathematical models of physical phenomena, and it can help us understand existing models. Many difficult concepts in fracture mechanics become relatively transparent when one considers the relevant dimensions of the problem. For example, dimensional analysis gives us a clue as to when a particular model, such as linear elastic fracture mechanics, is no longer valid.

Let us review the fundamental theorem of dimensional analysis and then look at a few simple applications to fracture mechanics.

1.5.1   THE BUCKINGHAM Π-THEOREM

The first step in building a mathematical model of a physical phenomenon is to identify all of the parameters that may influence the phenomenon. Assume that a problem, or at least an idealized version of it, can be described by the following set of scalar quantities: $\{u, W_1, W_2, \ldots, W_n\}$. The dimensions of all quantities in this set is denoted by $\{[u], [W_1], [W_2], \ldots, [W_n]\}$. Now suppose that we wish to express the first variable u as a function of the remaining parameters:

$$u = f(W_1, W_2, \ldots, W_n) \qquad (1.6)$$

Thus, the process of modeling the problem is reduced to finding a mathematical relationship that represents f as best as possible. We might accomplish this by performing a set of experiments in which we measure u while varying each W_i independently. The number of experiments can be greatly reduced, and the modeling processes simplified, through dimensional analysis. The first step is to identify all of the *fundamental dimensional units* (fdu's) in the problem: $\{L_1, L_2, \ldots, L_m\}$. For example, a typical mechanics problem may have $\{L_1 = \text{length}, L_2 = \text{mass}, L_3 = \text{time}\}$. We can express the dimensions of each quantity in our problem as the product of the powers of the fdu's; i.e., for any quantity X, we have

$$[X] = L_1^{a_1} L_2^{a_2}, \ldots, L_m^{a_m} \qquad (1.7)$$

The quantity X is dimensionless if $[X] = 1$.

In the set of Ws, we can identify m *primary quantities* that contain all of the fdu's in the problem. The remaining variables are secondary quantities, and their dimensions can be expressed in terms of the primary quantities:

$$[W_{m+j}] = [W_1]^{a_{m+j(1)}}, \ldots, [W_m]^{a_{m+j(m)}} \qquad (j = 1, 2, \ldots, n-m \qquad (1.8)$$

Thus, we can define a set of new quantities π_i that are dimensionless:

$$\pi_i = \frac{W_{m+j}}{W_1^{a_{m+j(1)}}, \ldots, W_m^{a_{m+j(m)}}} \qquad (1.9)$$

Similarly, the dimensions of u can be expressed in terms of the dimensions of the primary quantities:

$$[u] = [W_1]^{a_1}, \ldots, [W_m]^{a_r} \qquad (1.10)$$

and we can form the following dimensionless quantity:

$$\pi = \frac{u}{W_1^{a_1}, \ldots, W_m^{a_m}} \tag{1.11}$$

According to the Buckingham Π-theorem, π depends only on the other dimensionless groups:

$$\pi = F(\pi_1, \pi_2, \ldots, \pi_{n-m}) \tag{1.12}$$

This new function F is independent of the system of measurement units. Note that the number of quantities in F has been reduced from the old function by m, the number of fdu's. Thus dimensional analysis has reduced the degrees of freedom in our model, and we need vary only $n-m$ quantities in our experiments or computer simulations.

The Buckingham Π-theorem gives guidance on how to scale a problem to different sizes or to other systems of measurement units. Each dimensionless group (π_i) must be scaled in order to obtain equivalent conditions at two different scales. Suppose, for example, that we wish to perform wind tunnel tests on a model of a new airplane design. Dimensional analysis tells us that we should reduce all length dimensions in the same proportion; thus we would build a "scale" model of the airplane. The length dimensions of the plane are not the only important quantities in the problem, however. In order to model the aerodynamic behavior accurately, we would need to scale the wind velocity and the viscosity of the air in accordance with the reduced size of the airplane model. Modifying the viscosity of the air is not practical in most cases. In real wind tunnel tests, the size of the model is usually close enough to full scale that the errors introduced by not scaling viscosity are minor.

1.5.2 Dimensional Analysis in Fracture Mechanics

Dimensional analysis proves to be a very useful tool in fracture mechanics. Later chapters describe how dimensional arguments play a key role in developing mathematical descriptions for important phenomena. For now, let us explore a few simple examples.

Consider a series of cracked plates under a remote tensile stress σ^∞, as illustrated in Figure 1.13. Assume that each is a two-dimensional problem; that is, the thickness dimension does not enter into the problem. The first case, Figure 1.13(a), is an edge crack of length a in an elastic, semi-infinite plate. In this case infinite means that the plate width is much larger than the crack size. Suppose that we wish to know how one of the stress components σ_{ij} varies with position. We will adopt a polar coordinate system with the origin at the crack tip, as illustrated in Figure 1.9. A generalized functional relationship can be written as

$$\sigma_{ij} = f_1(\sigma^\infty, E, \nu, \sigma_{kl}, \varepsilon_{kl}, a, r, \theta) \tag{1.13}$$

where

    ν = Poisson's ratio
    σ_{kl} = other stress components
    ε_{kl} = all nonzero components of the strain tensor

We can eliminate σ_{kl} and ε_{kl} from f_1 by noting that for a linear elastic problem, strain is uniquely defined by stress through Hooke's law and the stress components at a point increase in proportion to one another. Let σ^∞ and a be the primary quantities. Invoking the Buckingham Π-theorem gives

$$\frac{\sigma_{ij}}{\sigma^\infty} = F_1\left(\frac{E}{\sigma^\infty}, \frac{r}{a}, \nu, \theta\right) \tag{1.14}$$

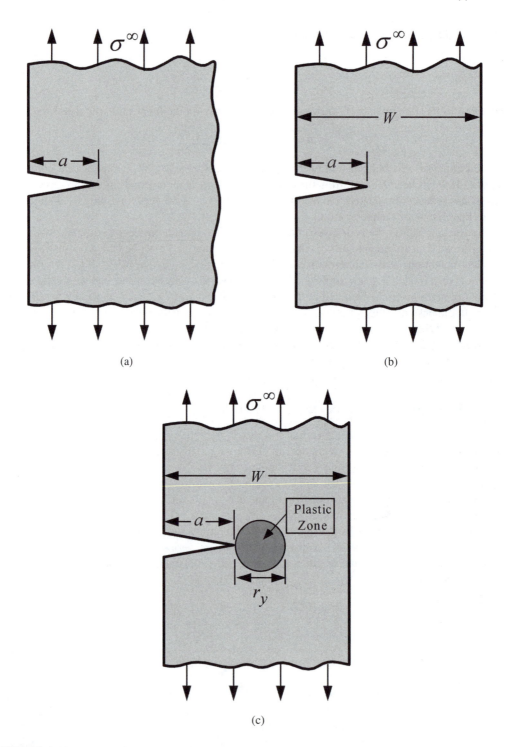

FIGURE 1.13 Edge-cracked plates subject to a remote tensile stress: (a) edge crack in a wide elastic plate, (b) edge crack in a finite width elastic plate, and (c) edge crack with a plastic zone at the crack tip.

When the plate width is finite (Figure 1.13(b)), an additional dimension is required to describe the problem:

$$\frac{\sigma_{ij}}{\sigma^{\infty}} = F_2\left(\frac{E}{\sigma^{\infty}}, \frac{r}{a}, \frac{W}{a}, \nu, \theta\right) \tag{1.15}$$

Thus, one might expect Equation (1.14) to give erroneous results when the crack extends across a significant fraction of the plate width. Consider a large plate and a small plate made of the same material (same E and ν), with the same a/W ratio, loaded to the same remote stress. The local stress at an angle θ from the crack plane in each plate would depend only on the r/a ratio, as long as both plates remained elastic.

When a plastic zone forms ahead of the crack tip (Figure 1.13(c)), the problem is complicated further. If we assume that the material does not strain harden, the yield strength is sufficient to define the flow properties. The stress field is given by

$$\frac{\sigma_{ij}}{\sigma} = F_3\left(\frac{E}{\sigma}, \frac{\sigma_{YS}}{\sigma}, \frac{r}{a}, \frac{W}{a}, \frac{r_y}{a}, \nu, \theta\right) \tag{1.16}$$

The first two functions, F_1 and F_2, correspond to LEFM, while F_3 is an elastic-plastic relationship. Thus, dimensional analysis tells us that LEFM is valid only when $r_y \ll a$ and $\sigma^{\infty} \ll \sigma_{YS}$. In Chapter 2, the same conclusion is reached through a somewhat more complicated argument.

REFERENCES

1. Duga, J.J., Fisher, W.H., Buxbaum, R.W., Rosenfield, A.R., Burh, A.R., Honton, E.J., and McMillan, S.C., "The Economic Effects of Fracture in the United States." NBS Special Publication 647-2, U.S. Department of Commerce, Washington, DC, March 1983.
2. Garwood, S.J., Private Communication, 1990.
3. Jones, R.E. and Bradley, W.L., "Failure Analysis of a Polyethylene Natural Gas Pipeline." *Forensic Engineering,* Vol. 1, 1987, pp. 47–59.
4. Jones, R.E. and Bradley, W.L., "Fracture Toughness Testing of Polyethylene Pipe Materials." ASTM STP 995, Vol. 1, American Society for Testing and Materials, Philadelphia, PA, 1989, pp. 447–456.
5. Shank, M.E., "A Critical Review of Brittle Failure in Carbon Plate Steel Structures Other than Ships." Ship Structure Committee Report SSC-65, National Academy of Science-National Research Council, Washington, DC, December 1953.
6. Love, A.E.H., *A Treatise on the Mathematical Theory of Elasticity.* Dover Publications, New York, 1944.
7. Griffith, A.A., "The Phenomena of Rupture and Flow in Solids." *Philosophical Transactions,* Series A, Vol. 221, 1920, pp. 163–198.
8. Inglis, C.E., "Stresses in a Plate Due to the Presence of Cracks and Sharp Corners." *Transactions of the Institute of Naval Architects,* Vol. 55, 1913, pp. 219–241.
9. Williams, M.L. and Ellinger, G.A., "Investigation of Stractural Failures of Welded Ships." Welding Journal, Vol. 32, 1953, pp. 498–528.
10. Irwin, G.R., "Fracture Dynamics." *Fracturing of Metals,* American Society for Metals, Cleveland, OH, 1948, pp. 147–166.
11. Orowan, E., "Fracture and Strength of Solids." *Reports on Progress in Physics,* Vol. XII, 1948, pp. 185–232.
12. Mott, N.F., "Fracture of Metals: Theoretical Considerations." *Engineering,* Vol. 165, 1948, pp. 16–18.
13. Irwin, G.R., "Onset of Fast Crack Propagation in High Strength Steel and Aluminum Alloys." *Sagamore Research Conference Proceedings,* Vol. 2, 1956, pp. 289–305.

14. Westergaard, H.M., "Bearing Pressures and Cracks." *Journal of Applied Mechanics,* Vol. 6, 1939, pp. 49–53.

15. Irwin, G.R., "Analysis of Stresses and Strains near the End of a Crack Traversing a Plate." *Journal of Applied Mechanics,* Vol. 24, 1957, pp. 361–364.

16. Williams, M.L., "On the Stress Distribution at the Base of a Stationary Crack." *Journal of Applied Mechanics,* Vol. 24, 1957, pp. 109–114.

17. Wells, A.A., "The Condition of Fast Fracture in Aluminum Alloys with Particular Reference to Comet Failures." British Welding Research Association Report, April 1955.

18. Winne, D.H. and Wundt, B.M., "Application of the Griffith-Irwin Theory of Crack Propagation to the Bursting Behavior of Disks, Including Analytical and Experimental Studies." *Transactions of the American Society of Mechanical Engineers,* Vol. 80, 1958, pp. 1643–1655.

19. Paris, P.C., Gomez, M.P., and Anderson, W.P., "A Rational Analytic Theory of Fatigue." *The Trend in Engineering,* Vol. 13, 1961, pp. 9–14.

20. Irwin, G.R., "Plastic Zone Near a Crack and Fracture Toughness." *Sagamore Research Conference Proceedings,* Vol. 4, Syracuse University Research Institute, Syracuse, NY, 1961, pp. 63–78.

21. Dugdale, D.S., "Yielding in Steel Sheets Containing Slits." *Journal of the Mechanics and Physics of Solids,* Vol. 8, 1960, pp. 100–104.

22. Barenblatt, G.I., "The Mathematical Theory of Equilibrium Cracks in Brittle Fracture." *Advances in Applied Mechanics,* Vol. VII, Academic Press, 1962, pp. 55–129.

23. Wells, A.A., "Unstable Crack Propagation in Metals: Cleavage and Fast Fracture." *Proceedings of the Crack Propagation Symposium,* Vol. 1, Paper 84, Cranfield, UK, 1961.

24. Rice, J.R., "A Path Independent Integral and the Approximate Analysis of Strain Concentration by Notches and Cracks." *Journal of Applied Mechanics,* Vol. 35, 1968, pp. 379–386.

25. Eshelby, J.D., "The Continuum Theory of Lattice Defects." *Solid State Physics,* Vol. 3, 1956.

26. Hutchinson, J.W., "Singular Behavior at the End of a Tensile Crack Tip in a Hardening Material." *Journal of the Mechanics and Physics of Solids,* Vol. 16, 1968, pp. 13–31.

27. Rice, J.R. and Rosengren, G.F., "Plane Strain Deformation near a Crack Tip in a Power-Law Hardening Material." *Journal of the Mechanics and Physics of Solids,* Vol. 16, 1968, pp. 1–12.

28. Begley, J.A. and Landes, J.D., "The *J*-Integral as a Fracture Criterion." ASTM STP 514, American Society for Testing and Materials, Philadelphia, PA, 1972, pp. 1–20.

29. E 813-81, "Standard Test Method for J_{Ic}, a Measure of Fracture Toughness." American Society for Testing and Materials, Philadelphia, PA, 1981.

30. Shih, C.F. and Hutchinson, J.W., "Fully Plastic Solutions and Large-Scale Yielding Estimates for Plane Stress Crack Problems." *Journal of Engineering Materials and Technology,* Vol. 98, 1976, pp. 289–295.

31. Kumar, V., German, M.D., and Shih, C.F., "An Engineering Approach for Elastic-Plastic Fracture Analysis." EPRI Report NP-1931, Electric Power Research Institute, Palo Alto, CA, 1981.

32. Burdekin, F.M. and Dawes, M.G., "Practical Use of Linear Elastic and Yielding Fracture Mechanics with Particular Reference to Pressure Vessels." *Proceedings of the Institute of Mechanical Engineers Conference,* London, May 1971, pp. 28–37.

33. Wells, A.A., "Application of Fracture Mechanics at and beyond General Yielding." *British Welding Journal,* Vol. 10, 1963, pp. 563–570.

34. Harrison, R.P., Loosemore, K., Milne, I., and Dowling, A.R., "Assessment of the Integrity of Structures Containing Defects." Central Electricity Generating Board Report R/H/R6-Rev 2, April 1980.

35. Shih, C.F., "Relationship between the *J*-Integral and the Crack Opening Displacement for Stationary and Extending Cracks." *Journal of the Mechanics and Physics of Solids,* Vol. 29, 1981, pp. 305–326.

Part II

Fundamental Concepts

2 Linear Elastic Fracture Mechanics

The concepts of fracture mechanics that were derived prior to 1960 are applicable only to materials that obey Hooke's law. Although corrections for small-scale plasticity were proposed as early as 1948, these analyses are restricted to structures whose global behavior is linear elastic.

Since 1960, fracture mechanics theories have been developed to account for various types of nonlinear material behavior (i.e., plasticity and viscoplasticity) as well as dynamic effects. All of these more recent results, however, are extensions of linear elastic fracture mechanics (LEFM). Thus a solid background in the fundamentals of LEFM is essential to an understanding of more advanced concepts in fracture mechanics.

This chapter describes both the energy and stress intensity approaches to linear fracture mechanics. The early work of Inglis and Griffith is summarized, followed by an introduction to the energy release rate and stress intensity parameters. The appendix at the end of this chapter includes mathematical derivations of several important results in LEFM.

Subsequent chapters also address linear elastic fracture mechanics. Chapter 7 and Chapter 8 discuss laboratory testing of linear elastic materials, Chapter 9 addresses application of LEFM to structures, Chapter 10 and chapter 11 apply LEFM to fatigue crack propagation and environmental cracking, respectively. Chapter 12 outlines numerical methods for computing stress intensity factor and energy release rate.

2.1 AN ATOMIC VIEW OF FRACTURE

A material fractures when sufficient stress and work are applied at the atomic level to break the bonds that hold atoms together. The bond strength is supplied by the attractive forces between atoms.

Figure 2.1 shows schematic plots of the potential energy and force vs. the separation distance between atoms. The equilibrium spacing occurs where the potential energy is at a minimum. A tensile force is required to increase the separation distance from the equilibrium value; this force must exceed the cohesive force to sever the bond completely. The bond energy is given by

$$E_b = \int_{x_o}^{\infty} P \, dx \qquad (2.1)$$

where x_o is the equilibrium spacing and P is the applied force.

It is possible to estimate the cohesive strength at the atomic level by idealizing the interatomic force-displacement relationship as one half of the period of a sine wave:

$$P = P_c \sin\left(\frac{\pi x}{\lambda}\right) \qquad (2.2a)$$

where the distance λ is defined in Figure 2.1. For the sake of simplicity, the origin is defined at x_o. For small displacements, the force-displacement relationship is linear:

$$P = P_c\left(\frac{\pi x}{\lambda}\right) \qquad (2.2b)$$

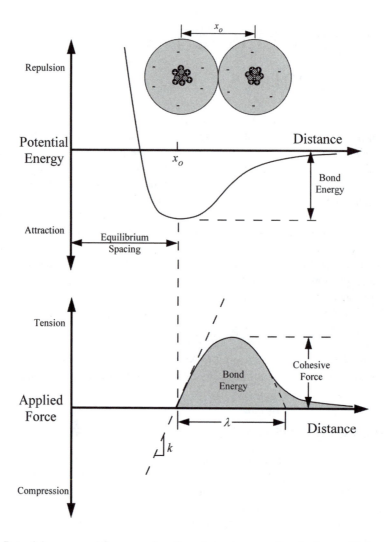

FIGURE 2.1 Potential energy and force as a function of atomic separation. At the equilibrium separation x_o the potential energy is minimized, and the attractive and repelling forces are balanced.

and the bond stiffness (i.e., the spring constant) is given by

$$k = P_c \left(\frac{\pi}{\lambda} \right) \tag{2.3}$$

Multiplying both sides of this equation by the number of bonds per unit area and the gage length, x_o, converts k to Young's modulus E and P_c to the cohesive stress σ_c. Solving for σ_c gives

$$\sigma_c = \frac{E\lambda}{\pi x_o} \tag{2.4}$$

or

$$\sigma_c \approx \frac{E}{\pi} \tag{2.5}$$

if λ is assumed to be approximately equal to the atomic spacing.

The surface energy can be estimated as follows:

$$\gamma_s = \frac{1}{2}\int_0^\lambda \sigma_c \sin\left(\frac{\pi x}{\lambda}\right)dx = \sigma_c \frac{\lambda}{\pi} \tag{2.6}$$

The surface energy per unit area, γ_s, is equal to one-half of the fracture energy because two surfaces are created when a material fractures. Substituting Equation (2.4) into Equation (2.6) and solving for σ_c gives

$$\sigma_c = \sqrt{\frac{E\gamma_s}{x_o}} \tag{2.7}$$

2.2 STRESS CONCENTRATION EFFECT OF FLAWS

The derivation in the previous section showed that the theoretical cohesive strength of a material is approximately E/π, but experimental fracture strengths for brittle materials are typically three or four orders of magnitude below this value. As discussed in Chapter 1, experiments by Leonardo da Vinci, Griffith, and others indicated that the discrepancy between the actual strengths of brittle materials and theoretical estimates was due to flaws in these materials. Fracture cannot occur unless the stress at the atomic level exceeds the cohesive strength of the material. Thus, the flaws must lower the global strength by magnifying the stress locally.

The first quantitative evidence for the stress concentration effect of flaws was provided by Inglis [1], who analyzed elliptical holes in flat plates. His analyses included an elliptical hole $2a$ long by $2b$ wide with an applied stress perpendicular to the major axis of the ellipse (see Figure 2.2). He assumed that the hole was not influenced by the plate boundary, i.e., the plate width >> $2a$ and the plate height >> $2b$. The stress at the tip of the major axis (Point A) is given by

$$\sigma_A = \sigma\left(1 + \frac{2a}{b}\right) \tag{2.8}$$

The ratio σ_A/σ is defined as the stress concentration factor k_t. When $a = b$, the hole is circular and $k_t = 3.0$, a well-known result that can be found in most strength-of-materials textbooks.

As the major axis, a, increases relative to b, the elliptical hole begins to take on the appearance of a sharp crack. For this case, Inglis found it more convenient to express Equation (2.8) in terms of the radius of curvature ρ:

$$\sigma_A = \sigma\left(1 + 2\sqrt{\frac{a}{\rho}}\right) \tag{2.9}$$

where

$$\rho = \frac{b^2}{a} \tag{2.10}$$

When $a >> b$, Equation (2.9) becomes

$$\sigma_A = 2\sigma\sqrt{\frac{a}{\rho}} \tag{2.11}$$

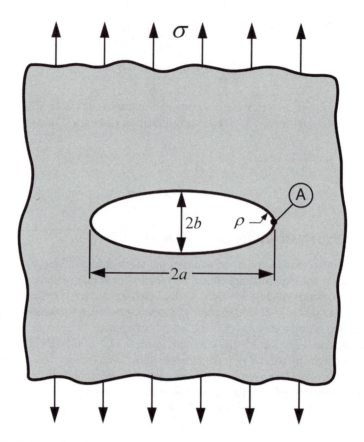

FIGURE 2.2 Elliptical hole in a flat plate.

Inglis showed that Equation (2.11) gives a good approximation for the stress concentration due to a notch that is not elliptical except at the tip.

Equation (2.11) predicts an infinite stress at the tip of an infinitely sharp crack, where $\rho = 0$. This result caused concern when it was first discovered, because no material is capable of withstanding infinite stress. A material that contains a sharp crack should theoretically fail upon the application of an infinitesimal load. The paradox of a sharp crack motivated Griffith [2] to develop a fracture theory based on energy rather than local stress (Section 2.3).

An infinitely sharp crack in a continuum is a mathematical abstraction that is not relevant to real materials, which are made of atoms. Metals, for instance, deform plastically, which causes an initially sharp crack to blunt. In the absence of plastic deformation, the minimum radius a crack tip can have is on the order of the atomic radius. By substituting $\rho = x_o$ into Equation (2.11), we obtain an estimate of the local stress concentration at the tip of an atomically sharp crack:

$$\sigma_A = 2\sigma\sqrt{\frac{a}{x_o}} \tag{2.12}$$

If it is assumed that fracture occurs when $\sigma_A = \sigma_c$, Equation (2.12) can be set equal to Equation (2.7), resulting in the following expression for the remote stress at failure:

$$\sigma_f = \left(\frac{E\gamma_s}{4a}\right)^{1/2} \tag{2.13}$$

Equation (2.13) must be viewed as a rough estimate of failure stress, because the continuum assumption upon which the Inglis analysis is based is not valid at the atomic level. However, Gehlen and Kanninen [3] obtained similar results from a numerical simulation of a crack in a two-dimensional lattice, where discrete "atoms" were connected by nonlinear springs:

$$\sigma_f = \alpha \left(\frac{E\gamma_s}{a} \right)^{1/2} \tag{2.14}$$

where α is a constant, on the order of unity, which depends slightly on the assumed atomic force-displacement law (Equation (2.2)).

2.3 THE GRIFFITH ENERGY BALANCE

According to the first law of thermodynamics, when a system goes from a nonequilibrium state to equilibrium, there is a net decrease in energy. In 1920, Griffith applied this idea to the formation of a crack [2]:

> It may be supposed, for the present purpose, that the crack is formed by the sudden annihilation of the tractions acting on its surface. At the instant following this operation, the strains, and therefore the potential energy under consideration, have their original values; but in general, the new state is not one of equilibrium. If it is not a state of equilibrium, then, by the theorem of minimum potential energy, the potential energy is reduced by the attainment of equilibrium; if it is a state of equilibrium, the energy does not change.

A crack can form (or an existing crack can grow) only if such a process causes the total energy to decrease or remain constant. Thus the critical conditions for fracture can be defined as the point where crack growth occurs under equilibrium conditions, with no net change in total energy.

Consider a plate subjected to a constant stress σ which contains a crack $2a$ long (Figure 2.3). Assume that the plate width $\gg 2a$ and that plane stress conditions prevail. (Note that the plates in Figure 2.2 and Figure 2.3 are identical when $a \gg b$). In order for this crack to increase in size, sufficient potential energy must be available in the plate to overcome the surface energy of the material. The Griffith energy balance for an incremental increase in the crack area $d\mathrm{A}$, under equilibrium conditions, can be expressed in the following way:

$$\frac{dE}{d\mathrm{A}} = \frac{d\Pi}{d\mathrm{A}} + \frac{dW_s}{d\mathrm{A}} = 0 \tag{2.15a}$$

or

$$-\frac{d\Pi}{d\mathrm{A}} = \frac{dW_s}{d\mathrm{A}} \tag{2.15b}$$

where

E = total energy

Π = potential energy supplied by the internal strain energy and external forces

W_s = work required to create new surfaces

For the cracked plate illustrated in Figure 2.3, Griffith used the stress analysis of Inglis [1] to show that

$$\Pi = \Pi_o - \frac{\pi\sigma^2 a^2 B}{E} \tag{2.16}$$

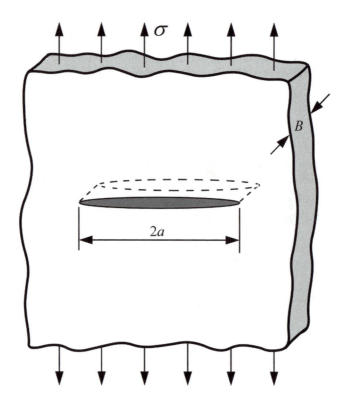

FIGURE 2.3 A through-thickness crack in an infinitely wide plate subjected to a remote tensile stress.

where Π_o is the potential energy of an uncracked plate and B is the plate thickness. Since the formation of a crack requires the creation of two surfaces, W_s is given by

$$W_s = 4aB\gamma_s \tag{2.17}$$

where γ_s is the surface energy of the material. Thus

$$-\frac{d\Pi}{dA} = \frac{\pi\sigma^2 a}{E} \tag{2.18a}$$

and

$$\frac{dW_s}{dA} = 2\gamma_s \tag{2.18b}$$

Equating Equation (2.18a) and Equation (2.18b) and solving for fracture stress gives

$$\sigma_f = \left(\frac{2E\gamma_s}{\pi a}\right)^{1/2} \tag{2.19}$$

It is important to note the distinction between *crack area* and *surface area*. The crack area is defined as the projected area of the crack ($2aB$ in the present example), but since a crack includes two matching surfaces, the surface area is $2A$.

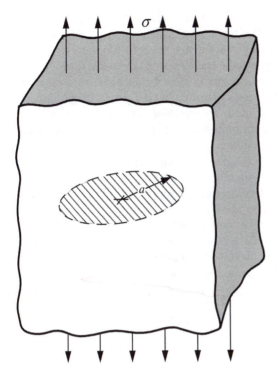

FIGURE 2.4 A penny-shaped (circular) crack embedded in a solid subjected to a remote tensile stress.

The Griffith approach can be applied to other crack shapes. For example, the fracture stress for a penny-shaped flaw embedded in the material (Figure 2.4) is given by

$$\sigma_f = \left(\frac{\pi E \gamma_s}{2(1-v^2)a} \right)^{1/2} \tag{2.20}$$

where a is the crack radius and v is Poisson's ratio.

2.3.1 COMPARISON WITH THE CRITICAL STRESS CRITERION

The Griffith model is based on a global energy balance: for fracture to occur, the energy stored in the structure must be sufficient to overcome the surface energy of the material. Since fracture involves the breaking of bonds, the stress on the atomic level must be equal to the cohesive stress. This local stress intensification can be provided by flaws in the material, as discussed in Section 2.2.

The similarity between Equation (2.13), Equation (2.14), and Equation (2.19) is obvious. Predictions of the global fracture stress from the Griffith approach and the local stress criterion differ by less than 40%. Thus, these two approaches are consistent with one another, at least in the case of a sharp crack in an ideally brittle solid.

An apparent contradiction emerges when the crack-tip radius is significantly greater than the atomic spacing. The change in the stored energy with crack formation (Equation (2.16)) is insensitive to the notch radius as long as $a \gg b$; thus, the Griffith model implies that the fracture stress is insensitive to ρ. According to the Inglis stress analysis, however, in order for σ_c to be attained at the tip of the notch, σ_f must vary with $1/\sqrt{\rho}$.

Consider a crack with $\rho = 5 \times 10^{-6}$ m. Such a crack would appear sharp under a light microscope, but ρ would be four orders of magnitude larger than the atomic spacing in a typical crystalline solid. Thus the local stress approach would predict a global fracture strength 100 times larger than the Griffith equation. The actual material behavior is somewhere between these extremes; fracture stress does depend on notch root radius, but not to the extent implied by the Inglis stress analysis.

The apparent discrepancy between the critical stress criterion and the energy criterion based on thermodynamics can be resolved by viewing fracture as a nucleation and growth process. When the global stress and crack size satisfy the Griffith energy criterion, there is sufficient thermodynamic driving force to grow the crack, but fracture must first be nucleated. This situation is analogous to the solidification of liquids. Water, for example, is in equilibrium with ice at 0°C, but the liquid–solid reaction requires ice crystals to be nucleated, usually on the surface of another solid (e.g., your car windshield on a January morning). When nucleation is suppressed, liquid water can be super cooled (at least momentarily) to as much as 30°C below the equilibrium freezing point.

Nucleation of fracture can come from a number of sources. For example, microscopic surface roughness at the tip of the flaw could produce sufficient local stress concentration to nucleate failure. Another possibility, illustrated in Figure 2.5, involves a sharp microcrack near the tip of a macroscopic flaw with a finite notch radius. The macroscopic crack magnifies the stress in the vicinity of the microcrack, which propagates when it satisfies the Griffith equation. The microcrack links with the large flaw, which then propagates if the Griffith criterion is satisfied globally. This type of mechanism controls cleavage fracture in ferritic steels, as discussed in Chapter 5.

2.3.2 Modified Griffith Equation

Equation (2.19) is valid only for ideally brittle solids. Griffith obtained a good agreement between Equation (2.19) and the experimental fracture strength of glass, but the Griffith equation severely underestimates the fracture strength of metals.

Irwin [4] and Orowan [5] independently modified the Griffith expression to account for materials that are capable of plastic flow. The revised expression is given by

$$\sigma_f = \left(\frac{2E(\gamma_s + \gamma_p)}{\pi a} \right)^{1/2}$$

(2.21)

where γ_p is the plastic work per unit area of surface created and is typically much larger than γ_s.

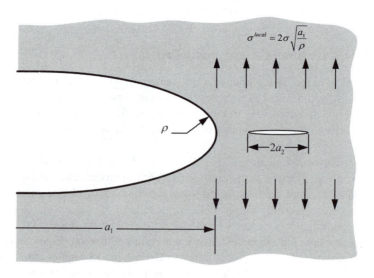

FIGURE 2.5 A sharp microcrack at the tip of a macroscopic crack.

In an ideally brittle solid, a crack can be formed merely by breaking atomic bonds; γ_s reflects the total energy of broken bonds in a unit area. When a crack propagates through a metal, however, a dislocation motion occurs in the vicinity of the crack tip, resulting in additional energy dissipation.

Although Irwin and Orowan originally derived Equation (2.21) for metals, it is possible to generalize the Griffith model to account for any type of energy dissipation:

$$\sigma_f = \left(\frac{2Ew_f}{\pi a}\right)^{1/2} \tag{2.22}$$

where w_f is the fracture energy, which could include plastic, viscoelastic, or viscoplastic effects, depending on the material. The fracture energy can also be influenced by crack meandering and branching, which increase the surface area. Figure 2.6 illustrates various types of material behavior and the corresponding fracture energy.

A word of caution is necessary when applying Equation (2.22) to materials that exhibit nonlinear deformation. The Griffith model, in particular Equation (2.16), applies only to linear elastic material behavior. Thus, the global behavior of the structure must be elastic. Any nonlinear effects, such as plasticity, must be confined to a small region near the crack tip. In addition, Equation (2.22) assumes that w_f is constant; in many ductile materials, the fracture energy increases with crack growth, as discussed in Section 2.5.

EXAMPLE 2.1

A flat plate made from a brittle material contains a macroscopic through-thickness crack with half length a_1 and notch tip radius ρ. A sharp penny-shaped microcrack with radius a_2 is located near the tip of the larger flaw, as illustrated in Figure 2.5. Estimate the minimum size of the microcrack required to cause failure in the plate when the Griffith equation is satisfied by the global stress and a_1.

Solution: The nominal stress at failure is obtained by substituting a_1 into Equation (2.19). The stress in the vicinity of the microcrack can be estimated from Equation (2.11), which is set equal to the Griffith criterion for the penny-shaped microcrack (Equation 2.20):

$$2\left(\frac{2E\gamma_s}{\pi a_1}\right)^{1/2}\sqrt{\frac{a_1}{\rho}} = \left(\frac{\pi E\gamma_s}{2(1-v^2)a_2}\right)^{1/2}$$

Solving for a_2 gives

$$a_2 = \frac{\pi^2 \rho}{16(1-v^2)}$$

for $v = 0.3$, $a_2 = 0.68\rho$. Thus the nucleating microcrack must be approximately the size of the macroscopic crack-tip radius.

This derivation contains a number of simplifying assumptions. The notch-tip stress computed from Equation (2.11) is assumed to act uniformly ahead of the notch, in the region of the microcrack; the actual stress would decay away from the notch tip. Also, this derivation neglects free boundary effects from the tip of the macroscopic notch.

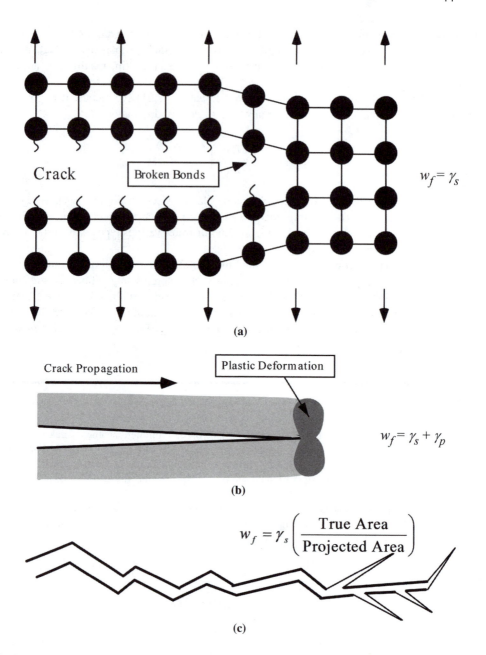

FIGURE 2.6 Crack propagation in various types of materials, with the corresponding fracture energy. (a) ideally brittle material, (b) quasi-brittle elastic-plastic material and, (c) brittle material with crack meandering and branching.

2.4 THE ENERGY RELEASE RATE

In 1956, Irwin [6] proposed an energy approach for fracture that is essentially equivalent to the Griffith model, except that Irwin's approach is in a form that is more convenient for solving engineering problems. Irwin defined an *energy release rate* $\mathcal{G}$, which is a measure of the energy available for an increment of crack extension:

$$\mathcal{G} = -\frac{d\Pi}{dA} \qquad (2.23)$$

The term *rate*, as it is used in this context, does not refer to a derivative with respect to time; $\mathcal{G}$ is the rate of change in potential energy with the crack area. Since $\mathcal{G}$ is obtained from the derivative of a potential, it is also called the *crack extension force* or the *crack driving force*. According to Equation (2.18a), the energy release rate for a wide plate in plane stress with a crack of length $2a$ (Figure 2.3) is given by

$$\mathcal{G} = \frac{\pi\sigma^2 a}{E} \tag{2.24}$$

Referring to the previous section, crack extension occurs when $\mathcal{G}$ reaches a critical value, i.e.,

$$\mathcal{G}_c = \frac{dW_s}{dA} = 2w_f \tag{2.25}$$

where $\mathcal{G}_c$ is a measure of the *fracture toughness* of the material.

The potential energy of an elastic body, Π, is defined as follows:

$$\Pi = U - F \tag{2.26}$$

where U is the strain energy stored in the body and F is the work done by external forces.

Consider a cracked plate that is dead loaded, as illustrated in Figure 2.7. Since the load is fixed at P, the structure is said to be *load controlled*. For this case

$$F = P\Delta$$

and

$$U = \int_0^\Delta P d\Delta = \frac{P\Delta}{2}$$

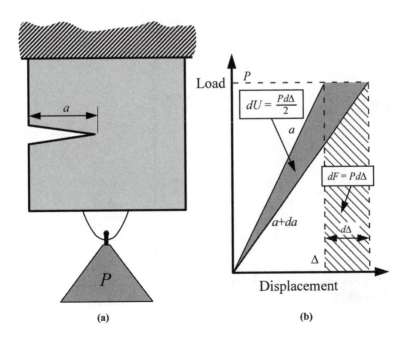

FIGURE 2.7 Cracked plate at a fixed load P.

Therefore

$$\Pi = -U$$

and

$$\mathcal{G} = \frac{1}{B}\left(\frac{dU}{da}\right)_P = \frac{P}{2B}\left(\frac{d\Delta}{da}\right)_P \tag{2.27}$$

When displacement is fixed (Figure 2.8), the plate is *displacement controlled*; $F = 0$ and $\Pi = U$. Thus

$$\mathcal{G} = -\frac{1}{B}\left(\frac{dU}{da}\right)_\Delta = -\frac{\Delta}{2B}\left(\frac{dP}{da}\right)_\Delta \tag{2.28}$$

It is convenient at this point to introduce the compliance, which is the inverse of the plate stiffness:

$$C = \frac{\Delta}{P} \tag{2.29}$$

By substituting Equation (2.29) into Equation (2.27) and Equation (2.28) it can be shown that

$$\mathcal{G} = \frac{P^2}{2B}\frac{dC}{da} \tag{2.30}$$

for both load control and displacement control. Therefore, the energy release rate, as defined in Equation (2.23), is the same for load control and displacement control. Also

$$\left(\frac{dU}{da}\right)_P = -\left(\frac{dU}{da}\right)_\Delta \tag{2.31}$$

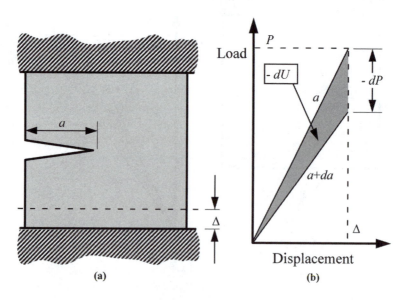

(a)   (b)

FIGURE 2.8 Cracked plate at a fixed displacement Δ.

Equation (2.31) is demonstrated graphically in Figure 2.7(b) and Figure 2.8(b). In load control, a crack extension *da* results in a net *increase* in strain energy because of the contribution of the external force *P*:

$$(dU)_P = Pd\Delta - \frac{Pd\Delta}{2} = \frac{Pd\Delta}{2}$$

When displacement is fixed, $dF = 0$ and the strain energy *decreases*:

$$(dU)_\Delta = \frac{\Delta dP}{2}$$

where *dP* is negative. As can be seen in Figure 2.7(b) and Figure 2.8(b), the absolute values of these energies differ by the amount $dPd\Delta/2$, which is negligible. Thus

$$(dU)_P = -(dU)_\Delta$$

for an increment of crack growth at a given *P* and Δ.

EXAMPLE 2.2

Determine the energy release rate for a double cantilever beam (DCB) specimen (Figure 2.9)

Solution: From beam theory

$$\frac{\Delta}{2} = \frac{Pa^3}{3EI} \qquad \text{where} \qquad I = \frac{Bh^3}{12}$$

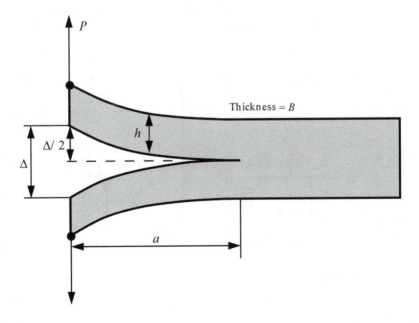

FIGURE 2.9 Double cantilever beam (DCB) specimen.

The elastic compliance is given by

$$C = \frac{\Delta}{P} = \frac{2\,a^3}{3\,E\,I}$$

Substituting C into Equation (2.30) gives

$$\mathcal{G} = \frac{P^2 a^2}{B\,E\,I} = \frac{12\,P^2\,a^2}{B^2\,h^3\,E}$$

2.5 INSTABILITY AND THE R CURVE

Crack extension occurs when $\mathcal{G} = 2w_f$; but crack growth may be stable or unstable, depending on how $\mathcal{G}$ and w_f vary with crack size. To illustrate stable and unstable behavior, it is convenient to replace $2w_f$ with R, the material resistance to crack extension. A plot of R vs. crack extension is called a *resistance curve* or *R curve*. The corresponding plot of $\mathcal{G}$ vs. crack extension is the *driving force curve*.

Consider a wide plate with a through crack of initial length $2a_o$ (Figure 2.3). At a fixed remote stress σ, the energy release rate varies linearly with crack size (Equation (2.24)). Figure 2.10 shows schematic driving force vs. R curves for two types of material behavior.

The first case, Figure 2.10(a), shows a flat R curve, where the material resistance is constant with crack growth. When the stress is σ_1, the crack is stable. Fracture occurs when the stress reaches σ_2; the crack propagation is unstable because the driving force increases with crack growth, but the material resistance remains constant.

Figure 2.10(b) illustrates a material with a rising R curve. The crack grows a small amount when the stress reaches σ_2, but cannot grow further unless the stress increases. When the stress is fixed at σ_2, the driving force increases at a slower rate than R. Stable crack growth continues as the stress increases to σ_3. Finally, when the stress reaches σ_4, the driving force curve is tangent to the R curve. The plate is unstable with further crack growth because the rate of change in the driving force exceeds the slope of the R curve.

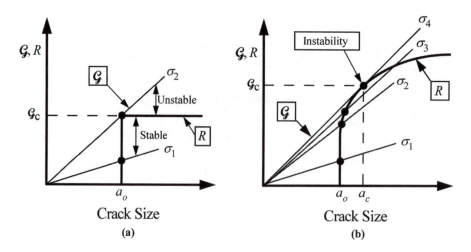

FIGURE 2.10 Schematic driving force vs. R curve diagrams (a) flat R curve and (b) rising R curve.

The conditions for *stable* crack growth can be expressed as follows:

$$\mathcal{G} = R \tag{2.32a}$$

and

$$\frac{d\mathcal{G}}{da} \leq \frac{dR}{da} \tag{2.32b}$$

Unstable crack growth occurs when

$$\frac{d\mathcal{G}}{da} > \frac{dR}{da} \tag{2.33}$$

When the resistance curve is flat, as in Figure 2.10(a), one can define a critical value of energy release rate $\mathcal{G}_c$, unambiguously. A material with a rising R curve, however, cannot be uniquely characterized with a single toughness value. According to Equation (2.33) a flawed structure fails when the driving force curve is tangent to the R curve, but this point of tangency depends on the shape of the driving force curve, which depends on the configuration of the structure. The driving force curve for the through crack configuration is linear, but $\mathcal{G}$ in the DCB specimen (Example 2.2) varies with a^2; these two configurations would have different $\mathcal{G}_c$ values for a given R curve.

Materials with rising R curves can be characterized by the value of $\mathcal{G}$ at the initiation of the crack growth. Although the initiation toughness is usually not sensitive to structural geometry, there are other problems with this measurement. It is virtually impossible to determine the precise moment of crack initiation in most materials. An engineering definition of initiation, analogous to the 0.2% offset yield strength in tensile tests, is usually required. Another limitation of initiation toughness is that it characterizes only the onset of crack growth; it provides no information on the shape of the R curve.

2.5.1 REASONS FOR THE R CURVE SHAPE

Some materials exhibit a rising R curve, while the R curve for other materials is flat. The shape of the R curve depends on the material behavior and, to a lesser extent, on the configuration of the cracked structure.

The R curve for an ideally brittle material is flat because the surface energy is an invariant material property. When nonlinear material behavior accompanies fracture, however, the R curve can take on a variety of shapes. For example, ductile fracture in metals usually results in a rising R curve; a plastic zone at the tip of the crack increases in size as the crack grows. The driving force must increase in such materials to maintain the crack growth. If the cracked body is infinite (i.e., if the plastic zone is small compared to the relevant dimensions of the body) the plastic zone size and R eventually reach steady-state values, and the R curve becomes flat with further growth (see Section 3.5.2).

Some materials can display a falling R curve. When a metal fails by cleavage, for example, the material resistance is provided by the surface energy and local plastic dissipation, as illustrated in Figure 2.6(b). The R curve would be relatively flat if the crack growth were stable. However, cleavage propagation is normally unstable; the material near the tip of the growing crack is subject to very high strain rates, which suppress plastic deformation. Thus, the resistance of a rapidly growing cleavage crack is less than the initial resistance at the onset of fracture.

The size and geometry of the cracked structure can exert some influence on the shape of the *R* curve. A crack in a thin sheet tends to produce a steeper *R* curve than a crack in a thick plate because there is a low degree of stress triaxiality at the crack tip in the thin sheet, while the material near the tip of the crack in the thick plate may be in plane strain. The *R* curve can also be affected if the growing crack approaches a free boundary in the structure. Thus, a wide plate may exhibit a somewhat different crack growth resistance behavior than a narrow plate of the same material.

Ideally, the *R* curve, as well as other measures of fracture toughness, should be a property only of the material and not depend on the size or shape of the cracked body. Much of fracture mechanics is predicated on the assumption that fracture toughness is a material property. Configurational effects can occur, however. A practitioner of fracture mechanics should be aware of these effects and their potential influence on the accuracy of an analysis. This issue is explored in detail in Section 2.10, Section 3.5, and Section 3.6.

2.5.2 LOAD CONTROL VS. DISPLACEMENT CONTROL

According to Equation (2.32) and Equation (2.33), the stability of crack growth depends on the rate of change in $\mathcal{G}$, i.e., the second derivative of potential energy. Although the driving force $\mathcal{G}$ is the same for both load control and displacement control, the *rate of change* of the driving force curve depends on how the structure is loaded.

Displacement control tends to be more stable than load control. With some configurations, the driving force actually decreases with crack growth in displacement control. A typical example is illustrated in Figure 2.11.

Referring to Figure 2.11, consider a cracked structure subjected to a load P_3 and a displacement Δ_3. If the structure is load controlled, it is at the point of instability where the driving force curve is tangent to the *R* curve. In displacement control, however, the structure is stable because the driving force decreases with crack growth; the displacement must be increased for further crack growth.

When an *R* curve is determined experimentally, the specimen is usually tested in displacement control, or as near to pure displacement control as is possible in the test machine. Since most of the common test specimen geometries exhibit falling driving force curves in displacement control,

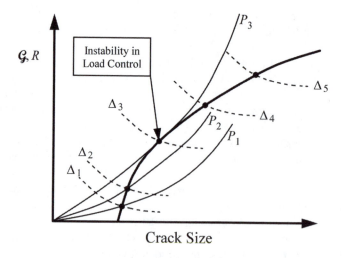

Crack Size

FIGURE 2.11 Schematic driving force/R curve diagram that compares load control and displacement control.

it is possible to obtain a significant amount of stable crack growth. If an instability occurs during the test, the R curve cannot be defined beyond the point of ultimate failure.

EXAMPLE 2.3

Evaluate the relative stability of a DCB specimen (Figure 2.9) in load control and displacement control.

Solution: From the result derived in Example 2.2, the slope of the driving force curve in load control is given by

$$\left(\frac{d\mathcal{G}}{da}\right)_P = \frac{2P^2 a}{BEI} = \frac{2\mathcal{G}}{a}$$

In order to evaluate displacement control, it is necessary to express $\mathcal{G}$ in terms of Δ and a. From beam theory, load is related to displacement as follows:

$$P = \frac{3\Delta EI}{2a^3}$$

Substituting the above equation into expression for energy release rate gives

$$\mathcal{G} = \frac{9\Delta^2 EI}{4Ba^4}$$

Thus

$$\left(\frac{d\mathcal{G}}{da}\right)_\Delta = -\frac{9\Delta^2 EI}{Ba^5} = -\frac{4\mathcal{G}}{a}$$

Therefore, the driving force increases with crack growth in load control and decreases in displacement control. For a flat R curve, crack growth in load control is always unstable, while displacement control is always stable.

2.5.3 STRUCTURES WITH FINITE COMPLIANCE

Most real structures are subject to conditions between load control and pure displacement control. This intermediate situation can be schematically represented by a spring in series with the flawed structure (Figure 2.12). The structure is fixed at a constant remote displacement Δ_T; the spring represents the system compliance C_m. Pure displacement control corresponds to an infinitely stiff spring, where $C_m = 0$. Load control (dead loading) implies an infinitely soft spring, i.e., $C_m = \infty$.

When the system compliance is finite, the point of fracture instability obviously lies somewhere between the extremes of pure load control and pure displacement control. However, determining the precise point of instability requires a rather complex analysis.

At the moment of instability, the following conditions are satisfied:

$$\mathcal{G} = R \tag{2.34a}$$

UNIVERSITY OF SHEFFIELD LIBRARY

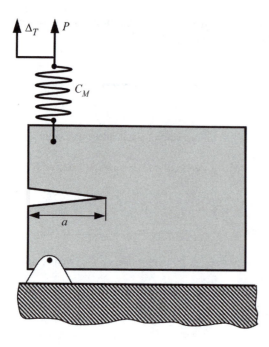

FIGURE 2.12 A cracked structure with finite compliance, represented schematically by a spring in series.

and

$$\left(\frac{d\mathcal{G}}{da}\right)_{\Delta_T} = \frac{dR}{da} \qquad (2.34b)$$

The left side of Equation (2.34b) is given by Hutchinson and Paris [7]

$$\left(\frac{d\mathcal{G}}{da}\right)_{\Delta_T} = \left(\frac{\partial\mathcal{G}}{\partial a}\right)_P - \left(\frac{\partial\mathcal{G}}{\partial P}\right)_a \left(\frac{\partial\Delta}{\partial a}\right)_P \left[C_m + \left(\frac{\partial\Delta}{\partial P}\right)_a \right]^{-1} \qquad (2.35)$$

Equation (2.35) is derived in Appendix 2.2.

2.6   STRESS ANALYSIS OF CRACKS

For certain cracked configurations subjected to external forces, it is possible to derive closed-form expressions for the stresses in the body, assuming isotropic linear elastic material behavior. Westergaard [8], Irwin [9], Sneddon [10], and Williams [11] were among the first to publish such solutions. If we define a polar coordinate axis with the origin at the crack tip (Figure 2.13), it can be shown that the stress field in any linear elastic cracked body is given by

$$\sigma_{ij} = \left(\frac{k}{\sqrt{r}}\right) f_{ij}(\theta) + \sum_{m=0}^{\infty} A_m r^{\frac{m}{2}} g_{ij}^{(m)}(\theta) \qquad (2.36)$$

where

  σ_{ij} = stress tensor
  r and θ are as defined in Figure 2.13
  k = constant
  f_{ij} = dimensionless function of θ in the leading term

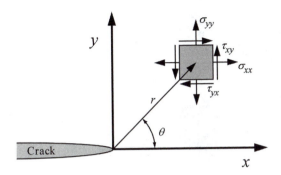

FIGURE 2.13 Definition of the coordinate axis ahead of a crack tip. The z direction is normal to the page.

For the higher-order terms, A_m is the amplitude and $g_{ij}^{(m)}$ is a dimensionless function of θ for the mth term. The higher-order terms depend on geometry, but the solution for any given configuration contains a leading term that is proportional to $1/\sqrt{r}$. As $r \to 0$, the leading term approaches infinity, but the other terms remain finite or approach zero. Thus, stress near the crack tip varies with $1/\sqrt{r}$, regardless of the configuration of the cracked body. It can also be shown that displacement near the crack tip varies with $\sqrt{r}$. Equation (2.36) describes a stress *singularity*, since stress is asymptotic to $r = 0$. The basis of this relationship is explored in more detail in Appendix 2.3.

There are three types of loading that a crack can experience, as Figure 2.14 illustrates. Mode I loading, where the principal load is applied normal to the crack plane, tends to open the crack. Mode II corresponds to in-plane shear loading and tends to slide one crack face with respect to the other. Mode III refers to out-of-plane shear. A cracked body can be loaded in any one of these modes, or a combination of two or three modes.

2.6.1 THE STRESS INTENSITY FACTOR

Each mode of loading produces the $1/\sqrt{r}$ singularity at the crack tip, but the proportionality constants k and f_{ij} depend on the mode. It is convenient at this point to replace k by the *stress intensity factor* K, where $K = k\sqrt{2\pi}$. The stress intensity factor is usually given a subscript to denote the mode of loading, i.e., K_I, K_{II}, or K_{III}. Thus, the stress fields ahead of a crack tip in an isotropic linear elastic

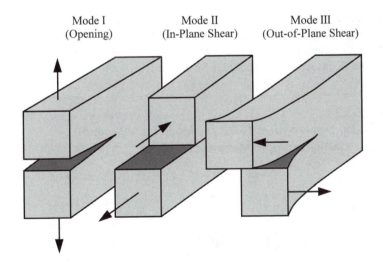

FIGURE 2.14 The three modes of loading that can be applied to a crack.

TABLE 2.1
Stress Fields Ahead of a Crack Tip for Mode I and Mode II in a Linear Elastic, Isotropic Material

	Mode I	Mode II
σ_{xx}	$\dfrac{K_I}{\sqrt{2\pi r}}\cos\left(\dfrac{\theta}{2}\right)\left[1-\sin\left(\dfrac{\theta}{2}\right)\sin\left(\dfrac{3\theta}{2}\right)\right]$	$-\dfrac{K_{II}}{\sqrt{2\pi r}}\sin\left(\dfrac{\theta}{2}\right)\left[2+\cos\left(\dfrac{\theta}{2}\right)\cos\left(\dfrac{3\theta}{2}\right)\right]$
σ_{yy}	$\dfrac{K_I}{\sqrt{2\pi r}}\cos\left(\dfrac{\theta}{2}\right)\left[1+\sin\left(\dfrac{\theta}{2}\right)\sin\left(\dfrac{3\theta}{2}\right)\right]$	$\dfrac{K_{II}}{\sqrt{2\pi r}}\sin\left(\dfrac{\theta}{2}\right)\cos\left(\dfrac{\theta}{2}\right)\cos\left(\dfrac{3\theta}{2}\right)$
τ_{xy}	$\dfrac{K_I}{\sqrt{2\pi r}}\cos\left(\dfrac{\theta}{2}\right)\sin\left(\dfrac{\theta}{2}\right)\cos\left(\dfrac{3\theta}{2}\right)$	$\dfrac{K_{II}}{\sqrt{2\pi r}}\cos\left(\dfrac{\theta}{2}\right)\left[1-\sin\left(\dfrac{\theta}{2}\right)\sin\left(\dfrac{3\theta}{2}\right)\right]$
σ_{zz}	0 (Plane stress) $\nu(\sigma_{xx}+\sigma_{yy})$ (Plane strain)	0 (Plane stress) $\nu(\sigma_{xx}+\sigma_{yy})$ (Plane strain)
τ_{xz}, τ_{yz}	0	0

Note: υ is Poisson's ratio.

material can be written as

$$\lim_{r\to 0}\sigma_{ij}^{(I)}=\frac{K_I}{\sqrt{2\pi r}}f_{ij}^{(I)}(\theta) \tag{2.37a}$$

$$\lim_{r\to 0}\sigma_{ij}^{(II)}=\frac{K_{II}}{\sqrt{2\pi r}}f_{ij}^{(II)}(\theta) \tag{2.37b}$$

$$\lim_{r\to 0}\sigma_{ij}^{(III)}=\frac{K_{III}}{\sqrt{2\pi r}}f_{ij}^{(III)}(\theta) \tag{2.37c}$$

for Modes I, II, and III, respectively. In a mixed-mode problem (i.e., when more than one loading mode is present), the individual contributions to a given stress component are additive:

$$\sigma_{ij}^{(total)}=\sigma_{ij}^{(I)}+\sigma_{ij}^{(II)}+\sigma_{ij}^{(III)} \tag{2.38}$$

Equation (2.38) stems from the principle of linear superposition.

Detailed expressions for the singular stress fields for Mode I and Mode II are given in Table 2.1. Displacement relationships for Mode I and Mode II are listed in Table 2.2. Table 2.3 lists the nonzero stress and displacement components for Mode III.

TABLE 2.2
Crack-Tip Displacement Fields for Mode I and Mode II (Linear Elastic, Isotropic Material)

	Mode I	Mode II
u_x	$\dfrac{K_I}{2\mu}\sqrt{\dfrac{r}{2\pi}}\cos\left(\dfrac{\theta}{2}\right)\left[\kappa-1+2\sin^2\left(\dfrac{\theta}{2}\right)\right]$	$\dfrac{K_{II}}{2\mu}\sqrt{\dfrac{r}{2\pi}}\sin\left(\dfrac{\theta}{2}\right)\left[\kappa+1+2\cos^2\left(\dfrac{\theta}{2}\right)\right]$
u_y	$\dfrac{K_I}{2\mu}\sqrt{\dfrac{r}{2\pi}}\sin\left(\dfrac{\theta}{2}\right)\left[\kappa+1-2\cos^2\left(\dfrac{\theta}{2}\right)\right]$	$-\dfrac{K_{II}}{2\mu}\sqrt{\dfrac{r}{2\pi}}\cos\left(\dfrac{\theta}{2}\right)\left[\kappa-1-2\sin^2\left(\dfrac{\theta}{2}\right)\right]$

Note: μ is the shear modulus. $\kappa = 3-4\nu$ (plane strain) and $\kappa=(3-\nu)/(1+\nu)$ (plane stress).

TABLE 2.3
Nonzero Stress and Displacement Components in Mode III
(Linear Elastic, Isotropic Material)

$$\tau_{xz} = -\frac{K_{III}}{\sqrt{2\pi r}} \sin\left(\frac{\theta}{2}\right)$$

$$\tau_{yz} = \frac{K_{III}}{\sqrt{2\pi r}} \cos\left(\frac{\theta}{2}\right)$$

$$u_z = \frac{2K_{III}}{\mu} \sqrt{\frac{r}{2\pi}} \sin\left(\frac{\theta}{2}\right)$$

Consider the Mode I singular field on the crack plane, where $\theta = 0$. According to Table 2.1, the stresses in the x and y direction are equal:

$$\sigma_{xx} = \sigma_{yy} = \frac{K_I}{\sqrt{2\pi r}} \tag{2.39}$$

When $\theta = 0$, the shear stress is zero, which means that the crack plane is a principal plane for pure Mode I loading. Figure 2.15 is a schematic plot of σ_{yy}, the stress normal to the crack plane vs. distance from the crack tip. Equation (2.39) is valid only near the crack tip, where the $1/\sqrt{r}$ singularity dominates the stress field. Stresses far from the crack tip are governed by the remote boundary conditions. For example, if the cracked structure is subjected to a uniform remote tensile stress, σ_{yy} approaches a constant value σ^∞. We can define a *singularity-dominated zone* as the region where the equations in Table 2.1 to Table 2.3 describe the crack-tip fields.

The stress intensity factor defines the amplitude of the crack-tip singularity. That is, stresses near the crack tip increase in proportion to K. Moreover, the stress intensity factor completely defines the crack tip conditions; if K is known, it is possible to solve for all components of stress, strain, and displacement as a function of r and θ. This single-parameter description of crack tip conditions turns out to be one of the most important concepts in fracture mechanics.

2.6.2 Relationship between K and Global Behavior

In order for the stress intensity factor to be useful, one must be able to determine K from remote loads and the geometry. Closed-form solutions for K have been derived for a number of simple

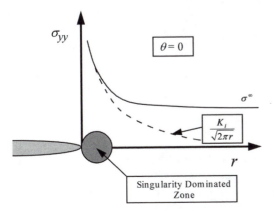

FIGURE 2.15 Stress normal to the crack plane in Mode I.

configurations. For more complex situations, the stress intensity factor can be estimated by experiment or numerical analysis (see Chapter 12).

One configuration for which a closed-form solution exists is a through crack in an infinite plate subjected to a remote tensile stress (Figure 2.3). Since the remote stress σ is perpendicular to the crack plane, the loading is pure Mode I. Linear elastic bodies must undergo proportional stressing, i.e., all the stress components at all locations increase in proportion to the remotely applied forces. Thus the crack-tip stresses must be proportional to the remote stress, and $K_I \propto \sigma$. According to Equation (2.37), stress intensity has units of $\text{stress} \bullet \sqrt{\text{length}}$. Since the only relevant length scale in Figure 2.3 is the crack size, the relationship between K_I and the global conditions must have the following form:

$$K_I = O\left(\sigma \sqrt{a}\right) \qquad (2.40)$$

The actual solution, which is derived in Appendix 2.3, is given by

$$K_I = \sigma \sqrt{\pi a} \qquad (2.41)$$

Thus the amplitude of the crack-tip singularity for this configuration is proportional to the remote stress and the square root of the crack size. The stress intensity factor for Mode II loading of the plate in Figure 2.3 can be obtained by replacing σ in Equation (2.41) by the remotely applied shear stress (see Figure 2.18 and Equation (2.43) below).

A related solution is that for a semi-infinite plate with an edge crack (Figure 2.16). Note that this configuration can be obtained by slicing the plate in Figure 2.3 through the middle of the crack. The stress intensity factor for the edge crack is given by

$$K_I = 1.12\sigma \sqrt{\pi a} \qquad (2.42)$$

which is similar to Equation (2.41). The 12% increase in K_I for the edge crack is caused by different boundary conditions at the free edge. As Figure 2.17 illustrates, the edge crack opens more because it is less restrained than the through crack, which forms an elliptical shape when loaded.

Consider a through crack in an infinite plate where the normal to the crack plane is oriented at an angle β with the stress axis (Figure 2.18(a)). If $\beta \neq 0$, the crack experiences combined Mode I and Mode II loading; $K_{III} = 0$ as long as the stress axis and the crack normal both lie in the plane of the plate. If we redefine the coordinate axis to coincide with the crack orientation (Figure 2.18(b)), we see that the applied stress can be resolved into normal and shear components. The stress normal to the crack plane, $\sigma_{y'y'}$, produces pure Mode I loading, while $\tau_{x'y'}$ applies Mode II loading to the crack. The stress intensity factors for the plate in Figure 2.18 can be inferred by relating $\sigma_{y'y'}$ and $\tau_{x'y'}$ to σ and β through Mohr's circle:

$$K_I = \sigma_{y'y'} \sqrt{\pi a}$$

$$= \sigma \cos^2(\beta) \sqrt{\pi a} \qquad (2.43a)$$

and

$$K_{II} = \tau_{x'y'} \sqrt{\pi a}$$

$$= \sigma \sin(\beta)\cos(\beta) \sqrt{\pi a} \qquad (2.43b)$$

Note that Equation (2.43) reduces to the pure Mode I solution when $\beta = 0$. The maximum K_{II} occurs at $\beta = 45°$, where the shear stress is also at a maximum. Section 2.11 addresses fracture under mixed mode conditions.

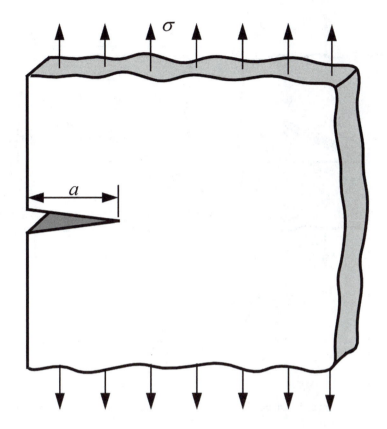

FIGURE 2.16 Edge crack in a semi-infinite plate subject to a remote tensile stress.

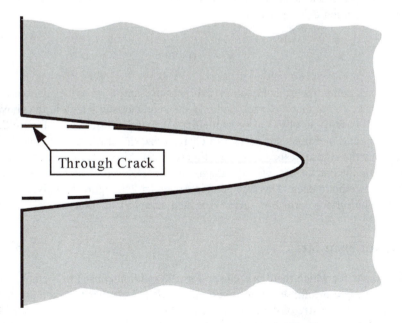

FIGURE 2.17 Comparison of crack-opening displacements for an edge crack and through crack. The edge crack opens wider at a given stress, resulting in a stress intensity that is 12% higher.

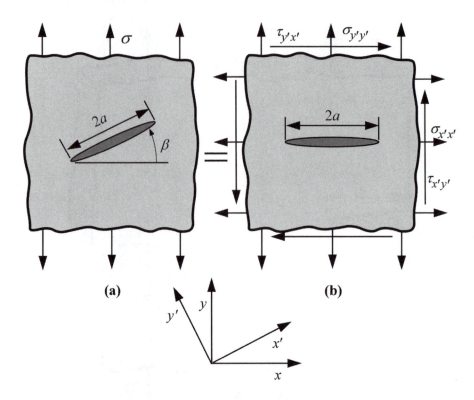

FIGURE 2.18 Through crack in an infinite plate for the general case where the principal stress is not perpendicular to the crack plane.

The penny-shaped crack in an infinite medium (Figure 2.4) is another configuration for which a closed-form K_I solution exists [11]:

$$K_I = \frac{2}{\pi} \sigma \sqrt{\pi a} \qquad (2.44)$$

where a is the crack radius. Note that Equation (2.44) has the same form as the previous relationships for a through crack, except that the crack radius is the characteristic length in the above equation. The more general case of an elliptical or semielliptical flaw is illustrated in Figure 2.19. In this instance, two length dimensions are needed to characterize the crack size: $2c$ and $2a$, the major and minor axes of the ellipse, respectively (see Figure 2.19). Also, when $a < c$, the stress intensity factor varies along the crack front, with the maximum K_I at $\phi = 90°$. The flaw shape parameter Q is obtained from an elliptic integral, as discussed in Appendix 2.4. Figure 2.19 gives an approximate solution for Q. The surface correction factor λ_s is also an approximation.

2.6.3 Effect of Finite Size

Most configurations for which there is a closed-form K solution consist of a crack with a simple shape (e.g., a rectangle or ellipse) in an infinite plate. Stated another way, the crack dimensions are small compared to the size of the plate; the crack-tip conditions are not influenced by external boundaries. As the crack size increases, or as the plate dimensions decrease, the outer boundaries begin to exert an influence on the crack tip. In such cases, a closed-form stress intensity solution is usually not possible.

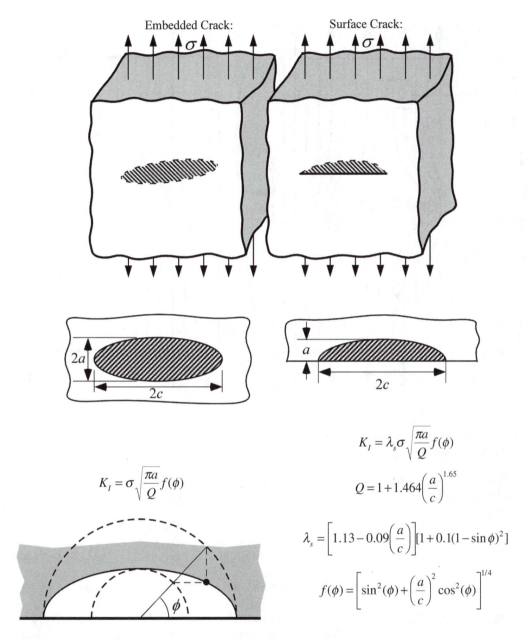

FIGURE 2.19 Mode I stress intensity factors for elliptical and semielliptical cracks. These solutions are valid only as long as the crack is small compared to the plate dimensions and $a \leq c$.

Consider a cracked plate subjected to a remote tensile stress. Figure 2.20 schematically illustrates the effect of finite width on the crack tip stress distribution, which is represented by lines of force; the local stress is proportional to the spacing between lines of force. Since a tensile stress cannot be transmitted through a crack, the lines of force are diverted around the crack, resulting in a local stress concentration. In the infinite plate, the line of force at a distance W from the crack centerline has force components in the x and y directions. If the plate width is restricted to $2W$, the x force must be zero on the free edge; this boundary condition causes the lines of force to be compressed, which results in a higher stress intensification at the crack tip.

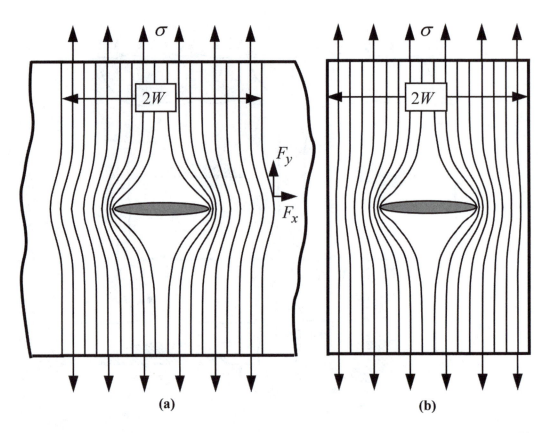

FIGURE 2.20 Stress concentration effects due to a through crack in finite and infinite width plates: (a) infinite plate and (b) finite plate.

One technique to approximate the finite width boundary condition is to assume a periodic array of collinear cracks in an infinite plate (Figure 2.21). The Mode I stress intensity factor for this situation is given by

$$K_I = \sigma\sqrt{\pi a}\left[\frac{2W}{\pi a}\tan\left(\frac{\pi a}{2W}\right)\right]^{1/2} \tag{2.45}$$

The stress intensity approaches the infinite plate value as a/W approaches zero; K_I is asymptotic to $a/W = 1$.

More accurate solutions for a through crack in a finite plate have been obtained from finite-element analysis; solutions of this type are usually fit to a polynomial expression. One such solution [12] is given by

$$K_I = \sigma\sqrt{\pi a}\left[\sec\left(\frac{\pi a}{2W}\right)^{1/2}\right]\left[1 - 0.025\left(\frac{a}{W}\right)^2 + 0.06\left(\frac{a}{w}\right)^4\right] \tag{2.46}$$

Figure 2.22 compares the finite width corrections in Equation (2.45) and Equation (2.46). The secant term (without the polynomial term) in Equation (2.46) is also plotted. Equation (2.45) agrees with the finite-element solution to within 7% for $a/W < 0.6$. The secant correction is much closer to the finite element solution; the error is less than 2% for $a/W < 0.9$. Thus, the polynomial term in Equation (2.46) contributes little and can be neglected in most cases.

Table 2.4 lists stress intensity solutions for several common configurations. These K_I solutions are plotted in Figure 2.23. Several handbooks devoted solely to stress intensity solutions have been published [12–14].

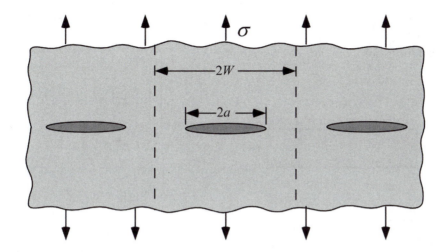

FIGURE 2.21 Collinear cracks in an infinite plate subject to remote tension.

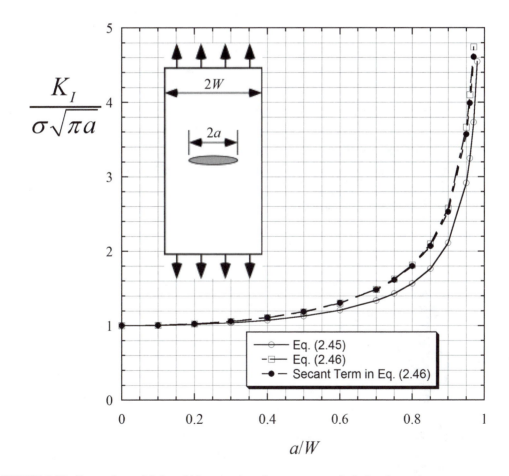

FIGURE 2.22 Comparison of finite width corrections for a center-cracked plate in tension.

TABLE 2.4
K_I Solutions for Common Test Specimens[a]

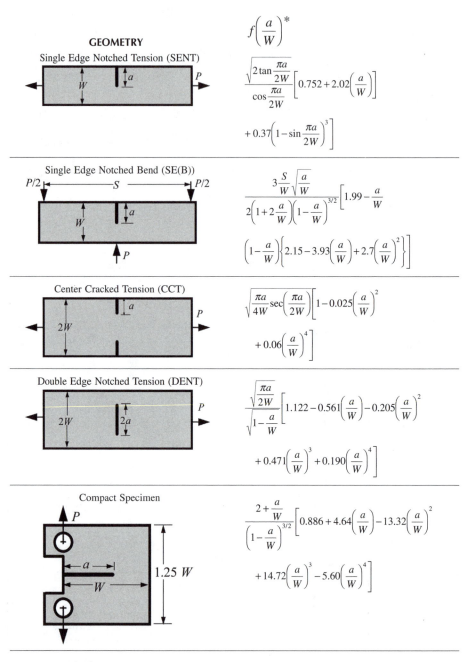

$$f\left(\frac{a}{W}\right)^*$$

GEOMETRY

Single Edge Notched Tension (SENT)

$$\frac{\sqrt{2\tan\dfrac{\pi a}{2W}}}{\cos\dfrac{\pi a}{2W}}\left[0.752+2.02\left(\frac{a}{W}\right)\right]$$

$$+0.37\left(1-\sin\frac{\pi a}{2W}\right)^3\right]$$

Single Edge Notched Bend (SE(B))

$$\frac{3\dfrac{S}{W}\sqrt{\dfrac{a}{W}}}{2\left(1+2\dfrac{a}{W}\right)\left(1-\dfrac{a}{W}\right)^{3/2}}\left[1.99-\frac{a}{W}\right.$$

$$\left(1-\frac{a}{W}\right)\left\{2.15-3.93\left(\frac{a}{W}\right)+2.7\left(\frac{a}{W}\right)^2\right\}\right]$$

Center Cracked Tension (CCT)

$$\sqrt{\frac{\pi a}{4W}\sec\left(\frac{\pi a}{2W}\right)}\left[1-0.025\left(\frac{a}{W}\right)^2\right.$$

$$+0.06\left(\frac{a}{W}\right)^4\right]$$

Double Edge Notched Tension (DENT)

$$\frac{\sqrt{\dfrac{\pi a}{2W}}}{\sqrt{1-\dfrac{a}{W}}}\left[1.122-0.561\left(\frac{a}{W}\right)-0.205\left(\frac{a}{W}\right)^2\right.$$

$$+0.471\left(\frac{a}{W}\right)^3+0.190\left(\frac{a}{W}\right)^4\right]$$

Compact Specimen

$$\frac{2+\dfrac{a}{W}}{\left(1-\dfrac{a}{W}\right)^{3/2}}\left[0.886+4.64\left(\frac{a}{W}\right)-13.32\left(\frac{a}{W}\right)^2\right.$$

$$+14.72\left(\frac{a}{W}\right)^3-5.60\left(\frac{a}{W}\right)^4\right]$$

$$^*K_I=\frac{P}{B\sqrt{W}}f\left(\frac{a}{W}\right)\quad\text{where }B\text{ is the specimen thickness.}$$

[a] Taken from Tada, H., Paris, P.C., and Irwin, G.R., *The Stress Analysis of Cracks Handbook*. 2nd Ed., Paris Productions, St. Louis, MO, 1985.

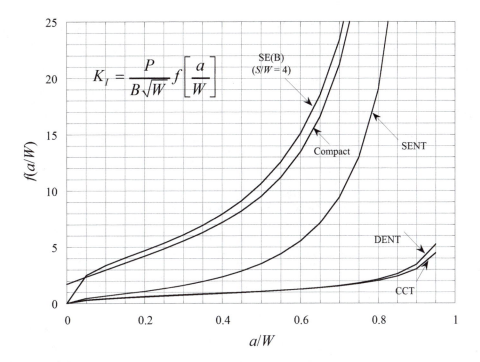

FIGURE 2.23 Plot of stress intensity solutions from Table 2.4.

Although stress intensity solutions are given in a variety of forms, K can always be related to the through crack (Figure 2.4) through the appropriate correction factor:

$$K_{(I,II,III)} = Y\sigma\sqrt{\pi a} \qquad (2.47)$$

where

σ = characteristic stress
a = characteristic crack dimension
Y = dimensionless constant that depends on the geometry and the mode of loading

EXAMPLE 2.4

Show that the K_I solution for the single edge notched tensile panel reduces to Equation (2.42) when $a \ll W$.

Solution: All of the K_I expressions in Table 2.4 are of the form:

$$K_I = \frac{P}{B\sqrt{W}} f\left(\frac{a}{w}\right)$$

where

P = applied force
B = plate thickness
$f(a/W)$ = dimensionless function

The above equation can be expressed in the form of Equation (2.47):

$$\frac{P}{B\sqrt{W}}\,f\!\left(\frac{a}{w}\right)=\frac{P}{BW}\,f\!\left(\frac{a}{w}\right)\sqrt{\frac{W}{\pi a}}\sqrt{\pi a}=Y\sigma\sqrt{\pi a}$$

where

$$Y=f\!\left(\frac{a}{W}\right)\sqrt{\frac{W}{\pi a}}$$

In the limit of a small flaw, the geometry correction factor in Table 2.4 becomes

$$\lim_{a/W\to 0}f\!\left(\frac{a}{W}\right)=\sqrt{\frac{\pi a}{W}}(0.752+0.37)$$

Thus,

$$\lim_{a/W\to 0}f(Y)=1.12$$

2.6.4 PRINCIPLE OF SUPERPOSITION

For linear elastic materials, individual components of stress, strain, and displacement are additive. For example, two normal stresses in the x direction imposed by different external forces can be added to obtain the total σ_{xx}, but a normal stress cannot be summed with a shear stress. Similarly, stress intensity factors are additive as long as the mode of loading is consistent. That is

$$K_I^{(\text{total})}=K_I^{(A)}+K_I^{(B)}+K_I^{(C)}$$

but

$$K_{(\text{total})}\neq K_I+K_{II}+K_{III}$$

In many instances, the principle of superposition allows stress intensity solutions for complex configurations to be built from simple cases for which the solutions are well established. Consider, for example, an edge-cracked panel (Table 2.4) subject to combined membrane (axial) loading P_m, and three-point bending P_b. Since both types of loading impose pure Mode I conditions, the K_I values can be added:

$$K_I^{(\text{total})}=K_I^{(\text{membrane})}+K_I^{(\text{bending})}$$

$$=\frac{1}{B\sqrt{W}}\left[P_m f_m\!\left(\frac{a}{W}\right)+P_b f_b\!\left(\frac{a}{W}\right)\right] \tag{2.48}$$

where f_m and f_b are the geometry correction factors for membrane and bending loading, respectively, listed in Table 2.4 and plotted in Figure 2.23.

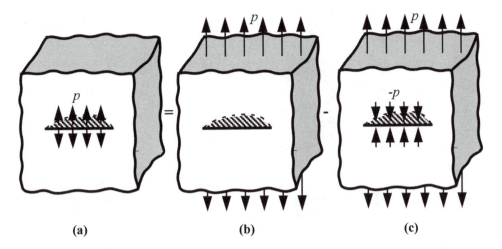

FIGURE 2.24 Determination of K_I for a semielliptical surface crack under internal pressure p by means of the principle of superposition.

EXAMPLE 2.5

Determine the stress intensity factor for a semielliptical surface crack subjected to an internal pressure p (Figure 2.24(a)).

Solution: The principle of superposition enables us to construct the solution from known cases. One relevant case is the semielliptical surface flaw under uniform remote tension p (Figure 2.24(b)). If we impose a uniform compressive stress $-p$ on the crack surface (Figure 2.24(c)), $K_I = 0$ because the crack faces close, and the plate behaves as if the crack were not present. The loading configuration of interest is obtained by subtracting the stresses in Figure 2.24(c) from those of Figure 2.24(b):

$$K_I^{(a)} = K_I^{(b)} - K_I^{(c)}$$

$$= \lambda_s p \sqrt{\frac{\pi a}{Q}} f(\phi) - 0 = \lambda_s p \sqrt{\frac{\pi a}{Q}} f(\phi)$$

Example 2.5 is a simple illustration of a more general concept, namely, stresses acting on the boundary (i.e., tractions) can be replaced with tractions that act on the crack face, such that the two loading configurations (boundary tractions vs. crack-face tractions) result in the same stress intensity factor. Consider an uncracked body subject to a boundary traction $P(x)$, as illustrated in Figure 2.25. This boundary traction results in a normal stress distribution $p(x)$ on Plane A-B. In order to confine the problem to Mode I, let us assume that no shear stresses act on Plane A-B. (This assumption is made only for the sake of simplicity; the basic principle can be applied to all three modes of loading.) Now assume that a crack that forms on Plane A-B and the boundary traction $P(x)$ remains fixed, as Figure 2.26(a) illustrates. If we remove the boundary traction and apply a traction $p(x)$ on the crack face (Figure 2.26(b)), the principle of superposition indicates that the applied K_I will be unchanged. That is

$$K_I^{(a)} = K_I^{(b)} - K_I^{(c)} = K_I^{(b)} \qquad \left(\text{since } K_I^{(c)} = 0\right)$$

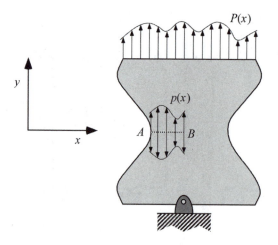

FIGURE 2.25 Uncracked body subject to an arbitrary boundary traction $P(x)$, which results in a normal stress distribution $p(x)$ acting on Plane A-B.

2.6.5 WEIGHT FUNCTIONS

When one performs an analysis to infer a stress intensity factor for a cracked body, the K value that is computed applies only to one particular set of boundary conditions; different loading conditions result in a different stress intensity factors for that geometry. It turns out, however, that the solution to one set of boundary conditions contains sufficient information to infer K for *any other* boundary conditions on that same geometry.

Consider two arbitrary loading conditions on an isotropic elastic cracked body in plane stress or plane strain. For now, we assume that both loadings are symmetric with respect to the crack plane, such that pure Mode I loading is achieved in each case. Suppose that we know the stress intensity factor for loading (1) and we wish to solve for $K_I^{(2)}$, the stress intensity factor for the second set of boundary conditions. Rice [15] showed that $K_I^{(1)}$ and $K_I^{(2)}$ are related as follows:

$$K_I^{(2)} = \frac{E'}{2K_I^{(1)}} \left[\int_\Gamma T_i \frac{\partial u_i^{(1)}}{\partial a} d\Gamma + \int_A F_i \frac{\partial u_i^{(1)}}{\partial a} dA \right] \tag{2.49}$$

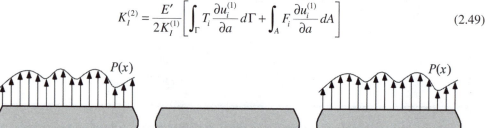

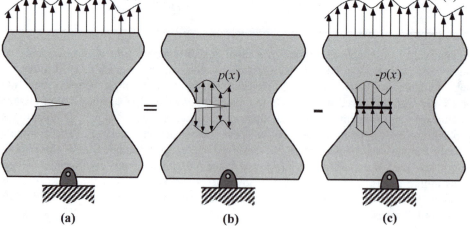

FIGURE 2.26 Application of superposition to replace a boundary traction $P(x)$ with a crack face traction $p(x)$ that results in the same K_I.

where Γ and A are the perimeter and area of the body, respectively, and u_i are the displacements in the x and y directions. Since loading systems (1) and (2) are arbitrary, it follows that $K_I^{(2)}$ cannot depend on $K_I^{(1)}$ and $u_i^{(1)}$. Therefore, the function

$$h(x_i) = \frac{E}{2K_I^{(1)}} \frac{\partial u_i^{(1)}}{\partial a} \tag{2.50}$$

where x_i represents the x and y coordinates, must be independent of the nature of loading system (1). Bueckner [16] derived a similar result to Equation (2.50) two years before Rice, and referred to h as a *weight function*.

Weight functions are first-order tensors that depend only on the geometry of the cracked body. Given the weight function for a particular configuration, it is possible to compute K_I from Equation (2.49) for any boundary condition. Moreover, the previous section invoked the principle of super-position to show that any loading configuration can be represented by appropriate tractions applied directly to the crack face. Thus K_I for a two-dimensional cracked body can be inferred from the following expression:

$$K_I = \int_{\Gamma_c} p(x)h(x)dx \tag{2.51}$$

where $p(x)$ is the crack face traction (equal to the normal stress acting on the crack plane when the body is uncracked) and Γ_c is the perimeter of the crack. The weight function $h(x)$ can be interpreted as the stress intensity resulting from a unit force applied to the crack face at x, and the above integral represents the superposition of the K_I values from discrete opening forces along the crack face.

EXAMPLE 2.6

Derive an expression for K_I for an arbitrary traction on the face of a through crack in an infinite plate.

Solution: We already know K_I for this configuration when a uniform tensile stress is applied:

$$K_I = \sigma\sqrt{\pi a}$$

where a is the half-crack length. From Equation (A2.43), the opening displacement of the crack faces in this case is given by

$$u_y = \pm \frac{2\sigma}{E'}\sqrt{x(2a-x)}$$

where the x-y coordinate axis is defined in Figure 2.27(a). Since the crack length is $2a$, we must differentiate u_y with respect to $2a$ rather than a:

$$\frac{\partial u_y}{\partial(2a)} = \pm \frac{2\sigma}{E}\sqrt{\frac{x}{2a-x}}$$

Thus, the weight function for this crack geometry is given by

$$h(x) = \pm \frac{1}{\sqrt{\pi a}}\sqrt{\frac{x}{2a-x}}$$

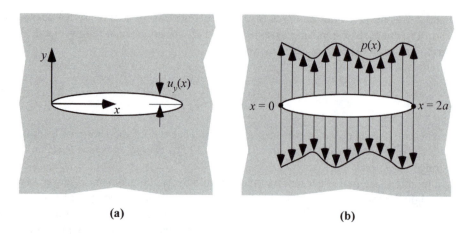

FIGURE 2.27 Through crack configuration analyzed in Example 2.6: (a) definition of coordinate axes and (b) arbitrary traction applied to crack faces.

If we apply a surface traction of $\pm p(x)$ on the crack faces, the Mode I stress intensity factor for the two crack tips is as follows:

$$K_{I(x=2a)} = \frac{1}{\sqrt{\pi a}} \int_0^{2a} p(x) \sqrt{\frac{x}{2a-x}}\, dx$$

$$K_{I(x=0)} = \frac{1}{\sqrt{\pi a}} \int_0^{2a} p(x) \sqrt{\frac{2a-x}{x}}\, dx$$

The weight function concept is not restricted to two-dimensional bodies, Mode I loading, or isotropic elastic materials. In their early work on weight functions, Rice [15] extended the theory to three dimensions, Bueckner [16] considered combined Mode I/II loading, and both allowed for anisotropy in the elastic properties. Subsequent researchers [17–22] have shown that the theory applies to all linear elastic bodies that contain an arbitrary number of cracks.

For mixed-mode problems, separate weight functions are required for each mode: h_I, h_{II}, and h_{III}. Since the stress intensity factors can vary along a three-dimensional crack front, the weight functions also vary along the crack front. That is

$$h_a = h_a(x_i, \eta) \qquad (2.52)$$

where $\alpha(= 1, 2, 3)$ indicates the mode of loading and η is the crack front position.

Given that any loading configuration in a cracked body can be represented by equivalent crack-face tractions, the general mixed-mode, three-dimensional formulation of the weight function approach can be expressed in the following form:

$$K_a(\eta) = \int_{S_c} T_i h_a(x_i, \eta)\, dS \qquad (2.53)$$

where T_i are the tractions assumed to act on the crack surface S_c.

See Chapter 9 for examples of practical applications of weight functions.

2.7 RELATIONSHIP BETWEEN K AND $\mathcal{G}$

Two parameters that describe the behavior of cracks have been introduced so far: the energy release rate and the stress intensity factor. The former parameter quantifies the net change in potential energy that accompanies an increment of crack extension; the latter quantity characterizes the stresses, strains, and displacements near the crack tip. The energy release rate describes global behavior, while K is a local parameter. For linear elastic materials, K and $\mathcal{G}$ are uniquely related.

For a through crack in an infinite plate subject to a uniform tensile stress (Figure 2.3), $\mathcal{G}$ and K_I are given by Equation (2.24) and Equation (2.41), respectively. Combining these two equations leads to the following relationship between $\mathcal{G}$ and K_I for plane stress:

$$\mathcal{G} = \frac{K_I^2}{E} \tag{2.54}$$

For plane strain conditions, E must be replaced by $E/(1 - v^2)$. To avoid writing separate expressions for plane stress and plane strain, the following notation will be adopted throughout this book:

$$E' = E \quad \text{for plane stress} \tag{2.55a}$$

and

$$E' = \frac{E}{1 - v^2} \quad \text{for plane strain} \tag{2.55b}$$

Thus the $\mathcal{G}$-K_I relationship for both plane stress and plane strain becomes

$$G = \frac{K_I^2}{E'} \tag{2.56}$$

Since Equation (2.24) and Equation (2.41) apply only to a through crack in an infinite plate, we have yet to prove that Equation (2.56) is a general relationship that applies to all configurations. Irwin [9] performed a crack closure analysis that provides such a proof. Irwin's analysis is presented below.

Consider a crack of initial length $a + \Delta a$ subject to Mode I loading, as illustrated in Figure 2.28(a). It is convenient in this case to place the origin a distance Δa behind the crack tip.

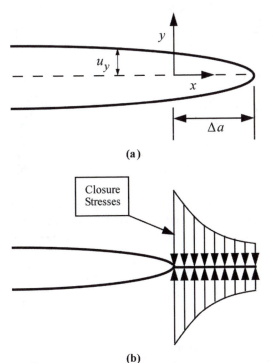

(a)

(b)

FIGURE 2.28 Application of closure stresses which shorten a crack by Δa.

Assume that the plate has a unit thickness. Let us now apply a compressive stress field to the crack faces between $x = 0$ and $x = \Delta a$ of sufficient magnitude to close the crack in this region. The work required to close the crack at the tip is related to the energy release rate:

$$\mathcal{G} = \lim_{\Delta a \to 0} \left(\frac{\Delta U}{\Delta a} \right)_{\text{fixed load}} \tag{2.57}$$

where ΔU is the work of crack closure, which is equal to the sum of contributions to work from $x = 0$ to $x = \Delta a$:

$$\Delta U = \int_{x=0}^{x=\Delta a} dU(x) \tag{2.58}$$

and the incremental work at x is equal to the area under the force-displacement curve:

$$dU(x) = 2 \times \frac{1}{2} F_y(x) u_y(x) = \sigma_{yy}(x) u_y(x) dx \tag{2.59}$$

The factor of 2 on the work is required because both crack faces are displaced an absolute distance $u_y(x)$. The crack-opening displacement u_y for Mode I is obtained from Table 2.2 by setting $\theta = \pi$.

$$u_y = \frac{(\kappa + 1) K_I(a + \Delta a)}{2\mu} \sqrt{\frac{\Delta a - x}{2\pi}} \tag{2.60}$$

where $K_I(a + \Delta a)$ denotes the stress intensity factor at the original crack tip. The normal stress required to close the crack is related to K_I for the shortened crack:

$$\sigma_{yy} = \frac{K_I(a)}{\sqrt{2\pi x}} \tag{2.61}$$

Combining Equation (2.57) to Equation (2.61) gives

$$\mathcal{G} = \lim_{\Delta a \to 0} \frac{(\kappa + 1) K_I(a) K_I(a + \Delta a)}{4\pi\mu\Delta a} \int_0^{\Delta a} \sqrt{\frac{\Delta a - x}{x}} dx$$

$$= \frac{(\kappa + 1) K_I^2}{8\mu} = \frac{K_I^2}{E'} \tag{2.62}$$

Thus, Equation (2.56) is a general relationship for Mode I. The above analysis can be repeated for other modes of loading; the relevant closure stress and displacement for Mode II are, respectively, τ_{yx} and u_x, and the corresponding quantities for Mode III are τ_{yz} and u_z. When all three modes of loading are present, the energy release rate is given by

$$\mathcal{G} = \frac{K_I^2}{E'} + \frac{K_{II}^2}{E'} + \frac{K_{III}^2}{2\mu} \tag{2.63}$$

Contributions to $\mathcal{G}$ from the three modes are additive because energy release rate, like energy, is a scalar quantity. Equation (2.63), however, assumes a self-similar crack growth, i.e., a planar crack is assumed to remain planar and maintain a constant shape as it grows. Such is usually not the case for mixed-mode fractures. See Section 2.11 for further discussion of energy release rate in mixed-mode problems.

2.8 CRACK-TIP PLASTICITY

Linear elastic stress analysis of sharp cracks predicts infinite stresses at the crack tip. In real materials, however, stresses at the crack tip are finite because the crack-tip radius must be finite (Section 2.2). Inelastic material deformation, such as plasticity in metals and crazing in polymers, leads to further relaxation of crack-tip stresses.

The elastic stress analysis becomes increasingly inaccurate as the inelastic region at the crack tip grows. Simple corrections to linear elastic fracture mechanics (LEFM) are available when moderate crack-tip yielding occurs. For more extensive yielding, one must apply alternative crack-tip parameters that take nonlinear material behavior into account (see Chapter 3).

The size of the crack-tip-yielding zone can be estimated by two methods: the Irwin approach, where the elastic stress analysis is used to estimate the elastic-plastic boundary, and the strip-yield model. Both approaches lead to simple corrections for crack-tip yielding. The term *plastic zone* usually applies to metals, but will be adopted here to describe inelastic crack-tip behavior in a more general sense. Differences in the yielding behavior between metals and polymers are discussed in Chapter 6.

2.8.1 THE IRWIN APPROACH

On the crack plane ($\theta = 0$), the normal stress σ_{yy} in a linear elastic material is given by Equation (2.39). As a first approximation, we can assume that the boundary between elastic and plastic behavior occurs when the stresses given by Equation (2.39) satisfy a yield criterion. For plane stress conditions, yielding occurs when $\sigma_{yy} = \sigma_{YS}$, the uniaxial yield strength of the material. Substituting yield strength into the left side of Equation (2.39) and solving for r gives a first-order estimate of the plastic zone size:

$$r_y = \frac{1}{2\pi}\left(\frac{K_I}{\sigma_{YS}}\right)^2 \tag{2.64}$$

If we neglect strain hardening, the stress distribution for $r = r_y$ can be represented by a horizontal line at $\sigma_{yy} = \sigma_{YS}$, as Figure 2.29 illustrates; the stress singularity is truncated by yielding at the crack tip.

The simple analysis in the preceding paragraph is not strictly correct because it was based on an elastic crack-tip solution. When yielding occurs, stresses must redistribute in order to satisfy equilibrium. The cross-hatched region in Figure 2.29 represents forces that would be present in an elastic material but cannot be carried in the elastic-plastic material because the stress cannot exceed the yield. The plastic zone must increase in size in order to accommodate these forces. A simple force balance leads to a second-order estimate of the plastic zone size r_p:

$$\sigma_{YS}r_p = \int_0^{r_y} \sigma_{yy}\,dr = \int_0^{r_y} \frac{K_I}{\sqrt{2\pi r}}\,dr \tag{2.65}$$

Integrating and solving for r_p gives

$$r_p = \frac{1}{\pi}\left(\frac{K_I}{\sigma_{YS}}\right)^2 \tag{2.66}$$

which is twice as large as r_y, the first-order estimate.

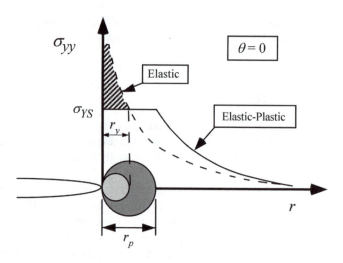

FIGURE 2.29 First-order and second-order estimates of plastic zone size (r_y and r_p, respectively). The cross-hatched area represents load that must be redistributed, resulting in a larger plastic zone.

Referring to Figure 2.29, the material in the plastic zone carries less stress than it would otherwise carry if the material remained elastic. Irwin [23] accounted for the softer material in the plastic zone by defining an effective crack length that is slightly longer than the actual crack size. The effective crack length is defined as the sum of the actual crack size and a plastic zone correction:

$$a_{eff} = a + r_y \tag{2.67}$$

where r_y for plane stress is given by Equation (2.64). In plane strain, yielding is suppressed by the triaxial stress state, and the Irwin plastic zone correction is smaller by a factor of 3:

$$r_y = \frac{1}{6\pi} \left(\frac{K_I}{\sigma_{YS}} \right)^2 \tag{2.68}$$

The effective stress intensity is obtained by inserting a_{eff} into the K expression for the geometry of interest:

$$K_{eff} = Y(a_{eff}) \sigma \sqrt{\pi a_{eff}} \tag{2.69}$$

Since the effective crack size is taken into account in the geometry correction factor Y, an iterative solution is usually required to solve for K_{eff}. That is, K is first determined in the absence of a plasticity correction; a first-order estimate of a_{eff} is then obtained from Equation (2.64) or Equation (2.68), which in turn is used to estimate K_{eff}. A new a_{eff} is computed from the K_{eff} estimate, and the process is repeated until successive K_{eff} estimates converge. Typically, no more than three or four iterations are required for reasonable convergence.

In certain cases, this iterative procedure is unnecessary because a closed-form solution is possible. For example, the effective Mode I stress intensity factor for a through crack in an infinite plate in plane stress is given by

$$K_{eff} = \frac{\sigma \sqrt{\pi a}}{\sqrt{1 - \frac{1}{2} \left(\frac{\sigma}{\sigma_{YS}} \right)^2}} \tag{2.70}$$

Elliptical and semielliptical flaws (Figure 2.20) also have an approximate closed-form plastic zone correction, provided the flaw is small compared to the plate dimensions. In the case of the embedded elliptical flaw, K_{eff} is given by

$$K_{eff} = \sigma \sqrt{\frac{\pi a}{Q_{eff}}} \left[\sin^2(\phi) + \left(\frac{a}{c}\right)^2 \cos^2(\phi) \right]^{1/4} \tag{2.71}$$

where Q_{eff} is the effective flaw shape parameter defined as

$$Q_{eff} = Q - 0.212 \left(\frac{\sigma}{\sigma_{YS}}\right)^2 \tag{2.72}$$

Equation (2.72) must be multiplied by a surface correction factor for a semielliptical surface flaw (see Figure 2.20).

One interpretation of the Irwin plastic zone adjustment is that of an *effective compliance*. Figure 2.30 compares the load-displacement behavior of a purely elastic cracked plate with that of a cracked plate with a plastic zone at the tip. The load-displacement curve for the latter case deviates from the purely elastic curve as the load increases. At a given load P, the displacement for the plate with a plastic zone is greater than that of the elastic plate; referring to Figure 2.30, $\Delta_2 > \Delta_1$. One can define an effective compliance as follows:

$$C_{eff} = \frac{\Delta_2}{P}$$

Therefore, a_{eff} in this instance can be viewed as the crack length that results in the compliance C_{eff} in a purely elastic material.

Finally, it should be noted that the author does not recommend using the Irwin plastic zone adjustment for practical applications. It was presented here primarily to provide a historical context to the development of both linear and nonlinear fracture mechanics. See Chapter 9 for recommended approaches for handling plasticity effects.

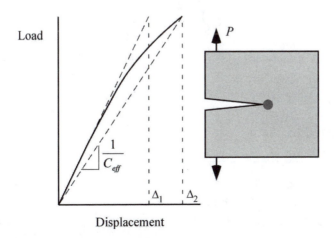

FIGURE 2.30 Definition of the effective compliance to account for crack-tip plasticity.

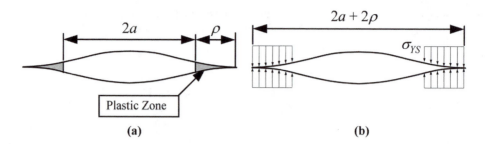

(a)   **(b)**

FIGURE 2.31 The strip-yield model. The plastic zone is modeled by yield magnitude compressive stresses at each crack tip (b).

2.8.2 THE STRIP-YIELD MODEL

The strip-yield model, which is illustrated in Figure 2.31, was first proposed by Dugdale [24] and Barenblatt [25]. They assumed a long, slender plastic zone at the crack tip in a nonhardening material in plane stress. These early analyses considered only a through crack in an infinite plate. The strip-yield plastic zone is modeled by assuming a crack of length $2a + 2\rho$, where ρ is the length of the plastic zone, with a closure stress equal to σ_{YS} applied at each crack tip (Figure 2.31(b)).

This model approximates elastic-plastic behavior by superimposing two elastic solutions: a through crack under remote tension and a through crack with closure stresses at the tip. Thus the strip-yield model is a classical application of the principle of superposition.

Since the stresses are finite in the strip-yield zone, there cannot be a stress singularity at the crack tip. Therefore, the leading term in the crack-tip field that varies with $1/\sqrt{r}$ (Equation (2.36)) must be zero. The plastic zone length ρ must be chosen such that the stress intensity factors from the remote tension and closure stress cancel one another.

The stress intensity due to the closure stress can be estimated by considering a normal force P applied to the crack at a distance x from the centerline of the crack (Figure 2.32). The stress intensities for the two crack tips are given by

$$K_{I(+a)} = \frac{P}{\sqrt{\pi a}} \sqrt{\frac{a+x}{a-x}}$$ (2.73a)

$$K_{I(-a)} = \frac{P}{\sqrt{\pi a}} \sqrt{\frac{a-x}{a+x}}$$ (2.73b)

assuming the plate is of unit thickness. The closure force at a point within the strip-yield zone is equal to

$$P = -\sigma_{YS} dx$$ (2.74)

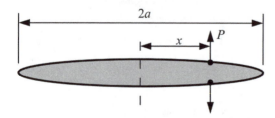

FIGURE 2.32 Crack-opening force applied at a distance x from the center-line.

Thus, the total stress intensity at each crack tip resulting from the closure stresses is obtained by replacing a with $a + \rho$ in Equation (2.73) and summing the contribution from both crack tips:

$$
K_{closure} = -\frac{\sigma_{YS}}{\sqrt{\pi(a+\rho)}} \int_a^{a+\rho} \left\{ \sqrt{\frac{a+\rho+x}{a+\rho-x}} + \sqrt{\frac{a+\rho-x}{a+\rho+x}} \right\} dx
$$

$$
= -2\sigma_{YS} \sqrt{\frac{a+\rho}{\pi}} \int_a^{a+\rho} \frac{dx}{\sqrt{(a+\rho)^2 - x^2}}
\tag{2.75}
$$

Solving this integral gives

$$
K_{closure} = -2\sigma_{YS} \sqrt{\frac{a+\rho}{\pi}} \cos^{-1}\left(\frac{a}{a+\rho}\right)
\tag{2.76}
$$

The stress intensity from the remote tensile stress, $K_\sigma = \sigma\sqrt{\pi(a+\rho)}$, must balance with $K_{closure}$. Therefore,

$$
\frac{a}{a+\rho} = \cos\left(\frac{\pi\sigma}{2\sigma_{YS}}\right)
\tag{2.77}
$$

Note that ρ approaches infinity as $\sigma \to \sigma_{YS}$. Let us explore the strip-yield model further by performing a Taylor series expansion on Equation (2.77):

$$
\frac{a}{a+\rho} = 1 - \frac{1}{2!}\left(\frac{\pi\sigma}{2\sigma_{YS}}\right)^2 + \frac{1}{4!}\left(\frac{\pi\sigma}{2\sigma_{YS}}\right)^4 - \frac{1}{6!}\left(\frac{\pi\sigma}{2\sigma_{YS}}\right)^6 + \cdots
\tag{2.78}
$$

Neglecting all but the first two terms and solving for the plastic zone size gives

$$
\rho = \frac{\pi^2\sigma^2 a}{8\sigma_{YS^2}} = \frac{\pi}{8}\left(\frac{K_I}{\sigma_{YS}}\right)^2
\tag{2.79}
$$

for $\sigma \ll \sigma_{YS}$. Note the similarity between Equation (2.79) and Equation (2.66); since $1/\pi = 0.318$ and $\pi/8 = 0.392$, the Irwin and strip-yield approaches predict similar plastic zone sizes.

One way to estimate the effective stress intensity with the strip-yield model is to set a_{eff} equal to $a + \rho$:

$$
K_{eff} = \sigma\sqrt{\pi a \sec\left(\frac{\pi\sigma}{2\sigma_{YS}}\right)}
\tag{2.80}
$$

However, Equation (2.80) tends to overestimate K_{eff}; the actual a_{eff} is somewhat less than $a + \rho$ because the strip-yield zone is loaded to σ_{YS}. Burdekin and Stone [26] obtained a more realistic estimate of K_{eff} for the strip-yield model

$$
K_{eff} = \sigma_{YS}\sqrt{\pi a}\left[\frac{8}{\pi^2}\ln\sec\left(\frac{\pi\sigma}{2\sigma_{YS}}\right)\right]^{1/2}
\tag{2.81}
$$

Refer to Appendix 3.1 for a derivation of Equation (2.81).

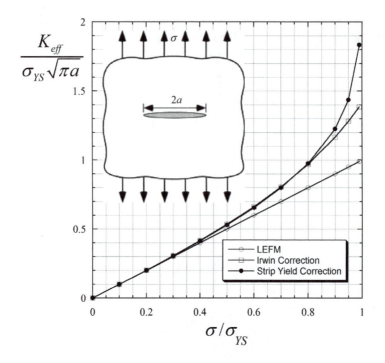

FIGURE 2.33 Comparison of plastic zone corrections for a through crack in plane strain.

2.8.3 Comparison of Plastic Zone Corrections

Figure 2.33 shows a comparison between a pure LEFM analysis (Equation (2.41)), the Irwin correction for plane stress (Equation (2.70)), and the strip-yield correction on stress intensity (Equation (2.81)). The effective stress intensity, nondimensionalized by $\sigma_{YS}\sqrt{\pi a}$, is plotted against the normalized stress. The LEFM analysis predicts a linear relationship between K and stress. Both the Irwin and strip-yield corrections deviate from the LEFM theory at stresses greater than $0.5\sigma_{YS}$. The two plasticity corrections agree with each other up to approximately $0.85\sigma_{YS}$. According to the strip-yield model, K_{eff} is infinite at yield; the strip-yield zone extends completely across the plate, which has reached its maximum load capacity.

The plastic zone shape predicted by the strip-yield model bears little resemblance to actual plastic zones in metals (see Section 2.8.4), but many polymers produce crack-tip craze zones that look very much like Figure 2.31. Thus, although Dugdale originally proposed the strip-yield model to account for yielding in thin steel sheets, this model is better suited to polymers (see Chapter 6).

In the 1970s, the strip-yield model was used to derive a practical methodology for assessing fracture in structural components. This approach is called the failure assessment diagram (FAD) and is described in Chapter 9.

2.8.4 Plastic Zone Shape

The estimates of plastic zone size that have been presented so far consider only the crack plane $\theta = 0$. It is possible to estimate the extent of plasticity at all angles by applying an appropriate yield criterion to the equations in Table 2.1 and Table 2.3. Consider the von Mises equation:

$$\sigma_e = \frac{1}{\sqrt{2}}\left[(\sigma_1 - \sigma_2)^2 + (\sigma_1 - \sigma_3)^2 + (\sigma_2 - \sigma_3)^2\right]^{1/2} \qquad (2.82)$$

where σ_e is the effective stress, and σ_1, σ_2, and σ_3 are the three principal normal stresses. According to the von Mises criterion, yielding occurs when $\sigma_e = \sigma_{YS}$, the uniaxial yield strength. For plane stress or plane strain conditions, the principal stresses can be computed from the two-dimensional Mohr's circle relationship:

$$\sigma_1, \sigma_2 = \frac{\sigma_{xx} + \sigma_{yy}}{2} \pm \left[\left(\frac{\sigma_{xx} - \sigma_{yy}}{2} \right)^2 + \tau_{xy^2} \right]^{1/2} \tag{2.83}$$

For plane stress $\sigma_3 = 0$, and $\sigma_3 = v(\sigma_1 + \sigma_2)$ for plane strain. Substituting the Mode I stress fields into Equation (2.83) gives

$$\sigma_1 = \frac{K_I}{\sqrt{2\pi r}} \cos\left(\frac{\theta}{2} \right) \left[1 + \sin\left(\frac{\theta}{2} \right) \right] \tag{2.84a}$$

$$\sigma_2 = \frac{K_I}{\sqrt{2\pi r}} \cos\left(\frac{\theta}{2} \right) \left[1 - \sin\left(\frac{\theta}{2} \right) \right] \tag{2.84b}$$

$$\sigma_3 = 0 \quad \text{(plane stress)}$$

$$= \frac{2vK_I}{\sqrt{2\pi r}} \cos\left(\frac{\theta}{2} \right) \quad \text{(plane strain)} \tag{2.84c}$$

By substituting Equation (2.84) into Equation (2.82), setting $\sigma_e = \sigma_{YS}$, and solving for r, we obtain estimates of the Mode I plastic zone radius as a function of θ:

$$r_y(\theta) = \frac{1}{4\pi} \left(\frac{K_I}{\sigma_{YS}} \right)^2 \left[1 + \cos\theta + \frac{3}{2} \sin^2\theta \right] \tag{2.85a}$$

for plane stress, and

$$r_y(\theta) = \frac{1}{4\pi} \left(\frac{K_I}{\sigma_{YS}} \right)^2 \left[(1 - 2v)^2 (1 + \cos\theta) + \frac{3}{2} \sin^2\theta \right] \tag{2.85b}$$

for plane strain. Equation (2.85a) and Equation (2.85b), which are plotted in Figure 2.34(a), define the approximate boundary between elastic and plastic behavior. The corresponding equations for Modes II and III are plotted in Figure 2.34(b) and Figure 2.34(c), respectively.

Note the significant difference in the size and shape of the Mode I plastic zones for plane stress and plane strain. The latter condition suppresses yielding, resulting in a smaller plastic zone for a given K_I value.

Equation (2.85a) and Equation (2.85b) are not strictly correct because they are based on a purely elastic analysis. Recall Figure 2.29, which schematically illustrates how crack-tip plasticity causes stress redistribution, which is not taken into account in Figure 2.34. The Irwin plasticity correction, which accounts for stress redistribution by means of an effective crack length, is also simplistic and not totally correct.

Figure 2.35 compares the plane strain plastic zone shape predicted from Equation (2.85b) with a detailed elastic-plastic crack-tip stress solution obtained from the finite element analysis. The latter, which was published by Dodds et al. [27], assumed a material with the following uniaxial stress-strain relationship:

$$\frac{\varepsilon}{\varepsilon_o} = \frac{\sigma}{\sigma_o} + \alpha \left(\frac{\sigma}{\sigma_o} \right)^n \tag{2.86}$$

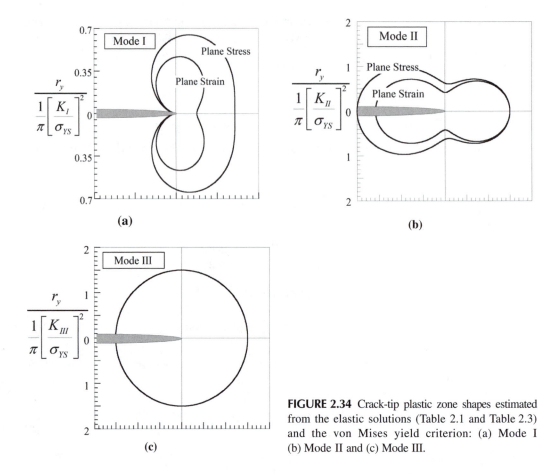

(a)   **(b)**

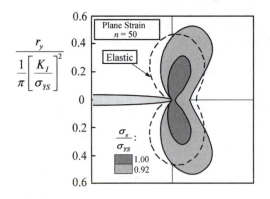

(c)

FIGURE 2.34 Crack-tip plastic zone shapes estimated from the elastic solutions (Table 2.1 and Table 2.3) and the von Mises yield criterion: (a) Mode I (b) Mode II and (c) Mode III.

where ε_o, σ_o, α, and n are material constants. We will examine the above relationship in more detail in Chapter 3. For now, it is sufficient to note that the exponent n characterizes the strain-hardening rate of a material. Dodds et al. analyzed materials with $n = 5$, 10, and 50, which corresponds to high, medium, and low strain-hardening, respectively. Figure 2.35 shows the contours of constant σ_e for $n = 50$. The definition of the elastic-plastic boundary is somewhat arbitrary, since materials that can be described by Equation (2.86) do not have a definite yield point. When the plastic zone boundary is defined at $\sigma_e = \sigma_{YS}$ (the 0.2% offset yield strength), the plane strain plastic zone is considerably smaller than predicted by Equation (2.85b). Defining the boundary at a slightly lower

FIGURE 2.35 Contours of constant effective stress in Mode I, obtained from finite element analysis.[a] The elastic-plastic boundary estimated from Equation (2.85a) is shown for comparison.

[a] Taken from Dodds, R.H., Jr., Anderson, T.L., and Kirk, M.T., *International Journal of Fracture,* Vol. 48, 1991.

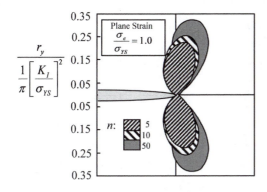

FIGURE 2.36 Effect of strain-hardening on the Mode I plastic zone; $n = 5$ corresponds to a high strain-hardening material, while $n = 50$ corresponds to very low hardening (cf. Equation (2.86)).

effective stress results in a much larger plastic zone. Given the difficulties of defining the plastic zone unambiguously with a detailed analysis, the estimates of the plastic zone size and shape from the elastic analysis (Figure 2.34) appear to be reasonable.

Figure 2.36 illustrates the effect of strain hardening on the plastic zone. A high strain-hardening rate results in a smaller plastic zone because the material inside of the plastic zone is capable of carrying higher stresses, and less stress redistribution is necessary.

2.9 K-CONTROLLED FRACTURE

Section 2.6.1 introduced the concept of the singularity-dominated zone and alluded to single-parameter characterization of crack-tip conditions. The stresses near the crack tip in a linear elastic material vary as $1/\sqrt{r}$; the stress intensity factor defines the amplitude of the singularity. Given the equations in Table 2.1 to Table 2.3, one can completely define the stresses, strains, and displacements in the singularity-dominated zone if the stress intensity factor is known. If we assume a material fails locally at some combination of stresses and strains, then crack extension must occur at a critical K value. This K_{crit} value, which is a measure of *fracture toughness*, is a material constant that is independent of the size and geometry of the cracked body. Since energy release rate is uniquely related to stress intensity (Section 2.7), $\mathcal{G}$ also provides a single-parameter description of crack-tip conditions, and $\mathcal{G}_c$ is an alternative measure of toughness.

The foregoing discussion does not consider plasticity or other types of nonlinear material behavior at the crack tip. Recall that the $1/\sqrt{r}$ singularity applies only to linear elastic materials. The equations in Table 2.1 to Table 2.3 do not describe the stress distribution inside the plastic zone. As discussed in Chapter 5 and Chapter 6, the microscopic events that lead to fracture in various materials generally occur well within the plastic zone (or damage zone, to use a more generic term). Thus, even if the plastic zone is very small, fracture may not nucleate in the singularity-dominated zone. This fact raises an important question: Is stress intensity a useful failure criterion in materials that exhibit inelastic deformation at the crack tip?

Under certain conditions, K still uniquely characterizes crack-tip conditions when a plastic zone is present. In such cases, K_{crit} is a geometry-independent material constant, as discussed below.

Consider a test specimen and structure loaded to the same K_I level, as illustrated in Figure 2.37. Assume that the plastic zone is small compared to all the length dimensions in the structure and test specimen. Let us construct a free-body diagram with a small region removed from the crack tip of each material. If this region is sufficiently small to be within the singularity-dominated zone, the stresses and displacements at the boundary are defined by the relationships in Table 2.1 and Table 2.2. The disc-shaped region in Figure 2.37 can be viewed as an independent problem. The imposition of the $1/\sqrt{r}$ singularity at the boundary results in a plastic zone at the crack tip. The size of the plastic zone and the stress distribution within the disc-shaped region are a function only of the boundary conditions and material properties. Therefore, even though we do not know the actual stress distribution in the plastic zone, we can argue that it is uniquely characterized by the

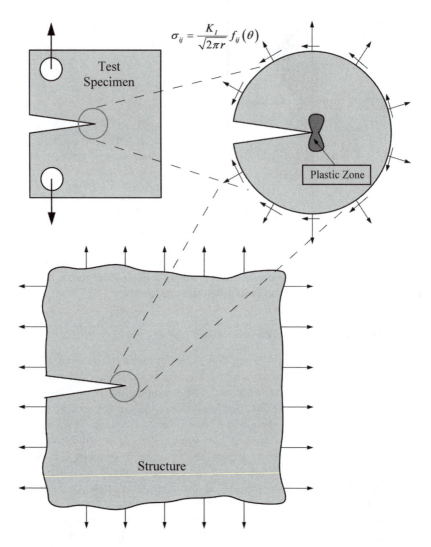

FIGURE 2.37 Schematic test specimen and structure loaded to the same stress intensity. The crack-tip conditions should be identical in both configurations as long as the plastic zone is small compared to all relevant dimensions. Thus, both will fail at the same critical K value.

boundary conditions, i.e., K_I characterizes crack-tip conditions even though the $1/\sqrt{r}$ singularity does not apply to the plastic zone. Since the structure and test specimen in Figure 2.37 are loaded to the same K_I value, the crack-tip conditions must be identical in the two configurations. Furthermore, as the load is increased, both configurations will fail at the same critical stress intensity, provided the plastic zone remains small in each case. Similarly, if both structures are loaded in fatigue at the same ΔK, the crack-growth rates will be similar as long as the cyclic plastic zone is embedded within the singularity-dominated zone in each case (see Chapter 10).

Figure 2.38 schematically illustrates the stress distributions in the structure and test specimen from the previous figure. In the singularity-dominated zone, a log-log plot of the stress distribution is linear with a slope of $-1/2$. Inside of the plastic zone, the stresses are lower than predicted by the elastic solution, but are identical for the two configurations. Outside of the singularity-dominated zone, higher-order terms become significant (Equation (2.36)) and the stress fields are different for the structure and test specimen. K does not uniquely characterize the magnitude of the higher-order terms.

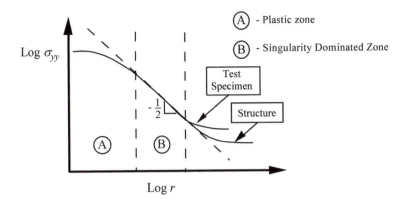

FIGURE 2.38 Crack-tip stress fields for the specimen and structure in Figure 2.37.

As the plastic zone increases in size, it eventually engulfs the singularity zone, and there is no longer a region where the stress varies as $1/\sqrt{r}$. In such cases, K no longer characterizes stresses near the crack tip. What this means in practical terms is that LEFM is not valid once the plastic zone size becomes large relative to key dimensions such as the crack size.

EXAMPLE 2.7

Estimate the relative size of the singularity-dominated zone ahead of a through crack in an infinite plate subject to remote uniaxial tension (Figure 2.3). The full solution for the stresses on the crack plane ($\theta = 0$) for this geometry are as follows (see Appendix 2.3.2):

$$\sigma_{yy} = \frac{\sigma(a+r)}{\sqrt{2ar+r^2}}$$

$$\sigma_{xx} = \frac{\sigma(a+r)}{\sqrt{2ar+r^2}} - \sigma$$

where σ is the remotely applied tensile stress. Also, estimate the value of K_I where the plane strain plastic zone engulfs the singularity-dominated zone.

Solution: As $r \to 0$ both of the above relationships reduce to the expected result:

$$\sigma_{yy} = \sigma_{xx} = \frac{\sigma\sqrt{a}}{\sqrt{2r}} = \frac{K_I}{\sqrt{2\pi r}}$$

Figure 2.39 is a plot of the ratio of the total stress to the singular stress given by the above equation. Note that the stress in the y direction is close to the singular limit to relatively large distances from the crack tip, but the x stress diverges considerably from the near-tip limit. When $r/a = 0.02$, the singularity approximation results in roughly a 2% underestimate of σ_{yy} and a 20% overestimate of σ_{xx}. Let us arbitrarily define this point as the limit of the singularity zone:

$$r_s \approx \frac{a}{50}$$

By setting the plane strain plastic zone size estimate (Equation (2.63)) equal to a/50, we obtain an estimate of the K_I value at which the singularity zone is engulfed by crack-tip plasticity:

$$a = \frac{50}{6\pi}\left(\frac{K_I^*}{\sigma_{YS}}\right)^2 = 2.65\left(\frac{K_I^*}{\sigma_{YS}}\right)^2$$

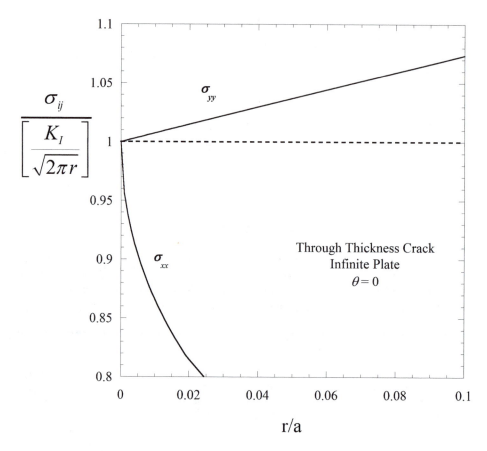

FIGURE 2.39 Ratio of actual stresses on the crack plane to the singularity limit in an infinite plate with a through-thickness crack (Example 2.7).

or

$$K_I^* = 0.35\,\sigma_{YS}\sqrt{\pi a}$$

Therefore, when the nominal stress exceeds approximately 35% of yield in this case, the accuracy of K_I as a crack-tip–characterizing parameter in this particular geometry is suspect.[1]

2.10 PLANE STRAIN FRACTURE: FACT VS. FICTION

In the 1960s, massive testing programs were undertaken by NASA and other organizations in an effort to develop experimental procedures for measuring fracture toughness in high strength materials. Among the variables that were considered in these studies were the dimensions of the test specimen.

The NASA data exhibited an apparent effect of specimen thickness on the critical stress intensity for fracture, K_{crit}. The explanation that was originally offered was that thin specimens are subject to plane stress loading at the crack tip, while thick specimens experience plane strain conditions. The biaxial stress state associated with plane stress, it was argued, results in a higher measured toughness than is observed in the same material when subject to a triaxial stress state. Failure in thin sections was referred to as "plane stress fracture," while the term "plane strain fracture" was

---

[1] The singularity zone is small in this geometry because of a significant transverse compressive stress. In cracked geometries loaded in bending, this transverse stress (also called the T stress) is near zero or slightly positive; consequently, the singularity zone is larger in these configurations. See Section 3.6 for a further discussion on the effect of the T stress.

applied to toughness tests on thick sections. This two-dimensional viewpoint, which is still prevalent in textbooks and published literature, is simplistic and misleading.

Much of the classical fracture mechanics theory is predicated on two-dimensional approximations. For example, the relationship between K_I and energy release rate (Equation (2.56)) is rigorously correct only for the special cases of plane stress and plane strain. There are cases where a two-dimensional model is appropriate, but there are other instances where a two-dimensional outlook gives a distorted view of reality. The relationship between specimen dimensions and apparent fracture toughness is an example of the latter.

In the 1960s, when "plane stress fracture" and "plane strain fracture" mechanisms were first postulated, a detailed three-dimensional analysis of the stress state in front of a crack was simply not possible. Today, three-dimensional finite element analyses of components with cracks are commonplace (Chapter 12). Advances in computer technology have significantly aided in our understanding of the behavior of material at the tip of a crack.

This section presents an updated perspective on the interrelationship between specimen dimensions, crack-tip triaxiality, and fracture toughness.

2.10.1 CRACK-TIP TRIAXIALITY

Consider a cracked plate with thickness B subject to in-plane loading, as illustrated in Figure 2.40. If there was no crack, the plate would be in a state of plane stress. Thus, regions of the plate that are sufficiently far from the crack tip must also be loaded in plane stress. Material near the crack tip is loaded to higher stresses than the surrounding material. Because of the large stress normal to the crack plane, the crack-tip material tries to contract in the x and z directions, but is prevented from doing so by the surrounding material. This constraint causes a triaxial state of stress near the crack-tip, as Figure 2.40 illustrates.

Figure 2.41 is a schematic plot of the stress parallel to the crack front, σ_{zz}, in the plate depicted in Figure 2.40. In the interior of the plate, the z stress, and therefore the level of triaxiality, is high. The stress state in this central region is essentially plane strain at distances from the crack tip that are small compared to the plate thickness. Near the free surface, the stress triaxiality is lower, but a state of pure plane stress exists only at the free surface.

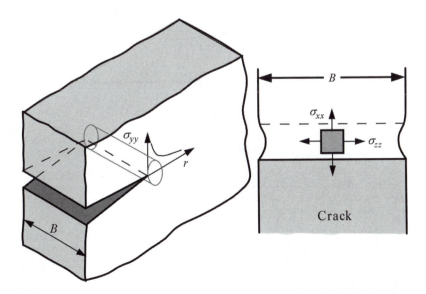

FIGURE 2.40 Three-dimensional deformation at the tip of a crack. The high normal stress at the crack tip causes material near the surface to contract, but material in the interior is constrained, resulting in a triaxial stress state.

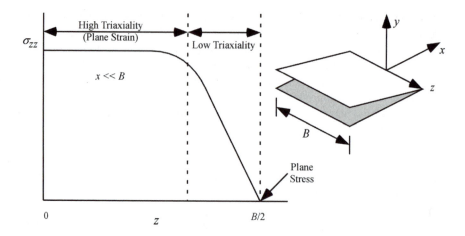

FIGURE 2.41 Schematic variation of transverse stress and strain through the thickness at a point near the crack tip.

Figure 2.42 is a plot of σ_{zz} as a function of z/B and x/B. These results were obtained from a three-dimensional elastic-plastic finite element analysis performed by Narasimhan and Rosakis [28]. Material near the crack tip experiences high triaxiality, but $\sigma_{zz} = 0$ when x is a significant fraction of the plate thickness. Therefore, plane stress conditions exist remote from the crack tip, but the stress state close to the crack tip is essentially plane strain in the interior of the plate.[2]

The stress state can have a significant effect on the fracture behavior of a given material. To illustrate this effect, consider a point on the crack plane ($\theta = 0$) just ahead of the crack-tip. According to Equation (2.39), $\sigma_{xx} = \sigma_{yy}$ under linear elastic conditions. If the stress state is plane

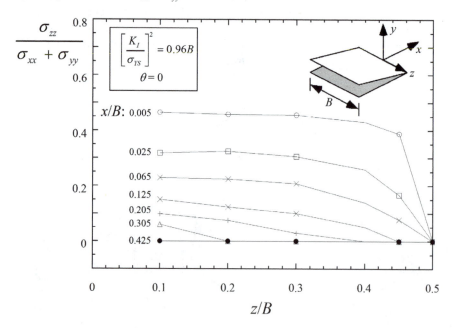

FIGURE 2.42 Transverse stress through the thickness as a function of distance from the crack tip [28].

---

[2] Under plane strain loading, the quantity $\sigma_{zz}/(\sigma_{xx} + \sigma_{yy})$ is equal to Poisson's ratio for elastic material behavior and is equal to 0.5 for incompressible plastic deformation. Therefore, the curve for $x/B = 0.005$ in Figure 2.42 is inside the plastic zone at $(KI/\sigma_{YS}) = 0.96B$.

stress, $\sigma_{zz} = 0$ by definition. Under plain strain conditions, $\sigma_{zz} = 2\nu\sigma_{yy}$. Substituting these stresses into the von Mises yield criterion (Equation (2.82)) leads to the following:

$$\sigma_{yy}(\text{at yield}) = \begin{cases} \sigma_{YS} & \text{(plane stress)} \\ 2.5\sigma_{YS} & \text{(plane strain)} \end{cases} \qquad (2.87)$$

assuming $\nu = 0.3$. Therefore, the triaxial stress state associated with plane strain leads to higher stresses in the plastic zone. For fracture mechanisms that are governed by normal stress, such as cleavage in metals (Section 5.2), the material will behave in a more brittle fashion when subjected to a triaxial stress state. Triaxial stresses also assist ductile fracture processes such as microvoid coalescence (Section 5.1).

2.10.2 EFFECT OF THICKNESS ON APPARENT FRACTURE TOUGHNESS

Figure 2.43 and Figure 2.44 show two sets of data that have commonly been used to illustrate thickness effects on fracture toughness [29]. The measured K_{crit} values decrease with specimen thickness until a plateau is reached, at which point the toughness appears to be relatively insensitive to further increase in thickness. This apparent asymptote in the toughness vs. thickness trend is designated by the symbol K_{Ic}, and is referred to as "plane strain fracture toughness" [30, 31]. A K_{Ic} value is purported to be a specimen-size–independent material property.[3]

In the past, the decreasing trend in K_{crit} with increasing thickness in Figure 2.43 and Figure 2.44 was attributed to a transition from plane stress to plane strain at the crack tip. Although this trend

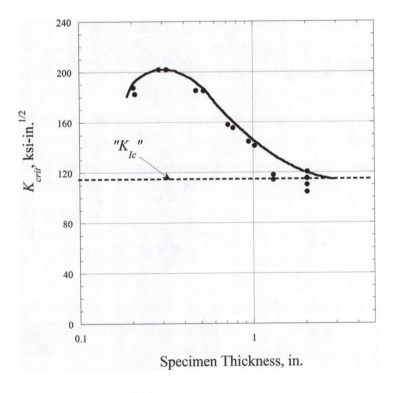

FIGURE 2.43 Variation of measured fracture toughness with specimen thickness for an unspecified alloy. Adapted from Barsom and Rolfe, *Fracture and Fatigue Control in Structures*. 2nd Ed., Prentice-Hall, Englewood Cliffs, NJ, 1987.

[3] In reality, fracture toughness, as it is defined standardized K_{Ic} test methods, does not usually exhibit a true asymptote with increasing specimen size. Refer to Section 7.2 for a detailed discussion of the K_{Ic} test.

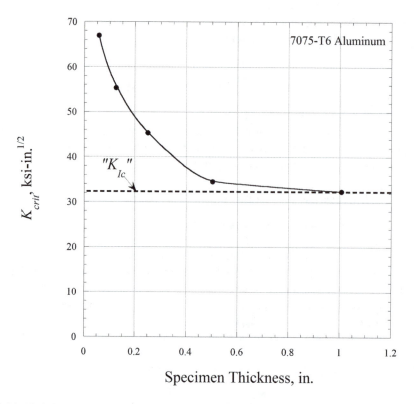

FIGURE 2.44 Variation of measured fracture toughness with specimen thickness for 7075-T6 Aluminum. Adapted from Barsom and Rolfe, *Fracture and Fatigue Control in Structures.* 2nd Ed., Prentice-Hall, Englewood Cliffs, NJ, 1987.

is related to the crack-tip stress state, the traditional plane stress–plane strain transition model is far too simplistic.

Plots such as Figure 2.43 and Figure 2.44, that show a decrease in apparent toughness with specimen thickness, generally correspond to materials in which the crack propagation is ductile (microvoid coalescence). In such tests, the crack "tunnels" through the center of the specimen. That is, the crack grows preferentially in the region of high triaxiality. Crack growth on the outer regions of the specimen lags behind, and occurs at a 45° angle to the applied load. The resulting fracture surface exhibits a flat region in the central region and 45° shear lips on the edges. Section 5.1 provides additional information on crack tunneling and the formation of shear lips. Figure 2.45

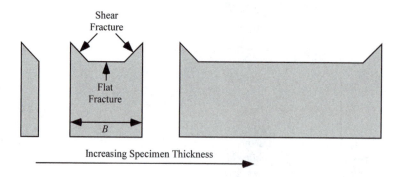

FIGURE 2.45 Effect of specimen thickness on fracture surface morphology for materials that exhibit ductile crack growth.

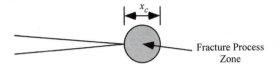

Fracture Process
Zone

FIGURE 2.46 Fracture process zone at the tip of a crack.

illustrates the fracture surface morphology for three specimen thicknesses. Fracture toughness tests on very thin plates or sheets typically result in a 45° shear fracture. At larger thicknesses, there is generally some mixture of shear fracture and flat fracture. The thickness effect on the apparent fracture toughness is due to the relative portions of flat and shear fracture. In the limit of a very thick specimen, the flat fracture mechanism dominates, and further increases in thickness have relatively little effect on the measured toughness.

The stress distribution depicted in Figure 2.41 is characteristic of all section thicknesses. A central plane strain region can exist even in thin sheet specimens, as long as the distance from the crack-tip is sufficiently small. Pure slant fracture occurs when the distance over which high triaxiality conditions exist is smaller than the fracture process zone, which is a function of micro-structural parameters such as inclusion spacing. Figure 2.46 illustrates the concept of a fracture process zone. The micromechanical processes that lead to ductile crack extensions occur over a finite distance x_c, which typically is much less than the plastic zone size. Figure 2.47 illustrates the effect of thickness on the crack-tip stress state at $x = x_c$. For very thin sections, plane strain conditions do not exist at $x = x_c$. As the thickness increases, the size of the plane strain zone increases relative to the low triaxiality zone near the free surfaces. It is this stress state that results in the varying fracture surface morphology depicted in Figure 2.45, which in turn leads to the apparent thickness dependence of fracture toughness shown in Figure 2.43 and Figure 2.44.

To summarize, the trends in Figure 2.43 and Figure 2.44 are not indicative of a transition from "plane stress fracture" to "plane strain fracture." Rather, this trend reflects the differing relative contributions of two distinct fracture mechanisms. In point of fact, there is no such thing as "plane stress fracture" except perhaps in very thin foil. There is nearly always some level of triaxiality along the crack front.

The shear lips, that cause the apparent thickness dependence of toughness in materials that exhibit ductile crack growth, are an artifact of the way in which K_{Ic} tests have been conducted since the early 1960s. More recent fracture toughness test methods typically use side-grooved specimens (Section 7.1). This specimen design eliminates shear lips and provides an accurate measure of the resistance of the material to flat, ductile fracture. The apparent thickness dependence of toughness observed in Figure 2.43 and Figure 2.44 would disappear if the same materials were tested with side-grooved specimens.

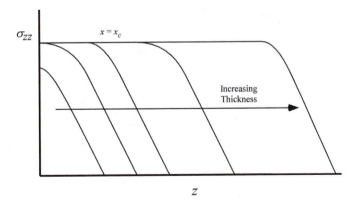

FIGURE 2.47 Effect of thickness on stress — the crack-tip stress state in the fracture process zone.

Fracture toughness specimens that fail by cleavage fracture usually do not form shear lips, so the trends in Figure 2.43 and Figure 2.44 do not apply to such data. Cleavage fracture toughness does exhibit a slight thickness-dependence due to weakest-link sampling effects. See Section 5.2 for a detailed discussion of this fracture mechanism.

2.10.3 Plastic Zone Effects

Section 2.9 outlines the conditions required for K-controlled fractures. The plastic zone must be embedded within an elastic singularity zone in order for K to characterize crack-tip conditions. Traditionally, the loss of K dominance with plastic zone growth has been lumped together with the purported transition from "plane strain fracture" to "plane stress fracture," as if these phenomena were synonymous. In fact, there is not a direct correspondence between the plastic zone size and the existence (or absence) of plane strain conditions near the crack tip. Three-dimensional elastic-plastic finite element analyses of standard laboratory fracture toughness specimens have shown that a high degree of triaxiality near the crack tip exists even when the entire cross-section has yielded. Although K is not valid as a characterizing parameter under fully plastic conditions, a single-parameter description of fracture toughness is still possible using the J integral, or crack-tip-opening displacement (Chapter 3).

Figure 2.48 shows the evolution of the Mode I plastic zone at mid-thickness in a plate containing an edge crack. These results were obtained from a three-dimensional elastic-plastic finite element analysis performed by Nakamura and Parks [32]. The plastic zone boundary is defined at $\sigma_e = \sigma_{YS}$ in this case. As the quantity $(K_I/\sigma_{YS})^2$ increases relative to plate thickness B, the plastic zone size increases, as one would expect. What is interesting about these results is the change in plastic zone shape. At low K_I values, the plastic zone has a typical plane strain shape, but evolves into a plane stress shape at higher K_I values (Figure 2.34(a)). This transition can be understood by referring to Figure 2.42. At distances from the crack tip on the order of half the plate thickness, $\sigma_{zz} = 0$. As a result, the plastic zone takes on a plane stress shape when it grows to approximately half the plate

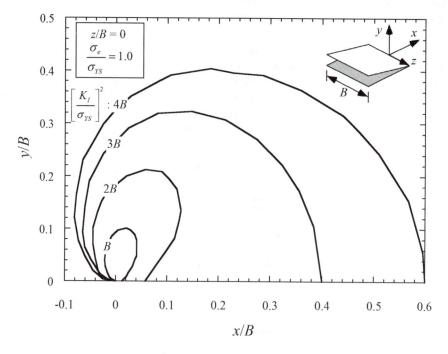

FIGURE 2.48 Effect of K_I, relative to thickness, of the plastic zone size and shape. Taken from Nakamura, T. and Parks, D.M., ASME AMD-91. American Society of Mechanical Engineers, New York, 1988.

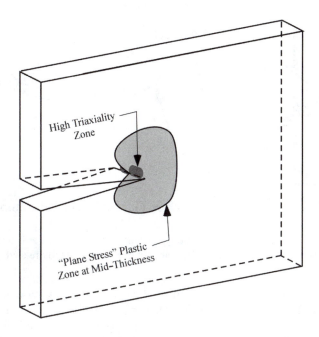

FIGURE 2.49 Cracked plate in which the plastic zone size is of the same order of magnitude as the plate thickness. The plastic zone at mid thickness has a plane stress shape, but there is a zone of high triaxiality close to the crack tip.

thickness. Although the stress state at the plastic zone boundary is plane stress, the material close to the crack tip is subject to a triaxial stress state. This is illustrated in Figure 2.49, which depicts a plastic zone in the center of an edge-cracked plate. Because the plastic zone size in Figure 2.49 is of the same order of magnitude as the plate thickness, the plastic zone has a plane stress shape. At the crack tip, however, there is a zone of high triaxiality. As stated above, the zone of high triaxiality at the crack tip can persist even in the presence of large-scale plasticity.

When performing laboratory K_{Ic} tests on standard specimens, such as those illustrated in Table 2.4, the following size requirements have been adopted [30, 31]:

$$a, B, (W - a) \geq 2.5 \left(\frac{K_{Ic}}{\sigma_{YS}} \right)^2 \qquad (2.88)$$

Recall that the quantity $(K_I/\sigma_{YS})^2$ is proportional to the plastic zone size. The minimum requirements on the crack length and ligament length $(W - a)$ are designed to ensure that the plastic zone is sufficiently small for fracture to be K-controlled. The thickness requirement, which is based on experimental data such as Figure 2.43 and Figure 2.44, is intended to ensure plane strain conditions along the crack front. As stated earlier, however, the apparent thickness dependence in fracture toughness is a result of the relative mixtures of flat fracture and shear fracture, and side grooves would eliminate this effect. The thickness requirement in Equation (2.88) is far more stringent than is necessary to ensure plane strain conditions along the majority of the crack front.

2.10.4 Implications for Cracks in Structures

A final, very important point is that the observed thickness dependence of fracture toughness in laboratory tests is usually not directly transferable to structural components. For example, Figure 2.50 schematically compares the crack-tip stress state of a laboratory fracture toughness

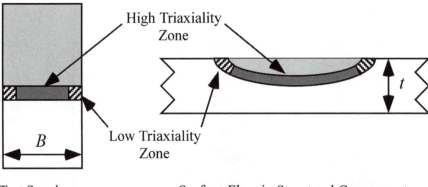

FIGURE 2.50 Schematic comparison of a laboratory specimen with a flaw in a structural component. In the latter case, the stress state and fracture morphology are not necessarily directly related to section thickness.

specimen with that of a surface crack in a structural component. In the case of the test specimen, the relative size of the high-triaxiality zone is directly related to the thickness. For the surface flaw, however, the size of the high-triaxiality zone is governed by the crack front length, which need not be related in any way to the section thickness. A standard laboratory specimen and a surface crack would not necessarily produce the same fracture morphology (e.g., the relative fractions of flat fracture vs. shear fracture). The observed trends in Figure 2.43 and Figure 2.44 cannot legitimately be used to predict the fracture behavior of structural components.

In 1967, Irwin et al. [33] developed a simple empirical relationship to describe the trends in Figure 2.43 and Figure 2.44, as well as other similar data sets:

$$K_{\text{crit}} = K_{Ic}\left(1 + 1.4\beta_{Ic}^{2}\right)^{1/2} \tag{2.89}$$

where K_{crit} is the measured fracture toughness for thinner sections, K_{Ic} is the presumed asymptotic fracture toughness for large section thicknesses, and

$$\beta_{Ic} = \frac{1}{B}\left(\frac{K_{Ic}}{\sigma_{YS}}\right)^{2} \tag{2.90}$$

Unfortunately, this empirical fit to a few data sets has been grossly misused over the years. Equation (2.89) has often been applied to structural components in an effort to account for the presumed improvement in toughness in thinner sections. As Figure 2.50 illustrates, however, the stress state at the tip of a surface crack is not directly related to the plate thickness.

Equation (2.89) is suitable only for edge-cracked laboratory specimens that exhibit shear lips on the fracture surface, as Figure (2.45) illustrates. It does not apply to flaws in structural components, nor does it apply to side-grooved laboratory specimens or specimens that fail by cleavage.

2.11 MIXED-MODE FRACTURE

When two or more modes of loading are present, Equation (2.63) indicates that the energy release rate contributions from each mode are additive. This equation assumes self-similar crack growth, however. Consider the angled crack problem depicted in Figure 2.18. Equation (2.63) gives the energy release rate for planar crack growth at an angle $90° − \beta$ from the applied stress.

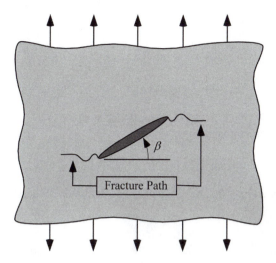

FIGURE 2.51 Typical propagation from an initial crack that is not orthogonal to the applied normal stress. The loading for the initial angled crack is a combination of Mode I and Mode II, but the crack tends to propagate normal to the applied stress, resulting in pure Mode I loading.

Figure 2.51 illustrates a more typical scenario for an angled crack. When fracture occurs, the crack tends to propagate orthogonal to the applied normal stress; i.e., the mixed-mode crack becomes a Mode I crack.

A propagating crack seeks the path of least resistance (or the path of maximum driving force) and need not be confined to its initial plane. If the material is isotropic and homogeneous, the crack will propagate in such a way as to maximize the energy release rate. What follows is an evaluation of the energy release rate as a function of propagation direction in mixed-mode problems. Only Mode I and Mode II are considered here, but the basic methodology can, in principle, be applied to a more general case where all three modes are present. This analysis is based on similar work in Refs. [34–36].

2.11.1 PROPAGATION OF AN ANGLED CRACK

We can generalize the angled through-thickness crack of Figure 2.18 to *any* planar crack oriented $90° - \beta$ from the applied normal stress. For uniaxial loading, the stress intensity factors for Mode I and Mode II are given by

$$K_I = K_{I(0)} \cos^2 \beta \qquad (2.91a)$$

$$K_{II} = K_{I(0)} \cos \beta \sin \beta \qquad (2.91b)$$

where $K_{I(0)}$ is the Mode I stress intensity when $\beta = 0$. The crack-tip stress fields (in polar coordinates) for the Mode I portion of the loading are given by

$$\sigma_{rr} = \frac{K_I}{\sqrt{2\pi r}} \left[\frac{5}{4} \cos \left(\frac{\theta}{2} \right) - \frac{1}{4} \cos \left(\frac{3\theta}{2} \right) \right] \qquad (2.92a)$$

$$\sigma_{\theta\theta} = \frac{K_I}{\sqrt{2\pi r}} \left[\frac{3}{4} \cos \left(\frac{\theta}{2} \right) + \frac{1}{4} \cos \left(\frac{3\theta}{2} \right) \right] \qquad (2.92b)$$

$$\tau_{r\theta} = \frac{K_I}{\sqrt{2\pi r}} \left[\frac{1}{4} \sin \left(\frac{\theta}{2} \right) + \frac{1}{4} \sin \left(\frac{3\theta}{2} \right) \right] \qquad (2.92c)$$

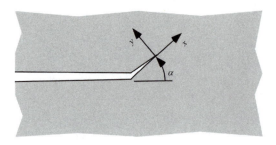

FIGURE 2.52 Infinitesimal kink at the tip of a macroscopic crack.

As stated earlier, these singular fields only apply as $r \to 0$. The singular stress fields for Mode II are given by

$$\sigma_{rr} = \frac{K_{II}}{\sqrt{2\pi r}}\left[-\frac{5}{4}\sin\left(\frac{\theta}{2}\right)+\frac{3}{4}\sin\left(\frac{3\theta}{2}\right)\right]$$ (2.93a)

$$\sigma_{\theta\theta} = \frac{K_{II}}{\sqrt{2\pi r}}\left[-\frac{3}{4}\sin\left(\frac{\theta}{2}\right)-\frac{3}{4}\sin\left(\frac{3\theta}{2}\right)\right]$$ (2.93b)

$$\tau_{r\theta} = \frac{K_{II}}{\sqrt{2\pi r}}\left[\frac{1}{4}\cos\left(\frac{\theta}{2}\right)+\frac{3}{4}\cos\left(\frac{3\theta}{2}\right)\right]$$ (2.93c)

Suppose that the crack in question forms an infinitesimal kink at an angle α from the plane of the crack, as Figure 2.52 illustrates. The local stress intensity factors at the tip of this kink differ from the nominal K values of the main crack. If we define a local x-y coordinate system at the tip of the kink and assume that Equation (2.92) and Equation (2.93) define the local stress fields, the local Mode I and Mode II stress intensity factors at the tip are obtained by summing the normal and shear stresses, respectively, at α:

$$k_I(\alpha) = \sigma_{yy}\sqrt{2\pi r} = C_{11}K_I + C_{12}K_{II}$$ (2.94a)

$$k_{II}(\alpha) = \tau_{xy}\sqrt{2\pi r} = C_{21}K_I + C_{22}K_{II}$$ (2.94b)

where k_I and k_{II} are the local stress intensity factors at the tip of the kink and K_I and K_{II} are the stress intensity factors for the main crack, which are given by Equation (2.91) for the tilted crack. The coefficients C_{ij} are given by

$$C_{11} = \frac{3}{4}\cos\left(\frac{\alpha}{2}\right)+\frac{1}{4}\cos\left(\frac{3\alpha}{2}\right)$$ (2.95a)

$$C_{12} = -\frac{3}{4}\left[\sin\left(\frac{\alpha}{2}\right)+\sin\left(\frac{3\alpha}{2}\right)\right]$$ (2.95b)

$$C_{21} = \frac{1}{4}\left[\sin\left(\frac{\alpha}{2}\right)+\sin\left(\frac{3\alpha}{2}\right)\right]$$ (2.95c)

$$C_{22} = \frac{1}{4}\cos\left(\frac{\alpha}{2}\right)+\frac{3}{4}\cos\left(\frac{3\alpha}{2}\right)$$ (2.95d)

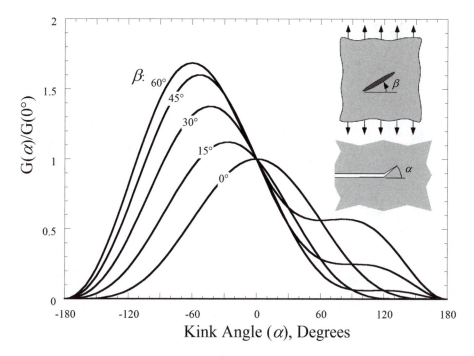

FIGURE 2.53 Local energy release rate at the tip of a kinked crack.

The energy release rate for the kinked crack is given by

$$\mathcal{G}(\alpha) = \frac{k_I^2(\alpha) + k_{II}^2(\alpha)}{E} \qquad (2.96)$$

Figure 2.53 is a plot of $\mathcal{G}(\alpha)$ normalized by $\mathcal{G}(\alpha = 0)$. The peak in $\mathcal{G}(\alpha)$ at each β corresponds to the point where k_I exhibits a maximum and $k_{II} = 0$. Thus, the maximum energy release rate is given by

$$\mathcal{G}_{\max} = \frac{k_I^2(\alpha^*)}{E} \qquad (2.97)$$

where α^* is the angle at which both $\mathcal{G}$ and k_I exhibit a maximum and $k_{II} = 0$. Crack growth in a homogeneous material should initiate along α^*.

Figure 2.54 shows the effect of β on the optimum propagation angle. The dashed line corresponds to propagation perpendicular to the remote principal stress. Note that the $\mathcal{G}_{\max}$ criterion implies an initial propagation plane that differs slightly from the normal to the remote stress.

2.11.2 EQUIVALENT MODE I CRACK

Let us now introduce an effective Mode I crack that results in the same stress intensity and energy release rate as a crack oriented at an angle β and propagating at an angle α^*:

$$K_I(a_{eq}) = k_I(\alpha^*, \beta, a) \qquad (2.98)$$

For the special case of a through-thickness crack in an infinite plate (Figure 2.18), Equation (2.98) becomes

$$\sigma \sqrt{\pi a_{eq}} = \sigma \sqrt{\pi a} \left[\cos^2 \beta C_{11}(\alpha^*) + \sin \beta \cos \beta C_{12}(\alpha^*) \right] \qquad (2.99)$$

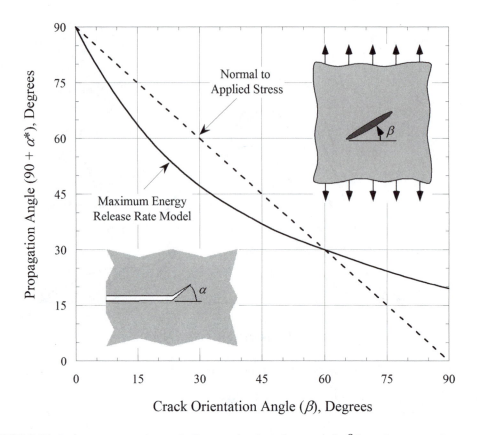

FIGURE 2.54 Optimum propagation angle for a crack oriented at an angle β from the stress axis.

Solving for a_{eq} gives

$$\frac{a_{eq}}{a} = \left[\cos^2 \beta C_{11}(\alpha^*) + \sin \beta \cos \beta C_{12}(\alpha^*)\right]^2 \qquad (2.100)$$

2.11.3 BIAXIAL LOADING

Figure 2.55 illustrates a cracked plate subject to principal stresses σ_1 and σ_2, where σ_1 is the greater of the two stresses; β is defined as the angle between the crack and the σ_1 plane. Applying superposition leads to the following expressions for K_I and K_{II}:

$$K_I = K_{I(0)}(\cos^2 \beta + B\sin^2 \beta) \qquad (2.101a)$$

$$K_{II} = K_{I(0)}(\sin \beta \cos \beta)(1 - B) \qquad (2.101b)$$

where B is the biaxiality ratio, defined as

$$B = \frac{\sigma_2}{\sigma_1} \qquad (2.102)$$

The local Mode I stress intensity for a kinked crack is obtained by substituting Equation (2.101) into Equation (2.94a):

$$k_I(\alpha) = K_{I(0)}\left[(\cos^2 \beta + B\sin^2 \beta)C_{11}(\alpha) + (\sin \beta \cos \beta)(1 - B)C_{12}(\alpha)\right] \qquad (2.103)$$

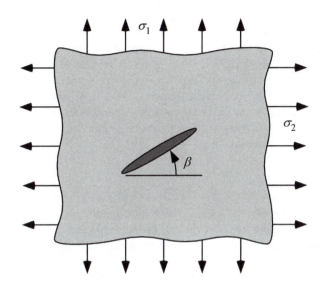

FIGURE 2.55 Cracked plane subject to a biaxial stress state.

The maximum local stress intensity factor and energy release rate occurs at the optimum propagation angle α^*, which depends on the biaxiality ratio. Figure 2.56 illustrates the effect of B and β on the propagation angle. Note that when $B > 0$ and $\beta = 90°$, propagation occurs in the crack plane ($\alpha^* = 0$), since the crack lies on a principal plane and is subject to pure Mode I loading.

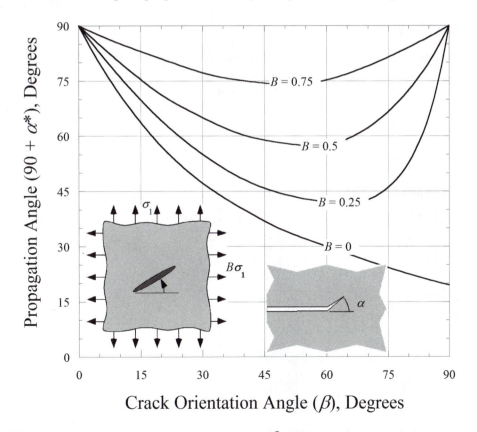

FIGURE 2.56 Optimum propagation angle as a function of β and biaxialty.

2.12 INTERACTION OF MULTIPLE CRACKS

The local stress field and crack driving force for a given flaw can be significantly affected by the presence of one or more neighboring cracks. Depending on the relative orientation of the neighboring cracks, the interaction can either magnify or diminish the stress intensity factor. An example of the former is an infinite array of coplanar cracks (Figure 2.21). The K_I solution for this configuration is given by Equation (2.45) and is plotted in Figure 2.22. When cracks are parallel to one another, K_I tends to decrease due to the interaction. The interaction of both coplanar and parallel cracks is discussed further in the following section.

2.12.1 COPLANAR CRACKS

Figure 2.57 illustrates two identical coplanar cracks in an infinite plate. The lines of force represent the relative stress-concentrating effect of the cracks. As the ligament between the cracks shrinks in size, the area through which the force must be transmitted decreases. Consequently, K_I is magnified for each crack as the two cracks approach one another.

Figure 2.58 is a plot of the K_I solution for the configuration in Figure 2.57. As one might expect, the crack tip closest to the neighboring crack experiences the greater magnification in K_I. The K_I solution at tip B increases asymptotically as $s \to 0$. At tip A, the solution approaches $\sqrt{2}$ as $s \to 0$ because the two cracks become a single crack with twice the original length of each crack.

Figure 2.57 and Figure 2.58 illustrate the general principle that multiple cracks in the same plane have the effect of magnifying K_I in one another.

2.12.2 PARALLEL CRACKS

Figure 2.59 illustrates two parallel cracks. In this case, the cracks tend to shield one another, which results in a decrease in K_I relative to the single crack case. Figure 2.60 shows the K_I solution for this geometry. This is indicative of the general case where two or more parallel cracks have a mutual shielding interaction when subject to Mode I loading. Consequently, multiple cracks that are parallel to one another are of less concern than multiple cracks in the same plane.

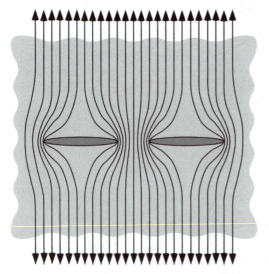

FIGURE 2.57 Coplanar cracks. Interaction between cracks results in a magnification of K_I.

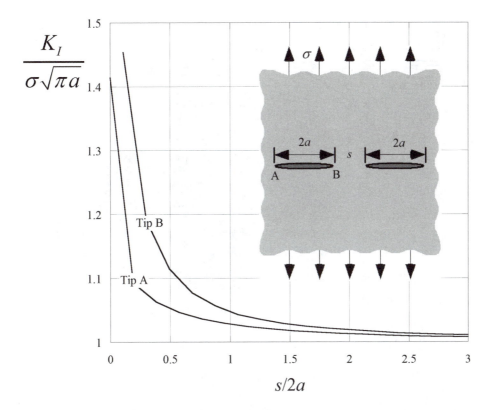

FIGURE 2.58 Interaction of two identical coplanar through-wall cracks in an infinite plate. Taken from Murakami, Y., *Stress Intensity Factors Handbook.* Pergamon Press, New York, 1987.

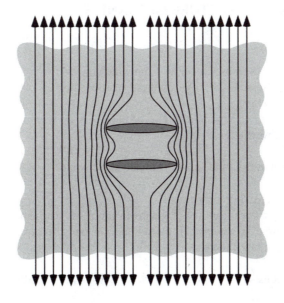

FIGURE 2.59 Parallel cracks. A mutual shielding effect reduces K_I in each crack.

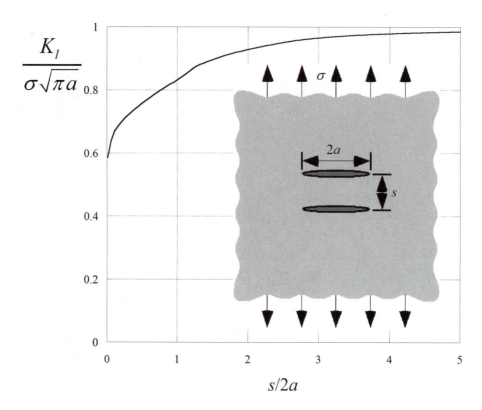

FIGURE 2.60 Interaction between two identical parallel through-wall cracks in an infinite plate. Taken from Murakami, Y., *Stress Intensity Factors Handbook*. Pergamon Press, New York, 1987.

APPENDIX 2: MATHEMATICAL FOUNDATIONS OF LINEAR ELASTIC FRACTURE MECHANICS

A2.1 PLANE ELASTICITY

This section catalogs the governing equations from which linear fracture mechanics is derived. The reader is encouraged to review the basis of these relationships by consulting one of the many textbooks on elasticity theory.[4]

The equations that follow are simplifications of more general relationships in elasticity and are subject to the following restrictions:

- Two-dimensional stress state (plane stress or plane strain)
- Isotropic material
- Quasistatic, isothermal deformation
- Absence of body forces from the problem (In problems where body forces are present, a solution can first be obtained in the absence of body forces, and then modified by superimposing the body forces.)

Imposing these restrictions simplifies crack problems considerably, and permits closed-form solutions in many cases.

---

[4] This appendix is intended only for more advanced readers, who have at least taken one graduate-level course in the theory of elasticity.

The governing equations of plane elasticity are given below for rectangular Cartesian coordinates. Section A2.1.2 lists the same relationships in terms of polar coordinates.

A2.1.1 Cartesian Coordinates

Strain-displacement relationships:

$$\varepsilon_{xx} = \frac{\partial u_x}{\partial x} \qquad \varepsilon_{yy} = \frac{\partial u_y}{\partial_y} \qquad \varepsilon_{xy} = \frac{1}{2}\left(\frac{\partial u_x}{\partial y} + \frac{\partial u_y}{\partial x}\right) \tag{A2.1}$$

where

x and y = horizontal and vertical coordinates

ε_{xx}, ε_{yy}, etc. = strain components

u_x and u_y = displacement components

Stress-strain relationships:

1. Plane strain

$$\sigma_{xx} = \frac{E}{(1+v)(1-2v)}[(1-v)\varepsilon_{xx} + v\varepsilon_{yy}] \tag{A2.2a}$$

$$\sigma_{yy} = \frac{E}{(1+v)(1-2v)}[(1-v)\varepsilon_{yy} + v\varepsilon_{xx}] \tag{A2.2b}$$

$$\tau_{xy} = 2\mu\varepsilon_{xy} = \frac{E}{1+v}\varepsilon_{xy} \tag{A2.2c}$$

$$\sigma_{zz} = v(\sigma_{xx} + \sigma_{yy}) \tag{A2.2d}$$

$$\varepsilon_{zz} = \varepsilon_{xz} = \varepsilon_{yz} = \tau_{xz} = \tau_{yz} = 0 \tag{A2.2e}$$

where

σ and τ = normal and shear stress components

E = Young's modulus

μ = shear modulus

v = Poisson's ratio

2. Plane stress

$$\sigma_{xx} = \frac{E}{1-v^2}[\varepsilon_{xx} + v\varepsilon_{yy}] \tag{A2.3a}$$

$$\sigma_{yy} = \frac{E}{1-v^2}[\varepsilon_{yy} + v\varepsilon_{xx}] \tag{A2.3b}$$

$$\tau_{xy} = 2\mu\varepsilon_{xy} = \frac{E}{1+v}\varepsilon_{xy} \tag{A2.3c}$$

$$\varepsilon_{zz} = \frac{-v}{1-v}(\varepsilon_{xx} + \varepsilon_{yy}) \tag{A2.3d}$$

$$\sigma_{zz} = \varepsilon_{xz} = \varepsilon_{yz} = \tau_{xz} = \tau_{yz} = 0 \tag{A2.3e}$$

Equilibrium equations:

$$\frac{\partial \sigma_{xx}}{\partial x} + \frac{\partial \tau_{xy}}{\partial y} = 0 \tag{A2.4a}$$

$$\frac{\partial \sigma_{yy}}{\partial y} + \frac{\partial \tau_{xy}}{\partial x} = 0 \tag{A2.4b}$$

Compatibility equation:

$$\nabla^2 (\sigma_{xx} + \sigma_{yy}) = 0 \tag{A2.5}$$

where

$$\nabla^2 = \frac{\partial^2}{\partial x^2} + \frac{\partial^2}{\partial y^2}$$

Airy stress function: For a two-dimensional continuous elastic medium, there exists a function $\Phi(x, y)$ from which the stresses can be derived:

$$\sigma_{xx} = \frac{\partial^2 \Phi}{\partial y^2} \tag{A2.6a}$$

$$\sigma_{yy} = \frac{\partial^2 \Phi}{\partial x^2} \tag{A2.6b}$$

$$\tau_{xy} = -\frac{\partial^2 \Phi}{\partial x \partial y} \tag{A2.6c}$$

where Φ is the Airy stress function. The equilibrium and compatibility equations are automatically satisfied if Φ has the following property:

$$\frac{\partial^4 \Phi}{\partial x^4} + 2\frac{\partial^4 \Phi}{\partial x^2 \partial y^2} + \frac{\partial^4 \Phi}{\partial y^4} = 0$$

or

$$\nabla^2 \nabla^2 \Phi = 0 \tag{A2.7}$$

A2.1.2   Polar Coordinates

Strain-displacement relationships:

$$\varepsilon_{rr} = \frac{\partial u_r}{\partial r} \tag{A2.8a}$$

$$\varepsilon_{\theta\theta} = \frac{u_r}{r} + \frac{1}{r}\frac{\partial u\theta}{\partial \theta} \tag{A2.8b}$$

$$\varepsilon_{r\theta} = \frac{1}{2}\left(\frac{1}{r}\frac{\partial u_r}{\partial \theta} + \frac{\partial u_\theta}{\partial r} - \frac{u_\theta}{r}\right) \tag{A2.8c}$$

where u_r and u_θ are the radial and tangential displacement components, respectively.

Stress-strain relationships: The stress-strain relationships in polar coordinates can be obtained by substituting r and θ for x and y in Equation (A2.2) and Equation (A2.3). For example, the radial stress is given by

$$\sigma_{rr} = \frac{E}{(1+v)(1-2v)}[(1-v)\varepsilon_{rr} + v\varepsilon_{\theta\theta}] \tag{A2.9a}$$

for plane strain, and

$$\sigma_{rr} = \frac{E}{1-v^2}[\varepsilon_{rr} + v\varepsilon_{\theta\theta}] \tag{A2.9b}$$

for plane stress.

Equilibrium equations:

$$\frac{\partial \sigma_{rr}}{\partial r} + \frac{1}{r}\frac{\partial \tau_{r\theta}}{\partial \theta} + \frac{\sigma_{rr} - \sigma_{\theta\theta}}{r} = 0 \tag{A2.10a}$$

$$\frac{1}{r}\frac{\partial \sigma_{\theta\theta}}{\partial \theta} + \frac{\partial \tau_{r\theta}}{\partial r} + \frac{2\tau_{r\theta}}{r} = 0 \tag{A2.10b}$$

Compatibility equation:

$$\nabla^2(\sigma_{rr} + \sigma_{\theta\theta}) = 0 \tag{A2.11}$$

where

$$\nabla^2 + \frac{\partial^2}{\partial R^2} + \frac{1}{r}\frac{\partial}{\partial r} + \frac{1}{r^2}\frac{\partial^2}{\partial \theta^2}$$

Airy stress function

$$\nabla^2\nabla^2\Phi = 0 \tag{A2.12}$$

where $\Phi = \Phi(r, \theta)$ and

$$\sigma_{rr} = \frac{1}{r^2}\frac{\partial^2\Phi}{\partial \theta^2} + \frac{1}{r}\frac{\partial\Phi}{\partial r} \tag{A2.13a}$$

$$\sigma_{\theta\theta} = \frac{\partial^2\Phi}{\partial r^2} \tag{A2.13b}$$

$$\tau_{r\theta} = -\frac{1}{r}\frac{\partial^2\Phi}{\partial r\partial\theta} + \frac{1}{r^2}\frac{\partial\Phi}{\partial\theta} \tag{A2.13c}$$

A2.2 CRACK GROWTH INSTABILITY ANALYSIS

Figure 2.12 schematically illustrates the general case of a cracked structure with finite system compliance C_M. The structure is held at a fixed remote displacement Δ_T given by

$$\Delta_T = \Delta + C_M P \tag{A2.14}$$

where Δ is the local load line displacement and P is the applied load. Differentiating Equation (A2.14) gives

$$d\Delta_T = \left(\frac{\partial\Delta}{\Delta a}\right)_P da + \left(\frac{\partial\Delta}{\partial P}\right)_a dP + C_M dP = 0 \qquad (A2.15)$$

assuming Δ depends only on load and crack length. We can make this same assumption about the energy release rate:

$$d\mathcal{G} = \left(\frac{\partial\mathcal{G}}{\partial a}\right)_P da + \left(\frac{\partial\mathcal{G}}{\partial P}\right)_a dP \qquad (A2.16)$$

Dividing both sides of Equation (A2.16) by da and fixing Δ_T yields

$$\left(\frac{d\mathcal{G}}{da}\right)_{\Delta_T} = \left(\frac{\partial\mathcal{G}}{\partial a}\right)_P + \left(\frac{\partial\mathcal{G}}{\partial P}\right)_a \left(\frac{dP}{da}\right)_{\Delta_T} \qquad (A2.17)$$

which, upon substitution of Equation (A2.15), leads to

$$\left(\frac{d\mathcal{G}}{da}\right)_{\Delta_T} = \left(\frac{\partial\mathcal{G}}{\partial a}\right)_P - \left(\frac{\partial\mathcal{G}}{\partial P}\right)_a \left(\frac{\partial\Delta}{\partial a}\right)_P \left[C_M + \left(\frac{\partial\Delta}{\partial P}\right)_a\right]^{-1} \qquad (A2.18)$$

A virtually identical expression for the J integral (Equation 3.52) can be derived by assuming J depends only on P and a, and expanding dJ into its partial derivatives.

Under dead-loading conditions, $C_M = \infty$, and all but the first term in Equation (A2.18) vanish. Conversely, $C_M = 0$ corresponds to an infinitely stiff system, and Equation (A2.18) reduces to the pure displacement control case.

A2.3 CRACK-TIP STRESS ANALYSIS

A variety of techniques are available for analyzing stresses in cracked bodies. This section focuses on two early approaches developed by Williams [11, 37] and Westergaard [8]. These two analyses are complementary; the Williams approach considers the local crack-tip fields under generalized in-plane loading, while Westergaard provided a means for connecting the local fields to global boundary conditions in certain configurations.

Space limitations preclude listing every minute step in each derivation. Moreover, stress, strain, and displacement distributions are not derived for all modes of loading. The derivations that follow serve as illustrative examples. The reader who is interested in further details should consult the original references.

A2.3.1 Generalized In-Plane Loading

Williams was the first to demonstrate the universal nature of the $1/\sqrt{r}$ singularity for elastic crack problems, although Inglis [1], Westergaard, and Sneddon [10] had earlier obtained this result in specific configurations. Williams actually began by considering stresses at the corner of a plate with various boundary conditions and included angles; a crack is a special case where the included angle of the plate corner is 2π and the surfaces are traction free (Figure A2.1).

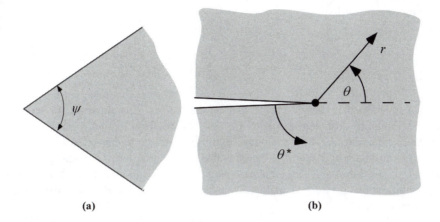

FIGURE A2.1 Plate corner configuration analyzed by Williams. A crack is formed when $\psi = 2\pi$: (a) plate corner with included angle ψ and (b) special case of a sharp crack. Taken from Williams, J.G. and Ewing, P.D., *International Journal of Fracture Mechanics,* Vol. 8, 1972.

For the configuration shown in Figure A2.1(b), Williams postulated the following stress function:

$$\Phi = r^{\lambda+1}\left[c_1 \sin(\lambda+1)\theta^* + c_2 \cos(\lambda+1)\theta^* + c_3 \sin(\lambda-1)\theta^* + c_4 \cos(\lambda-1)\theta^*\right]$$
$$= r^{\lambda+1}\Phi(\theta^*,\lambda) \tag{A2.19}$$

where c_1, c_2, c_3, and c_4 are constants, and θ^* is defined in Figure A2.1(b). Invoking Equation (A2.13) gives the following expressions for the stresses:

$$\sigma_{rr} = r^{\lambda-1}[F''(\theta^*) + (\lambda+1)F(\theta^*)] \tag{A2.20a}$$

$$\sigma_{\theta\theta} = r^{\lambda-1}[\lambda(\lambda+1)F(\theta^*)] \tag{A2.20b}$$

$$\tau_{r\theta} = r^{\lambda-1}[-\lambda F'(\theta^*)] \tag{A2.20c}$$

where the primes denote derivatives with respect to θ^*. Williams also showed that Equation (A2.19) implies that the displacements vary with r^λ. In order for displacements to be finite in all regions of the body, λ must be > 0. If the crack faces are traction free, $\sigma_{\theta\theta}(0) = \sigma_{\theta\theta}(2\pi) = \tau_{r\theta}(0) = \tau_{r\theta} = (2\pi) = 0$, which implies the following boundary conditions:

$$F(0) = F(2\pi) = F'(0) = F'(2\pi) = 0 \tag{A2.21}$$

Assuming the constants in Equation (A2.19) are nonzero in the most general case, the boundary conditions can be satisfied only when $\sin(2\pi\lambda) = 0$. Thus,

$$\lambda = \frac{n}{2}, \quad \text{where } n = 1, 2, 3, \dots$$

There are an infinite number of λ values that satisfy the boundary conditions; the most general solution to a crack problem, therefore, is a polynomial of the form

$$\Phi = \sum_{n=1}^{N}\left(r^{\frac{n}{2}+1}F\left(\theta^*,\frac{n}{2}\right)\right) \tag{A2.22}$$

and the stresses are given by

$$\sigma_{ij} = \frac{\Gamma_{ij}\left(\theta^*, -\frac{1}{2}\right)}{\sqrt{r}} + \sum_{m=0}^{M}\left(r^{m/2}\Gamma_{ij}(\theta^*, m)\right) \tag{A2.23}$$

where Γ is a function that depends on F and its derivatives. The order of the stress function polynomial, M, must be sufficient to model the stresses in all regions of the body. When $r \to 0$, the first term in Equation (A2.23) approaches infinity, while the higher-order terms remain finite (when $m = 0$) or approach zero (for $m > 0$). Thus the higher-order terms are negligible close to the crack-tip, and stress exhibits a $1/\sqrt{r}$ singularity. Note that this result was obtained without assuming a specific configuration. It can be concluded that the inverse square-root singularity is universal for cracks in isotropic elastic media.

A further evaluation of Equation (A2.19) and Equation (A2.20) with the appropriate boundary conditions reveals the precise nature of the function Γ. Recall that Equation (A2.19) contains four, as yet unspecified, constants; by applying Equation (A2.21), it is possible to eliminate two of these constants, resulting in

$$\Phi(r, \theta) = r^{n/2+1}\left\{c_3\left[\sin\left(\frac{n}{2}-1\right)\theta^* - \frac{n-2}{n+2}\sin\left(\frac{n}{2}+1\right)\theta^*\right] + c_4\left[\cos\left(\frac{n}{2}-1\right)\theta - \cos\left(\frac{n}{2}+1\right)\theta\right]\right\} \tag{A2.24}$$

for a given value of n. For crack problems, it is more convenient to express the stress function in terms of θ, the angle from the symmetry plane (Figure A2.1). Substituting $\theta = \theta^* - \pi$ into Equation (A2.24) yields, after some algebra, the following stress function for the first few values of n:

$$\Phi(r, \theta) = r^{3/2}\left[s_1\left(-\cos\frac{\theta}{2} - \frac{1}{3}\cos\frac{3\theta}{2}\right) + t_1\left(-\sin\frac{\theta}{2} - \sin\frac{3\theta}{2}\right)\right] + s_2 r^2[1 - \cos 2\theta] + O(r^{5/2}) + \cdots \tag{A2.25}$$

where s_i and t_i are constants to be defined. The stresses are given by

$$\sigma_{rr} = \frac{1}{4\sqrt{r}}\left\{s_1\left[-5\cos\frac{\theta}{2} + \cos\frac{3\theta}{2}\right] + t_1\left[-5\sin\frac{\theta}{2} + 3\sin\frac{3\theta}{2}\right]\right\} + 4s_2\cos^2\theta + O(r^{1/2}) + \cdots \tag{A2.26a}$$

$$\sigma_{\theta\theta} = \frac{1}{4\sqrt{r}}\left\{s_1\left[-3\cos\frac{\theta}{2} - \cos\frac{3\theta}{2}\right] + t_1\left[-3\sin\frac{\theta}{2} - 3\sin\frac{3\theta}{2}\right]\right\} + 4s_2\sin^2\theta + O(r^{1/2}) + \cdots \tag{A2.26b}$$

$$\tau_{r\theta} = \frac{1}{4\sqrt{r}}\left\{s_1\left[-\sin\frac{\theta}{2} - \sin\frac{3\theta}{2}\right] + t_1\left[\cos\frac{\theta}{2} + 3\cos\frac{3\theta}{2}\right]\right\} - 2s_2\sin 2\theta + O(r^{1/2}) + \cdots \tag{A2.26c}$$

Note that the constants s_i in the stress function (Equation (A2.25)) are multiplied by cosine terms while the t_i are multiplied by sine terms. Thus, the stress function contains symmetric and antisymmetric components, with respect to $\theta = 0$. When the loading is symmetric about $\theta = 0$, $t_i = 0$, while $s_i = 0$ for the special case of pure antisymmetric loading. Examples of symmetric loading include pure bending and pure tension; in both cases the principal stress is normal to the crack plane. Therefore, symmetric loading corresponds to Mode I (Figure 2.14); antisymmetric loading is produced by in-plane shear on the crack faces and corresponds to Mode II.

It is convenient in most cases to treat the symmetric and antisymmetric stresses separately. The constants s_1 and t_1 can be replaced by the Mode I and Mode II stress intensity factors, respectively:

$$s_1 = -\frac{K_I}{\sqrt{2\pi}} \tag{A2.27a}$$

$$t_1 = \frac{K_{II}}{\sqrt{2\pi}} \tag{A2.27b}$$

The crack-tip stress fields for symmetric (Mode I) loading (assuming the higher-order terms are negligible) are given by

$$\sigma_{rr} = \frac{K_I}{\sqrt{2\pi r}}\left[\frac{5}{4}\cos\left(\frac{\theta}{2}\right) - \frac{1}{4}\cos\left(\frac{3\theta}{2}\right)\right] \tag{A2.28a}$$

$$\sigma_{\theta\theta} = \frac{K_I}{\sqrt{2\pi r}}\left[\frac{3}{4}\cos\left(\frac{\theta}{2}\right) + \frac{1}{4}\cos\left(\frac{3\theta}{2}\right)\right] \tag{A2.28b}$$

$$\tau_{r\theta} = \frac{K_I}{\sqrt{2\pi r}}\left[\frac{1}{4}\sin\left(\frac{\theta}{2}\right) + \frac{1}{4}\sin\left(\frac{3\theta}{2}\right)\right] \tag{A2.28c}$$

The singular stress fields for Mode II are given by

$$\sigma_{rr} = \frac{K_{II}}{\sqrt{2\pi r}}\left[-\frac{5}{4}\sin\left(\frac{\theta}{2}\right) + \frac{3}{4}\sin\left(\frac{3\theta}{2}\right)\right] \tag{A2.29a}$$

$$\sigma_{\theta\theta} = \frac{K_{II}}{\sqrt{2\pi r}}\left[-\frac{3}{4}\sin\left(\frac{\theta}{2}\right) - \frac{3}{4}\sin\left(\frac{3\theta}{2}\right)\right] \tag{A2.29b}$$

$$\tau_{r\theta} = \frac{K_{II}}{\sqrt{2\pi r}}\left[\frac{1}{4}\cos\left(\frac{\theta}{2}\right) + \frac{3}{4}\cos\left(\frac{3\theta}{2}\right)\right] \tag{A2.29c}$$

The relationships in Table 2.1 can be obtained by converting Equation (A2.28) and Equation (A2.29) to Cartesian coordinates.

The stress intensity factor defines the amplitude of the crack-tip singularity; all the stress and strain components at points near the crack tip increase in proportion to K, provided the crack is stationary. The precise definition of the stress intensity factor is arbitrary, however; the constants s_1 and t_1 would serve equally well for characterizing the singularity. The accepted definition of stress intensity stems from the early work of Irwin [9], who quantified the amplitude of the Mode I singularity with $\sqrt{\mathcal{G}E}$, where $\mathcal{G}$ is the energy release rate. It turns out that the $\sqrt{\pi}$ in the denominators of Equation (A2.28) and Equation (A2.29) is superfluous (see Equation (A2.34)–(A2.36), but convention established over the last 35 years precludes redefining K in a more straightforward form.

Williams also derived relationships for radial and tangential displacements near the crack tip. We will postpone the evaluation of displacements until the next section, however, because the Westergaard approach for deriving displacements is somewhat more compact.

A2.3.2 The Westergaard Stress Function

Westergaard showed that a limited class of problems could be solved by introducing a complex stress function $Z(z)$, where $z = x + iy$ and $i = \sqrt{-1}$. The Westergaard stress function is related to the Airy stress function as follows:

$$\Phi = \operatorname{Re} \overline{\overline{Z}} + y \operatorname{Im} \overline{Z} \tag{A2.30}$$

where Re and Im denote real and imaginary parts of the function, respectively, and the bars over Z represent integrations with respect to z, i.e.,

$$\overline{\overline{Z}} = \frac{d\overline{\overline{Z}}}{dz} \quad \text{and} \quad \overline{Z} = \frac{d\overline{Z}}{dz}$$

Applying Equation (A2.6) gives

$$\sigma_{xx} = \operatorname{Re} Z - y \operatorname{Im} Z' \tag{A2.31a}$$

$$\sigma_{yy} = \operatorname{Re} Z + y \operatorname{Im} Z' \tag{A2.31b}$$

$$\tau_{xy} = -y \operatorname{Re} Z' \tag{A2.31c}$$

Note that the imaginary part of the stresses vanishes when $y = 0$. In addition, the shear stress vanishes when $y = 0$, implying that the crack plane is a principal plane. Thus, the stresses are symmetric about $\theta = 0$ and Equation (A2.31) implies Mode I loading.

The Westergaard stress function, in its original form, is suitable for solving a limited range of Mode I crack problems. Subsequent modifications [38–41] generalized the Westergaard approach to be applicable to a wider range of cracked configurations.

Consider a through crack in an infinite plate subject to biaxial remote tension (Figure A2.2). If the origin is defined at the center of the crack, the Westergaard stress function is given by

$$Z(z) = \frac{\sigma z}{\sqrt{z^2 - a^2}} \tag{A2.32}$$

where σ is the remote stress and a is the half-crack length, as defined in Figure A2.2. Consider the crack plane where $y = 0$. For $-a < x < a$, Z is pure imaginary, while Z is real for $|x| > |a|$. The normal stresses on the crack plane are given by

$$\sigma_{xx} = \sigma_{yy} = \operatorname{Re} Z = \frac{\sigma x}{\sqrt{x^2 - a^2}} \tag{A2.33}$$

Let us now consider the horizontal distance from each crack tip, $x^* = x - a$, Equation (A2.33) becomes

$$\sigma_{xx} = \sigma_{yy} = \frac{\sigma \sqrt{a}}{\sqrt{2x^*}} \tag{A2.34}$$

for $x^* \ll a$. Thus, the Westergaard approach leads to the expected inverse square-root singularity. One advantage of this analysis is that it relates the local stresses to the global stress and crack size. From Equation (A2.28), the stresses on the crack plane ($\theta = 0$) are given by

$$\sigma_{rr} = \sigma_{\theta\theta} = \sigma_{xx} = \sigma_{yy} = \frac{K_I}{\sqrt{2\pi x^*}} \tag{A2.35}$$

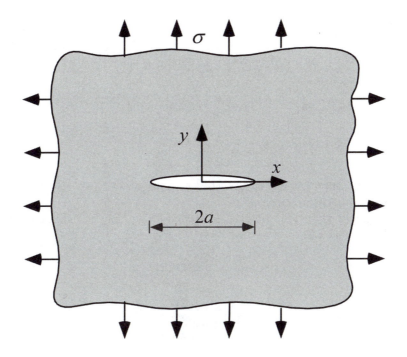

FIGURE A2.2 Through-thickness crack in an infinite plate loaded in biaxial tension.

Comparing Equation (A2.34) and Equation (A2.35) gives

$$K_I = \sigma\sqrt{\pi a} \qquad\qquad (A2.36)$$

for the configuration in Figure A2.2. Note that $\sqrt{\pi}$ appears in Equation (A2.36) because K was originally defined in terms of the energy release rate; an alternative definition of stress intensity might be

$$\sigma_{yy}(\theta = 0) = \frac{K_I^*}{\sqrt{2x^*}} \qquad \text{where} \qquad K_I^* = \sigma\sqrt{a}$$

for the plate in Figure A2.2.

Substituting Equation (A3.26) into Equation (A2.32) results in an expression of the Westergaard stress function in terms of K_I:

$$Z(z^*) = \frac{K_I}{\sqrt{2\pi z^*}} \qquad\qquad (A2.37)$$

where $z^* = z - a$. It is possible to solve for the singular stresses at other angles by making the following substitution in Equation (A2.37):

$$z^* = re^{i\theta}$$

where

$$r^2 = (x - a)^2 + y^2 \qquad \text{and} \qquad \theta = \tan^{-1}\left(\frac{y}{x - a}\right)$$

which leads to

$$\sigma_{xx} = \frac{K_I}{\sqrt{2\pi r}} \cos\left(\frac{\theta}{2}\right)\left[1 - \sin\left(\frac{\theta}{2}\right)\sin\left(\frac{3\theta}{2}\right)\right] \tag{A2.38a}$$

$$\sigma_{yy} = \frac{K_I}{\sqrt{2\pi r}} \cos\left(\frac{\theta}{2}\right)\left[1 + \sin\left(\frac{\theta}{2}\right)\sin\left(\frac{3\theta}{2}\right)\right] \tag{A2.38b}$$

$$\tau_{xy} = \frac{K_I}{\sqrt{2\pi r}} \cos\left(\frac{\theta}{2}\right)\sin\left(\frac{\theta}{2}\right) \tag{A2.38c}$$

assuming $r \gg a$. Equation (A2.38) is equivalent to Equation (A2.28), except that the latter is expressed in terms of polar coordinates.

Westergaard published the following stress function for an array of collinear cracks in a plate in biaxial tension (Figure 2.21):

$$Z(z) = \frac{\sigma}{\left\{1 - \left[\frac{\sin\left(\frac{\pi a}{2W}\right)}{\sin\left(\frac{\pi z}{2W}\right)}\right]^2\right\}^{1/2}} \tag{A2.39}$$

where a is the half-crack length and $2W$ is the spacing between the crack centers. The stress intensity for this case is given in Equation (2.45); early investigators used this solution to approximate the behavior of a center-cracked tensile panel with finite width.

Irwin [9] published stress functions for several additional configurations, including a pair of crack-opening forces located at a distance X from the crack center (Figure 2.32):

$$Z(z) = \frac{Pa}{p(z-X)z}\sqrt{\frac{1-(X/a)^2}{1-(a/z)^2}} \tag{A2.40}$$

where P is the applied force. When there are matching forces at $\pm X$, the appropriate stress function can be obtained by superposition:

$$Z(z) = \frac{2Pa}{\pi(z^2-X^2)}\sqrt{\frac{1-(X/a)^2}{1-(a/z)^2}} \tag{A2.41}$$

In each case, the stress function can be expressed in the form of Equation (A2.37) and the near-tip stresses are given by Equation (A2.38). This is not surprising, since all of the above cases are pure Mode I and the Williams analysis showed that the inverse square root singularity is universal.

For plane strain conditions, the in-plane displacements are related to the Westergaard stress function as follows:

$$u_x = \frac{1}{2\mu}[(1-2v)\operatorname{Re}\overline{Z} - y\operatorname{Im}Z] \tag{A2.42a}$$

$$u_y = \frac{1}{2\mu}[2(1-v)\operatorname{Im}\overline{Z} - y\operatorname{Re}Z] \tag{A2.42b}$$

For the plate in Figure A2.2, the crack-opening displacement is given by

$$2u_y = \frac{1-v}{\mu}\operatorname{Im}\overline{Z} = \frac{2(1-v^2)}{E}\operatorname{Im}\overline{Z} = \frac{4(1-v^2)\sigma}{E}\sqrt{a^2-x^2} \tag{A2.43a}$$

assuming plane strain, and

$$2u_y = \frac{4\sigma}{E}\sqrt{a^2 - x^2} \tag{A2.43b}$$

for plane stress. Equation (A2.43) predicts that a through crack forms an elliptical opening profile when subjected to tensile loading.

The near-tip displacements can be obtained by inserting Equation (A2.37) into Equation (A2.42):

$$u_x = \frac{K_I}{2\mu}\sqrt{\frac{r}{2\pi}}\cos\left(\frac{\theta}{2}\right)\left[\kappa - 1 + 2\sin^2\left(\frac{\theta}{2}\right)\right] \tag{A2.44a}$$

$$u_y = \frac{K_I}{2\mu}\sqrt{\frac{r}{2\pi}}\sin\left(\frac{\theta}{2}\right)\left[\kappa + 1 - 2\cos^2\left(\frac{\theta}{2}\right)\right] \tag{A2.44b}$$

for $r \ll a$, where

$$\kappa = 3 - 4v \quad \text{for plane strain} \tag{A2.45a}$$

and

$$k = \frac{3 - v}{1 + v} \quad \text{for plane stress} \tag{A2.45b}$$

Although the original Westergaard approach correctly describes the singular Mode I stresses in certain configurations, it is not sufficiently general to apply to all Mode I problems; this shortcoming has prompted various modifications to the Westergaard stress function. Irwin [38] noted that photoelastic fringe patterns observed by Wells and Post [42] on center-cracked panels did not match the shear strain contours predicted by the Westergaard solution. Irwin achieved a good agreement between theory and experiment by subtracting a uniform horizontal stress:

$$\sigma_{xx} = \mathrm{Re}\, Z - y\, \mathrm{Im}\, Z' - \sigma_{oxx} \tag{A2.46}$$

where σ_{oxx} depends on the remote stress. The other two stress components remain the same as in Equation (A2.31). Subsequent analyses have revealed that when a center-cracked panel is loaded in uniaxial tension, a transverse compressive stress develops in the plate. Thus, Irwin's modification to the Westergaard solution has a physical basis in the case of a center-cracked panel.[5] Equation (A2.46) has been used to interpret photoelastic fringe patterns in a variety of configurations.

Sih [39] provided a theoretical basis for the Irwin modification. A stress function for Mode I must lead to zero shear stress on the crack plane. Sih showed that the Westergaard function was more restrictive than it needed to be, and was thus unable to account for all situations. Sih generalized the Westergaard approach by applying a complex potential formulation for the Airy stress function [39]. He imposed the condition $\tau_{xy} = 0$ at $y = 0$, and showed that the stresses could be expressed in terms of a new function $\phi(z)$:

$$\sigma_{xx} = 2\,\mathrm{Re}\,\phi'(z) - 2y\,\mathrm{Im}\,\phi''(z) - A \tag{A2.47a}$$

$$\sigma_{yy} = 2\,\mathrm{Re}\,\phi'(z) + 2y\,\mathrm{Im}\,\phi''(z) + A \tag{A2.47b}$$

$$\tau_{xy} = 2y\,\mathrm{Re}\,\phi''(z) \tag{A2.47c}$$

---

[5] Recall that the stress function in Equation (A2.32) is strictly valid only for biaxial loading. Although this restriction was not imposed in Westergaard's original work, a transverse tensile stress is necessary in order to cancel with $-\sigma_{oxx}$. However, the transverse stresses, whether compressive or tensile, do not affect the singular term; thus the stress intensity factor is the same for uniaxial and biaxial tensile loading and is given by Equation (A2.36).

where A is a real constant. Equation (A2.47) is equivalent to the Irwin modification of the Westergaard approach if

$$2\phi'(z) = Z(z) - A \tag{A2.48}$$

Substituting Equation (A2.48) into Equation (A2.47) gives

$$\sigma_{xx} = \text{Re}\,Z - y\,\text{Im}\,Z' - 2A \tag{A2.49a}$$

$$\sigma_{yy} = \text{Re}\,Z + y\,\text{Im}\,Z' \tag{A2.49b}$$

$$\tau_{xy} = -y\,\text{Re}\,Z \tag{A2.49c}$$

Comparing Equation (A2.49) with Equation (A2.31) and Equation (A2.46), it is obvious that the Sih and Irwin modifications are equivalent, and $2A = \sigma_{oxx}$.

Sanford [41] showed that the Irwin-Sih approach is still too restrictive, and he proposed replacing A with a complex function $\eta(z)$:

$$2\phi'(z) = Z(z) - \eta(z) \tag{A2.50}$$

The modified stresses are given by

$$\sigma_{xx} = \text{Re}\,Z - y\,\text{Im}\,Z' + y\,\text{Im}\,\eta' - 2\,\text{Re}\,\eta \tag{A2.51a}$$

$$\sigma_{yy} = \text{Re}\,Z + y\,\text{Im}\,Z' + y\,\text{Im}\,\eta' \tag{A2.51b}$$

$$\tau_{xy} = -y\,\text{Re}\,Z + y\,\text{Re}\,\eta' + \text{Im}\,\eta \tag{A2.51c}$$

Equation (A2.51) represents the most general form of Westergaard-type stress functions. When $\eta(z) = $ a real constant for all z, Equation (A2.51) reduces to the Irwin-Sih approach, while Equation (A2.51) reduces to the original Westergaard solution when $\eta(z) = 0$ for all z.

The function η can be represented as a polynomial of the form

$$\eta(z) = \sum_{m=0}^{M} \alpha_m z^{m/2} \tag{A2.52}$$

Combining Equation (A2.37), Equation (A2.50), and Equation (A2.52) and defining the origin at the crack tip gives

$$2\phi' = \frac{K_I}{\sqrt{2\pi z}} - \sum_{m=0}^{M} \alpha_m z^{m/2} \tag{A2.53}$$

which is consistent with the Williams [11, 38] asymptotic expansion.

A2.4 ELLIPTICAL INTEGRAL OF THE SECOND KIND

The solution of stresses in the vicinity of elliptical and semielliptical cracks in elastic solids [10, 44] involves an elliptic integral of the second kind:

$$\Psi = \int_0^{\pi/2} \sqrt{1 - \frac{c^2 - a^2}{c^2} \sin^2 \phi}\, d\phi \tag{A2.54}$$

where $2c$ and $2a$ are the major and minor axes of the elliptical flaw, respectively. A series expansion of Equation (A2.54) gives

$$\Psi = \frac{\pi}{2}\left[1 - \frac{1}{4}\frac{c^2 - a^2}{c^2} - \frac{3}{64}\left(\frac{c^2 - a^2}{c^2}\right)^2 - \cdots \right] \qquad (A2.55)$$

Most stress intensity solutions for elliptical and semiellipical cracks in published literature are written in terms of a flaw shape parameter Q, which can be approximated by

$$Q = \Psi^2 \approx 1 + 1.464\left(\frac{a}{c}\right)^{1.65} \qquad (A2.56)$$

REFERENCES

1. Inglis, C.E., "Stresses in a Plate Due to the Presence of Cracks and Sharp Corners." *Transactions of the Institute of Naval Architects,* Vol. 55, 1913, pp. 219–241.
2. Griffith, A.A., "The Phenomena of Rupture and Flow in Solids." *Philosophical Transactions,* Series A, Vol. 221, 1920, pp. 163–198.
3. Gehlen, P.C. and Kanninen, M.F., "An Atomic Model for Cleavage Crack Propagation in Iron." *Inelastic Behavior of Solids,* McGraw-Hill, New York, 1970, pp. 587–603.
4. Irwin, G.R., "Fracture Dynamics." *Fracturing of Metals,* American Society for Metals, Cleveland, OH, 1948, pp. 147–166.
5. Orowan, E., "Fracture and Strength of Solids." *Reports on Progress in Physics,* Vol. XII, 1948, p. 185.
6. Irwin, G.R., "Onset of Fast Crack Propagation in High Strength Steel and Aluminum Alloys." *Sagamore Research Conference Proceedings,* Vol. 2, 1956, pp. 289–305.
7. Hutchinson, J.W. and Paris, P.C., "Stability Analysis of J-Controlled Crack Growth." ASTM STP 668, American Society for Testing and Materials, Philadelphia, PA, 1979, pp. 37–64.
8. Westergaard, H.M., "Bearing Pressures and Cracks." *Journal of Applied Mechanics,* Vol. 6, 1939, pp. 49–53.
9. Irwin, G.R., "Analysis of Stresses and Strains near the End of a Crack Traversing a Plate." *Journal of Applied Mechanics,* Vol. 24, 1957, pp. 361–364.
10. Sneddon, I.N., "The Distribution of Stress in the Neighbourhood of a Crack in an Elastic Solid." *Proceedings, Royal Society of London,* Vol. A-187, 1946, pp. 229–260.
11. Williams, M.L., "On the Stress Distribution at the Base of a Stationary Crack." *Journal of Applied Mechanics,* Vol. 24, 1957, pp. 109–114.
12. Tada, H., Paris, P.C., and Irwin, G.R., *The Stress Analysis of Cracks Handbook.* 2nd Ed., Paris Productions, St. Louis, MO, 1985.
13. Murakami, Y., *Stress Intensity Factors Handbook.* Pergamon Press, New York, 1987.
14. Rooke, D.P. and Cartwright, D.J., *Compendium of Stress Intensity Factors.* Her Majesty's Stationery Office, London, 1976.
15. Rice, J.R., "Some Remarks on Elastic Crack-Tip Stress Fields." *International Journal of Solids and Structures,* Vol. 8, 1972, pp. 751–758.
16. Bueckner, H.F., "A Novel Principle for the Computation of Stress Intensity Factors." *Zeitschrift für Angewandte Mathematik und Mechanik,* Vol. 50, 1970, pp. 529–545.
17. Rice, J.R., "Weight Function Theory for Three-Dimensional Elastic Crack Analysis." ASTM STP 1020, American Society for Testing and Materials, Philadelphia, PA, 1989, pp. 29–57.
18. Parks, D.M. and Kamentzky, E.M., "Weight Functions from Virtual Crack Extension." *International Journal for Numerical Methods in Engineering,* Vol. 14, 1979, pp. 1693–1706.
19. Vainshtok, V.A., "A Modified Virtual Crack Extension Method of the Weight Functions Calculation for Mixed Mode Fracture Problems." *International Journal of Fracture,* Vol. 19, 1982, pp. R9–R15.
20. Sha, G.T. and Yang, C.-T., "Weight Function Calculations for Mixed-Mode Fracture Problems with the Virtual Crack Extension Technique." *Engineering Fracture Mechanics.* Vol. 21, 1985, pp. 1119–1149.

21. Atluri, S.N. and Nishoika, T., "On Some Recent Advances in Computational Methods in the Mechanics of Fracture." *Advances in Fracture Research: Seventh International Conference on Fracture,* Pergamon Press, Oxford, 1989, pp. 1923–1969.

22. Sham, T.-L., "A Unified Finite Element Method for Determining Weight Functions in Two and Three Dimensions." *International Journal of Solids and Structures,* Vol. 23, 1987, pp. 1357–1372.

23. Irwin, G.R., "Plastic Zone Near a Crack and Fracture Toughness." *Sagamore Research Conference Proceedings,* Vol. 4, 1961 Syracuse University Research Institute, Syracuse NY pp. 63–78.

24. Dugdale, D.S., "Yielding in Steel Sheets Containing Slits." *Journal of the Mechanics and Physics of Solids,* Vol. 8, 1960, pp. 100–104.

25. Barenblatt, G.I., "The Mathematical Theory of Equilibrium Cracks in Brittle Fracture." *Advances in Applied Mechanics,* Vol. VII, Academic Press, 1962, NY pp. 55–129.

26. Burdekin, F.M. and Stone, D.E.W., "The Crack Opening Displacement Approach to Fracture Mechanics in Yielding Materials." *Journal of Strain Analysis,* Vol. 1, 1966, pp. 145–153.

27. Dodds, R.H., Jr., Anderson, T.L., and Kirk, M.T., "A Framework to Correlate a/W Effects on Elastic-Plastic Fracture Toughness (J_c)." *International Journal of Fracture,* Vol. 48, 1991, pp. 1–22.

28. Narasimhan, R. and Rosakis, A.J., "Three Dimensional Effects Near a Crack-Tip in a Ductile Three Point Bend Specimen - Part I: A Numerical Investigation." California Institute of Technology, Division of Engineering and Applied Science, Report SM 88-6, Pasadena, CA, January 1988.

29. Barsom, J.M. and Rolfe, S.T., *Fracture and Fatigue Control in Structures.* 2nd Ed., Prentice-Hall, Englewood Cliffs, NJ, 1987.

30. E 399-90, "Standard Test Method for Plane-Strain Fracture Toughness of Metallic Materials." American Society for Testing and Materials, Philadelphia, PA, 1990.

31. Brown, W.F., Jr. and Srawley, J.E., *Plane Strain Crack Toughness Testing of High Strength Metallic Materials.* ASTM STP 410, American Society for Testing and Materials, Philadelphia, PA, 1966.

32. Nakamura, T, and Parks, D.M., "Conditions of J-Dominance in Three-Dimensional Thin Cracked Plates." *Analytical, Numerical, and Experimental Aspects of Three-Dimensional Fracture Processes,* ASME AMD-91, American Society of Mechanical Engineers, New York, 1988, pp. 227–238.

33. Irwin, G.R., Kraft, J.M., Paris, P.C., and Wells, A.R., "Basic Aspects of Crack Growth and Fracture." NRL Report 6598, Naval Research Lab, Washington, DC, 1967.

34. Erdogan, F. and Sih, G.C., "On the Crack Extension in Plates under Plane Loading and Transverse Shear." *Journal of Basic Engineering,* Vol. 85, 1963, pp. 519–527.

35. Williams, J.G. and Ewing, P.D., "Fracture under Complex Stress—The Angled Crack Problem." *International Journal of Fracture Mechanics,* Vol. 8, 1972, pp. 441–446.

36. Cottrell, B. and Rice, J.R., "Slightly Curved or Kinked Cracks." *International Journal of Fracture,* Vol. 16, 1980, pp. 155–169.

37. Williams, M.L., "Stress Singularities Resulting from Various Boundary Conditions in Angular Corners of Plates in Extension." *Journal of Applied Mechanics,* Vol. 19, 1952, pp. 526–528.

38. Irwin, G.R., Discussion of Ref. 9, 1958.

39. Sih, G.C., "On the Westergaard Method of Crack Analysis." *International Journal of Fracture Mechanics,* Vol. 2, 1966, pp. 628–631.

40. Eftis, J. and Liebowitz, H., "On the Modified Westergaard Equations for Certain Plane Crack Problems." *International Journal of Fracture Mechanics,* Vol. 8, 1972, p. 383.

41. Sanford, R.J., "A Critical Re-Examination of the Westergaard Method for Solving Opening Mode Crack Problems." *Mechanics Research Communications,* Vol. 6, 1979, pp. 289–294.

42. Wells, A.A. and Post, D., "The Dynamic Stress Distribution Surrounding a Running Crack—A Photoelastic Analysis." *Proceedings of the Society for Experimental Stress Analysis,* Vol. 16, 1958, pp. 69–92.

43. Muskhelishvili, N.I., *Some Basic Problems in the Theory of Elasticity.* Noordhoff, Netherlands, 1953.

44. Green, A.E. and Sneddon, I.N., "The Distribution of Stress in the Neighbourhood of a Flat Elliptical Crack in an Elastic Solid." *Proceedings, Cambridge Philosophical Society,* Vol. 46, 1950, pp. 159–163.

3 Elastic-Plastic Fracture Mechanics

Linear elastic fracture mechanics (LEFM) is valid only as long as nonlinear material deformation is confined to a small region surrounding the crack tip. In many materials, it is virtually impossible to characterize the fracture behavior with LEFM, and an alternative fracture mechanics model is required.

Elastic-plastic fracture mechanics applies to materials that exhibit time-independent, nonlinear behavior (i.e., plastic deformation). Two elastic-plastic parameters are introduced in this chapter: the crack-tip-opening displacement (CTOD) and the *J* contour integral. Both parameters describe crack-tip conditions in elastic-plastic materials, and each can be used as a fracture criterion. Critical values of CTOD or *J* give nearly size-independent measures of fracture toughness, even for relatively large amounts of crack-tip plasticity. There are limits to the applicability of *J* and CTOD (Section 3.5 and Section 3.6), but these limits are much less restrictive than the validity requirements of LEFM.

3.1 CRACK-TIP-OPENING DISPLACEMENT

When Wells [1] attempted to measure K_{Ic} values in a number of structural steels, he found that these materials were too tough to be characterized by LEFM. This discovery brought both good news and bad news: High toughness is obviously desirable to designers and fabricators, but Wells' experiments indicated that the existing fracture mechanics theory was not applicable to an important class of materials. While examining fractured test specimens, Wells noticed that the crack faces had moved apart prior to fracture; plastic deformation had blunted an initially sharp crack, as illustrated in Figure 3.1. The degree of crack blunting increased in proportion to the toughness of the material. This observation led Wells to propose the opening at the crack tip as a measure of fracture toughness. Today, this parameter is known as CTOD.

In his original paper, Wells [1] performed an approximate analysis that related CTOD to the stress intensity factor in the limit of small-scale yielding. Consider a crack with a small plastic zone, as illustrated in Figure 3.2. Irwin [2] postulated that crack-tip plasticity makes the crack behave as if it were slightly longer (Section 2.8.1). Thus, we can estimate the CTOD by solving for the displacement at the physical crack tip, assuming an effective crack length of $a + r_y$. From Table 2.2, the displacement r_y behind the effective crack tip is given by

$$u_y = \frac{\kappa+1}{2\mu} K_I \sqrt{\frac{r_y}{2\pi}} = \frac{4}{E'} K_I \sqrt{\frac{r_y}{2\pi}} \tag{3.1}$$

where E' is the effective Young's modulus, as defined in Section 2.7. The Irwin plastic zone correction for plane stress is

$$r_y = \frac{1}{2\pi} \left(\frac{K_I}{\sigma_{YS}} \right)^2 \tag{3.2}$$

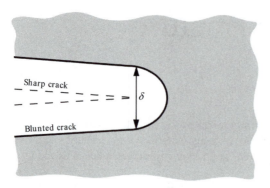

FIGURE 3.1 Crack-tip-opening displacement (CTOD). An initially sharp crack blunts with plastic deformation, resulting in a finite displacement (δ) at the crack tip.

Substituting Equation (3.2) into Equation (3.1) gives

$$\delta = 2u_y = \frac{4}{\pi} \frac{K_I^2}{\sigma_{YS} E} \tag{3.3}$$

where δ is the CTOD. Alternatively, CTOD can be related to the energy release rate by applying Equation (2.54):

$$\delta = \frac{4}{\pi} \frac{\mathcal{G}}{\sigma_{YS}} \tag{3.4}$$

Thus, in the limit of small-scale yielding, CTOD is related to $\mathcal{G}$ and K_I. Wells postulated that CTOD is an appropriate crack-tip-characterizing parameter when LEFM is no longer valid. This assumption was shown to be correct several years later when a unique relationship between CTOD and the J integral was established (Section 3.3).

The strip-yield model provides an alternate means for analyzing CTOD [3]. Recall Section 2.8.2, where the plastic zone was modeled by yield magnitude closure stresses. The size of the strip-yield zone was defined by the requirement of finite stresses at the crack tip. The CTOD can be defined as the crack-opening displacement at the end of the strip-yield zone, as Figure 3.3 illustrates. According to this definition, CTOD in a through crack in an infinite plate subject to a remote tensile stress (Figure 2.3) is given by [3]

$$\delta = \frac{8\sigma_{YS} a}{\pi E} \ln \sec\left(\frac{\pi}{2} \frac{\sigma}{\sigma_{YS}}\right) \tag{3.5}$$

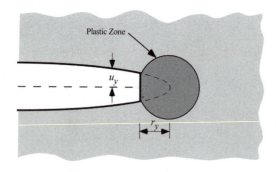

FIGURE 3.2 Estimation of CTOD from the displacement of the effective crack in the Irwin plastic zone correction.

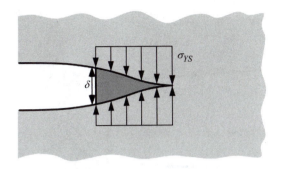

FIGURE 3.3 Estimation of CTOD from the strip-yield model. Taken from Burdekin, F.M. and Stone, D.E.W., "The Crack Opening Displacement Approach to Fracture Mechanics in Yielding Materials." *Journal of Strain Analysis,* Vol. 1, 1966, pp. 145–153.

Equation (3.5) is derived in Appendix 3.1. Series expansion of the "ln sec" term gives

$$\delta = \frac{8\sigma_{YS}a}{\pi E}\left[\frac{1}{2}\left(\frac{\pi}{2}\frac{\sigma}{\sigma_{YS}}\right)^2 + \frac{1}{12}\left(\frac{\pi}{2}\frac{\sigma}{\sigma_{YS}}\right)^4 + \cdots\right]$$

$$= \frac{K_I^2}{\sigma_{YS}E}\left[1 + \frac{1}{6}\left(\frac{\pi}{2}\frac{\sigma}{\sigma_{YS}}\right)^2 + \cdots\right]$$

(3.6)

Therefore, as $\sigma/\sigma_{YS} \to 0$,

$$\delta = \frac{K_I^2}{\sigma_{YS}E} = \frac{\mathcal{G}}{\sigma_{YS}}$$

(3.7)

which differs slightly from Equation (3.3).

The strip-yield model assumes plane stress conditions and a nonhardening material. The actual relationship between CTOD and K_I and $\mathcal{G}$ depends on stress state and strain hardening. The more general form of this relationship can be expressed as follows:

$$\delta = \frac{K_I^2}{m\sigma_{YS}E'} = \frac{\mathcal{G}}{m\sigma_{YS}}$$

(3.8)

where m is a dimensionless constant that is approximately 1.0 for plane stress and 2.0 for plane strain.

There are a number of alternative definitions of CTOD. The two most common definitions, which are illustrated in Figure 3.4, are the displacement at the original crack tip and the 90° intercept. The latter definition was suggested by Rice [4] and is commonly used to infer CTOD in finite element measurements. Note that these two definitions are equivalent if the crack blunts in a semicircle.

Most laboratory measurements of CTOD have been made on edge-cracked specimens loaded in three-point bending (see Table 2.4). Early experiments used a flat paddle-shaped gage that was inserted into the crack; as the crack opened, the paddle gage rotated, and an electronic signal was sent to an *x-y* plotter. This method was inaccurate, however, because it was difficult to reach the crack tip with the paddle gage. Today, the displacement *V* at the crack mouth is measured, and the CTOD is inferred by assuming the specimen halves are rigid and rotate about a hinge point, as illustrated in Figure 3.5. Referring to this figure, we can estimate CTOD from a similar triangles construction:

$$\frac{\delta}{r(W-a)} = \frac{V}{r(W-a)+a}$$

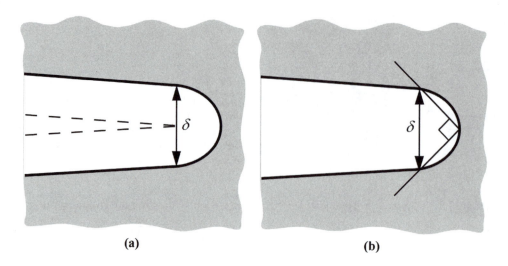

(a)   **(b)**

FIGURE 3.4 Alternative definitions of CTOD: (a) displacement at the original crack tip and (b) displacement at the intersection of a 90° vertex with the crack flanks.

Therefore

$$\delta = \frac{r(W-a)V}{r(W-a)+a} \tag{3.9}$$

where r is the rotational factor, a dimensionless constant between 0 and 1.

   The hinge model is inaccurate when displacements are primarily elastic. Consequently, standard methods for CTOD testing [5, 6] typically adopt a modified hinge model, in which displacements are separated into elastic and plastic components; the hinge assumption is applied only to plastic displacements. Figure 3.6 illustrates a typical load (P) vs. displacement (V) curve from a CTOD test. The shape of the load-displacement curve is similar to a stress-strain curve: It is initially linear but deviates from linearity with plastic deformation. At a given point on the curve, the displacement is separated into elastic and plastic components by constructing a line parallel to the elastic loading line.

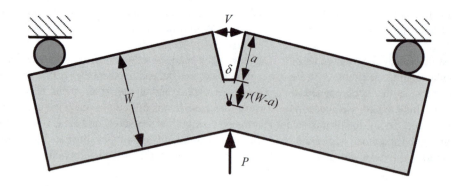

FIGURE 3.5 The hinge model for estimating CTOD from three-point bend specimens.

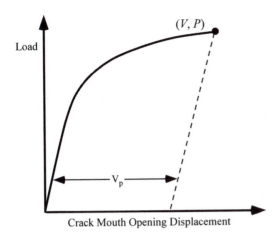

Load

(V, P)

V_p

Crack Mouth Opening Displacement

FIGURE 3.6 Determination of the plastic component of the crack-mouth-opening displacement.

The dashed line represents the path of unloading for this specimen, assuming the crack does not grow during the test. The CTOD in this specimen is estimated by

$$\delta = \delta_{el} + \delta_p = \frac{K_I^2}{m\sigma_{YS}E'} + \frac{r_p(W-a)V_p}{r_p(W-a)+a} \tag{3.10}$$

The subscripts el and p denote elastic and plastic components, respectively. The elastic stress intensity factor is computed by inserting the load and specimen dimensions into the appropriate expression in Table 2.4. The plastic rotational factor r_p is approximately 0.44 for typical materials and test specimens. Note that Equation (3.10) reduces to the small-scale yielding result (Equation (3.8)) for linear elastic conditions, but the hinge model dominates when $V \approx V_p$.

Further details of CTOD testing are given in Chapter 7. Chapter 9 outlines how CTOD is used in design.

3.2 THE J CONTOUR INTEGRAL

The J contour integral has enjoyed great success as a fracture characterizing parameter for nonlinear materials. By idealizing elastic-plastic deformation as nonlinear elastic, Rice [4] provided the basis for extending fracture mechanics methodology well beyond the validity limits of LEFM.

Figure 3.7 illustrates the uniaxial stress-strain behavior of elastic-plastic and nonlinear elastic materials. The loading behavior for the two materials is identical, but the material responses differ when each is unloaded. The elastic-plastic material follows a linear unloading path with the slope equal to Young's modulus, while the nonlinear elastic material unloads along the same path as it was loaded. There is a unique relationship between stress and strain in an elastic material, but a given strain in an elastic-plastic material can correspond to more than one stress value if the material is unloaded or cyclically loaded. Consequently, it is much easier to analyze an elastic material than a material that exhibits irreversible plasticity.

As long as the stresses in both materials in Figure 3.7 increase monotonically, the mechanical response of the two materials is identical. When the problem is generalized to three dimensions, it does not necessarily follow that the loading behavior of the nonlinear elastic and elastic-plastic materials is identical, but there are many instances where this is a good assumption (see Appendix 3.6). Thus an analysis that assumes nonlinear elastic behavior may be valid for an elastic-plastic material, provided no unloading occurs. The *deformation theory of plasticity*, which relates total strains to stresses in a material, is equivalent to nonlinear elasticity.

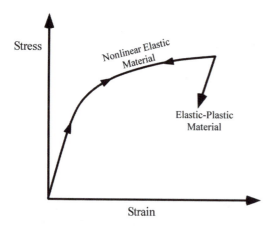

FIGURE 3.7 Schematic comparison of the stress-strain behavior of elastic-plastic and nonlinear elastic materials.

Rice [4] applied deformation plasticity (i.e., nonlinear elasticity) to the analysis of a crack in a nonlinear material. He showed that the nonlinear energy release rate J could be written as a path-independent line integral. Hutchinson [7] and Rice and Rosengren [8] also showed that J uniquely characterizes crack-tip stresses and strains in nonlinear materials. Thus the J integral can be viewed as both an energy parameter and a stress intensity parameter.

3.2.1 NONLINEAR ENERGY RELEASE RATE

Rice [4] presented a path-independent contour integral for the analysis of cracks. He then showed that the value of this integral, which he called J, is equal to the energy release rate in a nonlinear elastic body that contains a crack. In this section, however, the energy release rate interpretation is discussed first because it is closely related to concepts introduced in Chapter 2. The J contour integral is outlined in Section 3.2.2. Appendix 3.2 gives a mathematical proof, similar to what Rice [4] presented, that shows that this line integral is equivalent to the energy release rate in nonlinear elastic materials.

Equation (2.23) defines the energy release rate for linear materials. The same definition holds for nonlinear elastic materials, except that $\mathcal{G}$ is replaced by J:

$$J = -\frac{d\Pi}{dA} \tag{3.11}$$

where Π is the potential energy and A is the crack area. The potential energy is given by

$$\Pi = U - F \tag{3.12}$$

where U is the strain energy stored in the body and F is the work done by external forces. Consider a cracked plate which exhibits a nonlinear load-displacement curve, as illustrated in Figure 3.8. If the plate has unit thickness, $A = a$.[1] For load control

$$\Pi = U - P\Delta = -U^*$$

where U^* is the complimentary strain energy, defined as

$$U^* = \int_0^P \Delta\, dP \tag{3.13}$$

---

[1] It is important to remember that the energy release rate is defined in terms of the crack area, not crack length. Failure to recognize this can lead to errors and confusion when computing $\mathcal{G}$ or J for configurations other than edge cracks; examples include a through crack, where $dA = 2da$ (assuming unit thickness), and a penny-shaped crack, where $dA = 2\pi a\, da$.

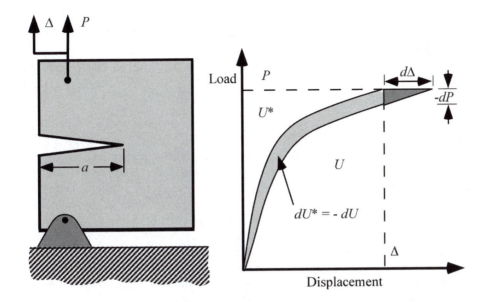

FIGURE 3.8 Nonlinear energy release rate.

Thus if the plate in Figure 3.8 is in load control, J is given by

$$J = \left(\frac{dU^*}{da}\right)_P \qquad (3.14)$$

If the crack advances at a fixed displacement, $F = 0$, and J is given by

$$J = -\left(\frac{dU}{da}\right)_\Delta \qquad (3.15)$$

According to Figure 3.8, dU^* for load control differs from $-dU$ for displacement control by the amount $\tfrac{1}{2}\,dP d\Delta$, which is vanishingly small compared to dU. Therefore, J for load control is equal to J for displacement control. Recall that we obtained this same result for $\mathcal{G}$ in Section 2.4.

By invoking the definitions for U and U^*, we can express J in terms of load and displacement:

$$J = \left(\frac{\partial}{\partial a}\int_0^P \Delta\, dP\right)_P$$

$$= \int_0^P \left(\frac{\partial \Delta}{\partial a}\right)_P dP \qquad (3.16)$$

or

$$J = -\left(\frac{\partial}{\partial a}\int_0^\Delta P\, d\Delta\right)_\Delta$$

$$= -\int_0^\Delta \left(\frac{\partial P}{\partial a}\right)_\Delta d\Delta \qquad (3.17)$$

Integrating Equation (3.17) by parts leads to a rigorous proof of what we have already inferred from Figure 3.8. That is, Equation (3.16) and Equation (3.17) are equal, and J is the same for fixed load and fixed grip conditions.

Thus, J is a more general version of the energy release rate. For the special case of a linear elastic material, $J = \mathcal{G}$. Also

$$J = \frac{K_I^2}{E'} \tag{3.18}$$

for linear elastic Mode I loading. For mixed mode loading refer to Equation (2.63).

A word of caution is necessary when applying J to elastic-plastic materials. The energy release rate is normally defined as the potential energy that is *released* from a structure when the crack grows in an elastic material. However, much of the strain energy absorbed by an elastic-plastic material is not recovered when the crack grows or the specimen is unloaded; a growing crack in an elastic-plastic material leaves a plastic wake (Figure 2.6(b)). Thus the energy release rate concept has a somewhat different interpretation for elastic-plastic materials. Rather than defining the energy released from the body when the crack grows, Equation (3.15) relates J to the difference in energy absorbed by specimens with neighboring crack sizes. This distinction is important only when the crack grows (Section 3.4.2). See Appendix 4.2 and Chapter 12 for a further discussion of the energy release rate concept.

The energy release rate definition of J is useful for elastic-plastic materials when applied in an appropriate manner. For example, Section 3.2.5 describes how Equations (3.15)–(3.17) can be exploited to measure J experimentally.

3.2.2 J AS A PATH-INDEPENDENT LINE INTEGRAL

Consider an arbitrary counterclockwise path (Γ) around the tip of a crack, as in Figure 3.9. The J integral is given by

$$J = \int_{\Gamma} \left(w\,dy - T_i \frac{\partial u_i}{\partial x}\,ds \right) \tag{3.19}$$

where
- w = strain energy density
- T_i = components of the traction vector
- u_i = displacement vector components
- ds = length increment along the contour Γ

The strain energy density is defined as

$$w = \int_0^{\varepsilon_{ij}} \sigma_{ij}\,d\varepsilon_{ij} \tag{3.20}$$

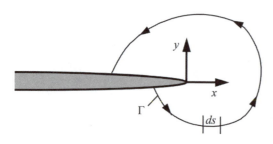

FIGURE 3.9 Arbitrary contour around the tip of a crack.

where σ_{ij} and ε_{ij} are the stress and strain tensors, respectively. The traction is a stress vector at a given point on the contour. That is, if we were to construct a free body diagram of the material inside of the contour, T_i would define the stresses acting at the boundaries. The components of the traction vector are given by

$$T_i = \sigma_{ij} n_j \tag{3.21}$$

where n_j are the components of the unit vector normal to Γ.

Rice [4] showed that the value of the J integral is independent of the path of integration around the crack. Thus J is called a *path-independent* integral. Appendix 3.2 demonstrates this path independence, and shows that Equation (3.19) is equal to the energy release rate.

3.2.3 J AS A STRESS INTENSITY PARAMETER

Hutchinson [7] and Rice and Rosengren [8] independently showed that J characterizes crack-tip conditions in a nonlinear elastic material. They each assumed a power law relationship between plastic strain and stress. If elastic strains are included, this relationship for uniaxial deformation is given by

$$\frac{\varepsilon}{\varepsilon_o} = \frac{\sigma}{\sigma_o} + \alpha \left(\frac{\sigma}{\sigma_o} \right)^n \tag{3.22}$$

where

σ_o = reference stress value that is usually equal to the yield strength
$\varepsilon_o = \sigma_o/E$
α = dimensionless constant
n = strain-hardening exponent.[2]

Equation (3.22) is known as the Ramberg-Osgood equation, and is widely used for curve-fitting stress-strain data. Hutchinson, Rice, and Rosengren showed that in order to remain path independent, stress–strain must vary as $1/r$ near the crack tip. At distances very close to the crack tip, well within the plastic zone, elastic strains are small in comparison to the total strain, and the stress-strain behavior reduces to a simple power law. These two conditions imply the following variation of stress and strain ahead of the crack tip:

$$\sigma_{ij} = k_1 \left(\frac{J}{r} \right)^{\frac{1}{n+1}} \tag{3.23a}$$

$$\varepsilon_{ij} = k_2 \left(\frac{J}{r} \right)^{\frac{n}{n+1}} \tag{3.23b}$$

where k_1 and k_2 are proportionality constants, which are defined more precisely below. For a linear elastic material, $n = 1$, and Equation (3.23) predicts a $1/\sqrt{r}$ singularity, which is consistent with LEFM theory.

---

[2] Although Equation (3.22) contains four material constants, there are only two fitting parameters. The choice of σ_o, which is arbitrary, defines ε_o; a linear regression is then performed on a log-log plot of stress vs. plastic strain to determine α and n.

The actual stress and strain distributions are obtained by applying the appropriate boundary conditions (see Appendix 3.4):

$$\sigma_{ij} = \sigma_o \left(\frac{EJ}{\alpha \sigma_o^2 I_n r} \right)^{\frac{1}{n+1}} \tilde{\sigma}_{ij}(n, \theta) \qquad (3.24a)$$

and

$$\varepsilon_{ij} = \frac{\alpha \sigma_o}{E} \left(\frac{EJ}{\alpha \sigma_o^2 I_n r} \right)^{\frac{n}{n+1}} \tilde{\varepsilon}_{ij}(n, \theta) \qquad (3.24b)$$

where I_n is an integration constant that depends on n, and $\tilde{\sigma}_{ij}$ and $\tilde{\varepsilon}_{ij}$ are the dimensionless functions of n and θ. These parameters also depend on the assumed stress state (i.e., plane stress or plane strain). Equation (3.24a) and Equation (3.24b) are called the HRR singularity, named after Hutchinson, Rice, and Rosengren [7, 8]. Figure 3.10 is a plot of I_n vs. n for plane stress and plane strain. Figure 3.11 shows the angular variation of $\tilde{\sigma}_{ij}(n, \theta)$ [7]. The stress components in Figure 3.11 are defined in terms of polar coordinates rather than x and y.

The J integral defines the amplitude of the HRR singularity, just as the stress intensity factor characterizes the amplitude of the linear elastic singularity. Thus J completely describes the conditions within the plastic zone. A structure in small-scale yielding has two singularity-dominated zones: one in the elastic region, where stress varies as $1/\sqrt{r}$, and one in the plastic zone where stress varies as $r^{-1/(n+1)}$. The latter often persists long after the linear elastic singularity zone has been destroyed by crack-tip plasticity.

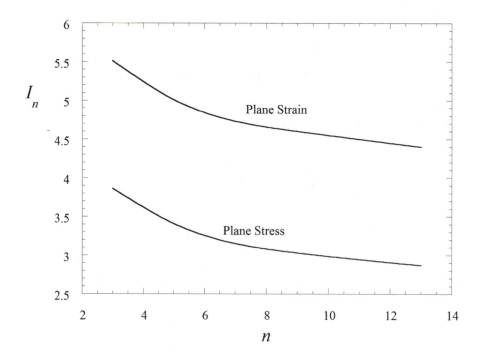

FIGURE 3.10 Effect of the strain-hardening exponent on the HRR integration constant.

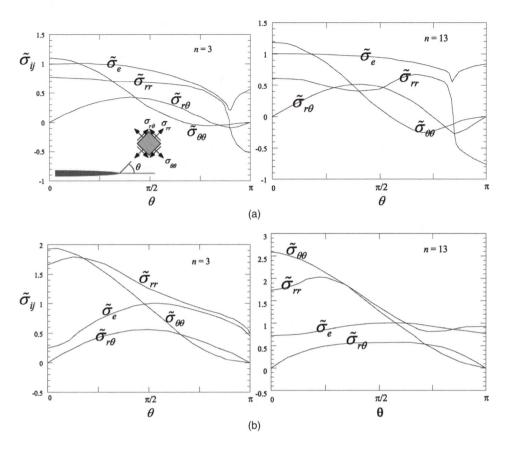

FIGURE 3.11 Angular variation of dimensionless stress for $n = 3$ and $n = 13$: (a) plane stress (b) plane strain. Taken from Hutchinson, J.W., "Singular Behavior at the End of a Tensile Crack Tip in a Hardening Material." *Journal of the Mechanics and Physics of Solids,* Vol. 16, 1968, pp. 13–31.

3.2.4 THE LARGE STRAIN ZONE

The HRR singularity contains the same apparent anomaly as the LEFM singularity; namely, both predict infinite stresses as $r \to 0$. However, the singular field does not persist all the way to the crack tip. The large strains at the crack tip cause the crack to blunt, which reduces the stress triaxiality locally. The blunted crack tip is a free surface; thus σ_{xx} must vanish at $r = 0$.

The analysis that leads to the HRR singularity does not consider the effect of the blunted crack tip on the stress fields, nor does it take account of the large strains that are present near the crack tip. This analysis is based on the small strain theory, which is the multi-axial equivalent of engineering strain in a tensile test. Small strain theory breaks down when strains are greater than ~0.10 (10%).

McMeeking and Parks [9] performed crack-tip finite element analyses that incorporated large strain theory and finite geometry changes. Some of their results are shown in Figure 3.12, which is a plot of stress normal to the crack plane vs. distance. The HRR singularity (Equation (3.24a)) is also shown on this plot. Note that both axes are nondimensionalized in such a way that both curves are invariant, as long as the plastic zone is small compared to the specimen dimensions.

The solid curve in Figure 3.12 reaches a peak when the ratio $x\sigma_o/J$ is approximately unity, and decreases as $x \to 0$. This distance corresponds approximately to twice the CTOD. The HRR singularity is invalid within this region, where the stresses are influenced by large strains and crack blunting.

The breakdown of the HRR solution at the crack tip leads to a similar question to one that was posed in Section 2.9: Is the J integral a useful fracture criterion when a blunting zone forms

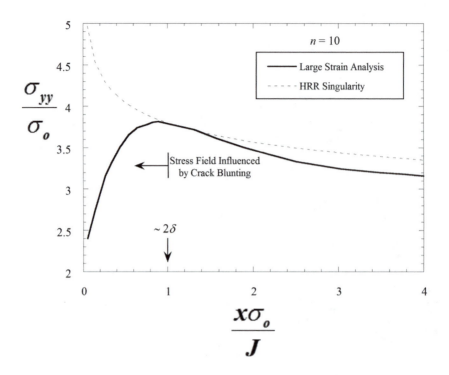

FIGURE 3.12 Large-strain crack-tip finite element results of McMeeking and Parks. Blunting causes the stresses to deviate from the HRR solution close to the crack tip. Taken from McMeeking, R.M. and Parks, D.M., "On Criteria for *J*-Dominance of Crack Tip Fields in Large-Scale Yielding." *Elastic Plastic Fracture* ASTM STP 668, American Society for Testing and Materials, Philadelphia, PA, 1979, pp. 175–194.

at the crack tip? The answer is also similar to the argument offered in Section 2.9. That is, as long as there is a region *surrounding* the crack tip that can be described by Equation (3.24), the *J* integral uniquely characterizes crack-tip conditions, and a critical value of *J* is a size-independent measure of fracture toughness. The question of *J* controlled fracture is explored further in Section 3.5.

3.2.5 Laboratory Measurement of *J*

When the material behavior is linear elastic, the calculation of *J* in a test specimen or structure is relatively straightforward because $J = \mathcal{G}$, and $\mathcal{G}$ is uniquely related to the stress intensity factor. The latter quantity can be computed from the load and crack size, assuming a *K* solution for that particular geometry is available.

Computing the *J* integral is somewhat more difficult when the material is nonlinear. The principle of superposition no longer applies, and *J* is not proportional to the applied load. Thus a simple relationship between *J*, load, and crack length is usually not available.

One option for determining *J* is to apply the line integral definition Equation (3.19) to the configuration of interest. Read [10] has measured the *J* integral in test panels by attaching an array of strain gages in a contour around the crack tip. Since *J* is path independent and the choice of contour is arbitrary, he selected a contour in such a way as to simplify the calculation of *J* as much as possible. This method can also be applied to finite element analysis, i.e., stresses, strains, and displacements can be determined along a contour and *J* can then be calculated according to Equation (3.19). However, the contour method for determining *J* is impractical in most cases. The instrumentation required for experimental measurements of the contour integral is highly cumbersome, and the contour method is also not very attractive in numerical analysis (see Chapter 12).

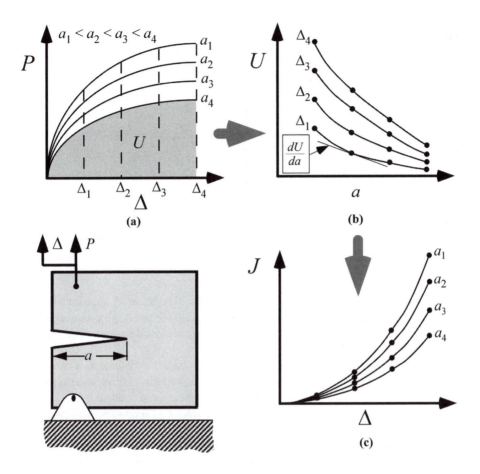

FIGURE 3.13 Schematic of early experimental measurements of J, performed by Landes and Begley. Taken from Begley, J.A. and Landes, J.D., "The J-Integral as a Fracture Criterion." ASTM STP 514, American Society for Testing and Materials, Philadelphia, PA, 1972, pp. 1–20; Landes, J.D. and Begley, J.A., "The Effect of Specimen Geometry on J_{Ic}." ASTM STP 514, American Society for Testing and Materials, Philadelphia, PA, 1972, pp. 24–29.

A much better method for determining J numerically is outlined in Chapter 12. More practical experimental approaches are developed below and are explored further in Chapter 7.

Landes and Begley [11, 12], who were among the first to measure J experimentally, invoked the energy release rate definition of J (Equation (3.11)). Figure 3.13 schematically illustrates their approach. They obtained a series of test specimens of the same size, geometry, and material and introduced cracks of various lengths.[3] They deformed each specimen and plotted load vs. displacement (Figure 3.13(a)). The area under a given curve is equal to U, the energy absorbed by the specimen. Landes and Begley plotted U vs. crack length at various fixed displacements (Fig. 3.13(b)). For an edge-cracked specimen of thickness B, the J integral is given by

$$J = -\frac{1}{B}\left(\frac{\partial U}{\partial a}\right)_{\Delta} \tag{3.25}$$

Thus J can be computed by determining the slope of the tangent to the curves in Figure 3.13(b). Applying Equation (3.25) leads to Figure 3.13(c), a plot of J vs. displacement at various crack

---

[3] See Chapter 7 for a description of fatigue-precracking procedures for test specimens.

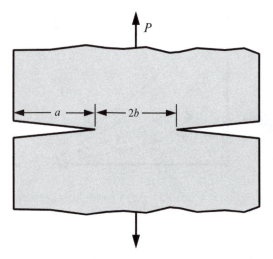

FIGURE 3.14 Double-edge-notched tension (DENT) panel.

lengths. The latter is a calibration curve, which only applies to the material, specimen size, specimen geometry, and temperature for which it was obtained. The Landes and Begley approach has obvious disadvantages, since multiple specimens must be tested and analyzed to determine J in a particular set of circumstances.

Rice et al. [13] showed that it was possible, in certain cases, to determine J directly from the load displacement curve of a single specimen. Their derivations of J relationships for several specimen configurations demonstrate the usefulness of dimensional analysis.[4]

Consider a double-edge-notched tension panel of unit thickness (Figure 3.14). Cracks of length a on opposite sides of the panel are separated by a ligament of length $2b$. For this configuration, $dA = 2da = -2db$ (see Footnote 1); Equation (3.16) must be modified accordingly:

$$J = \frac{1}{2}\int_0^P \left(\frac{\partial \Delta}{\partial a}\right)_P dP = -\frac{1}{2}\int_0^P \left(\frac{\partial \Delta}{\partial b}\right)_P dP \qquad (3.26)$$

In order to compute J from the above expression, it is necessary to determine the relationship between load, displacement, and panel dimensions. Assuming an isotropic material that obeys a Ramberg-Osgood stress-strain law (Equation (3.22)), the dimensional analysis gives the following functional relationship for displacement:

$$\Delta = b\Phi\left(\frac{P}{\sigma_o b};\frac{a}{b};\frac{\sigma_o}{E};\nu;\alpha;n\right) \qquad (3.27)$$

where Φ is a dimensionless function. For fixed material properties, we need only consider the load and specimen dimensions. For reasons described below, we can simplify the functional relationship for displacement by separating Δ into elastic and plastic components:

$$\Delta = \Delta_{el} + \Delta_p \qquad (3.28)$$

---

[4] See Section 1.5 for a review of the fundamentals of dimensional analysis.

Substituting Equation (3.28) into Equation (3.26) leads to a relationship for the elastic and plastic components of J:

$$J = -\frac{1}{2} \int_0^P \left[\left(\frac{\partial \Delta_{el}}{\partial b} \right)_P + \left(\frac{\partial \Delta_p}{\partial b} \right)_P \right] dP$$

$$= \frac{K_I^2}{E'} - \frac{1}{2} \int_0^P \left(\frac{\partial \Delta_p}{\partial b} \right)_P dP \tag{3.29}$$

where $E' = E$ for plane stress and $E' = E/(1 - v^2)$ for plane strain, as defined in Chapter 2. Thus we need only be concerned about plastic displacements because a solution for the elastic component of J is already available (Table 2.4). If plastic deformation is confined to the ligament between the crack tips (Figure 3.14(b)), we can assume that b is the only length dimension that influences Δ_p. That is a reasonable assumption, provided the panel is deeply notched so that the average stress in the ligament is substantially higher than the remote stress in the gross cross section. We can define a new function for Δ_p:

$$\Delta_p = bH\left(\frac{P}{b} \right) \tag{3.30}$$

Note that the net-section yielding assumption has eliminated the dependence on the a/b ratio. Taking a partial derivative with respect to the ligament length gives

$$\left(\frac{\partial \Delta_p}{\partial b} \right)_P = H\left(\frac{P}{b} \right) - H'\left(\frac{P}{b} \right) \frac{P}{b}$$

where H' denotes the first derivative of the function H. We can solve for H' by taking a partial derivative of Equation (3.30) with respect to load:

$$\left(\frac{\partial \Delta_p}{\partial P} \right)_b = H'\left(\frac{P}{b} \right)$$

Therefore

$$\left(\frac{\partial \Delta_p}{\partial b} \right)_P = \frac{1}{b} \left[\Delta_p - P\left(\frac{\partial \Delta_p}{\partial P} \right)_b \right] \tag{3.31}$$

Substituting Equation (3.31) into Equation (3.29) and integrating by parts gives

$$J = \frac{K_I^2}{E'} + \frac{1}{2b} \left[2 \int_0^{\Delta_p} P d\Delta_p - P\Delta_p \right] \tag{3.32}$$

Recall that we assumed a unit thickness at the beginning of this derivation. In general, the plastic term must be divided by the plate thickness; the term in square brackets, which depends on the load displacement curve, is normalized by the net cross-sectional area of the panel. The J integral has units of energy/area.

Another example from the Rice et al. article [13] is an edge-cracked plate in bending (Figure 3.15). In this case they chose to separate displacements along somewhat different lines from the previous problem. If the plate is subject to a bending moment M, it would displace by an angle Ω_{nc} if no crack were present, and an additional amount Ω_c when the plate is cracked.

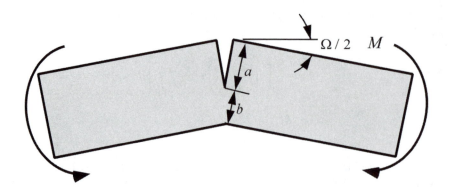

FIGURE 3.15 Edge-cracked plate in pure bending.

Thus the total angular displacement can be written as

$$\Omega = \Omega_{nc} + \Omega_c \qquad (3.33)$$

If the crack is deep, $\Omega_c >> \Omega_{nc}$. The energy absorbed by the plate is given by

$$U = \int_0^{\Omega} M \, d\Omega \qquad (3.34)$$

When we differentiate U with respect to the crack area in order to determine J, only Ω_c contributes to the energy release rate because Ω_{nc} is not a function of crack size, by definition. By analogy with Equation (3.16), J for the cracked plate in bending can be written as

$$J = \int_0^M \left(\frac{\partial \Omega_c}{\partial a} \right)_M dM = -\int_0^M \left(\frac{\partial \Omega_c}{\partial b} \right)_M dM \qquad (3.35)$$

If the material properties are fixed, dimensional analysis leads to

$$\Omega_c = F\left(\frac{M}{b^2} \right) \qquad (3.36)$$

assuming the ligament length is the only relevant length dimension, which is reasonable if the crack is deep. When Equation (3.36) is differentiated with respect to b and inserted into Equation (3.35), the resulting expression for J is as follows:

$$J = \frac{2}{b} \int_0^{\Omega_c} M \, d\Omega_c \qquad (3.37)$$

The decision to separate Ω into "crack" and "no-crack" components was somewhat arbitrary. The angular displacement could have been divided into elastic and plastic components as in the previous example. If the crack is relatively deep, Ω_{nc} should be entirely elastic, while Ω_c may contain both elastic and plastic contributions. Therefore, Equation (3.37) can be written as

$$J = \frac{2}{b} \left[\int_0^{\Omega_{c(el)}} M \, d\Omega_{c(el)} + \int_0^{\Omega_p} M \, d\Omega_p \right]$$

or

$$J = \frac{K_I^2}{E'} + \frac{2}{b} \int_0^{\Omega_p} M d\Omega_p \qquad (3.38)$$

Conversely, the prior analysis on the double-edged cracked plate in tension could have been written in terms of Δ_c and Δ_{nc}. Recall, however, that the dimensional analysis was simplified in each case (Equation (3.30) and Equation (3.36)) by assuming a negligible dependence on a/b. This turns out to be a reasonable assumption for plastic displacements in deeply notched DENT panels, but less so for elastic displacements. Thus while elastic and plastic displacements due to the crack can be combined to compute J in bending (Equation (3.37)), it is not advisable to do so for tensile loading. The relative accuracy and the limitations of Equation (3.32) and Equation (3.37) are evaluated in Chapter 9.

In general, the J integral for a variety of configurations can be written in the following form:

$$J = \frac{\eta U_c}{Bb} \qquad (3.39)$$

where η is a dimensionless constant. Note that Equation (3.39) contains the actual thickness, while the above derivations assumed a unit thickness for convenience. Equation (3.39) expresses J as the energy absorbed, divided by the cross-sectional area, times a dimensionless constant. For a deeply cracked plate in pure bending, $\eta = 2$. Equation (3.39) can be separated into elastic and plastic components:

$$\begin{aligned} J &= \frac{\eta_{el} U_{c(el)}}{Bb} + \frac{\eta_p U_p}{Bb} \\ &= \frac{K_I^2}{E'} + \frac{\eta_p U_p}{Bb} \end{aligned} \qquad (3.40)$$

EXAMPLE 3.1

Determine the plastic η factor for the DENT configuration, assuming the load-plastic displacement curve follows a power law:

$$P = C\Delta_p^N$$

Solution: The plastic energy absorbed by the specimen is given by

$$U_p = \int_0^{\Delta_p} \Delta_p^N d\Delta_p = \frac{C\Delta_p^{N+1}}{N+1} = \frac{P\Delta p}{N+1}$$

Comparing Equation (3.32) and Equation (3.40) and solving for η_p gives

$$\eta_p = \frac{P\Delta p \left(\dfrac{2}{N+1} - 1 \right)}{\dfrac{P\Delta p}{N+1}} = 1 - N$$

For a nonhardening material, $N = 0$ and $\eta_p = 1$.

3.3 RELATIONSHIPS BETWEEN *J* AND CTOD

For linear elastic conditions, the relationship between CTOD and $\mathcal{G}$ is given by Equation (3.8). Since $J = \mathcal{G}$ for linear elastic material behavior, these equations also describe the relationship between CTOD and J in the limit of small-scale yielding. That is

$$J = m\sigma_{YS}\delta \qquad (3.41)$$

where m is a dimensionless constant that depends on the stress state and material properties. It can be shown that Equation (3.41) applies well beyond the validity limits of LEFM.

Consider, for example, a strip-yield zone ahead of a crack tip, as illustrated in Figure 3.16. Recall (from Chapter 2) that the strip-yield zone is modeled by surface tractions along the crack face. Let us define a contour Γ along the boundary of this zone. If the damage zone is long and slender, i.e., if $\rho \gg \delta$, the first term in the J contour integral (Equation 3.19) vanishes because $dy = 0$. Since the only surface tractions within ρ are in the y direction, $n_y = 1$ and $n_x = n_z = 0$. Thus the J integral is given by

$$J = \int_\Gamma \sigma_{yy}\frac{\partial u_y}{\partial x}ds \qquad (3.42)$$

Let us define a new coordinate system with the origin at the tip of the strip-yield zone: $X = \rho - x$. For a fixed δ, σ_{yy} and u_y depend only on X, provided ρ is small compared to the in-plane dimensions of the cracked body. The J integral becomes

$$J = 2\int_0^\rho \sigma_{yy}(X)\left(\frac{du_y(X)}{dX}\right)dX$$
$$= \int_0^\delta \sigma_{yy}(\delta)d\delta \qquad (3.43)$$

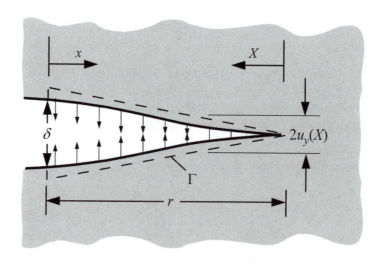

FIGURE 3.16 Contour along the boundary of the strip-yield zone ahead of a crack tip.

where $\delta = 2u_y$ $(X = \rho)$. Since the strip-yield model assumes $\sigma_{yy} = \sigma_{YS}$ within the plastic zone, the J-CTOD relationship is given by

$$J = \sigma_{YS}\delta \qquad (3.44)$$

Note the similarity between Equation (3.44) and Equation (3.7). The latter was derived from the strip-yield model by neglecting the higher-order terms in a series expansion; no such assumption was necessary to derive Equation (3.44). Thus the strip-yield model, which assumes plane stress conditions and a nonhardening material, predicts that $m = 1$ for both linear elastic and elastic-plastic conditions.

Shih [14] provided further evidence that a unique J-CTOD relationship applies well beyond the validity limits of LEFM. He evaluated the displacements at the crack tip implied by the HRR solution and related the displacement at the crack tip to J and flow properties. According to the HRR solution, the displacements near the crack tip are as follows:

$$u_i = \frac{\alpha\sigma_o}{E}\left(\frac{EJ}{\alpha\sigma_o^2 I_n r}\right)^{\frac{n}{n+1}} r\tilde{u}_i(\theta, n) \qquad (3.45)$$

where $\tilde{u}_i$ is a dimensionless function of θ and n, analogous to $\tilde{\sigma}_{ij}$ and $\tilde{\varepsilon}_{ij}$ (Equation (3.24)). Shih [14] invoked the 90° intercept definition of CTOD, as illustrated in Figure 3.4(b). This 90° intercept construction is examined further in Figure 3.17. The CTOD is obtained by evaluating u_x and u_y at $r = r^*$ and $\theta = \pi$:

$$\frac{\delta}{2} = u_y(r^*, \pi) = r^* - u_x(r^*, \pi) \qquad (3.46)$$

Substituting Equation (3.46) into Equation (3.45) and solving for r^* gives

$$r^* = \left(\frac{\alpha\sigma_o}{E}\right)^{1/n}\{\tilde{u}_x(\pi, n) + \tilde{u}_y(\pi, n)\}^{\frac{n+1}{n}}\frac{J}{\sigma_o I_n} \qquad (3.47)$$

Setting $\delta = 2u_y(r^*, \pi)$ leads to

$$\delta = \frac{d_n J}{\sigma_o} \qquad (3.48)$$

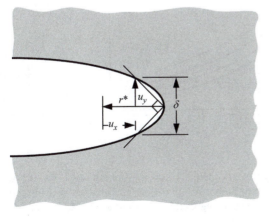

FIGURE 3.17 Estimation of CTOD from a 90° intercept construction and HRR displacements.

where d_n is a dimensionless constant, given by

$$d_n = \frac{2\tilde{u}_y(\pi,n)\left[\dfrac{\alpha\sigma_o}{E}\{\tilde{u}_x(\pi,n)+\tilde{u}_y(\pi,n)\}\right]^{1/n}}{I_n} \qquad (3.49)$$

Figure 3.18 shows plots of d_n for $\alpha = 1.0$, which exhibits a strong dependence on the strain-hardening exponent and a mild dependence on $\alpha\sigma_o/E$. A comparison of Equation (3.41) and

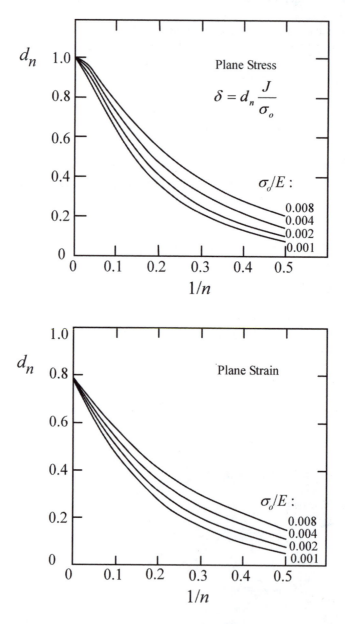

FIGURE 3.18 Predicted J-CTOD relationships for plane stress and plane strain, assuming $\alpha = 1$. For $\alpha \neq 1$, the above values should be multiplied by $\alpha^{1/n}$. Taken from Shih, C.F. "Relationship between the J-Integral and the Crack Opening Displacement for Stationary and Extending Cracks." *Journal of the Mechanics and Physics of Solids,* Vol. 29, 1981, pp. 305–326.

Equation (3.48) indicates that $d_n = 1/m$, assuming $\sigma_o = \sigma_{YS}$ (see Footnote 2). According to Figure 3.18(a), $d_n = 1.0$ for a nonhardening material ($n = \infty$) in plane stress, which agrees with the strip-yield model (Equation (3.44)).

The Shih analysis shows that there is a unique relationship between J and CTOD for a given material. Thus these two quantities are equally valid crack-tip-characterizing parameters for elastic-plastic materials. The fracture toughness of a material can be quantified either by a critical value of J or CTOD.

The above analysis contains an apparent inconsistency. Equation (3.48) is based on the HRR singularity, which does not account for large geometry changes at the crack tip. Figure 3.12 indicates that the stresses predicted by the HRR theory are inaccurate for $r < 2\delta$, but the Shih analysis uses the HRR solution to evaluate displacements well within the large strain region. Crack-tip finite element analyses [14], however, are in general agreement with Equation (3.48). Thus the displacement fields predicted from the HRR theory are reasonably accurate, despite the large plastic strains at the crack tip.

3.4 CRACK-GROWTH RESISTANCE CURVES

Many materials with high toughness do not fail catastrophically at a particular value of J or CTOD. Rather, these materials display a rising R curve, where J and CTOD increase with crack growth. In metals, a rising R curve is normally associated with the growth and coalescence of microvoids. See Chapter 5 for a discussion of microscopic fracture mechanisms in ductile metals.

Figure 3.19 schematically illustrates a typical J resistance curve for a ductile material. In the initial stages of deformation, the R curve is nearly vertical; there is a small amount of apparent crack growth due to blunting. As J increases, the material at the crack tip fails locally and the crack advances further. Because the R curve is rising, the initial crack growth is usually stable, but an instability can be encountered later, as discussed below.

One measure of fracture toughness J_{Ic} is defined near the initiation of stable crack growth. The precise point at which crack growth begins is usually ill-defined. Consequently, the definition of J_{Ic} is somewhat arbitrary, much like a 0.2% offset yield strength. The corresponding CTOD near the initiation of stable crack growth is denoted δ_i by U.S. and British testing standards. Chapter 7 describes experimental measurements of J_{Ic} and δ_i in more detail.

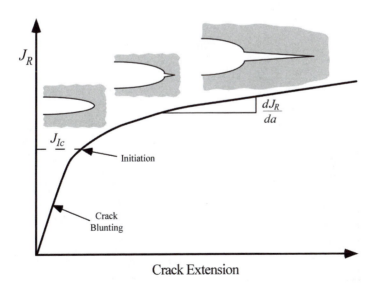

FIGURE 3.19 Schematic J resistance curve for a ductile material.

While initiation toughness provides some information about the fracture behavior of a ductile material, the entire R curve gives a more complete description. The slope of the R curve at a given amount of crack extension is indicative of the relative stability of the crack growth; a material with a steep R curve is less likely to experience unstable crack propagation. For J resistance curves, the slope is usually quantified by a dimensionless *tearing modulus*:

$$T_R = \frac{E}{\sigma_o^2} \frac{dJ_R}{da}$$

(3.49)

where the subscript R indicates a value of J on the resistance curve.

3.4.1 Stable and Unstable Crack Growth

The conditions that govern the stability of crack growth in elastic-plastic materials are virtually identical to the elastic case presented in Section 2.5. Instability occurs when the driving force curve is tangent to the R curve. As Figure 3.20 indicates, load control is usually less stable than displacement control. The conditions in most structures are somewhere between the extremes of load control and displacement control. The intermediate case can be represented by a spring in series with the structure, where remote displacement is fixed (Figure 2.12). Since the R curve slope has been represented by a dimensionless tearing modulus (Equation (3.49)), it is convenient to express the driving force in terms of an *applied tearing modulus*:

$$T_{app} = \frac{E}{\sigma_o^2} \left(\frac{dJ}{da} \right)_{\Delta_T}$$

(3.50)

where Δ_T is the total remote displacement defined as

$$\Delta_T = \Delta + C_m P$$

(3.51)

and C_m is the system compliance. The slope of the driving force curve for a fixed Δ_T is identical to the linear elastic case (Equation (2.35)), except that $\mathcal{G}$ is replaced by J:

$$\left(\frac{dJ}{da} \right)_{\Delta_T} = \left(\frac{\partial J}{\partial a} \right)_P - \left(\frac{\partial J}{\partial P} \right)_a \left(\frac{\partial \Delta}{\partial a} \right)_P \left[C_m + \left(\frac{\partial \Delta}{\partial P} \right)_a \right]^{-1}$$

(3.52)

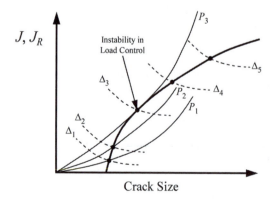

FIGURE 3.20 Schematic J driving force/R curve diagram which compares load control and displacement control.

For load control, $C_m = \infty$, and the second term in Equation (3.52) vanishes:

$$\left(\frac{dJ}{da}\right)_{\Delta_T} = \left(\frac{\partial J}{\partial a}\right)_P$$

For displacement control, $C_m = 0$, and $\Delta_T = \Delta$. Equation (3.52) is derived in Appendix 2.2 for the linear elastic case.

The conditions during stable crack growth can be expressed as follows:

$$J = J_R \qquad (3.53a)$$

and

$$T_{app} \leq T_R \qquad (3.53b)$$

Unstable crack propagation occurs when

$$T_{app} > T_R \qquad (3.54)$$

Chapter 9 gives practical guidance on assessing structural stability with Equation (3.50) to Equation (3.54). A simple example is presented below.

EXAMPLE 3.2

Derive an expression for the applied tearing modulus in the double cantilever beam (DCB) specimen with a spring in series (Figure 3.21), assuming linear elastic conditions.

Solution: From Example 2.1, we have the following relationships:

$$J = \mathcal{G} = \frac{P^2 a^2}{BEI} \quad \text{and} \quad \Delta = \mathcal{G} = \frac{2Pa^3}{3EI}$$

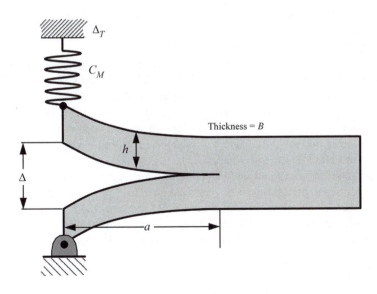

FIGURE 3.21 Double cantilever beam specimen with a spring in series.

Therefore, the relevant partial derivatives are given by

$$\left(\frac{\partial J}{\partial a}\right)_P = \frac{2P^2 a}{BEI}$$

$$\left(\frac{\partial J}{\partial P}\right)_a = \frac{2Pa^2}{BEI}$$

$$\left(\frac{\partial \Delta}{\partial a}\right)_P = \frac{2Pa^2}{EI}$$

$$\left(\frac{\partial \Delta}{\partial P}\right)_a = \frac{2a^3}{3EI}$$

Substituting the above relationships into Equation (3.50) and Equation (3.52) gives

$$T_{app} = \frac{2P^2 a}{\sigma_o^2 BI}\left\{1 - \frac{2a^3}{EI}\left[C_M + \frac{2a^3}{3EI}\right]^{-1}\right\}$$

As discussed in Section 2.5, the point of instability in a material with a rising R curve depends on the size and geometry of the cracked structure; a critical value of J at instability is not a material property if J increases with crack growth. It is usually assumed that the R curve, including the J_{Ic} value, is a material property, independent of the configuration. This is a reasonable assumption, within certain limitations.

3.4.2 Computing J for a Growing Crack

The geometry dependence of a J resistance curve is influenced by the way in which J is calculated. The equations derived in Section 3.2.5 are based on the pseudo energy release rate definition of J and are valid only for a stationary crack. There are various ways to compute J for a growing crack, including the deformation J and the far-field J, which are described below. The former method is typically used to obtain experimental J resistance curves.

Figure 3.22 illustrates the load-displacement behavior in a specimen with a growing crack. Recall that the J integral is based on a deformation plasticity (or nonlinear elastic) assumption for material behavior. Consider point A on the load-displacement curve in Figure 3.22. The crack has grown to a length a_1 from an initial length a_o. The cross-hatched area represents the energy that would be released if the material were elastic. In an elastic-plastic material, only the elastic portion of this energy is released; the remainder is dissipated in a plastic wake that forms behind the growing crack (see Figure 2.6(b) and Figure 3.25).

In an elastic material, all quantities, including strain energy, are independent of the loading history. The energy absorbed during crack growth in an elastic-plastic material, however, exhibits a history dependence. The dashed curve in Figure 3.22 represents the load-displacement behavior when the crack size is fixed at a_1. The area under this curve is the strain energy in an elastic material; this energy depends only on the current load and crack length:

$$U_D = U_D(P, a) = \left(\int_0^\Delta P\, d\Delta\right)_{a=a_1} \qquad (3.55)$$

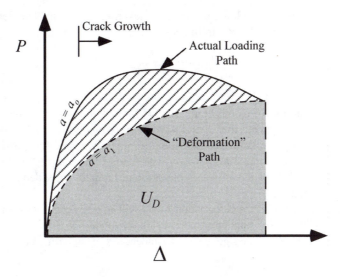

FIGURE 3.22 Schematic load-displacement curve for a specimen with a crack that grows to a_1 from an initial length a_o. U_D represents the strain energy in a nonlinear elastic material.

where the subscript D refers to the deformation theory. Thus the J integral for a nonlinear elastic body with a growing crack is given by

$$J_D = -\frac{1}{B}\left(\frac{\partial U_D}{\partial a}\right)_\Delta$$

$$= \frac{\eta U_D}{Bb} \tag{3.56a}$$

or

$$J_D = \frac{K_I^2}{E'} + \frac{\eta_p U_{D(p)}}{Bb} \tag{3.56b}$$

where b is the *current* ligament length. When the J integral for an elastic-plastic material is defined by Equation (3.56), the history dependence is removed and the energy release rate interpretation of J is restored. The *deformation J* is usually computed from Equation (3.56b) because no correction is required on the elastic term as long as K_I is determined from the current load and crack length. The calculation of $U_{D(p)}$ is usually performed incrementally, since the deformation theory load-displacement curve (Figure 3.22 and Equation (3.55)) depends on the crack size. Specific procedures for computing the deformation J are outlined in Chapter 7.

One can determine a far-field J from the contour integral definition of Equation (3.19), which may differ from J_D. For a deeply cracked bend specimen, Rice et al. [15] showed that the far-field J contour integral in a rigid, perfectly plastic material is given by

$$J_f = 0.73\sigma_o \int_0^\Omega b \, d\Omega \tag{3.57}$$

where the variation in b during the loading history is taken into account. The deformation theory leads to the following relationship for J in this specimen:

$$J_D = 0.73\sigma_o b\Omega \tag{3.58}$$

The two expressions are obviously identical when the crack is stationary.

　　Finite element calculations of Dodds et al. [16, 17] for a three-point bend specimen made from a strain hardening material indicate that J_f and J_D are approximately equal for moderate amounts of crack growth. The J integral obtained from a contour integration is path-dependent when a crack is growing in an elastic-plastic material, however, and tends to zero as the contour shrinks to the crack tip. See Appendix 4.2 for a theoretical explanation of the path dependence of J for a growing crack in an inelastic material.

　　There is no guarantee that either the deformation J_D or J_f will uniquely characterize crack-tip conditions for a growing crack. Without this single-parameter characterization, the J-R curve becomes geometry dependent. The issue of J validity and geometry dependence is explored in detail in Section 3.5 and Section 3.6.

3.5 J-CONTROLLED FRACTURE

The term *J-controlled fracture* corresponds to situations where J completely characterizes crack-tip conditions. In such cases, there is a unique relationship between J and CTOD (Section 3.3); thus J-controlled fracture implies CTOD-controlled fracture, and vice versa. Just as there are limits to LEFM, fracture mechanics analyses based on J and CTOD become suspect when there is excessive plasticity or significant crack growth. In such cases, fracture toughness and the J-CTOD relationship depend on the size and geometry of the structure or test specimen.

　　The required conditions for J-controlled fracture are discussed below. Fracture initiation from a stationary crack and stable crack growth are considered.

3.5.1 Stationary Cracks

Figure 3.23 schematically illustrates the effect of plasticity on the crack tip stresses; $\log(\sigma_{yy})$ is plotted against the normalized distance from the crack tip. The characteristic length scale L corresponds to the size of the structure; for example, L could represent the uncracked ligament length. Figure 3.23(a) shows the small-scale yielding case, where both K and J characterize crack-tip conditions. At a short distance from the crack tip, relative to L, the stress is proportional to $1/\sqrt{r}$; this area is called the *K-dominated region*. Assuming monotonic, quasistatic loading, a J-dominated region occurs in the plastic zone, where the elastic singularity no longer applies. Well inside of the plastic zone, the HRR solution is approximately valid and the stresses vary as $r^{-1/(n+1)}$. The finite strain region occurs within approximately 2δ from the crack tip, where large deformation invalidates the HRR theory. In small-scale yielding, K uniquely characterizes crack-tip conditions, despite the fact that the $1/\sqrt{r}$ singularity does not exist all the way to the crack tip. Similarly, J uniquely characterizes crack-tip conditions even though the deformation plasticity and small strain assumptions are invalid within the finite strain region.

　　Figure 3.23(b) illustrates elastic-plastic conditions, where J is still approximately valid, but there is no longer a K field. As the plastic zone increases in size (relative to L), the K-dominated zone disappears, but the J-dominated zone persists in some geometries. Thus although K has no meaning in this case, the J integral is still an appropriate fracture criterion. Since J dominance implies CTOD dominance, the latter parameter can also be applied in the elastic-plastic regime.

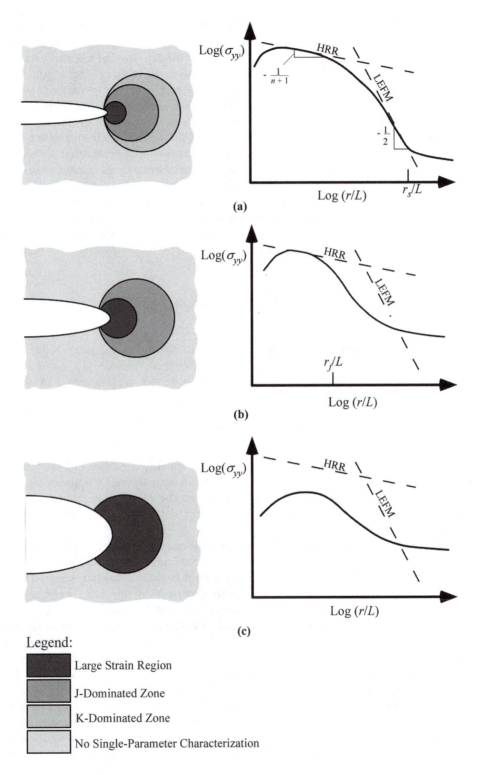

FIGURE 3.23 Effect of plasticity on the crack-tip stress fields: (a) small-scale yielding, (b) elastic-plastic conditions, and (c) large-scale yielding.

With large-scale yielding (Figure 3.23(c)), the size of the finite strain zone becomes significant relative to L, and there is no longer a region uniquely characterized by J. Single-parameter fracture mechanics is invalid in large-scale yielding, and critical J values exhibit a size and geometry dependence.

In certain configurations, the K and J zones are vanishingly small, and a single-parameter description is not possible except at very low loads. For example, a plate loaded in tension with a through-thickness crack is not amenable to a single-parameter description, either by K or J. Example 2.7 and Figure 2.39 indicate that the stress in the x direction in this geometry deviates significantly from the elastic singularity solution at small distances from the crack tip because of a compressive transverse (T) stress. Consequently the K-dominated zone is virtually nonexistent. The T stress influences stresses inside the plastic zone, so a highly negative T stress also invalidates a single-parameter description in terms of J. See Section 3.61 for further details about the T stress.

Recall Section 2.10.1, where a free-body diagram was constructed from a disk-shaped region removed from the crack tip of a structure loaded in small-scale yielding. Since the stresses on the boundary of this disk exhibit a $1/\sqrt{r}$ singularity, K_I uniquely defines the stresses and strains within the disk. For a given material,[5] dimensional analysis leads to the following functional relationship for the stress distribution within this region:

$$\frac{\sigma_{ij}}{\sigma_o} = F_{ij}\left(\frac{K_I^2}{\sigma_o^2 r}, \theta\right) \quad \text{for } 0 \leq r \leq r_s(\theta) \tag{3.59}$$

where r_s is the radius of the elastic singularity dominated zone, which may depend on θ. Note that the $1/\sqrt{r}$ singularity is a special case of F, which exhibits a different dependence on r within the plastic zone. Invoking the relationship between J and K_I for small-scale yielding (Equation 3.18) gives

$$\frac{\sigma_{ij}}{\sigma_o} = F_{ij}\left(\frac{E'J}{\sigma_o^2 r}, \theta\right) \quad \text{for } 0 \leq r \leq r_J(\theta) \tag{3.60}$$

where r_J is the radius of the J-dominated zone. The HRR singularity (Equation (3.24a)) is a special case of Equation (3.60), but stress exhibits a $r^{-1/(n+1)}$ dependence only over a limited range of r.

For small-scale yielding, $r_s = r_J$, but r_s vanishes when the plastic zone engulfs the elastic singularity dominated zone. The J-dominated zone usually persists longer than the elastic singularity zone, as Figure 3.23 illustrates.

It is important to emphasize that the J dominance at the crack tip does not require the existence of an HRR singularity. In fact, J dominance requires only that Equation (3.60) is valid in the *process zone* near the crack tip, where the microscopic events that lead to fracture occur. The HRR singularity is merely one possible solution to the more general requirement that J uniquely define crack-tip stresses and strains. The flow properties of most materials do not conform to the idealization of a Ramberg-Osgood power law, upon which the HRR analysis is based. Even in a Ramberg-Osgood material, the HRR singularity is valid over a limited range; large strain effects invalidate the HRR singularity close to the crack tip, and the computed stress lies below the HRR solution at greater distances. The latter effect can be understood by considering the analytical technique employed by Hutchinson [7], who represented the stress solution as an infinite series and showed that the leading term in the series was proportional to $r^{-1/(n+1)}$ (see Appendix 3.4). This singular term dominates as $r \to 0$; higher-order terms are significant for moderate values of r. When the computed stress field deviates from HRR, it still scales with $J/(\sigma_o r)$, as required by Equation (3.60). Thus J dominance does not necessarily imply agreement with the HRR fields.

---

[5] A complete statement of the functional relationship of σ_{ij} should include all material flow properties (e.g., α and n for a Ramberg-Osgood material). These quantities were omitted from Equation (3.59) and Equation (3.60) for the sake of clarity, since material properties are assumed to be fixed in this problem.

Equation (3.59) and Equation (3.60) gradually become invalid as specimen boundaries interact with the crack tip. We can apply dimensional arguments to infer when a single-parameter description of crack-tip conditions is suspect. As discussed in Chapter 2, the LEFM solution breaks down when the plastic-zone size is a significant fraction of in-plane dimensions. Moreover, the crack-tip conditions evolve from plane strain to plane stress as the plastic-zone size grows to a significant fraction of the thickness. The J integral becomes invalid as a crack-tip-characterizing parameter when the large strain region reaches a finite size relative to in-plane dimensions. Section 3.6 provides quantitative information on size effects.

3.5.2 *J*-CONTROLLED CRACK GROWTH

According to the dimensional argument in the previous section, J-controlled conditions exist at the tip of a stationary crack (loaded monotonically and quasistatically), provided the large strain region is small compared to the in-plane dimensions of the cracked body. Stable crack growth, however, introduces another length dimension, i.e., the change in the crack length from its original value. Thus J may not characterize crack-tip conditions when the crack growth is significant compared to the in-plane dimensions. Prior crack growth should not have any adverse effects in a purely elastic material, because the local crack-tip fields depend only on current conditions. However, prior history does influence the stresses and strains in an elastic-plastic material. Therefore, we might expect the J integral theory to break down when there is a combination of significant plasticity and crack growth. This heuristic argument based on dimensional analysis agrees with experiment and with more complex analyses.

Figure 3.24 illustrates crack growth under J-controlled conditions. The material behind the growing crack tip has unloaded elastically. Recall Figure 3.7, which compares the unloading behavior of nonlinear elastic and elastic-plastic materials; the material in the unloading region of Figure 3.24 obviously violates the assumptions of deformation plasticity. The material directly in front of the crack also violates the single-parameter assumption because the loading is highly nonproportional, i.e., the various stress components increase at different rates and some components actually decrease. In order for the crack growth to be J controlled, the elastic unloading and non-proportional plastic loading regions must be embedded within a zone of J dominance. When the crack grows out of the zone of J dominance, the measured R curve is no longer uniquely characterized by J.

In small-scale yielding, there is always a zone of J dominance because the crack-tip conditions are defined by the elastic stress intensity, which depends only on the current values of the load and crack size. The crack never grows out of the J-dominated zone as long as all the specimen boundaries are remote from the crack tip and the plastic zone.

Figure 3.25 illustrates three distinct stages of crack growth resistance behavior in small-scale yielding. During the initial stage the crack is essentially stationary; the finite slope of the R curve

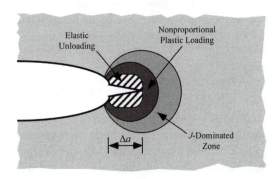

FIGURE 3.24 *J*-controlled crack growth.

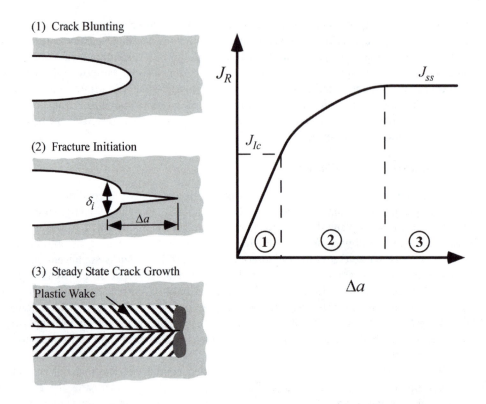

(1) Crack Blunting

(2) Fracture Initiation

(3) Steady State Crack Growth

Plastic Wake

FIGURE 3.25 Three stages of crack growth in an infinite body.

is caused by blunting. The crack-tip fields for Stage 1 are given by

$$\frac{\sigma_{ij}}{\sigma_o} = F_{ij}^{(1)}\left(\frac{E'J}{\sigma_o^2 r}, \theta\right) \tag{3.61}$$

which is a restatement of Equation (3.60). The crack begins to grow in Stage 2. The crack-tip stresses and strains are probably influenced by the original blunt crack tip during the early stages of crack growth. Dimensional analysis implies the following relationship:

$$\frac{\sigma_{ij}}{\sigma_o} = F_{ij}^{(2)}\left(\frac{E'J}{\sigma_o^2 r}, \theta, \frac{\Delta a}{\delta_i}\right) \tag{3.62}$$

where δ_i is the CTOD at initiation of stable tearing. When the crack grows well beyond the initial blunted tip, a steady-state condition is reached, where the local stresses and strains are independent of the extent of crack growth:

$$\frac{\sigma_{ij}}{\sigma_o} = F_{ij}^{(3)}\left(\frac{E'J}{\sigma_o^2 r}, \theta\right) \tag{3.63}$$

Although Equation (3.61) and Equation (3.63) would predict identical conditions in the elastic-singularity zone, the material in the plastic zone at the tip of a growing crack is likely to experience a different loading history from the material in the plastic zone of a blunting stationary crack; thus $F^{(1)} \neq F^{(3)}$ as $r \to 0$. During steady-state crack growth, a plastic zone of constant size sweeps through the material, leaving a plastic wake, as illustrated in Figure 3.25. The R curve is flat; J does not increase with crack extension, provided the material properties do not vary with position. Appendix 3.5.2 presents a formal mathematical argument for a flat R curve during steady-state growth; a heuristic explanation is given below.

If Equation (3.63) applies, J uniquely describes crack-tip conditions, independent of crack extension. If the material fails at some critical combination of stresses and strains, it follows that local failure at the crack tip must occur at a critical J value, as in the stationary crack case. This critical J value must remain constant with crack growth. A rising or falling R curve would imply that the local material properties varied with position.

The second stage in Figure 3.25 corresponds to the transition between the blunting of a stationary crack and crack growth under steady state conditions. A rising R curve is possible in Stage 2. For small-scale yielding conditions the R curve depends only on crack extension:

$$J_R = J_R(\Delta a) \tag{3.64}$$

That is, the J-R curve is a material property.

The steady-state limit is usually not observed in laboratory tests on ductile materials. In typical test specimens, the ligament is fully plastic during crack growth, thereby violating the small-scale yielding assumption. Moreover, the crack approaches a finite boundary while still in Stage 2 growth. Enormous specimens would be required to observe steady-state crack growth in tough materials.

3.6 CRACK-TIP CONSTRAINT UNDER LARGE-SCALE YIELDING

Under small-scale yielding conditions, a single parameter (e.g., K, J, or CTOD) characterizes crack-tip conditions and can be used as a geometry-independent fracture criterion. Single-parameter fracture mechanics breaks down in the presence of excessive plasticity, and the fracture toughness depends on the size and geometry of the test specimen.

McClintock [18] applied the slip line theory to estimate the stresses in a variety of configurations under plane strain, fully plastic conditions. Figure 3.26 summarizes some of these results. For small-scale yielding (Figure 3.26(a)), the maximum stress at the crack tip is approximately $3\sigma_o$ in a nonhardening material. According to the slip line analysis, a deeply notched double-edged notched tension (DENT) panel, illustrated in Figure 3.26(b), maintains a high level of triaxiality under fully plastic conditions, such that the crack-tip conditions are similar to the small-scale yielding case. An edge-cracked plate in bending (Figure 3.26(c)) exhibits slightly less stress elevation, with the maximum principal stress approximately $2.5\sigma_o$. A center-cracked panel in pure tension (Figure 3.26(d)) is incapable of maintaining significant triaxiality under fully plastic conditions.

The results in Figure 3.26 indicate that for a nonhardening material under fully yielded conditions, the stresses near the crack tip are not unique, but depend on geometry. Traditional fracture mechanics approaches recognize that the stress and strain fields remote from the crack tip may depend on geometry, but it is assumed that the near-tip fields have a similar form in all configurations that can be scaled by a single parameter. The single-parameter assumption is obviously not valid for nonhardening materials under fully plastic conditions, because the near-tip fields depend on the configuration. Fracture toughness, whether quantified by J, K, or CTOD, must also depend on the configuration.

The prospects for applying fracture mechanics in the presence of large-scale yielding are not quite as bleak as the McClintock analysis indicates. The configurational effects on the near-tip fields are much less severe when the material exhibits strain hardening. Moreover, single-parameter fracture mechanics may be approximately valid in the presence of significant plasticity, provided the specimen maintains a relatively high level of triaxiality. Both the DENT specimen and the edge-cracked plate in bending (SE(B)) apparently satisfy this requirement. Most laboratory measurements

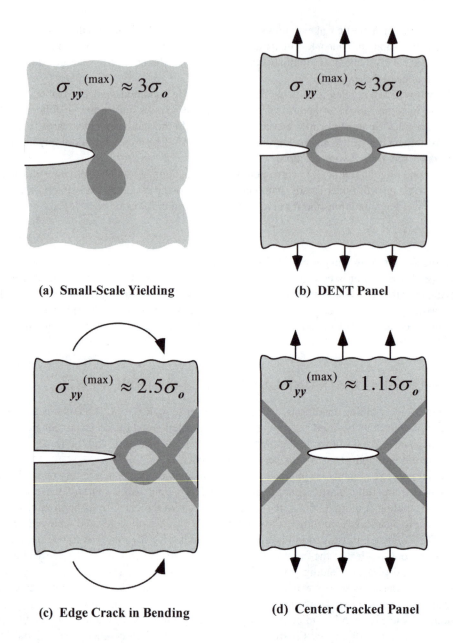

$$\sigma_{yy}^{(max)} \approx 3\sigma_o$$

(a) Small-Scale Yielding

$$\sigma_{yy}^{(max)} \approx 3\sigma_o$$

(b) DENT Panel

$$\sigma_{yy}^{(max)} \approx 2.5\sigma_o$$

(c) Edge Crack in Bending

$$\sigma_{yy}^{(max)} \approx 1.15\sigma_o$$

(d) Center Cracked Panel

FIGURE 3.26 Comparison of the plastic deformation pattern in small-scale yielding (a) with slip patterns under fully plastic conditions in three configurations. The estimated local stresses are based on the slip-line analyses of McClintock, and apply only to nonhardening materials. Taken from McClintock, F.A., "Plasticity Aspects of Fracture." *Fracture: An Advanced Treatise,* Vol. 3, Academic Press, New York, 1971, pp. 47–225.

of fracture toughness are performed with bend-type specimens, such as the compact and three-point bend geometries, because these specimens present the fewest experimental difficulties.

Figure 3.27 compares the cleavage-fracture toughness for bending and tensile loading. Although the scatter bands overlap, the average toughness for the single-edge-notched bend specimens is considerably lower than that of the center-cracked tension panels or the surface-cracked panels.

Crack depth and specimen size can also have an effect on fracture toughness, as Figure 3.28 illustrates. Note that the bend specimens with shallow cracks tend to have higher toughness than

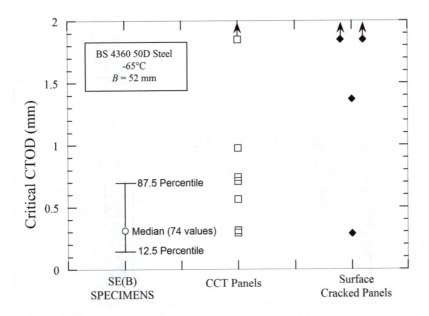

FIGURE 3.27 Critical CTOD values for cleavage fracture in bending and tensile loading for a low-alloy structural steel. Taken from Anderson, T.L., "Ductile and Brittle Fracture Analysis of Surface Flaws Using CTOD." *Experimental Mechanics*, 1988, pp. 188–193.

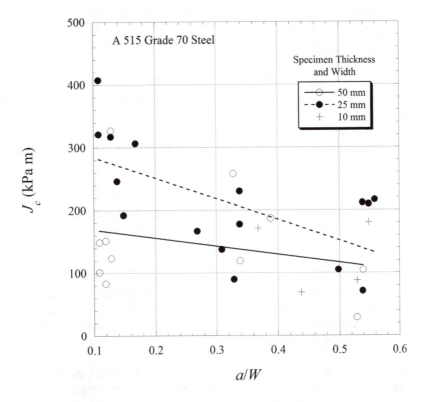

FIGURE 3.28 Critical *J* values for cleavage as a function of crack depth and specimen size of single-edge-notched bend (SE(B)) specimens. Taken from Kirk, M.T., Koppenhoefer, K.C., and Shih, C.F., "Effect of Constraint on Specimen Dimensions Needed to Obtain Structurally Relevant Toughness Measures." *Constraint Effects in Fracture*, ASTM STP 1171, American Society for Testing and Materials, Philadelphia, PA 1993, pp. 79–103.

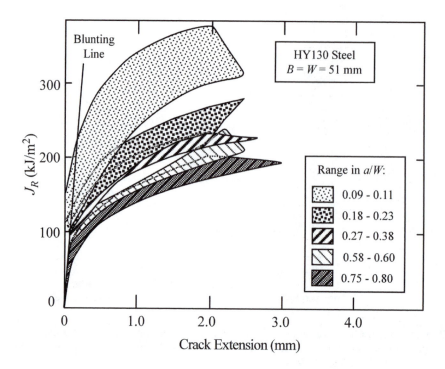

FIGURE 3.29 Effect of crack length/specimen width ratio on *J-R* curves for HY130 steel single-edge-notched bend (SE(B)) specimens. Taken from Towers, O.L. and Garwood, S.J., "Influence of Crack Depth on Resistance Curves for Three-Point Bend Specimens in HY130." ASTM STP 905, American Society for Testing and Materials, Philadelphia, PA, 1986, pp. 454–484.

deep-cracked specimens, and the specimens with 50 mm × 50 mm cross sections have a lower average toughness than smaller specimens with the same a/W ratio.

Figure 3.27 and Figure 3.28 illustrate the effect of specimen size and geometry on cleavage-fracture toughness. Specimen configuration can also influence the *R* curve of ductile materials. Figure 3.29 shows the effect of crack depth on crack growth resistance behavior. Note that the trend is the same as in Figure 3.28. Joyce and Link [21] measured *J-R* curves for several geometries and found that the initiation toughness J_{Ic} is relatively insensitive to geometry (Figure 3.30), but the tearing modulus, as defined in Equation (3.49), is a strong function of geometry (Figure 3.31). Configurations that have a high level of constraint under full plastic conditions, such as the compact and deep-notched SE(B) specimens, have low T_R values relative to low constraint geometries, such as single edge notched tension panels.

Note that the DENT specimens have the highest tearing modulus in Figure 3.31, but McClintock's slip-line analysis indicates that this configuration should have a high level of constraint under fully plastic conditions. Joyce and Link presented elastic-plastic finite element results for the DENT specimen that indicated significant constraint loss in this geometry,[6] which is consistent with the observed elevated tearing modulus. Thus the slip-line analysis apparently does not reflect the actual crack-tip conditions of this geometry.

A number of researchers have attempted to extend fracture mechanics theory beyond the limits of the single-parameter assumption. Most of these new approaches involve the introduction of a second parameter to characterize crack-tip conditions. Several such methodologies are described later.

---

[6] Joyce and Link quantified crack-tip constraint with the *T* and *Q* parameters, which are described in Section 3.6.1 and Section 3.6.2, respectively.

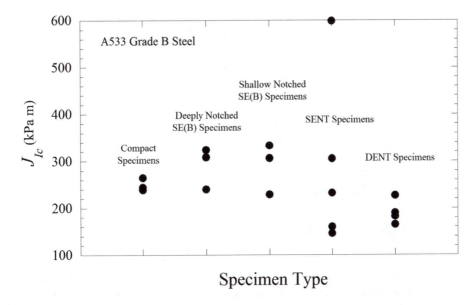

FIGURE 3.30 Effect of specimen geometry on critical J values for initiation of ductile tearing. Taken from Joyce, J.A. and Link, R.E., "Effect of Constraint on Upper Shelf Fracture Toughness." *Fracture Mechanics,* Vol. 26, ASTM STP 1256, American Society for Testing and Materials, Philadelphia, PA (in press).

3.6.1 THE ELASTIC T STRESS

Williams [23] showed that the crack-tip stress fields in an isotropic elastic material can be expressed as an infinite power series, where the leading term exhibits a $1/\sqrt{r}$ singularity, the second term is constant with r, the third term is proportional to $\sqrt{r}$, and so on. Classical fracture mechanics theory normally neglects all but the singular term, which results in a single-parameter description of the near-tip

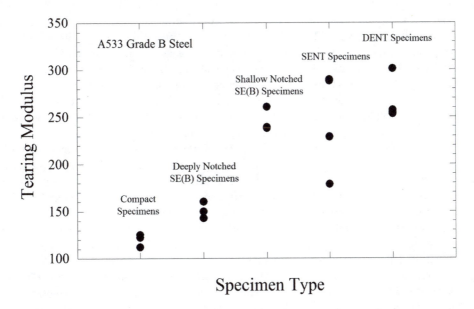

FIGURE 3.31 Effect of specimen geometry on tearing modulus at $\Delta a = 1$ mm. Taken from Towers, O.L. and Garwood, S.J., "Influence of Crack Depth on Resistance Curves for Three-Point Bend Specimens in HY130." ASTM STP 905, American Society for Testing and Materials, Philadelphia, PA, 1986, pp. 454–484.

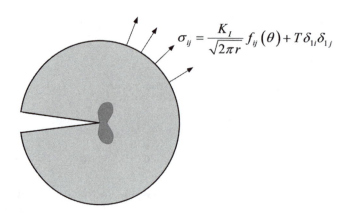

FIGURE 3.32 Modified boundary layer analysis. The first two terms of the Williams series are applied as boundary conditions.

fields (see Chapter 2). Although the third and higher terms in the Williams solution, which have positive exponents on r, vanish at the crack tip, the second (uniform) term remains finite. It turns out that this second term can have a profound effect on the plastic zone shape and the stresses deep inside the plastic zone [24, 25].

For a crack in an isotropic elastic material subject to plane strain Mode I loading, the first two terms of the Williams solution are as follows:

$$\sigma_{ij} = \frac{K_I}{\sqrt{2\pi r}} f_{ij}(\theta) + \begin{bmatrix} T & 0 & 0 \\ 0 & 0 & 0 \\ 0 & 0 & vT \end{bmatrix} \tag{3.65}$$

where T is a uniform stress in the x direction (which induces a stress vT in the z direction in plane strain).

We can assess the influence of the T stress by constructing a circular model that contains a crack, as illustrated in Figure 3.32. On the boundary of this model, let us apply in-plane tractions that correspond to Equation (3.65). A plastic zone develops at the crack tip, but its size must be small relative to the size of the model in order to ensure the validity of the boundary conditions, which are inferred from an elastic solution. This configuration, often referred to as a *modified boundary layer analysis*, simulates the near-tip conditions in an arbitrary geometry, provided the plasticity is well contained within the body. It is equivalent to removing a core region from the crack tip and constructing a free-body diagram, as in Figure 2.43.

Figure 3.33 is a plot of finite element results from a modified boundary layer analysis [26] that show the effect of the T stress on stresses deep inside the plastic zone. The special case of $T = 0$ corresponds to the small-scale yielding limit, where the plastic zone is a negligible fraction of the crack length and size of the body,[7] and the singular term uniquely defines the near-tip fields. The single-parameter description is rigorously correct only for $T = 0$. Note that negative T values cause a significant downward shift in the stress fields. Positive T values shift the stresses to above the small-scale yielding limit, but the effect is much less pronounced than it is for the negative T stress.

Note that the HRR solution does not match the $T = 0$ case. The stresses deep inside the plastic zone can be represented by a power series, where the HRR solution is the leading term. Figure 3.33 indicates that the higher-order plastic terms are not negligible when $T = 0$. A single-parameter description in terms of J is still valid, however, as discussed in Section 3.5.1.

---

[7] In this case, "body" refers to the global configuration, not the modified boundary layer model. A modified boundary layer model with $T = 0$ simulates an infinite body with an infinitely long crack.

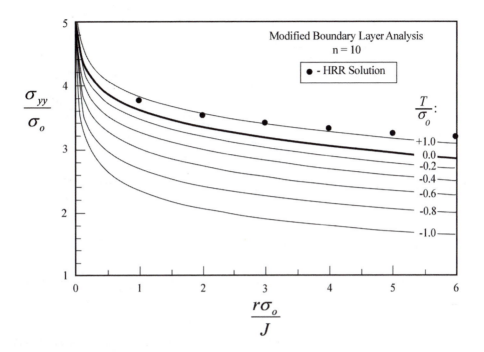

FIGURE 3.33 Stress fields obtained from modified boundary layer analysis. Taken from Kirk, M.T., Dodds, R.H., Jr., and Anderson, T.L., "Approximate Techniques for Predicting Size Effects on Cleavage Fracture Toughness." *Fracture Mechanics*, Vol. 24, ASTM STP 1207, American Society for Testing and Materials, Philadelphia, PA (in press).

In a cracked body subject to Mode I loading, the T stress, like K_I, scales with the applied load. The *biaxiality ratio* relates T to stress intensity:

$$\beta = \frac{T\sqrt{\pi a}}{K_I} \qquad (3.66)$$

For a through-thickness crack in an infinite plate subject to a remote normal stress (Figure 2.3), $\beta = -1$. Thus a remote stress σ induces a T stress of $-\sigma$ in the x direction. Recall Example 2.7, where a rough estimate of the elastic-singularity zone and plastic zone size led to the conclusion that K breaks down in this configuration when the applied stress exceeds 35% of the yield, which corresponds to $T/\sigma_o = -0.35$. From Figure 3.33, we see that such a T stress leads to a significant relaxation in crack-tip stresses, relative to the small-scale yielding case.

For laboratory specimens with K_I solutions of the form in Table 2.4, the T stress is given by

$$T = \frac{\beta P}{B\sqrt{\pi a W}} f\left(\frac{a}{W}\right) \qquad (3.67)$$

Figure 3.34 is a plot of β for several geometries. Note that β is positive for deeply notched SENT and SE(B) specimens, where the uncracked ligament is subject predominately to bending stresses. As discussed above, such configurations maintain a high level of constraint under fully plastic conditions. Thus a positive T stress in the elastic case generally leads to high constraint under fully plastic conditions, while geometries with negative T stress lose constraint rapidly with deformation.

The biaxiality ratio can be used as a qualitative index of the relative crack-tip constraint of various geometries. The T stress, combined with the modified boundary layer solution (Figure 3.33)

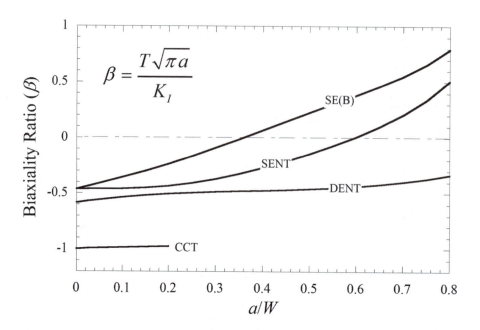

FIGURE 3.34 Biaxiality ratio for single-edge-notched bend, single-edge-notched tension, double-edge-notched tension, and center-cracked tension geometries.

can also be used *quantitatively* to estimate the crack-tip stress field in a particular geometry [26–28]. For a given load level, the T stress can be inferred from Equation (3.66) or Equation (3.67), and the corresponding crack-tip stress field for the same T stress can be estimated from the modified boundary layer solution with the same applied T. This methodology has limitations, however, because T is an *elastic* parameter. A T stress estimated from load through Equation (3.67) has no physical meaning under fully plastic conditions. Errors in stress fields inferred from T stress and the modified boundary layer solution increase with plastic deformation. This approximate procedure works fairly well for $|\beta| > 0.9$ but breaks down when $|\beta| < 0.4$ [26].

3.6.2 J-Q Theory

Assuming the small-strain theory, the crack-tip fields deep inside the plastic zone can be represented by a power series, where the HRR solution is the leading term. The higher-order terms can be grouped together into a *difference field*:

$$\sigma_{ij} = (\sigma_{ij})_{HRR} + (\sigma_{ij})_{Diff} \qquad (3.68a)$$

Alternatively, the difference field can be defined as the deviation from the $T = 0$ reference solution:

$$\sigma_{ij} = (\sigma_{ij})_{T=0} + (\sigma_{ij})_{Diff} \qquad (3.68b)$$

Note from Figure 3.33 that nonzero T stresses cause the near-tip field at $\theta = 0$ to shift up or down uniformly, i.e., the magnitude of the shift is constant with distance from the crack tip. O'Dowd and Shih [29, 30] observed that the difference field is relatively constant with both distance and angular position in the forward sector of the crack-tip region ($|\theta| \leq \pi/2$). Moreover, they noted that

$$(\sigma_{yy})_{Diff} \approx (\sigma_{xx})_{Diff} \gg (\sigma_{xy})_{Diff} \quad \text{for } |\theta| \leq \frac{\pi}{2}$$

Thus the difference field corresponds approximately to a uniform hydrostatic shift of the stress field in front of the crack tip. O'Dowd and Shih designated the amplitude of this approximate difference field by the letter Q. Equation (3.68b) then becomes

$$\sigma_{ij} \approx (\sigma_{ij})_{T=0} + Q\sigma_o\delta_{ij} \quad \left(|\theta| \leq \frac{\pi}{2}\right) \qquad (3.69)$$

where δ_{ij} is the Kronecker delta. The Q parameter can be inferred by subtracting the stress field for the $T = 0$ reference state from the stress field of interest. O'Dowd and Shih and most subsequent researchers defined Q as follows:

$$Q \equiv \frac{\sigma_{yy} - (\sigma_{yy})_{T=0}}{\sigma_o} \quad \text{at } \theta = 0 \quad \text{and} \quad \frac{r\sigma_o}{J} = 2 \qquad (3.70)$$

Referring to Figure 3.33, we see that Q is negative when T is negative. For the modified boundary layer solution, T and Q are uniquely related. Figure 3.35 is a plot of Q vs. T for a wide range of hardening exponents.

In a given cracked body, $Q = 0$ in the limit of small-scale yielding, but Q generally becomes increasingly negative with deformation. Figure 3.36 shows the evolution of Q for a deeply cracked bend (SENB) specimen and a center-cracked panel. Note that the SENB specimen stays close to the $Q = 0$ limit to fairly high deformation levels, but Q for the center-cracked panel becomes highly negative at relatively small J values.

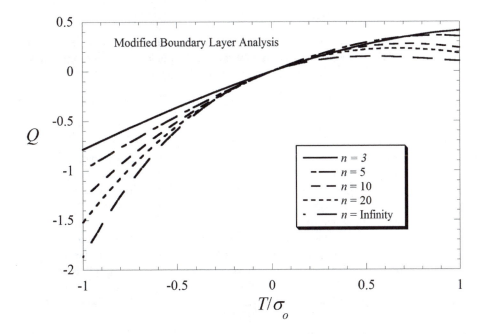

FIGURE 3.35 Relationship between Q and T as a function of strain-hardening exponent. Taken from O'Dowd, N.P. and Shih, C.F., "Family of Crack-Tip Fields Characterized by a Triaxiality Parameter–I. Structure of Fields." *Journal of the Mechanics and Physics of Solids,* Vol. 39, 1991, pp. 898–1015.

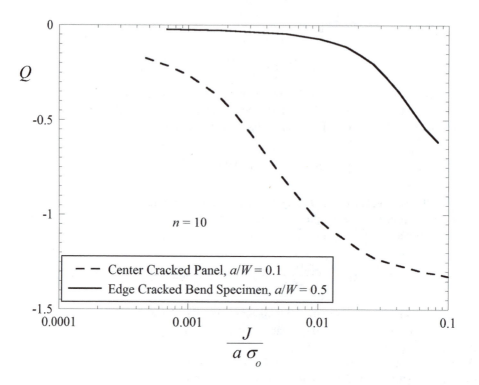

FIGURE 3.36 Evolution of the Q parameter with deformation in two geometries. Taken from O'Dowd, N.P. and Shih, C.F., "Family of Crack-Tip Fields Characterized by a Triaxiality Parameter–I. Structure of Fields." *Journal of the Mechanics and Physics of Solids,* Vol. 39, 1991, pp. 898–1015.

3.6.2.1 The *J-Q* Toughness Locus

Classical single-parameter fracture mechanics assumes that fracture toughness is a material constant. With the *J-Q* theory, however, an additional degree of freedom has been introduced, which implies that the critical *J* value for a given material depends on *Q*:

$$J_c = J_c(Q) \tag{3.71}$$

Thus fracture toughness is no longer viewed as a single value; rather, it is a *curve* that defines a critical locus of *J* and *Q* values.

Figure 3.37 is a plot of critical *J* values (for cleavage fracture) as a function of *Q* [29]. Although there is some scatter, the trend in Figure 3.37 is clear. The critical *J* increases as *Q* becomes more negative. This trend is consistent with Figures 3.27–3.31. That is, fracture toughness tends to increase as the constraint decreases. The *Q* parameter is a direct measure of the relative stress triaxiality (constraint) at the crack tip.

Since the *T* stress is also an indication of the level of crack-tip constraint, a *J-T* failure locus can be constructed [27, 28]. Such plots have similar trends to *J-Q* plots, but the ordering of data points sometimes differs. That is, the relative ranking of geometries can be influenced by whether the constraint is quantified by *T* or *Q*. Under well-contained yielding, *T* and *Q* are uniquely related (Figure 3.35), but the *T* stress loses its meaning for large-scale yielding. Thus, a *J-T* toughness locus is unreliable when significant yielding precedes fracture.

The single-parameter fracture mechanics theory assumes that toughness values obtained from laboratory specimens can be transferred to structural applications. Two-parameter approaches such as the *J-Q* theory imply that the laboratory specimen must match the constraint of the structure, i.e., the two geometries must have the same *Q* at failure in order for the respective J_c values to be

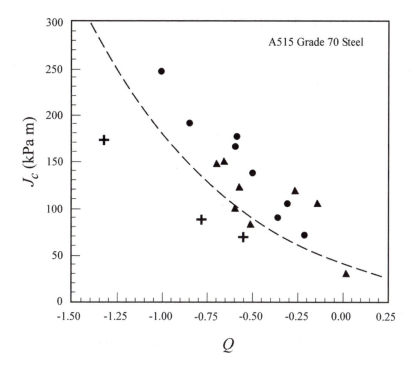

FIGURE 3.37 *J-Q* toughness locus for SE(B) specimens of A515 Grade 70 steel. Taken from Shih, C.F., O'Dowd, N.P., and Kirk, M.T., "A Framework for Quantifying Crack Tip Constraint." *Constraint Effects in Fracture*, ASTM STP 1171, American Society for Testing and Materials, Philadelphia, PA., 1993, pp. 2–20.

equal. Figure 3.38 illustrates the application of the *J-Q* approach to structures. The applied *J* vs. *Q* curve for the configuration of interest is obtained from finite element analysis and plotted with the *J-Q* toughness locus. Failure is predicted when the driving force curve passes through the toughness locus. Since toughness data are often scattered, however, there is not a single unambiguous crossover point. Rather, there is a range of possible J_c values for the structure.

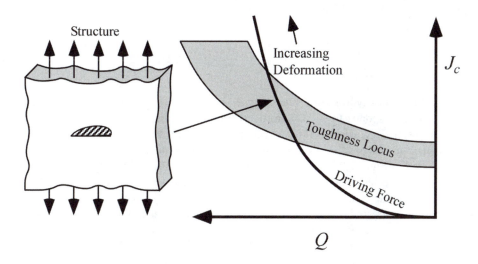

FIGURE 3.38 Application of a *J-Q* toughness locus. Failure occurs when the applied *J-Q* curve passes through the toughness locus.

3.6.2.2 Effect of Failure Mechanism on the J-Q Locus

The J-Q approach is descriptive but not predictive. That is, Q quantifies the crack-tip constraint, but it gives no indication as to the effect of constraint on fracture toughness. A J-Q failure locus, such as Figure 3.37, can be inferred from a series of experiments on a range of geometries. Alternatively, a micromechanical failure criterion can be invoked.

Consider, for example, the Ritchie-Knott-Rice (RKR) [32] model for cleavage fracture, which states that fracture occurs when a critical fracture stress σ_f is exceeded over a characteristic distance r_c. As an approximation, let us replace the $T = 0$ reference solution with the HRR field in Equation (3.69):

$$\sigma_{ij} \approx (\sigma_{ij})_{HRR} + Q\sigma_o\delta_{ij} \tag{3.72}$$

Setting the stress normal to the crack plane equal to σ_f and $r = r_c$, and relating the resulting equation to the $Q = 0$ limit leads to

$$\frac{\sigma_f}{\sigma_o} = \left(\frac{J_c}{\alpha\varepsilon_o\sigma_o I_n r_c}\right)^{\frac{1}{n+1}} \tilde{\sigma}_{yy}(n,0) + Q = \left(\frac{J_o}{\alpha\varepsilon_o\sigma_o I_n r_c}\right)^{\frac{1}{n+1}} \tilde{\sigma}_{yy}(n,0) \tag{3.73}$$

where J_o is the critical J value for the $Q = 0$ small-scale yielding limit. Rearranging gives

$$\frac{J_c}{J_o} = \left[1 - Q\left(\frac{\sigma_o}{\sigma_f}\right)\right]^{n+1} \tag{3.74}$$

which is a prediction of the J-Q toughness locus. Equation (3.74) predicts that toughness is highly sensitive to Q, since the quantity in brackets is raised to the $n + 1$ power.

The shape of the J-Q locus depends on the failure mechanism. Equation (3.74) refers to stress-controlled fracture, such as cleavage in metals, but strain-controlled fracture is less sensitive to the crack-tip constraint. A simple parametric study illustrates the influence of the local failure criterion.

Suppose that fracture occurs when a damage parameter Φ reaches a critical value r_c ahead of the crack tip, where Φ is given by

$$\Phi = \left(\frac{\sigma_m}{\sigma_o}\right)^{\gamma} \bar{\varepsilon}_{pl}^{1-\gamma} \quad (0 \leq \gamma \leq 1) \tag{3.75}$$

where σ_m is the mean (hydrostatic) stress and $\bar{\varepsilon}_{pl}$ is the equivalent plastic strain. When $\gamma = 1$, Equation (3.75) corresponds to stress-controlled fracture, similar to the RKR model. The other limit $\gamma = 0$ corresponds to strain-controlled failure. By varying γ and applying Equation (3.75) to the finite element results of O'Dowd and Shih [29, 30], we obtain a family of J-Q toughness loci, which are plotted in Figure 3.39. The J-Q locus for stress-controlled fracture is highly sensitive to constraint, as expected. For strain-controlled fracture, the locus has a slight negative slope, indicating that toughness *decreases* as constraint relaxes. As Q decreases (i.e., becomes more negative), crack-tip stresses relax, but the plastic *strain* fields at a given J value *increase* with constraint loss. Thus as constraint relaxes, a smaller J_c is required for failure for a purely strain-controlled mechanism. The predicted J_c is nearly constant for $\gamma = 0.5$. Microvoid growth in metals is governed by a combination of plastic strain and hydrostatic stress (see Chapter 5). Consequently, critical J values for the initiation of ductile crack growth are relatively insensitive to geometry, as Figure 3.30 indicates.

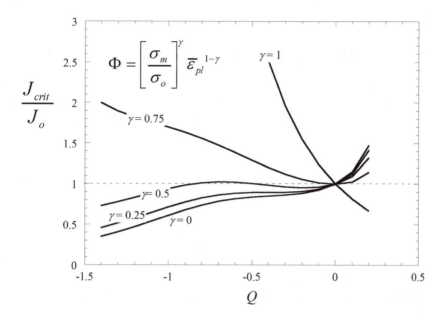

FIGURE 3.39 Effect of failure criterion on the *J-Q* locus. Fracture is assumed to occur when Φ reaches a critical value at a specific distance from the crack tip. Taken from Anderson, T.L., Vanaparthy, N.M.R., and Dodds, R.H., Jr., ''Predictions of Specimen Size Dependence on Fracture Toughness for Cleavage and Ductile Tearing.'' *Constraint Effects in Fracture,* ASTM STP 1171, American Society for Testing and Materials, Philadelphia, PA, 1993, pp. 473–491.

According to Figure 3.31, the slope of the *J* resistance curve is influenced by specimen configuration. However, the stress and strain fields ahead of a growing crack are different from the stationary crack case [16, 17], and the *J-Q* theory is not applicable to a growing crack.

3.6.3 SCALING MODEL FOR CLEAVAGE FRACTURE

Both the *J-Q* and *T* stress methodologies are based on continuum theory. As stated above, these approaches characterize the crack-tip fields but they cannot predict the effect of these fields on a material's fracture resistance. A micromechanical failure criterion must be introduced to relate crack-tip fields to fracture toughness. The RKR model provides a simple means for such predictions. Anderson and Dodds [34–36] have developed a somewhat more sophisticated model for cleavage, which is described below.

3.6.3.1 Failure Criterion

Cleavage initiation usually involves a local Griffith instability of a microcrack which forms from a microstructural feature such as a carbide or inclusion. The Griffith energy balance is satisfied when a critical stress is reached in the vicinity of the microcrack. The size and location of the critical microstructural feature dictate the fracture toughness; thus cleavage toughness is subject to considerable scatter. See Chapter 5 for a more detailed description of the micromechanisms of cleavage fracture.

The Griffith instability criterion implies fracture at a critical normal stress near the tip of the crack; the statistical sampling nature of cleavage initiation (i.e., the probability of finding a critical microstructural feature near the crack tip) suggests that the volume of the process zone is also important. Thus the probability of cleavage fracture in a cracked specimen can be expressed in the following general form:

$$F = F[V(\sigma_1)] \tag{3.76}$$

where

F = failure probability

σ_1 = maximum principal stress at a point

$V(\sigma_1)$ = cumulative volume sampled where the principal stress is $\geq \sigma_1$

For a specimen subjected to plane strain conditions along the crack front, $V = BA$, where B is the specimen thickness and A is the cumulative area on the x-y plane.

The J_o Parameter

For small-scale yielding, dimensional analysis shows that the principal stress ahead of the crack tip can be written as

$$\frac{\sigma_1}{\sigma_o} = f\left(\frac{J}{\sigma_o r}, \theta\right) \tag{3.77}$$

Equation (3.77) implies that the crack-tip stress fields depend only on J. When J dominance is lost, there is a relaxation in triaxiality; the principal stress at a fixed r and θ is less than the small-scale yielding value.

Equation (3.77) can be inverted to solve for the radius corresponding to a given stress and angle:

$$r(\sigma_1/\sigma_o, \theta) = \frac{J}{\sigma_o} g(\sigma_1/\sigma_o, \theta) \tag{3.78}$$

Solving for the area inside a specific principal stress contour gives

$$A(\sigma_1/\sigma_o) = \frac{J^2}{\sigma_o^2} h(\sigma_1/\sigma_o) \tag{3.79}$$

where

$$h(\sigma_1/\sigma_o) = \frac{1}{2}\int_{-\pi}^{\pi} g^2(\sigma_1/\sigma_o, \theta)d\theta \tag{3.80}$$

Thus, for a given stress, the area scales with J^2 in the case of small-scale yielding. Under large-scale yielding conditions, the test specimen or structure experiences a loss in constraint, and the area inside a given principal stress contour (at a given J value) is less than predicted from small-scale yielding:

$$A(\sigma_1/\sigma_o) = \phi \frac{J^2}{\sigma_o^2} h(\sigma_1/\sigma_o) \tag{3.81}$$

where ϕ is a constraint factor that is ≤ 1. Let us define an *effective J* in large-scale yielding that relates the area inside the principal stress contour to the small-scale yielding case:

$$A(\sigma_1/\sigma_o) = \frac{(J_o)^2}{\sigma_o^2} h(\sigma_1/\sigma_o) \tag{3.82}$$

where J_o is the effective small-scale yielding J, i.e., the value of J that would result in the area $A(\sigma_1/\sigma_o)$ if the structure were large relative to the plastic zone, and $T = Q = 0$. Therefore, the ratio of the applied J to the effective J is given by

$$\frac{J}{J_o} = \sqrt{\frac{1}{\phi}} \qquad (3.83)$$

The small-scale yielding J value (J_o) can be viewed as *the effective driving force for cleavage*, while J is the *apparent* driving force.

The J/J_o ratio quantifies the size dependence of cleavage fracture toughness. Consider, for example, a finite size test specimen that fails at $J_c = 200$ kPa m. If the J/J_o ratio were 2.0 in this case, a very large specimen made from the same material would fail at $J_c = 100$ kPa m. An equivalent toughness ratio in terms of the crack-tip-opening displacement (CTOD) can also be defined.

3.6.3.2 Three-Dimensional Effects

The constraint model described above considers only stressed *areas* in front of the crack tip. This model is incomplete, because it is the *volume* of material sampled ahead of the crack tip that controls the cleavage fracture. The stressed volume obviously scales with specimen thickness (or crack front length in the more general case). Moreover, the stressed volume is a function of the constraint parallel to the crack front; a higher constraint results in a larger volume, as is the case for in-plane constraint.

One way to treat three-dimensional constraint effects is to define an effective thickness based on an equivalent two-dimensional case. Consider a three-dimensional specimen that is loaded to a given J value. If we choose a principal stress value and construct contours at two-dimensional slices on the x-y plane, the area inside of these contours will vary along the crack front because the center of the specimen is more highly constrained than the free surface, as Figure 3.40 illustrates. The volume can be obtained by summing the areas in these two-dimensional contours. This volume can then be related to an equivalent two-dimensional specimen loaded to the same J value:

$$V = 2\int_0^{B/2} A(\sigma_1, z)\,dz = B_{eff} A_c(\sigma_1) \qquad (3.84)$$

where A_c is the area inside the σ_1 contour on the center plane of the three-dimensional specimen and B_{eff} is the effective thickness.

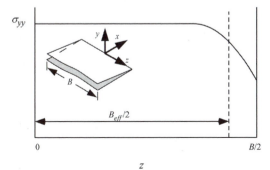

FIGURE 3.40 Schematic illustration of the effective thickness, B_{eff}.

The effective thickness influences the cleavage driving force through a sample volume effect: longer crack fronts have a higher probability of cleavage fracture because more volume is sampled along the crack front. This effect can be characterized by a three-parameter Weibull distribution (See Chapter 5):

$$F = 1 - \exp\left[-\frac{B}{B_o}\left(\frac{K_{JC} - K_{min}}{\Theta_K - K_{min}} \right)^4 \right]$$

(3.85)

where

B = thickness (or crack front length)
B_o = reference thickness
K_{min} = threshold toughness
Θ_K = 63rd percentile toughness when $B = B_o$

Consider two samples with effective crack front lengths B_1 and B_2. If a value of $K_{JC(1)}$ is measured for Specimen 1, the expected toughness for Specimen 2 can be inferred from Equation (3.85) by equating failure probabilities:

$$K_{JC(2)} = \left(\frac{B_1}{B_2} \right)^{1/4}\left(K_{JC(1)} - K_{min} \right) + K_{min}$$

(3.86)

Equation (3.86) is a statistical thickness adjustment that can be used to relate two sets of data with different thicknesses.

3.6.3.3 Application of the Model

As with the J-Q approach, the implementation of the scaling model requires detailed elastic-plastic finite element analysis of the configuration of interest. The principal stress contours must be constructed and the areas compared with the $T = 0$ reference solution obtained from a modified boundary layer analysis. The effective driving force J_o is then plotted against the applied J, as Figure 3.41 schematically illustrates. At low deformation levels, the J_o-J curves follow the 1:1 line, but deviate from the line with further deformation. When $J \approx J_o$, the crack-tip stress fields are close to the $Q = 0$ limit, and fracture toughness is not significantly influenced by specimen boundaries. At high deformation levels $J > J_o$ and the fracture toughness is artificially elevated by constraint loss. Constraint loss occurs more rapidly in specimens with shallow cracks, as Figure 3.28 illustrates. A specimen with $a/W = 0.15$ would tend to fail at a higher J_c value than a specimen with $a/W = 0.5$. Given the J_o-J curve, however, the J_c values for both specimens can be corrected to J_o, as Figure 3.41 illustrates.

Figure 3.42 is a nondimensional plot of J_o at the midplane vs. the average J through the thickness of SENB specimens with various W/B ratios [36]. These curves were inferred from a

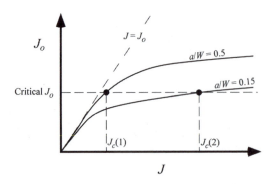

FIGURE 3.41 Schematic illustration of the scaling model. A specimen with $a/W = 0.15$ will fail at a higher J_c value than a specimen with $a/W = 0.5$, but both J_c values can be corrected down to the same critical J_o value.

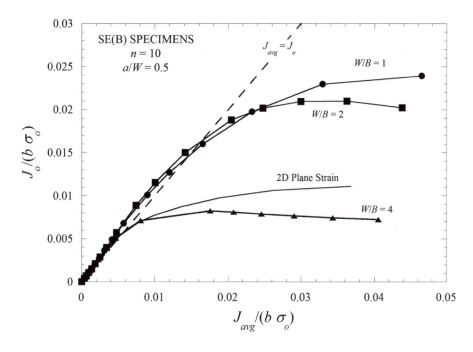

FIGURE 3.42 Effective driving force for cleavage J_o for deeply notched SENB specimens.

three-dimensional elastic-plastic analysis. The corresponding curve from a two-dimensional plane strain analysis is shown for comparison. Note that for $W/B = 1$ and 2, J_o at the midplane lies well above the plane-strain curve. For $W/B = 4$, J_o at the midplane follows the plane-strain curve initially, but falls below the plane-strain results at high deformation levels. The three-dimensional nature of the plastic deformation apparently results in a high level of constraint at the midplane when the uncracked ligament length is $\leq$ the specimen thickness.

Figure 3.43 is a plot of effective thickness B_{eff} as a function of deformation. The trends in this plot are consistent with Figure 3.42; namely, the constraint increases with decreasing W/B. Note that all three curves reach a plateau. Recall that B_{eff} is defined in such a way as to be a measure of the through-thickness relaxation of constraint, relative to the in-plane constraint at the midplane. At low deformation levels there is negligible relation at the midplane and $J \approx J_o$, but a through-thickness constraint relation occurs, resulting in a falling B_{eff}/B ratio. At high deformation levels, the B_{eff}/B ratio is essentially constant, indicating that the constraint relaxation is proportional in three dimensions. Figure 3.44 and Figure 3.45 show data that have been corrected with the scaling model.

3.6.4 LIMITATIONS OF TWO-PARAMETER FRACTURE MECHANICS

The T stress approach, J-Q theory, and the cleavage scaling model are examples of two-parameter fracture theories, where a second quantity (e.g., T, Q, or J_o) has been introduced to characterize the crack-tip environment. Thus these approaches assume that the crack-tip fields contain two degrees of freedom. When single-parameter fracture mechanics is valid, the crack-tip fields have only one degree of freedom. In such cases, any one of several parameters (e.g., J, K, or CTOD) will suffice to characterize the crack-tip conditions, provided the parameter can be defined unambiguously; K is a suitable characterizing parameter only when an elastic singularity zone exists ahead of the crack tip.[8] Similarly, the choice of a second parameter in

---

[8] An effective K can be inferred from J through Equation (3.18). Such a parameter has units of K, but it loses its meaning as the amplitude of the elastic singularity when such a singularity no longer exists.

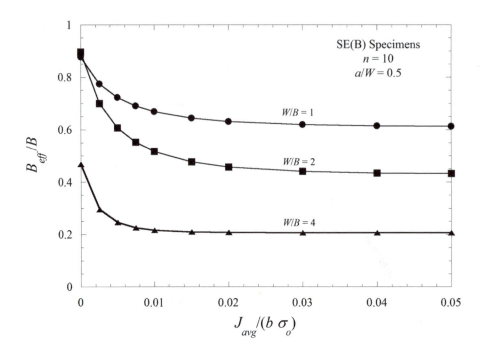

FIGURE 3.43 Effective thickness for deeply notched SENB specimens.

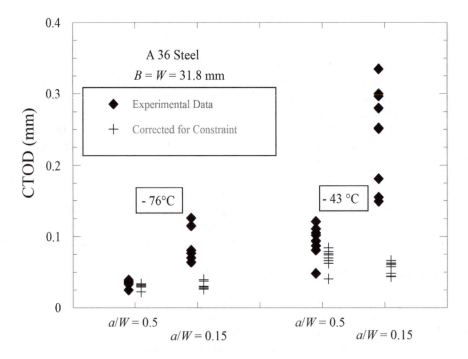

FIGURE 3.44 Fracture toughness data for a mild steel, corrected for constraint loss. Taken from Anderson, T.L. and Dodds, R.H., Jr., "Specimen Size Requirements for Fracture Toughness Testing in the Ductile-Brittle Transition Region." *Journal of Testing and Evaluation*, Vol. 19, 1991, pp. 123–134; Sorem, W.A., "The Effect of Specimen Size and Crack Depth on the Elastic-Plastic Fracture Toughness of a Low-Strength High-Strain Hardening Steel." Ph.D. Dissertation, The University of Kansas, Lawrence, KS, 1989.

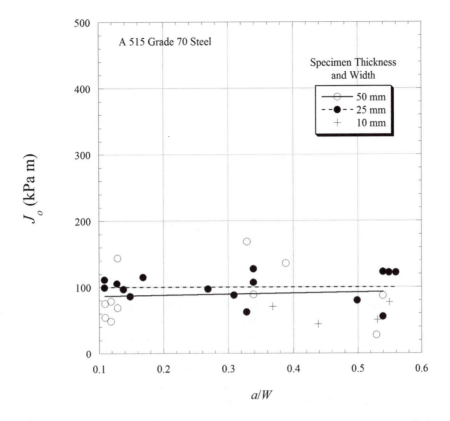

FIGURE 3.45 Experimental data from Figure 3.28 corrected for constraint loss. Taken from Anderson, T.L., Stienstra, D.I.A., and Dodds, R.H., Jr., "A Theoretical Framework for Addressing Fracture in the Ductile-Brittle Transition Region." *Fracture Mechanics,* Vol. 24, ASTM STP 1207, American Society for Testing and Materials, Philadelphia, PA (in press).

the case of two-parameter theory is mostly arbitrary, but the T stress has no physical meaning under large-scale yielding conditions.

Just as plastic flow invalidates single-parameter fracture mechanics in many geometries, two-parameter theories eventually break down with extensive deformation. If we look at the structure of the crack-tip fields in the plastic zone, we can evaluate the range of validity of both single- and two-parameter methodologies.

A number of investigators [39–43] have performed asymptotic analyses of the crack-tip fields for elastic-plastic materials. These analyses utilize deformation plasticity and small-strain theory. In the case of plane strain, these analyses assume incompressible strain. Consequently, asymptotic analyses are not valid close to the crack tip (in the large-strain zone) nor remote from the crack tip, where elastic strains are a significant fraction of the total strain. Despite these limitations, asymptotic analysis provides insights into the range of validity of both single- and two-parameter fracture theories.

In the case of a plane strain crack in a power-law-hardening material, asymptotic analysis leads to the following power series:

$$\sigma_{ij} = \sum_{k=1}^{\infty} A_k \left(\frac{J}{\alpha \varepsilon_o \sigma_o r} \right)^{s_k} \hat{\sigma}_{ij}^{(k)}(\theta, n) \tag{3.87}$$

The exponents s_k and the angular functions for each term in the series can be determined from the asymptotic analysis. The amplitudes for the first five terms are as follows:

$$\begin{bmatrix} A_1 \\ A_2 \\ A_3 \\ A_4 \\ A_5 \end{bmatrix} = \begin{bmatrix} (I_n)^{-\frac{1}{n+1}} \\ \text{(unspecified)} \\ \dfrac{A_2^2}{A_1} \\ \text{(unspecified)} \\ \dfrac{A_2^3}{A_1^2} \end{bmatrix}$$

The two unspecified coefficients A_2 and A_4 are governed by the far-field boundary conditions. The first five terms of the series have three degrees of freedom, where J, A_2, and A_4 are independent parameters. For low and moderate strain hardening materials, Crane [43] showed that a fourth independent parameter does not appear in the series for many terms. For example, when $n = 10$, the fourth independent coefficient appears in approximately the 100th term. Thus for all practical purposes, the crack-tip stress field inside the plastic zone has three degrees of freedom.

Since two-parameter theories assume two degrees of freedom, they cannot be rigorously correct in general. There are, however, situations where two-parameter approaches provide a good engineering approximation.

Consider the modified boundary layer model in Figure 3.32. Since the boundary conditions have only two degrees of freedom (K and T), the resulting stresses and strains inside the plastic zone must be two-parameter fields. Thus there must be a unique relationship between A_2 and A_4 in this case. That is

$$A_4 = A_4(A_2) \tag{3.88}$$

The two-parameter theory is approximately valid for other geometries to the extent that the crack-tip fields obey Equation (3.88). Figure 3.46 schematically illustrates the A_2-A_4 relationship. This relationship can be established by varying the boundary conditions on the modified boundary layer model. When a given cracked geometry is loaded, A_2 and A_4 initially will evolve in accordance with Equation (3.88) because the crack-tip conditions in the geometry of interest can be represented by the modified boundary layer model when the plastic zone is relatively small. Under large-scale yielding conditions, however, the A_2-A_4 relationship may deviate from the modified boundary layer solution, in which case the two-parameter theory is no longer valid.

Figure 3.47 is a schematic three-dimensional plot of J, A_2, and A_4. Single-parameter fracture mechanics can be represented by a vertical line, since A_2 and A_4 must be constant in this case. The two-parameter theory, where Equation (3.88) applies, can be represented by a surface in this three-dimensional space. The loading path for a cracked body initially follows the vertical

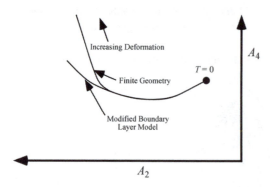

FIGURE 3.46 Schematic relationship between the two independent amplitudes in the asymptotic power series.

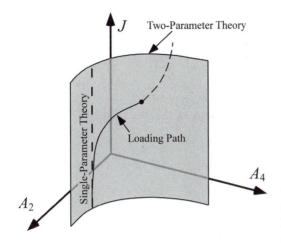

A_4

FIGURE 3.47 Single- and two-parameter assumptions in terms of the three independent variables in the elastic-plastic crack-tip field. The loading path initially lines in the two-parameter surface and then diverges, as indicated by the dashed line.

single-parameter line. When it deviates from this line, it may remain in the two-parameter surface for a time before diverging from the surface.

The loading path in J-A_2-A_4 space depends on geometry [43]. Low-constraint configurations such as the center-cracked panel and shallow notched bend specimens diverge from the single-parameter theory almost immediately, but follow Equation (3.88) to fairly high deformation levels. Deeply notched bend specimens maintain a high constraint to relatively high J values, but they do not follow Equation (3.88) when constraint loss eventually occurs. Thus low-constraint geometries should be treated with the two-parameter theory, and high-constraint geometries can be treated with the single-parameter theory in many cases. When high-constraint geometries violate the single-parameter assumption, however, the two-parameter theory is of little value.

APPENDIX 3:   MATHEMATICAL FOUNDATIONS OF ELASTIC-PLASTIC FRACTURE MECHANICS

A3.1   Determining CTOD from the Strip-Yield Model

Burdekin and Stone [3] applied the Westergaard [44] complex stress function approach to the strip-yield model. They derived an expression for CTOD by superimposing a stress function for closure forces on the crack faces in the strip-yield zone. Their result was similar to previous analyses based on the strip-yield model performed by Bilby et al. [45] and Smith [46].

Recall from Appendix 2.3 that the Westergaard approach expresses the in-plane stresses (in a limited number of cases) in terms of Z:

$$\sigma_{xx} = \text{Re}\,Z - y\,\text{Im}\,Z' \tag{A3.1a}$$

$$\sigma_{yy} = \text{Re}\,Z + y\,\text{Im}\,Z' \tag{A3.1b}$$

$$\tau_{xy} = -y\,\text{Re}\,Z' \tag{A3.1c}$$

where Z is an analytic function of the complex variable $z = x + iy$, and the prime denotes a first derivative with respect to z. By invoking the equations of elasticity for the plane problem, it can be shown that the displacement in the y direction is as follows:

$$u_y = \frac{1}{E}[2\,\text{Im}\,\bar{Z} - y(1+v)\,\text{Re}\,Z] \quad \text{for plane stress} \tag{A3.2a}$$

and

$$u_y = \frac{1}{E}[2(1-v^2)\operatorname{Im}\bar{Z} - y(1+v)\operatorname{Re}Z] \quad \text{for plane strain} \tag{A3.2b}$$

where $\bar{Z}$ is the integral of Z with respect to z, as discussed in Appendix 2. For a through crack of length $2a_1$ in an infinite plate under biaxial tensile stress σ, the Westergaard function is given by

$$Z = \frac{\sigma z}{\sqrt{z^2 - a_1^2}} \tag{A3.3}$$

where the origin is defined at the crack center.

The stress function for a pair of splitting forces P at $\pm x$ within a crack of length $2a_1$ (see Figure 2.32) is given by

$$Z = \frac{2Pz\sqrt{a_1^2 - x^2}}{\pi\sqrt{z^2 - a_1^2}(z^2 - a_1^2)} \tag{A3.4}$$

For a uniform compressive stress σ_{YS} along the crack surface between a and a_1 (Figure A3.1), the Westergaard stress function is obtained by substituting $P = -\sigma_{YS}dx$ into Equation (A3.4) and integrating:

$$Z = -\int_a^{a_1} \frac{2\sigma_{YS}z\sqrt{a_1^2 - x^2}}{\pi\sqrt{z^2 - a_1^2}\left(z^2 - a_1^2\right)}dx$$

$$\tag{A3.5}$$

$$= -\frac{2\sigma_{YS}}{\pi}\left[\frac{z}{\sqrt{z^2 - a_1^2}}\cos^{-1}\left(\frac{a}{a_1}\right) - \cot^{-1}\left(\frac{a}{z}\sqrt{\frac{z^2 - a_1^2}{a_1^2 - a^2}}\right)\right]$$

The stress functions of Equation (A3.3) and Equation (A3.5) can be superimposed, resulting in the strip-yield solution for the through crack. Recall from Section 2.8.2 that the size of the strip-yield

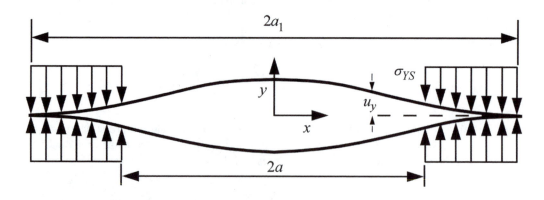

FIGURE A3.1 Strip-yield model for a through crack.

zone was chosen so that the stresses at the tip would be finite. Thus

$$k \equiv \frac{a}{a_1} = \cos\left(\frac{\pi\sigma}{\sigma_{YS}}\right)$$

(A3.6)

When Equation (A3.6) is substituted into Equation (A3.5) and Equation (A3.3) is superimposed, the first term in Equation (A3.5) cancel with Equation (A3.3), which leads to

$$Z = \frac{2\sigma_{YS}}{\pi}\left[\frac{k}{z}\sqrt{\frac{z^2 - a_1^2}{1 - k^2}}\right]$$

(A3.7)

Integrating Equation (A3.7) gives

$$\bar{Z} = \frac{2\sigma_{YS}}{\pi}[z\omega_1 - a\omega_2]$$

(A3.8)

where

$$\omega_1 = \cot^{-1}\frac{\sqrt{1 - \left(\frac{a_1}{a}\right)^2}}{\sqrt{\frac{1}{k^2} - 1}}$$

and

$$\omega_2 = \cot^{-1}\sqrt{\frac{z^2 - a_1^2}{1 - k^2}}$$

On the crack plane, $y = 0$ and the displacement in the y direction (Equation (A3.2)) reduces to

$$u_y = \frac{2}{E}\operatorname{Im}\bar{Z}$$

(A3.9)

for plane stress. Solving for the imaginary part of Equation (A3.8) gives

$$u_y = \frac{4\sigma_{YS}}{\pi E}\left[a\coth^{-1}\left(\frac{1}{a_1}\sqrt{\frac{a_1^2 - z^2}{1 - k^2}}\right) - z\coth^{-1}\left(\frac{k}{z}\sqrt{\frac{a_1^2 - z^2}{1 - k^2}}\right)\right]$$

for $|z| \leq a_1$. Setting $z = a$ leads to

$$\delta = 2u_y = \frac{8\sigma_{YS}a}{\pi E}\ln\left(\frac{1}{k}\right)$$

(A3.10)

which is identical to Equation (3.5).

Recall the *J*-CTOD relationship (Equation (3.44)) derived from the strip-yield model. Let us define an effective stress intensity for elastic-plastic conditions in terms of the *J* integral:

$$K_{eff} \equiv \sqrt{JE} \qquad (A3.11)$$

Combining Equation (3.44), Equation (A3.10) and Equation (A3.11) gives

$$K_{eff} = \sigma_{YS}\sqrt{\pi a}\left[\frac{8}{\pi^2}\ln \sec\left(\frac{\pi\sigma}{2\sigma_{YS}}\right)\right]^{1/2} \qquad (A3.12)$$

which is the strip-yield plastic zone correction given in Equation (2.81) and plotted in Figure 2.33. Thus the strip-yield correction to K_I is equivalent to a *J*-based approach for a nonhardening material in plane stress.

A3.2 THE *J* CONTOUR INTEGRAL

Rice [4] presented a mathematical proof of the path independence of the *J* contour integral. He began by evaluating *J* along a closed contour Γ^* (Figure A3.2):

$$J* = \int_{\Gamma^*}\left(w\,dy - T_i\frac{\partial u_i}{\partial x}\,ds\right) \qquad (A3.13)$$

where the various terms in this expression are defined in Section 3.2.2. Rice then invoked the divergence theorem to convert Equation (A3.13) into an area integral:

$$J* = \int_{A^*}\left[\frac{\partial w}{\partial x} - \frac{\partial}{\partial x_j}\left(\sigma_{ij}\frac{\partial u_i}{\partial x}\right)\right]dx\,dy \qquad (A3.14)$$

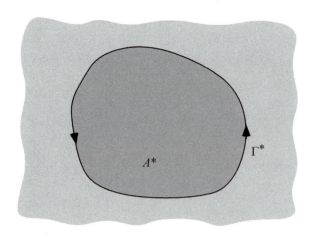

FIGURE A3.2 Closed contour Γ^* in a two-dimensional solid.

where A^* is the area enclosed by Γ^*. By invoking the definition of strain energy density given by Equation (3.20), we can evaluate the first term in square brackets in Equation (A3.14):

$$\frac{\partial w}{\partial x} = \frac{\partial w}{\partial \varepsilon_{ij}} \frac{\partial \varepsilon_{ij}}{\partial x} = \sigma_{ij} \frac{\partial \varepsilon_{ij}}{\partial x}$$

(A3.15)

Note that Equation (A3.15) applies only when w exhibits the properties of an elastic potential. Applying the strain-displacement relationship (for small strains) to Equation (A3.15) gives

$$\frac{\partial w}{\partial x} = \frac{1}{2} \sigma_{ij} \left[\frac{\partial}{\partial x} \left(\frac{\partial u_i}{\partial x_j} \right) + \frac{\partial}{\partial x} \left(\frac{\partial u_j}{\partial x_i} \right) \right]$$

$$= \sigma_{ij} \frac{\partial}{\partial x_j} \left(\frac{\partial u_i}{\partial x} \right)$$

(A3.16)

since $\sigma_{ij} = \sigma_{ji}$. Invoking the equilibrium condition:

$$\frac{\partial \sigma_{ij}}{\partial x_j} = 0$$

(A3.17)

leads to

$$\sigma_{ij} \frac{\partial}{\partial x_j} \left(\frac{\partial u_i}{\partial x} \right) = \frac{\partial}{\partial x_j} \left(\sigma_{ij} \frac{\partial u_i}{\partial x} \right)$$

(A3.18)

which is identical to the second term in square brackets in Equation (A3.14). Thus the integrand in Equation (A3.14) vanishes and $J = 0$ for any closed contour.

Consider now two arbitrary contours Γ_1 and Γ_2 around a crack tip, as illustrated in Figure A3.3. If Γ_1 and Γ_2 are connected by segments along the crack face (Γ_3 and Γ_4), a closed contour is formed. The total J along the closed contour is equal to the sum of contributions from each segment:

$$J = J_1 + J_2 + J_3 + J_4 = 0$$

(A3.19)

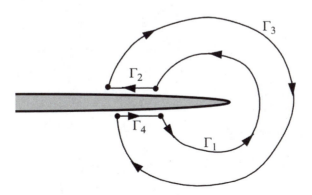

FIGURE A3.3 Two arbitrary contours Γ_1 and Γ_2 around the tip of a crack. When these contours are connected by Γ_3 and Γ_4, a closed contour is formed, and the total $J = 0$.

On the crack face, $T_i = dy = 0$. Thus, $J_3 = J_4 = 0$ and $J_1 = -J_2$. Therefore, any arbitrary (counterclockwise) path around a crack will yield the same value of J; J is path-independent.

A3.3 J As A Nonlinear Elastic Energy Release Rate

Consider a two-dimensional cracked body bounded by the curve Γ (Figure A3.4). Let A' denote the area of the body. The coordinate axis is attached to the crack tip. Under quasistatic conditions and in the absence of body forces, the potential energy is given by

$$\Pi = \int_{A'} w\, dA - \int_{\Gamma''} T_i u_i\, ds \qquad (A3.20)$$

where Γ^* is the portion of the contour on which the tractions are defined. Let us now consider the change in potential energy resulting from a virtual extension of the crack:

$$\frac{d\Pi}{da} = \int_{A'} \frac{dw}{da}\, dA - \int_{\Gamma'} T_i \frac{du_i}{da}\, ds \qquad (A3.21)$$

The line integration in Equation (A3.21) can be performed over the entire contour Γ because $du_i/da = 0$ over the region where displacements are specified; also, $dT_i/da = 0$ over the region the tractions are specified. When the crack grows, the coordinate axis moves. Thus a derivative with respect to crack length can be written as

$$\frac{d}{da} = \frac{\partial}{\partial a} + \frac{\partial x}{\partial a}\frac{\partial}{\partial x} = \frac{\partial}{\partial a} - \frac{\partial}{\partial x} \qquad (A3.22)$$

since $\partial x/\partial a = -1$. Applying this result to Equation (A3.21) gives

$$\frac{d\Pi}{da} = \int_{A'} \left(\frac{\partial w}{\partial a} - \frac{\partial w}{\partial x} \right) dA - \int_{\Gamma'} T_i \left(\frac{\partial u_i}{\partial a} - \frac{\partial u_i}{\partial x} \right) ds \qquad (A3.23)$$

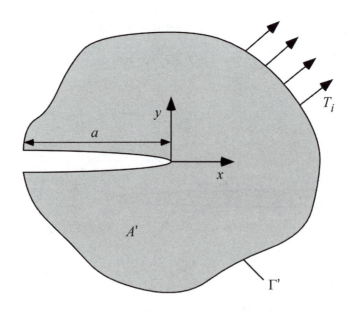

FIGURE A3.4 A two-dimensional cracked body bounded by the curve Γ'.

By applying the same assumptions as in Equation (A3.15) and Equation (A3.16), we obtain:

$$\frac{\partial w}{\partial a} = \frac{\partial w}{\partial \varepsilon_{ij}}\frac{\partial \varepsilon_{ij}}{\partial a} = \sigma_{ij}\frac{\partial}{\partial x_j}\left(\frac{\partial u_i}{\partial a}\right) \tag{A3.24}$$

Invoking the principle of virtual work gives

$$\int_{A'}\sigma_{ij}\frac{\partial}{\partial x_j}\left(\frac{\partial u_i}{\partial a}\right)dA = \int_{\Gamma'}T_i\frac{\partial u_i}{\partial a}ds \tag{A3.25}$$

which cancels with one of the terms in the line integral in Equation (A3.23), resulting in the following:

$$\frac{d\Pi}{da} = \int_{\Gamma'}T_i\frac{\partial u_i}{\partial x}ds - \int_{A'}\frac{\partial w}{\partial x}dA \tag{A3.26}$$

Applying the divergence theorem and multiplying both sides by −1 leads to

$$-\frac{d\Pi}{da} = \int_{\Gamma'}\left(wn_x - T_i\frac{\partial u_i}{\partial x}\right)ds$$

$$= \int_{\Gamma'}\left(wdy - T_i\frac{\partial u_i}{\partial x}ds\right) \tag{A3.26a}$$

since $n_x\,ds = dy$. Therefore, the J contour integral is equal to the energy release rate for a linear or nonlinear elastic material under quasistatic conditions.

A3.4 THE HRR SINGULARITY

Hutchinson [7] and Rice and Rosengren [8] independently evaluated the character of crack-tip stress fields in the case of power-law-hardening materials. Hutchinson evaluated both plane stress and plane strain, while Rice and Rosengren considered only plane-strain conditions. Both articles, which were published in the same issue of the *Journal of the Mechanics and Physics of Solids*, argued that stress times strain varies as $1/r$ near the crack tip, although only Hutchinson was able to provide a mathematical proof of this relationship.

The Hutchinson analysis is outlined below. Some of the details are omitted for brevity. We focus instead on his overall approach and the ramifications of this analysis.

Hutchinson began by defining a stress function Φ for the problem. The governing differential equation for deformation plasticity theory for a plane problem in a Ramberg-Osgood material is more complicated than the linear elastic case:

$$\Delta^4\Phi + \gamma(\Phi, \sigma_e, r, n, \alpha) = 0 \tag{A3.27}$$

where the function γ differs for plane stress and plane strain. For the Mode I crack problem, Hutchinson chose to represent Φ in terms of an asymptotic expansion in the following form:

$$\Phi = C_1(\theta)r^s + C_2(\theta)r^t + \cdots \tag{A3.28}$$

where C_1 and C_2 are constants that depend on θ, the angle from the crack plane. Equation (A3.28) is analogous to the Williams expansion for the linear elastic case (Appendix 2.3). If $s < t$, and t is less than all subsequent exponents on r, then the first term dominates as $r \rightarrow 0$. If the analysis is restricted to the region near the crack tip, then the stress function can be expressed as follows:

$$\Phi = \kappa \sigma_o r^s \tilde{\Phi}(\theta) \tag{A3.29}$$

where κ is the amplitude of the stress function and $\tilde{\Phi}$ is a dimensionless function of θ. Although Equation (A3.27) is different from the linear elastic case, the stresses can still be derived from Φ through Equation (A2.6) or Equation (A2.13). Thus the stresses, in polar coordinates, are given by

$$\sigma_{rr} = \kappa \sigma_o r^{s-2} \tilde{\sigma}_{rr}(\theta) = \kappa \sigma_o r^{s-2}(s\tilde{\Phi} + \tilde{\Phi}'')$$

$$\sigma_{\theta\theta} = \kappa \sigma_o r^{s-2} \tilde{\sigma}_{\theta\theta}(\theta) = \kappa \sigma_o r^{s-2} s(s-1)\tilde{\Phi}$$

$$\sigma_{r\theta} = \kappa \sigma_o r^{s-2} \tilde{\sigma}_{r\theta}(\theta) = \kappa \sigma_o r^{s-2}(1-s)\tilde{\Phi}' \tag{A3.30}$$

$$\sigma_e = \kappa \sigma_o r^{s-2} \tilde{\sigma}_e(\theta) = \kappa \sigma_o r^{s-2}\left(\tilde{\sigma}_{rr}^2 + \tilde{\sigma}_{\theta\theta}^2 - \tilde{\sigma}_{rr}\tilde{\sigma}_{\theta\theta} + 3\tilde{\sigma}_{r\theta}^2\right)^{1/2}$$

The boundary conditions for the crack problem are as follows:

$$\tilde{\Phi}(\pm\pi) = \tilde{\Phi}'(\pm\pi) = 0$$

In the region close to the crack tip where Equation (A3.29) applies, elastic strains are negligible compared to plastic strains; only the second term in Equation (A3.27) is relevant in this case. Hutchinson substituted the boundary conditions and Equation (A3.29) into Equation (A3.27) and obtained a nonlinear eigenvalue equation for s. He then solved this equation numerically for a range of n values. The numerical analysis indicated that s could be described quite accurately (for both plane stress and plane strain) by a simple formula:

$$s = \frac{2n+1}{n+1} \tag{A3.31}$$

which implies that the strain energy density varies as $1/r$ near the crack tip. This numerical analysis also yielded relative values for the angular functions $\tilde{\sigma}_{ij}$. The amplitude, however, cannot be obtained without connecting the near-tip analysis with the remote boundary conditions. The J contour integral provides a simple means for making this connection in the case of small-scale yielding. Moreover, by invoking the path-independent property of J, Hutchinson was able to obtain a direct proof of the validity of Equation (A3.31).

Consider two circular contours of radius r_1 and r_2 around the tip of a crack in small-scale yielding, as illustrated in Figure A3.5. Assume that r_1 is in the region described by the elastic singularity, while r_2 is well inside of the plastic zone, where the stresses are described by Equation (A3.30). When the stresses and displacements in Table 2.1 and Table 2.2 are inserted into Equation (A3.26), and the J integral is evaluated along r_1, one finds that $J = K_I^2/E'$ as expected from the previous section. Since the connection between K_I and the global boundary conditions is well established for a wide range of configurations, and J is path-independent, the near-tip problem for small-scale yielding can be solved by evaluating J at r_2 and relating J to the amplitude (κ).

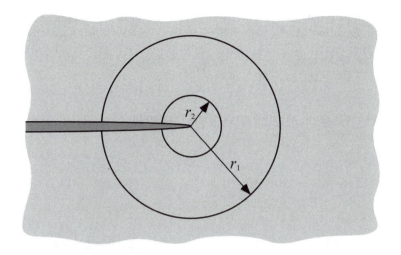

FIGURE A3.5 Two circular contours around the crack tip. r_1 is in the zone dominated by the elastic singularity, while r_2 is in the plastic zone where the leading term of the Hutchinson asymptotic expansion dominates.

Solving for the integrand in the J integral at r_2 leads to

$$w = \alpha \sigma_o \varepsilon_o \kappa^{n+1} \frac{n}{n+1} r^{(n+1)(s-2)} \tilde{\sigma}_e^{n+1} \tag{A3.32a}$$

and

$$T_i \frac{\partial u_i}{\partial x} = \alpha \sigma_o \varepsilon_o \kappa^{n+1} r^{(n+1)(s-2)} \{\sin \theta [\tilde{\sigma}_{rr}(\tilde{u}_\theta - \tilde{u}_r') - \tilde{\sigma}_{r\theta}(\tilde{u}_r + \tilde{u}_\theta')]$$
$$+ \cos \theta [n(s-2)+1][\tilde{\sigma}_{rr}\tilde{u}_r + \tilde{\sigma}_{\theta\theta}\tilde{u}_\theta]\} \tag{A3.32b}$$

where $\tilde{u}_r$ and $\tilde{u}_\theta$ are dimensionless displacements, defined by

$$u_r = \alpha \varepsilon_o \kappa^n r^{n(s-2)+1} \tilde{u}_r(\theta)$$
$$u_\theta = \alpha \varepsilon_o \kappa^n r^{n(s-2)+1} \tilde{u}_\theta(\theta) \tag{A3.33}$$

u_r and u_θ can be derived from the strain-displacement relationships. Evaluating the J integral at r_2 gives

$$J = \alpha \varepsilon_o \sigma_o \kappa^{n+1} r_2^{(n+1)(s-2)+1} I_n \tag{A3.34}$$

where I_n is an integration constant, given by

$$I_n = \int_{-\pi}^{+\pi} \left\{ \frac{n}{n+1} \tilde{\sigma}_e^{n+1} \cos \theta - \{\sin \theta [\tilde{\sigma}_{rr}(\tilde{u}_\theta - \tilde{u}') - \tilde{\sigma}_{r\theta}(\tilde{u}_r + \tilde{u}_\theta')] \right.$$
$$\left. + \cos \theta [n(s-2)+1](\tilde{\sigma}_{rr}\tilde{u}_r + \tilde{\sigma}_{\theta\theta}\tilde{u}_\theta)\} \right\} d\theta \tag{A3.35}$$

In order for J to be path independent, it cannot depend on r_2, which was defined arbitrarily. The radius vanishes in Equation (A3.34) only when $(n+1)(s-2)+1=0$ $(n + 1)(s - 2) + 1 = 0$, or

$$s = \frac{2n+1}{n+1}$$

which is identical to the result obtained numerically (Equation (A3.31)). Thus the amplitude of the stress function is given by

$$\kappa = \left(\frac{J}{\alpha \sigma_o \varepsilon_o I_n} \right)^{\frac{1}{n+1}} \tag{A3.36}$$

Substituting Equation (A3.36) into Equation (A3.30) yields the familiar form of the HRR singularity:

$$\sigma_{ij} = \sigma_o \left(\frac{EJ}{\alpha \sigma_o^2 I_n r} \right)^{\frac{1}{n+1}} \tilde{\sigma}_{ij}(n, \theta) \tag{A3.37}$$

since $\varepsilon_o = \sigma_o/E$. The integration constant is plotted in Figure 3.10 for both plane stress and plane strain, while Figure 3.11 shows the angular variation of $\tilde{\sigma}_{ij}$ for $n = 3$ and $n = 13$.

Rice and Rosengren [8] obtained essentially identical results to Hutchinson (for plane strain), although they approached the problem in a somewhat different manner. Rice and Rosengren began with a heuristic argument for the $1/r$ variation of strain energy density, and then introduced an Airy stress function of the form of Equation (A3.29) with the exponent on r given by Equation (A3.31). They computed stresses, strains, and displacements in the vicinity of the crack tip by applying the appropriate boundary conditions.

The HRR singularity was an important result because it established J as a stress amplitude parameter within the plastic zone, where the linear elastic solution is invalid. The analyses of Hutchinson, Rice, and Rosengren demonstrated that the stresses in the plastic zone are much higher in plane strain than in plane stress. This is consistent with the simplistic analysis in Section 2.10.1.

One must bear in mind the limitations of the HRR solution. Since the singularity is merely the leading term in an asymptotic expansion, and elastic strains are assumed to be negligible, this solution dominates only near the crack tip, well within the plastic zone. For very small r values, however, the HRR solution is invalid because it neglects finite geometry changes at the crack tip. When the HRR singularity dominates, the loading is proportional, which implies a single-parameter description of crack-tip fields. When the higher-order terms in the series are significant, the loading is often nonproportional and a single-parameter description may no longer be possible (see Section 3.6).

A3.5 ANALYSIS OF STABLE CRACK GROWTH IN SMALL-SCALE YIELDING

A3.5.1 The Rice-Drugan-Sham Analysis

Rice, Drugan, and Sham (RDS) [15] performed an asymptotic analysis of a growing crack in an elastic-plastic solid in small-scale yielding. They assumed crack extension at a constant crack-opening angle, and predicted the shape of J resistance curves. They also speculated about the effect of large-scale yielding on the crack growth resistance behavior.

Small-Scale Yielding

Rice et al. analyzed the local stresses and displacements at a growing crack by modifying the classical Prandtl slip-line field to account for elastic unloading behind the crack tip. They assumed small-scale yielding conditions and a nonhardening material; the details of the derivation are omitted for brevity. The RDS crack growth analysis resulted in the following expression:

$$\dot{\delta} = \alpha \frac{\dot{J}}{\sigma_o} + \beta \frac{\sigma_o}{E} \dot{a} \ln\left(\frac{R}{r}\right) \quad \text{for } r \to 0 \tag{A3.38}$$

where

$\dot{\delta}$ = rate of crack-opening displacement at a distance R behind the crack tip
$\dot{J}$ = rate of change in the J integral
$\dot{a}$ = crack-growth rate
α, β, and R = constants[9]

The asymptotic analysis indicated that $\beta = 5.083$ for $v = 0.3$ and $\beta = 4.385$ for $v = 0.5$. The other constants, α and R, could not be inferred from the asymptotic analysis. Rice et al. [15] performed an elastic-plastic finite element analysis of a growing crack and found that R, which has units of length, scales approximately with the plastic zone size, and can be estimated by

$$R = \frac{\lambda E J}{\sigma_o^2} \quad \text{where } \lambda \approx 0.2 \tag{A3.39}$$

The dimensionless constant α can be estimated by considering a stationary crack ($\dot{a} = 0$):

$$\delta = \alpha \frac{J}{\sigma_o} \tag{A3.40}$$

Referring to Equation (3.48), α obviously equals d_n when δ is defined by the 90° intercept method. The finite element analysis performed by Rice et al. indicated that α for a growing crack is nearly equal to the stationary crack case.

Rice et al. performed an asymptotic integration of (Equation (A3.38)) for the case where the crack length increases continuously with J, which led to

$$\delta = \frac{\alpha r}{\sigma_o} \frac{dJ}{da} + \beta r \frac{\sigma_o}{E} \ln\left(\frac{eR}{r}\right) \tag{A3.41}$$

where δ, in this case, is the crack-opening displacement at a distance r from the crack tip, and $e (= 2.718)$ is the natural logarithm base. Equation (A3.41) can be rearranged to solve for the nondimensional tearing modulus:

$$T \equiv \frac{E}{\sigma_o^2} \frac{dJ}{da} = \frac{E\delta}{\alpha \sigma_o r} - \frac{\beta}{\alpha} \ln\left(\frac{eR}{r}\right) \tag{A3.42}$$

---

[9] The constant α in the RDS analysis should not be confused with the dimensionless constant in the Ramberg-Osgood relationship (Equation (3.22)), for which the same symbol is used.

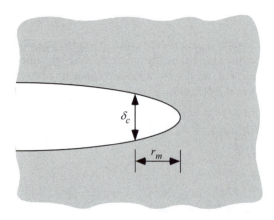

FIGURE A3.6 The RDS crack growth criterion. The crack is assumed to extend with a constant opening displacement δ_c at distance r_m behind the crack tip. This criterion corresponds approximately to crack extension at a constant crack-tip-opening angle (CTOA).

Rice et al. proposed a failure criterion that corresponds approximately to crack extension at a constant crack-tip-opening angle (CTOA). Since $d\delta/dr = \infty$ at the crack tip, CTOA is undefined, but an approximate CTOA can be inferred a finite distance from the tip. Figure A3.6 illustrates the RDS crack-growth criterion. They postulated that crack growth occurs at a critical crack-opening displacement δ_c at a distance r_m behind the crack tip. That is

$$\frac{\delta_c}{r_m} = \frac{\alpha}{\sigma_o}\frac{dJ}{da} + \beta\frac{\sigma_o}{E}\ln\left(\frac{eR}{r_m}\right) = \text{constant} \tag{A3.43}$$

Rice et al. found that it was possible to define the micromechanical failure parameters δ_c and r_m in terms of global parameters that are easy to obtain experimentally. Setting $J = J_{Ic}$ and combining Equation (A3.39), Equation (A3.42), and Equation (A3.43) gives

$$T_o = \frac{E\delta_c}{\alpha\sigma_o r_m} - \frac{\beta}{\alpha}\ln\left(\frac{e\lambda EJ_{Ic}}{r_m\sigma_o^2}\right) \tag{A3.44}$$

where T_o is the initial tearing modulus. Thus for $J > J_{Ic}$, the tearing modulus is given by

$$T = T_o - \frac{\beta}{\alpha}\ln\left(\frac{J}{J_{Ic}}\right) \tag{A3.45}$$

Rice et al. computed normalized R curves (J/J_{Ic} vs. $\Delta a/R$) for a range of T_o values and found that $T = T_o$ in the early stages of crack growth, but the R curve slope decreases until the steady-state plateau is reached. The steady-state J can easily be inferred from Equation (A3.45) by setting $T = 0$:

$$J_{ss} = J_{IC}\exp\left(\frac{\alpha T_o}{\beta}\right) \tag{A3.46}$$

Large-Scale Yielding

Although the RDS analysis was derived for small-scale yielding conditions, Rice et al. speculated that the form of Equation (A3.38) might also be valid for fully plastic conditions. The numerical values of some of the constants, however, probably differ for the large-scale yielding case.

The most important difference between small-scale yielding and fully plastic conditions is the value of R. Rice et al. argued that R would no longer scale with the plastic zone size, but should be proportional to the ligament length. They made a rough estimate of $R \sim b/4$ for the fully plastic case.

The constant α depends on crack-tip triaxiality and thus may differ for small-scale yielding and fully yielded conditions. For highly constrained configurations, such as bend specimens, α for the two cases should be similar.

In small-scale yielding, the definition of J is unambiguous, since it is related to the elastic stress intensity factor. The J integral for a growing crack under fully plastic conditions can be computed in a number of ways, however, and not all definitions of J are appropriate in the large-scale yielding version of Equation (A3.38).

Assume that the crack-growth resistance behavior is to be characterized by a J-like parameter J_x. Assuming J_x depends on the crack length and displacement, the rate of change in J_x should be linearly related to the displacement rate and $\dot{a}$:

$$\dot{J}_x = \xi \dot{\Delta} + \chi \dot{a} \tag{A3.47}$$

where ξ and χ are functions of displacement and crack length. Substituting Equation (A3.47) into Equation (A3.38) gives

$$\dot{\delta} = \frac{\alpha}{\sigma_o} \xi \dot{\Delta} + \left[\beta \frac{\sigma_o}{E} \ln\left(\frac{R}{r}\right) + \frac{\alpha}{\sigma_o} \chi \right] \dot{a} \tag{A3.48}$$

In the limit of a rigid ideally plastic material, $\sigma_o/E = 0$. Also, the local crack-opening rate must be proportional to the global displacement rate for a rigid ideally plastic material:

$$\dot{\delta} = \psi \dot{\Delta} \tag{A3.49}$$

Therefore, the term in square brackets in Equation (A3.48) must vanish, which implies that $\chi = 0$, at least in the limit of a rigid ideally plastic material. Thus, in order for the RDS model to apply to large-scale yielding, the rate of change in the J-like parameter must not depend on the crack-growth rate:

$$\dot{J}_x \neq \dot{J}_x(\dot{a}) \tag{A3.50}$$

Rice et al. showed that neither the deformation theory J nor the far-field J satisfy Equation (A3.50) for all configurations.

Satisfying Equation (A3.50) does not necessarily imply that a J_x-R curve is geometry independent. The RDS model suggests that a resistance curve obtained from a fully yielded specimen will not, in general, agree with the small-scale yielding R curve for the same material. Assuming $R = b/4$ for the fully plastic case, the RDS model predicts the following tearing modulus:

$$T = \left[T_o - \frac{\beta}{\alpha_{ssy}} \ln\left(\frac{b/4}{\lambda E J_{IC}/\sigma_o^2} \right) \right] \frac{\alpha_{ssy}}{\alpha_{fy}} \tag{A3.51}$$

where the subscripts *ssy* and *fy* denote small-scale yielding and fully yielded conditions, respectively. According to Equation (A3.51), the crack-growth-resistance curve under fully yielded conditions has a constant initial slope, but this slope is not equal to T_o (the initial tearing modulus in small-scale yielding) unless $\alpha_{fy} = \alpha_{ssy}$ and $b = 4\lambda\, EJ_{Ic}/\sigma_o^2$. Equation (A3.51) does not predict a steady-state limit where $T = 0$; rather this relationship predicts that T actually increases as the ligament becomes smaller.

The foregoing analysis implies that crack-growth-resistance curves obtained from specimens with fully yielded ligaments are suspect. One should exercise extreme caution when applying experimental results from small specimens to predict the behavior of large structures.

A3.5.2 Steady State Crack Growth

The RDS analysis, which assumed a local failure criterion based on crack-opening angle, indicated that crack growth in small-scale yielding reaches a steady state, where $dJ/da \to 0$. The derivation that follows shows that the steady-state limit is a general result for small-scale yielding; the R curve must eventually reach a plateau in an infinite body, regardless of the failure mechanism.

Generalized Damage Integral

Consider a material element a small distance from a crack tip, as illustrated in Figure A3.7. This material element will fail when it is deformed beyond its capacity. The crack will grow as consecutive material elements at the tip fail. Let us define a generalized damage integral Θ, which characterizes the severity of loading at the crack tip:

$$\Theta = \int_0^{\varepsilon_{eq}} \Omega(\sigma_{ij}, \varepsilon_{ij}) d\varepsilon_{eq} \qquad (A3.52)$$

where ε_{eq} is the equivalent (von Mises) plastic strain and Ω is a function of the stress and strain tensors (σ_{ij} and ε_{ij}, respectively). The above integral is sufficiently general that it can depend on the current values of all stress and strain components, as well as the entire deformation history. Referring to Figure A3.7, the material element will fail at a critical value of Θ. At the moment of crack initiation or during crack extension, the material near the crack tip will be close to the point of failure. At a distance r^* from the crack tip, where r^* is arbitrarily small, we can assume that $\Theta = \Theta_c$.

The precise form of the damage integral depends on the micromechanism of fracture. For example, a modified Rice and Tracey [47] model for ductile hole growth (see Chapter 5) can be used to characterize ductile fracture in metals:

$$\Theta = \ln\left(\frac{R}{R_o}\right) = 0.283 \int_0^{\varepsilon_{eq}} \exp\left(\frac{1.5\sigma_m}{\sigma_e}\right) d\varepsilon_{eq} \qquad (A3.53)$$

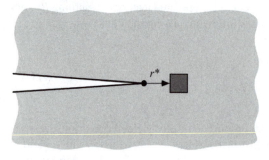

FIGURE A3.7 Material point a distance r^* from the crack tip.

where

R = void radius

R_o = initial radius

σ_m = mean (hydrostatic) stress

σ_e = effective (von Mises) stress

Failure, in this case, is assumed when the void radius reaches a critical value.

Stable Crack Growth

Consider an infinite body[10] that contains a crack that is growing in a stable, self-similar, and quasistatic manner. If the crack has grown well beyond the initial blunted tip, dimensional analysis indicates that the local stresses and strains are uniquely characterized by the far-field J integral, as stated in Equation (3.69). In light of this single-parameter condition, the integrand of Equation (A3.52) becomes

$$\Omega = \Omega\left(\frac{J}{\sigma_o r}, \theta\right) \qquad (A3.54)$$

We can restrict this analysis to $\theta = 0$ by assuming that the material on the crack plane fails during Mode I crack growth. For a given material point on the crack plane, r decreases as the crack grows, and the plastic strain increases. If strain increases monotonically as this material point approaches the crack tip, Equation (A3.54) permits writing Ω as a function of the von Mises strain:

$$\Omega = \Omega(\varepsilon_{eq}) \qquad (A3.55)$$

Therefore the local failure criterion is given by

$$\Theta_c = \int_0^{\varepsilon^*} \Omega(\varepsilon_{eq}) d\varepsilon_{eq} \qquad (A3.56)$$

where ε^* is the critical strain (i.e., the von Mises strain at $r = r^*$). Since the integrand is a function only of ε_{eq}, the integration path is the same for all material points ahead of the crack tip, and ε^* is constant during crack growth. That is, the equivalent plastic strain at r^* will always equal ε^* when the crack is growing. Based on Equation (3.69) and Equation (A3.54), ε^* is a function only of r^* and the applied J:

$$\varepsilon^* = \varepsilon^*(J, r^*) \qquad (A3.57)$$

Solving for the differential of ε^* gives

$$d\varepsilon^* = \frac{\partial \varepsilon^*}{\partial J} dJ + \frac{\partial \varepsilon^*}{\partial r^*} dr^* \qquad (A3.58)$$

Since ε^* and r^* are both fixed, $d\varepsilon^* = dr^* = dJ = 0$.

---

[10] In practical terms "infinite" means that external boundaries are sufficiently far from the crack tip so that the plastic zone is embedded within an elastic singularity zone.

Thus the J integral remains constant during crack extension ($dJ/da = 0$) when Equation (3.69) is satisfied. Steady-state crack growth is usually not observed experimentally because large-scale yielding in finite-sized specimens precludes characterizing a growing crack with J. Also, a significant amount of crack growth may be required before a steady state is reached (Figure 3.25); the crack tip in a typical laboratory specimen approaches a free boundary well before the crack growth is sufficient to be unaffected by the initial blunted tip.

A3.6 Notes on the Applicability of Deformation Plasticity to Crack Problems

Since elastic-plastic fracture mechanics is based on the deformation plasticity theory, it may be instructive to take a closer look at this theory and assess its validity for crack problems.

Let us begin with the plastic portion of the Ramberg-Osgood equation for uniaxial deformation, which can be expressed in the following form:

$$\varepsilon_p = \alpha \left(\frac{\sigma}{\sigma_o} \right)^{n-1} \frac{\sigma}{E} \tag{A3.59}$$

Differentiating Equation (A3.59) gives

$$d\varepsilon_p = \alpha n \left(\frac{\sigma}{\sigma_o} \right)^{n-2} \frac{\sigma}{E} \frac{d\sigma}{\sigma_o} \tag{A3.60}$$

for an increment of plastic strain. For the remainder of this section, the subscript on strain is suppressed for brevity; only plastic strains are considered, unless stated otherwise.

Equation (A3.59) and Equation (A3.60) represent the deformation and incremental flow theories, respectively, for uniaxial deformation in a Ramberg-Osgood material. In this simple case, there is no difference between the incremental and deformation theories, provided no unloading occurs. Equation (A3.60) can obviously be integrated to obtain Equation (A3.59). Stress is uniquely related to strain when both increase monotonically. It does not necessarily follow that deformation and incremental theories are equivalent in the case of three-dimensional monotonic loading, but there are many cases where this is a good assumption.

Equation (A3.59) can be generalized to three dimensions by assuming deformation plasticity and isotropic hardening:

$$\varepsilon_{ij} = \frac{3}{2} \alpha \left(\frac{\sigma_e}{\sigma_o} \right)^{n-1} \frac{S_{ij}}{E} \tag{A3.61}$$

where σ_e is the effective (von Mises) stress and S_{ij} is the deviatoric component of the stress tensor, defined by

$$S_{ij} = \sigma_{ij} - \frac{1}{3} \sigma_{kk} \delta_{ij} \tag{A3.62}$$

where δ_{ij} is the Kronecker delta. Equation (A3.61) is the deformation theory *flow rule* for a Ramberg-Osgood material. The corresponding flow rule for incremental plasticity theory is given by

$$d\varepsilon_{ij} = \frac{3}{2} \alpha n \left(\frac{\sigma_e}{\sigma_o} \right)^{n-2} \frac{S_{ij}}{E} \frac{d\sigma_e}{\sigma_o} \tag{A3.63}$$

By comparing Equation (A3.61) and Equation (A3.63), one sees that the deformation and incremental theories of plasticity coincide only if the latter equation can be integrated to obtain the former. If the deviatoric stress components are proportional to the effective stress:

$$S_{ij} = \omega_{ij} \sigma_e \qquad (A3.64)$$

where ω_{ij} is a constant tensor that does not depend on strain, then the integration of Equation (A3.63) results in Equation (A3.61). Thus deformation and incremental theories of plasticity are identical when the loading is proportional in the deviatoric stresses. Note that the *total* stress components need not be proportional in order for the two theories to coincide; the flow rule is not influenced by the hydrostatic portion of the stress tensor.

Proportional loading of the deviatoric components does not necessarily mean that deformation plasticity theory is rigorously correct; it merely implies that the deformation theory is no more objectionable than the incremental theory. Classical plasticity theory, whether based on incremental strain or total deformation, contains simplifying assumptions about material behavior. Both Equation (A3.61) and Equation (A3.63) assume that the yield surface expands symmetrically and that its radius does not depend on hydrostatic stress. For monotonic loading ahead of a crack in a metal, these assumptions are probably reasonable; the assumed hardening law is of little consequence for monotonic loading, and hydrostatic stress effects on the yield surface are relatively small for most metals.

Budiansky [48] showed that the deformation theory is still acceptable when there are modest deviations from proportionality. Low work-hardening materials are the least sensitive to nonproportional loading.

Since most of classical fracture mechanics assumes either plane stress or plane strain, it is useful to examine plastic deformation in the two-dimensional case, and determine under what conditions the requirement of proportional deviatoric stresses is at least approximately satisfied. Consider, for example, plane strain. When elastic strains are negligible, the in-plane deviatoric normal stresses are given by

$$S_{xx} = \frac{\sigma_{xx} - \sigma_{yy}}{2} \quad \text{and} \quad S_{yy} = \frac{\sigma_{yy} - \sigma_{xx}}{2} \qquad (A3.65)$$

assuming incompressible plastic deformation, where $\sigma_{zz} = (\sigma_{xx} + \sigma_{yy})/2$. The expression for von Mises stress in plane strain reduces to

$$\sigma_e = \frac{1}{\sqrt{2}} \left[3S_{xx}^2 + 6S_{xy}^2 \right]^{1/2} \qquad (A3.66)$$

where $S_{xy} = \tau_{xy}$. Alternatively, σ_e can be written in terms of principal normal stresses:

$$\sigma_e = \frac{\sqrt{3}}{2} [\sigma_1 - \sigma_2] \quad \text{where } \sigma_1 > \sigma_2$$

$$= \sqrt{3} S_1 \qquad (A3.67)$$

Therefore, the principal deviatoric stresses are proportional to σ_e in the case of plane strain. It can easily be shown that the same is true for plane stress. If the principal axes are fixed, S_{xx}, S_{yy}, and

S_{xy} must also be proportional to σ_e. If, however, the principal axes rotate during deformation, the deviatoric stress components defined by a fixed coordinate system will not increase in proportion to one another.

In the case of Mode I loading of a crack, τ_{xy} is always zero on the crack plane, implying that the principal directions on the crack plane are always parallel to the x-y-z coordinate axes. Thus, the deformation and incremental plasticity theories should be equally valid on the crack plane, well inside the plastic zone (where elastic strains are negligible). At finite angles from the crack plane, the principal axes may rotate with deformation, which will produce nonproportional deviatoric stresses. If this effect is small, the deformation plasticity theory should be adequate to analyze stresses and strains near the crack tip in either plane stress or plane strain.

The validity of the deformation plasticity theory does not automatically guarantee that the crack-tip conditions can be characterized by a single parameter, such as J or K. Single-parameter fracture mechanics requires that the *total* stress components be proportional near the crack tip,[11] a much more severe restriction. Proportional total stresses imply that the deviatoric stresses are proportional, but the reverse is not necessarily true. In both the linear elastic case (Appendix 2.3) and the nonlinear case (Appendix 3.4) the stresses near the crack tip were derived from a stress function of the form

$$\Phi = \kappa r^s f(\theta) \qquad\qquad (A3.68)$$

where κ is a constant. The form of Equation (A3.68) guarantees that all stress components are proportional to κ, and thus proportional to one another. Therefore any monotonic function of κ uniquely characterizes the stress fields in the region where Equation (A3.68) is valid. Nonproportional loading automatically invalidates Equation (A3.68) and the single-parameter description that it implies.

As stated earlier, the deviatoric stresses are proportional on the crack plane, well within the plastic zone. However, the hydrostatic stress may not be proportional to σ_e. For example, the loading is highly nonproportional in the large-strain region, as Figure 3.12 indicates. Consider a material point at a distance x from the crack tip, where x is in the current large-strain region. At earlier stages of deformation the loading on this point was proportional, but σ_{yy} reached a peak when the ratio $x \sigma_o/J$ was approximately unity, and the normal stress decreased with subsequent deformation. Thus, the most recent loading on this point was nonproportional, but the deviatoric stresses are still proportional to σ_e.

When the crack grows, the material behind the crack tip unloads elastically and the deformation plasticity theory is no longer valid. The deformation theory is also suspect near the elastic-plastic boundary. Equation (A3.65) to Equation (A3.67) were derived assuming the elastic strains were negligible, which implies $\sigma_{zz} = 0.5(\sigma_{xx} + \sigma_{yy})$ in plane strain. At the onset of yielding, however, $\sigma_{zz} = v\,(\sigma_{xx} + \sigma_{yy})$, and the proportionality constants between σ_e and the deviatoric stress components are different than for the fully plastic case. Thus when elastic and plastic strains are of comparable magnitude, the deviatoric stresses are nonproportional, as ω_{ij} (Equation (A3.64)) varies from its elastic value to the fully plastic limit. The errors in the deformation theory that may arise from the transition from elastic to plastic behavior should not be appreciable in crack problems, because the strain gradient ahead of the crack tip is relatively steep, and the transition zone is small.

---

[11] The proportional loading region need not extend all the way to the crack tip, but the nonproportional zone at the tip must be embedded within the proportional zone in order for a single loading parameter to characterize crack-tip conditions.

REFERENCES

1. Wells, A.A., "Unstable Crack Propagation in Metals: Cleavage and Fast Fracture." *Proceedings of the Crack Propagation Symposium,* Vol. 1, Paper 84, Cranfield, UK, 1961.
2. Irwin, G.R., "Plastic Zone Near a Crack and Fracture Toughness." *Sagamore Research Conference Proceedings,* Vol. 4, 1961, pp. 63–78.
3. Burdekin, F.M. and Stone, D.E.W., "The Crack Opening Displacement Approach to Fracture Mechanics in Yielding Materials." *Journal of Strain Analysis,* Vol. 1, 1966, pp. 145–153.
4. Rice, J.R., "A Path Independent Integral and the Approximate Analysis of Strain Concentration by Notches and Cracks." *Journal of Applied Mechanics*, Vol. 35, 1968, pp. 379–386.
5. BS 7448: Part 1: 1991, "Fracture Mechanics Toughness Tests, Part 1, Method for Determination of K_{IC}, Critical CTOD and Critical J Values of Metallic Materials," The British Standards Institution, London, 1991.
6. E 1290-99, "Standard Test Method for Crack Tip Opening Displacement (CTOD) Fracture Toughness Measurement." American Society for Testing and Materials, Philadelphia, PA, 1999.
7. Hutchinson, J.W., "Singular Behavior at the End of a Tensile Crack Tip in a Hardening Material." *Journal of the Mechanics and Physics of Solids,* Vol. 16, 1968, pp. 13–31.
8. Rice, J.R. and Rosengren, G.F., "Plane Strain Deformation near a Crack Tip in a Power-Law Hardening Material." *Journal of the Mechanics and Physics of Solids,* Vol. 16, 1968, pp. 1–12.
9. McMeeking, R.M. and Parks, D.M., "On Criteria for J-Dominance of Crack Tip Fields in Large-Scale Yielding." *Elastic Plastic Fracture*, ASTM STP 668, American Society for Testing and Materials, Philadelphia, PA, 1979, pp. 175–194.
10. Read, D.T., "Applied J-Integral in HY130 Tensile Panels and Implications for Fitness for Service Assessment." Report NBSIR 82-1670, National Bureau of Standards, Boulder, CO, 1982.
11. Begley, J.A. and Landes, J.D., "The J-Integral as a Fracture Criterion." ASTM STP 514, American Society for Testing and Materials, Philadelphia, PA, 1972, pp. 1–20.
12. Landes, J.D. and Begley, J.A., "The Effect of Specimen Geometry on J_{Ic}." ASTM STP 514, American Society for Testing and Materials, Philadelphia, PA, 1972, pp. 24–29.
13. Rice, J.R., Paris, P.C., and Merkle, J.G., "Some Further Results of J-Integral Analysis and Estimates." ASTM STP 536, American Society for Testing and Materials, Philadelphia, PA, 1973, pp. 231–245.
14. Shih, C.F. "Relationship between the J-Integral and the Crack Opening Displacement for Stationary and Extending Cracks." *Journal of the Mechanics and Physics of Solids,* Vol. 29, 1981, pp. 305–326.
15. Rice, J.R., Drugan, W.J., and Sham, T.-L., "Elastic-Plastic Analysis of Growing Cracks." ASTM STP 700, American Society for Testing and Materials, Philadelphia, PA, 1980, pp. 189–221.
16. Dodds, R.H., Jr. and Tang, M., "Numerical Techniques to Model Ductile Crack Growth in Fracture Test Specimens." *Engineering Fracture Mechanics,* Vol. 46, 1993, pp. 246–253.
17. Dodds, R.H., Jr., Tang, M., and Anderson, T.L. "Effects of Prior Ductile Tearing on Cleavage Fracture Toughness in the Transition Region." *Second Symposium on Constraint Effects in Fracture,* American Society for Testing and Materials, Philadelphia, PA (in press).
18. McClintock, F.A., "Plasticity Aspects of Fracture." *Fracture: An Advanced Treatise*, Vol. 3, Academic Press, New York, 1971, pp. 47–225.
19. Anderson, T.L., "Ductile and Brittle Fracture Analysis of Surface Flaws Using CTOD." *Experimental Mechanics,* 1988, pp. 188–193.
20. Kirk, M.T., Koppenhoefer, K.C., and Shih, C.F., "Effect of Constraint on Specimen Dimensions Needed to Obtain Structurally Relevant Toughness Measures." *Constraint Effects in Fracture,* ASTM STP 1171, American Society for Testing and Materials, Philadelphia, PA 1993, pp. 79–103.
21. Joyce, J.A. and Link, R.E., "Effect of Constraint on Upper Shelf Fracture Toughness." *Fracture Mechanics,* Vol. 26, ASTM STP 1256, American Society for Testing and Materials, Philadelphia, PA (in press).
22. Towers, O.L. and Garwood, S.J., "Influence of Crack Depth on Resistance Curves for Three-Point Bend Specimens in HY130." ASTM STP 905, American Society for Testing and Materials, Philadelphia, PA, 1986, pp. 454–484.
23. Williams, M.L., "On the Stress Distribution at the Base of a Stationary Crack." *Journal of Applied Mechanics,* Vol. 24, 1957, pp. 109–114.
24. Bilby, B.A., Cardew, G.E., Goldthorpe, M.R., and Howard, I.C., "A Finite Element Investigation of the Effects of Specimen Geometry on the Fields of Stress and Strain at the Tips of Stationary Cracks." *Size Effects in Fracture*, Institute of Mechanical Engineers, London, 1986, pp. 37–46.

25. Betegon, C. and Hancock, J.W., "Two Parameter Characterization of Elastic-Plastic Crack Tip Fields." *Journal of Applied Mechanics*, Vol. 58, 1991, pp. 104–110.

26. Kirk, M.T., Dodds, R.H., Jr., and Anderson, T.L., "Approximate Techniques for Predicting Size Effects on Cleavage Fracture Toughness." *Fracture Mechanics*, Vol. 24, ASTM STP 1207, American Society for Testing and Materials, Philadelphia, PA (in press).

27. Hancock, J.W., Reuter, W.G., and Parks, D.M., "Constraint and Toughness Parameterized by *T*." *Constraint Effects in Fracture*, ASTM STP 1171, American Society for Testing and Materials, Philadelphia, PA, 1993, pp. 21–40.

28. Sumpter, J.D.G., "An Experimental Investigation of the *T* Stress Approach." *Constraint Effects in Fracture*, ASTM STP 1171, American Society for Testing and Materials, Philadelphia, PA, 1993, pp. 492–502.

29. O'Dowd, N.P. and Shih, C.F., "Family of Crack-Tip Fields Characterized by a Triaxiality Parameter–I. Structure of Fields." *Journal of the Mechanics and Physics of Solids,* Vol. 39, 1991, pp. 898–1015.

30. O'Dowd, N.P. and Shih, C.F., "Family of Crack-Tip Fields Characterized by a Triaxiality Parameter–II. Fracture Applications." *Journal of the Mechanics and Physics of Solids*, Vol. 40, 1992, pp. 939–963.

31. Shih, C.F., O'Dowd, N.P., and Kirk, M.T., "A Framework for Quantifying Crack Tip Constraint." *Constraint Effects in Fracture*, ASTM STP 1171, American Society for Testing and Materials, Philadelphia, PA., 1993, pp. 2–20.

32. Ritchie, R.O., Knott, J.F., and Rice, J.R., "On the Relationship between Critical Tensile Stress and Fracture Toughness in Mild Steel." *Journal of the Mechanics and Physics of Solids*, Vol. 21, 1973, pp. 395–410.

33. Anderson, T.L., Vanaparthy, N.M.R., and Dodds, R.H., Jr., "Predictions of Specimen Size Dependence on Fracture Toughness for Cleavage and Ductile Tearing." *Constraint Effects in Fracture*, ASTM STP 1171, American Society for Testing and Materials, Philadelphia, PA, 1993, pp. 473–491.

34. Anderson, T.L. and Dodds, R.H., Jr., "Specimen Size Requirements for Fracture Toughness Testing in the Ductile-Brittle Transition Region." *Journal of Testing and Evaluation*, Vol. 19, 1991, pp. 123–134.

35. Dodds, R.H., Jr., Anderson, T.L., and Kirk, M.T., "A Framework to Correlate *a/W* Effects on Elastic-Plastic Fracture Toughness (J_c)." *International Journal of Fracture,* Vol. 48, 1991, pp. 1–22.

36. Anderson, T.L. and Dodds, R.H., Jr., "An Experimental and Numerical Investigation of Specimen Size Requirements for Cleavage Fracture Toughness." NUREG/CR-6272, Nuclear Regulatory Commission, Washington, DC (in press).

37. Sorem, W.A., "The Effect of Specimen Size and Crack Depth on the Elastic-Plastic Fracture Toughness of a Low-Strength High-Strain Hardening Steel." Ph.D. Dissertation, The University of Kansas, Lawrence, KS, 1989.

38. Anderson, T.L., Stienstra, D.I.A., and Dodds, R.H., Jr., "A Theoretical Framework for Addressing Fracture in the Ductile-Brittle Transition Region." *Fracture Mechanics,* Vol. 24, ASTM STP 1207, American Society for Testing and Materials, Philadelphia, PA (in press).

39. Li, W.C. and Wang, T.C., "Higher-Order Asymptotic Field of Tensile Plane Strain Nonlinear Crack Problems." *Scientia, Sinica (Series A),* Vol. 29, 1986, pp. 941–955.

40. Sharma, S.M. and Aravas, N., "Determination of Higher-Order Terms in Asymptotic Elastoplastic Crack Tip Solutions." *Journal of the Mechanics and Physics of Solids,* Vol. 39, 1991, pp. 1043–1072.

41. Yang, S., Chao, Y.J, and Sutton, M.A., "Higher Order Asymptotic Crack Tip Fields in a Power Law Hardening Material." *Engineering Fracture Mechanics,* Vol. 45, 1993, pp. 1–20.

42. Xia, L., Wang, T.C., and Shih, C.F., "Higher Order Analysis of Crack-Tip Fields in Elastic-Plastic Power-Law Hardening Materials." *Journal of the Mechanics and Physics of Solids,* Vol. 41, 1993, pp. 665–687.

43. Crane, D.L., "Deformation Limits on Two-Parameter Fracture Mechanics in Terms of Higher Order Asymptotics." Ph.D. Dissertation, Texas A&M University, College Station, TX, 1994.

44. Westergaard, H.M., "Bearing Pressures and Cracks." *Journal of Applied Mechanics*, Vol. 6, 1939, pp. 49–53.

45. Bilby, B.A., Cottrell, A.H., and Swindon, K.H., "The Spread of Plastic Yield from a Notch." *Proceedings, Royal Society of London,* Vol. A-272, 1963, pp. 304–314.

46. Smith, E., "The Spread of Plasticity from a Crack: an Approach Based on the Solution of a Pair of Dual Integral Equations." CEGB Research Laboratories, Lab. Note No. RD/L/M31/62, 1962.

47. Rice, J.R. and Tracey, D.M., *Journal of the Mechanics and Physics of Solids,* Vol. 17, 1969, pp. 201–217.

48. Budiansky, B., "A Reassessment of Deformation Theories of Plasticity." *Journal of Applied Mechanics,* Vol. 81, 1959, pp. 259–264.

4 Dynamic and Time-Dependent Fracture

In certain fracture problems, time is an important variable. At high loading rates, for example, inertia effects and material rate dependence can be significant. Metals and ceramics also exhibit rate-dependent deformation (creep) at temperatures that are close to the melting point of the material. The mechanical behavior of polymers is highly sensitive to strain rate, particularly above the glass transition temperature. In each of these cases, linear elastic and elastic-plastic fracture mechanics, which assume quasistatic, rate-independent deformation, are inadequate.

Early fracture mechanics researchers considered dynamic effects, but only for the special case of linear elastic material behavior. More recently, fracture mechanics has been extended to include time-dependent material behavior such as viscoplasticity and viscoelasticity. Most of these newer approaches are based on generalizations of the J contour integral.

This chapter gives an overview of time-dependent fracture mechanics. The treatment of this subject is far from exhaustive, but should serve as an introduction to a complex and rapidly developing field. The reader is encouraged to consult the published literature for a further background.

4.1 DYNAMIC FRACTURE AND CRACK ARREST

As any undergraduate engineering student knows, dynamics is more difficult than statics. Problems become more complicated when the equations of equilibrium are replaced by the equations of motion.

In the most general case, dynamic fracture mechanics contains three complicating features that are not present in LEFM and elastic-plastic fracture mechanics: inertia forces, rate-dependent material behavior, and reflected stress waves. Inertia effects are important when the load changes abruptly or the crack grows rapidly; a portion of the work that is applied to the specimen is converted to kinetic energy. Most metals are not sensitive to moderate variations in strain rate near ambient temperature, but the flow stress can increase appreciably when the strain rate increases by several orders of magnitude. The effect of rapid loading is even more pronounced in rate-sensitive materials such as polymers. When the load changes abruptly or the crack grows rapidly, stress waves propagate through the material and reflect off free surfaces, such as the specimen boundaries and the crack plane. Reflecting stress waves influence the local crack-tip stress and strain fields which, in turn, affect the fracture behavior.

In certain problems, one or more of the above effects can be ignored. If all three effects are neglected, the problem reduces to the quasistatic case.

The dynamic version of LEFM is termed *elastodynamic fracture mechanics*, where nonlinear material behavior is neglected, but inertia forces and reflected stress waves are incorporated when necessary. The theoretical framework of elastodynamic fracture mechanics is fairly well established, and practical applications of this approach are becoming more common. Extensive reviews of this subject have been published by Freund [1–5], Kanninen and Poplar [6], Rose [7], and others. Elastodynamic fracture mechanics has limitations, but is approximately valid in many cases. When the plastic zone is restricted to a small region near the crack tip in a dynamic problem, the stress-intensity approach, with some modifications, is still applicable.

Dynamic fracture analyses that incorporate nonlinear, time-dependent material behavior are a relatively recent innovation. A number of researchers have generalized the J integral to account for inertia and viscoplasticity [8–13].

There are two major classes of dynamic fracture problems: (1) fracture initiation as a result of rapid loading, and (2) rapid propagation of a crack. In the latter case, the crack propagation may initiate either by quasistatic or rapid application of a load; the crack may arrest after some amount of unstable propagation. Dynamic initiation, propagation, and crack arrest are discussed later in this chapter.

4.1.1 RAPID LOADING OF A STATIONARY CRACK

Rapid loading of a structure can come from a number of sources, but most often occurs as the result of impact with a second object (e.g., a ship colliding with an offshore platform or a missile striking its target). Impact loading is often applied in laboratory tests when a high strain rate is desired. The Charpy test [14], where a pendulum dropped from a fixed height fractures a notched specimen, is probably the most common dynamic mechanical test. Dynamic loading of a fracture mechanics specimen can be achieved through impact loading [15, 16], a controlled explosion near the specimen [17], or servo-hydraulic testing machines that are specially designed to impart high displacement rates. Chapter 7 describes some of the practical aspects of high rate fracture testing.

Figure 4.1 schematically illustrates a typical load-time response for dynamic loading. The load tends to increase with time, but oscillates at a particular frequency that depends on specimen geometry and material properties. Note that the loading rate is finite, i.e., a finite time is required to reach a particular load. The amplitude of the oscillations decreases with time, as kinetic energy is dissipated by the specimen. Thus, inertia effects are most significant at short times, and are minimal after sufficiently long times, where the behavior is essentially quasistatic.

Determining a fracture characterizing parameter, such as the stress-intensity factor or the J integral, for rapid loading can be very difficult. Consider the case where the plastic zone is confined to a small region surrounding the crack tip. The near-tip stress fields for high rate Mode I loading are given by Equation 4.1.

$$\sigma_{ij} = \frac{K_I(t)}{\sqrt{2\pi r}}$$

(4.1)

where (t) denotes a function of time. The angular functions f_{ij} are identical to the quasistatic case and are given in Table 2.1. The stress-intensity factor, which characterizes the amplitude of the elastic singularity, varies erratically in the early stages of loading. Reflecting stress waves that pass through the specimen constructively and destructively interfere with one another, resulting in a highly complex time-dependent stress distribution. The instantaneous K_I depends on the magnitude of the discrete stress waves that pass through the crack-tip region at that particular moment in time. When the discrete waves are significant, it is not possible to infer K_I from the remote loads.

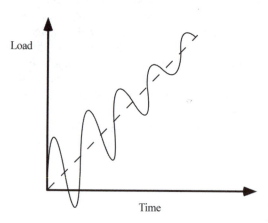

Load

Time

FIGURE 4.1 Schematic load-time response of a rapidly loaded structure.

Recent work by Nakamura et al. [18, 19] quantified inertia effects in laboratory specimens and showed that these effects can be neglected in many cases. They observed that the behavior of a dynamically loaded specimen can be characterized by a short-time response, dominated by discrete waves, and a long-time response that is essentially quasistatic. At intermediate times, global inertia effects are significant but local oscillations at the crack are small, because kinetic energy is absorbed by the plastic zone. To distinguish short-time response from long-time response, Nakamura et al. defined a *transition time* t_τ when the kinetic energy and the deformation energy (the energy absorbed by the specimen) are equal. Inertia effects dominate prior to the transition time, but the deformation energy dominates at times significantly greater than t_τ. In the latter case, a *J*-dominated field should exist near the crack tip and quasistatic relationships can be used to infer *J* from global load and displacement.

Since it is not possible to measure kinetic and deformation energies separately during a fracture mechanics experiment, Nakamura et al. developed a simple model to estimate the kinetic energy and the transition time in a three-point bend specimen (Figure 4.2). This model was based on the Bernoulli-Euler beam theory and assumed that the kinetic energy at early times was dominated by the elastic response of the specimen. Incorporating the known relationship between the load-line displacement and the strain energy in a three-point bend specimen leads to an approximate relationship for the ratio of kinetic to deformation energy:

$$\frac{E_k}{U} = \left(\Lambda \frac{W\dot{\Delta}(t)}{c_o \Delta(t)} \right)^2 \tag{4.2}$$

where
  E_k = kinetic energy
  U = deformation energy
  W = specimen width
  Δ = load line displacement
  $\dot{\Delta}$ = displacement rate
  c_o = longitudinal wave speed (i.e., the speed of sound) in a one-dimensional bar
  Λ = geometry factor, which for the bend specimen is given by

$$\Lambda = \sqrt{\frac{SBEC}{W}} \tag{4.3}$$

where S is the span of the specimen.

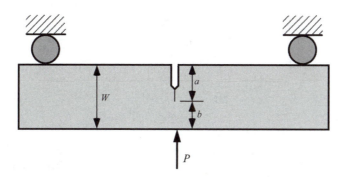

FIGURE 4.2 Three-point bend specimen.

The advantage of Equation (4.2) is that the displacement and displacement rate can be measured experimentally. The transition time is defined at the moment in the test when the ratio $E_k/U = 1$. In order to obtain an explicit expression for t_τ, it is convenient to introduce a dimensionless displacement coefficient D:

$$D = \frac{t\dot{\Delta}(t)}{\Delta(t)}\bigg|_{t_\tau}$$

(4.4)

If, for example, the displacement varies with time as a power law: $\Delta = \beta t\gamma$, then $D = \gamma$. Combining Equation (4.2) and Equation (4.4) and setting $E_k/U = 1$ leads to

$$t_\tau = D\Lambda \frac{W}{c_o}$$

(4.5)

Nakamura et al. [18, 19] performed a dynamic finite element analysis on a three-point bend specimen in order to evaluate the accuracy of Equation (4.2) and Equation (4.5). Figure 4.3 compares the E_k/U ratio computed from a finite element analysis with that determined from the experiment and Equation (4.2). The horizontal axis is a dimensionless time scale, and c_1 is the longitudinal wave speed in an unbounded solid. The ratio W/c_1 is an estimate of the time required for a stress wave to traverse the width of the specimen. Based on Equation (4.2) and the experiment, $t_\tau c_1/W \approx 28$ (or $t_\tau c_o/H \approx 24$), while the finite element analysis estimated $t_\tau c_1/W \approx 27$. Thus the simple model agrees quite well with a more detailed analysis.

The simple model was based on the global kinetic energy and did not consider discrete stress waves. Thus the model is only valid after stress waves have traversed the width of the specimen several times. This limitation does not affect the analysis of the transition time, since stress waves have made approximately 27 passes when t_τ is reached. Note, in Figure 4.3, that the simple model agrees very well with the finite element analysis when $tc_1/W > 20$.

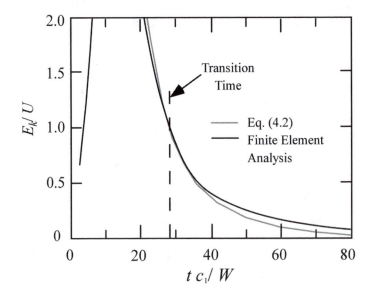

FIGURE 4.3 Ratio of kinetic to stress work energy in a dynamically loaded three-point bend specimen. Taken from Nakamura, T., Shih, C.F., and Freund, L.B., "Three-Dimensional Transient Analysis of a Dynamically Loaded Three-Point-Bend Ductile Fracture Specimen." ASTM STP 995, Vol. I, American Society for Testing and Materials, Philadelphia, PA, 1989, pp. 217–241.

When $t \gg t_\tau$, inertia effects are negligible and quasistatic models should apply to the problem. Consequently, the J integral for a deeply cracked bend specimen at long times can be estimated by

$$J_{dc} = \frac{2}{Bb} \int_0^{\Omega(t^*)} M(t) d\Omega(t) \qquad (4.6)$$

where
    B = plate thickness
    b = uncracked ligament length
    M = applied moment on the ligament
    Ω = angle of rotation
    t^* = current time

Equation (4.6), which was originally published by Rice et al. [20], is derived in Section 3.2.5.

Nakamura et al. [19] performed a three-dimensional dynamic elastic-plastic finite element analysis on a three-point bend specimen in order to determine the range of applicability of Equation (4.6). They evaluated a dynamic J integral (see Section 4.1.3) at various thickness positions and observed a through-thickness variation of J that is similar to Figure 3.36. They computed a nominal J that averaged the through-thickness variations and compared this value with J_{dc}. The results of this exercise are plotted in Figure 4.4. At short times, the average dynamic J is significantly lower than the J computed from the quasistatic relationship. For $t > 2t_\tau$, the J_{dc}/J_{ave} reaches a constant value that is slightly greater than 1. The modest discrepancy between J_{dc} and J_{ave} at long times is probably due to three-dimensional effects rather than dynamic effects (Equation (4.6) is essentially a two-dimensional formula).

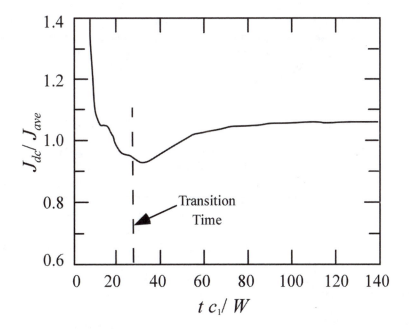

FIGURE 4.4 Ratio of J computed from Equation (4.6) to the through-thickness average J computed from a three-dimensional dynamic finite element analysis.

According to Figure 4.4, Equation (4.6) provides a good estimate of J in a high-rate test at times greater than approximately twice the transition time. It follows that if the fracture initiation occurs after $2t_\tau$, the critical value of J obtained from Equation (4.6) is a measure of the fracture toughness for high-rate loading. If small-scale yielding assumptions apply, the critical J can be converted to an equivalent K_{Ic} through Equation (3.18).

Given the difficulties associated with defining a fracture parameter in the presence of inertia forces and reflected stress waves, it is obviously preferable to apply Equation (4.6) whenever possible. For a three-point bend specimen with $W = 50$ mm, the transition time is approximately 300 μs [19]. Thus the quasistatic formula can be applied as long as fracture occurs after ~600 μs. This requirement is relatively easy to meet in impact tests on ductile materials [15, 16]. For more brittle materials, the transition-time requirement can be met by decreasing the displacement rate or the width of the specimen.

The transition-time concept can be applied to other configurations by adjusting the geometry factor in Equation (4.2). Duffy and Shih [17] have applied this approach to dynamic fracture toughness measurement in notched round bars. Small round bars have proved to be suitable for the dynamic testing of brittle materials such as ceramics, where the transition time must be small.

If the effects of inertia and reflected stress waves can be eliminated, one is left with the rate-dependent material response. The transition-time approach allows material rate effects to be quantified independent of inertia effects. High strain rates tend to elevate the flow stress of the material. The effect of flow stress on fracture toughness depends on the failure mechanism. High strain rates tend to decrease cleavage resistance, which is stress controlled. Materials whose fracture mechanisms are strain controlled often see an increase in toughness at high loading rates because more energy is required to reach a given strain value.

Figure 4.5 shows the fracture toughness data for a structural steel at three loading rates [21]. The critical K_I values were determined from quasistatic relationships. For a given loading rate, the fracture toughness increases rapidly with temperature at the onset of the ductile-brittle transition. Note that increasing the loading rate has the effect of shifting the transition to higher temperatures. Thus, at a constant temperature, fracture toughness is highly sensitive to strain rate.

The effect of the loading rate on the fracture behavior of a structural steel on the upper shelf of toughness is illustrated in Figure 4.6. In this instance, the strain rate has the opposite effect from Figure 4.5, because ductile fracture of metals is primarily strain controlled. The J integral at a given amount of crack extension is elevated by high strain rates.

4.1.2 RAPID CRACK PROPAGATION AND ARREST

When the driving force for crack extension exceeds the material resistance, the structure is unstable, and rapid crack propagation occurs. Figure 4.7 illustrates a simple case, where the (quasistatic) energy release rate increases linearly with the crack length and the material resistance is constant. Since the first law of thermodynamics must be obeyed even by an unstable system, the excess energy, denoted by the shaded area in Figure 4.7, does not simply disappear, but is converted into kinetic energy. The magnitude of the kinetic energy dictates the crack speed.

In the quasistatic case, a crack is stable if the driving force is less than or equal to the material resistance. Similarly, if the energy available for an incremental extension of a rapidly propagating crack falls below the material resistance, the crack arrests. Figure 4.8 illustrates a simplified scenario for crack arrest. Suppose that cleavage fracture initiates when $K_I = K_{Ic}$. The resistance encountered by a rapidly propagating cleavage crack is less than for cleavage initiation, because plastic deformation at the moving crack tip is suppressed by the high local strain rates. If the structure has a falling driving force curve, it eventually crosses the resistance curve. Arrest does not occur at this point, however, because the structure contains kinetic energy that can be converted to fracture

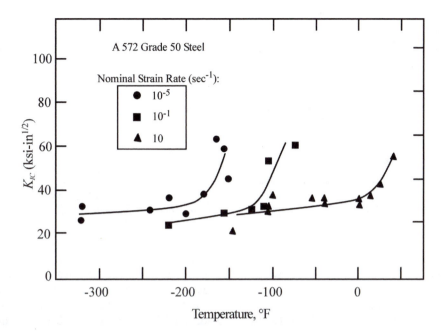

FIGURE 4.5 Effect of loading rate on the cleavage fracture toughness of a structural steel. Taken from Barsom, J.M., "Development of the AASHTO Fracture Toughness Requirements for Bridge Steels." *Engineering Fracture Mechanics,* Vol. 7, 1975, pp. 605–618.

energy. Arrest occurs below the resistance curve, after most of the available energy has been dissipated. The apparent arrest toughness K_{Ia} is less than the true material resistance K_{IA}. The difference between K_{Ia} and K_{IA} is governed by the kinetic energy created during crack propagation; K_{IA} is a material property, but K_{Ia} depends on geometry.

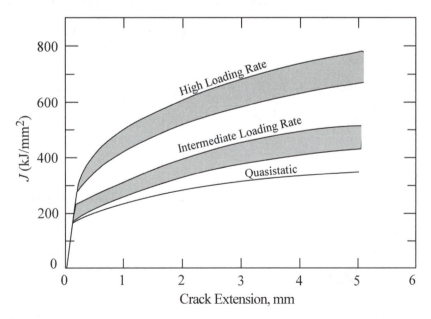

FIGURE 4.6 Effect of loading rate on the *J-R* curve behavior of HY80 steel. Taken from Joyce, J.A. and Hacket, E.M., "Dynamic *J-R* Curve Testing of a High Strength Steel Using the Multispecimen and Key Curve Techniques." ASTM STP 905, American Society for Testing and Materials, Philadelphia, PA, 1984, pp. 741–774.

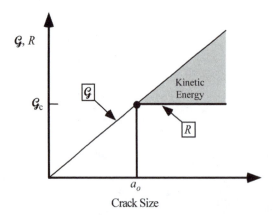

FIGURE 4.7 Unstable crack propagation, which results in the generation of kinetic energy.

Figure 4.7 and Figure 4.8 compare material resistance with *quasistatic* driving force curves. That is, these curves represent K_I and $\mathcal{G}$ values computed with the procedures described in Chapter 2. Early researchers [22–26] realized that the crack-driving force should incorporate the effect of kinetic energy. The Griffith-Irwin energy balance (Section 2.3 and Section 2.4) can be modified to include kinetic energy, resulting in a dynamic definition of the energy release rate:

$$\mathcal{G}(t) = \frac{dF}{dA} - \frac{dU}{dA} - \frac{dE_k}{dA} \tag{4.7}$$

where F is the work done by external forces and A is the crack area. Equation (4.7) is consistent with the original Griffith approach, which is based on the first law of thermodynamics. The kinetic energy must be included in a general statement of the first law; Griffith implicitly assumed that $E_k = 0$.

4.1.2.1 Crack Speed

Mott [22] applied dimensional analysis to a propagating crack in order to estimate the relationship between kinetic energy and crack speed. For a through crack of length $2a$ in an infinite plate in tension, the displacements must be proportional to the crack size, since a is the only relevant length

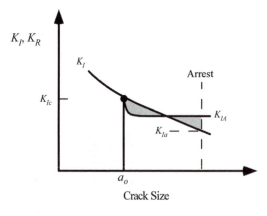

FIGURE 4.8 Unstable crack propagation and arrest with a falling driving-force curve. The apparent arrest toughness K_{Ia} is slightly below the true material resistance K_{IA} due to excess kinetic energy.

dimension. Assuming the plate is elastic, the displacements must also be proportional to the nominal applied strain; thus

$$u_x = \alpha_x a \frac{\sigma}{E} \quad \text{and} \quad u_y = \alpha_y a \frac{\sigma}{E} \tag{4.8}$$

where α_x and α_y are dimensionless constants. (Note that quantitative estimates for α_x and α_y near the crack tip in the quasistatic case can be obtained by applying the relationships in Table 2.2.) The kinetic energy is equal to half the mass times the velocity squared. Therefore, E_k for the cracked plate (assuming a unit thickness) is given by

$$E_k = \frac{1}{2}\rho a^2 V^2 \left(\frac{\sigma}{E}\right)^2 \iint (\alpha_x^2 + \alpha_y^2)\, dxdy \tag{4.9}$$

where ρ is the mass density of the material and $V(=\dot{a})$ is the crack speed. Assuming the integrand depends only on position,[1] E_k can be written in the following form:

$$E_k = \frac{1}{2}k\rho a^2 V^2 \left(\frac{\sigma}{E}\right)^2 \tag{4.10}$$

where k is a constant. Applying the modified Griffith energy balance (Equation (4.7)) gives

$$\mathcal{G}(t) = \frac{1}{2}\frac{d}{da}\left[\frac{\pi\sigma^2 a^2}{E} - \frac{k}{2}\rho a^2 V^2 \left(\frac{\sigma}{E}\right)^2\right] = 2w_f \tag{4.11}$$

where w_f is the work of fracture, defined in Chapter 2; in the limit of an ideally brittle material, $w_f = \gamma_s$, the surface energy. Note that Equation (4.11) assumes a flat R curve (constant w_f). At initiation, the kinetic energy term is not present, and the initial crack length a_o can be inferred from Equation (2.22):

$$a_o = \frac{2Ew_f}{\pi\sigma^2} \tag{4.12}$$

Substituting Equation (4.12) into Equation (4.11) and solving for V leads to

$$V = \sqrt{\frac{2\pi}{k}}\,c_o\left(1 - \frac{a_o}{a}\right) \tag{4.13}$$

where $c_o = \sqrt{E/\rho}$, the speed of sound for one-dimensional wave propagation. Mott [22] actually obtained a somewhat different relationship from Equation (4.13), because he solved Equation (4.11) by making the erroneous assumption that $dV/da = 0$. Dulaney and Brace [27] and Berry [28] later corrected the Mott analysis and derived Equation (4.13).

Roberts and Wells [29] obtained an estimate for k by applying the Westergaard stress function (Appendix 2.3) for this configuration. After making a few assumptions, they showed that $\sqrt{2\pi/k} \approx 0.38$.

---

[1] In a rigorous dynamic analysis, α_x and α_y and thus k depend on the crack speed.

According to Equation (4.13) and the Roberts and Wells analysis, the crack speed reaches a limiting value of $0.38c_o$ when $a \gg a_o$. This estimate compares favorably with measured crack speeds in metals, which typically range from 0.2 to $0.4c_o$ [30].

Freund [2–4] performed a more detailed numerical analysis of a dynamically propagating crack in an infinite body and obtained the following relationship:

$$V = c_r\left(1 - \frac{a_o}{a}\right) \tag{4.14}$$

where c_r is the Raleigh (surface) wave speed. For Poisson's ratio = 0.3, the c_r/c_o ratio = 0.57. Thus the Freund analysis predicts a larger limiting crack speed than the Roberts and Wells analysis. The limiting crack speed in Equation (4.14) can be argued on physical grounds [26]. For the special case where $w_f = 0$, a propagating crack is merely a disturbance on a free surface that must move at the Raleigh wave velocity. In both Equation (4.13) and Equation (4.14), the limiting velocity is independent of the fracture energy; thus the maximum crack speed should be c_r for all w_f.

Experimentally observed crack speeds do not usually reach c_r. Both the simple analysis that resulted in Equation (4.13) and Freund's more detailed dynamic analysis assumed that the fracture energy does not depend on crack length or crack speed. The material resistance actually increases with crack speed, as discussed below. The good agreement between experimental crack velocities and the Roberts and Wells estimate of $0.38c_o$ is largely coincidental.

4.1.2.2 Elastodynamic Crack-Tip Parameters

The governing equation for Mode I crack propagation under elastodynamic conditions can be written as

$$K_I(t) = K_{ID}(V) \tag{4.15}$$

where K_I is the instantaneous stress intensity and K_{ID} is the material resistance to crack propagation, which depends on the crack velocity. In general, $K_I(t)$ is not equal to the static stress-intensity factor, as defined in Chapter 2. A number of researchers [8–10, 31–33] have obtained a relationship for the dynamic stress intensity of the form

$$K_I(t) = k(V)K_I(0) \tag{4.16}$$

where k is a universal function of crack speed and $K_I(0)$ is the static stress-intensity factor. The function $k(V) = 1.0$ when $V = 0$, and decreases to zero as V approaches the Raleigh wave velocity. An approximate expression for k was obtained by Rose [34]:

$$k(V) \approx \left(1 - \frac{V}{c_r}\right)\sqrt{1 - hV} \tag{4.17}$$

where h is a function of the elastic wave speeds and can be approximated by

$$h \approx \frac{2}{c_1}\left(\frac{c_2}{c_r}\right)^2\left[1 - \left(\frac{c_2}{c_1}\right)\right]^2 \tag{4.18}$$

where c_1 and c_2 are the longitudinal and shear wave speeds, respectively.

Equation (4.16) is valid only at short times or in infinite bodies. This relationship neglects reflected stress waves, which can have a significant effect on the local crack-tip fields. Since the crack speed is proportional to the wave speed, Equation (4.16) is valid as long as the length of crack propagation $(a - a_o)$ is small compared to the specimen dimensions, because reflecting stress waves will not have had time to reach the crack tip (Example 4.1). In finite specimens where stress waves reflect back to the propagating crack tip, the dynamic stress intensity must be determined experimentally or numerically on a case-by-case basis.

EXAMPLE 4.1

Rapid crack propagation initiates in a deeply notched specimen with an initial ligament b_o (Figure 4.9). Assuming the average crack speed = 0.2 c_1, estimate how far the crack will propagate before it encounters a reflected longitudinal wave.

Solution: At the moment the crack encounters the first reflected wave, the crack has traveled a distance Δa, while the wave has traveled $2b_o - \Delta a$. Equating the travel times gives

$$= \frac{\Delta a}{0.2c_1} = \frac{2b_o - \Delta a}{c_1}$$

thus,

$$\Delta a = \frac{b_o}{3}$$

Equation (4.16) is valid in this case as long as the crack extension is less than $b_o/3$ and the plastic zone is small compared to b_o.

For an infinite body or short times, Freund [10] showed that the dynamic energy release rate could be expressed in the following form:

$$\mathcal{G}(t) = g(V)\mathcal{G}(0) \tag{4.19}$$

where g is a universal function of crack speed that can be approximated by

$$g(V) \approx 1 - \frac{V}{c_r} \tag{4.20}$$

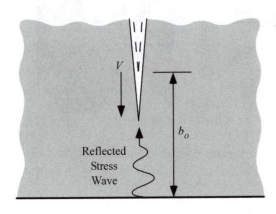

FIGURE 4.9 Propagating crack encountering a reflected stress wave.

Combining Equation (4.16)–(4.20) with Equation (2.51) gives

$$\mathcal{G}(t) = A(V)\frac{K_I^2(t)}{E'} \tag{4.21}$$

where

$$A(V) \approx \left[\left(1 - \frac{V}{c_r}\right)(1 - hV)\right]^{-1} \tag{4.22}$$

Thus the relationship between K_I and $\mathcal{G}$ depends on crack speed. A more accurate (and more complicated) relationship for $A(V)$ is given in Appendix 4.1.

When the plastic zone ahead of the propagating crack is small, $K_I(t)$ uniquely defines the crack-tip stress, strain, and displacement fields, but the angular dependence of these quantities is different from the quasistatic case. For example, the stresses in the elastic singularity zone are given by [32, 33, 35]

$$\sigma_{ij} = \frac{K_I(t)}{\sqrt{2\pi r}} f_{ij}(\theta, V) \tag{4.23}$$

The function f_{ij} reduces to the quasistatic case (Table 2.1) when $V = 0$. Appendix 4.1 outlines the derivation of Equation (4.23) and gives specific relationships for f_{ij} in the case of rapid crack propagation. The displacement functions also display an angular dependence that varies with V. Consequently, α_x and α_y in Equation (4.9) must depend on crack velocity as well as position, and the Mott analysis is not rigorously correct for dynamic crack propagation.

4.1.2.3 Dynamic Toughness

As Equation (4.15) indicates, the dynamic stress intensity is equal to K_{ID}, the dynamic material resistance, which depends on crack speed. This equality permits experimental measurements of K_{ID}.

Dynamic propagation toughness can be measured as a function of crack speed by means of high-speed photography and optical methods, such as photoelasticity [36, 37] and the method of caustics [38]. Figure 4.10 shows photoelastic fringe patterns for dynamic crack propagation in Homalite 100 [37]. Each fringe corresponds to a contour of maximum shear stress. Sanford and Dally [36] describe procedures for inferring stress intensity from photoelastic patterns.

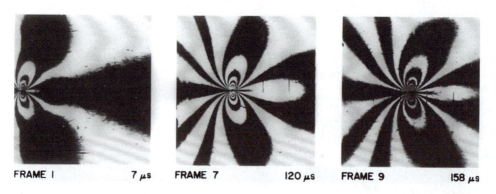

FIGURE 4.10 Photoelastic fringe patterns for a rapidly propagating crack in Homalite 100. Photograph provided by R. Chona. Taken from Chona, R., Irwin, G.R., and Shukla, A., "Two and Three Parameter Representation of Crack Tip Stress Fields." *Journal of Strain Analysis,* Vol. 17, 1982, pp. 79–86.

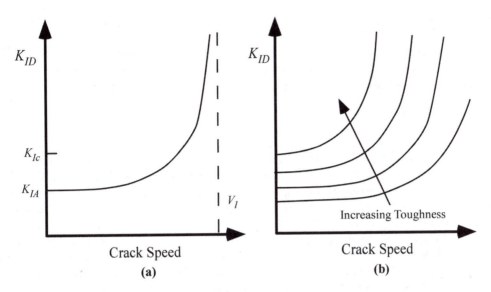

FIGURE 4.11 Schematic K_{ID}-crack speed curves: (a) effect of crack speed on K_{ID} and (b) effect of material toughness.

Figure 4.11 illustrates the typical variation of K_{ID} with crack speed. At low speeds, K_{ID} is relatively insensitive to V, but K_{ID} increases asymptotically as V approaches a limiting value. Figure 4.12 shows K_{ID} data for 4340 steel published by Rosakis and Freund [39].

In the limit of $V = 0$, $K_{ID} = K_{IA}$, the arrest toughness of the material. In general, $K_{IA} < K_{Ic}$, the quasistatic initiation toughness. When a stationary crack in an elastic-plastic material is loaded monotonically, the crack-tip blunts and a plastic zone forms. A propagating crack, however, tends to be sharper and has a smaller plastic zone than a stationary crack. Consequently, more energy is required to initiate fracture from a stationary crack than is required to maintain the propagation of a sharp crack.

The crack-speed dependence of K_{ID} can be represented by an empirical equation of the form

$$K_{ID} = \frac{K_{IA}}{1 - \left(\dfrac{V}{V_I}\right)^m} \tag{4.24}$$

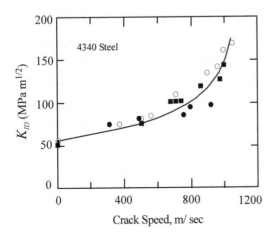

FIGURE 4.12 Experimental K_{ID} vs. crack speed data for 4340 steel. Taken from Rosakis, A.J. and Freund, L.B., "Optical Measurement of the Plane Strain Concentration at a Crack Tip in a Ductile Steel Plate." *Journal of Engineering Materials Technology,* Vol. 104, 1982, pp. 115–120.

where V_l is the limiting crack speed in the material and m is an experimentally determined constant. As Figure 4.11(b) illustrates, K_{IA} increases and V_l decreases with increasing material toughness. The trends in Figure 4.11(a) and Figure 4.11(b) have not only been observed experimentally, but have also been obtained by numerical simulation [40, 41]. The upturn in propagation toughness at high crack speeds is apparently caused by local inertia forces in the plastic zone.

4.1.2.4 Crack Arrest

Equation (4.15) defines the conditions for rapid crack advance. If, however, $K_I(t)$ falls below the minimum K_{ID} value for a finite length of time, propagation cannot continue, and the crack arrests. There are a number of situations that might lead to crack arrest. Figure 4.8 illustrates one possibility: If the driving force decreases with crack extension, it may eventually be less than the material resistance. Arrest is also possible when the material resistance increases with crack extension. For example, a crack that initiates in a brittle region of a structure, such as a weld, may arrest when it reaches a material with higher toughness. A temperature gradient in a material that exhibits a ductile-brittle transition is another case where the toughness can increase with position: A crack may initiate in a cold region of the structure and arrest when it encounters warmer material with a higher toughness. An example of this latter scenario is a pressurized thermal shock event in a nuclear pressure vessel [42].

In many instances, it is not possible to guarantee with absolute certainty that an unstable fracture will not initiate in a structure. Transient loads, for example, may occur unexpectedly. In such instances crack arrest can be the second line of defense. Thus, the crack arrest toughness K_{IA} is an important material property.

Based on Equation (4.16), one can argue that $K_I(t)$ at arrest is equivalent to the quasistatic value, since $V = 0$. Thus it should be possible to infer K_{IA} from a quasistatic calculation based on the load and crack length at arrest. This quasistatic approach to arrest is actually quite common, and is acceptable in many practical situations. Chapter 7 describes a standardized test method for measuring crack-arrest toughness that is based on quasistatic assumptions.

However, the quasistatic arrest approach must be used with caution. Recall that Equation (4.16) is valid only for infinite structures or short crack jumps, where reflected stress waves do not have sufficient time to return to the crack tip. When reflected stress wave effects are significant, Equation (4.16) is no longer valid, and a quasistatic analysis tends to give misleading estimates of the arrest toughness. Quasistatic estimates of arrest toughness are sometimes given the designation K_{Ia}; for short crack jumps $K_{Ia} = K_{IA}$.

The effect of stress waves on the apparent arrest toughness K_{Ia} was demonstrated dramatically by Kalthoff et al. [43], who performed dynamic propagation and arrest experiments on wedge-loaded double cantilever beam (DCB) specimens. Recall from Example 2.3 that the DCB specimen exhibits a falling driving-force curve in displacement control. Kalthoff et al. varied the K_I at initiation by varying the notch-root radius. When the crack was sharp, fracture initiated slightly above K_{IA} and arrested after a short crack jump; the length of crack jump increased with the notch-tip radius.

Figure 4.13 is a plot of the Kalthoff et al. results. For the shortest crack jump, the true arrest toughness and the apparent quasistatic value coincide, as expected. As the length of crack jump increases, the discrepancy between the true arrest and the quasistatic estimate increases, with $K_{IA} > K_{Ia}$. Note that K_{IA} appears to be a material constant but K_{Ia} varies with the length of crack propagation. Also note that the dynamic stress intensity during crack growth is considerably different from the quasistatic estimate of K_I. Kobayashi et al. [44] obtained similar results.

A short time after arrest, the applied stress intensity reaches K_{Ia}, the quasistatic value. Figure 4.14 shows the variation of K_I after arrest in one of the Kalthoff et al. experiments. When the crack arrests, $K_I = K_{IA}$, which is greater than K_{Ia}. Figure 4.14 shows that the specimen "rings down" to K_{Ia} after ~2000 μs. The quasistatic value, however, is not indicative of the true material-arrest properties.

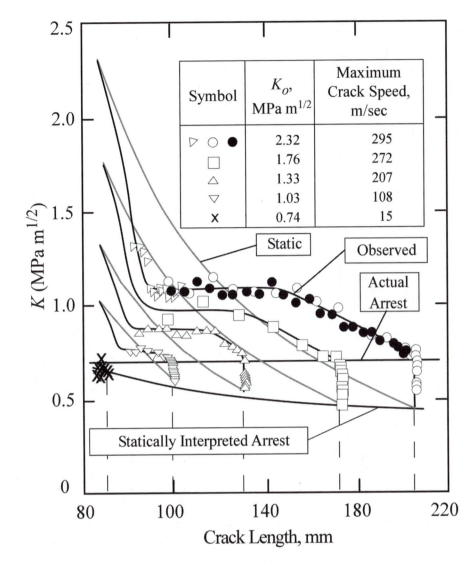

FIGURE 4.13 Crack arrest experiments on wedge-loaded DCB Araldite B specimens. The statically interpreted arrest toughness underestimates the true K_{IA} of the material; this effect is most pronounced for long crack jumps. Taken from Kalthoff, J.F., Beinart, J., and Winkler, S., "Measurement of Dynamic Stress Intensity Factors for Fast Running and Arresting Cracks in Double-Cantilever Beam Specimens." ASTM STP 627, American Society for Testing and Materials, Philadelphia, PA, 1977, pp. 161–176.

Recall the schematic in Figure 4.8, where it was argued that arrest, when quantified by the quasistatic stress intensity, would occur below the true arrest toughness K_{IA}, because of the kinetic energy in the specimen. This argument is a slight oversimplification, but it leads to the correct qualitative conclusion.

The DCB specimen provides an extreme example of reflected stress wave effects; the specimen design is such that stress waves can traverse the width of the specimen and return to the crack tip in a very short time. In many structures, the quasistatic approach is approximately valid, even for relatively long crack jumps. In any case, K_{Ia} gives a lower bound estimate of K_{IA}, and thus is conservative in most instances.

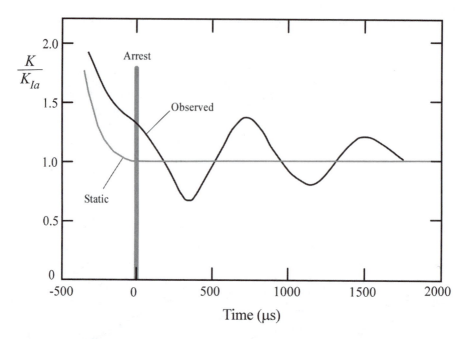

FIGURE 4.14 Comparison of dynamic measurements of stress intensity with static calculations for a wedge loaded DCB Araldite B specimen. Taken from Kalthoff, J.F., Beinart, J., and Winkler, S., "Measurement of Dynamic Stress Intensity Factors for Fast Running and Arresting Cracks in Double-Cantilever Beam Specimens." ASTM STP 627, American Society for Testing and Materials, Philadelphia, PA, 1977, pp. 161–176.

4.1.3 DYNAMIC CONTOUR INTEGRALS

The original formulation of the J contour integral is equivalent to the nonlinear elastic energy release rate for quasistatic deformation. By invoking a more general definition of energy release rate, it is possible to incorporate dynamic effects and time-dependent material behavior into the J integral.

The energy release rate is usually defined as the energy released from the body per unit crack advance. A more precise definition [11] involves the work input into the crack tip. Consider a vanishingly small contour Γ around the tip of a crack in a two-dimensional solid (Figure 4.15). The energy release rate is equal to the energy flux into the crack tip, divided by the crack speed:

$$J = \frac{\mathcal{F}}{V} \qquad (4.25)$$

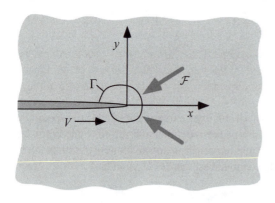

FIGURE 4.15 Energy flux into a small contour at the tip of a propagating crack.

where $\mathcal{F}$ is the energy flux into the area bounded by Γ. The generalized energy release rate, including inertia effects, is given by

$$J = \lim_{\Gamma \to 0} \int_{\Gamma} \left[(w + T)dy - \sigma_{ij} n_j \frac{\partial u_i}{\partial x} ds \right]$$ (4.26)

where w and T are the stress work and kinetic energy densities defined as

$$w = \int_0^{\varepsilon_{ij}} \sigma_{ij} d\varepsilon_{ij}$$ (4.27)

and

$$T = \frac{1}{2} \rho \frac{\partial u_i}{\partial t} \frac{\partial u_i}{\partial t}$$ (4.28)

Equation (4.26) has been published in a variety of forms by several researchers [8–12]. Appendix 4.2 gives a derivation of this relationship.

Equation (4.26) is valid for time-dependent as well as history-dependent material behavior. When evaluating J for a time-dependent material, it may be convenient to express w in the following form:

$$w = \int_{t_o}^{t} \sigma_{ij} \dot{\varepsilon}_{ij} dt$$ (4.29)

where $\dot{\varepsilon}_{ij}$ is the strain rate.

Unlike the conventional J integral, the contour in Equation (4.26) cannot be chosen arbitrarily. Consider, for example, a dynamically loaded cracked body with stress waves reflecting off free surfaces. If the integral in Equation (4.26) were computed at two arbitrary contours a finite distance from the crack tip and a stress wave passed through one contour but not the other, the values of these integrals would normally be different for the two contours. Thus, the generalized J integral is not path independent, except in the immediate vicinity of the crack tip. If, however, $T = 0$ at all points in the body, the integrand in Equation (4.26) reduces to the form of the original J integral. In the latter case, the path-independent property of J is restored if w displays the property of an elastic potential (see Appendix 4.2).

The form of Equation (4.26) is not very convenient for numerical calculations, since it is extremely difficult to obtain adequate numerical precision from a contour integration very close to the crack tip. Fortunately, Equation (4.26) can be expressed in a variety of other forms that are more conducive to numerical analysis. The energy release rate can also be generalized to three dimensions. The results in Figure 4.3 and Figure 4.4 are obtained from a finite element analysis that utilized alternate forms of Equation (4.26). Chapter 12 discusses the numerical calculations of J for both quasistatic and dynamic loading.

4.2 CREEP CRACK GROWTH

Components that operate at high temperatures relative to the melting point of the material may fail by the slow and stable extension of a macroscopic crack. Traditional approaches to design in the creep regime apply only when creep and material damage are uniformly distributed. Time-dependent fracture mechanics approaches are required when creep failure is controlled by a dominant crack in the structure.

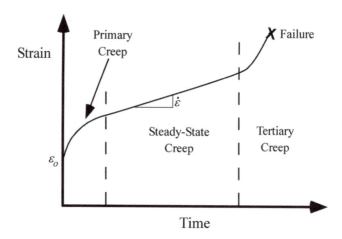

FIGURE 4.16 Schematic creep behavior of a material subject to a constant stress.

Figure 4.16 illustrates the typical creep response of a material subject to constant stress. Deformation at high temperatures can be divided into four regimes: instantaneous (elastic) strain, primary creep, secondary (steady state) creep, and tertiary creep. The elastic strain occurs immediately upon application of the load. As discussed in the previous section on dynamic fracture, the elastic stress-strain response of a material is not instantaneous (i.e., it is limited by the speed of sound in the material), but it can be viewed as such in creep problems, where the time scale is usually measured in hours. Primary creep dominates at short times after the application of the load; the strain rate decreases with time, as the material strain hardens. In the secondary creep stage, the deformation reaches a steady state, where strain hardening and strain softening are balanced; the creep rate is constant in the secondary stage. In the tertiary stage, the creep rate accelerates, as the material approaches ultimate failure. Microscopic failure mechanisms, such as grain boundary cavitation, nucleate in this final stage of creep.

During the growth of a macroscopic crack at high temperatures, all four types of creep response can occur simultaneously in the most general case (Figure 4.17). The material at the tip of a growing crack is in the tertiary stage of creep, since the material is obviously failing locally. The material may be elastic remote from the crack tip, and in the primary and secondary stages of creep at moderate distances from the tip.

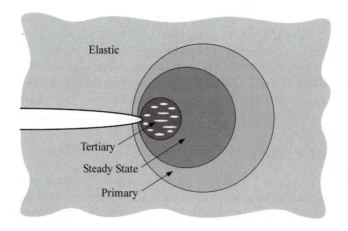

FIGURE 4.17 Creep zones at the tip of a crack.

Most analytical treatments of creep crack growth assume limiting cases, where one or more of these regimes are not present or are confined to a small portion of the component. If, for example, the component is predominantly elastic, and the creep zone is confined to a small region near the crack tip, the crack growth can be characterized by the stress-intensity factor. In the other extreme, when the component deforms globally in steady-state creep, elastic strains and tertiary creep can be disregarded. A parameter that applies to the latter case is described below, followed by a brief discussion of approaches that consider the transition from elastic to steady-state creep behavior.

4.2.1 THE C^* INTEGRAL

A formal fracture mechanics approach to creep crack growth was developed soon after the J integral was established as an elastic-plastic fracture parameter. Landes and Begley [45], Ohji et al. [46], and Nikbin et al. [47] independently proposed what became known as the C^* integral to characterize crack growth in a material undergoing steady-state creep. They applied Hoff's analogy [48], which states that if there exists a nonlinear elastic body that obeys the relationship $\varepsilon_{ij} = f(\sigma_{ij})$ and a viscous body that is characterized by $\dot{\varepsilon}_{ij} = f(\sigma_{ij})$, where the function of stress is the same for both, then both bodies develop identical stress distributions when the same load is applied. Hoff's analogy can be applied to steady-state creep, since the creep rate is a function only of the applied stress.

The C^* integral is defined by replacing strains with strain rates, and displacements with displacement rates in the J contour integral:

$$C^* = \int_{\Gamma} \left(\dot{w}dy - \sigma_{ij}n_j \frac{\partial \dot{u}_i}{\partial x} ds \right) \tag{4.30}$$

where $\dot{w}$ is the stress work rate (power) density, defined as

$$\dot{w} = \int_0^{\dot{\varepsilon}_{kl}} \sigma_{ij}d\dot{\varepsilon}_{ij} \tag{4.31}$$

Hoff's analogy implies that the C^* integral is path independent, because J is path independent. Also, if secondary creep follows a power law:

$$\dot{\varepsilon}_{ij} = A\sigma_{ij}^n \tag{4.32}$$

where A and n are material constants, then it is possible to define an HRR-type singularity for stresses and strain rates near the crack tip:

$$\sigma_{ij} = \left(\frac{C^*}{AI_n r} \right)^{\frac{1}{n+1}} \tilde{\sigma}_{ij}(n,\theta) \tag{4.33a}$$

and

$$\dot{\varepsilon}_{ij} = \left(\frac{C^*}{AI_n r} \right)^{\frac{n}{n+1}} \tilde{\varepsilon}_{ij}(n,\theta) \tag{4.33b}$$

where the constants I_n, $\tilde{\sigma}_{ij}$, and $\tilde{\varepsilon}_{ij}$ are identical to the corresponding parameters in the HRR relationship (Equation (3.24)). Note that in the present case, n is a creep exponent rather than a strain-hardening exponent.

Just as the J integral characterizes the crack-tip fields in an elastic or elastic-plastic material, the C^* integral uniquely defines crack-tip conditions in a viscous material. Thus the time-dependent crack growth rate in a viscous material should depend only on the value of C^*. Experimental studies [45–49] have shown that creep crack growth rates correlate very well with C^*, provided steady-state creep is the dominant deformation mechanism in the specimen. Figure 4.18 shows typical creep crack growth data. Note that the crack growth rate follows a power law:

$$\dot{a} = \gamma (C^*)^m \tag{4.34}$$

where γ and m are material constants. In many materials, $m \approx n/(n + 1)$, a result that is predicted by grain boundary cavitation models [49].

Experimental measurements of C^* take advantage of analogies with the J integral. Recall that J is usually measured by invoking the energy release rate definition:

$$J = -\frac{1}{B}\left(\frac{\partial}{\partial a}\int_0^\Delta P d\Delta \right)_\Delta \tag{4.35}$$

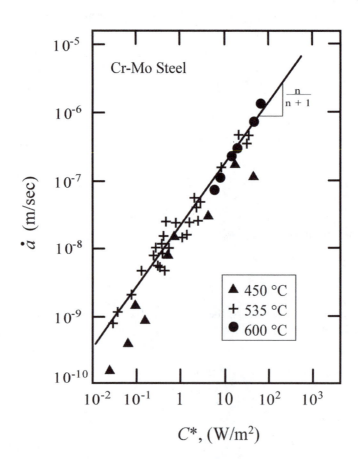

FIGURE 4.18 Creep crack growth data in a Cr-Mo Steel at three temperatures. Taken from Riedel, H., "Creep Crack Growth." ASTM STP 1020, American Society for Testing and Materials, Philadelphia, PA, 1989, pp. 101–126.

where P is the applied load and Δ is the load-line displacement. Similarly, C^* can be defined in terms of a power release rate:

$$C^* = -\frac{1}{B}\left(\frac{\partial}{\partial a}\int_0^\Delta Pd\Delta\right)_{\dot{\Delta}} \qquad (4.36)$$

The J integral can be related to the energy absorbed by a laboratory specimen, divided by the ligament area:[2]

$$J = \frac{\eta}{Bb}\int_0^\Delta Pd\Delta \qquad (4.37)$$

where η is a dimensionless constant that depends on geometry. Therefore, C^* is given by

$$C^* = \frac{\eta}{Bb}\int_0^{\dot{\Delta}} Pd\dot{\Delta} \qquad (4.38)$$

For a material that creeps according to a power law (Equation (4.32)), the displacement rate is proportional to P^n, assuming global creep in the specimen. In this case, Equation (4.38) reduces to

$$C^* = \frac{n}{n+1}\frac{\eta}{Bb}P\dot{\Delta} \qquad (4.39)$$

The geometry factor η has been determined for a variety of test specimens. For example, $\eta = 2.0$ for a deeply notched bend specimen (Equation (3.37) and Equation (4.6)).

4.2.2 SHORT-TIME VS. LONG-TIME BEHAVIOR

The C^* parameter applies only to crack growth in the presence of global steady-state creep. Stated another way, C^* applies to long-time behavior, as discussed below.

Consider a stationary crack in a material that is susceptible to creep deformation. If a remote load is applied to the cracked body, the material responds almost immediately with the corresponding elastic strain distribution. Assuming the loading is pure Mode I, the stresses and strains exhibit a $1/\sqrt{r}$ singularity near the crack tip and are uniquely defined by K_I. However, large-scale creep deformation does not occur immediately. Soon after the load is applied, a small creep zone, analogous to a plastic zone, forms at the crack tip. The crack-tip conditions can be characterized by K_I as long as the creep zone is embedded within the singularity dominated zone. The creep zone grows with time, eventually invalidating K_I as a crack-tip parameter. At long times, the creep zone spreads throughout the entire structure.

When the crack grows with time, the behavior of the structure depends on the crack growth rate relative to the creep rate. In brittle materials, the crack growth rate is so fast that it overtakes the creep zone; crack growth can be characterized by K_I because the creep zone at the tip of the

---

[2] The load-line displacement Δ in Equation (4.37)–(4.39) corresponds to the portion of the displacement due to the presence of the crack, as discussed in Section 3.2.5. This distinction is not necessary in Equation (4.35) and Equation (4.36), because the displacement component attributed to the uncracked configuration vanishes when differentiated with respect to a.

growing crack remains small. At the other extreme, if the crack growth is sufficiently slow that the creep zone spreads throughout the structure, C^* is the appropriate characterizing parameter.

Riedel and Rice [50] analyzed the transition from short-time elastic behavior to long-time viscous behavior. They assumed a simplified stress-strain rate law that neglects primary creep:

$$\dot{\varepsilon} = \frac{\dot{\sigma}}{E} + A\sigma^n \qquad (4.40)$$

for uniaxial tension. If a load is suddenly applied and then held constant, a creep zone gradually develops in an elastic singularity zone, as discussed earlier. Riedel and Rice argued that the stresses well within the creep zone can be described by

$$\sigma_{ij} = \left(\frac{C(t)}{AI_n r}\right)^{\frac{1}{n+1}} \tilde{\sigma}_{ij}(n, \theta) \qquad (4.41)$$

where $C(t)$ is a parameter that characterizes the amplitude of the local stress singularity in the creep zone; $C(t)$ varies with time and is equal to C^* in the limit of long-time behavior. If the remote load is fixed, the stresses in the creep zone relax with time, as creep strain accumulates in the crack-tip region. For small-scale creep conditions, $C(t)$ decays as $1/t$ according to the following relationship:

$$C(t) = \frac{K_I^2(1-v^2)}{(n+1)Et} \qquad (4.42)$$

The approximate size of the creep zone is given by

$$r_c(\theta, t) = \frac{1}{2\pi} \left(\frac{K_I}{E}\right)^2 \left[\frac{(n+1)AI_n E^n t}{2\pi(1-v^2)}\right]^{\frac{2}{n-1}} \tilde{r}_c(\theta, n) \qquad (4.43)$$

At $\theta = 90°$, $\tilde{r}_c$ is a maximum and ranges from 0.2 to 0.5, depending on n. As r_c increases in size, $C(t)$ approaches the steady-state value C^*. Riedel and Rice defined a characteristic time for the transition from short-time to long-time behavior:

$$t_1 = \frac{K_I^2(1-v^2)}{(n+1)EC^*} \qquad (4.44a)$$

or

$$t_1 = \frac{J}{(n+1)C^*} \qquad (4.44b)$$

When significant crack growth occurs over time scales much less than t_1, the behavior can be characterized by K_I, while C^* is the appropriate parameter when significant crack growth requires times $\gg t_1$. Based on a finite element analysis, Ehlers and Riedel [51] suggested the following simple formula to interpolate between small-scale creep and extensive creep (short- and long-time behavior, respectively):

$$C(t) \approx C^* \left(\frac{t_1}{t} + 1\right) \qquad (4.45)$$

Note the similarity to the transition time concept in dynamic fracture (Section 4.1.1). In both instances, a transition time characterizes the interaction between two competing phenomena.

4.2.2.1 The C_t Parameter

Unlike K_I and C^*, a direct experimental measurement of $C(t)$ under transient conditions is usually not possible. Consequently, Saxena [52] defined an alternate parameter C_t which was originally intended as an approximation of $C(t)$. The advantage of C_t is that it can be measured relatively easily.

Saxena began by separating global displacement into instantaneous elastic and time-dependent creep components:

$$\Delta = \Delta_e + \Delta_t \tag{4.46}$$

The creep displacement Δ_t increases with time as the creep zone grows. Also, if load is fixed, $\dot{\Delta}_t = \dot{\Delta}$. The C_t parameter is defined as the creep component of the power release rate:

$$C_t = -\frac{1}{B}\left(\frac{\partial}{\partial a}\int_0^{\dot{\Delta}_t} P\, d\dot{\Delta}_t\right)_{\dot{\Delta}_t} \tag{4.47}$$

Note the similarity between Equation (4.36) and Equation (4.47).

For small-scale creep conditions, Saxena defined an effective crack length, analogous to the Irwin plastic zone correction described in Chapter 2:

$$a_{eff} = a + \beta r_c \tag{4.48}$$

where $\beta \approx \frac{1}{3}$ and r_c is defined at $\theta = 90°$. The displacement due to the creep zone is given by

$$\Delta_t = \Delta - \Delta_e = P\frac{dC}{da}\beta r_c \tag{4.49}$$

where C is the elastic compliance, defined in Chapter 2. Saxena showed that the small-scale creep limit for C_t can be expressed as follows:

$$(C_t)_{ssc} = \left(\frac{f'\left(\frac{a}{W}\right)}{f\left(\frac{a}{W}\right)}\right)\frac{P\dot{\Delta}_t}{BW} \tag{4.50}$$

where $f(a/W)$ is the geometry correction factor for Mode I stress intensity (see Table 2.4):

$$f\left(\frac{a}{W}\right) = \frac{K_I B\sqrt{W}}{P}$$

and f' is the first derivative of f. Equation (4.50) predicts that $(C_t)_{ssc}$ is proportional to K_I^4; thus C_t does not coincide with $C(t)$ in the limit of small-scale creep (Equation (4.42)).

Saxena proposed the following interpolation between small-scale creep and extensive creep:

$$C_t = (C_t)_{ssc}\left(1 - \frac{\dot{\Delta}}{\dot{\Delta}_t}\right) + C^* \tag{4.51}$$

where C^* is determined from Equation (4.38) using the *total* displacement rate. In the limit of long-time behavior, $C^*/C_t = 1.0$, but this ratio is less than unity for small-scale creep and transient behavior.

Bassani et al. [53] applied the C_t parameter to experimental data with various C^*/C_t ratios and found that C_t characterized crack growth rates much better than C^* or K_I. They state that C_t, when defined by Equation (4.50) and Equation (4.51), characterizes experimental data better than $C(t)$, as defined by Riedel's approximation (Equation (4.45)).

Although C_t was originally intended as an approximation of $C(t)$, it has become clear that these two parameters are distinct from one another. The $C(t)$ parameter characterizes the stresses ahead of a stationary crack, while C_t is related to the rate of expansion of the creep zone. The latter quantity appears to be better suited to materials that experience relatively rapid creep crack growth. Both parameters approach C^* in the limit of steady-state creep.

4.2.2.2 Primary Creep

The analyses introduced so far do not consider primary creep. Referring to Figure 4.17, which depicts the most general case, the outer ring of the creep zone is in the primary stage of creep. Primary creep may have an appreciable effect on the crack growth behavior if the size of the primary zone is significant.

Recently, researchers have begun to develop crack growth analyses that include the effects of primary creep. One such approach [54] considers a strain-hardening model for the primary creep deformation, resulting in the following expression for total strain rate:

$$\dot{\varepsilon} = \frac{\sigma}{E} + A_1 \sigma^n + A_2 \sigma^{m(1+p)} \varepsilon^{-p} \tag{4.51}$$

Riedel [54] introduced a new parameter C_h^* which is the primary creep analog to C^*. The characteristic time that defines the transition from primary to secondary creep is defined as

$$t_2 = \left(\frac{C_h^*}{(1+p)C^*} \right)^{\frac{p+1}{p}} \tag{4.52}$$

The stresses within the steady-state creep zone are still defined by Equation (4.41), but the interpolation scheme for $C(t)$ is modified when primary creep strains are present [54]:

$$C(t) \approx \left[\frac{t_1}{t} + \left(\frac{t_2}{t} \right)^{\frac{p+1}{p}} + 1 \right] C^* \tag{4.53}$$

Equation (4.53) has been applied to experimental data in a limited number of cases. This relationship appears to give a better description of experimental data than Equation (4.45), where the primary term is omitted.

Chun-Pok and McDowell [55] have incorporated the effects of primary creep into the estimation of the C_t parameter.

4.3 VISCOELASTIC FRACTURE MECHANICS

Polymeric materials have seen increasing service in structural applications in recent years. Consequently, the fracture resistance of these materials has become an important consideration. Much of the fracture mechanics methodology that was developed for metals is not directly transferable to polymers, however, because the latter behave in a viscoelastic manner.

Theoretical fracture mechanics analyses that incorporate viscoelastic material response are relatively new, and practical applications of viscoelastic fracture mechanics are rare, as of this writing. Most current applications to polymers utilize conventional, time-independent fracture mechanics methodology (see Chapter 6 and Chapter 8). Approaches that incorporate time dependence should become more widespread, however, as the methodology is developed further and is validated experimentally.

This section introduces viscoelastic fracture mechanics and outlines a number of recent advances in this area. The work of Schapery [56–61] is emphasized, because he has formulated the most complete theoretical framework, and his approach is related to the J and C^* integrals, which were introduced earlier in this text.

4.3.1 LINEAR VISCOELASTICITY

Viscoelasticity is perhaps the most general (and complex) type of time-dependent material response. From a continuum mechanics viewpoint, viscoplastic creep in metals is actually a special case of viscoelastic material behavior. While creep in metals is generally considered permanent deformation, the strains can recover with time in viscoelastic materials. In the case of polymers, time-dependent deformation and recovery is a direct result of their molecular structure, as discussed in Chapter 6.

Let us introduce the subject by considering linear viscoelastic material behavior. In this case, *linear* implies that the material meets two conditions: superposition and proportionality. The first condition requires that stresses and strains at time t be additive. For example, consider two uniaxial strains ε_1 and ε_2, at time t, and the corresponding stresses $\sigma(\varepsilon_1)$ and $\sigma(\varepsilon_2)$. Superposition implies

$$\sigma[\varepsilon_1(t)] + \sigma[\varepsilon_2(t)] = \sigma[\varepsilon_1(t) + \varepsilon_2(t)] \qquad (4.54)$$

If each stress is multiplied by a constant, the proportionality condition gives

$$\lambda_1 \sigma[\varepsilon_1(t)] + \lambda_2 \sigma[\varepsilon_2(t)] = \sigma[\lambda_1 \varepsilon_1(t) + \lambda_2 \varepsilon_2(t)] \qquad (4.55)$$

If a uniaxial constant stress creep test is performed on a linear viscoelastic material, such that $\sigma = 0$ for $t < 0$ and $\sigma = \sigma_o$ for $t > 0$, the strain increases with time according to

$$\varepsilon(t) = D(t)\sigma_o \qquad (4.56)$$

where $D(t)$ is the creep compliance. The loading in this case can be represented more compactly as $\sigma_o H(t)$, where $H(t)$ is the Heaviside step function, defined as

$$H(t) \equiv \begin{cases} 0 & \text{for } t < 0 \\ 1 & \text{for } t > 0 \end{cases}$$

In the case of a constant uniaxial strain, i.e., $\varepsilon = \varepsilon_o H(t)$, the stress is given by

$$\sigma(t) = E(t)\varepsilon_o \qquad (4.57)$$

where $E(t)$ is the relaxation modulus. When ε_o is positive, the stress relaxes with time. Figure 4.19 schematically illustrates creep at a constant stress, and stress relaxation at a fixed strain.

When stress and strain both vary, the entire deformation history must be taken into account. The strain at time t is obtained by summing the strain increments from earlier times.

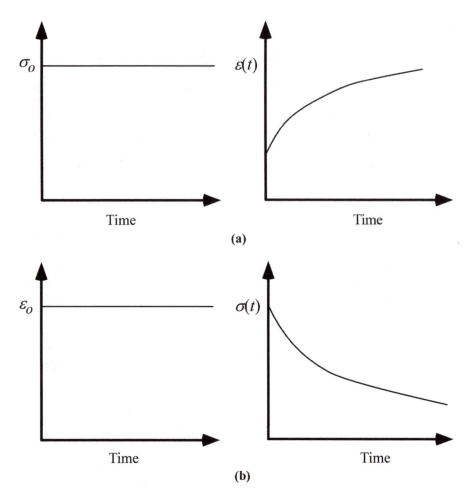

FIGURE 4.19 Schematic uniaxial viscoelastic deformation: (a) creep at a constant stress and (b) stress relaxation at a constant strain.

The incremental strain at time τ, where $0 < \tau < t$, that results from an incremental stress $d\sigma\, H(t - \tau)$ is given by

$$d\varepsilon(\tau) = D(t - \tau)d\sigma(\tau) \qquad (4.58)$$

Integrating this expression with respect to time t gives

$$\varepsilon(t) = \int_0^t D(t - \tau)\frac{d\sigma(\tau)}{d\tau}\,d\tau \qquad (4.59)$$

where it is assumed that $\varepsilon = \sigma = 0$ at $t = 0$. In order to allow for a discontinuous change in stress at $t = 0$, the lower integration limit is assumed to be 0^-, an infinitesimal time before $t = 0$. Relationships such as Equation (4.59) are called *hereditary integrals* because the conditions at time t depend on prior history. The corresponding hereditary integral for stress is given by the inverse of Equation (4.59):

$$\sigma(t) = \int_0^t E(t - \tau)\frac{d\varepsilon(\tau)}{d\tau}\,d\tau \qquad (4.60)$$

By performing a Laplace transform on Equation (4.59) and Equation (4.60), it can be shown that the creep compliance and the relaxation modulus are related as follows:

$$\int_{\tau_o}^{t} E(t-\tau)\frac{dD(\tau-\tau_o)}{d\tau}d\tau = H(t-\tau_o) \tag{4.61}$$

For deformation in three dimensions, the generalized hereditary integral for strain is given by

$$\varepsilon_{ij}(t) = \int_{0}^{t} D_{ijkl}\left(t-\tau\frac{d\sigma_{kl}(\tau)}{d\tau}\right)d\tau \tag{4.62}$$

but symmetry considerations reduce the number of independent creep compliance constants. In the case of a linear viscoelastic isotropic material, there are two independent constants, and the mechanical behavior can be described by $E(t)$ or $D(t)$, which are uniquely related, plus $v_c(t)$, the Poisson's ratio for creep.

Following an approach developed by Schapery [59], it is possible to define a pseudo-elastic strain, which for uniaxial conditions is given by

$$\varepsilon^e(t) = \frac{\sigma(t)}{E_R} \tag{4.63}$$

where E_R is a reference modulus. Substituting Equation (4.63) into Equation (4.59) gives

$$\varepsilon(t) = E_R \int_{0}^{t} D(t-\tau)\frac{d\varepsilon^e(\tau)}{d\tau}d\tau \tag{4.64}$$

The pseudo-strains in three dimensions are related to the stress tensor through Hooke's law, assuming isotropic material behavior:

$$\varepsilon_{ij}^e = E_R^{-1}[(1+v)\sigma_{ij} - v\sigma_{kk}\delta_{ij}] \tag{4.65}$$

where δ_{ij} is the Kronecker delta, and the standard convention of summation on repeated indices is followed. If $v_c = v = $ constant with time, it can be shown that the three-dimensional generalization of Equation (4.64) is given by

$$\varepsilon_{ij}(t) = E_R \int_{0}^{t} D(t-\tau)\frac{d\varepsilon_{ij}^e(\tau)}{d\tau}d\tau \tag{4.66}$$

and the inverse of Equation (4.66) is as follows:

$$\varepsilon_{ij}^e(t) = E_R^{-1} \int_{0}^{t} E(t-\tau)\frac{d\varepsilon_{ij}(\tau)}{d\tau}d\tau \tag{4.67}$$

The advantage of introducing pseudo-strains is that they can be related to stresses through Hooke's law. Thus, if a linear elastic solution is known for a particular geometry, it is possible to determine the corresponding linear viscoelastic solution through a hereditary integral. Given two identical configurations, one made from a linear elastic material and the other made from a linear viscoelastic material, the stresses in both bodies must be identical, and the strains are related through

Equation (4.66) or Equation (4.67), provided both configurations are subject to the same applied loads. This is a special case of a *correspondence principle*, which is discussed in more detail below; note the similarity to Hoff's analogy for elastic and viscous materials (Section 4.2).

4.3.2 THE VISCOELASTIC *J* INTEGRAL

4.3.2.1 Constitutive Equations

Schapery [59] developed a generalized *J* integral that is applicable to a wide range of viscoelastic materials. He began by assuming a nonlinear viscoelastic constitutive equation in the form of a hereditary integral:

$$\varepsilon_{ij}(t) = E_R \int_0^t D(t-\tau, t) \frac{\partial \varepsilon_{ij}^e(\tau)}{\partial \tau} d\tau \tag{4.68}$$

where the lower integration limit is taken as 0^-. The pseudo-elastic strain ε_{ij}^e is related to stress through a linear or nonlinear elastic constitutive law. The similarity between Equation (4.66) and Equation (4.68) is obvious, but the latter relationship also applies to certain types of nonlinear viscoelastic behavior. The creep compliance $D(t)$ has a somewhat different interpretation for the nonlinear case.

The pseudo-strain tensor and reference modulus in Equation (4.68) are analogous to the linear case. In the previous section, these quantities were introduced to relate a linear viscoelastic problem to a reference elastic problem. This idea is generalized in the present case, where the nonlinear viscoelastic behavior is related to a reference nonlinear elastic problem through a correspondence principle, as discussed below.

The inverse of Equation (4.68) is given by

$$\varepsilon_{ij}^e(t) = E_R^{-1} \int_0^t E(t-\tau, t) \frac{\partial \varepsilon_{ij}(\tau)}{\partial \tau} d\tau \tag{4.69}$$

Since hereditary integrals of the form of Equation (4.68) and Equation (4.69) are used extensively in the remainder of this discussion, it is convenient to introduce an abbreviated notation:

$$\{Ddf\} \equiv E_R \int_0^t D(t-\tau, t) \frac{\partial f}{\partial \tau} d\tau \tag{4.70a}$$

and

$$\{Edf\} \equiv E_R^{-1} \int_0^t E(t-\tau, t) \frac{\partial f}{\partial \tau} d\tau \tag{4.70b}$$

where *f* is a function of time. In each case, it is assumed that integration begins at 0^-. Thus Equation (4.68) and Equation (4.69) become, respectively,

$$\varepsilon_{ij}(t) = \left\{ Dd\varepsilon_{ij}^e \right\} \quad \text{and} \quad \varepsilon_{ij}^e = \{Ed\varepsilon_{ij}\}$$

4.3.2.2 Correspondence Principle

Consider two bodies with the same instantaneous geometry, where one material is elastic and the other is viscoelastic and is described by Equation (4.68). Assume that at time *t*, a surface traction $T_i = \sigma_{ij} n_j$ is applied to both configurations along the outer boundaries. If the stresses and strains in the elastic body are σ_{ij}^e and ε_{ij}^e, respectively, while the corresponding quantities in the viscoelastic body are σ_{ij} and ε_{ij}, the stresses, strains, and displacements are related as follows [59]:

$$\sigma_{ij} = \sigma_{ij}^e, \qquad \varepsilon_{ij} = \left\{ Dd\varepsilon_{ij}^e \right\}, \qquad u_i = \left\{ Ddu_i^e \right\} \tag{4.71}$$

Equation (4.71) defines a correspondence principle, introduced by Schapery [59], which allows the solution to a viscoelastic problem to be inferred from a reference elastic solution. This correspondence principle stems from the fact that the stresses in both bodies must satisfy equilibrium, and the strains must satisfy compatibility requirements in both cases. Also, the stresses are equal on the boundaries by definition:

$$T_i = \sigma_{ij} n_j = \sigma_{ij}^e n$$

Schapery [59] gives a rigorous proof of Equation (4.71) for viscoelastic materials that satisfy Equation (4.68).

Applications of correspondence principles in viscoelasticity, where the viscoelastic solution is related to a corresponding elastic solution, usually involve performing a Laplace transform on a hereditary integral in the form of Equation (4.62), which contains actual stresses and strains. The introduction of pseudo-quantities makes the connection between viscoelastic and elastic solutions more straightforward.

4.3.2.3 Generalized J Integral

The correspondence principle in Equation (4.71) makes it possible to define a generalized time-dependent J integral by forming an analogy with the nonlinear elastic case:

$$J_v = \int_\Gamma \left(w^e dy - \sigma_{ij} n_j \frac{\partial u_i^e}{\partial x} ds \right) \tag{4.72}$$

where w^e is the pseudo-strain energy density:

$$w^e = \int \sigma_{ij} d\varepsilon_{ij}^e \tag{4.73}$$

The stresses in Equation (4.72) are the actual values in the body, but the strains and displacements are pseudo-elastic values. The actual strains and displacements are given by Equation (4.71). Conversely, if ε_{ij} and u_i are known, J_v can be determined by computing pseudo-values, which are inserted into Equation (4.73). The pseudo-strains and displacements are given by

$$\varepsilon_{ij}^e = \{ Ed\varepsilon_{ij} \} \quad \text{and} \quad u_i^e = \{ Edu_i \} \tag{4.74}$$

Consider a simple example, where the material exhibits steady-state creep at $t > t_o$. The hereditary integrals for strain and displacement reduce to

$$\varepsilon_{ij}^e = \dot{\varepsilon}_{ij} \quad \text{and} \quad u_i^e = \dot{u}_i$$

By inserting the above results into Equation (4.73), we see that $J_v = C^*$. Thus C^* is a special case of J_v. The latter parameter is capable of taking account of a wide range of time-dependent material behavior, and includes viscous creep as a special case.

Near the tip of the crack, the stresses and pseudo-strains are characterized by J_v through an HRR-type relationship in the form of Equation (4.33). The viscoelastic J can also be determined through a pseudo-energy release rate:

$$J_v = -\frac{1}{B}\left(\frac{\partial}{\partial a}\int_0^{\Delta^e} Pd\Delta^e\right)_{\Delta^e} \tag{4.75}$$

where Δ^e is the pseudo-displacement in the loading direction, which is related to the actual displacement by

$$\Delta = \{Dd\Delta^e\} \tag{4.76}$$

Finally, for Mode I loading of a linear viscoelastic material in plane strain, J_v is related to the stress-intensity factor as follows:

$$J_v = \frac{K_I^2(1-v^2)}{E_R} \tag{4.77}$$

The stress-intensity factor is related to specimen geometry, applied loads, and crack dimensions through the standard equations outlined in Chapter 2.

4.3.2.4 Crack Initiation and Growth

When characterizing crack initiation and growth, it is useful to relate J_v to physical parameters such as CTOD and fracture work, which can be used as local failure criteria. Schapery [59] derived simplified relationships between these parameters by assuming a strip-yield-type failure zone ahead of the crack tip, where a closure stress σ_m acts over ρ, as illustrated in Figure 4.20. While the

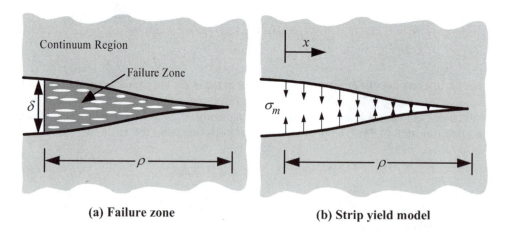

(a) Failure zone   (b) Strip yield model

FIGURE 4.20 Failure zone at the crack tip in a viscoelastic material. This zone is modeled by surface tractions within $0 < x < \rho$.

material in the failure zone may be severely damaged and contain voids and other discontinuities, it is assumed that the surrounding material can be treated as a continuum. If σ_m does not vary with x, applying Equation (3.44) gives

$$J_v = \sigma_m \delta^e \qquad (4.78)$$

where δ^e is the pseudo-crack-tip-opening displacement, which is related to the actual CTOD through a hereditary integral of the form of Equation (4.77). Thus the CTOD is given by

$$\delta = \{Dd(J_v/\sigma_m)\} \qquad (4.79)$$

Although σ_m was assumed to be independent of x at time t, Equation (4.79) permits σ_m to vary with time. The CTOD can be utilized as a local failure criterion: If crack initiation occurs at δ_i, the J_v at initiation can be inferred from Equation (4.79). If δ_i is assumed to be constant, the critical J_v would, in general, depend on the strain rate. A more general version of Equation (4.79) can be derived by allowing σ_m to vary with x.

An alternative local failure criterion is the fracture work w_f. Equating the work input to the crack tip to the energy required to advance the crack tip by da results in the following energy balance at initiation:

$$\int_0^{\delta_i} \sigma_m d\delta = 2w_f \qquad (4.80)$$

assuming unit thickness and Mode I loading. This energy balance can also be written in terms of a time integral:

$$\int_0^{\tau_i} \sigma_m \frac{\partial \delta}{\partial t} dt = 2w_f \qquad (4.81)$$

Inserting Equation (4.79) into Equation (4.81) gives

$$\int_0^{\tau_i} \sigma_m \frac{\partial\{Dd(J_v/\sigma_m)\}}{\partial t} dt = 2w_f \qquad (4.82)$$

If σ_m is independent of time, it cancels out of Equation (4.82), which then simplifies to

$$E_R \int_0^{\tau_i} D(t_i - \tau, t_i) \frac{\partial J_v}{\partial \tau} d\tau = 2w_f \qquad (4.83)$$

For an elastic material, $D = E_R^{-1}$ and $J_v = 2w_f$. If the failure zone is viscoelastic and the surrounding continuum is elastic, J_v may vary with time. If the surrounding continuum is viscous, $D = (t - \tau)/(t_v E_R)$, where t_v is a constant with units of time. Inserting this latter result into Equation (4.83) and integrating by parts gives

$$t_v^{-1} \int_0^{t_i} J_v dt = 2w_f \qquad (4.84)$$

4.3.3 Transition from Linear to Nonlinear Behavior

Typical polymers are linear viscoelastic at low stresses and nonlinear at high stresses. A specimen that contains a crack may have a zone of nonlinearity at the crack tip, analogous to a plastic zone, which is surrounded by linear viscoelastic material. The approach described in the previous section applies only when one type of behavior (linear or nonlinear) dominates.

Schapery [61] has modified the J_v concept to cover the transition from small-stress to large-stress behavior. He introduced a modified constitutive equation, where the strain is given by the sum of two hereditary integrals: one corresponding to linear viscoelastic strains and the other describing nonlinear strains. For the latter term, he assumed power-law viscoelasticity. For the case of uniaxial constant tensile stress σ_o the creep strain in this modified model is given by

$$\varepsilon(t) = E_R D(t) \left(\frac{\sigma_o}{\sigma_{ref}} \right)^n + D_L(t)\sigma_o \qquad (4.85)$$

where D and D_L are the nonlinear and linear creep compliance, respectively, and σ_{ref} is a reference stress.

At low stresses and short times, the second term in Equation (4.85) dominates, while the nonlinear term dominates at high stresses or long times. In the case of a viscoelastic body with a stationary crack at a fixed load, the nonlinear zone is initially small but normally increases with time, until the behavior is predominantly nonlinear. Thus there is a direct analogy between the present case and the transition from elastic to viscous behavior described in Section 4.2.

Close to the crack tip, but outside of the failure zone, the stresses are related to a pseudo-strain through a power law:

$$\varepsilon^e = \left(\frac{\sigma_o}{\sigma_{ref}} \right)^n \qquad (4.86)$$

In the region dominated by Equation (4.86), the stresses are characterized by J_v, regardless of whether the global behavior is linear or nonlinear:

$$\sigma_{ij} = \sigma_{ref} \left(\frac{J_v}{\sigma_{ref} I_n r} \right)^{\frac{1}{n+1}} \tilde{\sigma}_{ij}(n, \theta) \qquad (4.87)$$

If the global behavior is linear, there is a second singularity further away from the crack tip:

$$\sigma_{ij} = \frac{K_I}{\sqrt{2\pi r}} f_{ij}(\theta) \qquad (4.88)$$

Let us define a pseudo-strain tensor that, when inserted into the path-independent integral of Equation (4.72), yields a value J_L. Also suppose that this pseudo-strain tensor is related to the stress tensor by means of linear and power-law pseudo-complementary strain energy density functions (w_{cl} and w_{cn}, respectively):

$$\varepsilon_{ij}^{eL} = \frac{\partial}{\partial \sigma_{ij}} (f w_{cn} + w_{cl}) \qquad (4.89)$$

where $f(t)$ is an as yet unspecified aging function, and the complementary strain energy density is defined by

$$w_c = \int \varepsilon_{ij}^{eL} d\sigma_{ij}$$

For uniaxial deformation, Equation (4.89) reduces to

$$\varepsilon^{eL} = f \left(\frac{\sigma}{\sigma_{ref}} \right)^n + \frac{\sigma}{E_R} \qquad (4.90)$$

Comparing Equation (4.85) and Equation (4.90), it can be seen that

$$f = \frac{D(t)}{D_L(t)} \qquad \text{if} \qquad \varepsilon^{eL} \equiv \frac{\varepsilon(t)}{E_R D_L(t)} \qquad \text{for constant stress creep.}$$

The latter relationship for pseudo-strain agrees with the conventional definition in the limit of linear behavior.

Let us now consider the case where the inner and outer singularities, Equation (4.87) and Equation (4.88), exist simultaneously. For the outer singularity, the second term in Equation (4.90) dominates, the stresses are given by Equation (4.88), and J_L is related to K_I as follows:

$$J_L = \frac{K_I^2(1-v^2)}{E_R} \qquad (4.91)$$

Closer to the crack tip, the stresses are characterized by J_v through Equation (4.87), but J_L is not necessarily equal to J_v, because f appears in the first term of the modified constitutive relationship (Equation (4.90)), but not in Equation (4.86). These two definitions of J coincide if σ_{ref} in Equation (4.90) is replaced with $\sigma_{ref} f^{1/n}$. Thus, the near-tip singularity in terms of J_L is given by

$$\sigma_{ij} = \sigma_{ref} \left(\frac{J_L}{f \sigma_{ref} I_n r} \right)^{\frac{1}{n+1}} \tilde{\sigma}_{ij}(n, \theta) \qquad (4.92)$$

therefore,

$$J_v = \frac{J_L}{f} \qquad (4.93)$$

Schapery showed that $f = 1$ in the limit of purely linear behavior; thus J_L is the limiting value of J_v when the nonlinear zone is negligible. The function f is indicative of the extent of nonlinearity. In most cases, f increases with time, until J_v reaches J_n, the limiting value when the specimen is dominated by nonlinear viscoelasticity. Schapery also confirmed that

$$f = \frac{D(t)}{D_L(t)} \qquad (4.94)$$

for small-scale nonlinearity. Equation (4.93) and Equation (4.94) provide a reasonable description of the transition to nonlinear behavior. Schapery defined a transition time by setting $J_v = J_n$ in Equation (4.93):

$$J_n = \frac{J_L}{f(t_\tau)} \tag{4.95a}$$

or

$$t_\tau = f^{-1}\left(\frac{J_L}{J_n}\right) \tag{4.95b}$$

For the special case of linear behavior followed by viscous creep, Equation (4.95b) becomes

$$t_\tau = \frac{J_L}{(n+1)C^*} \tag{4.96}$$

which is identical to the transition time defined by Riedel and Rice [50].

APPENDIX 4:   DYNAMIC FRACTURE ANALYSIS

A4.1   ELASTODYNAMIC CRACK TIP FIELDS

Rice [31], Sih [35], and Irwin [62] each derived expressions for the stresses ahead of a crack propagating at a constant speed. They found that the moving crack retained the $1/\sqrt{r}$ singularity, but that the angular dependence of the stresses, strains, and displacements depends on crack speed. Freund and Clifton [32] and Nilsson [33] later showed that the solution for a constant speed crack was valid in general; the near-tip quantities depend only on instantaneous crack speed. The following derivation presents the more general case, where the crack speed is allowed to vary.

For dynamic problems, the equations of equilibrium are replaced by the equations of motion, which, in the absence of body forces, are given by

$$\frac{\partial \sigma_{ji}}{\partial x_j} = \rho \ddot{u}_i \tag{A4.1}$$

where x_j denotes the orthogonal coordinates and each dot indicates a time derivative. For quasistatic problems, the term on the right side of Equation (A4.1) vanishes. For a linear elastic material, it is possible to write the equations of motion in terms of displacements and elastic constants by invoking the strain-displacement and stress-strain relationships:

$$\mu \frac{\partial^2 u_i}{\partial x_j^2} + (\lambda + \mu) \frac{\partial^2 u_j}{\partial x_i \partial x_j} = \rho \ddot{u}_i \tag{A4.2}$$

where μ and λ are the Lame' constants; μ is the shear modulus and

$$\lambda = \frac{2\mu v}{1 - 2v}$$

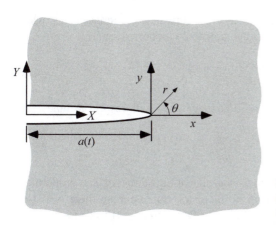

FIGURE A4.1 Definition of coordinate axes for a rapidly propagating crack. The X, Y axes are fixed in space and the x, y axes are attached to the crack tip.

Consider rapid crack propagation in a body subject to plane strain loading. Let us define a fixed coordinate axis X-Y with an origin on the crack plane at $a(t) = 0$, as illustrated in Figure A4.1. It is convenient at this point to introduce two displacement potentials, defined by

$$u_X = \frac{\partial \psi_1}{\partial X} + \frac{\partial \psi_2}{\partial Y}, \qquad u_Y = \frac{\partial \psi_1}{\partial Y} - \frac{\partial \psi_2}{\partial X} \tag{A4.3}$$

Substituting Equation (A4.3) into Equation (A4.2) leads to

$$\frac{\partial^2 \psi_1}{\partial X^2} + \frac{\partial^2 \psi_1}{\partial Y^2} = \frac{1}{c_1^2} \ddot{\psi}_1 \tag{A4.4a}$$

and

$$\frac{\partial^2 \psi_2}{\partial X^2} + \frac{\partial^2 \psi_2}{\partial Y^2} = \frac{1}{c_2^2} \ddot{\psi}_2 \tag{A4.4b}$$

since the wave speeds are given by

$$c_1^2 = \frac{\lambda + \mu}{\rho}, \qquad c_2^2 = \frac{\mu}{\rho}$$

for plane strain. Thus ψ_1 and ψ_2 are the longitudinal and shear wave potentials, respectively. The stresses can be written in terms of ψ_1 and ψ_2 by invoking Equation (A2.1) and Equation (A2.2):

$$\sigma_{XX} + \sigma_{YY} = 2(\lambda + \mu)\left(\frac{\partial^2 \psi_1}{\partial X^2} + \frac{\partial^2 \psi_1}{\partial Y^2} \right) \tag{A4.5a}$$

$$\sigma_{XX} - \sigma_{YY} = 2\mu\left(\frac{\partial^2 \psi_1}{\partial X^2} - \frac{\partial^2 \psi_1}{\partial Y^2} + 2\frac{\partial^2 \psi_2}{\partial X \partial Y} \right) \tag{A4.5b}$$

$$\tau_{XY} = \mu\left(\frac{\partial^2 \psi_2}{\partial Y^2} - \frac{\partial^2 \psi_2}{\partial X^2} + 2\frac{\partial^2 \psi_1}{\partial X \partial Y} \right) \tag{A4.5c}$$

Let us now introduce a moving coordinate system (x, y) attached to the crack tip, where $x = X - a(t)$ and $y = Y$. The rate of change of each wave potential can be written as

$$\frac{d\psi_i}{dt} = \frac{\partial \psi_i}{\partial t} - V \frac{\partial \psi_i}{\partial x} \quad (i = 1, 2) \tag{A4.6}$$

where $V(= -dx/dt)$ is the crack speed. Differentiating Equation (A4.6) with respect to time gives

$$\ddot{\psi}_i = V^2 \frac{\partial^2 \psi_i}{\partial x^2} - 2V \frac{\partial^2 \psi_i}{\partial x \partial t} + \frac{\partial^2 \psi_i}{\partial t^2} - \dot{V} \frac{\partial \psi_i}{\partial x} \tag{A4.7}$$

According to Equation (A4.5) the first term on the right-hand side of Equation (A4.7) is proportional to the stress tensor. This term should dominate close to the crack tip, assuming there is a stress singularity. Substituting the first term of Equation (A4.7) into Equation (A4.4) leads to

$$\beta_1^2 \frac{\partial^2 \psi_1}{\partial x^2} + \frac{\partial^2 \psi_1}{\partial y^2} = 0 \tag{A4.8a}$$

and

$$\beta_2^2 \frac{\partial^2 \psi_2}{\partial x^2} + \frac{\partial^2 \psi_2}{\partial y^2} = 0 \tag{A4.8b}$$

where

$$\beta_1^2 = 1 - \left(\frac{V}{c_1}\right)^2 \quad \text{and} \quad \beta_2^2 = 1 - \left(\frac{V}{c_2}\right)^2$$

Note that the governing equations depend only on instantaneous crack speed; the term that contains crack acceleration in Equation (A4.7) is negligible near the crack tip.

If we scale y by defining new coordinates, $y_1 = \beta_1 y$ and $y_2 = \beta_2 y$, Equation (A4.8) becomes the Laplace equation. Freund and Clifton [32] applied a complex variable method to solve Equation (A4.8). The general solutions to the wave potentials are as follows:

$$\psi_1 = \text{Re}[F(z_1)] \tag{A4.9}$$

and

$$\psi_2 = \text{Im}[G(z_2)]$$

where F and G are as yet unspecified complex functions, $z_1 = x + iy_1$ and $z_2 = x + iy_2$.

The boundary conditions are the same as for a stationary crack: $\sigma_{yy} = \tau_{xy} = 0$ on the crack surfaces. Freund and Clifton showed that these boundary conditions can be expressed in terms of second derivatives for F and G at $y = 0$ and $x < 0$:

$$(1 + \beta_2^2)[F''(x)_+ + F''(x)_-] + 2\beta_2[G''(x)_+ + G''(x)_-] = 0 \tag{A4.10a}$$

$$2\beta_1[F''(x)_+ - F''(x)_-] + (1 + \beta_1^2)[G''(x)_+ - G''(x)_-] = 0 \tag{A4.10b}$$

where the subscripts + and − correspond to the upper and lower crack surfaces, respectively. The following functions satisfy the boundary conditions and lead to integrable strain energy density and finite displacement at the crack tip:

$$F''(z_1) = \frac{C}{\sqrt{z_1}} \quad G''(z_2) = \frac{-2\beta_2 C}{\left(1+\beta_2^2\right)\sqrt{z_2}} \tag{A4.11}$$

where C is a constant. Making the substitution $z_1 = r_1 e^{i\theta_1}$ and $z_2 = r_2 e^{i\theta_2}$ leads to the following expressions for the Mode I crack-tip stress fields:

$$\sigma_{xx} = \frac{K_I(t)}{\sqrt{2\pi r}} \frac{1+\beta_2^2}{D(t)} \left[\left(1+2\beta_1^2 - \beta_2^2\right)\cos\left(\frac{\theta_1}{2}\right)\sqrt{\frac{r}{r_1}} - \frac{4\beta_1\beta_2}{1+\beta_2^2}\cos\left(\frac{\theta_2}{2}\right)\sqrt{\frac{r}{r_2}} \right] \tag{A4.12a}$$

$$\sigma_{yy} = \frac{K_I(t)}{\sqrt{2\pi r}} \frac{1+\beta_2^2}{D(t)} \left[-\left(1+\beta_2^2\right)\cos\left(\frac{\theta_1}{2}\right)\sqrt{\frac{r}{r_1}} + \frac{4\beta_1\beta_2}{1+\beta_2^2}\cos\left(\frac{\theta_2}{2}\right)\sqrt{\frac{r}{r_2}} \right] \tag{A4.12b}$$

$$\tau_{xy} = \frac{K_I(t)}{\sqrt{2\pi r}} \frac{2\beta_1\left(1+\beta_2^2\right)}{D(t)} \left[\sin\left(\frac{\theta_1}{2}\right)\sqrt{\frac{r}{r_1}} - \sin\left(\frac{\theta_2}{2}\right)\sqrt{\frac{r}{r_2}} \right] \tag{A4.12c}$$

where

$$D(t) = 4\beta_1\beta_2 - \left(1+\beta_2^2\right)^2$$

Equation (A4.12) reduces to the quasistatic relationship (Table 2.1) when $V = 0$.

Craggs [25] and Freund [10] obtained the following relationship between $K_I(t)$ and the energy release rate for crack propagation at a constant speed:

$$\mathcal{G} = A(V)\frac{K_I^2(1-v^2)}{E} \tag{A4.13}$$

for plane strain, where

$$A(V) = \frac{V^2\beta_1}{(1-v)c_2^2 D(t)}$$

It can be shown that $\lim_{V\to 0} A = 1$, and Equation (A4.13) reduces to the quasistatic result. Equation (A4.13) can be derived by substituting the dynamic crack-tip solution (Equation (A4.12) and the corresponding relationships for strain and displacement) into the generalized contour integral given by Equation (4.26).

The derivation that led to Equation (A4.12) implies that Equation (A4.13) is a general relationship that applies to accelerating cracks as well as constant speed cracks.

A4.2 DERIVATION OF THE GENERALIZED ENERGY RELEASE RATE

Equation (4.26) will now be derived. The approach closely follows that of Moran and Shih [11], who applied a general balance law to derive a variety of contour integrals, including the energy release rate. Other authors [8–10] have derived equivalent expressions using slightly different approaches.

Beginning with the equation of motion, Equation (A4.1), taking an inner product of both sides with displacement rate $\dot{u}_i$ and rearranging gives

$$\frac{\partial(\sigma_{ji}\dot{u}_i)}{\partial x_j} = \rho\ddot{u}_i + \sigma_{ji}\frac{\partial(\dot{u}_i)}{\partial x_j}$$

$$= \dot{T} + \dot{w} \tag{A4.14}$$

where T and w are the kinetic energy and stress work densities, respectively, as defined in Equation (4.27) to Equation (4.29). Equation (A4.14) is a general balance law that applies to all material behavior. Integrating this relationship over an arbitrary volume, and applying the divergence and transport theorems gives

$$\int_{\partial V} \sigma_{ji}\dot{u}_i m_j dS = \frac{d}{dt}\int_V (w+T)dV - \int_{\partial V} (w+T)V_j m_j dS \tag{A4.15}$$

where

ς = volume

m_j = outward normal to the surface ∂V

V_i = instantaneous velocity of ∂V

Consider now the special case of a crack in a two-dimensional body, where the crack is propagating along the x axis and the origin is attached to the crack tip (Figure A4.2). Let us define a contour C_o fixed in space, that contains the propagating crack and bounds the area A. The crack tip is surrounded by a small contour Γ that is fixed in size and moves with the crack. The balance law in Equation (A4.15) becomes

$$\int_{C_o} \sigma_{ji}\dot{u}_i m_j dC = \frac{d}{dt}\int_A (w+T)dA - \int_\Gamma [(w+T)V\delta_{1j} + \sigma_{ji}\dot{u}_i]m_j d\Gamma \tag{A4.16}$$

where V is the crack speed. The integral on the left side of Equation (A4.16) is the rate at which energy is input into the body. The first term on the right side of this relationship is the rate of increase in the internal energy in the body. Consequently, the second integral on the right side of

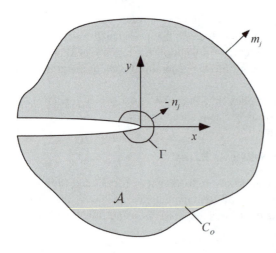

FIGURE A4.2 Conventions for the energy balance for a propagating crack. The outer contour C_o is fixed in space, and the inner contour Γ and the x, y axes are attached to the moving crack tip.

Equation (A4.16) corresponds to the rate at which energy is lost from the body due to flux through Γ. By defining $n_j = -m_j$ on Γ, we obtain the following expression for the energy flux into Γ:

$$F(\Gamma) = \int_\Gamma [(w + T)V\delta_{1j} + \sigma_{ji}\dot{u}_i]n_j d\Gamma \qquad (A4.17)$$

In the limit of a vanishingly small contour, the flux is independent of the shape of Γ. Thus, the energy flux to the crack tip is given by

$$F = \lim_{\Gamma \to 0} \int_\Gamma [(w + T)V\delta_{1j} + \sigma_{ji}\dot{u}_i]n_j d\Gamma \qquad (A4.18)$$

In an increment of time dt, the crack extends by $da = Vdt$ and the energy expended is $\Phi\, dt$. Thus, the energy release rate is given by

$$J = \frac{F}{V} \qquad (A4.19)$$

Substituting Equation (A4.18) into Equation (A4.19) will yield a generalized expression for the J integral. First, however, we must express the displacement rate in terms of crack speed. By analogy with Equation (A4.6), the displacement rate can be written as

$$\dot{u}_i = -V\frac{\partial u_i}{\partial x} + \frac{\partial u_i}{\partial t} \qquad (A4.20)$$

Under steady-state conditions, the second term in Equation (A4.20) vanishes; the displacement at a fixed distance from the propagating crack tip remains constant. Close to the crack tip, the displacement changes rapidly with position (at a fixed time) and the first term in Equation (A4.20) dominates in all cases. Thus, the J integral is given by

$$\begin{aligned} J &= \lim_{\Gamma \to 0} \int_\Gamma \left[(w + T)\delta_{1j} - \sigma_{ji}\frac{\partial u_i}{\partial x}\right]n_j d\Gamma \\ &= \lim_{\Gamma \to 0} \int_\Gamma \left[(w + T)dy - \sigma_{ji}n_j\frac{\partial u_i}{\partial x} d\Gamma\right] \end{aligned} \qquad (A4.21)$$

Equation (A4.21) applies to all types of material response (e.g., elastic, plastic, viscoplastic, and viscoelastic behavior), because it was derived from a generalized energy balance.[3] In the special case of an elastic material (linear or nonlinear), w is the strain energy density, which displays the properties of an elastic potential:

$$\sigma_{ij} = \frac{\partial w}{\partial \varepsilon_{ij}} \qquad (A4.22)$$

Recall from Appendix 3 that Equation (A4.22) is necessary to demonstrate the path independence of J in the quasistatic case. In general, Equation (A4.21) is not path independent except in a local

---

[3] Since the divergence and transport theorems were invoked, there is an inherent assumption that the material behaves as a continuum with smoothly varying displacement fields.

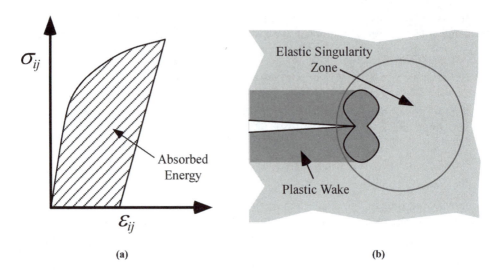

FIGURE A4.3 Crack growth in small-scale yielding. The plastic wake, which forms behind the growing crack, dissipates energy: (a) growing crack and (b) schematic stress-strain curve for material in the plastic wake.

region near the crack tip. For an elastic material, however, J is path independent in the dynamic case when the crack propagation is steady state $(\partial u_i/\partial t = 0)$ [8].

Although Equation (A4.22) is, in principle, applicable to all types of material response, special care must be taken when J is evaluated for a growing crack. Figure A4.3(a) illustrates a growing crack under small-scale yielding conditions. A small plastic zone (or process zone) is embedded within an elastic singularity zone. The plastic zone leaves behind a wake as it sweeps through the material. Unrecoverable work is performed on the material inside the plastic wake, as Figure A4.3(b) illustrates. The work necessary to form the plastic wake comes from the energy flux into the contour Γ. In an ideally elastic body, the energy flux is released from the body through the crack tip, but in an elastic-plastic material, the majority of this energy is dissipated in the wake.

Recall the modified Griffith model (Section 2.3.2), where the work required to increase the crack area a unit amount is equal to $2(\gamma_s + \gamma_p)$, where γ_s is the surface energy and γ_p is the plastic work. The latter term corresponds to the energy dissipated in the plastic wake (Figure 2.6(b)).

The energy release rate computed from Equation (A4.21) must therefore be interpreted as the energy flow to the plastic zone and plastic wake, rather than to the crack tip. That is, Γ cannot shrink to zero; rather, the contour must have a small, but finite radius. The J integral is path independent as long as Γ is defined within the elastic singularity zone, but J becomes path dependent when the contour is taken inside the plastic zone. In the limit, as Γ shrinks to the crack tip, the computed energy release rate would approach zero (in a continuum analysis), since the calculation would exclude the work dissipated by the plastic wake. The actual energy flow to the crack tip is not zero, since a portion of the energy is required to break bonds at the tip. In all but the most brittle materials, however, the bond energy (γ_s) is a small fraction of the total fracture energy.

As long as the plastic zone or process zone is embedded within an elastic singularity, the energy release rate can be defined unambiguously for a growing crack. In large-scale yielding conditions, however, J is path dependent. *Consequently, an unambiguous definition of energy release rate does not exist for a crack growing in an elastic-plastic or fully plastic body.* Recall from Chapter 3 that there are several definitions of J for growing cracks. The so-called deformation J, which is based on a pseudo-energy release rate concept, is the most common methodology. The deformation J is not, in general, equal to the J integral inferred from a contour integration.

REFERENCES

1. Freund, L.B., "Dynamic Crack Propagation." *The Mechanics of Fracture,* American Society of Mechanical Engineers, New York, 1976, pp. 105–134.
2. Freund, L.B., "Crack Propagation in an Elastic Solid Subjected to General Loading—I. Constant Rate of Extension." *Journal of the Mechanics and Physics of Solids,* Vol. 20, 1972, pp. 129–140.
3. Freund, L.B., "Crack Propagation in an Elastic Solid Subjected to General Loading—II. Non-Uniform Rate of Extension." *Journal of the Mechanics and Physics of Solids,* Vol. 20, 1972, pp. 141–152.
4. Freund, L.B., "Crack Propagation in an Elastic Solid Subjected to General Loading—III. Stress Wave Loading." *Journal of the Mechanics and Physics of Solids,* Vol. 21, 1973, pp. 47–61.
5. Freund, L.B., *Dynamic Fracture Mechanics,* Cambridge University Press, Cambridge, UK, 1990.
6. Kanninen, M.F. and Poplar, C.H., *Advanced Fracture Mechanics,* Oxford University Press, New York, 1985.
7. Rose, L.R.F., "Recent Theoretical and Experimental Results on Fast Brittle Fracture." *International Journal of Fracture,* Vol. 12, 1976, pp. 799–813.
8. Atkinson, C. and Eshlby, J.D., "The Flow of Energy into the Tip of a Moving Crack." *International Journal of Fracture Mechanics,* Vol. 4, 1968, pp. 3–8.
9. Sih, G.C., "Dynamic Aspects of Crack Propagation." *Inelastic Behavior of Solids,* McGraw-Hill, New York, 1970, pp. 607–633.
10. Freund, L.B., "Energy Flux into the Tip of an Extending Crack in an Elastic Solid." *Journal of Elasticity,* Vol. 2, 1972, pp. 341–349.
11. Moran, B. and Shih, C.F., "A General Treatment of Crack Tip Contour Integrals." *International Journal of Fracture,* Vol. 35, 1987, pp. 295–310.
12. Atluri, S.N., "Path-Independent Integrals in Finite Elasticity and Inelasticity, with Body Forces, Inertia, and Arbitrary Crack Face Conditions." *Engineering Fracture Mechanics,* Vol. 16, 1982, pp. 341–369.
13. Kishimoto, K., Aoki, S., and Sakata, M., "On the Path-Independent J Integral." *Engineering Fracture Mechanics,* Vol. 13, 1980, pp. 841–850.
14. E 23-88, "Standard Test Methods for Notched Bar Impact Testing of Metallic Materials." American Society for Testing and Materials, Philadelphia, PA, 1988.
15. Joyce, J.A. and Hacket, E.M., "Dynamic J-R Curve Testing of a High Strength Steel Using the Multispecimen and Key Curve Techniques." ASTM STP 905, American Society for Testing and Materials, Philadelphia, PA, 1984, pp. 741–774.
16. Joyce, J.A. and Hacket, E.M., "An Advanced Procedure for J-R Curve Testing Using a Drop Tower." ASTM STP 995, American Society for Testing and Materials, Philadelphia, PA, 1989, 298–317.
17. Duffy, J. and Shih, C.F., "Dynamic Fracture Toughness Measurement Methods for Brittle and Ductile Materials." *Advances in Fracture Research: Seventh International Conference on Fracture,* Pergamon Press, Oxford, 1989, pp. 633–642.
18. Nakamura, T., Shih, C.F., and Freund, L.B., "Analysis of a Dynamically Loaded Three-Point-Bend Ductile Fracture Specimen." *Engineering Fracture Mechanics,* Vol. 25, 1986, pp. 323–339.
19. Nakamura, T., Shih, C.F., and Freund, L.B., "Three-Dimensional Transient Analysis of a Dynamically Loaded Three-Point-Bend Ductile Fracture Specimen." ASTM STP 995, Vol. I, American Society for Testing and Materials, Philadelphia, PA, 1989, pp. 217–241.
20. Rice, J.R., Paris, P.C., and Merkle, J.G., "Some Further Results of J-Integral Analysis and Estimates." ASTM STP 536, American Society for Testing and Materials, Philadelphia, PA, 1973, pp. 231–245.
21. Barsom, J.M., "Development of the AASHTO Fracture Toughness Requirements for Bridge Steels." *Engineering Fracture Mechanics,* Vol. 7, 1975, pp. 605–618.
22. Mott, N.F., "Fracture of Metals: Theoretical Considerations." *Engineering,* Vol. 165, 1948, pp. 16–18.
23. Yoffe, E.H., "The Moving Griffith Crack." *Philosophical Magazine,* Vol. 42, 1951, pp. 739–750.
24. Broberg, K.B., "The Propagation of a Brittle Crack." *Arkvik for Fysik,* Vol 18, 1960, pp. 159–192.
25. Craggs, J.W., "On the Propagation of a Crack in an Elastic-Brittle Material." *Journal of the Mechanics and Physics of Solids,* Vol. 8, 1960, pp. 66–75.
26. Stroh, A.N., "A Simple Model of a Propagating Crack." *Journal of the Mechanics and Physics of Solids,* Vol. 8, 1960, pp. 119–122.
27. Dulaney, E.N. and Brace, W.F., "Velocity Behavior of a Growing Crack." *Journal of Applied Physics,* Vol. 31, 1960, pp. 2233–2236.

28. Berry, J.P., "Some Kinetic Considerations of the Griffith Criterion for Fracture." *Journal of the Mechanics and Physics of Solids,* Vol. 8, 1960, pp. 194–216.

29. Roberts, D.K. and Wells, A.A., "The Velocity of Brittle Fracture." *Engineering,* Vol. 178, 1954, pp. 820–821.

30. Bluhm, J.I., "Fracture Arrest." *Fracture: An Advanced Treatise,* Vol. V, Academic Press, New York, 1969.

31. Rice, J.R., "Mathematical Analysis in the Mechanics of Fracture." *Fracture: An Advanced Treatise,* Vol. II, Academic Press, New York, 1968, p. 191.

32. Freund, L.B. and Clifton, R.J., "On the Uniqueness of Plane Elastodynamic Solutions for Running Cracks." *Journal of Elasticity,* Vol. 4, 1974, pp. 293–299.

33. Nilsson, F., "A Note on the Stress Singularity at a Non-Uniformly Moving Crack Tip." *Journal of Elasticity,* Vol. 4, 1974, pp. 293–299.

34. Rose, L.R.F., "An Approximate (Wiener-Hopf) Kernel for Dynamic Crack Problems in Linear Elasticity and Viscoelasticity." *Proceedings, Royal Society of London,* Vol. A-349, 1976, pp. 497–521.

35. Sih, G.C., "Some Elastodynamic Problems of Cracks." *International Journal of Fracture Mechanics,* Vol. 4, 1968, pp. 51–68.

36. Sanford, R.J. and Dally, J.W., "A General Method for Determining Mixed-Mode Stress Intensity Factors from Isochromatic Fringe Patterns." *Engineering Fracture Mechanics,* Vol. 11, 1979, pp. 621–633.

37. Chona, R., Irwin, G.R., and Shukla, A., "Two and Three Parameter Representation of Crack Tip Stress Fields." *Journal of Strain Analysis,* Vol. 17, 1982, pp. 79–86.

38. Kalthoff, J.F., Beinart, J., Winkler, S., and Klemm, W., "Experimental Analysis of Dynamic Effects in Different Crack Arrest Test Specimens." ASTM STP 711, American Society for Testing and Materials, Philadelphia, PA, 1980, pp. 109–127.

39. Rosakis, A.J. and Freund, L.B., "Optical Measurement of the Plane Strain Concentration at a Crack Tip in a Ductile Steel Plate." *Journal of Engineering Materials Technology,* Vol. 104, 1982, pp. 115–120.

40. Freund, L.B. and Douglas, A.S., "The Influence of Inertia on Elastic-Plastic Antiplane Shear Crack Growth." *Journal of the Mechanics and Physics of Solids,* Vol. 30, 1982, pp. 59–74.

41. Freund, L.B., "Results on the Influence of Crack-Tip Plasticity During Dynamic Crack Growth." ASTM STP 1020, American Society for Testing and Materials, Philadelphia, PA, 1989, pp. 84–97.

42. Corwin, W.R., "Heavy Section Steel Technology Program Semiannual Progress Report for April-September 1987." U.S. Nuclear Regulatory Commission Report NUREG/CR-4219, Vol. 4, No. 2, 1987.

43. Kalthoff, J.F., Beinart, J., and Winkler, S., "Measurement of Dynamic Stress Intensity Factors for Fast Running and Arresting Cracks in Double-Cantilever Beam Specimens." ASTM STP 627, American Society for Testing and Materials, Philadelphia, PA, 1977, pp. 161–176.

44. Kobayashi, A.S., Seo, K.K., Jou, J.Y., and Urabe, Y., "A Dynamic Analysis of Modified Compact Tension Specimens Using Homolite-100 and Polycarbonate Plates." *Experimental Mechanics,* Vol. 20, 1980, pp. 73–79.

45. Landes, J.D. and Begley, J.A., "A Fracture Mechanics Approach to Creep Crack Growth." ASTM STP 590, American Society for Testing and Materials, Philadelphia, PA, 1976, pp. 128–148.

46. Ohji, K., Ogura, K., and Kubo, S., "Creep Crack Propagation Rate in SUS 304 Stainless Steel and Interpretation in Terms of Modified J-Integral." *Transactions, Japanese Society of Mechanical Engineers,* Vol. 42, 1976, pp. 350–358.

47. Nikbin, K.M., Webster, G.A., and Turner, C.E., ASTM STP 601, "Relevance of Nonlinear Fracture Mechanics to Creep Crack Growth." American Society for Testing and Materials, Philadelphia, PA, 1976, pp. 47–62.

48. Hoff, N.J., "Approximate Analysis of Structures in the Presence of Moderately Large Creep Deformations." *Quarterly of Applied Mathematics,* Vol. 12, 1954, pp. 49–55.

49. Riedel, H., "Creep Crack Growth." ASTM STP 1020, American Society for Testing and Materials, Philadelphia, PA, 1989, pp. 101–126.

50. Riedel, H. and Rice, J.R., "Tensile Cracks in Creeping Solids." ASTM STP 700, American Society for Testing and Materials, Philadelphia, PA, 1980, pp. 112–130.

51. Ehlers, R. and Riedel, H., "A Finite Element Analysis of Creep Deformation in a Specimen Containing a Macroscopic Crack." *Proceedings, 5th International Conference on Fracture,* Pergamon Press, Oxford, 1981, pp. 691–698.

52. Saxena, A., "Creep Crack Growth under Non-Steady-State Conditions." ASTM STP 905, American Society for Testing and Materials, Philadelphia, PA, 1986, pp. 185–201.

53. Bassani, J.L., Hawk, D.E., and Saxena, A., "Evaluation of the C_t Parameter for Characterizing Creep Crack Growth Rate in the Transient Regime." ASTM STP 995, Vol. I, American Society for Testing and Materials, Philadelphia, PA, 1990, pp. 112–130.

54. Riedel, H., "Creep Deformation at Crack Tips in Elastic-Viscoplastic Solids." *Journal of the Mechanics and Physics of Solids,* Vol. 29, 1981, pp. 35–49.

55. Chun-Pok, L. and McDowell, D.L., "Inclusion of Primary Creep in the Estimation of the C_t Parameter." *International Journal of Fracture,* Vol. 46, 1990, pp. 81–104.

56. Schapery, R.A., "A Theory of Crack Initiation and Growth in Viscoelastic Media—I. Theoretical Development." *International Journal of Fracture,* Vol. 11, 1975, pp. 141–159.

57. Schapery, R.A., "A Theory of Crack Initiation and Growth in Viscoelastic Media—II. Approximate Methods of Analysis." *International Journal of Fracture,* Vol. 11, 1975, pp. 369–388.

58. Schapery, R.A., "A Theory of Crack Initiation and Growth in Viscoelastic Media—III. Analysis of Continuous Growth." *International Journal of Fracture,* Vol. 11, 1975, pp. 549–562.

59. Schapery, R.A., "Correspondence Principles and a Generalized J Integral for Large Deformation and Fracture Analysis of Viscoelastic Media." *International Journal of Fracture,* Vol. 25, 1984, pp. 195–223.

60. Schapery, R.A., "Time-Dependent Fracture: Continuum Aspects of Crack Growth." *Encyclopedia of Materials Science and Engineering,* Pergamon Press, Oxford, 1986, pp. 5043–5054.

61. Schapery, R.A., "On Some Path Independent Integrals and Their Use in Fracture of Nonlinear Viscoelastic Media." *International Journal of Fracture,* Vol. 42, 1990, pp. 189–207.

62. Irwin, G.R., "Constant Speed Semi-Infinite Tensile Crack Opened by a Line Force." Lehigh University Memorandum, 1967.

Part III

Material Behavior

Chapter 5 and Chapter 6 give an overview of the micromechanisms of fracture in various material systems. This subject is of obvious importance to materials scientists, because an understanding of microstructural events that lead to fracture is essential to the development of materials with optimum toughness. Those who approach fracture from a solid mechanics viewpoint, however, often sidestep microstructural issues and consider only continuum models.

In certain cases, classical fracture mechanics provides some justification for disregarding microscopic failure mechanisms. Just as it is not necessary to understand dislocation theory to apply tensile data to design, it *may* not be necessary to consider the microscopic details of fracture when applying fracture mechanics on a global scale. When a single parameter (i.e., K, J, or crack-tip-opening displacement (CTOD)) uniquely characterizes crack-tip conditions, a critical value of this parameter is a material constant that is transferable from a test specimen to a structure made from the same material (see Section 2.9 and Section 3.5). A laboratory specimen and a flawed structure experience identical crack-tip conditions at failure when the single parameter assumption is valid, and it is not necessary to delve into the details of microscopic failure to characterize global fracture.

The situation becomes considerably more complicated when the single-parameter assumption ceases to be valid. A fracture toughness test on a small-scale laboratory specimen is no longer a reliable indicator of how a large structure will behave. The fracture toughness of the structure and test specimen are likely to be different, and the two configurations may even fail by different mechanisms. A number of researchers are currently attempting to develop alternatives to single parameter fracture mechanics (see Section 3.6). Such approaches cannot succeed with continuum theory alone, but must also consider microscopic fracture mechanisms. Thus, the next two chapters should be of equal value to materials scientists and solid mechanicians.

5 Fracture Mechanisms in Metals

Figure 5.1 schematically illustrates three of the most common fracture mechanisms in metals and alloys. (A fourth mechanism, fatigue, is discussed in Chapter 10.) Ductile materials (Figure 5.1(a)) usually fail as the result of nucleation, growth, and the coalescence of microscopic voids that initiate at inclusions and second-phase particles. Cleavage fracture (Figure 5.1(b)) involves separation along specific crystallographic planes. Note that the fracture path is transgranular. Although cleavage is often called brittle fracture, it can be preceded by large-scale plasticity and ductile crack growth. Intergranular fracture (Figure 5.1(c)), as its name implies, occurs when the grain boundaries are the preferred fracture path in the material.

5.1 DUCTILE FRACTURE

Figure 5.2 schematically illustrates the uniaxial tensile behavior in a ductile metal. The material eventually reaches an instability point, where strain hardening cannot keep pace with the loss in the cross-sectional area, and a necked region forms beyond the maximum load. In very high purity materials, the tensile specimen may neck down to a sharp point, resulting in extremely large local plastic strains and nearly 100% reduction in area. Materials that contain impurities, however, fail at much lower strains. Microvoids nucleate at inclusions and second-phase particles; the voids grow together to form a macroscopic flaw, which leads to fracture.

The commonly observed stages in ductile fracture [1–5] are as follows:

1. Formation of a free surface at an inclusion or second-phase particle by either interface decohesion or particle cracking.
2. Growth of the void around the particle, by means of plastic strain and hydrostatic stress.
3. Coalescence of the growing void with adjacent voids.

In materials where the second-phase particles and inclusions are well-bonded to the matrix, void nucleation is often the critical step; fracture occurs soon after the voids form. When void nucleation occurs with little difficulty, the fracture properties are controlled by the growth and coalescence of voids; the growing voids reach a critical size, relative to their spacing, and a local plastic instability develops between voids, resulting in failure.

5.1.1 VOID NUCLEATION

A void forms around a second-phase particle or inclusion when sufficient stress is applied to break the interfacial bonds between the particle and the matrix. A number of models for estimating void nucleation stress have been published, some of which are based on continuum theory [6, 7] while others incorporate dislocation-particle interactions [8, 9]. The latter models are required for particles <1 μm in diameter.

The most widely used continuum model for void nucleation is due to Argon et al. [6]. They argued that the interfacial stress at a cylindrical particle is approximately equal to the sum of the mean (hydrostatic) stress and the effective (von Mises) stress. The decohesion stress is defined as a critical combination of these two stresses:

$$\sigma_c = \sigma_e + \sigma_m \qquad (5.1)$$

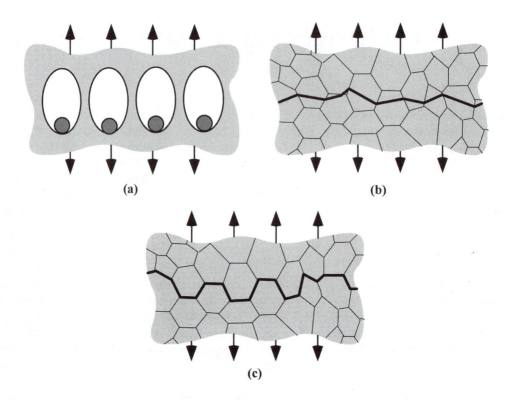

FIGURE 5.1 Three micromechanisms of fracture in metals: (a) ductile fracture, (b) cleavage, and (c) inter-granular fracture.

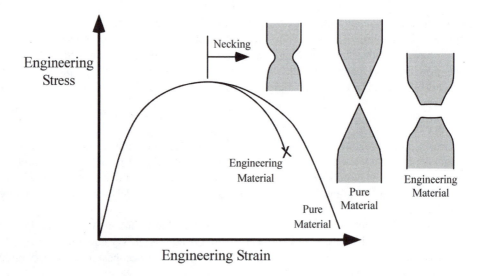

FIGURE 5.2 Uniaxial tensile deformation of ductile materials.

where σ_e is the effective stress, given by

$$\sigma_e = \frac{1}{\sqrt{2}}\left[(\sigma_1 - \sigma_2)^2 + (\sigma_1 - \sigma_3)^2 + (\sigma_3 - \sigma_2)^2\right]^{1/2} \tag{5.2}$$

σ_m is the mean stress, defined as

$$\sigma_m = \frac{\sigma_1 + \sigma_2 + \sigma_3}{3} \tag{5.3}$$

and σ_1, σ_2, and σ_3 are the principal normal stresses. According to the Argon et al. model, the nucleation strain decreases as the hydrostatic stress increases. That is, void nucleation occurs more readily in a triaxial tensile stress field, a result that is consistent with experimental observations.

The Beremin research group in France [7] applied the Argon et al. criterion to experimental data for a carbon manganese steel, but found that the following semiempirical relationship gave better predictions of void nucleation at MnS inclusions that were elongated in the rolling direction:

$$\sigma_c = \sigma_m + C(\sigma_e - \sigma_{YS}) \tag{5.4}$$

where σ_{YS} is the yield strength and C is a fitting parameter that is approximately 1.6 for longitudinal loading and 0.6 for loading transverse to the rolling direction.

Goods and Brown [9] have developed a dislocation model for void nucleation at submicron particles. They estimated that dislocations near the particle elevate the stress at the interface by the following amount:

$$\Delta\sigma_d = 5.4\alpha\mu\sqrt{\frac{\varepsilon_1 b}{r}} \tag{5.5}$$

where

α = constant that ranges from 0.14 to 0.33
μ = shear modulus
ε_1 = maximum remote normal strain
b = magnitude of Burger's vector
r = particle radius

The total maximum interface stress is equal to the maximum principal stress plus $\Delta\sigma_d$. Void nucleation occurs when the sum of these stresses reaches a critical value:

$$\sigma_c = \Delta\sigma_d + \sigma_1 \tag{5.6}$$

An alternative but equivalent expression can be obtained by separating σ_1 into deviatoric and hydrostatic components:

$$\sigma_c = \Delta\sigma_d + S_1 + \sigma_m \tag{5.7}$$

where S_1 is the maximum deviatoric stress.

The Goods and Brown dislocation model indicates that the local stress concentration increases with decreasing particle size; void nucleation is more difficult with larger particles. The continuum models (Equation (5.1) and Equation (5.4)), which apply to particles with $r > 1$ μm, imply that σ_c is independent of particle size.

Experimental observations usually differ from both continuum and dislocation models, in that void nucleation tends to occur more readily at large particles [10]. Recall, however, that these models only considered nucleation by particle-matrix debonding. Voids can also be nucleated when particles crack. Larger particles are more likely to crack in the presence of plastic strain, because they are more likely to contain small defects that can act like Griffith cracks (see Section 5.2). In addition, large nonmetallic inclusions, such as oxides and sulfides, are often damaged during fabrication; some of these particles may be cracked or debonded prior to plastic deformation, making void nucleation relatively easy. Further research is obviously needed to develop void nucleation models that are more in line with experiments.

5.1.2 Void Growth and Coalescence

Once voids form, further plastic strain and hydrostatic stress cause the voids to grow and eventually coalesce. Figure 5.3 and Figure 5.4 are scanning electron microscope (SEM) fractographs that show dimpled fracture surfaces that are typical of microvoid coalescence. Figure 5.4 shows an inclusion that nucleated a void.

Figure 5.5 schematically illustrates the growth and coalescence of microvoids. If the initial volume fraction of voids is low (<10%), each void can be assumed to grow independently; upon further growth, neighboring voids interact. Plastic strain is concentrated along a sheet of voids, and local necking instabilities develop. The orientation of the fracture path depends on the stress state [11].

Many materials contain a bimodal or trimodal distribution of particles. For example, a precipitation-hardened aluminum alloy may contain relatively large intermetallic particles, together with a fine dispersion of submicron second-phase precipitates. These alloys also contain micron-size dispersoid particles for grain refinement. Voids form much more readily in the inclusions, but the smaller particles can contribute in certain cases. Bimodal particle distributions can lead to so-called shear fracture surfaces, as described in the next paragraph.

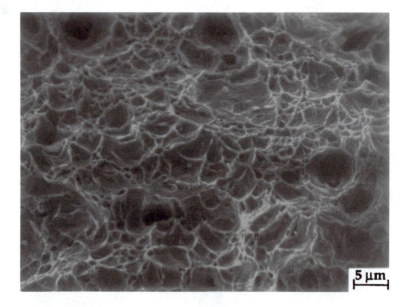

FIGURE 5.3 Scanning electron microscope (SEM) fractograph which shows ductile fracture in a low carbon steel. Photograph courtesy of Mr. Sun Yongqi.

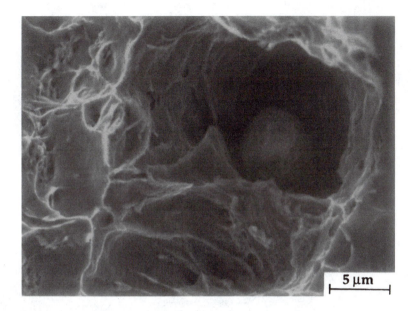

FIGURE 5.4 High magnification fractograph of the steel ductile fracture surface. Note the spherical inclusion which nucleated a microvoid. Photograph courtesy of Mr. Sun Yongqi.

Figure 5.6 illustrates the formation of the "cup and cone" fracture surface that is commonly observed in uniaxial tensile tests. The neck produces a triaxial stress state in the center of the specimen, which promotes void nucleation and growth in the larger particles. Upon further strain, the voids coalesce, resulting in a penny-shaped flaw. The outer ring of the specimen contains relatively few voids, because the hydrostatic stress is lower than in the center. The penny-shaped flaw produces deformation bands at 45° from the tensile axis. This concentration of strain provides sufficient plasticity to nucleate voids in the smaller more numerous particles. Since the small particles are closely spaced, an instability occurs soon after these smaller voids form, resulting in the total fracture of the specimen and the cup and cone appearance of the matching surfaces. The central region of the fracture surface has a fibrous appearance at low magnifications, but the outer region is relatively smooth. Because the latter surface is oriented 45° from the tensile axis and there is little evidence (at low magnifications) of microvoid coalescence, many refer to this type of surface as "shear fracture." The 45° angle between the fracture plane and the applied stress results in a combined Mode I/Mode II loading.

Figure 5.7 is a photograph of the cross-section of a fractured tensile specimen; note the high concentration of microvoids in the center of the necked region, compared with the edges of the necked region.

Figure 5.8 shows SEM fractographs of a cup and cone fracture surface. The central portion of the specimen exhibits a typical dimpled appearance, but the outer region appears to be relatively smooth, particularly at low magnification (Figure 5.8(a)). At somewhat higher magnification (Figure 5.8(b)), a few widely spaced voids are evident in the outer region. Figure 5.9 shows a representative fractograph at higher magnification of the 45° shear surface. Note the dimpled appearance, which is characteristic of microvoid coalescence. The average void size and spacing, however, are much smaller than in the central region of the specimen.

There are a number of mathematical models for void growth and coalescence. The two most widely referenced models were published by Rice and Tracey [12] and Gurson [13]. The latter approach was actually based on the work of Berg [14], but it is commonly known as the Gurson model. The Gurson model has subsequently been modified by Tvergaard and others [15–18].

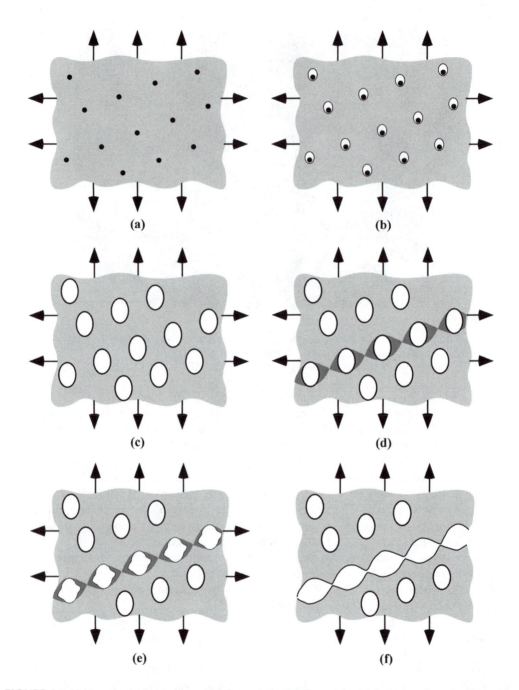

FIGURE 5.5 Void nucleation, growth, and coalescence in ductile metals: (a) inclusions in a ductile matrix, (b) void nucleation, (c) void growth, (d) strain localization between voids, (e) necking between voids, and (f) void coalescence and fracture.

Rice and Tracey considered a single void in an infinite solid, as illustrated in Figure 5.10. The void is subject to remote normal stresses σ_1, σ_2, σ_3, and remote normal strain rates $\dot{\varepsilon}_1$, $\dot{\varepsilon}_2$, $\dot{\varepsilon}_3$. The initial void is assumed to be spherical, but it becomes ellipsoidal as it deforms. Rice and Tracey analyzed both rigid plastic material behavior and linear strain hardening. They showed that the rate

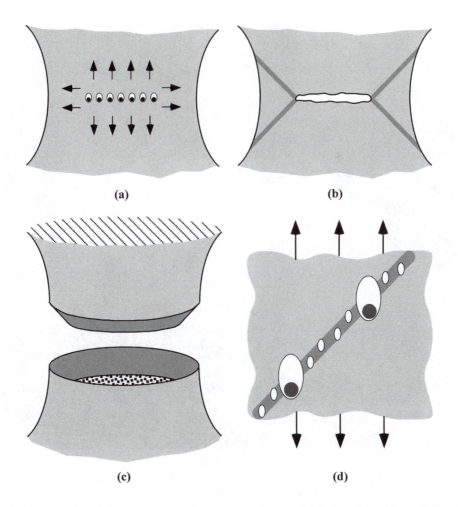

FIGURE 5.6 Formation of the cup and cone fracture surface in uniaxial tension: (a) void growth in a triaxial stress state, (b) crack and deformation band formation, (c) nucleation at smaller particles along the deformation bands, and (d) cup and cone fracture.

of change of radius in each principal direction has the form:

$$\dot{R}_i = \left[(1+G)\dot{\varepsilon}_i + \sqrt{\frac{2}{3}\dot{\varepsilon}_j \dot{\varepsilon}_j} D \right] R_o \quad (i, j = 1, 2, 3) \tag{5.8}$$

where D and G are constants that depend on stress state and strain hardening, and R_o is the radius of the initial spherical void. The standard notation, where repeated indices imply summation, is followed here. Invoking the incompressibility condition $(\dot{\varepsilon}_1 + \dot{\varepsilon}_2 + \dot{\varepsilon}_3 = 0)$ reduces the number of independent principal strain rates to two. Rice and Tracey chose to express $\dot{\varepsilon}_2$ and $\dot{\varepsilon}_3$ in terms of $\dot{\varepsilon}_1$ and a second parameter:

$$\dot{\varepsilon}_2 = \frac{-2\phi}{3+\phi}\dot{\varepsilon}_1 \tag{5.9a}$$

$$\dot{\varepsilon}_3 = \frac{\phi-3}{3+\phi}\dot{\varepsilon}_1 \tag{5.9b}$$

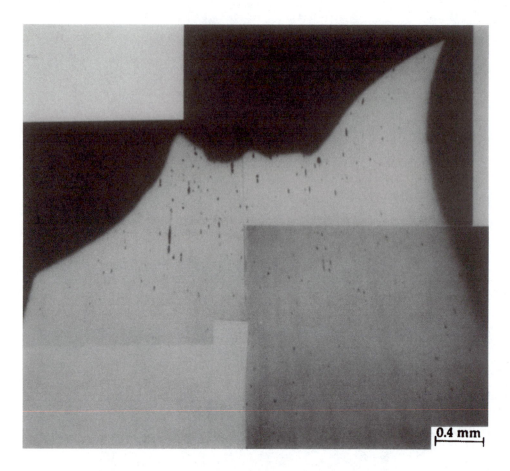

FIGURE 5.7 Metallographic cross-section (unetched) of a ruptured austenitic stainless steel tensile specimen (The dark areas in the necked region are microvoids). Photograph courtesy of P.T. Purtscher.

where

$$\phi = -\frac{3\dot{\varepsilon}_2}{\dot{\varepsilon}_1 - \dot{\varepsilon}_3}$$

Substituting Equation (5.9) into Equation (5.8) and making a few simplifying assumptions leads to the following expressions for the radial displacements of the ellipsoidal void:

$$R_1 = \left(A + \frac{(3+\phi)}{2\sqrt{\phi^2+3}} B \right) R_o \qquad (5.10a)$$

$$R_2 = \left(A - \frac{\phi B}{\sqrt{\phi^2+3}} \right) R_o \qquad (5.10b)$$

$$R_3 = \left(A + \frac{(\phi-3)B}{2\sqrt{\phi^2+3}} \right) R_o \qquad (5.10c)$$

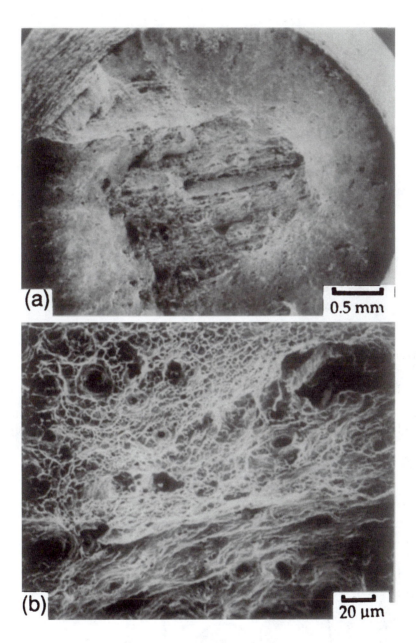

FIGURE 5.8 Cup and cone fracture in an austenitic stainless steel. Photographs courtesy of P.T. Purtscher. Taken from Purtscher, P.T., "Micromechanisms of Ductile Fracture and Fracture Toughness in a High Strength Austenitic Stainless Steel." Ph.D. Dissertation, Colorado School of Mines, Golden, CO, 1990.

where

$$A = \exp\left(\frac{2\sqrt{\phi^2+3}}{(3+\phi)} D\varepsilon_1\right)$$

$$B = \frac{(1+F)(A-1)}{D}$$

and ε_1 is the total strain, integrated from the undeformed configuration to the current state.

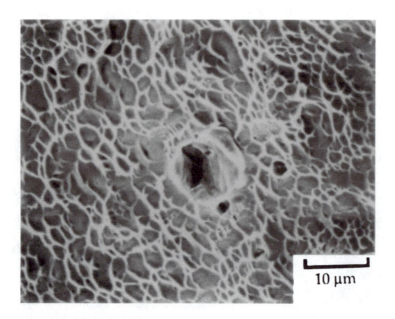

FIGURE 5.9 High-magnification fractograph of the "shear" region of a cup and cone fracture surface in austenitic stainless steel. Photograph courtesy of P.T. Purtscher. Taken from Purtscher, P.T., "Micromechanisms of Ductile Fracture and Fracture Toughness in a High Strength Austenitic Stainless Steel." Ph.D. Dissertation, Colorado School of Mines, Golden, CO, 1990.

Rice and Tracey solved Equation (5.10) for a variety of stress states and found that the void growth in all cases could be approximated by the following semiempirical relationship:

$$\ln\left(\frac{\overline{R}}{R_o}\right) = 0.283 \int_0^{\varepsilon_{eq}} \exp\left(\frac{1.5\sigma_m}{\sigma_{YS}}\right) d\varepsilon_{eq}^p \qquad (5.11)$$

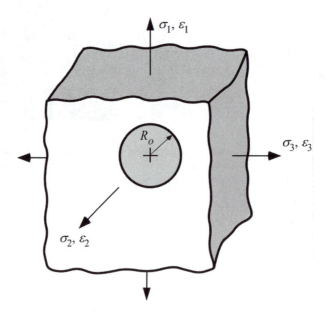

FIGURE 5.10 Spherical void in a solid, subject to a triaxial stress state.

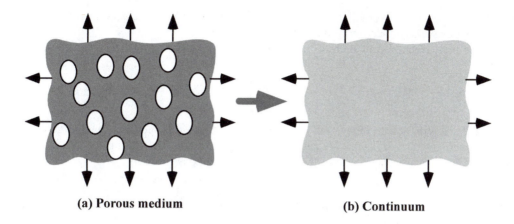

(a) Porous medium   **(b) Continuum**

FIGURE 5.11 The continuum assumption for modeling a porous medium. The material is assumed to be homogeneous, and the effect of the voids is averaged through the solid.

where $\bar{R} = (R_1 + R_2 + R_3)/3$ and ε_{eq}^p is the equivalent (von Mises) plastic strain. Subsequent investigators found that Equation (5.11) could be approximately modified for strain hardening by replacing the yield strength with σ_e, the effective stress [20].

Since the Rice and Tracey model is based on a single void, it does not take account of interactions between voids, nor does it predict ultimate failure. A separate failure criterion must be applied to characterize microvoid coalescence. For example, one might assume that fracture occurs when the nominal void radius reaches a critical value.

A model originally developed by Gurson [13] and later modified by Tvergaard [15, 16] analyzes the plastic flow in a porous medium by assuming that the material behaves as a continuum. Voids appear in the model indirectly through their influence on the global flow behavior. The effect of the voids is averaged through the material, which is assumed to be continuous and homogeneous (Figure 5.11). The main difference between the Gurson-Tvergaard (GT) model and classical plasticity is that the yield surface in the former exhibits a hydrostatic stress dependence, while classical plasticity assumes that yielding is independent of hydrostatic stress. This modification to conventional plasticity theory has the effect of introducing a strain-softening term.

Unlike the Rice and Tracey model, the GT model contains a failure criterion. Ductile fracture is assumed to occur as the result of a plastic instability that produces a band of localized deformation. Such an instability occurs more readily in a GT material because of the strain softening induced by hydrostatic stress. However, because the model does not consider discrete voids, it is unable to predict the necking instability between voids.

The model derives from a rigid-plastic limit analysis of a solid having a smeared volume fraction f of voids approximated by a homogeneous spherical body containing a spherical void. The yield surface and flow potential g is given by

$$g(\sigma_e, \sigma_m, \bar{\sigma}, f) = \left(\frac{\sigma_e}{\bar{\sigma}}\right)^2 + 2q_1 f \cosh\left(\frac{3q_2\sigma_m}{2\bar{\sigma}}\right) - (1 + q_3 f^2) = 0 \qquad (5.12)$$

where

  σ_e = macroscopic von Mises stress
  σ_m = macroscopic mean stress
  $\bar{\sigma}$ = flow stress for the matrix material of the cell
  f = current void fraction

Values of $q_1 = 1.5$, $q_2 = 1.0$, and $q_3 = q_1^2$ are typically used for metals. Setting $f = 0$ recovers the von Mises yield surface for an incompressible material.

In the most recent formulation of the GT model [17], the void growth rate has the following form:

$$\dot{f} = (1-f)\dot{\varepsilon}_{kk}^p + \Lambda \dot{\varepsilon}_{eq}^p \qquad (5.13)$$

where the first term defines the growth rate of the preexisting voids and the second term quantifies the contribution of new voids that are nucleated with plastic strain. The scaling coefficient Λ, which is applied to the plastic strain rate of the cell matrix material, is given by

$$\Lambda = \frac{f_N}{s_N \sqrt{2\pi}} \exp\left[-\frac{1}{2}\left(\frac{\varepsilon_{eq}^p - \varepsilon_N}{s_N}\right)^2 \right] \qquad (5.14)$$

In the above expression, the plastic strain range at nucleation of new voids follows a normal distribution with a mean value ε_N, a standard deviation s_N, and a volume fraction of void nucleating particles f_N. Introducing void nucleation into the GT model results in additional fitting parameters (ε_N, s_N, and f_N). Moreover, this void nucleation expression is not consistent with the models presented in Section 5.1.1 because the former implies that nucleation does not depend on hydrostatic stress. Further research is obviously necessary to sort out these inconsistencies.

In the absence of the void nucleation term in Equation (5.13), the key input parameters to the GT model are the initial void fraction f_o and the critical void fraction f_c. In real materials, voids grow very rapidly when the void fraction exceeds 10 to 20%. Equation (5.12) does not adequately capture the final stage of rapid void growth and coalescence. Consequently, failure is often assumed when the void fracture reaches a critical value f_c. This is a reasonable assumption because failure in real materials is very abrupt with little additional macroscopic strain once the void fraction exceeds f_c. A typical assumption for carbon steels is $f_c = 0.15$. The initial void fraction f_o is normally used as a fitting parameter to experimental data.

Tvergaard and Needleman [18] have attempted to model void coalescence by replacing f with an effective void volume fraction f^*:

$$f^* = \begin{cases} f & \text{for} \quad f \le f_c \\ f_c - \dfrac{f_u^* - f_c}{f_F - f_c}(f - f_c) & \text{for} \quad f > f_c \end{cases} \qquad (5.15)$$

where f_c, f_u^*, and f_F are fitting parameters. The effect of hydrostatic stress is amplified when $f > f_c$, which accelerates the onset of a plastic instability. As a practical matter, it is probably sufficient to assume failure when f exceeds f_c. The marginal benefit of applying Equation (5.15) is offset by the need to define the additional fitting parameters f_u^* and f_F. Consequently, Equation (5.15) has fallen out of favor in recent years.

Thomason [11] developed a simple limit load model for internal necking between microvoids. This model states that failure occurs when the net section stress between voids reaches a critical value $\sigma_{n(c)}$. Figure 5.12 illustrates a two-dimensional case, where cylindrical voids are growing in a material subject to plane strain loading ($\varepsilon_3 = 0$). If the in-plane dimensions of the voids are $2a$ and $2b$, and the spacing between voids is $2d$, the row of voids illustrated in Figure 5.12 is stable if

$$\sigma_{n(c)} \frac{d}{d+b} > \sigma_1 \qquad (5.16a)$$

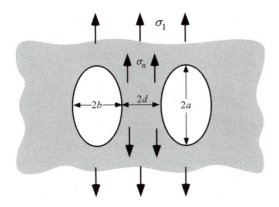

FIGURE 5.12 The limit-load model for void instability. Failure is assumed to occur when the net section stress between voids reaches a critical value.

and fracture occurs when

$$\sigma_{n(c)}\frac{d}{d+b}=\sigma_1 \tag{5.16b}$$

where σ_1 is the maximum remote principal stress. As with Equation (5.15), the Thomason model is of limited practical value. The void interactions leading to ductile failure are far too complex to be captured by a simple area reduction model. Moreover, the final stage in failure is very abrupt, as discussed earlier. Once the void fraction reaches 10 to 20%, failure occurs with only a minimal increase in the nominal strain.

5.1.3 DUCTILE CRACK GROWTH

Figure 5.13 schematically illustrates microvoid initiation, growth, and coalescence at the tip of a preexisting crack. As the cracked structure is loaded, local strains and stresses at the crack tip become sufficient to nucleate voids. These voids grow as the crack blunts, and they eventually link with the main crack. As this process continues, the crack grows.

Figure 5.14 is a plot of stress and strain near the tip of a blunted crack [21]. The strain exhibits a singularity near the crack tip, but the stress reaches a peak at approximately two times the crack-tip-opening displacement (CTOD).[1] In most materials, the triaxiality ahead of the crack tip provides sufficient stress elevation for void nucleation; thus the growth and coalescence of microvoids are usually the critical steps in ductile crack growth. Nucleation typically occurs when a particle is $\sim 2\delta$ from the crack tip, while most of the void growth occurs much closer to the crack tip, relative to CTOD. (Note that although a void remains approximately fixed in absolute space, its distance from the crack tip, relative to CTOD, decreases as the crack blunts; the *absolute* distance from the crack tip also decreases as the crack grows.)

Ductile crack growth is usually stable because it produces a rising resistance curve, at least during the early stages of crack growth. Stable crack growth and R curves are discussed in detail in Chapter 3 and Chapter 7.

When an edge crack in a plate grows by microvoid coalescence, the crack exhibits a *tunneling* effect, where it grows faster at the center of the plate, due to the higher stress triaxiality.

---

[1] Finite element analysis and slip line analysis of blunted crack tips predict a stress singularity very close to the crack tip (~0.1 CTOD), but it is not clear whether or not this actually occurs in real materials because the continuum assumptions break down at such fine scales.

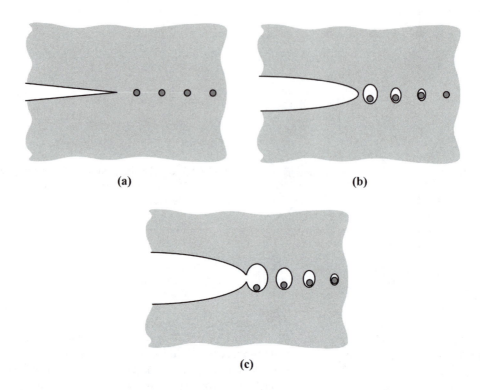

FIGURE 5.13 Mechanism for ductile crack growth: (a) initial state, (b) void growth at the crack tip, and, (c) coalescence of voids with the crack tip.

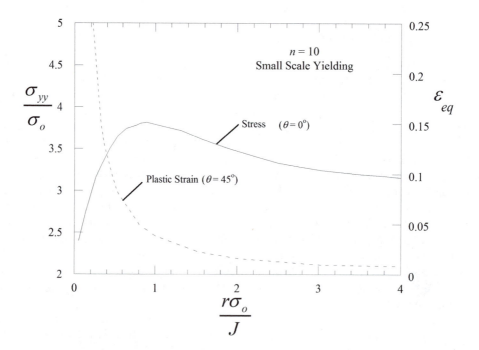

FIGURE 5.14 Stress and strain ahead of a blunted crack tip, determined by finite element analysis. Taken from McMeeking, R.M. and Parks, D.M., "On Criteria for J-Dominance of Crack-Tip Fields in Large-Scale Yielding." ASTM STP 668, American Society for Testing and Materials, Philadelphia, PA, 1979, pp. 175–194.

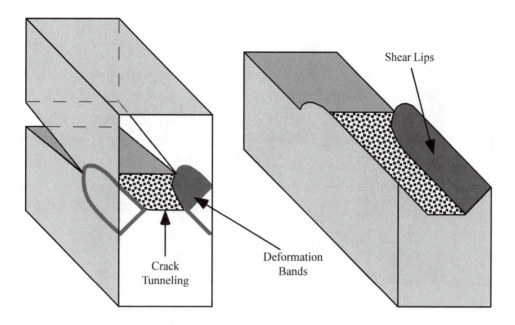

FIGURE 5.15 Ductile growth of an edge crack. The shear lips are produced by the same mechanism as the cup and cone in uniaxial tension (Figure 5.7).

The through-thickness variation of triaxiality also produces *shear lips*, where the crack growth near the free surface occurs at a 45° angle from the maximum principal stress, as illustrated in Figure 5.15. The shear lips are very similar to the cup and cone features in fractured tensile specimens. The growing crack at the center of the plate produces deformation bands that nucleate voids in small particles (Figure 5.6).

The high-triaxiality crack growth at the center of a plate appears to be relatively flat, but closer examination reveals a more complex structure. For a crack subject to plane strain Mode I loading, the maximum plastic strain occurs at 45° from the crack plane, as illustrated in Figure 5.16(a). On a local level, this angle is the preferred path for void coalescence, but global constraints require that the crack propagation remain in its original plane. One way to reconcile these competing requirements is for the crack to grow in a zigzag pattern (Figure 5.16(b)), such that the crack appears flat on a global scale, but oriented ±45° from the crack propagation direction when viewed at higher magnification. This zigzag pattern is often observed in ductile materials [22, 23]. Figure 5.17 shows a metallographic cross-section of a growing crack that exhibits this behavior.

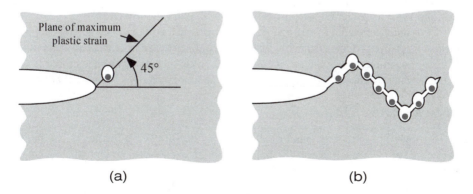

(a)   (b)

FIGURE 5.16. Ductile crack growth in a 45° zigzag pattern.

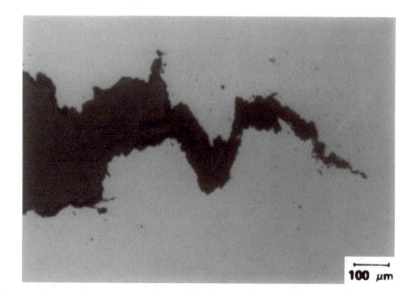

FIGURE 5.17 Optical micrograph (unetched) of ductile crack growth in an A 710 high-strength low-alloy steel. Photograph courtesy of J.P. Gudas. Taken from McMeeking, R.M. and Parks, D.M., "On Criteria for J-Dominance of Crack-Tip Fields in Large-Scale Yielding." ASTM STP 668, American Society for Testing and Materials, Philadelphia, PA, 1979, pp. 175–194.

5.2 CLEAVAGE

Cleavage fracture can be defined as the rapid propagation of a crack along a particular crystallographic plane. Cleavage may be brittle, but it can be preceded by large-scale plastic flow and ductile crack growth (see Section 5.3). The preferred cleavage planes are those with the lowest packing density, since fewer bonds must be broken and the spacing between planes is greater. In the case of body-centered cubic (BCC) materials, cleavage occurs on {100} planes. The fracture path is transgranular in polycrystalline materials, as Figure 5.1(b) illustrates. The propagating crack changes direction each time it crosses a grain boundary; the crack seeks the most favorably oriented cleavage plane in each grain. The nominal orientation of the cleavage crack is perpendicular to the maximum principal stress.

Cleavage is most likely when the plastic flow is restricted. Face-centered cubic (FCC) metals are usually not susceptible to cleavage because there are ample slip systems for ductile behavior at all temperatures. At low temperatures, BCC metals fail by cleavage because there are a limited number of active slip systems. Polycrystalline hexagonal close-packed (HCP) metals, which have only three slip systems per grain, are also susceptible to cleavage fracture.

This section and Section 5.3 focus on ferritic steel, because it is the most technologically important (and the most extensively studied) material that is subject to cleavage fracture. This class of materials has a BCC crystal structure, which undergoes a ductile-brittle transition with decreasing temperature. Many of the mechanisms described below also operate in other material systems that fail by cleavage.

5.2.1 FRACTOGRAPHY

Figure 5.18 shows SEM fractographs of cleavage fracture in a low-alloy steel. The multifaceted surface is typical of cleavage in a polycrystalline material; each facet corresponds to a single grain. The "river patterns" on each facet are also typical of cleavage fracture. These markings are so named because multiple lines converge to a single line, much like tributaries to a river.

Figure 5.19 illustrates how river patterns are formed. A propagating cleavage crack encounters a grain boundary, where the nearest cleavage plane in the adjoining grain is oriented at a finite

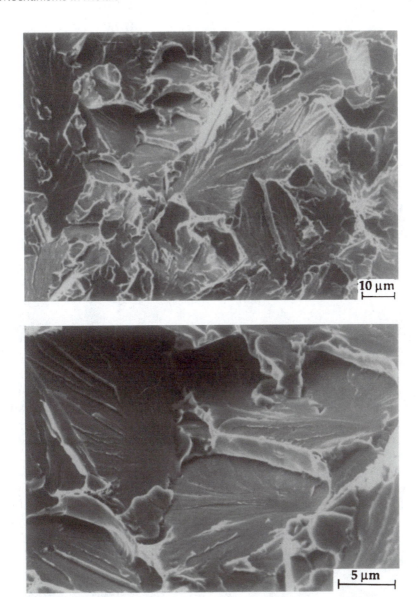

FIGURE 5.18 SEM fractographs of cleavage in an A 508 Class 3 alloy. Photographs courtesy of Mr. Sun Yongqi.

twist angle from the current cleavage plane. Initially, the crack accommodates the twist mismatch by forming on several parallel planes. As the multiple cracks propagate, they are joined by tearing between planes. Since this process consumes more energy than the crack propagation on a single plane, there is a tendency for the multiple cracks to converge into a single crack. Thus, the direction of crack propagation can be inferred from the river patterns. Figure 5.20 shows a fractograph of river patterns in a low-alloy steel, where tearing between parallel cleavage planes is evident.

5.2.2 MECHANISMS OF CLEAVAGE INITIATION

Since cleavage involves breaking bonds, the local stress must be sufficient to overcome the cohesive strength of the material. In Chapter 2, we learned that the theoretical fracture strength of a crystalline

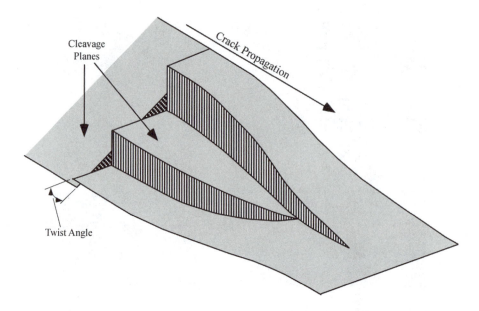

FIGURE 5.19 Formation of river patterns, as a result of a cleavage crack crossing a twist boundary between grains.

solid is approximately E/π. Figure 5.14, however, indicates that the maximum stress achieved ahead of the crack tip is three to four times the yield strength. For a steel with $\sigma_{YS} = 400$ MPa and $E = 210,000$ MPa, the cohesive strength would be ~50 times higher than the maximum stress achieved ahead of the crack tip. Thus, a macroscopic crack provides insufficient stress concentration to exceed the bond strength.

In order for cleavage to initiate, there must be a local discontinuity ahead of the macroscopic crack that is sufficient to exceed the bond strength. A sharp microcrack is one way to provide sufficient local stress concentration. Cottrell [24] postulated that microcracks form at intersecting

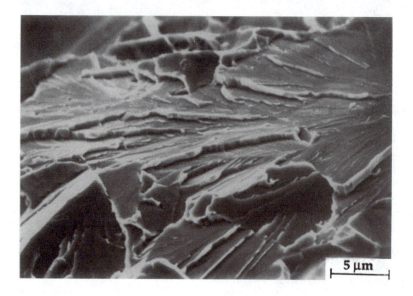

FIGURE 5.20 River patterns in an A 508 Class 3 steel. Note the tearing (light areas) between parallel cleavage planes. Photograph courtesy of Mr. Sun Yongqi.

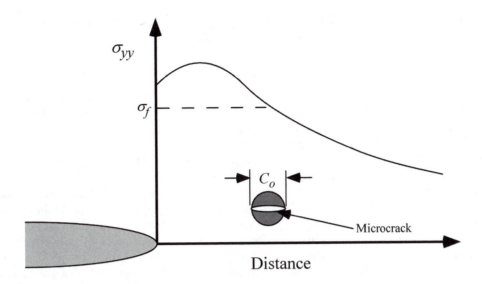

FIGURE 5.21 Initiation of cleavage at a microcrack that forms in a second-phase particle ahead of a macroscopic crack.

slip planes by means of dislocation interaction. A far more common mechanism for microcrack formation in steels, however, involves inclusions and second-phase particles [1, 25, 26].

Figure 5.21 illustrates the mechanism of cleavage nucleation in ferritic steels. The macroscopic crack provides a local stress and strain concentration. A second-phase particle, such as a carbide or inclusion, cracks because of the plastic strain in the surrounding matrix. At this point the microcrack can be treated as a Griffith crack (Section 2.3). If the stress ahead of the macroscopic crack is sufficient, the microcrack propagates into the ferrite matrix, causing failure by cleavage. For example, if the particle is spherical and it produces a penny-shaped crack, the fracture stress is given by

$$\sigma_f = \left(\frac{\pi E \gamma_p}{(1-v^2)C_o} \right)^{1/2}$$

(5.17)

where γ_p is the plastic work required to create a unit area of fracture surface in the ferrite and C_o is the particle diameter. It is assumed that $\gamma_p \gg \gamma_s$, where γ_s is the surface energy (c.f. Equation (2.21)). Note that the stress ahead of the macrocrack is treated as a remote stress in this case.

Consider the hypothetical material described earlier, where $\sigma_{YS} = 400$ MPa and $E = 210,000$ MPa. Knott [1] has estimated $\gamma_p = 14$ J/m^2 for ferrite. Setting $\sigma_f = 3$ σ_{YS} and solving for critical particle diameter yields $C_o = 7.05$ μm. Thus the Griffith criterion can be satisfied with relatively small particles.

The nature of the microstructural feature that nucleates cleavage depends on the alloy and heat treatment. In mild steels, cleavage usually initiates at grain boundary carbides [1, 25, 26]. In quenched and tempered alloy steels, the critical feature is usually either a spherical carbide or an inclusion [1, 27]. Various models have been developed to explain the relationship between cleavage fracture stress and microstructure; most of these models resulted in expressions similar to Equation (5.18). Smith [26] proposed a model for cleavage fracture that considers stress concentration due to a dislocation pile-up at a grain boundary carbide. The resulting failure criterion is as follows:

$$C_o \sigma_f^2 + k_y^2 \left[\frac{1}{2} + \frac{2\tau_i \sqrt{C_o}}{\pi k_y} \right]^2 = \frac{4E\gamma_p}{\pi(1-v^2)}$$

(5.18)

where C_o, in this case, is the carbide thickness, and τ_i and k_y are the friction stress and pile-up constant, respectively, as defined in the Hall-Petch equation:

$$\tau_y = \tau_i + k_y d^{-1/2} \tag{5.19}$$

where τ_y is the yield strength in shear. The second term on the left side of Equation (5.18) contains the dislocation contribution to cleavage initiation. If this term is removed, Equation (5.18) reduces to the Griffith relationship for a grain boundary microcrack.

Figure 5.22 shows SEM fractographs that give examples of cleavage initiation from a grain boundary carbide (a) and an inclusion at the interior of a grain (b). In both cases, the fracture origin was located by following river patterns on the fracture surface.

Susceptibility to cleavage fracture is enhanced by almost any factor that increases the yield strength, such as low temperature, a triaxial stress state, radiation damage, high strain rate, and strain aging. Grain size refinement not only increases the yield strength but also increases σ_f. There are a number of reasons for the grain size effect. In mild steels, a decrease in grain size implies an increase in the grain boundary area, which leads to smaller grain boundary carbides and an increase in σ_f. In fine-grained steels, the critical event may be the propagation of the microcrack across the first grain boundary it encounters. In such cases, the Griffith model implies the following expression for fracture stress:

$$\sigma_f = \left(\frac{\pi E \gamma_{gb}}{(1-v^2)d} \right)^{1/2} \tag{5.20}$$

where γ_{gb} is the plastic work per unit area required to propagate into the adjoining grains. Since there tends to be a high degree of mismatch between grains in a polycrystalline material, $\gamma_{gb} > \gamma_p$. Equation (5.20) assumes an equiaxed grain structure. For martensitic and bainitic microstructures, Dolby and Knott [28] derived a modified expression for σ_f based on the packet diameter.

In some cases cleavage nucleates, but total fracture of the specimen or structure does not occur. Figure 5.23 illustrates three examples of unsuccessful cleavage events. Part (a) shows a microcrack that has arrested at the particle–matrix interface. The particle cracks due to strain in the matrix, but the crack is unable to propagate because the applied stress is less than the required fracture stress. This microcrack does not reinitiate because subsequent deformation and dislocation motion in the matrix causes the crack to blunt. Microcracks must remain sharp in order for the stress on the atomic level to exceed the cohesive strength of the material. If a microcrack in a particle propagates into the ferrite matrix, it may arrest at the grain boundary, as illustrated in Figure 5.23(b). This corresponds to a case where Equation (5.20) governs cleavage. Even if a crack successfully propagates into the surrounding grains, it may still arrest if there is a steep stress gradient ahead of the macroscopic crack (Figure 5.23(c)). This tends to occur at low applied K_I values. Locally, the stress is sufficient to satisfy Equation (5.18) and Equation (5.20) but there is insufficient global driving force to continue the crack propagation. Figure 5.24 shows an example of arrested cleavage cracks in front of a macroscopic crack in a spheroidized 1008 steel [29].

5.2.3 MATHEMATICAL MODELS OF CLEAVAGE FRACTURE TOUGHNESS

A difficulty emerges when trying to predict fracture toughness from Equation (5.18) to Equation (5.20). The maximum stress ahead of a macroscopic crack occurs at approximately 2δ from the crack tip, but the absolute value of this stress is constant in small-scale yielding (Figure 5.14); the distance from the crack tip at which this stress occurs increases with increasing K, J, and δ. Thus, if attaining a critical fracture stress were a sufficient condition for cleavage fracture, the material might fail upon the application of an infinitesimal load, because the stresses would be high near the crack tip. Since ferritic materials have finite toughness, the attainment of a critical stress ahead of the crack tip is apparently necessary but not sufficient.

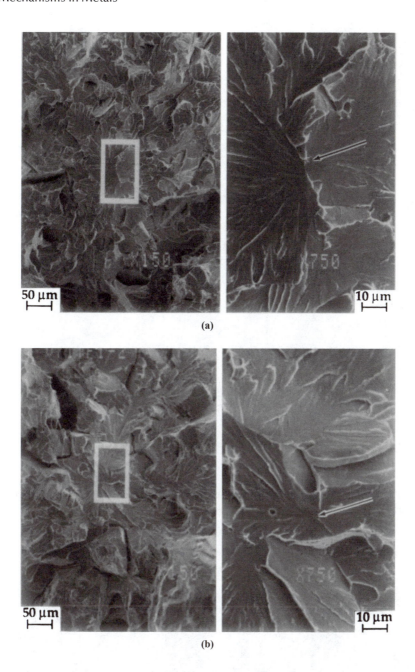

FIGURE 5.22 SEM fractographs of cleavage initiation in an A 508 Class 3 alloy: (a) initiation at a grain boundary carbide and (b) initiation at an inclusion near the center of a grain. Photographs courtesy of M.T. Miglin.

Ritchie, Knott, and Rice (RKR) [30] introduced a simple model to relate fracture stress to fracture toughness, and to explain why steels did not spontaneously fracture upon the application of a minimal load. They postulated that cleavage failure occurs when the stress ahead of the crack tip exceeds σ_f over a characteristic distance, as illustrated in Figure 5.25. They inferred σ_f in a mild steel from blunt-notched four-point bend specimens and measured K_{Ic} with conventional fracture toughness specimens. They inferred the crack-tip stress field from a finite element solution published by Rice and Tracey [31]. They found that the characteristic distance was equal to two grain diameters

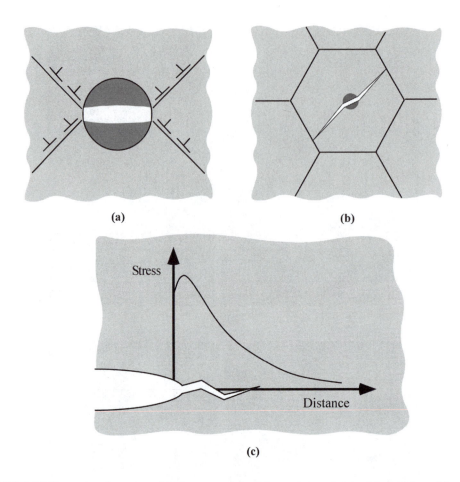

FIGURE 5.23 Examples of unsuccessful cleavage events: (a) arrest at particle/matrix interface, (b) arrest at a grain boundary, and (c) arrest due to a steep stress gradient.

FIGURE 5.24 Arrested cleavage cracks ahead of a macroscopic crack in a spherodized 1008 steel. Taken from Lin, T., Evans, A.G., and Ritchie, R.O., "Statistical Model of Brittle Fracture by Transgranular Cleavage." *Journal of the Mechanics and Physics of Solids,* Vol. 34, 1986, pp. 477–496.

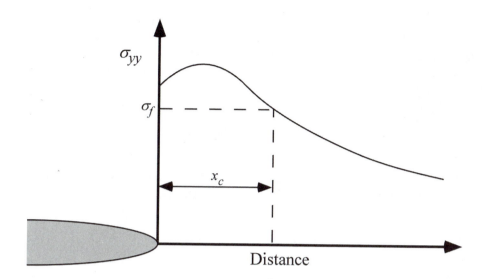

FIGURE 5.25 The Ritchie-Knott-Rice model for cleavage fracture. Failure is assumed to occur when the fracture stress is exceeded over a characteristic distance. Taken from Ritchie, R.O., Knott, J.F., and Rice, J.R., "On the Relationship between Critical Tensile Stress and Fracture Toughness in Mild Steel." *Journal of the Mechanics and Physics of Solids*, Vol. 21, 1973, pp. 395–410.

for the material they tested. Ritchie et al. argued that if fracture initiates in a grain boundary carbide and propagates into a ferrite grain, the stress must be sufficient to propagate the cleavage crack across the opposite grain boundary and into the next grain; thus σ_f must be exceeded over one or two grain diameters. Subsequent investigations [27, 32, 33], however, revealed no consistent relationship between the critical distance and grain size.

Curry and Knott [34] provided a statistical explanation for the RKR critical distance. A finite volume of material must be sampled ahead of the crack tip in order to find a particle that is sufficiently large to nucleate cleavage. Thus, a critical sample volume over which $\sigma_{yy} \geq \sigma_f$ is required for failure. The critical volume, which can be related to a critical distance, depends on the average spacing of cleavage nucleation sites.

The statistical argument also explains why cleavage fracture toughness data tend to be widely scattered. Two nominally identical specimens made from the same material may display vastly different toughness values because the location of the critical fracture-triggering particle is random. If one specimen samples a large fracture-triggering particle near the crack tip, while the fracture trigger in the other specimen is further from the crack tip, the latter specimen will display a higher fracture toughness, because a higher load is required to elevate the stress at the particle to a critical value. The statistical nature of fracture also leads to an apparent thickness effect on toughness. A thicker specimen is more likely to sample a large fracture trigger along the crack front, and therefore will have a lower toughness than a thin specimen, on average [36–38].

The Curry and Knott approach was followed by more formal statistical models for cleavage [29, 37–40]. These models all treated cleavage as a weakest link phenomenon, where the probability of failure is equal to the probability of sampling at least one critical fracture-triggering particle. For a volume of material V, with ρ critical particles per unit volume, the probability of failure can be inferred from the Poisson distribution:

$$F = 1 - \exp(-\rho V) \tag{5.21}$$

The second term is the probability of finding zero critical particles in V, so F is the probability of sampling one or more critical particles. The Poisson distribution can be derived from the binomial distribution by assuming that ρ is small and V is large, an assumption that is easily satisfied in the present problem.[2] The critical particle size depends on stress, as Equation (5.17), Equation (5.18), and Equation (5.20) indicate. In cases where stress varies with position, such as near the tip of a crack, ρ must also vary with position. In such instances, the failure probability must be integrated over individual volume elements ahead of the crack tip:

$$F = 1 - \exp\left[-\int_V \rho dV \right] \qquad (5.22)$$

Consider a uniformly stressed sample of volume V_o. A two-parameter Weibull distribution [41] can be used to represent the statistical variation of the fracture stress:

$$F = 1 - \exp\left[-\left(\frac{\sigma_f}{\sigma_u} \right)^m \right] \qquad (5.23)$$

where m and σ_u are material constants. Comparing Equation (5.21) and Equation (5.23), we see that

$$\rho = \frac{1}{V_o}\left(\frac{\sigma_f}{\sigma_u} \right) \qquad (5.24)$$

When stress varies with position, the cumulative fracture probability can be expressed in terms of an integral of maximum principal stress over volume, which can then be equated to an equivalent fracture stress:

$$F = 1 - \exp\left[-\frac{1}{V_o}\int_{V_f}\left(\frac{\sigma_1}{\sigma_u} \right)^m dV \right] = 1 - \exp\left[-\left(\frac{\sigma_w}{\sigma_u} \right)^m \right] \qquad (5.25)$$

where
  V_f = volume of the fracture process zone in the geometry of interest
  V_o = reference volume
  σ_w = Weibull stress [41]

The Weibull stress can be viewed as the equivalent fracture stress for uniform loading on a sample with volume V_o. Solving for Weibull stress gives

$$\sigma_w = \left[\frac{1}{V_o}\int_{V_f} \sigma_1^m dV \right]^{1/m} \qquad (5.26)$$

The fracture process zone is defined as the region where the stresses are sufficiently high for a realistic probability of cleavage.

    Let us now consider the special case of a cracked body subject to applied loading. Assuming ρ depends only on the local stress field, and the crack-tip conditions are uniquely defined by K or J,

---

[2] For a detailed discussion of the Poisson assumption, consult any textbook on probability and statistics.

it can be shown (Appendix 5) that the critical values of K and J follow a characteristic distribution when failure is controlled by a weakest link mechanism:[3]

$$F = 1 - \exp\left[-\left(\frac{K_{Ic}}{\Theta_K}\right)^4\right] \tag{5.27a}$$

or

$$F = 1 - \exp\left[-\left(\frac{J_c}{\Theta_J}\right)^2\right] \tag{5.27b}$$

where Θ_K and Θ_J are the material properties that depend on microstructure and temperature. Note that Equation (5.22a) and Equation (5.22b) have the form of a two-parameter Weibull distribution. The Weibull-shape parameter, which is sometimes called the Weibull slope, is equal to 4.0 for K_{Ic} data and (because of the relationship between K and J) 2.0 for J_c values for cleavage. The Weibull-scale parameters Θ_K and Θ_J are the 63rd percentile values of K_{Ic} and J_c, respectively. If Θ_K or Θ_J is known, the entire fracture toughness distribution can be inferred from Equation (5.27a) or Equation (5.27b).

The prediction of a fracture toughness distribution that follows a two-parameter Weibull function with a known slope is an important result. The Weibull slope is a measure of the relative scatter; a prior knowledge of the Weibull slope enables the relative scatter to be predicted *a priori*, as Example 5.1 illustrates.

EXAMPLE 5.1

Determine the relative size of the 90% confidence bounds of K_{Ic} and J_c data, assuming Equation (5.27a) and Equation (5.27b) describe the respective distributions.

Solution: The median, 5% lower bound, and 95% upper bound values are obtained by setting $F = 0.5$, 0.05, and 0.95, respectively, in Equation (5.27a) and Equation (5.27b). Both equations have the form:

$$F = 1 - \exp(-\lambda)$$

Solving for λ at each probability level gives

$$\lambda_{0.50} = 0.693, \quad \lambda_{0.05} = 0.0513, \quad \lambda_{0.95} = 2.996$$

The width of the 90% confidence band in K_{Ic} data, normalized by the median, is given by

$$\frac{K_{0.95} - K_{0.05}}{K_{0.50}} = \frac{(2.996)^{0.25} - (0.0513)^{0.25}}{(0.693)^{0.25}} = 0.920$$

and the relative width of the J_c scatter band is

$$\frac{J_{0.95} - J_{0.05}}{J_{0.50}} = \frac{\sqrt{2.996} - \sqrt{0.0513}}{\sqrt{0.693}} = 1.81$$

Note that Θ_K and Θ_J cancel out of the above results and the relative scatter depends only on the Weibull slope.

---

[3] Equation (5.27a) and Equation (5.27b) apply only when the thickness (i.e., the crack front length) is fixed. The weakest link model predicts a thickness effect, which is described in Appendix 5.2 but is omitted here for brevity.

There are two major problems with the weakest link model that leads to Equation (5.27a) and Equation (5.27b). First, these equations predict zero as the minimum toughness in the distribution. Intuition suggests that such a prediction is incorrect, and more formal arguments can be made for a nonzero threshold toughness. A crack cannot propagate in a material unless there is sufficient energy available to break bonds and perform plastic work. If the material is a polycrystal, additional work must be performed when the crack crosses randomly oriented grains. Thus, one can make an estimate of threshold toughness in terms of energy release rate:

$$\mathcal{G}_{c(min)} \approx 2\gamma_p \phi \qquad (5.28)$$

where ϕ is a grain misorientation factor. If the global driving force is less than $\mathcal{G}_{c(min)}$, the crack cannot propagate. The threshold toughness can also be viewed as a crack arrest value: a crack cannot propagate if $K_I < K_{IA}$.

A second problem with Equation (5.27a) and Equation (5.27b) is that they tend to overpredict the experimental scatter. That is, scatter in experimental cleavage fracture toughness data is usually less severe than predicted by the weakest link model.

According to the weakest link model, failure is controlled by the initiation of cleavage in the ferrite as the result of the cracking of a critical particle, i.e., a particle that satisfies Equation (5.17) or Equation (5.18). While weakest link initiation is necessary, it is apparently not sufficient for total failure. A cleavage crack, once initiated, must have a sufficient driving force to propagate. Recall Figure 5.22, which gives examples of unsuccessful cleavage events.

Both problems, threshold toughness and scatter, can be addressed by incorporating a conditional probability of propagation into the statistical model [42, 43]. Figure 5.26 is a probability tree for cleavage initiation and propagation. When a flawed structure is subject to an applied K, a microcrack may or may not initiate, depending on the temperature as well as the location of the eligible cleavage triggers. The initiation of cleavage cracks should be governed by a weakest link mechanism, because the process involves searching for a large enough trigger to propagate a microcrack into the first ferrite grain.

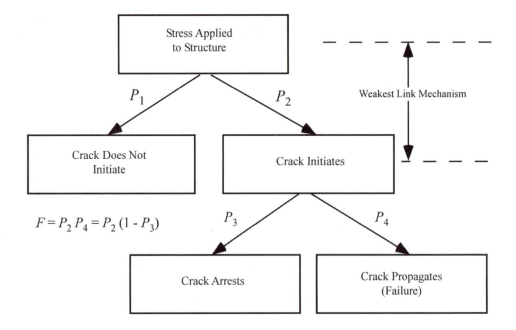

FIGURE 5.26 Probability tree for cleavage initiation and propagation.

Once cleavage initiates, the crack may either propagate in an unstable fashion or arrest, as in Figure 5.23. Initiation is governed by the local stress at the critical particle, while propagation is controlled by the orientation of the neighboring grains and the global driving force. The overall probability of failure is equal to the probability of initiation times the conditional probability of propagation.

This model assumes that if a microcrack arrests, it does not contribute to subsequent failure. This is a reasonable assumption, since only a rapidly propagating crack is sufficiently sharp to give the stress intensification necessary to break bonds. Once a microcrack arrests, it is blunted by local plastic flow.

Consider the case where the conditional probability of propagation is a step function:

$$P_{pr} = \begin{cases} 0, & K_I < K_o \\ 1, & K_I \geq K_o \end{cases}$$

That is, assume that all cracks arrest when $K_I < K_o$ and that a crack propagates if $K_I \geq K_o$ at the time of initiation. This assumption implies that the material has a crack-arrest toughness that is single valued. It can be shown (see Appendix 5.2) that such a material exhibits the following fracture toughness distribution on K values:

$$F = 1 - \exp\left\{ -\left[\left(\frac{K_{IC}}{\Theta_K} \right)^4 - \left(\frac{K_o}{\Theta_K} \right)^4 \right] \right\} \qquad \text{for} \qquad K_I > K_o \qquad (5.29a)$$

$$F = 0 \qquad \text{for} \qquad K_I \leq K_o \qquad (5.29b)$$

Equation (5.24) is a *truncated Weibull* distribution; Θ_K can no longer be interpreted as the 63rd percentile K_{Ic} value. Note that a threshold has been introduced, which removes one of the shortcomings of the weakest link model. Equation (5.24) also exhibits less scatter than the two-parameter distribution (Equation 5.22a), thereby removing the other objection to the weakest link model.

The threshold is obvious in Equation (5.29), but the reduction in relative scatter is less so. The latter effect can be understood by considering the limiting cases of Equation (5.24). If $K_o/\Theta_K \gg 1$, there are ample initiation sites for cleavage, but the microcracks cannot propagate unless $K_I > K_o$. Once K_I exceeds K_o, the next microcrack to initiate will cause total failure. Since initiation events are frequent in this case, K_{Ic} values will be clustered near K_o, and the scatter will be minimal. On the other hand, if $K_o/\Theta_K \ll 1$, Equation (5.29) reduces to the weakest link case. Thus the relative scatter decreases as K_o/Θ_K increases.

Equation (5.24) is an oversimplification, because it assumes a single-valued crack-arrest toughness. In reality, there is undoubtedly some degree of randomness associated with microscopic crack arrest. When a cleavage crack initiates in a single ferrite grain, the probability of propagation into the surrounding grains depends in part on their relative orientation; a high degree of mismatch increases the likelihood of arrest at the grain boundary. Anderson et al. [43] performed a probabilistic simulation of microcrack propagation and arrest in a polycrystalline solid. Initiation in a single grain ahead of the crack tip was assumed, and the tilt and twist angles at the surrounding grains were allowed to vary randomly (within the geometric constraints imposed by assuming {100} cleavage planes). An energy-based propagation criterion, suggested by the work of Gell and Smith [44], was applied. The conditional probability of propagation was estimated over a range of applied K_I values. The results fit an offset power law expression:

$$P_{pr} = \alpha (K_I - K_o)^\beta \qquad (5.30)$$

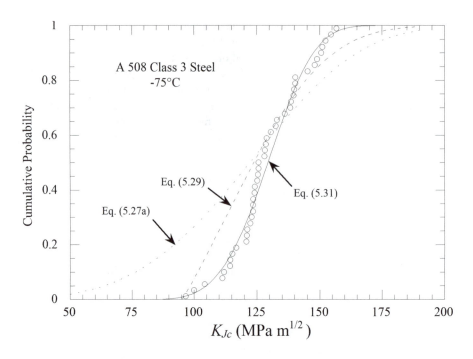

FIGURE 5.27 Cleavage fracture toughness data for an A 508 Class 3 steel at −75°C. The data have been fit to various statistical distributions. Taken from Anderson, T.L., Stienstra, D.I.A., and Dodds, R.H., Jr., "A Theoretical Framework for Addressing Fracture in the Ductile-Brittle Transition Region." *Fracture Mechanics: 24th Volume,* ASTM STP 1207, American Society for Testing and Materials, Philadelphia, PA (in press).

where α and β are material constants. Incorporating Equation (5.30) into the overall probability analysis leads to a complicated distribution function that is very difficult to apply to experimental data (see Appendix 5.2). Stienstra and Anderson found, however, that this new function could be approximated by a three-parameter Weibull distribution:

$$F = 1 - \exp\left[-\left(\frac{K_{JC} - K_{\min}}{\Theta_K - K_{\min}}\right)^4\right] \quad (5.31)$$

where $K_{\min}$ is the Weibull location parameter.

Figure 5.27 shows experimental cleavage fracture toughness data for a low-alloy steel. Critical J values measured experimentally were converted to equivalent K_{Ic} data. The data were corrected for constraint loss through an analysis developed by Anderson and Dodds [45] (see Section 3.6.1). Equation (5.27a), Equation (5.29), and Equation (5.31) were fit to the experimental data. The three-parameter Weibull distribution obviously gives the best fit. The weakest link model (Equation (5.27a)) overestimates the scatter, while the truncated Weibull distribution does not follow the data in the lower tail, presumably because the assumption of a single-valued arrest toughness is incorrect.

Wallin [46] has extended the Anderson et al. [45] model to show that the three-parameter Weibull distribution of Equation (5.31) implies a probability of propagation of the following form:

$$P_{pr} = A\left(1 - \frac{K_{\min}}{K_I}\right)^3 \quad (5.32)$$

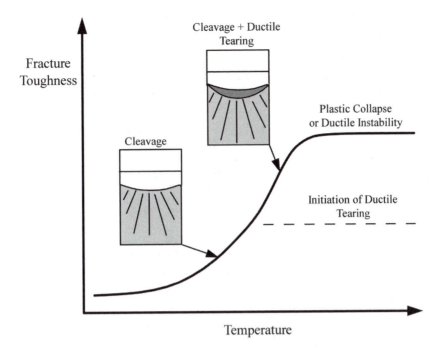

FIGURE 5.28 The ductile-brittle transition in ferritic steels. The fracture mechanism changes from cleavage to microvoid coalescence as the temperature increases.

5.3 THE DUCTILE-BRITTLE TRANSITION

The fracture toughness of ferritic steels can change drastically over a small temperature range, as Figure 5.28 illustrates. At low temperatures, steel is brittle and fails by cleavage. At high temperatures, the material is ductile and fails by microvoid coalescence. Ductile fracture initiates at a particular toughness value, as indicated by the dashed line in Figure 5.28. The crack grows as the load is increased. Eventually, the specimen fails by plastic collapse or tearing instability. In the transition region between ductile and brittle behavior, both micromechanisms of fracture can occur in the same specimen. In the lower transition region, the fracture mechanism is pure cleavage, but the toughness increases rapidly with temperature as cleavage becomes more diffi-cult. In the upper transition region, a crack initiates by microvoid coalescence but ultimate failure occurs by cleavage. On initial loading in the upper transition region, cleavage does not occur because there are no critical particles near the crack tip. As the crack grows by ductile tearing, however, more material is sampled. Eventually, the growing crack samples a critical particle and cleavage occurs. Because the fracture toughness in the transition region is governed by these statistical sampling effects, the data tend to be highly scattered. Wallin [47] has developed a statistical model for the transition region that incorporates the effect of prior ductile tearing on the cleavage probability.

Recent work by Heerens and Read [27] demonstrates the statistical sampling nature of cleavage fracture in the transition region. They performed a large number of fracture toughness tests on a quenched and tempered alloy steel at several temperatures in the transition region. As expected, the data at a given temperature were highly scattered. Some specimens failed without significant stable crack growth, while other specimens sustained high levels of ductile tearing prior to cleavage. Heerens and Read examined the fracture surface of each specimen to determine

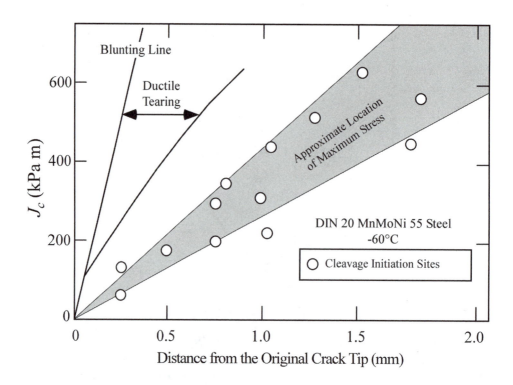

FIGURE 5.29 Relationship between cleavage fracture toughness and the distance between the crack tip and the cleavage trigger. Taken from Heerens, J. and Read, D.T., "Fracture Behavior of a Pressure Vessel Steel in the Ductile-to-Brittle Transition Region." NISTIR 88-3099, National Institute for Standards and Technology, Boulder, CO, 1988.

the site of cleavage initiation. The measured distance from the initiation site to the original crack tip correlated very well with the measured fracture toughness. In specimens that exhibited low toughness, this distance was small; a critical nucleus was available near the crack tip. In specimens that exhibited high toughness, there were no critical particles near the crack tip; the crack had to grow and sample additional material before a critical cleavage nucleus was found. Figure 5.29 is a plot of fracture toughness vs. the critical distance r_c, which Heerens and Read measured from the fracture surface; r_c is defined as the distance from the fatigue crack tip to the cleavage initiation site. The resistance curve for ductile crack growth is also shown in this plot. In every case, cleavage initiated near the location of the maximum tensile stress (c.f. Figure 5.14). Similar fractographic studies by Watanabe et al. [33] and Rosenfield and Shetty [48] also revealed a correlation between J_c, Δa, and r_c.

Cleavage propagation in the upper transition region often displays isolated islands of ductile fracture [23, 49]. When specimens with arrested macroscopic cleavage cracks are studied metallographically, unbroken ligaments are sometimes discovered behind the arrested crack tip. These two observations imply that a propagating cleavage crack in the upper transition region encounters barriers, such as highly misoriented grains or particles, through which the crack cannot propagate. The crack is diverted around these obstacles, leaving isolated unbroken ligaments in its wake. As the crack propagation continues, and the crack faces open, the ligaments that are well-behind the crack tip rupture. Figure 5.30 schematically illustrates this postulated mechanism. The energy required to rupture the ductile ligaments may provide the majority of the propagation resistance a cleavage crack experiences. The concentration of ductile ligaments on a fracture surface increases with temperature [49], which may explain why crack-arrest toughness (K_{Ia}) exhibits a steep brittle-ductile transition, much like K_{Ic} and J_c.

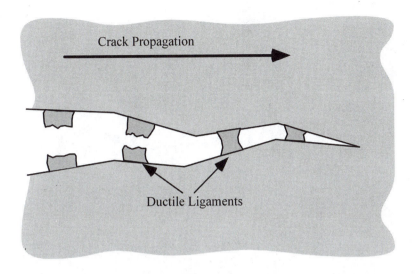

FIGURE 5.30 Schematic illustration of cleavage crack propagation in the ductile-brittle transition region. Ductile ligaments rupture behind the crack tip, resulting in increased propagation resistance.

5.4 INTERGRANULAR FRACTURE

In most cases metals do not fail along grain boundaries. Ductile metals usually fail by the coalescence of voids formed at inclusions and second-phase particles, while brittle metals typically fail by transgranular cleavage. Under special circumstances, however, cracks can form and propagate along grain boundaries.

There is no single mechanism for intergranular fracture. Rather, there are a variety of situations that can lead to cracking on grain boundaries, including:

1. Precipitation of a brittle phase on the grain boundary.
2. Environmental assisted cracking.
3. Intergranular corrosion.
4. Grain boundary cavitation and cracking at high temperatures.

Space limitations preclude discussing each of these mechanisms in detail. A brief description of the intergranular cracking mechanisms is given below. Chapter 11 provides an overview of environmental cracking.

Brittle phases can be deposited on the grain boundaries of steel through improper tempering [50]. Tempered martensite embrittlement, which results from tempering near 350°C, and temper embrittlement, which occurs when an alloy steel is tempered at ~550°C, both apparently involve the segregation of impurities such as phosphorous and sulfur, to prior austenite grain boundaries. These thin layers of impurity atoms are not resolvable on the fracture surface, but can be detected with surface analysis techniques such as Auger electron spectroscopy. Segregation of aluminum nitride particles on grain boundaries during solidification is a common embrittlement mechanism in cast steels [50]. Aluminum nitride, if present in sufficient quantity, can also contribute to the degradation of toughness resulting from temper embrittlement in wrought alloys.

Environmental cracking can occur in a range of material/environment combinations. In some cases, crack propagation is directly attributable to a corrosion reaction at the crack tip; that is, the material at the crack tip selectively corrodes. In other instances, crack propagation is due to hydrogen embrittlement. In both cases, high stresses at the crack tip promote crack propagation. These

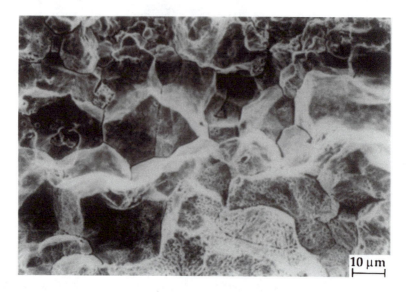

FIGURE 5.31 Intergranular fracture in a steel ammonia tank. Photograph courtesy of W.L. Bradley.

environmental cracking mechanisms are discussed in detail in Chapter 11. Figure 5.31 shows an intergranular fracture surface in a steel weld that was in contact with an ammonia environment.

Intergranular corrosion involves the preferential attack of the grain boundaries, as opposed to general corrosion, where the material is dissolved relatively uniformly across the surface. Intergranular attack is different from environmental assisted cracking, in that an applied stress is not necessary for grain boundary corrosion.

At high temperatures, grain boundaries are weak relative to the matrix, and a significant portion of creep deformation is accommodated by grain-boundary sliding. In such cases void nucleation and growth (at second-phase particles) is concentrated at the crack boundaries, and cracks form as grain boundary cavities grow and coalesce. Grain-boundary cavitation is the dominant mechanism of creep crack growth in metals [51], and it can be characterized with time-dependent parameters such as the C^* integral (see Chapter 4).

.

APPENDIX 5:   STATISTICAL MODELING OF CLEAVAGE FRACTURE

When one assumes that fracture occurs by a weakest link mechanism under J-controlled conditions, it is possible to derive a closed-form expression for the fracture-toughness distribution. When weakest link initiation is necessary but not sufficient for cleavage fracture, the problem becomes somewhat more complicated, but it is still possible to describe the cleavage process mathematically.

A5.1   WEAKEST LINK FRACTURE

As discussed in Section 5.2, the weakest link model for cleavage assumes that failure occurs when at least one critical fracture-triggering particle is sampled by the crack tip. Equation (5.22) describes the failure probability in this case.[4] Since cleavage is stress controlled, the

---

[4] It turns out that Equation (5.22) is valid even when the Poisson assumption is not applicable [40]; the quantity ρ is not the microcrack density in such cases but ρ is *uniquely related* to microcrack density. Thus, the derivation of the fracture-toughness distribution presented in this section does not hinge on the Poisson assumption.

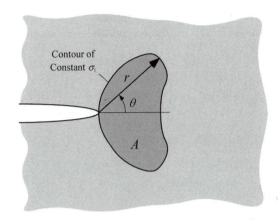

FIGURE A5.1 Definition of r, θ, and area for a principal stress contour.

microcrack density (i.e., the number of critical microcracks per unit volume) should depend only on the maximum principal stress.[5]

$$\rho = \rho(\sigma_1) \tag{A5.1}$$

This quantity must be integrated over the volume ahead of the crack tip. In order to perform this integration, it is necessary to relate the crack-tip stresses to the volume sampled at each stress level.

Recall Section 3.5, where dimensional analysis indicated that the stresses ahead of the crack tip in the limit of small-scale yielding are given by

$$\frac{\sigma_1}{\sigma_o} = f\left(\frac{J}{r\sigma_o}, \theta\right) \tag{A5.2}$$

assuming Young's modulus is fixed in the material and thus does not need to be included in the dimensional analysis. Equation (A5.2) can be inverted to solve for the distance ahead of the crack tip (at a given angle) which corresponds to a particular stress value:

$$r(\sigma_1/\sigma_o, \theta) = \frac{J}{\sigma_o} g(\sigma_1/\sigma_o, \theta) \tag{A5.3}$$

By fixing σ_1 and varying θ from $-\pi$ to $+\pi$, we can construct a contour of constant principal stress, as illustrated in Figure A5.1. The area inside this contour is given by

$$A(\sigma_1/\sigma_o) = \frac{J^2}{\sigma_o^2} h(\sigma_1/\sigma_o) \tag{A5.4}$$

where h is a dimensionless integration constant:

$$h(\sigma_1/\sigma_o) = \frac{1}{2} \int_{-\pi}^{+\pi} g(\sigma_1/\sigma_o, \theta) d\theta \tag{A5.5}$$

---

[5] Although this derivation assumes that the maximum principal stress at a point controls the incremental cleavage probability, the same basic result can be obtained by inserting any stress component into Equation (A5.1). For example, one might assume that the tangential stress $\sigma_{\theta\theta}$ governs cleavage.

For plane strain conditions in an edge-cracked test specimen, the volume sampled at a given stress value is simply BA, where B is the specimen thickness. Therefore, the incremental volume at a fixed J and σ_o is given by

$$dV(\sigma_1) = \frac{BJ^2}{\sigma_o^2} \frac{\partial h}{\partial \sigma_1} d\sigma_1 \qquad (A5.6)$$

Inserting Equation (A5.1) and Equation (A5.6) into Equation (5.21) gives

$$F = 1 - \exp\left(-\frac{BJ^2}{\sigma_o^2} \int_{\sigma_u}^{\sigma_{max}} \rho(\sigma_1) \frac{\partial h}{\partial \sigma_1} d\sigma_1\right) \qquad (A5.7)$$

here σ_{max} is the peak value of stress that occurs ahead of the crack tip and σ_u is the threshold fracture stress, which corresponds to the largest fracture-triggering particle the material is likely to contain.

Note that J appears outside of the integral in Equation (A5.7). By setting $J = J_c$ in Equation (A5.7), we obtain an expression for the statistical distribution of critical J values, which can be written in the following form:

$$F = 1 - \exp\left[-\frac{B}{B_o}\left(\frac{J_c}{\Theta_J}\right)^2\right] \qquad (A5.8)$$

where B_o is a reference thickness. When $B = B_o$, Θ_J is the 63rd percentile J_c value. Equation (A5.8) has the form of a two-parameter Weibull distribution, as discussed in Section 5.2.3. Invoking the relationship between K and J for small-scale yielding gives

$$F = 1 - \exp\left[-\frac{B}{B_o}\left(\frac{K_{IC}}{\Theta_K}\right)^4\right] \qquad (A5.9)$$

Equation (A5.8) and Equation (A5.9) both predict a thickness effect on toughness. The average toughness is proportional to $1/\sqrt{B}$ for critical J values and $B^{-0.25}$ for K_{Ic} data. The average toughness does not increase indefinitely with thickness, however. There are limits to the validity of the weakest link model, as discussed in the next section.

All of the above relationships are valid only when weakest link failure occurs under J-controlled conditions, i.e., the single-parameter assumption must apply. When the constraint relaxes, critical J values no longer follow a Weibull distribution with a specific slope, but the effective small-scale yielding J values, J_o (see Section 3.6.3), follow Equation (A5.8) if a weakest link mechanism controls failure. Actual J_c values would be more scattered than J_o values, however, because the ratio J/J_o increases with J.

A5.2 INCORPORATING A CONDITIONAL PROBABILITY
OF PROPAGATION

In many materials, the weakest link initiation of cleavage appears to be necessary but not sufficient. Figure 5.26 schematically illustrates a probability tree for cleavage initiation and propagation. This diagram is a slight oversimplification, because the cumulative failure probability must be computed incrementally.

Modifying the statistical cleavage model to account for propagation requires that the probability be expressed in terms of a *hazard function* [52], which defines the instantaneous risk of fracture.

For a random variable T, the hazard function $H(T)$, and the cumulative probability are related as follows:

$$F = 1 - \exp\left(-\int_{T_o}^{T} H(T)dT\right) \qquad (A5.10)$$

where T_o is the minimum value of T. By comparing Equation (A5.17) and Equation (A5.18), it can easily be shown that the hazard function for the weakest link initiation, in terms of stress intensity, is given by

$$H(K) = \frac{4K^3}{\Theta_K^4} \qquad (A5.11)$$

assuming $B = B_o$. The hazard function for *total failure* is equal to Equation (A5.11) times the conditional probability of failure:

$$H(K) = P_{pr}\frac{4K^3}{\Theta_K^4} \qquad (A5.12)$$

Thus, the overall probability of failure is given by

$$F = 1 - \exp\left(-\int_0^K P_{pr}\frac{4K^3}{\Theta_K^4}dK\right) \qquad (A5.13)$$

Consider the case where P_{pr} is a constant, i.e., it does not depend on the applied K. Suppose, for example, that half of the carbides of a critical size have a favorable orientation with respect to a cleavage plane in a ferrite grain. The failure probability becomes

$$F = 1 - \exp\left[-0.5\left(\frac{K}{\Theta_K}\right)^4\right] \qquad (A5.14)$$

In this instance, the finite propagation probability merely shifts the 63rd percentile toughness to a higher value:

$$\Theta_{K^*} = 2^{0.25}\Theta_K = 1.19\Theta_K$$

The shape of the distribution is unchanged, and the fracture process still follows a weakest link model. In this case, the weak link is defined as a particle that is greater than the critical size *that is also oriented favorably.*

Deviations from the weakest link distribution occur when P_{pr} depends on the applied K. If the conditional probability of propagation is a step function

$$P_{pr} = \begin{cases} 0, & K_I < K_o \\ 1, & K_I \geq K_o \end{cases}$$

the fracture-toughness distribution becomes a truncated Weibull (Equation (5.24)); failure can occur only when $K > K_o$. The introduction of a threshold toughness also reduces the relative scatter, as discussed in Section 5.2.3.

Equation (5.24) implies that the arrest toughness is single valued; a microcrack always propagates above K_o, but always arrests at or below K_o. Experimental data, however, indicate that arrest can occur over a range of K values. The data in Figure 5.27 exhibit a sigmoidal shape, while the truncated Weibull is nearly linear near the threshold.

A computer simulation of cleavage propagation in a polycrystalline material [42, 43] resulted in a prediction of P_{pr} as a function of the applied K; these results fit an offset power law expression (Equation (5.25)). The absolute values obtained from the simulation are questionable, but the predicted trend is reasonable. Inserting Equation (5.25) into Equation (A5.13) gives

$$F = 1 - \exp\left(-\int_{K_o}^{K} \alpha(K - K_o)^\beta \frac{4K^3}{\Theta_K^4} \, dK \right) \qquad (A5.15)$$

The integral in Equation (A5.15) has a closed-form solution, but it is rather lengthy. The above distribution exhibits a sigmoidal shape, much like the experimental data in Figure 5.27. Unfortunately, it is very difficult to fit experimental data to Equation (A5.22). Note that there are four fitting parameters in this distribution: α, β, K_o, and Θ_K. Even with fewer unknown parameters, the form of Equation (A5.15) is not conducive to curve-fitting because it cannot be linearized.

Equation (A5.15) can be approximated with a conventional three-parameter Weibull distribution with the slope fixed at 4 (Equation (5.26)). The latter expression also gives a reasonably good fit of experimental data (Figure 5.27). The three-parameter Weibull distribution is sufficiently flexible to model a wide range of behavior. The advantage of Equation (5.26) is that there are only two parameters to fit (the Weibull-shape parameter is fixed at 4.0) and it can be linearized. Wallin [46] has shown that Equation (5.26) is rigorously correct if P_{pr} is given by Equation (5.32).

REFERENCES

1. Knott, J.F., "Micromechanisms of Fracture and the Fracture Toughness of Engineering Alloys." *Fracture 1977, Proceedings of the Fourth International Conference on Fracture* (ICF4), Waterloo, Canada, Vol. 1, 1977, pp. 61–91.
2. Knott, J.F., "Effects of Microstructure and Stress-State on Ductile Fracture in Metallic Alloys." In: K. Salama, et al. *Advances in Fracture Research, Proceedings of the Seventh International Conference on Fracture (ICF7)*. Pergamon Press, Oxford, 1989, pp. 125–138.
3. Wilsforf, H.G.F., "The Ductile Fracture of Metals: A Microstructural Viewpoint." *Materials Science and Engineering*, Vol. 59, 1983, pp. 1–19.
4. Garrison, W.M., Jr. and Moody, N.R., "Ductile Fracture." *Journal of the Physics and Chemistry of Solids*, Vol. 48, 1987, pp. 1035–1074.
5. Knott, J.F., "Micromechanisms of Fibrous Crack Extension in Engineering Alloys." *Metal Science*, Vol. 14, 1980, pp. 327–336.
6. Argon, A.S., Im, J., and Safoglu, R., "Cavity Formation from Inclusions in Ductile Fracture." *Metallurgical Transactions*, Vol. 6A, 1975, pp. 825–837.
7. Beremin, F.M., "Cavity Formation from Inclusions in Ductile Fracture of A 508 Steel." *Metallurgical Transactions*, Vol. 12A, 1981, pp. 723–731.
8. Brown, L.M. and Stobbs, W.M., "The Work-Hardening of Copper-Silica vs. Equilibrium Plastic Relaxation by Secondary Dislocations." *Philosophical Magazine*, 1976, Vol. 34, pp. 351–372.
9. Goods, S.H. and Brown, L.M., "The Nucleation of Cavities by Plastic Deformation." *Acta Metallurgica*, Vol. 27, 1979, pp. 1–15.
10. Van Stone, R.H., Cox, T.B., Low, J.R., Jr., and Psioda, P.A., "Microstructural Aspects of Fracture by Dimpled Rupture." *International Metallurgical Reviews*, Vol. 30, 1985, pp. 157–179.
11. Thomason, P.F., *Ductile Fracture of Metals*. Pergamon Press, Oxford, 1990.
12. Rice, J.R. and Tracey, D.M., "On the Ductile Enlargement of Voids in Triaxial Stress Fields." *Journal of the Mechanics and Physics of Solids*, Vol. 17, 1969, pp. 201–217.

13. Gurson, A.L., "Continuum Theory of Ductile Rupture by Void Nucleation and Growth: Part 1—Yield Criteria and Flow Rules for Porous Ductile Media." *Journal of Engineering Materials and Technology,* Vol. 99, 1977, pp. 2–15.

14. Berg, C.A., "Plastic Dilation and Void Interaction." *Inelastic Behavior of Solids.* McGraw-Hill, New York, 1970, pp. 171–210.

15. Tvergaard, V., "On Localization in Ductile Materials Containing Spherical Voids." *International Journal of Fracture,* Vol. 18, 1982, pp. 237–252.

16. Tvegaard, V., "Material Failure by Void Growth to Coalescence." *Advances in Applied Mechanics,* Vol. 27, 1990, pp. 83–151.

17. Chu, C.C. and Needleman, A., "Void Nucleation Effects in Biaxially Stretched Sheets." *Journal of Engineering Materials and Technology,* Vol. 102, 1980, pp. 249–256.

18. Tvergaard, V. and Needleman, A., "Analysis of the Cup-Cone Fracture in a Round Tensile Bar." *Acta Metallurgica,* Vol. 32, 1984, pp. 157–169.

19. Purtscher, P.T., "Micromechanisms of Ductile Fracture and Fracture Toughness in a High Strength Austenitic Stainless Steel." Ph.D. Dissertation, Colorado School of Mines, Golden, CO, 1990.

20. d'Escata, Y. and Devaux, J.C., "Numerical Study of Initiation, Stable Crack Growth, and Maximum Load with a Ductile Fracture Criterion Based on the Growth of Holes." ASTM STP 668, American Society for Testing and Materials, Philadelphia, PA, 1979, pp. 229–248.

21. McMeeking, R.M. and Parks, D.M., "On Criteria for J-Dominance of Crack-Tip Fields in Large-Scale Yielding." ASTM STP 668, American Society for Testing and Materials, Philadelphia, PA, 1979, pp. 175–194.

22. Beachem, C.D. and Yoder, G.R., "Elastic-Plastic Fracture by Homogeneous Microvoid Coalescence Tearing Along Alternating Shear Planes." *Metallurgical Transactions,* Vol. 4A, 1973, pp. 1145–1153.

23. Gudas, J.P., "Micromechanisms of Fracture and Crack Arrest in Two High Strength Steels." Ph.D. Dissertation, Johns Hopkins University, Baltimore, MD, 1985.

24. Cottrell, A.H., "Theory of Brittle Fracture in Steel and Similar Metals." *Transactions of the ASME,* Vol. 212, 1958, pp. 192–203.

25. McMahon, C.J., Jr and Cohen, M., "Initiation of Cleavage in Polycrystalline Iron." *Acta Metallurgica,* Vol. 13, 1965, pp. 591–604.

26. Smith, E., "The Nucleation and Growth of Cleavage Microcracks in Mild Steel." *Proceedings of the Conference on the Physical Basis of Fracture,* Institute of Physics and Physics Society, 1966, pp. 36–46.

27. Heerens, J. and Read, D.T., "Fracture Behavior of a Pressure Vessel Steel in the Ductile-to-Brittle Transition Region." NISTIR 88-3099, National Institute for Standards and Technology, Boulder, CO, 1988.

28. Dolby, R.E. and Knott, J.F., "Toughness of Martensitic and Martensitic-Bainitic Microstructures with Particular Reference to Heat-Affected Zones." *Journal of the Iron and Steel Institute,* Vol. 210, 1972, pp. 857–865.

29. Lin, T., Evans, A.G., and Ritchie, R.O., "Statistical Model of Brittle Fracture by Transgranular Cleavage." *Journal of the Mechanics and Physics of Solids,* Vol. 34, 1986, pp. 477–496.

30. Ritchie, R.O., Knott, J.F., and Rice, J.R., "On the Relationship between Critical Tensile Stress and Fracture Toughness in Mild Steel." *Journal of the Mechanics and Physics of Solids,* Vol. 21, 1973, pp. 395–410.

31. Rice, J.R. and Tracey, D.M., "Computational Fracture Mechanics." *Numerical Computer Methods in Structural Mechanics,* Academic Press, New York, 1973, pp. 585–623.

32. Curry, D.A. and Knott, J.F., "Effects of Microstructure on Cleavage Fracture Stress in Steel." *Metal Science,* Vol. 12, 1978, pp. 511–514.

33. Watanabe, J., Iwadate, T., Tanaka, Y., Yokoboro, T., and Ando, K., "Fracture Toughness in the Transition Region." *Engineering Fracture Mechanics,* Vol. 28, 1987, pp. 589–600.

34. Curry, D.A. and Knott, J.F., "Effect of Microstructure on Cleavage Fracture Toughness in Mild Steel." *Metal Science,* Vol. 13, 1979, pp. 341–345

35. Landes, J.D. and Shaffer, D.H., "Statistical Characterization of Fracture in the Transition Region." ASTM STP 700, American Society for Testing and Materials, Philadelphia, PA, 1980, pp. 368–372.

36. Anderson, T.L. and Williams, S., "Assessing the Dominant Mechanism for Size Effects in the Ductile-to-Brittle Transition Region." ASTM STP 905, American Society for Testing and Materials, Philadelphia, PA, 1986, pp. 715–740.

37. Anderson, T.L. and Stienstra, D., "A Model to Predict the Sources and Magnitude of Scatter in Toughness Data in the Transition Region." *Journal of Testing and Evaluation,* Vol. 17, 1989, pp. 46–53.

38. Evans, A.G., "Statistical Aspects of Cleavage Fracture in Steel." *Metallurgical Transactions,* Vol. 14A, 1983, pp. 1349–1355.

39. Wallin, K., Saario, T., and Törrönen, K., "Statistical Model for Carbide Induced Brittle Fracture in Steel." *Metal Science,* Vol. 18, 1984, pp. 13–16.

40. Beremin, F.M., "A Local Criterion for Cleavage Fracture of a Nuclear Pressure Vessel Steel." *Metallurgical Transactions,* Vol. 14A, 1983, pp. 2277–2287.

41. Weibull, W., "A Statistical Distribution Function of Wide Applicability." *Journal of Applied Mechanics,* Vol. 18, 1953, pp. 293–297.

42. Stienstra, D.I.A., "Stochastic Micromechanical Modeling of Cleavage Fracture in the Ductile-Brittle Transition Region." Ph.D. Dissertation, Texas A&M University, College Station, TX, 1990.

43. Anderson, T.L., Stienstra, D.I.A., and Dodds, R.H., Jr., "A Theoretical Framework for Addressing Fracture in the Ductile-Brittle Transition Region." *Fracture Mechanics: 24th Volume,* ASTM STP 1207, American Society for Testing and Materials, Philadelphia, PA (in press).

44. Gell, M. and Smith, E., "The Propagation of Cracks Through Grain Boundaries in Polycrystalline 3% Silicon-Iron." *Acta Metallurgica,* Vol. 15, 1967, pp. 253–258.

45. Anderson, T.L. and Dodds, R.H., Jr., "Specimen Size Requirements for Fracture Toughness Testing in the Ductile-Brittle Transition Region." *Journal of Testing and Evaluation,* Vol. 19, 1991, pp. 123–134.

46. Wallin, K. "Microscopic Nature of Brittle Fracture." *Journal de Physique,* Vol. 3, 1993, pp. 575–584.

47. Wallin, K., "Fracture Toughness Testing in the Ductile-Brittle Transition Region." In: K. Salarna, et al. (eds.), *Advances in Fracture Research, Proceedings of the Seventh International Conference on Fracture (ICF7)* Pergamon Press, Oxford, 1989, pp. 267–276.

48. Rosenfield, A.R. and Shetty, D.K., "Cleavage Fracture in the Ductile-Brittle Transition Region." ASTM STP 856, American Society for Testing and Materials, Philadelphia, PA, 1985, pp. 196–209.

49. Hoagland, R.G., Rosenfield, A.R., and Hahn, G.T., "Mechanisms of Fast Fracture and Arrest in Steels." *Metallurgical Transactions,* Vol. 3, 1972, pp. 123–136.

50. Krauss, G., *Principles of Heat Treatment of Steel.* American Society for Metals, Metals Park, OH, 1980.

51. Riedel, H., *Creep Crack Growth.* ASTM STP 1020, American Society for Testing and Materials, Philadelphia, PA, 1989, pp. 101–126.

52. Bain, L.J., *Statistical Analysis of Reliability and Life-Testing Models.* Marcel Dekker, New York, 1978

6 Fracture Mechanisms in Nonmetals

Traditional structural metals such as steel and aluminum are being replaced with plastics, ceramics, and composites in a number of applications. Engineering plastics have a number of advantages, including low cost, ease of fabrication, and corrosion resistance. Ceramics provide superior wear resistance and creep strength. Composites offer high strength/weight ratios, and enable engineers to design materials with specific elastic and thermal properties. Traditional nonmetallic materials such as concrete continue to see widespread use.

Nonmetals, like metals, are not immune to fracture. Recall from Chapter 1, the example of the pinch clamping of a polyethylene pipe that led to time-dependent fracture. The so-called high-toughness ceramics that have been developed in recent years (Section 6.2) have lower toughness than even the most brittle steels. Relatively minor impact (e.g., an airplane mechanic accidentally dropping his wrench on a wing) can cause microscale damage in a composite material, which can adversely affect the subsequent performance. The lack of ductility of concrete (relative to steel) limits its range of application.

Compared with the fracture of metals, research into the fracture behavior of nonmetals is in its infancy. Much of the necessary theoretical framework is not yet fully developed for nonmetals, and there are many instances where fracture mechanics concepts that apply to metals have been misapplied to other materials.

This chapter gives a brief overview of the current state of understanding of fracture and failure mechanisms in selected nonmetallic structural materials. Although the coverage of the subject is far from complete, this chapter should enable the reader to gain an appreciation of the diverse fracture behavior that various materials can exhibit. The references listed at the end of the chapter provide a wealth of information to those who desire a more in-depth understanding of a particular material system. The reader should also refer to Chapter 8, which describes the current methods for fracture toughness measurements in nonmetallic materials.

Section 6.1 outlines the molecular structure and mechanical properties of polymeric materials, and describes how these properties influence the fracture behavior. This section also includes a discussion of the fracture mechanisms in polymer matrix composites. Section 6.2 considers fracture in ceramic materials, including the newest generation of ceramic composites. Section 6.3 addresses fracture in concrete and rock.

This chapter does not specifically address metal matrix composites, but these materials have many features in common with polymer and ceramic matrix composites [1]. Also, the metal matrix in these materials should exhibit the fracture mechanisms described in Chapter 5.

6.1 ENGINEERING PLASTICS

The fracture behavior of polymeric materials has only recently become a major concern, as engineering plastics have begun to appear in critical structural applications. In most consumer products made from polymers (e.g., toys, garbage bags, ice chests, lawn furniture, etc.), fracture may be an annoyance, but it is not a significant safety issue. Fracture in plastic natural gas piping systems or aircraft wings, however, can have dire consequences.

Several books devoted solely to the fracture and fatigue of plastics have been published in recent years [2–6]. These references proved invaluable to the author in preparing Chapter 6 and Chapter 8.

Let us begin the discussion of fracture in plastics by reviewing some of the basic principles of polymeric materials.

6.1.1 STRUCTURE AND PROPERTIES OF POLYMERS

A polymer is defined as the union of two or more compounds called mers. The *degree of polymerization* is a measure of the number of these units in a given molecule. Typical engineering plastics consist of very long chains, with the degree of polymerization on the order of several thousand.

Consider polyethylene, a polymer with a relatively simple molecular structure. The building block in this case is ethylene (C_2H_4), which consists of two carbon atoms joined by a double bond, with two hydrogen atoms attached to each carbon atom. If sufficient energy is applied to this compound, the double bond can be broken, resulting in two free radicals that can react with other ethylene groups:

$$
\begin{array}{ccc}
\text{H} \quad \text{H} & & \text{H} \, \text{H} \\
| \quad\ | & & | \ | \\
\text{C} = \text{C} + \text{ energy} & \rightarrow & -\text{C}\,\text{C}- \\
| \quad\ | & & | \ | \\
\text{H} \quad \text{H} & & \text{H}\,\text{H}
\end{array}
$$

The degree of polymerization (i.e., the length of the chain) can be controlled by the heat input, catalyst, as well as reagents that may be added to aid the polymerization process.

6.1.1.1 Molecular Weight

The molecular weight is a measure of the length of a polymer chain. Since there is typically a distribution of molecule sizes in a polymer sample, it is convenient to quantify an average molecular weight, which can be defined in one of two ways. The *number average* molecular weight is the total weight divided by the number of molecules:

$$
\overline{M}_n = \frac{\sum_{i=1}^{n} N_i M_i}{\sum_{i=1}^{n} N_i}
\tag{6.1}
$$

where N_i is the number of molecules with molecular weight M_i. The number average molecular weight attaches equal importance to all molecules, while the *weight average* molecular weight reflects the actual average weight of molecules by placing additional emphasis on the larger molecules:

$$
\overline{M}_w = \frac{\sum_{i=1}^{n} N_i M_i^2}{\sum_{i=1}^{n} N_i M_i}
\tag{6.2}
$$

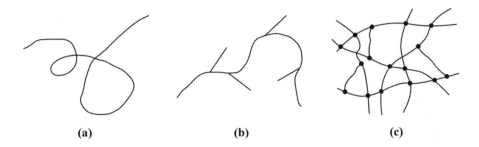

FIGURE 6.1 Three types of polymer chains: (a) linear polymer, (b) branched polymer, and (c) cross-linked polymer.

These two measures of molecular weight are obviously identical if all the molecules in the sample are of the same size, but the number average is usually lower than the weight average molecular weight. The *polydispersity* is defined as the ratio of these two quantities:

$$PD = \frac{\overline{M}_w}{\overline{M}_n} \tag{6.3}$$

A narrow distribution of molecular weights implies a PD close to 1, while PD can be greater than 20 in materials with broadly distributed molecule sizes. Both measures of molecular weight, as well as the PD, influence the mechanical properties of a polymer.

6.1.1.2 Molecular Structure

The structure of polymer chains also has a significant effect on the mechanical properties. Figure 6.1 illustrates three general classifications of polymer chains: linear, branched, and cross-linked. *Linear polymers* are not actual straight lines; rather, the carbon atoms in a linear molecule form a single continuous path from one end of the chain to the other. A *branched polymer* molecule, as the name suggests, contains a series of smaller chains that branch off from a main "backbone." A *cross-linked polymer* consists of a network structure rather than linear chains. A highly cross-linked structure is typical of *thermoset polymers*, while *thermoplastics* consist of linear and branched chains. *Elastomers* typically have lightly cross-linked structures and are capable of large elastic strains.

Epoxies are the most common example of thermoset polymers. Typically, two compounds that are in the liquid state at ambient temperature are mixed together to form an epoxy resin, which solidifies into a cross-linked lattice upon curing. This process is irreversible; a thermoset cannot be formed into another shape once it solidifies.

Thermomechanical processes in thermoplastics are reversible, because these materials do not form cross-linked networks. Thermoplastics become viscous upon heating, where they can be formed into the desired shape.

6.1.1.3 Crystalline and Amorphous Polymers

Polymer chains can be packed tightly together in a regular pattern, or they can form random entanglements. Materials that display the former configuration are called crystalline polymers, while the disordered state corresponds to amorphous (glassy) polymers. Figure 6.2 schematically illustrates crystalline and amorphous arrangements of polymer molecules.

The term *crystalline* does not have the same meaning for polymers as for metals and ceramics. A crystal structure in a metal or ceramic is a regular array of atoms with three-dimensional symmetry; all the atoms in a crystal have identical surroundings (except atoms that are adjacent to

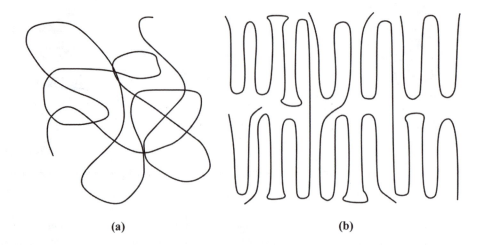

FIGURE 6.2 Amorphous and crystalline polymers: (a) amorphous polymer and (b) crystalline polymer.

a defect, such as a dislocation or vacancy). The degree of symmetry in a crystalline polymer, however, is much lower, as Figure 6.2 illustrates.

Figure 6.3 illustrates the volume-temperature relationships in crystalline and amorphous thermoplastics. As a crystalline polymer cools from the liquid state, an abrupt decrease in volume occurs at the melting temperature T_m, and the molecular chains pack efficiently in response to the thermodynamic drive to order into a crystalline state. The volume discontinuity at T_m resembles the behavior of crystalline metals and ceramics. An amorphous polymer bypasses T_m upon cooling, and remains in a viscous state until it reaches the glass transition temperature T_g, at which time the relative motion of the molecules becomes restricted. An amorphous polymer contains more free volume than the same material in the crystalline state, and thus has a lower density. The glass transition temperature is sensitive to the cooling rate; rapid heating or cooling tends to increase T_g, as Figure 6.3 indicates.

Semicrystalline polymers contain both crystalline and glassy regions. The relative fraction of each state depends on a number of factors, including the molecular structure and cooling rate. Slow cooling provides more time for the molecules to arrange themselves in an equilibrium crystal structure.

6.1.1.4 Viscoelastic Behavior

Polymers exhibit rate-dependent viscoelastic deformation, which is a direct result of their molecular structure. Figure 6.4 gives a simplified view of viscoelastic behavior on the molecular level. Two neighboring molecules, or different segments of a single molecule that is folded back upon itself, experience weak attractive forces called *Van der Waals bonds*. These secondary bonds resist any

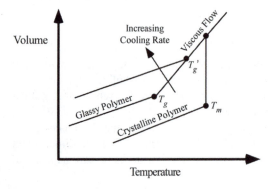

FIGURE 6.3 Volume-temperature relationships for amorphous (glassy) and crystalline polymers.

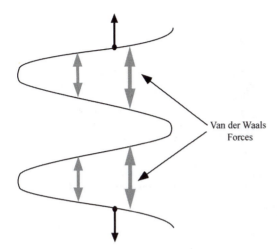

FIGURE 6.4 Schematic deformation of a polymer chain. Secondary Van der Waals bonds between chain segments resist forces that try to extend the molecule.

external force that attempts to pull the molecules apart. The elastic modulus of a typical polymer is significantly lower than Young's modulus for metals and ceramics, because the Van der Waals bonds are much weaker than primary bonds. Deforming a polymer requires cooperative motion among molecules. The material is relatively compliant if the imposed strain rate is sufficiently low to provide molecules sufficient time to move. At faster strain rates, however, the forced molecular motion produces friction, and a higher stress is required to deform the material. If the load is removed, the material attempts to return to its original shape, but molecular entanglements prevent instantaneous elastic recovery. If the strain is sufficiently large, yielding mechanisms occur, such as crazing and shear deformation (see Section 6.6.2), and much of the induced strain is essentially permanent.

Section 4.3 introduced the relaxation modulus $E(t)$, and the creep compliance $D(t)$, which describe the time-dependent response of viscoelastic materials. The relaxation modulus and creep compliance can be obtained experimentally by fixing the strain and stress, respectively:

$$E(t) = \frac{\sigma(t)}{\varepsilon_o}, \qquad D(t) = \frac{\varepsilon(t)}{\sigma_o} \qquad (6.4)$$

See Figure 4.19 for a schematic illustration of stress relaxation and creep experiments. For linear viscoelastic materials,[1] $E(t)$ and $D(t)$ are related through a hereditary integral (Equation (4.61)).

Figure 6.5(a) is a plot of the relaxation modulus vs. temperature at a fixed time for a thermoplastic. Below T_g, the modulus is relatively high, as molecular motion is restricted. At around T_g, the modulus (at a fixed time) decreases rapidly, and the polymer exhibits a "leathery" behavior. At higher temperatures, the modulus reaches a lower plateau, and the polymer is in a rubbery state. Natural and synthetic rubbers are merely materials whose glass transition temperature is below room temperature.[2] If the temperature is sufficiently high, linear polymers lose virtually all their load-carrying capacity and behave like a viscous fluid. Highly cross-linked polymers, however, maintain a modulus plateau.

---

[1] Linear viscoelastic materials do not, in general, have linear stress-strain curves (since the modulus is time dependent), but display other characteristics of linear elasticity such as superposition. See Section 4.3.1 for a definition of linear viscoelasticity.

[2] To demonstrate the temperature dependence of viscoelastic behavior, try blowing up a balloon after it has been in a freezer for an hour.

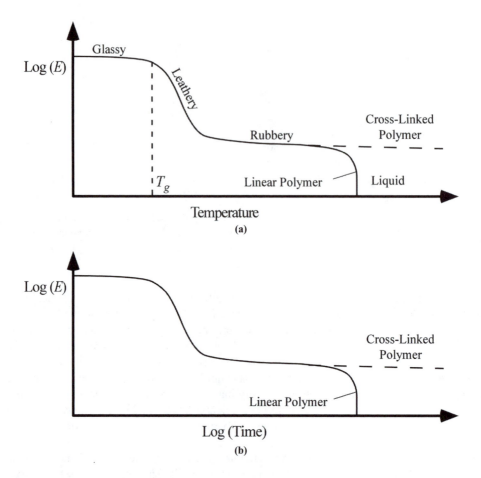

FIGURE 6.5 Effect of temperature and time on the modulus of an amorphous polymer: (a) modulus versus temperature at a fixed time and (b) modulus versus time at a fixed temperature.

Figure 6.5(b) shows a curve with the same characteristic shape as Figure 6.5(a), but with a fixed temperature and varying time. At short times, the polymer is glassy, but exhibits leathery, rubbery, and liquid behavior at sufficiently long times. Of course, *short time* and *long time* are relative terms that depend on temperature. A polymer significantly below T_g might remain in a glassy state during the time frame of a stress relaxation test, while a polymer well above T_g may pass through this state so rapidly that the glassy behavior cannot be detected.

The equivalence between high temperature and long times (i.e., the time-temperature superposition principle) led Williams, Landel, and Ferry [7] to develop a semiempirical equation that collapses data at different times onto a single modulus-temperature master curve. They defined a time shift factor a_T as follows:

$$\log a_T = \log \frac{t_T}{t_{T_o}} = \frac{C_1(T - T_o)}{C_2 + T - T_o} \tag{6.5}$$

where

t_T and t_{T_o} = times to reach a specific modulus at temperatures T and T_o, respectively

T_o = reference temperature (usually defined at T_g)

C_1 and C_2 = fitting parameters that depend on material properties.

Equation (6.5), which is known as the WLF relationship, typically is valid in the range $T_g < T < T_g + 100°C$. Readers familiar with creep in metals may recognize an analogy with the Larson-Miller parameter [8], which assumes a time-temperature equivalence for creep rupture.

6.1.1.5 Mechanical Analogs

Simple mechanical models are useful for understanding the viscoelastic response of polymers. Three such models are illustrated in Figure 6.6. The Maxwell model (Figure 6.6(a)) consists of a spring and a dashpot in series, where a dashpot is a moving piston in a cylinder of viscous fluid. The Voigt model (Figure 6.6(b)) contains a spring and a dashpot in parallel. Figure 6.6(c) shows a combined Maxwell-Voigt model. In each case, the stress-strain response in the spring is instantaneous:

$$\varepsilon = \frac{\sigma}{E} \tag{6.6}$$

while the dashpot response is time dependent:

$$\dot{\varepsilon} = \frac{\sigma}{\eta} \tag{6.7}$$

where $\dot{\varepsilon}$ is the strain rate and η is the fluid viscosity in the dashpot. The temperature dependence of η can be described by an Arrhenius rate equation:

$$\eta = \eta_o \, e^{\left(\frac{Q}{RT}\right)} \tag{6.8}$$

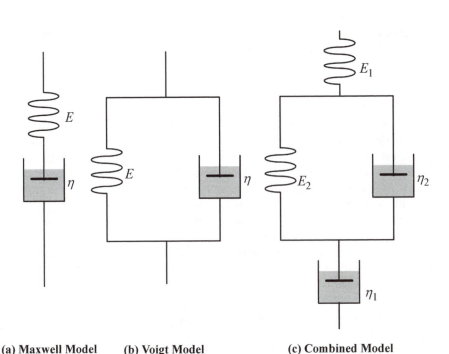

(a) Maxwell Model   **(b) Voigt Model**   **(c) Combined Model**

FIGURE 6.6 Mechanical analogs for viscoelastic deformation in polymers.

where

     Q = activation energy for viscous flow (which may depend on temperature)

     T = absolute temperature

     R = gas constant (= 8.314 J/(mole K))

In the Maxwell model, the stresses in the spring and dashpot are equal, and the strains are additive. Therefore,

$$\dot{\varepsilon} = \frac{\sigma}{\eta} + \frac{1}{E}\frac{d\sigma}{dt} \tag{6.9}$$

For a stress-relaxation experiment (Figure 4.19(b)), the strain is fixed at ε_o, and $\dot{\varepsilon} = 0$. Integrating stress with respect to time for this case leads to

$$\sigma(t) = \sigma_o\, e^{-t/t_R} \tag{6.10}$$

where σ_o is the stress at $t = 0$, and $t_R + \eta/E$ is the relaxation time.

When the spring and dashpot are in parallel (the Voigt model) the strains are equal and the stresses are additive:

$$\sigma(t) = E\varepsilon + \eta\dot{\varepsilon} \tag{6.11}$$

For a constant stress creep test, Equation (6.11) can be integrated to give

$$\varepsilon(t) = \frac{\sigma_o}{E}(1 - e^{-t/t_R}) \tag{6.12}$$

Note that the limiting value of creep strain in this model is σ_o/E, which corresponds to zero stress on the dashpot. If the stress is removed, the strain recovers with time:

$$\varepsilon(t) = \varepsilon_o\, e^{-t/t_R} \tag{6.13}$$

where ε_o is the strain at $t = 0$, and zero time is defined at the moment the load is removed.

Neither model describes all types of viscoelastic response. For example, the Maxwell model does not account for viscoelastic recovery, because the strain in the dashpot is not reversed when the stress is removed. The Voigt model cannot be applied to the stress relaxation case, because when strain is fixed in Equation (6.11), all of the stress is carried by the spring; the problem reduces to simple static loading, where both stress and strain remain constant.

If we combine the two models, however, we obtain a more realistic and versatile model of viscoelastic behavior. Figure 6.6(c) illustrates the combined Maxwell-Voigt model. In this case, the strains in the Maxwell and Voigt contributions are additive, and the stress carried by the Maxwell spring and dashpot is divided between the Voigt spring and dashpot. For a constant stress creep test, combining Equation (6.9) and Equation (6.13) gives

$$\varepsilon(t) = \frac{\sigma_o}{E_1} + \frac{\sigma_o}{E_2}(1 - e^{-t/t_{R(2)}}) + \frac{\sigma_o t}{\eta_1} \tag{6.14}$$

All three models are oversimplifications of actual polymer behavior, but are useful for approximating different types of viscoelastic response.

6.1.2 YIELDING AND FRACTURE IN POLYMERS

In metals, fracture and yielding are competing failure mechanisms. Brittle fracture occurs in materials in which yielding is difficult. Ductile metals, by definition, experience extensive plastic deformation before they eventually fracture. Low temperatures, high strain rates, and triaxial tensile stresses tend to suppress yielding and favor brittle fracture.

From a global point of view, the foregoing also applies to polymers, but the microscopic details of yielding and fracture in plastics are different from metals. Polymers do not contain crystallographic planes, dislocations, and grain boundaries; rather, they consist of long molecular chains. Section 2.1 states that fracture on the atomic level involves breaking bonds, and polymers are no exception. A complicating feature for polymers, however, is that two types of bond govern the mechanical response: the covalent bonds between carbon atoms and the secondary van der Waals forces between molecule segments. Ultimate fracture normally requires breaking the latter, but the secondary bonds often play a major role in the deformation mechanisms that lead to fracture.

The factors that govern the toughness and ductility of polymers include the strain rate, temperature, and molecular structure. At high rates or low temperatures (relative to T_g) polymers tend to be brittle, because there is insufficient time for the material to respond to stress with large-scale viscoelastic deformation or yielding. Highly cross-linked polymers are also incapable of large-scale viscoelastic deformation. The mechanism illustrated in Figure 6.4, where molecular chains overcome van der Waals forces, does not apply to cross-linked polymers; *primary* bonds between chain segments must be broken for these materials to deform.

6.1.2.1 Chain Scission and Disentanglement

Fracture, by definition, involves material separation, which normally implies severing bonds. In the case of polymers, fracture on the atomic level is called *chain scission.*

Recall from Chapter 2 that the theoretical bond strength in most materials is several orders of magnitude larger than the measured fracture stresses, but crack-like flaws can produce significant local stress concentrations. Another factor that aids chain scission in polymers is that molecules are not stressed uniformly. When a stress is applied to a polymer sample, certain chain segments carry a disproportionate amount of load, which can be sufficient to exceed the bond strength. The degree of nonuniformity in stress is more pronounced in amorphous polymers, while the limited degree of symmetry in crystalline polymers tends to distribute stress more evenly.

Free radicals form when covalent bonds in polymers are severed. Consequently, chain scission can be detected experimentally by means of electron spin resonance (ESR) and infrared spectroscopy [9, 10].

In some cases, fracture occurs by *chain disentanglement*, where molecules separate from one another intact. The likelihood of chain disentanglement depends on the length of molecules and the degree to which they are interwoven.[3]

Chain scission can occur at relatively low strains in cross-linked or highly aligned polymers, but the mechanical response of isotropic polymers with low cross-link density is governed by secondary bonds at low strains. At high strains, many polymers yield before fracture, as discussed below.

6.1.2.2 Shear Yielding and Crazing

Most polymers, like metals, yield at sufficiently high stresses. While metals yield by dislocation motion along slip planes, polymers can exhibit either shear yielding or crazing.

---

[3] An analogy that should be familiar to most Americans is the process of disentangling Christmas tree lights that have been stored in a box for a year. For those who are not acquainted with this holiday ritual, a similar example is a large mass of tangled strands of string; pulling on a single strand will either free it (chain disentanglement) or cause it to break (chain scission).

Shear yielding in polymers resembles plastic flow in metals, at least from a continuum mechanics viewpoint. Molecules slide with respect to one another when subjected to a critical shear stress. Shear-yielding criteria can either be based on the maximum shear stress or the octahedral shear stress [11, 12]:

$$\tau_{max} = \tau_o - \mu_s \sigma_m \qquad (6.15a)$$

or

$$\tau_{oct} = \tau_o - \mu_s \sigma_m \qquad (6.15b)$$

where σ_m is the hydrostatic stress and μ_s is a material constant that characterizes the sensitivity of the yield behavior to σ_m. When $\mu_s = 0$, Equation (6.15a) and Equation (6.15b) reduce to the Tresca and von Mises yield criteria, respectively.

Glassy polymers subject to tensile loading often yield by crazing, which is a highly localized deformation that leads to cavitation (void formation) and strains on the order of 100% [13, 14]. On the macroscopic level, crazing appears as a *stress-whitened region*, due to a low refractive index. The craze zone usually forms perpendicular to the maximum principal normal stress.

Figure 6.7 illustrates the mechanism for crazing in homogeneous glassy polymers. At sufficiently high strains, molecular chains form aligned packets called fibrils. Microvoids form between the fibrils due to an incompatibility of strains in neighboring fibrils. The aligned structure enables the fibrils to carry very high stresses relative to the undeformed amorphous state because covalent bonds are much stronger and stiffer than the secondary bonds. The fibrils elongate by incorporating additional material, as Figure 6.7 illustrates. Figure 6.8 shows an SEM fractograph of a craze zone.

Oxborough and Bowden [15] proposed the following craze criterion:

$$\varepsilon_1 = \frac{\beta(t, T)}{E} + \frac{\gamma(t, T)}{3\sigma_m} \qquad (6.16)$$

where ε_1 is the maximum principal normal strain, and β and γ are parameters that are time and temperature dependent. According to this model, the critical strain for crazing decreases with increasing modulus and hydrostatic stress.

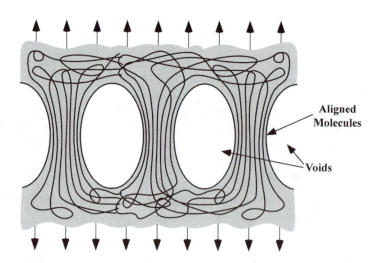

Aligned Molecules

Voids

FIGURE 6.7 Craze formation in glassy polymers. Voids form between fibrils, which are bundles of aligned molecular chains. The craze zone grows by drawing additional material into the fibrils.

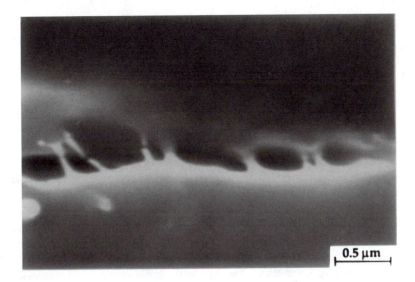

FIGURE 6.8 Craze zone in polypropylene. Photograph provided by M. Cayard.

Fracture occurs in a craze zone when individual fibrils rupture. This process can be unstable if, when a fibril fails, the redistributed stress is sufficient to rupture one or more neighboring fibrils. Fracture in a craze zone usually initiates from inorganic dust particles that are entrapped in the polymer [16]. There are a number of ways to neutralize the detrimental effects of these impurities, including the addition of soft second-phase particles.

Crazing and shear yielding are competing mechanisms; the dominant yielding behavior depends on the molecular structure, stress state, and temperature. A large hydrostatic tensile component in the stress tensor is conducive to crazing, while shear yielding favors a large deviatoric stress component. Each yielding mechanism displays a different temperature dependence; thus the dominant mechanism may change with temperature.

6.1.2.3 Crack-Tip Behavior

As with metals, a yield zone typically forms at the tip of a crack in polymers. In the case of shear yielding, the damage zone resembles the plastic zone in metals, because slip in metals and shear in polymers are governed by similar yield criteria. Craze yielding, however, produces a Dugdale-type strip-yield zone ahead of the crack tip. Of the two yielding mechanisms in polymers, crazing is somewhat more likely ahead of a crack tip, because of the triaxial tensile stress state. Shear yielding, however, can occur at crack tips in some materials, depending on the temperature and specimen geometry [17].

Figure 6.9 illustrates a craze zone ahead of a crack tip. If the craze zone is small compared to the specimen dimensions,[4] we can estimate its length ρ from the Dugdale-Barenblatt [18, 19] strip-yield model:

$$\rho = \frac{\pi}{8}\left(\frac{K_I}{\sigma_c}\right)^2 \qquad (6.17)$$

---

[4] Another implicit assumption of Equation (6.17) is that the global material behavior is linear elastic or linear viscoelastic. Chapter 8 discusses the requirements for the validity of the stress-intensity factor in polymers.

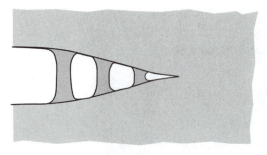

FIGURE 6.9 Schematic crack-tip craze zone.

which is a restatement of Equation (2.74), except that we have replaced the yield strength with σ_c, the crazing stress. Figure 6.10 is a photograph of a crack-tip craze zone [17], which exhibits a typical stress-whitening appearance.

The crack advances when the fibrils at the trailing edge of the craze rupture. In other words, cavities in the craze zone coalesce with the crack tip. Figure 6.11 is an SEM fractograph of the surface of a polypropylene fracture toughness specimen that has experienced craze-crack growth. Note the similarity to fracture surfaces for microvoid coalescence in metals (Figure 5.3 and Figure 5.8).

Craze-crack growth can either be stable or unstable, depending on the relative toughness of the material. Some polymers with intermediate toughness exhibit sporadic, so-called *stick/slip* crack growth: at a critical crack-tip-opening displacement, the entire craze zone ruptures, the crack arrests, and the craze zone reforms at the new crack tip [3]. Stick/slip crack growth can also occur in materials that exhibit shear yield zones.

6.1.2.4 Rubber Toughening

As stated earlier, the rupture of fibrils in a craze zone can lead to unstable crack propagation. Fracture initiates at inorganic dust particles in the polymer when the stress exceeds a critical value. It is possible to increase the toughness of a polymer by lowering the crazing stress to well below the critical fracture stress.

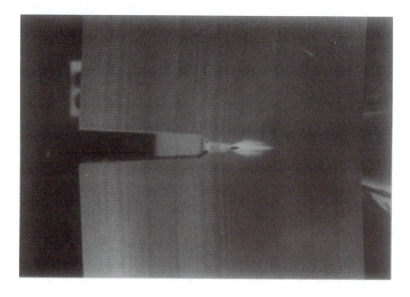

FIGURE 6.10 Stress-whitened zone ahead of a crack tip, which indicates crazing. Photograph provided by M. Cayard.

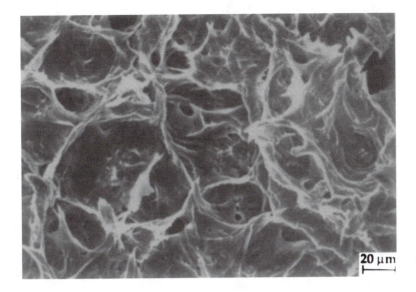

FIGURE 6.11 Fracture surface of craze-crack growth in polypropylene. Photograph provided by Mr. Sun Yongqi.

The addition of rubbery second-phase particles to a polymer matrix significantly increases toughness by making craze initiation easier [16]. The low-modulus particles provide sites for void nucleation, thereby lowering the stress required for craze formation. The detrimental effect of the dust particles is largely negated because the stress in the fibrils tends to be well below that required for fracture. Figure 6.12 is an SEM fractograph that shows crack growth in a rubber-toughened polymer. Note the high concentration of voids, compared to the fracture surface in Figure 6.11.

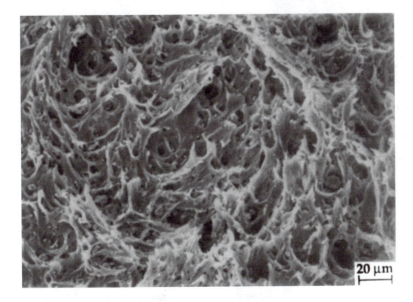

FIGURE 6.12 Fracture surface of a rubber-toughened polyvinyl chloride (PVC). Note the high concentration of microvoids. Photograph provided by Mr. Sun Yongqi.

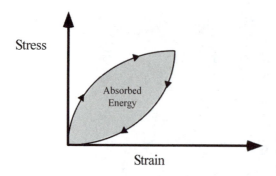

FIGURE 6.13 Cyclic stress-strain curve in a viscoelastic material. Hysteresis results in absorbed energy, which is converted to heat.

Of course there is a trade-off with rubber toughening, in that the increase in toughness and ductility comes at the expense of yield strength. A similar trade-off between toughness and strength often occurs in metals and alloys.

6.1.2.5 Fatigue

Time-dependent crack growth in the presence of cyclic stresses is a problem in virtually all material systems. Two mechanisms control fatigue in polymers: chain scission and hysteresis heating [5].

Crack growth by chain scission occurs in brittle systems, where crack-tip yielding is limited. A finite number of bonds are broken during each stress cycle, and measurable crack advance takes place after sufficient cycles.

Tougher materials exhibit significant viscoelastic deformation and yielding at the crack tip. Figure 6.13 illustrates the stress-strain behavior of a viscoelastic material for a single load-unload cycle. Unlike elastic materials, where the unloading and loading paths coincide and the strain energy is recovered, a viscoelastic material displays a hysteresis loop in the stress-strain curve; the area inside this loop represents the energy that remains in the material after it is unloaded. When a viscoelastic material is subject to multiple stress cycles, a significant amount of work is performed on the material. Much of this work is converted to heat, and the temperature in the material rises. The crack-tip region in a polymer subject to cyclic loading may rise to well above T_g, resulting in local melting and viscous flow of the material. The rate of crack growth depends on the temperature at the crack tip, which is governed by the loading frequency and the rate of heat conduction away from the crack tip. Fatigue crack growth data from small laboratory coupons may not be applicable to structural components because heat transfer properties depend on the size and geometry of the sample.

6.1.3 FIBER-REINFORCED PLASTICS

This section focuses on the fracture behavior of continuous fiber-reinforced plastics, as opposed to other types of polymer composites. The latter materials tend to be isotropic on the macroscopic scale, and their behavior is often similar to homogeneous materials. Continuous fiber-reinforced plastics, however, have orthotropic mechanical properties that lead to unique failure mechanisms such as delamination and microbuckling.

The combination of two or more materials can lead to a third material with highly desirable properties. Precipitation-hardened aluminum alloys and rubber-toughened plastics are examples of materials whose properties are superior to those of the parent constituents. While these materials form "naturally" through careful control of chemical composition and thermal treatments, the manufacture of *composite materials* normally involves a somewhat more heavy-handed human intervention. The constituents of a composite material are usually combined on a macroscopic scale through physical rather than chemical means [20]. The distinction between composites and multiphase materials is somewhat arbitrary, since many of the same strengthening mechanisms operate in both classes of material.

Composite materials usually consist of a matrix and a reinforcing constituent. The matrix is often soft and ductile compared to the reinforcement, but this is not always the case (see Section 6.2). Various types of reinforcement are possible, including continuous fibers, chopped fibers, whiskers, flakes, and particulates [20].

When a polymer matrix is combined with a strong, high-modulus reinforcement, the resulting material can have superior strength/weight and stiffness/weight ratios compared to steel and aluminum. Continuous fiber-reinforced plastics tend to give the best overall performance (compared to other types of polymer composites), but can also exhibit troubling fracture and damage behavior. Consequently, these materials have been the subject of extensive research over the past 40 years.

A variety of fiber-reinforced polymer composites are commercially available. The matrix material is usually a thermoset polymer (i.e., an epoxy), although thermoplastic composites have become increasingly popular in recent years. Two of the most common fiber materials are carbon, in the form of graphite, and aramid (also known by the trade name, Kevlar),[5] which is a high-modulus polymer. Polymers reinforced by continuous graphite or Kevlar fibers are intended for high-performance applications such as fighter planes, while *fiberglass* is an example of a polymer composite that appears in more down-to-earth applications. The latter material consists of randomly oriented chopped glass fibers in a thermoset matrix.

Figure 6.14 illustrates the structure of a fiber-reinforced composite. Consider a single ply (Figure 6.14(a)). The material has high strength and stiffness in the fiber direction, but has relatively poor mechanical properties when loaded transverse to the fibers. In the latter case, the strength and stiffness are controlled by the properties of the matrix. When the composite is subject to biaxial loading, several plies with differing fiber orientations can be bonded to form a laminated composite (Figure 6.14(b)). The individual plies interact to produce complex elastic properties in the laminate. The desired elastic response can be achieved through the appropriate choice of the fiber and matrix material, the fiber volume, and the lay-up sequence of the plies. The fundamentals of orthotropic elasticity and laminate theory are well established [21].

6.1.3.1 Overview of Failure Mechanisms

Many have attempted to apply fracture mechanics to fiber-reinforced composites, and have met with mixed success. Conventional fracture mechanics methodology assumes a single dominant crack that grows in a self-similar fashion, i.e., the crack increases in size (either through stable or unstable growth), but its shape and orientation remain the same. Fracture of a fiber-reinforced composite, however, is often controlled by numerous microcracks distributed throughout the material, rather than a single macroscopic crack. There are situations where fracture mechanics is appropriate for composites, but it is important to recognize the limitations of theories that were intended for homogeneous materials.

Figure 6.15 illustrates various failure mechanisms in fiber-reinforced composites. One advantage of composite materials is that fracture seldom occurs catastrophically without warning, but tends to be progressive, with subcritical damage widely dispersed through the material. Tensile loading (Figure 6.15 (a)) can produce matrix cracking, fiber bridging, fiber rupture, fiber pullout, and fiber/matrix debonding. Ultimate tensile failure of a fiber-reinforced composite often involves several of these mechanisms. Out-of-plane stresses can lead to delamination (Figure 6.15 (b)) because the fibers do not contribute significantly to the strength in this direction. Compressive loading can produce the microbuckling of fibers (Figure 6.15(c)); since the polymer matrix is soft compared to the fibers, the fibers are unstable in compression. Compressive loading can also lead to macroscopic delamination buckling (Figure 6.15(d)), particularly if the material contains a preexisting delaminated region.

---

[5] Kevlar is a trademark of the E.I. Dupont Company.

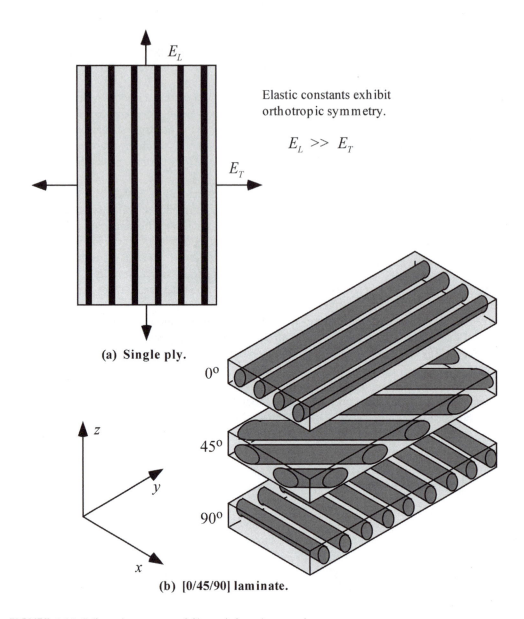

FIGURE 6.14 Schematic structure of fiber-reinforced composites.

6.1.3.2 Delamination

Out-of-plane tensile stresses can cause failure between plies, as Figure 6.15(b) illustrates. Stresses that lead to delamination could result from the structural geometry, such as if two composite panels are joined in a "T" configuration. Out-of-plane stresses, however, also arise from an unexpected source. Mismatch in Poisson's ratios between plies results in shear stresses in the x-y plane near the ply interface. These shear stresses produce a bending moment that is balanced by a stress in the z direction. For some lay-up sequences, substantial out-of-plane tensile stresses occur at the edge of the panel, which can lead to the formation of a delamination crack. Figure 6.16 shows a computed σ_z distribution for a particular lay-up [22].

Although the assumption of a self-similar growth of a dominant crack often does not apply to the failure of composite materials, such an assumption is appropriate in the case of delamination. Consequently, fracture mechanics has been very successful in characterizing this failure mechanism.

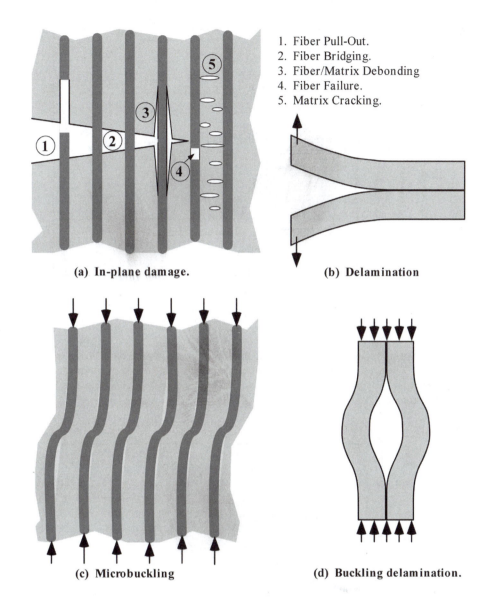

1. Fiber Pull-Out.
2. Fiber Bridging.
3. Fiber/Matrix Debonding
4. Fiber Failure.
5. Matrix Cracking.

(a) In-plane damage.   (b) Delamination

(c) Microbuckling   (d) Buckling delamination.

FIGURE 6.15 Examples of damage and fracture mechanisms in fiber-reinforced composites.

Delamination can occur in both Mode I and Mode II. The interlaminar fracture toughness, which is usually characterized by a critical energy release rate (see Chapter 8), is related to the fracture toughness of the matrix material. The matrix and composite toughness are seldom equal, however, due to the influence of the fibers in the latter.

Figure 6.17 is a compilation of $\mathcal{G}_{IC}$ values for various matrix materials, compared with the interlaminar toughness of the corresponding composite [23]. For brittle thermosets, the composite has a higher toughness than the neat resin, but the effect is reversed for high toughness matrices. Attempts to increase the composite toughness through tougher resins have yielded disappointing results; only a fraction of the toughness of a high ductility matrix is transferred to the composite.

Let us first consider the reasons for the high relative toughness of composites with brittle matrices. Figure 6.18 shows the fracture surface in a composite specimen with a brittle epoxy resin. The crack followed the fibers, implying that fiber/matrix debonding was the crack growth mechanism in this case. The fracture surface has a "corrugated roof" appearance; more surface area was

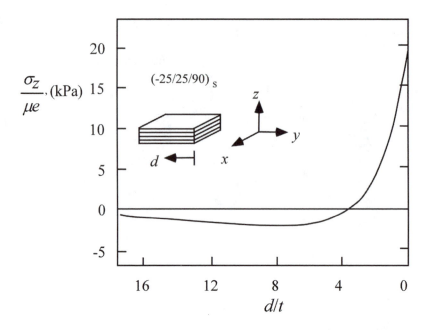

FIGURE 6.16 Out-of-plane stress at mid-thickness in a composite laminate, normalized by the remotely applied strain (μe represents microstrain). The distance from the free edge, d, is normalized by the ply thickness t. Taken from Wang, A.S.D., "An Overview of the Delamination Problem in Structural Composites." *Key Engineering Materials,* Vol. 37, 1989, pp. 1–20.

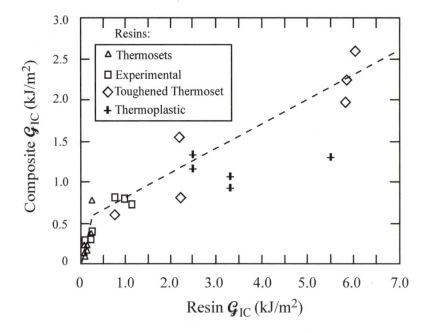

FIGURE 6.17 Compilation of interlaminar fracture toughness data, compared with the toughness of the corresponding neat resin. Taken from Hunston, D. and Dehl, R., "The Role of Polymer Toughness in Matrix Dominated Composite Fracture." Paper EM87-355, Society of Manufacturing Engineers, Deerborn, MI, 1987.

FIGURE 6.18 Fracture surface resulting from Mode I delamination of a graphite-epoxy composite with a brittle resin. Photograph provided by W.L. Bradley. Taken from Bradley, W.L., "Understanding the Translation of Neat Resin Toughness into Delamination Toughness in Composites." *Key Engineering Materials,* Vol. 37, 1989, pp. 161–198.

created in the composite experiment, which apparently resulted in higher fracture energy. Another contributing factor in the composite toughness in this case is fiber bridging. In some instances, the crack grows around a fiber, which then bridges the crack faces, and adds resistance to further crack growth.

With respect to the fracture of tough matrices, one possible explanation for the lower relative toughness of the composite is that the latter is limited by the fiber/matrix bond, which is weaker than the matrix material. Experimental observations, however, indicate that fiber constraint is a more likely explanation [24]. In high toughness polymers, a shear or craze damage zone forms ahead of the crack-tip. If the toughness is sufficient for the size of the damage zone to exceed the fiber spacing, the fibers restrain the crack-tip yielding, resulting in a smaller zone than in the neat resin. The smaller damage zone leads to a lower fracture energy between plies.

Delamination in Mode II loading is possible, but $\mathcal{G}_{IIC}$ is typically 2 to 10 times higher than the corresponding $\mathcal{G}_{IC}$ [24]. The largest disparity between Mode I and Mode II interlaminar toughness occurs in brittle matrices.

In-situ fracture experiments in an SEM enable one to view the fracture process during delamination; [24–26]. Long, slender damage zones containing numerous microcracks form ahead of the crack tip during Mode II loading. Figure 6.19 shows a sequence of SEM fractographs of a Mode II damage zone ahead of an interlaminar crack in a brittle resin; the same region was photographed at different damage states. Note that the microcracks are oriented approximately 45° from the main crack, which is subject to Mode II shear. Thus the microcracks are oriented perpendicular to the maximum normal stress and are actually Mode I cracks. As the loading progresses, these microcracks coalesce with the main crack tip. The high relative toughness in Mode II results from energy dissipation in this damage zone.

In more ductile matrices, the appearance of the Mode II damage zone is similar to the Mode I case, and the difference between $\mathcal{G}_{IC}$ and $\mathcal{G}_{IIC}$ is not as large as for brittle matrices [24].

6.1.3.3 Compressive Failure

High-modulus fibers provide excellent strength and stiffness in tension, but are of limited value for compressive loading. According to the Euler buckling equation, a column of length L with a

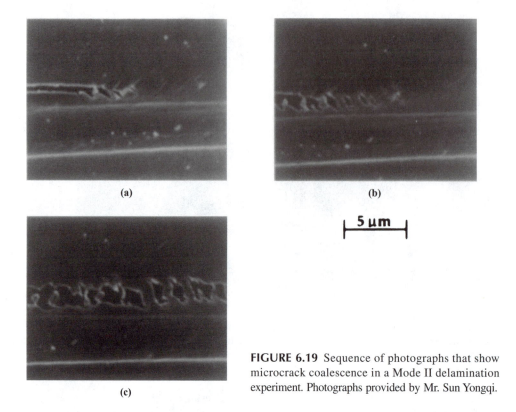

FIGURE 6.19 Sequence of photographs that show microcrack coalescence in a Mode II delamination experiment. Photographs provided by Mr. Sun Yongqi.

cross-section moment of inertia I, subject to a compressive force P, becomes unstable when

$$P \geq \frac{\pi^2 E I}{L^2}$$

(6.18)

assuming the loading is applied on the central axis of the column and the ends are unrestrained. Thus a long, slender fiber has very little load-carrying capacity in compression.

Equation (6.18) is much too pessimistic for composites, because the fibers are supported by matrix material. Early attempts [27] to model fiber buckling in composites incorporated an elastic foundation into the Euler buckling analysis, as Figure 6.20 illustrates. This led to the following compressive failure criterion for unidirectional composites:

$$\sigma_c = \mu_{LT} + \pi^2 E_f V_f \left(\frac{r}{L}\right)^2$$

(6.19)

where μ_{LT} is the longitudinal-transverse shear modulus of the matrix and E_f is Young's modulus of the fibers. This model overpredicts the actual compressive strength of composites by a factor of ~4.

One problem with Equation (6.19) is that it assumes that the response of the material remains elastic; matrix yielding is likely for large lateral displacements of fibers. Another shortcoming of this simple model is that it considers global fiber instability, while fiber buckling is a local phenomenon; microscopic *kink bands* form, usually at a free edge, and propagate across the panel [28, 29].[6] Figure 6.21 is a photograph of local fiber buckling in a graphite-epoxy composite.

---

[6] The long, slender appearance of the kink bands led several investigators [28, 29] to apply the Dugdale-Barenblatt strip-yield model to the problem. This model has been moderately successful in quantifying the size of the compressive damage zones.

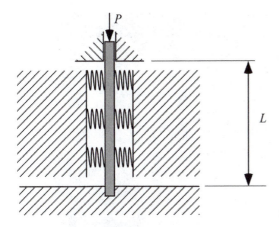

FIGURE 6.20 Compressive loading of a column that is supported laterally by an elastic foundation.

An additional complication in real composites is *fiber waviness*. Fibers are seldom perfectly straight; rather they tend to have a sine wave-like profile, as Figure 6.22 illustrates [30]. Such a configuration is less stable in compression than a straight column.

Recent investigators [30–32] have incorporated the effects of matrix nonlinearity and fiber waviness into failure models. Most failure models are based on continuum theory and thus do not address the localized nature of microbuckling. Guynn [32], however, has performed detailed numerical simulations of compression loading of fibers in a nonlinear matrix.

Microbuckling is not the only mechanism for compressive failure. Figure 6.15(d) illustrates buckling delamination, which is a macroscopic instability. This type of failure is common in composites that have been subject to impact damage, which produces microcracks and delamination flaws in the material. Delamination buckling induces Mode I loading, which causes the delamination flaw to propagate at sufficiently high loads. This delamination growth can be characterized with

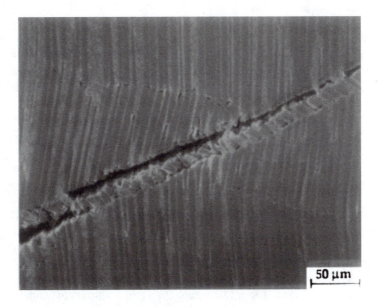

FIGURE 6.21 Kink band formation in a graphite-epoxy composite. Photograph provided by E.G. Guynn. Taken from Guynn, E.G., "Experimental Observations and Finite Element Analysis of the Initiation of Fiber Microbuckling in Notched Composite Laminates." Ph.D. Dissertation, Texas A&M University, College Station, TX, 1990.

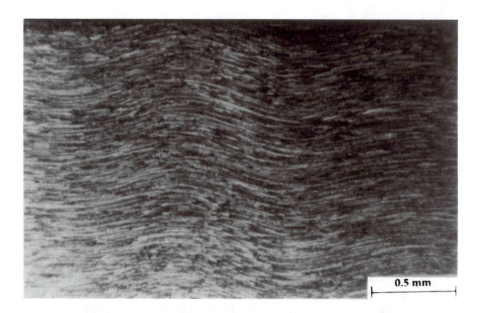

FIGURE 6.22 Fiber waviness in a graphite-epoxy composite. Photograph provided by A.L. Highsmith. Taken from Highsmith, A.L. and Davis, J., "The Effects of Fiber Waviness on the Compressive Response of Fiber-Reinforced Composite Materials." Progress Report for NASA Research Grant NAG-1-659, NASA Langley Research Center, Hampton, VA, 1990.

fracture mechanics methodology [33]. A *compression after impact* test is a common screening criterion for assessing the ability of a material to withstand impact loading without sustaining significant damage.

6.1.3.4 Notch Strength

The strength of a composite laminate that contains a hole or a notch is less than the unnotched strength because of the local stress concentration effect. A circular hole in an isotropic plate has a stress concentration factor (SCF) of 3.0, and the SCF can be much higher for an elliptical notch (Section 2.2). If a composite panel with a circular hole fails when the maximum stress reaches a critical value, the strength should be independent of the hole size, since the SCF does not depend on radius. Actual strength measurements, however, indicate a hole size effect, where strength decreases with increasing hole size [34].

Figure 6.23 illustrates the elastic stress distributions ahead of a large hole and a small hole. Although the peak stress is the same for both holes, the stress concentration effects of the large hole act over a wider distance. Thus the *volume* over which the stress acts appears to be important.

Whitney and Nuismer [35] proposed a simple model for notch strength, where failure is assumed to occur when the stress exceeds the unnotched strength over a critical distance.[7] This distance is a fitting parameter that must be obtained by experiment. Subsequent modifications to this model, including the work of Pipes et. al. [36], yielded additional fitting parameters, but did not result in a better understanding of the failure mechanisms.

Figure 6.24 shows the effect of notch length on the strength panels that contain elliptical center notches [34]. These experimental data actually apply to a boron-aluminum composite, but polymer composites exhibit a similar trend. The simple Whitney and Nuismer criterion gives a reasonably good fit of the data in this case.

---

[7] Note the similarity to the Ritchie-Knott-Rice model for cleavage fracture (Chapter 5).

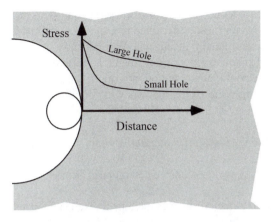

FIGURE 6.23 Effect of hole size on local stress distribution.

Some researchers [37] have applied fracture mechanics concepts to the failure of composites panels that contain holes and notches. They assume failure at a critical K, which is usually modified with a plastic zone correction to account for subcritical damage. Some of these models are capable of fitting experimental data such as that in Figure 6.24, because the plastic zone correction is an adjustable parameter. The physical basis of these models is dubious, however. Fracture mechanics formalism gives these models the illusion of rigor, but they have no more theoretical basis than the simple strength-of-materials approaches such as the Whitney-Nuismer criterion.

That linear elastic fracture mechanics is invalid for circular holes and blunt notches in composites should be self evident, since the LEFM theory assumes sharp cracks. If, however, a sharp slit is introduced into a composite panel (Figure 6.25), the validity (or lack of validity) of fracture mechanics is less obvious. This issue is explored below.

Recall Chapter 2, which introduced the concept of a singularity zone, where the stress and strain vary as $1/\sqrt{\pi}$ from the crack tip. Outside of the singularity zone, higher-order terms, which

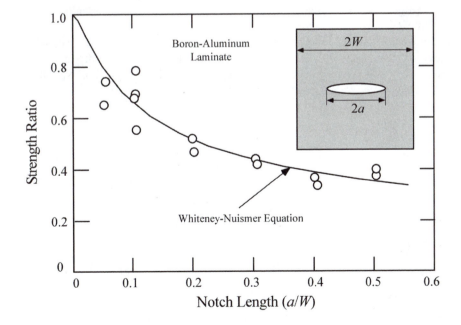

FIGURE 6.24 Strength of center-notched composite laminates, relative to the unnotched strength. Taken from Awerbuch, J. and Madhukar, M.S., "Notched Strength of Composite Laminates: Predictions and Experiments—A Review." *Journal of Reinforced Plastics and Composites,* Vol. 4, 1985, pp. 3–159.

are geometry dependent, become significant. For K to define uniquely the crack-tip conditions and be a valid failure criterion, all nonlinear material behavior must be confined to a small region inside the singularity zone. This theory is based entirely on continuum mechanics. While metals, plastics, and ceramics are often heterogeneous, the scale of microstructural constituents is normally small compared to the size of the singularity zone; thus the continuum assumption is approximately valid.

For LEFM to be valid for a sharp crack in a composite panel, the following conditions must be met:

1. The fiber spacing must be small compared to the size of the singularity zone. Otherwise, the continuum assumption is invalid.
2. Nonlinear damage must be confined to a small region within the singularity zone.

Harris and Morris [38] showed that K characterizes the *onset* of damage in cracked specimens, but not ultimate failure, because the damage spreads throughout the specimen before failure, and K no longer has any meaning. Figure 6.25 illustrates a typical damage zone in a specimen with a sharp macroscopic notch. The damage, which includes fiber/matrix debonding and matrix cracking, actually propagates perpendicular to the macrocrack. Thus the crack does not grow in a self-similar fashion.

One of the most significant shortcomings of tests on composite specimens with narrow slits is that defects of this type do not occur naturally in fiber-reinforced composites; therefore, the geometry in Figure 6.25 is of limited practical concern. Holes and blunt notches may be unavoidable in a design, but a competent design engineer would not be foolish enough to include a sharp notch in a load-bearing member of a structure.

6.1.3.5 Fatigue Damage

Cyclic loading of composite panels produces essentially the same type of damage as monotonic loading. Fiber rupture, matrix cracking, fiber/matrix debonding, and delamination all occur in response to fatigue loading. Fatigue damage reduces the strength and modulus of a composite laminate, and eventually leads to total failure.

Figure 6.26 and Figure 6.27 show the effect of cyclic stresses on the residual strength and modulus of graphite/epoxy laminates [39]. Both strength and modulus decrease rapidly after relatively few cycles, but remain approximately constant up to around 80% of the fatigue life. Near the end of the fatigue life, strength and modulus decrease further.

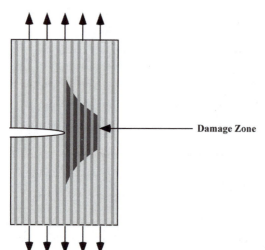

Damage Zone

FIGURE 6.25 Sharp notch artificially introduced into a composite panel.

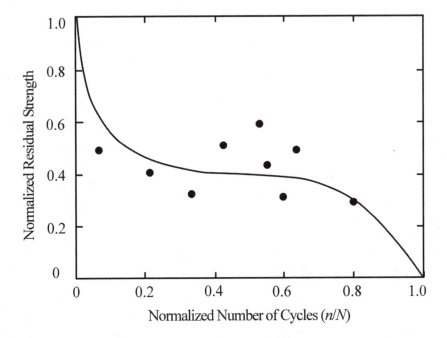

FIGURE 6.26 Residual strength after fatigue damage in a graphite-epoxy laminate. Taken from Charewicz, A. and Daniel, I.M., "Damage Mechanisms and Accumulation in Graphite/Epoxy Laminates." ASTM STP 907, American Society for Testing and Materials, Philadelphia, PA, 1986, pp. 274–297.

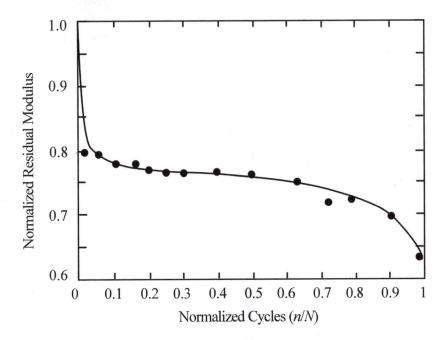

FIGURE 6.27 Residual modulus after fatigue damage in a graphite-epoxy laminate. Taken from Charewicz, A. and Daniel, I.M., "Damage Mechanisms and Accumulation in Graphite/Epoxy Laminates." ASTM STP 907, American Society for Testing and Materials, Philadelphia, PA, 1986, pp. 274–297.

6.2   CERAMICS AND CERAMIC COMPOSITES

A number of technological initiatives have been proposed whose implementation depends on achieving major advances in materials technology. For example, the National Aerospace Plane will re-enter the earth's atmosphere at speeds of up to Mach 25, creating extremes of both temperature and stress. Also, the Advanced Turbine Technology Applications Program (ATTAP) has the stated goal of developing heat engines that have a service life of 3000 h at 1350°C. Additional applications are on the horizon that will require materials that can perform at temperatures in excess of 2000°C. All metals, including cobalt-based superalloys, are inadequate at these temperatures. Only ceramics possess adequate creep resistance above 1000°C.

Ceramic materials include oxides, carbides, sulfides, and intermetallic compounds, which are joined either by covalent or ionic bonds. Most ceramics are crystalline but, unlike metals, they do not have close-packed planes on which dislocation motion can occur. Therefore, ceramic materials tend to be very brittle compared to metals.

Typical ceramics have very high melting temperatures, which explains their good creep properties. Also, many of these materials have superior wear resistance, and have been used for bearings and machine tools. Most ceramics, however, are too brittle for critical load-bearing applications. Consequently, a vast amount of research has been devoted to improving the toughness of ceramics.

Most traditional ceramics are monolithic (single phase) and have very low fracture toughness. Because they do not yield, monolithic ceramics behave as ideally brittle materials (Figure 2.6(a)), and a propagating crack need only overcome the surface energy of the material. The new generation of ceramics, however, includes multiphase materials and ceramic composites that have vastly improved toughness. Under certain conditions, two brittle solids can be combined to produce a material that is significantly tougher than either parent material.

The micromechanisms that lead to improved fracture resistance in modern ceramics include microcrack toughening, transformation toughening, ductile phase toughening, fiber toughening, and whisker toughening. Table 6.1 lists the dominant toughening mechanism in several materials, along with the typical fracture toughness values [40]. Fiber toughening, the most effective mechanism, produces toughness values around 20 MPa$\sqrt{\text{m}}$, which is below the lower shelf toughness of steels but is significantly higher than most ceramics.

TABLE 6.1
Ceramics with Enhanced Toughness [40]

Toughening Mechanism	Material	Maximum Toughness, MPa$\sqrt{\text{m}}$
Fiber reinforced	LAS/SiC	~20
	Glass/C	~20
	SiC/SiC	~20
Whisker reinforced	Al_2O_3/SiC(0.2)	10
	Si_3N_4/SiC(0.2)	14
Ductile network	Al_2O_3/Al(0.2)	12
	B_4C/Al(0.2)	14
	WC/Co(0.2)	20
Transformation toughened	PSZ	18
	TZP	16
	ZTA	10
Microcrack toughened	ZTA	7
	Si_3N_4/SiC	7

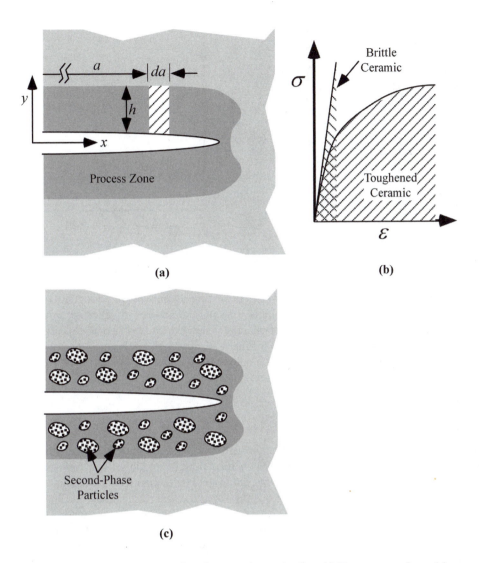

FIGURE 6.28 The process-zone mechanism for ceramic toughening. (a) Process zone formed by growing crack (b) Schematic stress-strain behavior (c) Nonlinear deformation of second-phase particles

Evans [40] divides the toughening mechanisms for ceramics into two categories: process zone formation and bridging. Both mechanisms involve energy dissipation at the crack tip. A third mechanism, crack deflection, elevates toughness by increasing the area of the fracture surface (Figure 2.6(c)).

The process zone mechanism for toughening is illustrated in Figure 6.28. Consider a material that forms a process zone at the crack tip (Figure 6.28(a)). When this crack propagates, it leaves a wake behind the crack tip. The critical energy release rate for propagation is equal to the work required to propagate the crack from a to $a + da$, divided by da:

$$\mathcal{G}_R = 2\int_0^h \left[\int_0^{\varepsilon_{ij}} \sigma_{ij}\, d\varepsilon_{ij} \right] dy + 2\gamma_s \qquad (6.20)$$

where h is the half width of the process zone and γ_s is the surface energy. The integral in the square brackets is the strain-energy density, which is simply the area under the stress-strain curve in the

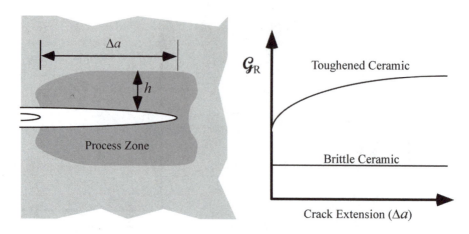

FIGURE 6.29 The process-zone toughening mechanism usually results in a rising R curve.

case of uniaxial loading. Figure 6.28(b) compares the stress-strain curve of brittle and toughened ceramics. The latter material is capable of higher strains, and absorbs more energy prior to failure.

Many toughened ceramics contain second-phase particles that are capable of nonlinear deformation, and are primarily responsible for the elevated toughness. Figure 6.28(c) illustrates the process zone for such a material. Assuming the particles provide all of the energy dissipation in the process zone, and the strain-energy density in this region does not depend on y, the fracture toughness is given by

$$\mathcal{G}_R = 2hf \int_0^{\varepsilon_{ij}} \sigma_{ij} d\varepsilon_{ij} + 2\gamma s \qquad (6.21)$$

where f is the volume fraction of second-phase particles. Thus the toughness is controlled by the width of the process zone, the concentration of second-phase particles, and the area under the stress-strain curve. Recall the delamination of composites with tough resins (Section 6.1.3), where the fracture toughness of the composite was not as great as the neat resin because the fibers restricted the size of the process zone (h).

The process zone mechanism often results in a rising R curve, as Figure 6.29 illustrates. The material resistance increases with crack growth, as the width of the processes zone grows. Eventually, h and $\mathcal{G}_R$ reach steady-state values.

Figure 6.30 illustrates the crack bridging mechanism, where the propagating crack leaves fibers or second-phase particles intact. The unbroken fibers or particles exert a traction force on the crack faces, much like the Dugdale-Barenblatt strip-yield model [18, 19]. The fibers eventually rupture when the stress reaches a critical value. According to Equation (3.42) and Equation (3.43), the

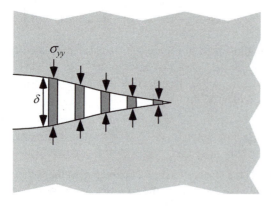

FIGURE 6.30 The fiber bridging mechanism for ceramic toughening.

critical energy release rate for crack propagation is given by

$$\mathcal{G}_c = J_c = f \int_0^{\delta_c} \sigma_{yy} d\delta \tag{6.22}$$

The sections that follow outline several specific toughening mechanisms in modern ceramics.

6.2.1 Microcrack Toughening

Although the formation of cracks in a material is generally considered deleterious, microcracking can sometimes lead to improved toughness. Consider a material sample of volume V that forms N microcracks when subject to a particular stress. If these cracks are penny shaped with an average radius a, the total work required to form these microcracks is equal to the surface energy times the total area created:

$$W_c = 2N\pi a^2 \gamma_s \tag{6.23}$$

The formation of microcracks releases the strain energy from the sample, which results in an increase in compliance. If this change in compliance is gradual, as existing microcracks grow and new cracks form, a nonlinear stress-strain curve results. The change in strain-energy density due to the microcrack formation is given by

$$\Delta w = 2\rho\pi a^2 \gamma_s \tag{6.24}$$

where $\rho \equiv N/V$ is the microcrack density. For a macroscopic crack that produces a process zone of microcracks, the increment of toughening due to microcrack formation can be inferred by inserting Equation (6.24) into Equation (6.21).

A major problem with the above scenario is that *stable* microcrack growth does not usually occur in a brittle solid. Preexisting flaws in the material remain stationary until they satisfy the Griffith criterion, at which time they become unstable. Stable crack advance normally requires either a rising R curve, where the fracture work w_f (Figure 2.6) increases with crack extension, or physical barriers in the material that inhibit crack growth. Stable microcracking occurs in concrete because aggregates act as crack arresters (see Section 6.3).

Certain multiphase ceramics have the potential for microcrack toughening. Figure 6.31 schematically illustrates this toughening mechanism [40]. Second-phase particles often are subject to residual stress due to thermal expansion mismatch or transformation. If the residual stress in the particle is tensile and the local stress in the matrix is compressive,[8] the particle cracks. If the signs on the stresses are reversed, the matrix material cracks at the interface. In both cases there is a residual opening of the microcracks, which leads to an increase in volume in the sample. Figure 6.31 illustrates the stress-strain response of such a material. The material begins to crack at a critical stress σ_c, and the stress-strain curve becomes nonlinear, due to a combination of compliance increase and dilatational strain. If the material is unloaded prior to total failure, the relative contributions of dilatational effects (residual microcrack opening) and modulus effects (due to the release of strain energy) are readily apparent.

A number of multiphase ceramic materials exhibit trends in toughness with particle size and temperature that are consistent with the microcracking mechanism, but this phenomenon has been directly observed only in aluminum oxide toughened with monoclinic zirconium dioxide [41].

---

[8] The residual stresses in the matrix and particle must balance in order to satisfy equilibrium.

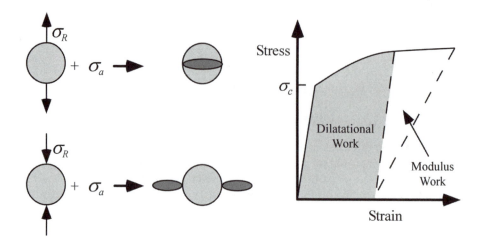

FIGURE 6.31 The microcrack toughening mechanism. The formation of microcracks in or near second-phase particles results in release of strain energy (modulus work) and residual microcrack opening (dilatational work). Taken from Evans, A.G., "The New High Toughness Ceramics." ASTM STP 1020, American Society for Testing and Materials, Philadelphia, PA, 1989, pp. 267–291.

This mechanism is relatively ineffective, as Table 6.1 indicates. Moreover, the degree of microcrack toughening is temperature dependent. Thermal mismatch and the resulting residual stresses tend to be lower at elevated temperatures, which implies less dilatational strain. Also, lower residual stresses may not prevent the microcracks from becoming unstable and propagating through the particle–matrix interface.

6.2.2 TRANSFORMATION TOUGHENING

Some ceramic materials experience a stress-induced martensitic transformation that results in shear deformation and a volume change (i.e., a dilatational strain). Ceramics that contain second-phase particles that transform often have improved toughness. Zirconium dioxide (ZrO_2) is the most widely studied material that exhibits a stress-induced martensitic transformation [42].

Figure 6.32 illustrates the typical stress-strain behavior for a martensitic transformation [42]. At a critical stress, the material transforms, resulting in both dilatational and shear strains. Figure 6.33(a) shows a crack-tip process zone, where second-phase particles have transformed.

The toughening mechanism for such a material can be explained in terms of the work argument: energy dissipation in the process zone results in higher toughness. An alternative explanation is that of *crack-tip shielding*, where the transformation lowers the local crack driving force [42]. Figure 6.33(b) shows the stress distribution ahead of the crack with a transformed process zone.

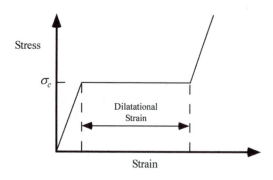

FIGURE 6.32 Schematic stress-strain response of a material that exhibits a martensitic transformation at a critical stress.

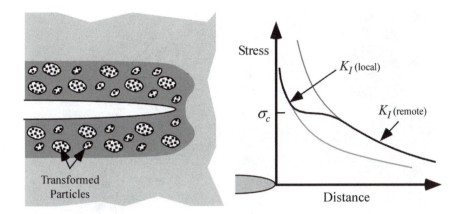

FIGURE 6.33 The martensitic toughening mechanism. Transformation of particles near the crack-tip results in a nonlinear process-zone and crack-tip shielding.

Outside of this zone, the stress field is defined by the global stress intensity, but the stress field in the process zone is lower, due to dilatational effects. The crack-tip work and shielding explanations are consistent with one another; more work is required for crack extension when the local stresses are reduced. Crack-tip shielding due to the martensitic transformation is analogous to the stress redistribution that accompanies plastic zone formation in metals (Chapter 2).

The transformation stress and the dilatational strain are temperature dependent. These quantities influence the size of the process zone h, and the strain-energy density within this zone. Consequently, the effectiveness of the transformation-toughening mechanism also depends on temperature. Below M_s, the martensite start temperature, the transformation occurs spontaneously, and the transformation stress is essentially zero. Thermally transformed martensite does not cause crack-tip shielding, however [42]. Above M_s, the transformation stress increases with temperature. When this stress becomes sufficiently large, the transformation-toughening mechanism is no longer effective.

6.2.3 DUCTILE PHASE TOUGHENING

Ceramics alloyed with ductile particles exhibit both bridging and process-zone toughening, as Figure 6.34 illustrates. Plastic deformation of the particles in the process zone contributes toughness, as does the ductile rupture of the particles that intersect the crack plane. Figure 6.35 is an SEM fractograph of bridging zones in Al_2O_3 reinforced with aluminum [40]. Residual stresses in the particles can also add to the material's toughness. The magnitude of the bridging and process-zone toughening depends on the volume fraction and flow properties of the particles. The process-zone toughening also depends on the particle size, with small particles giving the highest toughness [40].

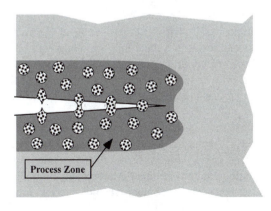

FIGURE 6.34 Ductile phase toughening. Ductile second-phase particles increase the ceramic toughness by plastic deformation in the process zone, as well as by a bridging mechanism.

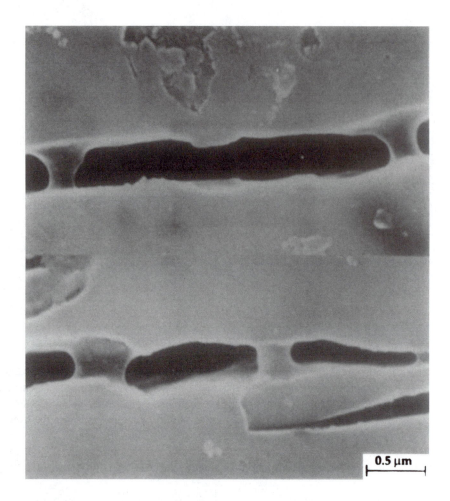

FIGURE 6.35 Ductile-phase bridging in Al$_2$O$_3$/Al. Photograph provided by A.G. Evans. Taken from Evans, A.G., "The New High Toughness Ceramics." ASTM STP 1020, American Society for Testing and Materials, Philadelphia, PA, 1989, pp. 267–291.

This toughening mechanism is temperature dependent, since the flow properties of the metal particles vary with temperature. Ductile phase ceramics are obviously inappropriate for applications above the melting temperature of the metal particles.

6.2.4 FIBER AND WHISKER TOUGHENING

One of the most interesting features of ceramic composites is that the combination of a brittle ceramic matrix with brittle ceramic fibers or whiskers can result in a material with relatively high toughness (Table 6.1). The secret to the high toughness of ceramic composites lies in the bond between the matrix and the fibers or whiskers. Having a *brittle* interface leads to higher toughness than a strong interface. Thus ceramic composites defy intuition: a brittle matrix bonded to a brittle fiber by a brittle interface results in a tough material.

A weak interface between the matrix and the reinforcing material aids the bridging mechanism. When a matrix crack encounters a fiber/matrix interface, this interface experiences Mode II loading; debonding occurs if the fracture energy of the interface is low (Figure 6.36(a)). If the extent of debonding is sufficient, the matrix crack bypasses the fiber, leaving it intact. Mathematical models [43] of fiber/matrix debonding predict crack bridging when the interfacial fracture energy is an

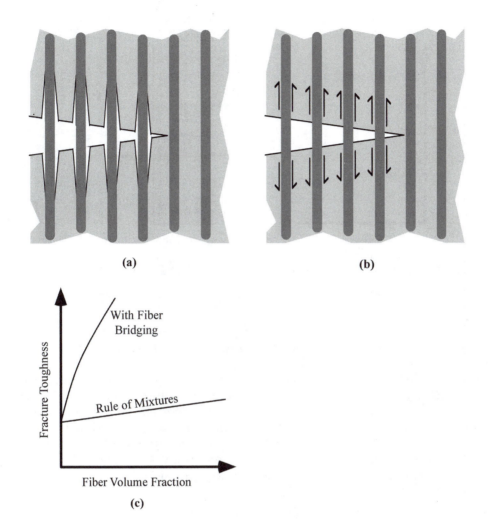

FIGURE 6.36 Fiber bridging in ceramic composites. Mathematical models treat bridging either in terms of fiber/matrix debonding or frictional sliding. This mechanism provides composite toughness well in excess of that predicted by the rule of mixtures. (a) Fiber/matrix debonding, (b) frictional sliding along interfaces, and (c) effect of bridging on toughness.

order of magnitude smaller than the matrix toughness. If the interfacial bond is strong, matrix cracks propagate through the fiber, and the composite toughness obeys a rule of mixtures; but bridging increases the composite toughness (Figure 6.36(c)).

An alternate model [43–45] for bridging in fiber-reinforced ceramics assumes that the fibers are not bonded, but that friction between the fibers and the matrix restricts the crack opening (Figure 6.36(b)). The model that considers Mode II debonding [43] neglects friction effects, and predicts that the length of the debond controls the crack opening.

Both models predict steady-state cracking, where the matrix cracks at a constant stress that does not depend on the initial flaw distribution in the matrix. Experimental data support the steady-state cracking theory. Because the cracking stress is independent of flaw size, fracture toughness measurements (e.g., K_{Ic} and $\mathcal{G}_c$) have little or no meaning.

Figure 6.37 illustrates the stress-strain behavior of a fiber-reinforced ceramic. The behavior is linear elastic up to σ_c, the steady-state cracking stress in the matrix. Once the matrix has cracked,

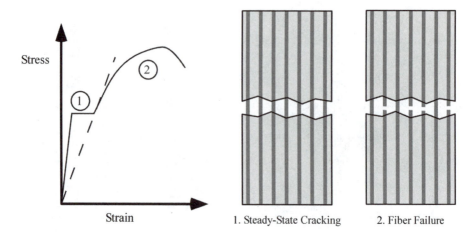

FIGURE 6.37 Stress-strain behavior of fiber-reinforced ceramic composites.

the load is carried by the fibers. The fibers do not fail simultaneously, because the fiber strength is subject to statistical variability [46]. Consequently, the material exhibits quasi ductility, where damage accumulates gradually until final failure.

Not only is fiber bridging the most effective toughening mechanism for ceramics (Table 6.1), it is also effective at high temperatures [47, 48]. Consequently applications that require load-bearing capability at temperatures above 1000°C will undoubtedly utilize fiber-reinforced ceramics.

Whisker-reinforced ceramics possess reasonably high toughness, although whisker reinforcement is not as effective as continuous fibers. The primary failure mechanism in whisker composites appears to be bridging [49]; crack deflection also adds an increment of toughness. Figure 6.38 is a micrograph that illustrates crack bridging in an Al_2O_3 ceramic reinforced by SiC whiskers.

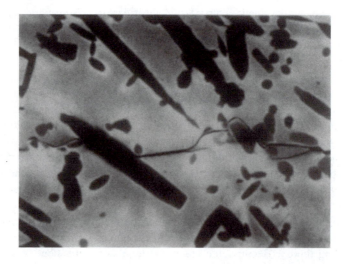

FIGURE 6.38 Crack bridging in Al_2O_3 reinforced with SiC whiskers. Photograph provided by A.G. Evans. Taken from Evans, A.G., "The New High Toughness Ceramics." ASTM STP 1020, American Society for Testing and Materials, Philadelphia, PA, 1989, pp. 267–291.

6.3 CONCRETE AND ROCK

Although concrete and rock are often considered brittle, they are actually *quasi-brittle* materials that are tougher than most of the so-called advanced ceramics. In fact, much of the research on toughening mechanisms in ceramics is aimed at trying to make ceramic composites behave more like concrete.

Concrete and rock derive their toughness from subcritical cracking that precedes ultimate failure. This subcritical damage results in a nonlinear stress-strain response and *R*-curve behavior.

A traditional strength-of-materials approach to designing with concrete has proved inadequate because the fracture strength is often size dependent [50]. This size dependence is due to the fact that nonlinear deformation in these materials is caused by subcritical cracking rather than plasticity. Initial attempts to apply fracture mechanics to concrete were unsuccessful because these early approaches were based on linear elastic fracture mechanics (LEFM) and failed to take account of the process zones that form in front of macroscopic cracks.

This section gives a brief overview of the mechanisms and models of fracture in concrete and rock. Although most of the experimental and analytical work has been directed at concrete as opposed to rock, due to the obvious technological importance of the former, rock and concrete behave in a similar manner. The remainder of this section will refer primarily to concrete, with the implicit understanding that most observations and models also apply to geologic materials.

Figure 6.39 schematically illustrates the formation of a fracture process zone in concrete, together with two idealizations of the process zone. Microcracks form ahead of a macroscopic crack, which consists of a bridged zone directly behind the tip and a traction-free zone further behind the tip. The bridging is a result of the weak interface between the aggregates and the matrix. Recall Section 6.2.4, where it was stated that fiber bridging, which occurs when the fiber-matrix bonds are weak, is the most effective toughening mechanism in ceramic composites. The process zone can be modeled as a region of strain softening (Figure 6.40(b)) or as a longer crack that is subject to closure tractions (Figure 6.40(c)). The latter is a slight modification to the Dugdale-Barenblatt strip-yield model.

Figure 6.40 illustrates the typical tensile response of concrete. After a small degree of nonlinearity caused by microcracking, the material reaches its tensile strength σ_t and then strain softens. Once σ_t is reached, the subsequent damage is concentrated in a local fracture zone. Virtually all of the displacement following the maximum stress is due to the damage zone. Note that Figure 6.40 shows a schematic stress-*displacement* curve rather than a stress-strain curve. The latter is inappropriate because nominal strain measured over the entire specimen is a function of gage length.

There are a number of models for fracture in concrete, but the one that is most widely referenced is the so-called *fictitious crack model* of Hillerborg [51, 52]. This model, which has also been called a *cohesive zone model*, is merely an application of the Dugdale-Barenblatt approach. The Hillerborg model assumes that the stress displacement behavior (σ-δ) observed in the damage zone of a tensile specimen is a material property. Figure 6.41(a) shows a schematic stress-displacement curve, and 6.41(b) illustrates the idealization of the damage zone ahead of a growing crack.

At the tip of the traction-free crack, the damage zone reaches a critical displacement δ_c. The tractions are zero at this point, but are equal to the tensile strength σ_t, at the tip of the damage zone (Figure 6.39(c)). Assuming that the closure stress σ and opening displacement δ are uniquely related, the critical energy release rate for crack growth is given by

$$\mathcal{G}_c = \int_0^{\delta_c} \sigma \, d\delta \tag{6.25}$$

which is virtually identical to Equation (3.43) and Equation (6.22).

The key assumption of the Hillerborg model that the σ-δ relationship is a unique material property is not strictly correct in most cases because process zones produced during the fracture

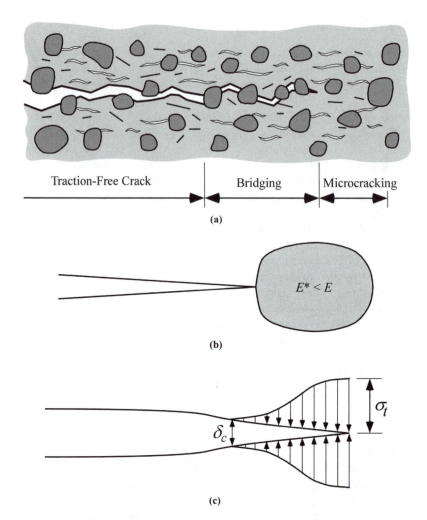

(a)

(b)

(c)

FIGURE 6.39 Schematic illustration of crack growth in concrete, together with two simplified models: (a) crack growth in concrete, (b) process zone idealized as a zone of strain softening, and (c) process zone idealized by closure tractions.

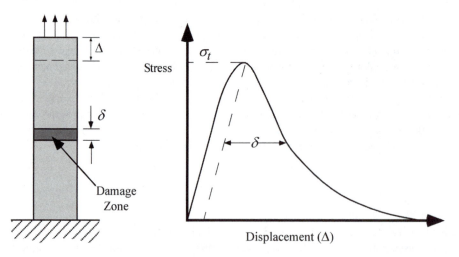

FIGURE 6.40 Typical tensile response of concrete.

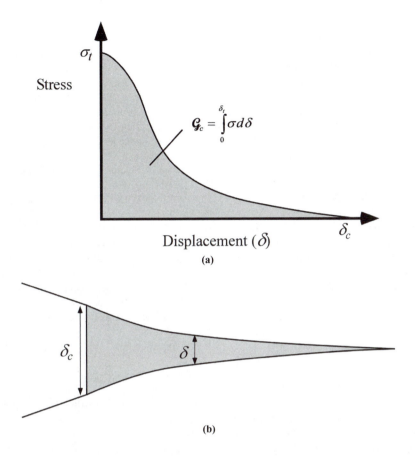

$$\mathcal{G}_c = \int_0^{\delta_t} \sigma d\delta$$

FIGURE 6.41 The "fictitious crack" model for concrete: (a) schematic stress-displacement response and (b) damage zone ahead of a growing crack. Taken from Hillerborg, A., Modeer, M., and Petersson, P.E., "Analysis of Crack Formation and Crack Growth in Concrete by Means of Fracture Mechanics and Finite Elements." *Cement and Concrete Research,* Vol. 6, 1976, pp. 773–782. Hillerborg, A., "Application of the Fictitious Crack Model to Different Materials." *International Journal of Fracture,* Vol. 51, 1991, pp. 95–102.

of concrete are often quite large, and the interaction between the process zone and free boundaries can influence the behavior. Consequently, $\mathcal{G}_c$ is not a material property in general, but can depend on specimen size. Fracture toughness results from small-scale tests tend to be lower than values obtained from larger samples.

The phenomena of size effects in concrete fracture has been the subject of considerable study [50, 53–56]. Some of the apparent size effects can be attributed to inappropriate data reduction methods. For example, if fracture toughness is computed by substituting the load at fracture into a linear elastic K or $\mathcal{G}$ relationship, the resulting value will be size dependent, because the nonlinearity due to the process zone has been neglected. Such an approach is analogous to applying linear elastic K equations to metal specimens that exhibit significant plasticity prior to failure. Even approaches that account for the process zone exhibit size dependence, however [54]. A fracture parameter that uniquely characterizes these materials would be of great benefit.

REFERENCES

1. Johnson, W.S. (ed.), *Metal Matrix Composites: Testing, Analysis, and Failure Modes.* ASTM STP 1032, American Society for Testing and Materials, Philadelphia, PA, 1989.
2. Williams, J.G., *Fracture Mechanics of Polymers.* Ellis Horwood Ltd., Chichester, UK., 1984.

3. Kinloch, A.J. and Young, R.J., *Fracture Behavior of Polymers,* Elsevier London, 1983.

4. Brostow, W. and Corneliussen, R.D. (eds.), *Failure of Plastics.* Hanser Publishers, Munich, 1986.

5. Hertzberg, R.W. and Manson, J.A., *Fatigue of Engineering Plastics.* Academic Press, New York, 1980.

6. Moore, D.R., Pavan, A., and Williams, J.G. (eds.), *Fracture Mechanics Testing Methods for Polymers, Adhesives and Composites.* Elsevier, Oxford, 2001.

7. Ferry, J.D., Landel, R.F., and Williams, M.L., "Extensions of the Rouse Theory of Viscoelastic Properties to Undiluted Linear Polymers." *Journal of Applied Physics,* Vol. 26, 1955, pp. 359–362.

8. Larson, F.R. and Miller, J., "A Time-Temperature Relationship for Rupture and Creep Stresses." *Transactions of the American Society for Mechanical Engineers,* Vol. 74, 1952, pp. 765–775.

9. Kausch, H.H., *Polymer Fracture.* Springer-Verlag, Heidelberg, 1978.

10. Zhurkov, S.N. and Korsukov, V.E., "Atomic Mechanism of Fracture of Solid Polymers." *Journal of Polymer Science: Polymer Physics Edition,* Vol. 12, 1974, pp. 385–398.

11. Ward, I.M., *Mechanical Properties of Solid Polymers.* John Wiley & Sons, New York, 1971.

12. Sternstein, S.S. and Ongchin, L., "Yield Criteria for Plastic Deformation of Glassy High Polymers in General Stress Fields." *American Chemical Society, Polymer Preprints,* Vol. 10, 1969, pp. 1117–1124.

13. Bucknall, C.B., *Toughened Plastics.* Applied Science Publishers, London, 1977.

14. Donald, A.M. and Kramer, E.J., "Effect of Molecular Entanglements on Craze Microstructure in Glassy Polymers." *Journal of Polymer Science: Polymer Physics Edition,* Vol. 27, 1982, pp. 899–909.

15. Oxborough, R.J. and Bowden, P.B., "A General Critical-Strain Criterion for Crazing in Amorphous Glassy Polymers." *Philosophical Magazine,* Vol. 28, 1973, pp. 547–559.

16. Argon, A.S., "The Role of Heterogeneities in Fracture." ASTM STP 1020, American Society for Testing and Materials, Philadelphia, PA, 1989, pp. 127–148.

17. Cayard, M., "Fracture Toughness Testing of Polymeric Materials." Ph.D. Dissertation, Texas A&M University, College Station, TX, 1990.

18. Dugdale, D.S., "Yielding in Steel Sheets Containing Slits." *Journal of the Mechanics and Physics of Solids,* Vol. 8, 1960, pp. 100–104.

19. Barenblatt, G.I., "The Mathematical Theory of Equilibrium Cracks in Brittle Fracture." *Advances in Applied Mechanics,* Vol. VII, 1962, pp. 55–129.

20. *Engineered Materials Handbook, Volume 1: Composites.* ASM International, Metals Park, OH, 1987.

21. Vinson, J.R. and Sierakowski, R.L., *The Behavior of Structures Composed of Composite Materials.* Marinus Nijhoff, Dordrecht, The Netherlands, 1987.

22. Wang, A.S.D., "An Overview of the Delamination Problem in Structural Composites." *Key Engineering Materials,* Vol. 37, 1989, pp. 1–20.

23. Hunston, D. and Dehl, R., "The Role of Polymer Toughness in Matrix Dominated Composite Fracture." Paper EM87-355, Society of Manufacturing Engineers, Deerborn, MI, 1987.

24. Bradley, W.L., "Understanding the Translation of Neat Resin Toughness into Delamination Toughness in Composites." *Key Engineering Materials,* Vol. 37, 1989, pp. 161–198.

25. Jordan, W.M. and Bradley, W.L., "Micromechanisms of Fracture in Toughened Graphite-Epoxy Laminates." ASTM STP 937, American Society for Testing and Materials, Philadelphia, PA, 1987, pp. 95–114.

26. Hibbs, M.F., Tse, M.K., and Bradley, W.L., "Interlaminar Fracture Toughness and Real-Time Fracture Mechanisms of Some Toughened Graphite/Epoxy Composites." ASTM STP 937, American Society for Testing and Materials, Philadelphia, PA, 1987, pp. 115–130.

27. Rosen, B.W., "Mechanics of Composite Strengthening." *Fiber Composite Materials,* American Society for Metals, Metals Park, OH, 1965, pp. 37–75.

28. Guynn, E.G., Bradley, W.L., and Elber, W., "Micromechanics of Compression Failures in Open Hole Composite Laminates." ASTM STP 1012, American Society for Testing and Materials, Philadelphia, PA, 1989, pp. 118–136.

29. Soutis, C., Fleck, N.A., and Smith, P.A., "Failure Prediction Technique for Compression Loaded Carbon Fibre-Epoxy Laminate with Open Holes." *Submitted to Journal of Composite Materials,* 1990.

30. Highsmith, A.L. and Davis, J., "The Effects of Fiber Waviness on the Compressive Response of Fiber-Reinforced Composite Materials." Progress Report for NASA Research Grant NAG-1-659, NASA Langley Research Center, Hampton, VA, 1990.

31. Wang, A.S.D., "A Non-Linear Microbuckling Model Predicting the Compressive Strength of Unidirectional Composites." ASME Paper 78-WA/Aero-1, American Society for Mechanical Engineers, New York, 1978.

32. Guynn, E.G., "Experimental Observations and Finite Element Analysis of the Initiation of Fiber Microbuckling in Notched Composite Laminates." Ph.D. Dissertation, Texas A&M University, College Station, TX, 1990.

33. Whitcomb, J.D., "Finite Element Analysis of Instability Related Delamination Growth." *Journal of Composite Materials,* Vol. 15, 1981, pp. 403–425.

34. Awerbuch, J. and Madhukar, M.S., "Notched Strength of Composite Laminates: Predictions and Experiments—A Review." *Journal of Reinforced Plastics and Composites,* Vol. 4, 1985, pp. 3–159.

35. Whitney, J.M. and Nuismer, R.J., "Stress Fracture Criteria for Laminated Composites Containing Stress Concentrations." *Journal of Composite Materials,* Vol. 8, 1974, pp. 253–265.

36. Pipes, R.B., Wetherhold, R.C., and Gillespie, J.W., Jr., "Notched Strength of Composite Materials." *Journal of Composite Materials,* Vol. 12, 1979, pp. 148–160.

37. Waddoups, M.E., Eisenmann, J.R., and Kaminski, B.E., "Macroscopic Fracture Mechanics of Advanced Composite Materials." *Journal of Composite Materials,* Vol. 5, 1971, pp. 446–454.

38. Harris, C.E. and Morris, D.H., "A Comparison of the Fracture Behavior of Thick Laminated Composites Utilizing Compact Tension, Three-Point Bend, and Center-Cracked Tension Specimens." ASTM STP 905, American Society for Testing and Materials, Philadelphia, PA, 1986, pp. 124–135.

39. Charewicz, A. and Daniel, I.M., "Damage Mechanisms and Accumulation in Graphite/Epoxy Laminates." ASTM STP 907, American Society for Testing and Materials, Philadelphia, PA, 1986, pp. 274–297.

40. Evans, A.G., "The New High Toughness Ceramics." ASTM STP 1020, American Society for Testing and Materials, Philadelphia, PA, 1989, pp. 267–291.

41. Hutchinson, J.W., "Crack Tip Shielding by Micro Cracking in Brittle Solids," *Acta Metallurgica,* Vol. 35, 1987, pp. 1605–1619.

42. A.G. Evans (ed.), *Fracture in Ceramic Materials: Toughening Mechanisms, Machining Damage, Shock.* Noyes Publications, Park Ridge, NJ, 1984.

43. Budiansky, B., Hutchinson, J.W., and Evans, A.G., "Matrix Fracture in Fiber-Reinforced Ceramics." *Journal of the Mechanics and Physics of Solids,* Vol. 34, 1986, pp. 167–189.

44. Aveston, J., Cooper G.A., and Kelly, A., *The Properties of Fiber Composites,* IPC Science and Technology Press, Guildford, UK, 1971, pp. 15–26.

45. Marshall, D.B., Cox, B.N., and Evans, A.G., "The Mechanics of Matrix Cracking in Brittle-Matrix Fiber Composites." *Acta Metallurgica,* Vol. 33, 1985, pp. 2013–2021.

46. Marshall, D.B. and Ritter, J.E., "Reliability of Advanced Structural Ceramics and Ceramic Matrix Composites—A Review." *Ceramic Bulletin,* Vol. 68, 1987, pp. 309–317.

47. Mah, T., Mendiratta, M.G., Katz, A.P., Ruh, R., and Mazsiyasni, K.S., "Room Temperature Mechanical Behavior of Fiber-Reinforced Ceramic Composites." *Journal of the American Ceramic Society,* Vol. 68, 1985, pp. C27–C30.

48. Mah, T., Mendiratta, M.G., Katz, A.P., Ruh, R., and Mazsiyasni, K.S., "High-Temperature Mechanical Behavior of Fiber-Reinforced Glass-Ceramic-Matrix Composites." *Journal of the American Ceramic Society,* Vol. 68, 1985, pp. C248–C251.

49. Ruhle, M., Dalgleish, B.J., and Evans, A.G., "On the Toughening of Ceramics by Whiskers." *Scripta Metallurgica,* Vol. 21, pp. 681–686.

50. Bazant, Z.P., "Size Effect in Blunt Fracture: Concrete, Rock, Metal." *Journal of Engineering Mechanics,* Vol. 110, 1984, pp. 518–535.

51. Hillerborg, A., Modeer, M., and Petersson, P.E., "Analysis of Crack Formation and Crack Growth in Concrete by Means of Fracture Mechanics and Finite Elements." *Cement and Concrete Research,* Vol. 6, 1976, pp. 773–782.

52. Hillerborg, A., "Application of the Fictitious Crack Model to Different Materials." *International Journal of Fracture,* Vol. 51, 1991, pp. 95–102.

53. Bazant, Z.P. and Kazemi, M.T., "Determination of Fracture Energy, Process Zone Length and Brittleness Number from Size Effect, with Application to Rock and Concrete." *International Journal of Fracture,* Vol. 44, 1990, pp. 111–131.

54. Bazant, Z.P. and Kazemi, M.T., "Size Dependence of Concrete Fracture Energy Determined by RILEM Work-of-Fracture Method." *International Journal of Fracture,* Vol. 51, 1991, pp. 121–138.

55. Planas, J. and Elices, M., "Nonlinear Fracture of Cohesive Materials." *International Journal of Fracture,* Vol. 51, 1991, pp. 139–157.

56. Mazars, J., Pijaudier-Cabot, G., and Saourdis, C., "Size Effect and Continuous Damage in Cementitious Materials." *International Journal of Fracture,* Vol. 51, 1991, pp. 159–173.

Part IV

Applications

7 Fracture Toughness Testing of Metals

A fracture toughness test measures the resistance of a material to crack extension. Such a test may yield either a single value of fracture toughness or a resistance curve, where a toughness parameter such as K, J, or CTOD is plotted against the crack extension. A single toughness value is usually sufficient to describe a test that fails by cleavage, because this fracture mechanism is typically unstable. The situation is similar to the schematic in Figure 2.10(a), which illustrates a material with a flat R curve. Cleavage fracture actually has a falling resistance curve, as Figure 4.8 illustrates. Crack growth by microvoid coalescence, however, usually yields a rising R curve, such as that shown in Figure 2.10(b); ductile crack growth can be stable, at least initially. When ductile crack growth initiates in a test specimen, that specimen seldom fails immediately. Therefore, one can quantify ductile fracture resistance either by the initiation value or by the entire R curve.

A variety of organizations throughout the world publish standardized procedures for fracture toughness measurements, including the American Society for Testing and Materials (ASTM), the British Standards Institution (BSI), the International Institute of Standards (ISO), and the Japan Society of Mechanical Engineers (JSME). The first standards for K and J testing were developed by ASTM in 1970 and 1981, respectively, while BSI published the first CTOD test method in 1979.

Existing fracture toughness standards include procedures for K_{Ic}, K-R curve, J_{Ic}, J-R curve, CTOD, and K_{Ia} testing. This chapter focuses primarily on ASTM standards, since they are the most widely used throughout the world. Standards produced by other organizations, however, are broadly consistent with the ASTM procedures, and usually differ only in minute details. The existing standards are continually evolving, as the technology improves and more experience is gained.

The reader should not rely on this chapter alone for guidance on conducting fracture toughness tests, but should consult the relevant standards. Also, the reader is strongly encouraged to review Chapter 2 and Chapter 3 in order to gain an understanding of the fundamental basis of K, J, and CTOD, as well as the limitations of these parameters.

7.1 GENERAL CONSIDERATIONS

Virtually all fracture toughness tests have several common features. The design of test specimens is similar in each of the standards, and the orientation of the specimen relative to symmetry directions in the material is always an important consideration. The cracks in test specimens are introduced by fatigue in each case, although the requirements for fatigue loads varies from one standard to the next. The basic instrumentation required to measure load and displacement is common to virtually all fracture mechanics tests, but some tests require additional instrumentation to monitor crack growth.

7.1.1 SPECIMEN CONFIGURATIONS

There are five types of specimens that are permitted in ASTM standards that characterize fracture initiation and crack growth, although no single standard allows all five configurations, and the design of a particular specimen type may vary between standards. The configurations that are currently standardized include the compact specimen, the single-edge-notched bend (SE(B)) geometry, the arc-shaped specimen, the disk specimen, and the middle tension (MT) panel. Figure 7.1 shows a drawing of each specimen type.

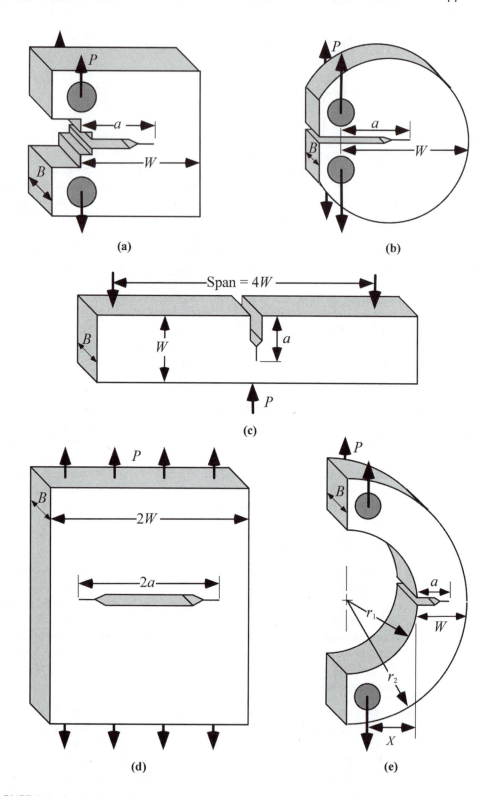

FIGURE 7.1 Standardized fracture mechanics test specimens: (a) compact specimen, (b) disk-shaped compact specimen, (c) single-edge-notched bend SE(B) specimen, (d) middle tension (MT) specimen, and (e) arc-shaped specimen.

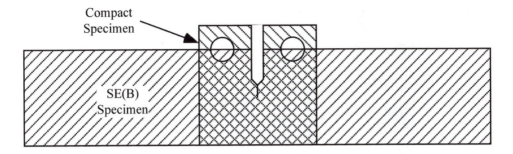

FIGURE 7.2 Comparison of the profiles of compact and SE(B) specimens with the same in-plane characteristic dimensions (W and a).

An additional configuration, the compact crack-arrest specimen, is used for K_{Ia} measurements and is described in Section 7.6. Specimens for qualitative toughness measurements, such as Charpy and drop-weight tests, are discussed in Section 7.9. Chevron-notched specimens, which are applied to brittle materials, are discussed in Chapter 8.

Each specimen configuration has three important characteristic dimensions: the crack length (a), the thickness (B) and the width (W). In most cases, $W = 2B$ and $a/W \approx 0.5$, but there are exceptions, which are discussed later in this chapter.

There are a number of specimen configurations that are used in research, but have yet to be standardized. Some of the more common nonstandard configurations include the single-edge-notched tensile panel, the double-edge-notched tensile panel, the axisymmetric-notched bar, and the double cantilever beam specimen.

The vast majority of fracture toughness tests are performed on either compact or SE(B) specimens. Figure 7.2 illustrates the profiles of these two specimen types, assuming the same characteristic dimensions (B, W, a). The compact geometry obviously consumes less material, but this specimen requires extra material in the width direction, due to the holes. If one is testing a plate material or a forging, the compact specimen is more economical, but the SE(B) configuration may be preferable for weldment testing, because less weld metal is consumed in some orientations (Section 7.7).

The compact specimen is pin-loaded by special clevises, as illustrated in Figure 7.3. Compact specimens are usually machined in a limited number of sizes because a separate test fixture must be fabricated for each specimen size. Specimen size is usually scaled geometrically; standard sizes include $\frac{1}{2}$ T, 1T, 2T, and 4T, where the nomenclature refers to the thickness in inches.[1] For example, a standard 1T compact specimen has the dimensions $B = 1$ in. (25.4 mm) and $W = 2$ in. (50.8 mm). Although ASTM has adopted SI units as their standard, the above nomenclature for compact specimen sizes persists.

The SE(B) specimen is more flexible with respect to size. The standard loading span for SE(B) specimens is $4W$. If the fixture is designed properly, the span can be adjusted continuously to any value that is within its capacity. Thus, SE(B) specimens with a wide range of dimensions can be tested with a single fixture. An apparatus for three-point bend testing is shown in Figure 7.4.

7.1.2 SPECIMEN ORIENTATION

Engineering materials are seldom homogeneous and isotropic. Microstructure, and thus, mechanical properties, are often sensitive to direction. The sensitivity to orientation is particularly pronounced in fracture toughness measurements, because a microstructure with a preferred orientation may contain planes of weakness, where crack propagation is relatively easy. Since specimen orientation

---

[1] An exception to this interpretation of the nomenclature occurs in thin sheet specimens, as discussed in Section 7.3.

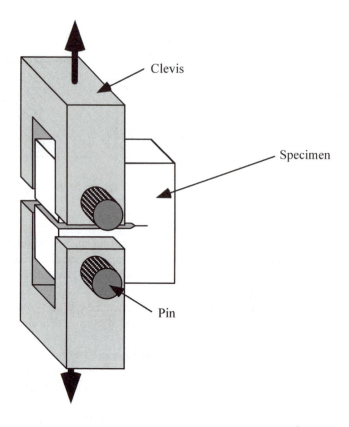

FIGURE 7.3 Apparatus for testing compact specimens.

FIGURE 7.4 Three-point bending apparatus for testing SE(B) specimens.

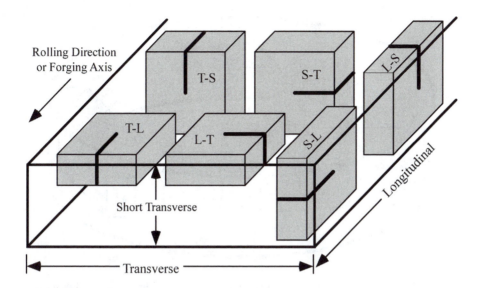

FIGURE 7.5 ASTM notation for specimens extracted from rolled plate and forgings. Taken from E 1823-96, "Standard Terminology Relating to Fatigue Fracture Testing." American Society for Testing and Materials, Philadelphia, PA, 1996 (Reapproved 2002).

is such an important variable in fracture toughness measurements, all ASTM fracture testing standards require that the orientation be reported along with the measured toughness; ASTM has adopted a notation for this purpose [1].

Figure 7.5 illustrates the ASTM notation for fracture specimens extracted from a rolled plate or forging. When the specimen is aligned with the axes of symmetry in the plate, there are six possible orientations. The letters L, T, and S denote the longitudinal, transverse, and short transverse directions, respectively, relative to the rolling direction or forging axis. Note that two letters are required to identify the orientation of a fracture mechanics specimen; the first letter indicates the direction of the principal tensile stress, which is always perpendicular to the crack plane in Mode I tests, and the second letter denotes the direction of crack propagation. For example, the *L-T* orientation corresponds to loading in the longitudinal direction and crack propagation in the transverse direction.

A similar notation applies to round bars and hollow cylinders, as Figure 7.6 illustrates. The symmetry directions in this case are circumferential, radial, and longitudinal (C, R, and L, respectively).

Ideally, one should measure the toughness of a material in several orientations, but this is often not practical. When choosing an appropriate specimen orientation, one should bear in mind the purpose of the test, as well as the geometrical constraints imposed by the material. A low toughness orientation, where the crack propagates in the rolling direction (*T-L* or *S-L*), should be adopted for general material characterization or screening. When the purpose of the test is to simulate conditions in a flawed structure, however, the crack orientation should match that of the structural flaw. Geometrical constraints may preclude testing some configurations; the *S-L* and *S-T* orientations, for example, are practical only in thick sections. The *T-S* and *L-S* orientations may limit the size of the compact specimen that can be extracted from a rolled plate.

7.1.3 Fatigue Precracking

Fracture mechanics theory applies to cracks that are infinitely sharp prior to loading. While laboratory specimens invariably fall short of this ideal, it is possible to introduce cracks that are sufficiently sharp for practical purposes. The most efficient way to produce such a crack is through cyclic loading.

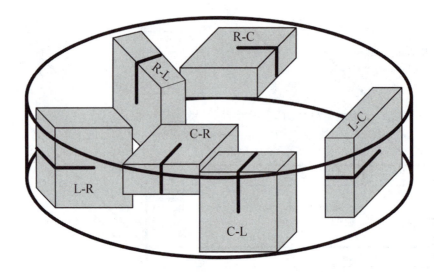

FIGURE 7.6 ASTM notation for specimens extracted from disks and hollow cylinders. Taken from E 1823-96, "Standard Terminology Relating to Fatigue Fracture Testing." American Society for Testing and Materials, Philadelphia, PA, 1996 (Reapproved 2002).

Figure 7.7 illustrates the precracking procedure in a typical specimen, where a fatigue crack initiates at the tip of a machined notch and grows to the desired size through careful control of the cyclic loads. Modern servo-hydraulic test machines can be programmed to produce sinusoidal loading, as well as a variety of other wave forms. Dedicated fatigue precracking machines that cycle at a high frequency are also available.

The fatigue crack must be introduced in such a way as not to adversely influence the toughness value that is to be measured. Cyclic loading produces a crack of finite radius with a small plastic zone at the tip, which contains strain-hardened material and a complicated residual stress distribution (see Chapter 10). In order for a fracture toughness measurement to reflect the true material properties, the fatigue crack must satisfy the following conditions:

- The crack-tip radius at failure must be much larger than the initial radius of the fatigue crack.
- The plastic zone produced during fatigue cracking must be small compared to the plastic zone at fracture.

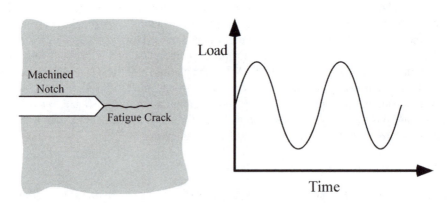

FIGURE 7.7 Fatigue precracking a fracture mechanics specimen. A fatigue crack is introduced at the tip of a machined notch by means of cyclic loading.

Each of the various fracture testing standards contains restrictions on fatigue loads, which are designed to satisfy the above requirements. The precise guidelines depend on the nature of the test. In K_{Ic} tests, for example, the maximum K during fatigue loading must be no greater than a particular fraction of K_{Ic}. In J and CTOD tests, where the test specimen is typically fully plastic at failure, the maximum fatigue load is defined as a fraction of the load at ligament yielding. Of course one can always perform fatigue precracking well below the allowable loads in order to gain additional assurance of the validity of the results, but the time required to produce the crack (i.e., the number of cycles) increases rapidly with decreasing fatigue loads.

7.1.4 INSTRUMENTATION

At a minimum, the applied load and a characteristic displacement on the specimen must be measured during a fracture toughness test. Additional instrumentation is applied to some specimens in order to monitor the crack growth or to measure more than one displacement.

Measuring load during a conventional fracture toughness test is relatively straightforward, since nearly all test machines are equipped with load cells. The most common displacement transducer in fracture mechanics tests is the clip gage, which is illustrated in Figure 7.8. The clip gage, which attaches to the mouth of the crack, consists of four resistance-strain gages bonded to a pair of cantilever beams. Deflection of the beams results in a change in voltage across the strain gages, which varies linearly with displacement. A clip gage must be attached to sharp knife edges in order to ensure that the ends of each beam are free to rotate. The knife edges can either be machined into the specimen or attached to the specimen at the crack mouth.

A *linear variable differential transformer* (LVDT) provides an alternative means for inferring displacements in fracture toughness tests. Figure 7.9 schematically illustrates the underlying principle of an LVDT. A steel rod is placed inside a hollow cylinder that contains a pair of tightly wound coils of wire. When a current passes through the first coil, the core becomes magnetized and induces a voltage in the second core. When the rod moves, the voltage drop in the second coil changes; the change in voltage varies linearly with displacement of the rod. The LVDT is useful for measuring displacements on a test specimen at locations other than the crack mouth.

The *potential drop* technique utilizes a voltage change to infer the crack growth, as illustrated in Figure 7.10. If a constant current passes through the uncracked ligament of a test specimen, the voltage must increase as the crack grows because the electrical resistance increases and the net

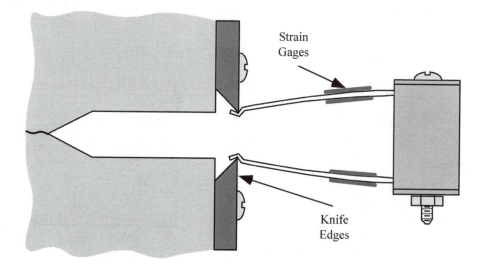

Strain Gages

Knife Edges

FIGURE 7.8. Measurement of the crack-mouth-opening displacement with a clip gage.

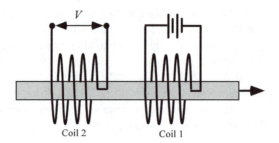

FIGURE 7.9 Schematic of a linear variable differential transformer (LVDT). Electric current in the first coil induces a magnetic field, which produces a voltage in the second coil. Displacement of the central core causes a variation in the output voltage.

cross-sectional area decreases. The potential drop method can use either DC or AC current. See Refs. [2] and [3] for examples of this technique.

The disadvantage of the potential drop technique is that it requires additional instrumentation. The *unloading compliance* technique [4], however, allows the crack growth to be inferred from the load and displacement transducers that are part of any standard fracture mechanics test. A specimen can be partially unloaded at various points during the test in order to measure the elastic compliance, which can be related to the crack length. Section 7.4 describes the unloading compliance technique in more detail.

In some cases it is necessary to measure more than one displacement on a test specimen. For example, one may want to measure both the crack-mouth-opening displacement (CMOD) and the displacement along the loading axis. A compact specimen can be designed such that the load line displacement and the CMOD are identical, but these two displacements do not coincide in an SE(B) specimen. Figure 7.11 illustrates simultaneous CMOD and load line displacement measurement in an SE(B) specimen. The CMOD is inferred from a clip gage attached to knife edges; the knife edge height must be taken into account when computing the relevant toughness parameter (see Section 7.5). The load-line displacement can be inferred by a number of methods, including the comparison bar technique [5, 6] that is illustrated in Figure 7.11. A bar is attached to the specimen at two points that are aligned with the outer loading points. The outer coil of an LVDT is attached to the comparison bar, which remains fixed during deformation, while the central rod is free to move as the specimen deflects.

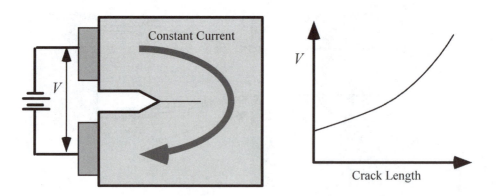

FIGURE 7.10 Potential drop method for monitoring crack growth. As the crack grows and the net cross-sectional area decreases, the effective resistance increases, resulting in an increase in voltage (V).

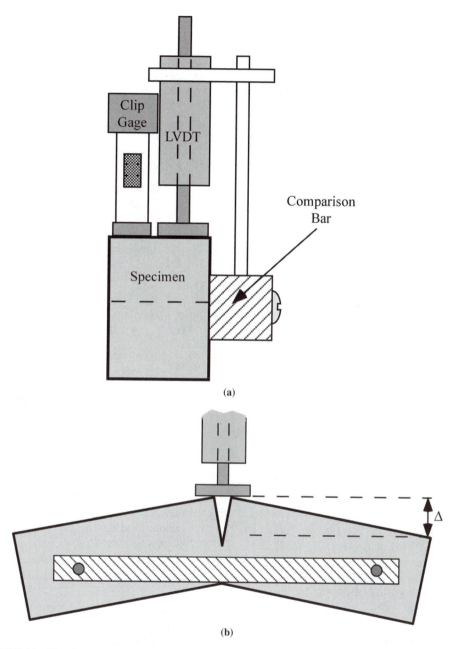

FIGURE 7.11 Simultaneous measurement of crack-mouth-opening displacement (CMOD) and load-line displacement on an SE(B) specimen. The CMOD is inferred from a clip gage attached to knife edges, while the load-line displacement can be determined from a comparison bar arrangement; the bar and outer coil of the LVDT remain fixed, while the inner rod moves with the specimen.

7.1.5 SIDE GROOVING

In certain cases, grooves are machined into the sides of a fracture toughness specimens [7], as Figure 7.12 illustrates. The primary purpose of side grooving is to maintain a straight crack front during an *R*-curve test. A specimen without side grooves is subject to crack tunneling and shear lip formation (Figure 5.15) because the material near the outer surfaces is in a state of low-stress triaxiality. Side grooves remove the low triaxiality zone and, if done properly, lead to relatively straight crack fronts.

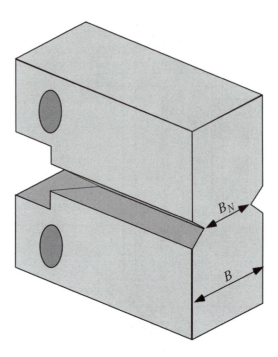

FIGURE 7.12 Side grooves in a fracture mechanics test specimen.

Typical side-grooved fracture toughness specimens have a net thickness that is approximately 80% of the gross thickness. If the side grooves are too deep, they produce lateral singularities, which cause the crack to grow more rapidly at the outer edges.

7.2 K_{Ic} TESTING

When a material behaves in a linear elastic manner prior to failure, such that the plastic zone is small compared to the specimen dimensions, a critical value of the Mode I stress-intensity factor K_{Ic} may be an appropriate fracture parameter. The first standardized test method for K_{Ic} testing, ASTM E 399 [8], was originally published in 1970. This standard has undergone a number of revisions over the years, but the key provisions have remained largely unchanged. Other K_{Ic} testing standards have been published throughout the world, including the British Standard 5447 [9], but are generally based on ASTM E 399.

In ASTM E 399 and similar test methods, K_{Ic} is referred to as "plane strain fracture toughness." This phrase actually appears in the title of ASTM E 399. In the 1960s, it was postulated that small specimens or thin sections fail under plane stress conditions, and that "plane strain fracture" occurs in thick sections. The ASTM E 399 test method reflects this viewpoint. Over the years, it has been taken as an indisputable fact that toughness decreases with increasing specimen size until a plateau is reached. Specimen size requirements in ASTM E 399 are intended to ensure that K_{Ic} measurements correspond to the supposed plane strain plateau.

There are a number of serious problems with ASTM E 399 and its underlying assumptions. Section 2.10 in Chapter 2 reexamines the conventional wisdom with respect to so-called "plane stress fracture" and "plane strain fracture." Three-dimensional finite element analysis of the stress state at the tip of a crack has revealed that the traditional view of the effect of specimen size on fracture toughness is simplistic and misleading. In addition, it can be shown that the E 399 test method results in a size dependence in the apparent K_{Ic} that is the opposite of what conventional wisdom suggests.

The existing K_{Ic} test procedure is outlined below. This is followed by a discussion of the limitations and pitfalls of the current approach.

7.2.1 ASTM E 399

The title of ASTM E 399, "*Standard Test Method for Plane Strain Fracture Toughness of Metallic Materials,*" is somewhat misleading. As discussed in Section 2.10, the specimen size requirements in this standard are far more stringent than they need to be to ensure predominately plane strain conditions at the crack tip. The real key to a K-based test method is ensuring that the specimen fractures under nominally linear elastic conditions. That is, the plastic zone must be small compared to the specimen cross section. Consequently, the important specimen dimensions to ensure a valid K test are the crack length a and the ligament length $W - a$, not the thickness B. Keeping these issues in mind, let us explore the current E 399 test method.

Four specimen configurations are permitted by the current version of E 399: the compact, SE(B), arc-shaped, and disk-shaped specimens. Specimens for K_{Ic} tests are usually fabricated with the width W equal to twice the thickness B. They are fatigue precracked so that the crack length/width ratio (a/W) lies between 0.45 and 0.55. Thus, the specimen design is such that all of the key dimensions, a, B, and $W-a$, are approximately equal. This design results in the efficient use of material, since the standard requires that each of these dimensions must be large compared to the plastic zone.

Most standardized mechanical tests (fracture toughness and otherwise) lead to valid results as long as the technician follows all of the procedures outlined in the standard. The K_{Ic} test, however, often produces invalid results through no fault of the technician. If the plastic zone at fracture is too large, it is not possible to obtain a valid K_{Ic}, regardless of how skilled the technician is.

Because of the strict size requirements, ASTM E 399 recommends that the user perform a preliminary validity check to determine the appropriate specimen dimensions. The size requirements for a valid K_{Ic} are as follows:

$$B, a \geq 2.5 \left(\frac{K_{IC}}{\sigma_{YS}} \right)^2, \quad 0.45 \leq a/W \leq 0.55$$

In order to determine the required specimen dimensions, the user must make a rough estimate of the anticipated K_{Ic} for the material. Such an estimate can come from data for similar materials. If such data are not available, the ASTM standard provides a table of recommended thicknesses for various strength levels. Although there is a tendency for toughness to decrease with increasing strength, there is not a unique relationship between K_{Ic} and σ_{YS} in metals. Thus, the strength-thickness table in E 399 should be used only when better data are not available.

During the initial stages of fatigue precracking, the peak value of stress intensity in a single cycle K_{max} should be no larger than $0.8 K_{Ic}$, according to ASTM E 399. As the crack approaches its final size, K_{max} should be less than $0.6 K_{Ic}$. If the specimen is fatigued at one temperature (T_1) and tested at a different temperature (T_2), the final K_{max} must be $\leq 0.6 (\sigma_{YS(1)}/\sigma_{YS(2)}) K_{Ic}$. The fatigue load requirements are less stringent at initiation because the final crack tip is remote from any damaged material that is produced in the early part of precracking. The maximum stress intensity during fatigue must always be less than K_{Ic}, however, in order to avoid the premature failure of the specimen.

Of course, one must know K_{Ic} to determine the maximum allowable fatigue loads. The user must specify fatigue loads based on the anticipated toughness of the material. If he or she is conservative and selects low loads, precracking could take a very long time. On the other hand, if precracking is conducted at high loads, the user risks an invalid result, in which case the specimen and the technician's time are wasted.

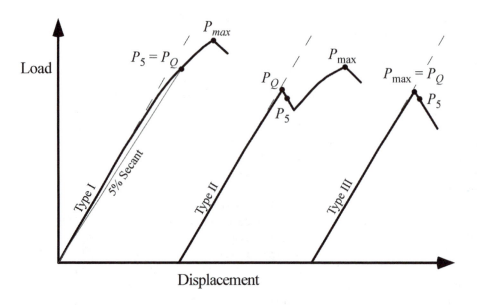

FIGURE 7.13 Three types of load-displacement behavior in a K_{Ic} test.

When a precracked test specimen is loaded to failure, load and displacement are monitored. Three types of load-displacement curves are shown in Figure 7.13. The critical load P_Q is defined in one of several ways, depending on the type of curve. One must construct a 5% secant line (i.e., a line from the origin with a slope equal to 95% of the initial elastic loading slope) to determine P_5. In the case of Type I behavior, the load-displacement curve is smooth and it deviates slightly from linearity before reaching a maximum load P_{max}. This nonlinearity can be caused by plasticity, subcritical crack growth, or both. For a Type I curve, $P_Q = P_5$. With a Type II curve, a small amount of unstable crack growth (i.e., a pop-in) occurs before the curve deviates from linearity by 5%. In this case P_Q is defined at the pop-in. A specimen that exhibits Type III behavior fails completely before achieving 5% nonlinearity. In such cases, $P_Q = P_{max}$.

The crack length must be measured from the fracture surface. Since there is a tendency for the crack depth to vary through the thickness, the crack length is defined as the average of three evenly spaced measurements. Once P_Q and the crack length are determined, a provisional fracture toughness K_Q is computed from the following relationship:

$$K_Q = \frac{P_Q}{B\sqrt{W}} f(a/W) \tag{7.1}$$

where $f(a/W)$ is a dimensionless function of a/W. This function is given in polynomial form in the E 399 standard for the four specimen types. Individual values of $f(a/W)$ are also tabulated in ASTM E 399. See Table 2.4 and Appendix 7 for K solutions for a variety of configurations.

The K_Q value computed from Equation (7.1) is a valid K_{Ic} result only if all validity requirements in the standard are met, including

$$0.45 \leq a/W \leq 0.55 \tag{7.2a}$$

$$B, a \geq 2.5 \left(\frac{K_Q}{\sigma_{YS}} \right)^2 \tag{7.2b}$$

$$P_{max} \leq 1.10 P_Q \tag{7.2c}$$

Additional validity requirements include the restrictions on the fatigue load mentioned earlier, as well as limits on the fatigue crack curvature. If the test meets all of the requirements of ASTM E 399, then $K_Q = K_{Ic}$.

Because the size requirements of ASTM E 399 are very stringent, it is very difficult and sometimes impossible to measure a valid K_{Ic} in most structural materials, as Example 7.1 and Example 7.2 illustrate. A material must either be relatively brittle or the test specimen must be very large for linear elastic fracture mechanics to be valid. In low- and medium-strength structural steels, valid K_{Ic} tests are normally possible only on the lower shelf of toughness; in the ductile-brittle transition and the upper shelf, elastic-plastic parameters such as the J integral and CTOD are required to characterize fracture.

Because of the strict validity requirements, the K_{Ic} test is of limited value to structural metals. The toughness and thickness of most materials precludes a valid K_{Ic} result. If, however, a valid K_{Ic} test can be measured on a given material, it is probably too brittle for most structural applications.

When attempting to measure fracture toughness using ASTM E 399, one runs the risk of invalid results due to the stringent size requirements. Once a result is declared invalid, E 399 offers no recourse for deriving useful information from the test. A more recent ASTM standard, E 1820 [4], provides an alterative test methodology that permits valid fracture toughness estimates from supposedly invalid K tests. ASTM E 1820 is a generalized test method for fracture toughness measurement that combines K, J, and CTOD parameters in a single standard. This standard provides a single test method for all three parameters and then offers a choice of post-test analysis procedures that pertain to a range of material behavior. If a test specimen exhibits too much plasticity to compute a valid K_{Ic}, the fracture toughness of the material can be characterized by J or CTOD. The calculation procedure and size requirements for K_{Ic} are essentially the same in E 399 and E 1820, but the latter relaxes specimen geometry requirements somewhat. For example, E 1820 permits side-grooved specimens in K_{Ic} tests. The standard BS 7448: Part 1 [10] is the British equivalent of ASTM E 1820.

EXAMPLE 7.1

Consider a structural steel with σ_{YS} = 350 MPa (51 ksi). Estimate the specimen dimensions required for a valid K_{Ic} test. Assume that this material is on the upper shelf of toughness, where typical K_{Ic} values for the initiation of microvoid coalescence in these materials are around 200 MPa$\sqrt{m}$.

Solution: Inserting the yield strength and estimated toughness into Equation (7.2a) gives

$$B, a = 2.5 \left(\frac{200 \text{ MPa}\sqrt{m}}{350 \text{ MPa}} \right)^2 = 0.816 \text{ m}(32.1 \text{ in.})$$

Since $a/W \approx 0.5$, W = 1.63 m (64.2 in.)! Thus a very large specimen would be required for a valid K_{Ic} test. Materials are seldom available in such thicknesses. Even if a sufficiently large section thickness were fabricated, testing such a large specimen would not be practical; machining would be prohibitively expensive, and a special testing machine with a high load capacity would be needed.

EXAMPLE 7.2

Suppose that the material in Example 7.1 is fabricated in 25 mm (1 in.) thick plate. Estimate the largest valid K_{Ic} that can be measured on such a specimen.

Solution: For the *L-T* or *T-L* orientation, a test specimen with a standard design could be no larger than $B = a = 25$ mm and $W = 50$ mm. Inserting these dimensions and the yield strength into Equation (7.2a) and solving for K_{Ic} gives

$$K_{Ic} = 350 \text{ MPa}\sqrt{\frac{0.025 \text{ m}}{2.5}} = 35 \text{ MPa}\sqrt{\text{m}}$$

Figure 4.5 shows fracture toughness data for A 572 Grade 50 steel. Note that the toughness level computed above corresponds to the lower shelf in this material. Thus, valid K_{Ic} tests on this material would be possible only at low temperatures, where the material is too brittle for most structural applications.

7.2.2 SHORTCOMINGS OF E 399 AND SIMILAR STANDARDS

Section 2.10 in Chapter 2 reexamines the conventional wisdom regarding the effect of plate thickness on apparent fracture toughness. The 1960s-vintage data that led to the hypothesis of a transition from "plane stress fracture" to "plane strain fracture" consisted almost entirely of materials that fail by microvoid coalescence (see Chapter 5). As Figure 2.45 illustrates, the observed thickness effect on fracture toughness is due to a competition between two fracture morphologies: slant fracture, which occurs on a 45° plane, and flat fracture, where the fracture plane is normal to the applied stress.

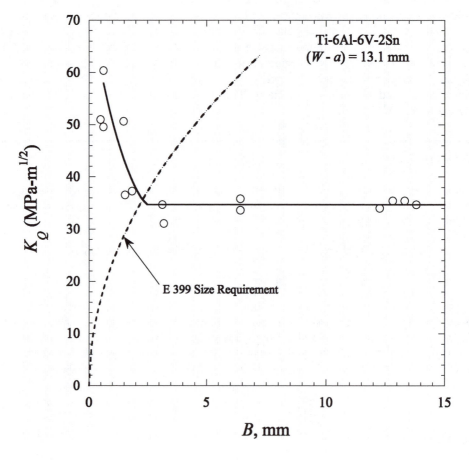

FIGURE 7.14   Effect of specimen thickness on apparent fracture toughness in a titanium alloy. Taken from Jones, M.H. and Brown, W.F., Jr., "The Influence of Crack Length and Thickness in Plane Strain Fracture Toughness Testing." ASTM STP 463, American Society for Testing and Materials, Philadelphia, PA, 1970, pp. 63–101.

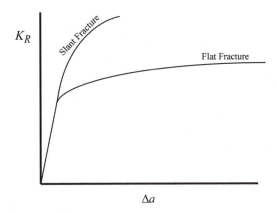

FIGURE 7.15 Effect of fracture morphology on the resistance to ductile tearing. Slant fracture results in a steeper R curve than a flat fracture.

In thinner specimens, the apparent fracture toughness is higher because slant fracture dominates. Figure 7.14 shows the effect of specimen thickness on K_Q in a titanium alloy [11]. In this case, the ligament length was fixed while thickness was varied. The measured toughness reaches a plateau value in this material when $B/(W - a) > 0.2$.

When crack extension occurs by ductile tearing (microvoid coalescence), the fracture toughness is characterized by a rising R curve. As Figure 7.15 illustrates, the R curve for slant fracture is significantly steeper than for flat fracture. The effective R curve for a specimen that experiences both morphologies will fall somewhere between these extremes. The relative amount of slant vs. flat fracture affects the K_Q value, as measured in accordance with the E 399 procedure. A side-grooved specimen (Figure 7.12) eliminates the shear lips and enables the R curve for flat fracture to be determined. ASTM E 399 does not permit side-grooved specimens, but E 1820 does allow side grooves in K_{Ic} tests.

Although shear lips can be eliminated by the proper use of side grooves, there is another serious problem with the E 399 test procedure when it is applied to a material that exhibits a rising R curve. Figure 7.16 is a plot of K_Q vs. ligament length for a high-strength aluminum alloy [12, 13]. Note that the measured toughness actually *increases* with specimen size, which runs counter to the conventional wisdom. This size effect on apparent toughness can be understood by examining how K_Q is measured. In a Type I load-displacement curve (Figure 7.13), which is typical of crack growth by ductile tearing, P_Q is defined as the point where the curve deviates from linearity by 5%. If the nonlinearity is due predominately to the crack growth, a 5% deviation from linearity corresponds approximately to the crack extension through 2% of the ligament. The specimen size effect in Figure 7.16 results from measuring K_Q at different amounts of absolute crack growth.

Figure 7.17 illustrates the inherent size effect in K_Q when it is based on a 2% relative crack growth. If the R curve is insensitive to the specimen size, specimens with larger ligaments will result in higher K_Q values because the measuring point is further up the R curve. The relative steepness of the R curve governs the magnitude of the K_Q size effect in materials that experience ductile crack extension. The E 399 procedure will result in size-independent toughness values only if the R curve is flat.

The original authors of E 399 were aware of the potential for size effects such as in Figure 7.16, although they may not have fully understood the reasons for this behavior. In an effort to address this issue, they set a maximum of 1.10 on the P_{max}/P_Q ratio (Equation (7.2c)). This additional restriction has been somewhat effective in reducing the size effect because it excludes materials with a steep R curve, as illustrated below.

Consider a material with an R curve that follows a power-law expression:

$$K_R = A(\Delta a)^m \tag{7.3}$$

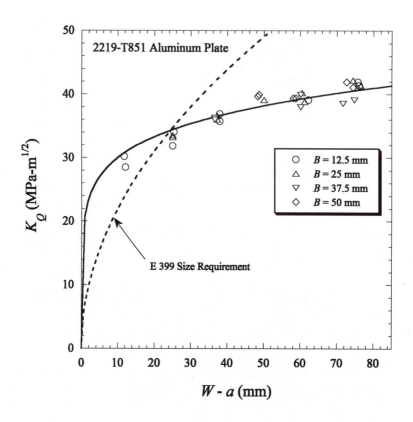

FIGURE 7.16 Effect of ligament length on apparent fracture toughness in an aluminum alloy. Taken from Kaufman, J.G. and Nelson, F.G., "More on Specimen Size Effects in Fracture Toughness Testing." ASTM STP 559, American Society for Testing and Materials, Philadelphia, PA, 1973, pp. 74–98. Wallin, K., "Critical Assessment of the Standard E 399." ASTM STP 1461, American Society for Testing and Materials, Philadelphia, PA, 2004.

where A and m are material constants. Using the above expression, and assuming that nonlinearity in the load-displacement curve is due entirely to crack growth, it is possible to compute the P_{max}/P_Q ratio for standard test specimens. Figure 7.18 shows the results of such an exercise for the compact specimen [13]. The load ratio is a function of the exponent m, but is insensitive to a/W. Moreover, this analysis predicts that the computed P_{max}/P_Q ratio is completely independent of the absolute

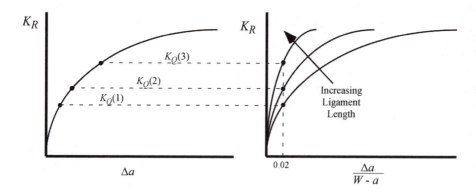

FIGURE 7.17 Schematic illustration of the inherent size effect in K_Q, as it is defined by the ASTM E 399 test method.

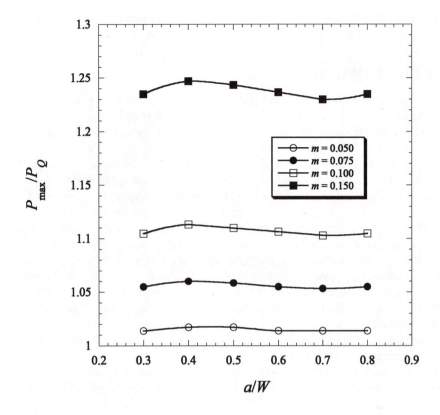

FIGURE 7.18 Theoretical calculation of the P_{max}/P_Q ratio, assuming all nonlinearity is due to crack growth and the R curve follows Equation (7.3). Taken from Wallin, K., "Critical Assessment of the Standard E 399." ASTM STP 1461, American Society for Testing and Materials, Philadelphia, PA, 2004.

specimen width W. The restriction of Equation (7.2c) is satisfied only for $m < 0.10$. Therefore, ASTM E 399 is applicable only to materials with relatively flat R curves, irrespective of specimen size.[2]

For materials that fail by cleavage fracture, the nonlinearity in the load-displacement curve is due to plastic zone formation rather than stable crack growth. In such cases, the 5% secant method is totally inappropriate for estimating the fracture toughness. Rather, the load at which unstable crack extension occurs should be used to compute the toughness. If the nonlinearity in the load-displacement curve is small, the fracture load can be substituted into Equation (7.1) to compute K_Q. If there is a significant nonlinearity due to plastic deformation, the toughness should be characterized by J or CTOD. A key advantage to generalized test methods such as ASTM E 1820 and BS 7448 is that they allow for both linear elastic and elastic-plastic material behavior.

As of this writing, the ASTM Committee E08 on fatigue and fracture is considering a complete overhaul of the E 399 standard to address the problems described earlier. Wallin [13] has outlined a proposed framework for an improved E 399 test method. He states that K_Q should be based on a fixed amount of absolute crack growth rather than 2% of the ligament. Given that such a

---

[2] This analysis is based on the assumption of a power-law R curve. Real materials do not exhibit an indefinitely rising R curve, however. When ductile crack extension occurs in small-scale yielding, the material resistance eventually reaches a steady state, as discussed in Section 3.5.2. Therefore, it is theoretically possible to test a specimen of sufficient size to meet the E 399 requirements, even if the initial R curve slope is relatively steep. However, since specimen size is limited by practical considerations, the current version of E 399 is suitable only for materials with relatively flat R curves.

modification would eliminate the size effect illustrated in Figure 7.16 and Figure 7.17, Wallin argues that the restriction on P_{max}/P_Q in Equation (7.2c) would no longer be necessary.

7.3 *K-R* CURVE TESTING

As discussed in the previous section, materials that fail by microvoid coalescence usually exhibit a rising R curve. The ASTM E 399 test method measures a single point on the R curve. This method contains an inherent size dependence on apparent toughness because the point on the R curve at which K_Q is defined is a function of the ligament length, as Figure 7.17 illustrates.

An alternative to measuring a single toughness value is determining the entire R curve for materials that exhibit ductile crack extension. The ASTM Standard E 561 [14] outlines a procedure for determining K vs. crack growth curves in such materials. Unlike ASTM E 399, the $K-R$ standard does not contain a minimum thickness requirement, and thus can be applied to thin sheets. This standard, however, is appropriate only when the plastic zone is small compared to the in-plane dimensions of the test specimen. This test method is often applied to high-strength sheet materials.

As Figure 7.15 illustrates, thin sheets generally have a steeper R curve than thick sections because the slant fracture morphology dominates in the former. There is a common misconception about the effect of section thickness on the shape of the R curve. A number of published articles and textbooks imply that thick sections, corresponding to so-called plane strain fracture, exhibit a single value of fracture toughness (K_{Ic}), while the same material in a thin section displays a rising R curve. The latter is often mistakenly referred to as "plane stress fracture." (Refer to Section 2.10 for a detailed discussion of the fallacies of the traditional "plane stress" and "plane strain" descriptions of crack-tip conditions.) The section thickness has an effect on the crack-tip stress state and the fracture morphology, which in turn affects the slope of the R curve (Figure 7.15). However, a material that fails by microvoid coalescence usually has a rising R curve even for flat fracture under predominately plane strain conditions. The only instance where a thin section might exhibit a rising R curve while a thick section of the same material has a flat R curve (and a single-valued toughness) is where the difference in crack-tip triaxiality causes a fracture mode change from ductile tearing to cleavage in thin and thick sections, respectively.

Figure 7.19 illustrates a typical $K-R$ curve in a predominantly linear elastic material. The R curve is initially very steep, as little or no crack growth occurs with increasing K_I. As the crack begins to grow, K increases with the crack growth until a steady state is reached, where the R curve becomes flat (see Section 3.5 and Appendix 3.5). It is possible to define a critical stress intensity K_c where the driving force is tangent to the R curve. This instability point is not a material property, however, because the point of tangency depends on the shape of the driving force curve, which is

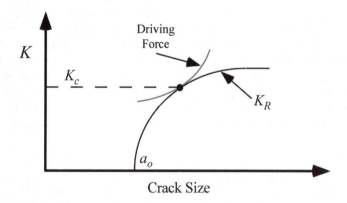

FIGURE 7.19 Schematic $K-R$ curve; K_c occurs at the point of tangency between the driving force and R curve.

governed by the size and geometry of the cracked body. In a laboratory specimen under load control, for example, K_c would correspond to P_{max} in a Type I load-displacement curve (Figure 7.13). Such a K_c value would exhibit a size dependence similar to that observed for K_Q based on a 2% crack growth criterion, as Figure 7.17 illustrates. Consequently, K_c values obtained from laboratory specimens are not usually transferable to structures.

7.3.1 SPECIMEN DESIGN

The ASTM standard for K-R curve testing [14] permits three configurations of test specimens: the middle tension (MT) geometry, the conventional compact specimen, and a wedge-loaded compact specimen. The latter configuration, which is similar to the compact crack-arrest specimen discussed in Section 7.6, is the most stable of the three specimen types, and thus is suitable for materials with relatively flat R curves.

Since this test method is often applied to thin sheets, specimens do not usually have the conventional geometry, with the width equal to twice the thickness. The specimen thickness is normally fixed by the sheet thickness, and the width is governed by the anticipated toughness of the material, as well as the available test fixtures.

A modified nomenclature is applied to thin-sheet compact specimens. For example, a specimen with $W = 50$ mm (2 in.) is designated as a *1T plan* specimen, since the in-plane dimensions correspond to the conventional 1T compact geometry. Standard fixtures can be used to test thin-sheet compact specimens, provided the specimens are fitted with spacers, as illustrated in Figure 7.20.

One problem with thin sheet fracture toughness testing is that the specimens are subject to out-of-plane buckling, which leads to combined Mode I–Mode III loading of the crack. Consequently, an antibuckling device should be fitted to the specimen. Figure 7.20 illustrates a typical antibuckling fixture for thin-sheet compact specimens. Plates on either side of the specimen prevent out-of-plane displacements. These plates should not be bolted too tightly together, because loads applied by the test machine should be carried by the specimen rather than the antibuckling plates. Some type of lubricant (e.g., Teflon sheet) is usually required to allow the specimen to slide freely through the two plates during the test.

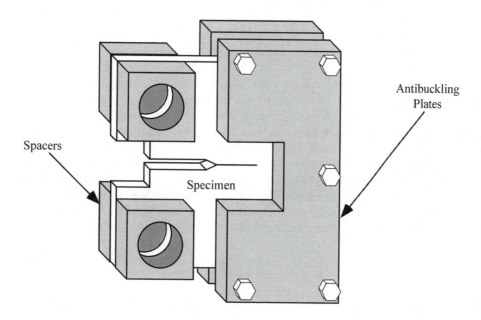

FIGURE 7.20 Antibuckling fixtures for testing thin compact specimens.

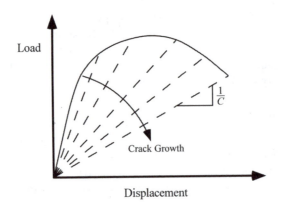

FIGURE 7.21 Load-displacement curve for crack growth in the absence of plasticity.

7.3.2 EXPERIMENTAL MEASUREMENT OF *K-R* CURVES

The ASTM Standard E 561 outlines a number of alternative methods for computing both K_I and the crack extension in an R curve test; the most appropriate approach depends on the relative size of the plastic zone. Let us first consider the special case of negligible plasticity, which exhibits a load-displacement behavior that is illustrated in Figure 7.21. As the crack grows, the load-displacement curve deviates from its initial linear shape because the compliance continuously changes. If the specimen were unloaded prior to fracture, the curve would return to the origin, as the dashed lines indicate. The compliance at any point during the test is equal to the displacement divided by the load. The instantaneous crack length can be inferred from the compliance through relationships that are given in the ASTM standard. See Appendix 7 for compliance-crack length equations for a variety of configurations. The crack length can also be measured optically during tests on thin sheets, where there is negligible through-thickness variation of crack length. The instantaneous stress intensity is related to the current values of load and crack length:

$$K_I = \frac{P}{B\sqrt{W}} f(a/W) \tag{7.4}$$

Consider now the case where a plastic zone forms ahead of the growing crack. The nonlinearity in the load-displacement curve is caused by a combination of crack growth and plasticity, as Figure 7.22 illustrates. If the specimen is unloaded prior to fracture, the load-displacement curve does not return to the origin; crack-tip plasticity produces a finite amount of permanent deformation in the specimen. The physical crack length can be determined optically or from unloading compliance, where the specimen is partially unloaded, the elastic compliance is measured, and the crack length is inferred from compliance. The stress intensity should be corrected for plasticity effects by determining an effective crack length. The ASTM standard suggests two alternative approaches for computing a_{eff}: the Irwin plastic zone correction and the secant method. According to the Irwin approach (Section 2.8.1), the effective crack length for plane stress is given by

$$a_{eff} = a + \frac{1}{2\pi} \left(\frac{K}{\sigma_{YS}} \right)^2 \tag{7.5}$$

The secant method consists of determining an effective crack size from the effective compliance, which is equal to the total displacement divided by the load (Figure 7.22). The effective stress-intensity

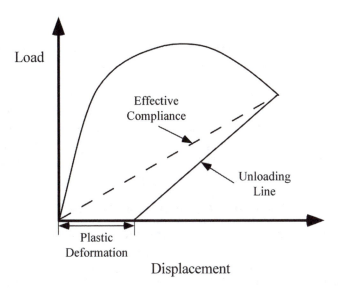

FIGURE 7.22 Load-displacement curve for crack growth with plasticity.

factor for both methods is computed from the load and the effective crack length:

$$K_{eff} = \frac{P}{B\sqrt{W}} f(a_{eff}/W) \tag{7.6}$$

   The Irwin correction requires an iterative calculation, where a first-order estimate of a_{eff} is used to estimate K_{eff}, which is inserted into Equation (7.5) to obtain a new a_{eff}; the process is repeated until the K_{eff} estimates converge.
   The choice of plasticity correction is left largely up to the user. When the plastic zone is small, ASTM E 561 suggests that the Irwin correction is acceptable, but recommends applying the secant approach when the crack-tip plasticity is more extensive. Experimental data typically display less size dependence when the stress intensity is determined by the secant method [15].
   The ASTM K-R curve standard requires that the stress intensity be plotted against the *effective* crack extension (Δa_{eff}). This practice is inconsistent with the J_{Ic} and J-R curve approaches (Section 7.4), where J is plotted against the *physical* crack extension. The estimate of the instability point K_c should not be sensitive to the way in which the crack growth is quantified, particularly when both the driving force and resistance curves are computed with a consistent definition of Δa.
   The ASTM E 561 standard does not contain requirements on the specimen size or the maximum allowable crack extension; thus there is no guarantee that a K-R curve produced according to this standard will be a geometry-independent material property. The in-plane dimensions must be large compared to the plastic zone in order for LEFM to be valid. Also, the growing crack must be remote from all external boundaries.
   Unfortunately, the size dependence of R curves in high strength sheet materials has yet to be quantified, so it is not possible to recommend specific size and crack growth limits for this type of testing. The user must be aware of the potential for size dependence in K-R curves. The application of the secant approach reduces but does not eliminate the size dependence. The user should test wide specimens whenever possible in order to ensure that the laboratory test is indicative of the structure under consideration.

7.4 *J* TESTING OF METALS

The current ASTM standard that covers *J*-integral testing is E 1820 [4]. This standard is actually a generalized fracture toughness standard, as it also covers K_{Ic} and CTOD tests. The British Standard BS 7448: Part 1 [10] is equivalent in scope to ASTM E 1820.

ASTM E 1820 has two alternative methods for *J* tests: the basic procedure and the resistance curve procedure. The basic procedure entails monotonically loading the specimen to failure or to a particular displacement, depending on the material behavior. The resistance curve procedure requires that the crack growth be monitored during the test. The *J* integral is calculated incrementally in the resistance curve procedure. The basic procedure can be used to measure *J* at fracture instability or near the onset of ductile crack extension. The latter toughness value is designated by the symbol J_{Ic}.

7.4.1 THE BASIC TEST PROCEDURE AND J_{Ic} MEASUREMENTS

Measuring toughness near the onset of ductile crack extension J_{Ic} requires the determination of a *J* resistance curve. If the basic procedure is used to generate such a resistance curve, the *J* values on the *R* curve may be subject to error because they have not been corrected for crack growth. This is of little consequence when measuring J_{Ic}, however, because the purpose of the *R* curve in this instance is to extrapolate back to a *J* value where Δa is small and a crack growth correction is not necessary. If a *J-R* curve is to be used in a tearing instability analysis (see Chapter 9), the test procedure in Section 7.4.2 should be applied.

Because crack growth is not monitored as a part of the basic test procedure, a multiple-specimen technique is normally required to obtain a *J-R* curve. In such cases, a series of nominally identical specimens are loaded to various levels and then unloaded. Different amounts of crack growth occur in the various specimens. The crack growth in each sample is marked by heat tinting or fatigue cracking after the test. Each specimen is then broken open and the crack extension is measured.

In addition to measuring crack growth, a *J* value must be computed for each specimen in order to generate the *R* curve. For estimation purposes, it is convenient to divide *J* into elastic and plastic components:

$$J = J_{el} + J_{pl} \tag{7.7}$$

The elastic *J* is computed from the elastic stress intensity:

$$J_{el} = \frac{K^2(1-v^2)}{E} \tag{7.8}$$

where *K* is inferred from the load and crack size through Equation (7.4). If, however, side-grooved specimens are used, the expression for *K* is modified:

$$K = \frac{P}{\sqrt{BB_N W}} f(a/W) \tag{7.9}$$

where *B* is the gross thickness and B_N is the net thickness (Figure 7.12). The basic procedure in ASTM E 1820 includes a simplified method for computing J_{pl} from the plastic area under the load-displacement curve:[3]

$$J_{pl} = \frac{\eta A_{pl}}{B_N b_o} \tag{7.10}$$

---

[3] Since *J* is defined in terms of the energy absorbed divided by the net cross-sectional area, B_N appears in the denominator. For nonside-grooved specimens $B_N = B$.

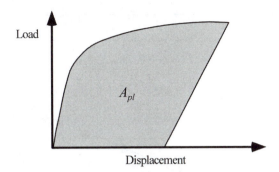

FIGURE 7.23 Plastic energy absorbed by a test specimen during a J_{Ic} test.

where η is a dimensionless constant, A_{pl} is the plastic area under the load-displacement curve (see Figure 7.23), and b_o is the initial ligament length. For an SE(B) specimen,

$$\eta = 2.0 \tag{7.11a}$$

and for a compact specimen,

$$\eta = 2 + 0.522\, b_o / W \tag{7.11b}$$

Recall from Section 3.2.5 that Equation (7.10) was derived from the energy release rate definition of J.

Note that Equation (7.10) and Equation (7.11b) do not correct J for crack growth, but are based on the initial crack length. The resistance curve procedure described in Section 7.4.2, in which J is computed incrementally with updated values of crack length and ligament length, can also be applied. This more elaborate procedure is usually not necessary for J_{Ic} measurements, however, because the crack growth is insignificant at the point on the R curve where J_{Ic} is measured. In the limit of a stationary crack, both formulas give identical results.

The ASTM procedure for computing J_Q, a provisional J_{Ic}, from the R curve is illustrated in Figure 7.24. Exclusion lines are drawn at crack extension (Δa) values of 0.15 and 1.5 mm. These lines have a slope of $M\sigma_Y$, where σ_Y is the flow stress, defined as the average of the yield and tensile strengths. The slope of the exclusion lines is intended to represent the component of crack extension that is due to crack blunting, as opposed to ductile tearing. The value of M can be determined experimentally, or a default value of 2 can be used. A horizontal exclusion line is defined at a maximum value of J:

$$J_{max} = \frac{b_o \sigma_Y}{15} \tag{7.12}$$

All data that fall within the exclusion limits are fit to a power-law expression:

$$J = C_1 (\Delta a)^{C_2} \tag{7.13}$$

The J_Q is defined as the intersection between Equation (7.13) and a 0.2-mm offset line. If all other validity criteria are met, $J_Q = J_{Ic}$ as long as the following size requirements are satisfied:

$$B, b_o \geq \frac{25 J_Q}{\sigma_Y} \tag{7.14}$$

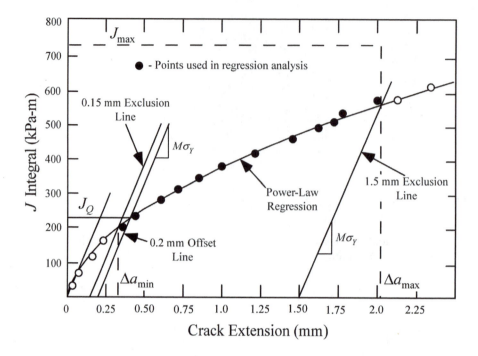

FIGURE 7.24 Determination of J_Q from a *J-R* curve. Taken from E 1820-01, "Standard Test Method for Measurement of Fracture Toughness." American Society for Testing and Materials, Philadelphia, PA, 2001.

EXAMPLE 7.3

Estimate the specimen size requirements for a valid J_{Ic} test on the material in Example 7.1. Assume $\sigma_{TS} = 450$ MPa and $E = 207,000$ MPa.

Solution: First we must convert the K_{Ic} value in Example 7.1 to an equivalent J_{Ic}:

$$J_{Ic} = \frac{K_{Ic}^2(1-v^2)}{E} = \frac{(200 \text{ MPa}\sqrt{m})^2(1-0.3^2)}{207,000 \text{ MPa}} = 0.176 \text{ MPa-m}$$

Substituting the above result into Equation (7.14) gives

$$B, b_o \geq \frac{(25)(0.176 \text{ MPa-m})}{400 \text{ MPa}} = 0.0110 \text{ m} = 11.0 \text{ mm} (0.433 \text{ in.})$$

which is nearly two orders of magnitude lower than the specimen dimension that ASTM E 399 requires for this material. Thus, the J_{Ic} size requirements are much more lenient than the K_{Ic} requirements.

7.4.2 *J-R* CURVE TESTING

The resistance curve test method in ASTM E 1820 requires that crack growth be monitored throughout the test. One disadvantage of this test method is that additional instrumentation is required. However, this complication is more than offset by the fact that the *J-R* curve can be obtained from a single specimen. Determining a *J-R* curve with the basic method requires tests on multiple specimens.

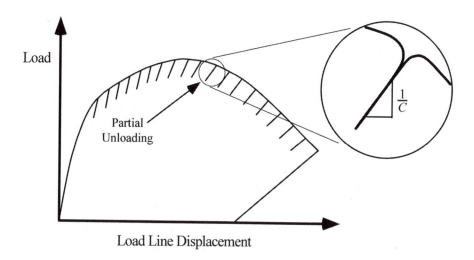

Load

Partial
Unloading

$\frac{1}{C}$

Load Line Displacement

FIGURE 7.25 The unloading compliance method for monitoring crack growth.

The most common single-specimen test technique is the unloading compliance method, which is illustrated in Figure 7.25. The crack length is computed at regular intervals during the test by partially unloading the specimen and measuring the compliance. As the crack grows, the specimen becomes more compliant (less stiff). The various J testing standards provide polynomial expressions that relate a/W to compliance. Table A7.3 in Appendix 7 lists these compliance equations for bend and compact specimens. The ASTM standard requires relatively deep cracks ($0.50 \leq a/W < 0.70$) because the unloading compliance technique is less sensitive for $a/W < 0.5$.

An alternative single-specimen test method is the potential drop procedure (Figure 7.10) in which crack growth is monitored through the change in electrical resistance that accompanies a loss in cross-sectional area.

A third option for monitoring crack growth during a J test is the normalization method [4, 16], which entails inferring the crack growth from the load-displacement curve. A specimen in which the crack is growing goes through a maximum load plateau followed by a decrease in load, but the load-displacement curve would continually rise in the absence of crack growth. The normalization method is particularly useful for high loading rates, where techniques such as unloading compliance are not possible.

When determining the J resistance curve for a given material, the specimens should be side-grooved to avoid shear lips and crack tunneling. Proper side-grooving will also produce relatively uniform ductile crack extension along the crack front.

There are a number of ways to compute J for a growing crack, as outlined in Section 3.4.2. The ASTM procedure for J-R curve testing utilizes the *deformation theory* definition of J, which corresponds to the rate of energy dissipation by the growing crack (i.e., the energy release rate). Recall Figure 3.22, which contrasts the actual loading path with the "deformation" path. The deformation J is related to the area under the load-displacement curve for a stationary crack, rather than the area under the actual load-displacement curve, where the crack length varies (see Equation (3.55) and Equation (3.56)).

Since the crack length changes continuously during a J-R curve test, the J integral must be calculated incrementally. For unloading compliance tests, the most logical time to update the J value is at each unloading point, where the crack length is also updated. Consider a J test with n measuring points. For a given measuring point i, where $1 \leq i \leq n$, the elastic and plastic components

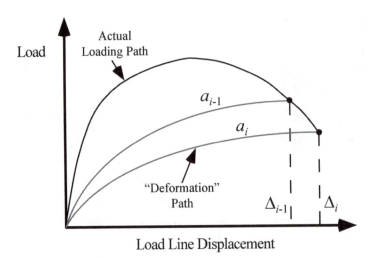

FIGURE 7.26 Schematic load-displacement curve for a *J-R* curve test.

of *J* can be estimated from the following expressions (see Figure 7.26):

$$J_{el(i)} = \frac{K_{(i)}^2(1-v^2)}{E}$$ (7.15a)

$$J_{pl(i)} = \left[J_{pl(i-1)} + \left(\frac{\eta_{i-1}}{B_N b_{i-1}} \right) \frac{(P_i + P_{i-1})\left(\Delta_{i(pl)} - \Delta_{i-1(pl)}\right)}{2} \right]$$

$$\times \left[1 - \gamma_{i-1} \frac{a_i - a_{i-1}}{b_{i-1}} \right]$$ (7.15b)

where $\Delta_{i(pl)}$ is the plastic load-line displacement, $\gamma_i = 1.0$ for SE(B) specimens and $\gamma_i = 1 + 0.76$ b_i/W for compact specimens, η_i is as defined in Equation (7.11), except that b_o is replaced by b_i, the instantaneous ligament length. The instantaneous *K* is related to P_i and a_i/W through Equation (7.9).

ASTM E 1820 has the following limits on *J* and crack extension relative to specimen size:

$$B, b_o \geq \frac{20J_{max}}{\sigma_Y}$$ (7.16)

and

$$\Delta a_{max} \leq 0.25 b_o$$ (7.17)

Figure 7.27 shows a typical *J-R* curve with the ASTM validity limits. The portion of the *J-R* curve that falls outside these limits is considered invalid.

7.4.3 CRITICAL *J* VALUES FOR UNSTABLE FRACTURE

Earlier *J* testing standards were restricted only to materials that exhibit ductile crack extension and a rising resistance curve. ASTM E 1820, however, also covers tests that terminate in an unstable

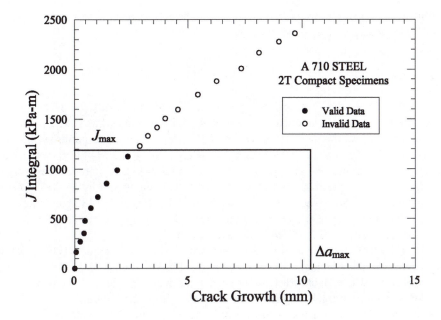

FIGURE 7.27 *J-R* curve for A 710 steel [17]. In this case, the data exceed the maximum *J* (Equation (7.16)) before the crack growth limit (Equation (7.17)).

fracture event, such as cleavage in steels. Either the basic or resistance curve test procedure may be applied to materials that exhibit unstable fracture.

The value of *J* at the point of unstable fracture is designated J_{Qc}, a provisional fracture toughness. If certain validity criterion are met, $J_{Qc} = J_c$. The specimen size requirement for J_c is as follows:

$$B, b_o \geq \frac{\lambda J_{Qc}}{\sigma_Y} \qquad (7.18)$$

where λ is a dimensionless constant. If the material is a ferritic steel with tensile properties within a certain range, $\lambda = 50$. Otherwise, $\lambda = 100$. The second validity criterion for J_c is the maximum allowable stable crack extension prior to stable fracture:

$$\Delta a < 0.2 \text{ mm} + J_Q/(M\sigma_Y) \qquad (7.19)$$

where $M\sigma_Y$ is the slope of the blunting line, as illustrated in Figure 7.24. Unless the blunting line slope has been experimentally determined for the material, *M* is normally assumed to equal 2.

If a test exhibits a significant stable crack growth prior to final fracture, such that Equation (7.19) is not satisfied, the *J* value at fracture instability is designated as J_u. There are no specimen size requirements for J_u. If the resistance curve test method is used such that crack growth is monitored prior to ultimate failure, it may be possible to construct a *J-R* curve and compute J_{lc}.

The size requirement for J_c is intended to ensure that the specimen has sufficient crack-tip triaxiality, and that further increases in size would not significantly affect the triaxiality. However, J_c values may still exhibit specimen size dependence. In ferritic steels, for example, there is a statistical size effect on fracture toughness because the probability of cleavage fracture is related to the length of the crack front. An ASTM toughness test method specifically for ferritic steels in the ductile-brittle transition range, E 1921 [18], addresses this phenomenon. See Section 7.8 for more information about ASTM E 1921.

The sensitivity of J_u values to specimen size has not been fully quantified. Consequently, ASTM warns that such values may be size dependent.

7.5 CTOD TESTING

The first CTOD test standard was published in Great Britain in 1979 [19]. Several years later, ASTM published E 1290, an American version of the CTOD standard. ASTM E 1290 has been revised several times, and the most recent version (as of this writing) was published in 2002 [20]. The original British CTOD test standard has been superceded by BS 7448 [10], which combines K, J, and CTOD testing into a single standard. ASTM E 1820 [4] also combined these three crack-tip parameters into a single testing standard, but E 1290 is still maintained by the ASTM Committee E08 on Fatigue and Fracture. The CTOD test methods in E 1290 and E 1820 are similar, but the latter standard includes provisions for generating a CTOD resistance curve. The discussion in this section focuses primarily on the ASTM E 1820 test method.

ASTM E 1820 includes both a basic and resistance curve procedure for CTOD, much like the J test methodology in this standard. The test method in E 1290 is comparable to the basic procedure. The basic procedure, where stable crack growth is not considered in the analysis, is described next. This is followed by a description of the CTOD resistance curve procedure.

Experimental CTOD estimates are made by separating the CTOD into elastic and plastic components, similar to J tests. The elastic CTOD is obtained from the elastic K:

$$\delta_{el} = \frac{K^2(1-v^2)}{2\sigma_{YS}E} \tag{7.20}$$

The elastic K is related to applied load through Equation (7.4). The above relationship assumes that $d_n = 0.5$ for linear elastic conditions (Equation (3.48)). The plastic component of CTOD is obtained by assuming that the test specimen rotates about a plastic hinge. This concept is illustrated in Figure 7.28 for an SE(B) specimen. The plastic displacement at the crack mouth, V_p, is related to the plastic CTOD through a similar triangles construction:

$$\delta_{pl} = \frac{r_p(W-a_o)V_p}{r_p(W-a_o)+a_o+z} \tag{7.21}$$

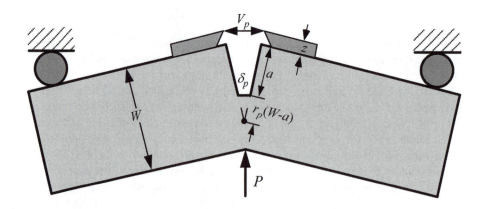

FIGURE 7.28 Hinge model for plastic displacements in an SE(B) specimen.

where r_p is the plastic rotational factor, a constant between 0 and 1 that defines the relative position of the apparent hinge point. The mouth-opening displacement is measured with a clip gage. In the case of an SE(B) specimen, knife edges are often attached in order to hold the clip gage, so that Equation (7.21) must take account of the knife-edge height z. The compact specimen is usually designed such that $z = 0$. The plastic component of V is obtained from the load-displacement curve by constructing a line parallel to the elastic loading line, as illustrated in Figure 3.6. The plastic rotational factor is given by

$$r_p = 0.44 \tag{7.22a}$$

for the SE(B) specimen and

$$r_p = 0.4 \left\{ 1 + 2 \left[\left(\frac{a_o}{b_o} \right)^2 + \frac{a_o}{b_o} + 0.5 \right]^{1/2} - 2 \left[\frac{a_o}{b_o} + 0.5 \right] \right\} \tag{7.22b}$$

for the compact specimen. The original British standard for CTOD tests, BS 5762:1979, applied only to SE(B) specimens and specified $r_p = 0.40$.

The crack-mouth-opening displacement V on an SE(B) specimen is not the same as the load-line displacement Δ. The latter displacement measurement is required for J estimation because A_{pl} in Figure 7.23 represents the plastic energy absorbed by the specimen. The CTOD standard utilizes V_p because this displacement is easier to measure in SE(B) specimens. If r_p is known, however, it is possible to infer J from a P-V curve or CTOD from a P-Δ curve [5, 6]. The compact specimen simplifies matters somewhat because $V = \Delta$ as long as $z = 0$.

When using the resistance curve procedure in ASTM E 1820, the CTOD equation must continually be updated with the current crack size. For the ith measurement point in the test, CTOD is given by

$$\delta_{(i)} = \frac{K_{(i)}^2 (1 - \nu^2)}{2 \sigma_{YS} E} + \frac{\left[r_{p(i)} (W - a_i) + \Delta a_i \right] V_{p(i)}}{\left[r_{p(i)} (W - a_i) + a_i + z \right]} \tag{7.23}$$

where a_i is the current crack length and $\Delta a_i = a_i - a_o$. The stress-intensity factor $K_{(i)}$ is computed using the current crack length. For compact specimens, the rotational factor must also be continually updated to account for crack growth:

$$r_{p(i)} = 0.4 \left\{ 1 + 2 \left[\left(\frac{a_i}{b_i} \right)^2 + \frac{a_i}{b_i} + 0.5 \right]^{1/2} - 2 \left[\frac{a_i}{b_i} + 0.5 \right] \right\} \tag{7.24}$$

Equation (7.23) and Equation (7.24) do not appear in ASTM E 1290.

When the material exhibits ductile crack extension, the above procedure can be used to generate a δ-R curve. The CTOD near the onset of ductile crack extension, δ_{Ic}, can be inferred from the resistance curve in a manner very similar to the J_{Ic} measurement (Figure 7.24). The slope of the blunting line and exclusion lines for a δ-R curve is M_δ, where M_δ is assumed to equal 1.4 unless an experimental measurement of the blunting line slope has been made. The provisional initiation toughness δ_Q is defined at the intersection between the δ-R curve, defined by a power-law fit within the exclusion lines, and a 0.2-mm offset line with a slope of M_δ. The size requirement for a valid δ_{Ic} is given by

$$b_o \geq 35\delta_Q \tag{7.25}$$

The basic procedure can be used to measure δ_{Ic}, but multiple specimens are required to define a $\delta\text{-}R$ curve. ASTM E 1290 does not provide a procedure to measure δ_{Ic}.

For tests that terminate in unstable fracture, such as cleavage in ferritic steels, the symbols δ_c and δ_u are assigned to the resulting CTOD values. The appropriate symbol depends on whether or not the fracture instability was preceded by significant stable crack extension, as is the case with J_c and J_u measurements in E 1820. When assessing toughness in accordance to ASTM E 1820, the δ_u label applies when $\Delta a \geq 0.2$ mm $+ \delta_{Qu}/M_\delta$. In ASTM E 1290, fracture instability toughness is designated as δ_u when $\Delta a \geq 0.2$ mm. When $\Delta a < 0.2$ mm $+ \delta_{Qu}/M_\delta$, ASTM E 1820 imposes the following size requirement on δ_c:

$$B, b_o \geq 300\delta_{Qc} \tag{7.26}$$

This size requirement is approximately two to four times more stringent than the requirement on J_c (Equation (7.19)). It is likely that a future revision of ASTM E 1820 will relax the CTOD size requirement to make it consistent with the J_c size requirement. ASTM E 1290 does not have a size requirement on δ_c.

When applying the basic test method to a ductile material that does not exhibit fracture instability, it is usually not possible to infer a $\delta\text{-}R$ curve or δ_{Ic}, but the value of CTOD at the maximum load plateau, δ_m, can be reported. This maximum load point is the result of competition between strain hardening, which causes the load to increase with deformation, and ductile crack growth, which reduces the cross section of the specimen. The CTOD at maximum load in a ductile material gives a relative indication of toughness, but δ_m values are highly dependent on specimen size.

Figure 7.29 is a series of schematic load-displacement curves that illustrate a variety of failure scenarios. Curve (a) illustrates a test that results in a δ_c value; unstable fracture occurs at P_c. Figure 7.29(b) corresponds to a δ_u result, where ductile tearing precedes unstable fracture. The ductile crack growth

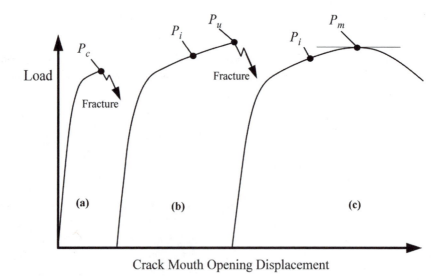

FIGURE 7.29 Various types of load-displacement curves from CTOD tests.

initiates at P_i. A test on a fully ductile material, such as steel on the upper shelf, produces a load-displacement curve like Figure 7.29(c); a maximum load plateau occurs at P_m. The specimen is still stable after maximum load if the test is performed in displacement control. Three types of CTOD result, δ_c, δ_u, and δ_m, are mutually exclusive, i.e., they cannot occur in the same test.

As Figure 7.29 illustrates, there is usually no detectable change in the load-displacement curve at the onset of ductile crack extension. The only deviation in the load-displacement behavior is the reduced rate of increase in load as the crack grows. The maximum load plateau (Figure 7.29(c)) occurs when the rate of strain hardening is exactly balanced by the rate of decrease in the cross section. However, the initiation of crack growth cannot be detected from the load-displacement curve because the loss of cross section is gradual. Thus δ_{Ic} must be determined from an R curve.

7.6 DYNAMIC AND CRACK-ARREST TOUGHNESS

When a material is subject to a rapidly applied load or a rapidly propagating crack, the response of that material may be drastically different from the quasistatic case. When rapid loading or unstable crack propagation are likely to occur in practice, it is important to duplicate these conditions when measuring material properties in the laboratory.

The dynamic fracture toughness and the crack-arrest toughness are two important material properties for many applications. The dynamic fracture toughness is a measure of the resistance of a material to crack propagation under rapid loading, while the crack-arrest toughness quantifies the ability of a material to stop a rapidly propagating crack. In the latter case, the crack may *initiate* under either dynamic or quasistatic conditions, but unstable *propagation* is generally a dynamic phenomenon.

Dynamic fracture problems are often complicated by inertia effects, material rate dependence, and reflected stress waves. One or more of these effects can be neglected in some cases, however. Refer to Chapter 4 for an additional discussion on this subject.

7.6.1 RAPID LOADING IN FRACTURE TESTING

Some testing standards, including ASTM E 399 [8] and E 1820 [4] include annexes for fracture toughness testing at high loading rates. This type of testing is more difficult than conventional fracture toughness measurements, and requires considerably more instrumentation.

High loading rates can be achieved in the laboratory by a number of means, including a drop tower, a high-rate testing machine, and explosive loading. With a drop tower, the load is imparted to the specimen through the force of gravity; a crosshead with a known weight is dropped onto the specimen from a specific height. A pendulum device such as a Charpy-testing machine is a variation of this principle. Some servo-hydraulic machines are capable of high displacement rates. While conventional testing machines are *closed loop*, where the hydraulic fluid circulates through the system, high-rate machines are *open loop*, where a single burst of hydraulic pressure is released over a short time interval. For moderately high displacement rates, a closed-loop machine may be adequate. Explosive loading involves setting off a controlled charge that sends stress waves through the specimen [21].

The dynamic loads resulting from impact are often inferred from an instrumented tup. Alternatively, strain gages can be mounted directly on the specimen; the output can be calibrated for load measurements, provided the gages are placed in a region of the specimen that remains elastic during the test. Crosshead displacements can be measured directly through an optical device mounted to the cross head. If this instrumentation is not available, a load-time curve can be converted to a load-displacement curve through momentum transfer relationships.

Certain applications require more advanced optical techniques, such as photoelasticity [22, 23] and the method of caustics [24]. These procedures provide more detailed information about the deformation of the specimen, but are also more complicated than global measurements of load and displacement.

Because high-rate fracture tests typically last only a few milliseconds, conventional data acquisition tools are inadequate. A storage oscilloscope has traditionally been required to capture data in a high-rate test; when a computer data acquisition system was used, the data were downloaded from the oscilloscope after the test. The newest generation of data acquisition cards for PCs removes the need for this two-step process. These cards are capable of collecting data at high rates, and enable the computer to simulate the functions of an oscilloscope.

Inertia effects can severely complicate the measurement of the relevant fracture parameters. The stress-intensity factor and J integral cannot be inferred from global loads and displacements when there is a significant kinetic energy component. Optical methods such as photoelasticity and caustics are necessary to measure J and K in such cases.

The transition time concept [25,26], which was introduced in Chapter 4, removes much of the complexity associated with J and K determination in high-rate tests. Recall that the transition time t_τ is defined as the time at which the kinetic energy and deformation energy are approximately equal. At times much less than t_τ, inertia effects dominate, while inertia is negligible at times significantly greater than t_τ. The latter case corresponds to essentially quasistatic conditions, where conventional equations for J and K apply. According to Figure 4.4, the quasistatic equation for J, based on the global load-displacement curve, is accurate at times greater that $2t_\tau$. Thus, if the critical fracture event occurs after $2t_\tau$, the toughness can be inferred from the conventional quasistatic relationships. For drop tower tests on ductile materials, the transition time requirement is relatively easy to meet [27,28]. For brittle materials (which fail sooner) or higher loading rates, the transition time can be shortened through specimen design.

7.6.2 K_{Ia} Measurements

In order to measure arrest toughness in a laboratory specimen, one must create conditions under which a crack initiates, propagates in an unstable manner, and then arrests. Unstable propagation followed by arrest can be achieved either through a rising R curve or a falling driving force curve. In the former case, a temperature gradient across a steel specimen produces the desired result; fracture can be initiated on the cold side of the specimen, where toughness is low, and propagate into warmer material where arrest is likely. A falling driving force can be obtained by loading the specimen in displacement control, as Example 2.3 illustrates.

The Robertson crack-arrest test [29] was one of the earliest applications of the temperature-gradient approach. This test is only qualitative, however, since the arrest temperature, rather than K_{Ia}, is determined from this test. The temperature at which a crack arrests in the Robertson specimen is only indicative of the relative arrest toughness of the material; designing above this temperature does not guarantee crack arrest under all loading conditions. The drop weight test; developed by Pellini (see Section 7.9) is another qualitative arrest test that yields a critical temperature. In this case, however, arrest is accomplished through a falling driving force.

While most crack-arrest tests are performed on small laboratory specimens, a limited number of experiments have been performed on larger configurations in order to validate the small-scale data. An extreme example of large-scale testing is the wide plate crack-arrest experiments conducted at the National Institute of Standards and Technology (NIST)[4] in Gaithersburg, Maryland [30]. Figure 7.30 shows a photograph of the NIST testing machine and one of the crack-arrest specimens. This specimen, which is a single-edge notched tensile panel, is 10 m long by 1 m wide. A temperature gradient is applied across the width, such that the initial crack is at the cold end. The specimen is then loaded until unstable cleavage occurs. These specimens are heavily instrumented, so that a variety of information can be inferred from each test. The crack-arrest toughness values measured from these tests are in broad agreement with small-scale specimen data.

---

[4] NIST was formerly known as the National Bureau of Standards (NBS), which explains the initials on either end of the specimen in Figure 7.30.

FIGURE 7.30 Photograph of a wide plate crack arrest test performed at NIST. Photograph provided by J.G. Merkle. Taken from Naus, D.J., Nanstad, R.K., Bass, B.R., Merkle, J.G., Pugh, C.E., Corwin, W.R., and Robinson, G.C., "Crack-Arrest Behavior in SEN Wide Plates of Quenched and Tempered A 533 Grade B Steel Tested under Nonisothermal Conditions." NUREG/CR-4930, U.S. Nuclear Regulatory Commission, Washington, DC, 1987.

In 1988, ASTM published a standard for crack-arrest testing, E 1221 [31]. This standard outlines a test procedure that is considerably more modest than the NIST experiments. A side-grooved compact crack-arrest specimen is wedge loaded until unstable fracture occurs. Because the specimen is held at a constant crack-mouth-opening displacement, the running crack experiences a falling K field. The crack-arrest toughness K_{Ia} is determined from the mouth-opening displacement and the arrested crack length.

The test specimen and loading apparatus for K_{Ia} testing are illustrated in Figure 7.31 and Figure 7.32. In most cases, a starter notch is placed in a brittle weld bead in order to facilitate

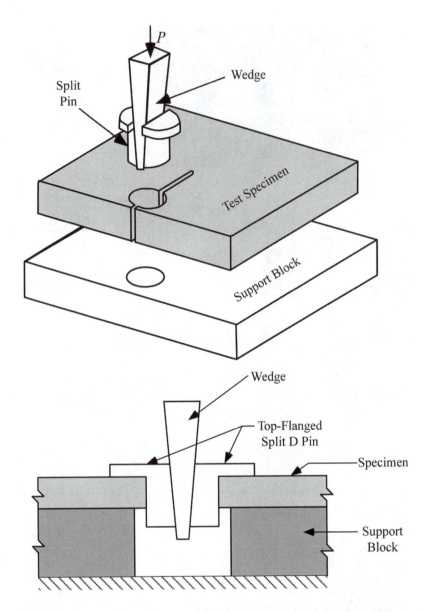

FIGURE 7.31 Apparatus for K_{Ia} tests. Taken from E 1221-96, "Standard Method for Determining Plane-Strain Crack-Arrest Toughness, K_{Ia}, of Ferritic Steels." American Society for Testing and Materials, Philadelphia, PA, 1996 (Reapproved 2002).

fracture initiation. A wedge is driven through a split pin that imparts a displacement to the specimen. A clip gage measures the displacement at the crack mouth (Figure 7.33).

Since the load normal to the crack plane is not measured in these tests, the stress intensity must be inferred from the clip-gage displacement. The estimation of K is complicated, however, by extraneous displacements, such as seating of the wedge/pin assembly. Also, local yielding can occur near the starter notch prior to fracture initiation. The ASTM standard outlines a cyclic loading procedure for identifying these displacements; Figure 7.33 shows a schematic load-displacement curve that illustrates this method. The specimen is first loaded to a predetermined displacement and, assuming the crack has not initiated, the specimen is unloaded. The displacement at zero load is assumed to represent the effects of fixture seating, and this component is subtracted from the

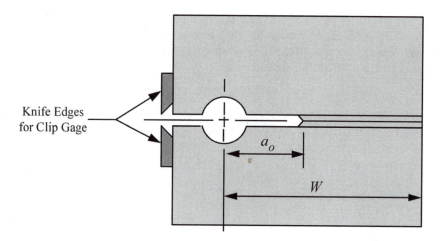

FIGURE 7.32 Side-grooved compact crack-arrest specimen. Taken from E 1221-96, "Standard Method for Determining Plane-Strain Crack-Arrest Toughness, K_{Ia}, of Ferritic Steels." American Society for Testing and Materials, Philadelphia, PA, 1996 (Reapproved 2002).

total displacement when stress intensity is computed. The specimen is reloaded to a somewhat higher displacement and then unloaded; this process continues until fracture initiates. The zero load offset displacements that occur after the first cycle can be considered to be due to notch-tip plasticity. The correct way to treat this displacement component in K calculations is unclear at present. Once the crack propagates through the plastic zone, the plastic displacement is largely recovered (i.e., converted into an elastic displacement), and thus may contribute to the driving force. It is not known whether or not there is sufficient time for this displacement component to exert an influence on the running crack. The ASTM standard takes the middle ground on this question, and requires that *half* of the plastic offset be included in the stress-intensity calculations.

After the test, the specimen should be heat tinted at 250–350°C for 10–90 min to mark the crack propagation. When the specimen is broken open, the arrested crack length can then be measured on the fracture surface. The critical stress intensity at initiation K_o is computed from the initial crack size and the critical clip-gage displacement. The provisional arrest toughness K_a is calculated from the *final* crack size, assuming a constant displacement. These calculations assume quasistatic conditions. As discussed in Chapter 4, this assumption can lead to underestimates of arrest toughness. The ASTM standard, however, cites experimental evidence that implies that the errors introduced by a quasistatic assumption are small in this case [32,33].

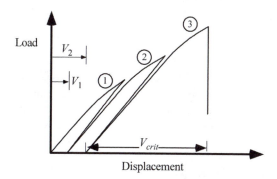

FIGURE 7.33 Schematic load-displacement curve for a K_{Ia} test [31], where V_1 and V_2 are zero load offset displacement. When computing V_{crit}, all of the first offset and half of the subsequent offsets are subtracted from the total displacement.

In order for the test to be valid, the crack propagation and arrest should occur under predominantly plane strain linear elastic conditions. The following validity requirements in ASTM E 1221 are designed to ensure that the plastic zone is small compared to specimen dimensions, and that the crack jump length is within acceptable limits:

$$W - a_a \geq 0.15W \tag{7.27a}$$

$$W - a_a \geq 1.25 \left(\frac{K_a}{\sigma_{Yd}} \right)^2 \tag{7.27b}$$

$$B \geq 1.0 \left(\frac{K_a}{\sigma_{Yd}} \right)^2 \tag{7.27c}$$

$$a_a - a_o \geq \frac{1}{2\pi} \left(\frac{K_a}{\sigma_{YS}} \right)^2 \tag{7.27d}$$

where
  a_a = arrested crack length
  a_o = initial crack length
  σ_{Yd} = assumed dynamic yield strength

which the ASTM standard specifies at 205 MPa (30 ksi) above the quasistatic value. Since unstable crack propagation results in very high strain rates, the recommended estimate of σ_{Yd} is probably very conservative.

If the above validity requirements are satisfied and all other provisions of ASTM E 1221 are followed, $K_a = K_{Ia}$.

7.7 FRACTURE TESTING OF WELDMENTS

All of the test methods discussed so far are suitable for specimens extracted from uniform sections of homogeneous material. Welded joints, however, have decidedly heterogeneous microstructures and, in many cases, irregular shapes. Weldments also contain complex residual stress distributions. Most existing fracture toughness testing standards do not address the special problems associated with weldment testing. An exception is Part 2 of British Standard 7448 [34]. This test method reflects practical experience that has been gained over the years [35–37].

The factors that make weldment testing difficult (i.e., heterogeneous microstructures, irregular shapes, and residual stresses) also tend to increase the risk of brittle fracture in welded structures. Thus, one cannot simply evaluate the regions of a structure where ASTM testing standards apply, and ignore the fracture properties of weldments.

When performing fracture toughness tests on weldments, a number of factors need special consideration. Specimen design and fabrication are more difficult because of the irregular shapes and curved surfaces associated with some welded joints. The heterogeneous microstructure of typical weldments requires special attention to the location of the notch in the test specimen. Residual stresses make fatigue precracking of weldment specimens more difficult. After the test, a weldment may be sectioned and examined metallographically to determine whether or not the fatigue crack sampled the intended microstructure.

7.7.1 SPECIMEN DESIGN AND FABRICATION

The underlying philosophy of the British Standards test procedure on specimen design and fabrication is that the specimen thickness should be as close to the section thickness as possible. Larger specimens tend to produce more crack-tip constraint, and hence lower toughness (see Chapter 2

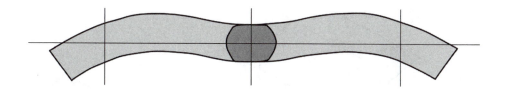

FIGURE 7.34 The gull-wing configuration for weldment specimens with excessive curvature. Taken from Dawes, M.G., Pisarski, H.G., and Squirrell, H.G., "Fracture Mechanics Tests on Welded Joints." ASTM STP 995, American Society for Testing and Materials, Philadelphia, PA, 1989, pp. II-191–II-213.

and Chapter 3). Achieving nearly full-thickness weldment often requires sacrifices in other areas. For example, if a specimen is to be extracted from a curved section such as a pipe, one can either produce a sub-size rectangular specimen that meets the tolerances of the existing ASTM standards, or a full-thickness specimen that is curved.

If curvature or distortion of a weldment is excessive, the specimen can be straightened by bending on either side of the notch to produce a "gull wing" configuration, which is illustrated in Figure 7.34. The bending must be performed so that the three loading points (in an SE(B) specimen) are aligned.

The fabrication of either a compact or SE(B) weldment specimen is possible, but the SE(B) specimen is preferable in most cases. Although the compact specimen consumes less material (for a given B and W) in parent metal tests, it requires more weld metal in a through-thickness orientation (L-T or T-L) than an SE(B) specimen (Figure 7.2). It is impractical to use a compact geometry for surface-notched specimens (T-S or L-S); such a specimen would be greatly undersized with the standard $B \times 2B$ geometry.

7.7.2 NOTCH LOCATION AND ORIENTATION

Weldments have a highly heterogeneous microstructure. Fracture toughness can vary considerably over relatively short distances. Thus, it is important to take great care in locating the fatigue crack in the correct region. If the fracture toughness test is designed to simulate an actual structural flaw, the fatigue crack must sample the same microstructure as the flaw. For a weld procedure qualification or a general assessment of a weldment's fracture toughness, location of the crack in the most brittle region may be desirable, but it is difficult to know in advance which region of the weld has the lowest toughness. In typical C–Mn structural steels, low toughness is usually associated with the coarse-grained heat-affected zone (HAZ) and the intercritically reheated HAZ. A microhardness survey can help identify low toughness regions because high hardness is often coincident with brittle behavior. The safest approach is to perform fracture toughness tests on a variety of regions in a weldment.

Once the microstructure of interest is identified, a notch orientation must be selected. The two most common alternatives are a through-thickness notch and a surface notch, which are illustrated in Figure 7.35. Since full-thickness specimens are desired, the surface-notched specimen should be a square section ($B \times B$), while the through-thickness notch will usually be in a rectangular ($B \times 2B$) specimen.

For weld metal testing, the through-thickness orientation is usually preferable because a variety of regions in the weld are sampled. However, there may be cases where the surface-notched specimen is the most suitable for testing the weld metal. For example, a surface notch can sample a particular region of the weld metal, such as the root or cap, or the notch can be located in a particular microstructure, such as unrefined weld metal.

Notch location in the HAZ often depends on the type of weldment. If welds are produced solely for mechanical testing, for example, as part of a weld procedure qualification or a research program,

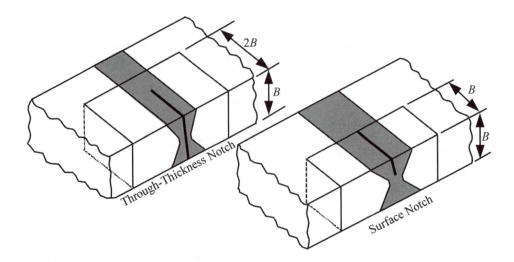

FIGURE 7.35 Notch orientation in weldment specimens. (a) through-thickness notch and (b) surface notch. Taken from Dawes, M.G., Pisarski, H.G., and Squirrell, H.G., "Fracture Mechanics Tests on Welded Joints." ASTM STP 995, American Society for Testing and Materials, Philadelphia, PA, 1989, pp. II-191–II-213.

the welded joint can be designed to facilitate HAZ testing. Figure 7.36 illustrates the K and half-K preparations, which simulate double-V and single-V welds, respectively. The plates should be tilted when these weldments are made, to have the same angle of attack for the electrode as in an actual single- or double-V joint. For fracture toughness testing, a through-thickness notch is placed in the straight side of the K or half-K HAZ.

In many instances, fracture toughness testing must be performed on an actual production weldment, where the joint geometry is governed by the structural design. In such cases, a surface notch is often necessary for the crack to sample sufficient HAZ material. The measured toughness is sensitive to the volume of HAZ material sampled by the crack tip because of the weakest-link nature of cleavage fracture (see Chapter 5).

Another application of the surface-notched orientation is the simulation of structural flaws. Figure 7.37 illustrates HAZ flaws in a structural weld and a surface-notched fracture toughness specimen that models one of the flaws.

Figure 7.37 demonstrates the advantages of allowing a range of a/W ratios in surface-notched specimens. A shallow notch is often required to locate a crack in the desired region, but most existing ASTM standards do not allow a/W ratios less than 0.45. Shallow-notched fracture toughness specimens tend to have lower constraint than deeply cracked specimens, as

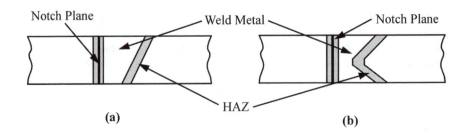

FIGURE 7.36 Special weld joint designs for fracture toughness testing of the heat-affected zone (HAZ). Taken from Dawes, M.G., Pisarski, H.G., and Squirrell, H.G., "Fracture Mechanics Tests on Welded Joints." ASTM STP 995, American Society for Testing and Materials, Philadelphia, PA, 1989, pp. II-191–II-213.

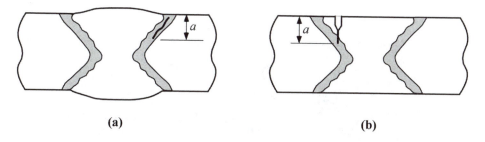

(a) **(b)**

FIGURE 7.37 Test specimen with notch orientation and depth that matches a flaw in a structure. (a) weldment with a flaw in the HAZ and (b) test specimen with simulated structural flaw. Taken from Dawes, M.G., Pisarski, H.G., and Squirrell, H.G., "Fracture Mechanics Tests on Welded Joints." ASTM STP 995, American Society for Testing and Materials, Philadelphia, PA, 1989, pp. II-191–II-213.

Figure 3.28 and Figure 3.44, illustrate. Thus, there is a conflict between the need to simulate a structural condition and the traditional fracture mechanics approach, where a toughness value is supposed to be a size-independent material property. One way to resolve this conflict is through constraint corrections, such as that applied to the data in Figure 3.44 and Figure 3.45.

7.7.3 FATIGUE PRECRACKING

Weldments that have not been stress relieved typically contain complex residual stress distributions that interfere with fatigue precracking of fracture toughness specimens. Tensile residual stresses accelerate fatigue crack initiation and growth, but compressive stresses retard fatigue. Since residual stresses vary through the cross section, fatigue crack fronts in as-welded samples are typically very nonuniform.

Towers and Dawes [38] evaluated the various methods for producing straight fatigue cracks in welded specimens, including reverse bending, high R ratio, and local compression.

The first method bends the specimen in the opposite direction to the normal loading configuration to produce residual tensile stresses along the crack front that counterbalance the compressive stresses. Although this technique gives some improvement, it does not usually produce acceptable fatigue crack fronts.

The R ratio in fatigue cracking is the ratio of the minimum stress to the maximum. A high R ratio minimizes the effect of residual stresses on fatigue, but also tends to increase the apparent toughness of the specimen. In addition, fatigue precracking at a high R ratio takes much longer than precracking at $R = 0.1$, the recommended R ratio of the various ASTM fracture-testing standards.

The only method that Towers and Dawes evaluated that produced consistently straight fatigue cracks was local compression, where the ligament is compressed to produce nominally 1% plastic strain through the thickness, mechanically relieving the residual stresses. However, local compression can reduce the toughness slightly. Towers and Dawes concluded that the benefits of local compression outweigh the disadvantages, particularly in the absence of a viable alternative.

7.7.4 POSTTEST ANALYSIS

Correct placement of a fatigue crack in weld metal is usually not difficult because this region is relatively homogeneous. The microstructure in the HAZ, however, can change dramatically over very small distances. Correct placement of a fatigue crack in the HAZ is often accomplished by

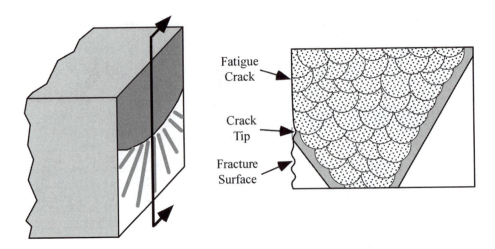

FIGURE 7.38 Posttest sectioning of a weldment fracture toughness specimen to identify the microstructure that caused fracture.

trial and error. Because fatigue cracks are usually slightly bowed, the precise location of the crack tip in the center of a specimen cannot be inferred from observations on the surface of the specimen. Thus, HAZ fracture toughness specimens should be examined metallographically after the test to determine the microstructure that initiated fracture. In certain cases, a posttest examination may be required in weld metal specimens.

Figure 7.38 illustrates a procedure for sectioning surface-notched and through-thickness-notched specimens [35]. First, the origin of the fracture must be located by the chevron markings on the fracture surface. After marking the origin with a small spot of paint, the specimen is sectioned perpendicular to the fracture surface and examined metallographically. The specimen should be sectioned slightly to one side of the origin and polished down to the initiation site. The spot of paint appears on the polished specimen when the origin is reached.

The API document RP2Z [37] outlines a posttest analysis of HAZ specimens, which is more detailed and cumbersome than the procedure outlined above. In addition to sectioning the specimen, the amount of coarse-grained material at the crack tip must be quantified. For the test to be valid, at least 15% of the crack front must be in the coarse-grained HAZ. The purpose of this procedure is to prequalify steels with respect to HAZ toughness, identifying those that produce low HAZ toughness so that they can be rejected before fabrication.

7.8 TESTING AND ANALYSIS OF STEELS IN THE DUCTILE-BRITTLE TRANSITION REGION

Chapter 5 described the micromechanisms of cleavage fracture, and indicated that cleavage toughness data tend to be highly scattered, especially in the transition region. Because of this substantial scatter, data should be treated statistically rather than deterministically. That is, a given steel does not have a single value of toughness at a particular temperature in the transition region; rather, the material has a toughness *distribution*. Testing numerous specimens to obtain a statistical distribution can be expensive and time consuming. Fortunately, a methodology has been developed that greatly simplifies this process for ferritic steels. A relatively new ASTM standard for the ductile-brittle transition region, E 1921 [18], implements this methodology.

Research over the past 20 years into the fracture of ferritic steels in the ductile-brittle transition region has led to two important conclusions:

1. Scatter in fracture toughness data in the transition region follows a characteristic statistical distribution that is the same for all ferritic steels.
2. The shape of the fracture toughness vs. temperature curve in the transition range is virtually identical for all ferritic steels. The only difference between steels is the absolute position of this curve on the temperature axis.

The term *Fracture Toughness Master Curve* [39–41] was coined to describe these two characteristics of steel. ASTM E 1921 outlines a fracture toughness test method that is based on the master curve concept.

Due to the idiosyncrasies of the micromechanism of cleavage fracture (Chapter 5), the fracture toughness at a fixed temperature in the transition region follows a 3-parameter Weibull distribution with a slope of 4:

$$F = 1 - \exp\left[-\frac{B}{25}\left(\frac{K_{Jc} - K_{min}}{K_o - K_{min}} \right)^4 \right] \tag{7.28}$$

where

F = cumulative probability

B = specimen thickness in millimeters

K_{Jc} = fracture toughness (critical J converted to the equivalent critical K) in MPa $\sqrt{m}$

K_{min} = threshold toughness

K_o = Weibull mean toughness, which corresponds to the 63rd percentile toughness for a 25-mm (1 in.) thick specimen

ASTM E 1920 fixes K_{min} at 20 MPa $\sqrt{m}$ (18.2 ksi $\sqrt{in}$.). Therefore, Equation (7.28) has only one unspecified parameter K_o. A statistical distribution that contains only one parameter can be fit with a relatively small sample size.

The thickness dependence in Equation (7.28) stems from statistical sampling effects. Cleavage fracture occurs by a weakest-link mechanism, and toughness tends to decrease with increasing crack front length.[5] Toughness data for various thicknesses can be converted to equivalent values for a 25-mm thick (i.e., standard 1T) specimen using the following relationship:

$$K_{Jc(1T)} = K_{min} + \left(K_{Jc(B)} - K_{min} \right)\left(\frac{B}{25} \right)^{1/4} \tag{7.29}$$

Once toughness values at a fixed temperature have been converted to 1T equivalent values using Equation (7.29), the Weibull mean toughness K_o can be computed as follows:

$$K_o = \left[\sum_{i=1}^{N} \frac{\left(K_{Jc(i)} - K_{min} \right)^4}{N} \right]^{1/4} + K_{min} \tag{7.30}$$

where N is the number of valid tests in the data set. The median toughness for a 1T specimen at a fixed temperature is related to the Weibull mean as follows:

$$K_{Jc(med)} = K_{min} + (K_o - K_{min})[\ln(2)]^{1/4} \tag{7.31}$$

---

[5] When assessing a flaw in a structural component, B is the crack front length, which generally is not equal to the section thickness.

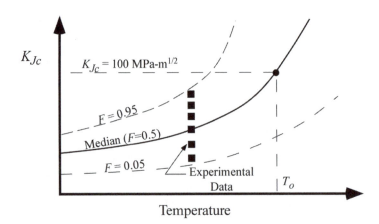

FIGURE 7.39 Fracture toughness master curve. Taken from E 1921-03, "Standard Test Method for Determination of Reference Temperature, T_o, for Ferritic Steels in the Transition Range." American Society for Testing and Materials, Philadelphia, PA, 2003.

According to the master curve model, the temperature dependence of median toughness in the ductile-brittle transition region is given by

$$K_{Jc(median)} = 30 + 70 \exp[0.019(T - T_o)] \tag{7.32}$$

where T_o is a reference transition temperature in °C and the units of K_{Jc} are $MPa\sqrt{m}$. At $T = T_o$, the median fracture toughness = 100 $MPa\sqrt{m}$. Once T_o is known for a given material, the fracture toughness distribution can be inferred as a function of temperature through Equation (7.28) and Equation (7.32).

The first step in determining T_o is to perform replicate fracture toughness tests at a constant temperature. ASTM E 1921 recommends at least six such tests. Next, the data are converted to equivalent 1T values using Equation (7.29). These data are then put into Equation (7.30) to determine K_o at the test temperature. The median toughness at this temperature is computed from Equation (7.31). Finally, T_o is computed by rearranging Equation (7.32):

$$T_o = T - \left(\frac{1}{0.019}\right) \ln\left[\frac{K_{Jc(median)} - 30}{70}\right] \tag{7.33}$$

ASTM E 1921 provides an alternative method for determining T_o when K_{Jc} data are obtained at multiple temperatures.

Figure 7.39 schematically illustrates the fracture toughness master curve for a particular steel. By combining Equation (7.28) and Equation (7.32), it is possible to infer median, upper-bound and lower-bound toughness as a function of temperature.

The Master Curve approach works best in the ductile-brittle transition region. It may not fit data in the lower shelf very well, and it is totally unsuitable for the upper shelf. Equation (7.32) increases without bound with increasing temperature, and thus does not model the upper shelf.

7.9 QUALITATIVE TOUGHNESS TESTS

Before the development of formal fracture mechanics methodology, engineers realized the importance of material toughness in avoiding brittle fracture. In 1901, a French scientist named G. Charpy developed a pendulum test that measured the energy of separation in notched metallic specimens. This energy was believed to be indicative of the resistance of the material to brittle fracture.

An investigation of the Liberty ship failures during World War II revealed that fracture was much more likely in steels with Charpy energy less than 20 J (15 ft-lb).

During the 1950s, when Irwin and his colleagues at the Naval Research Laboratory (NRL) were formulating the principles of linear elastic fracture mechanics, a metallurgist at NRL named W.S. Pellini developed the drop weight test, a qualitative measure of crack-arrest toughness.

Both the Charpy test and the Pellini drop weight test are still widely applied today to structural materials. ASTM has standardized the drop weight tests, as well as a number of related approaches, including the Izod, drop weight tear and dynamic tear tests [42–45] (see Section 7.9.1). Although these tests lack the mathematical rigor and predictive capabilities of fracture mechanics methods, these approaches provide a qualitative indication of material toughness. The advantage of these qualitative methods is that they are cheaper and easier to perform than fracture mechanics tests. These tests are suitable for material screening and quality control, but are not reliable indicators of structural integrity.

7.9.1 CHARPY AND IZOD IMPACT TEST

The ASTM Standard E 23 [42] covers Charpy and Izod testing. These tests both involve impacting a small notched bar with a pendulum and measuring the fracture energy. The Charpy specimen is a simple notched beam that is impacted in three-point bending, while the Izod specimen is a cantilever beam that is fixed at one end and impacted at the other. Figure 7.40 illustrates both types of tests.

Charpy and Izod specimens are relatively small, and thus do not consume much material. The standard cross section of both specimens is 10 mm × 10 mm, and the lengths are 55 and 75 mm for Charpy and Izod specimens, respectively.

The pendulum device provides a simple but elegant method for quantifying the fracture energy. As Figure 7.41 illustrates, the pendulum is released from a height y_1 and swings through the specimen to a height y_2. Assuming negligible friction and aerodynamic drag, the energy absorbed by the specimen is equal to the height difference times the weight of the pendulum. A simple mechanical device on the Charpy machine converts the height difference to a direct read-out of absorbed energy.

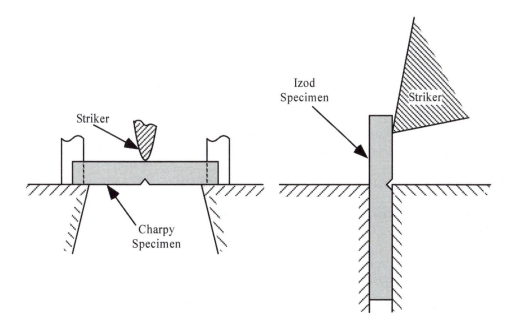

FIGURE 7.40 Charpy and Izod notched impact tests. Taken from E 23-02a, "Standard Test Methods for Notched Bar Impact Testing of Metallic Materials." American Society for Testing and Materials, Philadelphia, PA, 2002.

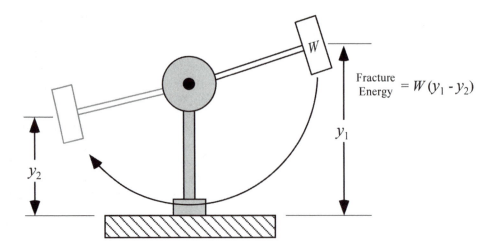

$$\text{Fracture Energy} = W\,(y_1 - y_2)$$

FIGURE 7.41 A pendulum device for impact testing. The energy absorbed by the specimen is equal to the weight of the crosshead, times the difference in height before and after impact.

A number of investigators [46–52] have attempted to correlate Charpy energy to fracture toughness parameters such as K_{Ic}. Some of these empirical correlations seem to work reasonably well, but most correlations are often unreliable. There are several important differences between the Charpy test and fracture mechanics tests that preclude simple relationships between the qualitative and quantitative measures of toughness. The Charpy test contains a blunt notch, while fracture mechanics specimens have sharp fatigue cracks. The Charpy specimen is subsize, and thus has low constraint. In addition, the Charpy specimen experiences impact loading, while most fracture toughness tests are conducted under quasistatic conditions.

It is possible to obtain quantitative information from fatigue precracked Charpy specimens, provided the tup (i.e., the striker) is instrumented [53,54]. Such an experiment is essentially a miniature dynamic fracture toughness test.

7.9.2 Drop Weight Test

The ASTM standard E 208 [43] outlines the procedure for performing the Pellini drop weight test. A plate specimen with a starter notch in a brittle weld bead is impacted in three-point bending. A cleavage crack initiates in the weld bead and runs into the parent metal. If the material is sufficiently tough, the crack arrests, otherwise the specimen fractures completely.

Figure 7.42 illustrates the drop weight specimen and the testing fixture. The crosshead drops onto the specimen, causing it to deflect a predetermined amount. The fixture is designed with a deflection stop, which limits the displacement in the specimen. A crack initiates at the starter notch and either propagates or arrests, depending on the temperature and material properties. A "break" result is recorded when the running crack reaches at least one specimen edge. A "no-break" result is recorded if the crack arrests in the parent metal. Figure 7.43 gives examples of break and no-break results.

A nil-ductility transition temperature (NDTT) is obtained by performing drop weight tests over a range of temperatures, in 5°C or 10°F increments. When a no-break result is recorded, the temperature is decreased for the next test; the test temperature is increased when a specimen fails. When break and no-break results are obtained at adjoining temperatures, a second test is performed at the no-break temperature. If this specimen fails, a test is performed at one temperature increment (5°C or 10°F) higher. The process is repeated until two no-break results are obtained at one

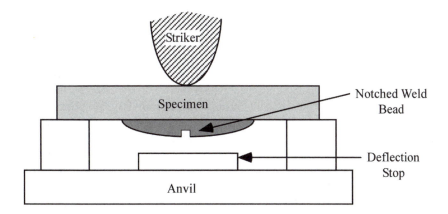

FIGURE 7.42 Apparatus for drop weight testing according to ASTM E 208-87. Taken from E 208-95a, "Standard Test Method for Conducting Drop-Weight Test to Determine Nil-Ductility Transition Temperature of Ferritic Steels." American Society for Testing and Materials, Philadelphia, PA, 1995 (Reapproved 2000).

temperature. The NDTT is defined as 5°C or 10°F below the lowest temperature where two no-breaks are recorded.

The nil-ductility transition temperature gives a qualitative estimate of the ability of a material to arrest a running crack. Arrest in structures is more likely to occur if the service temperature is above NDTT, but structures above NDTT are not immune to brittle fracture.

The ship building industry in the U.S. currently uses the drop weight test to qualify steels for ship hulls. The nuclear power industry relies primarily on quantitative fracture mechanics methodology, but uses the NDTT to index fracture toughness data for different heats of steel.

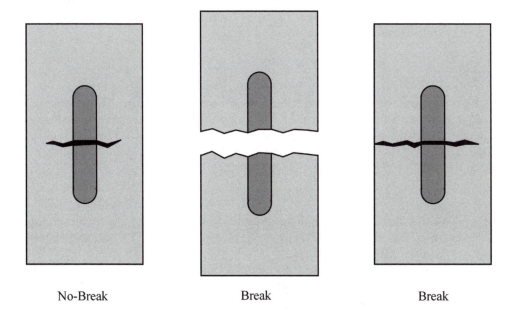

FIGURE 7.43 Examples of break and no-break behavior in drop weight tests. A break is recorded when the crack reaches at least one edge of the specimen.

7.9.3 Drop Weight Tear and Dynamic Tear Tests

Drop weight tear and dynamic tear tests are similar to the Charpy test, except that the former are performed on large specimens. The ASTM standards E 604 [45] and E 436 [44] cover drop weight tear and dynamic tear tests, respectively. Both test methods utilize three-point bend specimens that are impacted in a drop tower or pendulum machine.

Drop weight tear specimens are 41 mm (1.6 in.) wide, 16 mm (0.625 in.) thick, and are loaded over a span of 165 mm (6.5 in.). These specimens contain a sharp-machined notch. A 0.13-mm (0.010 in.)-deep indentation is made at the tip of this notch. The fracture energy is measured in this test, much like the Charpy and Izod tests. Since drop weight specimens are significantly larger than Charpy specimens, the fracture energy is much greater, and the capacity of the testing machine must be scaled accordingly. If a pendulum machine is used, the energy can be determined in the same manner as in the Charpy and Izod tests. A drop test must be instrumented, however, because only a portion of the potential energy is absorbed by the specimen; the remainder is transmitted through the foundation of the drop tower.

The dynamic tear test quantifies the toughness of steel through the appearance of the fracture surface. In the ductile-brittle transition region, a dynamic test produces a mixture of cleavage fracture and microvoid coalescence; the relative amount of each depends on the test temperature. The percent "shear" on the fracture surface is reported in dynamic tear tests, where the so-called shear fracture is actually microvoid coalescence (Chapter 5). Dynamic tear specimens are 76 mm (3 in.) wide, 305 mm (12 in.) long, and are loaded over a span of 254 mm (10 in.). The specimen thickness is equal to the thickness of the plate under consideration. The notch is pressed into the specimen by indentation.

APPENDIX 7: STRESS INTENSITY, COMPLIANCE, AND LIMIT LOAD SOLUTIONS FOR LABORATORY SPECIMENS

Figure A7.1 illustrates a variety of test specimen geometries and loading types. The stress-intensity solutions for all of these configurations are given in Table A7.1. Most of the K_I solutions in Table A7.1 have the following form:

$$K_1 = \frac{P}{B\sqrt{W}} f\left(\frac{a}{W}\right) \tag{A7.1}$$

where P is the applied load and $f(a/w)$ is a dimensionless geometry function. The specimen dimensions, B, W, and a are defined in Figure A7.1. The exception to the above form is the edge-cracked plate in pure bending (Figure A7.1h), which is written in terms of the applied bending moment.

Table A7.2 lists solutions for the load-line compliance for some of the configurations in Figure A7.1. Note that the load-line displacement has two components:

$$\Delta = \Delta_{nc} + \Delta_c \tag{A7.2}$$

where Δ_{nc} is the load-line displacement in the absence of a crack and Δ_c is the additional displacement due to the presence of the crack. For the compact specimen, $\Delta_{nc} = 0$ because the load-line displacement is measured at the crack mouth.

Table A7.3 provides polynomial expressions that enable crack length to be inferred from compliance measurements for compact and bend specimens. Expressions such as these are necessary for resistance curve measurement by means of the unloading compliance method (Figure 7.25).

Table A7.4 lists limit load solutions for several specimen types. These expressions are necessary for determining the required load capacity of the test machine when measuring fracture toughness in elastic-plastic regime.

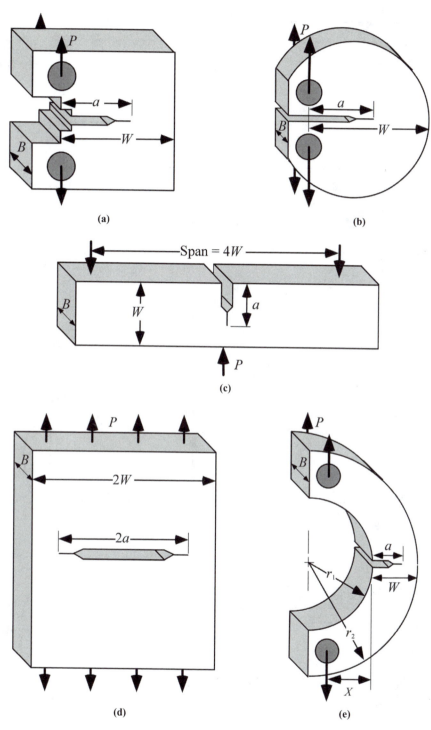

FIGURE A7.1 Common test specimen geometries: (a) compact specimen, (b) disk-shaped compact specimen, (c) single-edge-notched bend (SE(B)) specimen, (d) middle tension (MT) panel, (e) arc-shaped specimen, (f) single-edge-notched tension specimen, (g) double-edge-notched tension panel, (h) edge-cracked plate in pure bending, and (i) edge-cracked plate in combined bending and tension.

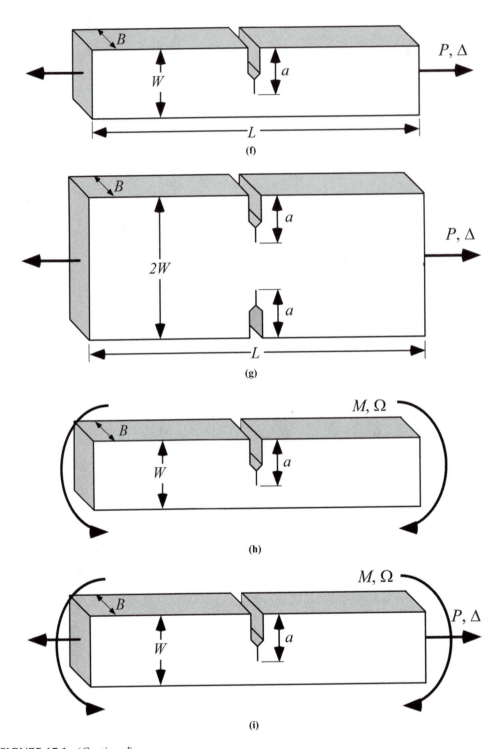

FIGURE A7.1 (*Continued*)

TABLE A7.1
Nondimensional K_I Solutions for Through-Thickness Cracks in Flat Plates [8, 55][a]

$$f\left(\frac{a}{W}\right) = \frac{K_I B \sqrt{W}}{P}$$

(a) Compact specimen

$$\frac{2+\dfrac{a}{W}}{\left(1-\dfrac{a}{W}\right)^{3/2}}\left[0.886 + 4.64\left(\frac{a}{W}\right) - 13.32\left(\frac{a}{W}\right)^2 + 14.72\left(\frac{a}{W}\right)^3 - 5.60\left(\frac{a}{W}\right)^4\right]$$

(b) Disk-shaped compact specimen

$$\frac{2+\dfrac{a}{W}}{\left(1-\dfrac{a}{W}\right)^{3/2}}\left[0.76 + 4.8\left(\frac{a}{W}\right) - 11.58\left(\frac{a}{W}\right)^2 + 11.43\left(\frac{a}{W}\right)^3 - 4.08\left(\frac{a}{W}\right)^4\right]$$

(c) Single-edge-notched bend specimen loaded in three-point bending

$$\frac{3\dfrac{S}{W}\sqrt{\dfrac{a}{W}}}{2\left(1+2\dfrac{a}{W}\right)\left(1-\dfrac{a}{W}\right)^{3/2}}\left[1.99 - \frac{a}{W}\left(1-\frac{a}{W}\right)\left\{2.15 - 3.93\left(\frac{a}{W}\right)+2.7\left(\frac{a}{W}\right)^2\right\}\right]$$

(d) Middle tension (MT) panel

$$\sqrt{\frac{\pi a}{4W}}\sec\frac{\pi a}{2W}\left[1-0.025\left(\frac{a}{W}\right)^2 + 0.06\left(\frac{a}{W}\right)^4\right]$$

(e) Arc shaped specimen

$$\left(\frac{3X}{W}+1.9+1.1\frac{a}{W}\right)\left[1+0.25\left(1-\frac{a}{W}\right)^2\left(1-\frac{r_1}{r_2}\right)\right]g\left(\frac{a}{W}\right)$$

where

$$g\left(\frac{a}{W}\right) = \frac{\sqrt{\dfrac{a}{W}}}{\left(1-\dfrac{a}{W}\right)^{3/2}}\left[3.74 - 6.30\left(\frac{a}{W}\right)+6.32\left(\frac{a}{W}\right)^2 - 2.43\left(\frac{a}{W}\right)^3\right]$$

$$f\left(\frac{a}{W}\right) = \frac{K_I B \sqrt{W}}{P}$$

(f) Single-edge-notched tension panel

$$\frac{\sqrt{2\tan\dfrac{\pi a}{2W}}}{\cos\dfrac{\pi a}{2W}}\left[0.752 + 2.02\left(\frac{a}{W}\right)+0.37\left(1-\sin\frac{\pi a}{2W}\right)^3\right]$$

(g) Double-edge-notched tension (DENT) panel

$$\frac{\sqrt{\dfrac{\pi a}{2W}}}{\sqrt{1-\dfrac{a}{W}}}\left[1.122 - 0.561\left(\frac{a}{W}\right) - 0.205\left(\frac{a}{W}\right)^2 + 0.471\left(\frac{a}{W}\right)^3 + 0.190\left(\frac{a}{W}\right)^4\right]$$

(Continued)

TABLE A7.1
(*Continued*)

(h) Edge-cracked plate subject to pure bending

$$f\left(\frac{a}{W}\right) = \frac{K_I BW^{3/2}}{M} = \frac{6\sqrt{2\tan\left(\frac{\pi a}{2W}\right)}}{\cos\left(\frac{\pi a}{2W}\right)}\left[0.923 + 0.199\left\{1 - \sin\left(\frac{\pi a}{2W}\right)\right\}^4\right]$$

(i) Edge-cracked plate subject to combined bending and tension

$$K_I = \frac{1}{B\sqrt{W}}\left[Pf_t\left(\frac{a}{W}\right) + \frac{M}{W}f_b\left(\frac{a}{W}\right)\right]$$

where f_t and f_b are given above in (f) and (h), respectively.

[a] See Figure A7.1 for a definition of the dimensions for each configuration.

TABLE A7.2
Nondimensional Load Line Compliance Solutions for Through-Thickness Cracks in Flat Plates [55][a,b]

Nondimensional Compliance: $z_{LL} = \dfrac{\Delta B E}{P}$ where $\Delta = \Delta_c + \Delta_{nc}$

(a) Compact specimen

$$\left(\frac{1+\frac{a}{W}}{1-\frac{a}{W}}\right)^2\left[2.163 + 12.219\left(\frac{a}{W}\right) - 20.065\left(\frac{a}{W}\right)^2 - 0.9925\left(\frac{a}{W}\right)^3 + 20.609\left(\frac{a}{W}\right)^4 - 9.9314\left(\frac{a}{W}\right)^5\right]$$

(b) Single-edge-notched bend (SENB) specimen loaded in three-point bending

$$\frac{S^3(1-v^2)}{W^3}\left[0.25 + 0.6\left(\frac{W}{S}\right)^2(1+v)\right] + 1.5\left(\frac{S}{W}\right)^2\left(\frac{\frac{a}{W}}{1-\frac{a}{W}}\right)^2\left[5.58 - 19.57\left(\frac{a}{W}\right) + 36.82\left(\frac{a}{W}\right)^2 - 34.94\left(\frac{a}{W}\right)^3 + 12.77\left(\frac{a}{W}\right)^4\right]$$

(c) Middle tension (MT) panel

$$\frac{L}{2W} + \left(\frac{a}{W}\right)\left[-1.071 + 0.250\left(\frac{a}{W}\right) - 0.357\left(\frac{a}{W}\right)^2 + 0.121\left(\frac{a}{W}\right)^3 - 0.047\left(\frac{a}{W}\right)^4 + 0.008\left(\frac{a}{W}\right)^5 - \frac{1.071}{\left(\frac{a}{W}\right)}\ln\left(1-\frac{a}{W}\right)\right]$$

(d) Single-edge-notched tension specimen

$$\frac{L}{W} + \frac{4\left(\frac{a}{W}\right)^2}{\left(1-\frac{a}{W}\right)^2}\left\{0.99 - \frac{a}{W}\left(1-\frac{a}{W}\right)\left[1.3 - 1.2\left(\frac{a}{W}\right) + 0.7\left(\frac{a}{W}\right)^2\right]\right\}$$

(e) Double-edge-notched tension panel

$$\frac{L}{2W} + \frac{4}{\pi}\left\{0.0629 - 0.0610\left[\cos\left(\frac{\pi a}{2W}\right)\right]^4 - 0.0019\left[\cos\left(\frac{\pi a}{2W}\right)\right]^8 + \ln\left[\sec\left(\frac{\pi a}{2W}\right)\right]\right\}$$

[a] For side-grooved specimens, B should be replaced by an effective thickness:

$$B_e = -\frac{(B-B_N)^2}{B}\quad\text{where }B_N\text{ is the net thickness.}$$

[b] See Figure A7.1 for a definition of the dimensions for each configuration.

TABLE A7.3
Crack Length-Compliance Relationships for Compact and Three-Point Bend Specimens [4].[a]

Compact specimen

$$\frac{a}{W} = 1.00196 - 4.06319U_{LL} + 11.242U_{LL}^2 - 106.043U_{LL}^3 + 464.335U_{LL}^4 - 650.677U_{LL}^5$$

where

$$U_{LL} = \frac{1}{\sqrt{Z_{LL}} + 1} \quad \text{and} \quad Z_{LL} = \frac{BE\Delta}{P}$$

Single-edge-notched bend specimen loaded in three-point bending

$$\frac{a}{W} = 0.999748 - 3.9504U_V + 2.9821U_V^2 - 3.21408U_V^3 + 51.51564U_V^4 - 113.031U_V^5$$

where

$$U_V = \frac{1}{\sqrt{\frac{4Z_V W}{S}} + 1} \quad Z_V = \frac{BEV}{P} \quad \text{and } V \text{ is the crack-mouth-opening displacement.}$$

[a] For side-grooved specimens, B should be replaced by an effective thickness:

$$B_e = B - \frac{(B - B_N)^2}{B}$$

where B_N is the net thickness.

TABLE A7.4
Limit Load Solutions for Through-Thickness Cracks in Flat Plates [56, 57][a,b]

(a) Compact specimen

$$P_L = 1.455\eta Bb\sigma_Y \qquad P_L = 1.455\eta Bb\sigma_Y \quad \text{(plane strain)}$$

$$P_L = 1.072\eta Bb\sigma_Y \quad \text{(plane stress)}$$

where

$$\eta = \sqrt{\left(\frac{2a}{b}\right)^2 + \frac{4a}{b} + 2} - \left(\frac{2a}{b} + 1\right)$$

(b) Single-edge-notched bend (SE(B)) specimen loaded in three-point bending

$$P_L = \frac{1.455 Bb^2 \sigma_Y}{S} \quad \text{(plane stress)}$$

$$P_L = \frac{1.072 Bb^2 \sigma_Y}{S} \quad \text{(plane stress)}$$

(c) Middle tension (MT) panel

$$P_L = \frac{4}{\sqrt{3}} Bb\sigma_Y \quad \text{(plane strain)}$$

$$P_L = 2Bb\sigma_Y \quad \text{(plane stress)}$$

(d) Single-edge-notched-tension (SENT) specimen

$$P_L = 1.455\eta Bb\sigma_Y \quad \text{(plane strain)}$$

$$P_L = 1.072\eta Bb\sigma_Y \quad \text{(plane stress)}$$

where

$$\eta = \sqrt{1 + \left(\frac{a}{b}\right)^2} - \frac{a}{b}$$

TABLE A7.4
(Continued)

(e) Double-edge-notched tension (DENT) panel

$$P_L = \left(0.72 + 1.82 \frac{b}{W} \right) Bb\sigma_Y \quad \text{(plane strain)}$$

$$P_L = \frac{4}{\sqrt{3}} Bb\sigma_Y \quad \text{(plane stress)}$$

Edge crack subject to combined bending and tension.

$$P_L = \frac{2Bb\sigma_Y}{\sqrt{3}} \left[-\left(2\lambda + \frac{a}{W} \right) + \sqrt{\left(2\lambda + \frac{a}{W} \right)^2 + \left(\frac{a}{b} \right)^2} \right] \quad \text{(plane strain)}$$

$$P_L = Bb\sigma_Y \left[-\left(2\lambda + \frac{a}{W} \right) + \sqrt{\left(2\lambda + \frac{a}{W} \right)^2 + \left(\frac{a}{b} \right)^2} \right] \quad \text{(plane stress)}$$

where

$$\lambda = \frac{M}{PW}$$

[a] The flow stress σ_Y is normally taken as the average of σ_{YS} and σ_{TS}.
[b] See Figure A7.1 for a definition of the dimensions for each configuration.

REFERENCES

1. E 1823-96, "Standard Terminology Relating to Fatigue Fracture Testing." American Society for Testing and Materials, Philadelphia, PA, 1996 (Reapproved 2002).
2. Baker, A., "A DC Potential Drop Procedure for Crack Initiation and R Curve Measurements During Ductile Fracture Tests." ASTM STP 856, American Society for Testing and Materials, Philadelphia, PA, 1985, pp. 394–410.
3. Schwalbe, K.-H., Hellmann, D., Heerens, J., Knaack, J., and Muller-Roos, J., "Measurement of Stable Crack Growth Including Detection of Initiation of Growth Using the DC Potential Drop and the Partial Unloading Methods." ASTM STP 856, American Society for Testing and Materials, Philadelphia, PA, 1985, pp. 338–362.
4. E 1820-01, "Standard Test Method for Measurement of Fracture Toughness." American Society for Testing and Materials, Philadelphia, PA, 2001.
5. Dawes, M.G., "Elastic-Plastic Fracture Toughness Based on the COD and J-Contour Integral Concepts." ASTM STP 668, American Society for Testing and Materials, Philadelphia, PA, 1979, pp. 306–333.
6. Anderson, T.L., McHenry, H.I., and Dawes, M.G., "Elastic-Plastic Fracture Toughness Testing with Single Edge Notched Bend Specimens." ASTM STP 856, American Society for Testing and Materials, Philadelphia, PA, 1985, pp. 210–229.
7. Andrews, W.R. and Shih, C.F., "Thickness and Side-Groove Effects on J- and δ-Resistance Curves for A533-B Steel at 93°C." ASTM STP 668, American Society for Testing and Materials, Philadelphia, PA, 1979, pp. 426–450.
8. E 399-90, "Standard Test Method for Plane-Strain Fracture Toughness of Metallic Materials." American Society for Testing and Materials, Philadelphia, PA, 1990 (Reapproved 1997).
9. BS 5447, "Methods of Testing for Plane Strain Fracture Toughness (K_{Ic}) of Metallic Materials." British Standards Institution, London, 1974.
10. BS 7448: Part 1, "Fracture Mechanics Toughness Tests, Part 1, Method for Determination of K_{IC}, Critical CTOD and Critical J Values of Metallic Materials." British Standards Institution, London, 1991.

11. Jones, M.H. and Brown, W.F., Jr., "The Influence of Crack Length and Thickness in Plane Strain Fracture Toughness Testing." ASTM STP 463, American Society for Testing and Materials, Philadelphia, PA, 1970, pp. 63–101.

12. Kaufman, J.G. and Nelson, F.G., "More on Specimen Size Effects in Fracture Toughness Testing." ASTM STP 559, American Society for Testing and Materials, Philadelphia, PA, 1973, pp. 74–98.

13. Wallin, K., "Critical Assessment of the Standard E 399." ASTM STP 1461, American Society for Testing and Materials, Philadelphia, PA, 2004.

14. E 561-98, "Standard Practice for R-Curve Determination." American Society for Testing and Materials, Philadelphia, PA, 1998.

15. Stricklin, L.L., "Geometry Dependence of Crack Growth Resistance Curves in Thin Sheet Aluminum Alloys." Master of Science Thesis, Texas A&M University, College Station, TX, 1988.

16. Landes, J.D., Zhou, Z., Lee, K., and Herrera, R., "Normalization Method for Developing *J-R* Curves with the LMN Function." *Journal of Testing and Evaluation,* Vol. 19, 1991, pp. 305–311.

17. Joyce, J.A. and Hackett, E.M., "Development of an Engineering Definition of the Extent of *J* Singularity Controlled Crack Growth." NUREG/CR-5238, U.S. Nuclear Regulatory Commission, Washington, DC, 1989.

18. E 1921-03, "Standard Test Method for Determination of Reference Temperature, T_o, for Ferritic Steels in the Transition Range." American Society for Testing and Materials, Philadelphia, PA, 2003.

19. BS 5762, *Methods for Crack Opening Displacement (COD) Testing.* British Standards Institution, London, 1979.

20. E 1290-02, "Standard Test Method for Crack Tip Opening Displacement Fracture Toughness Measurement." American Society for Testing and Materials, Philadelphia, PA, 2002.

21. Duffy, J. and Shih, C.F., "Dynamic Fracture Toughness Measurements for Brittle and Ductile Materials." *Advances in Fracture Research: Seventh International Conference on Fracture.* Pergamon Press, Oxford, 1989, pp. 633–642.

22. Sanford, R.J. and Dally, J.W., "A General Method for Determining Mixed-Mode Stress Intensity Factors from Isochromatic Fringe Patterns." *Engineering Fracture Mechanics,* Vol. 11, 1979, pp. 621–633.

23. Chona, R., Irwin, G.R., and Shukla, A., "Two and Three Parameter Representation of Crack Tip Stress Fields." *Journal of Strain Analysis,* Vol. 17, 1982, pp. 79–86.

24. Kalthoff, J.F., Beinart, J., Winkler, S., and Klemm, W., "Experimental Analysis of Dynamic Effects in Different Crack Arrest Test Specimens." ASTM STP 711, American Society for Testing and Materials, Philadelphia, PA, 1980, pp. 109–127.

25. Nakamura, T., Shih, C.F., and Freund, L.B., "Analysis of a Dynamically Loaded Three-Point-Bend Ductile Fracture Specimen." *Engineering Fracture Mechanics,* Vol. 25, 1986, pp. 323–339.

26. Nakamura, T., Shih, C.F., and Freund, L.B., "Three-Dimensional Transient Analysis of a Dynamically Loaded Three-Point-Bend Ductile Fracture Specimen." ASTM STP 995, American Society for Testing and Materials, Philadelphia, PA, Vol. I, 1989, pp. 217–241.

27. Joyce, J.A. and Hacket, E.M., "Dynamic *J-R* Curve Testing of a High Strength Steel Using the Multispecimen and Key Curve Techniques." ASTM STP 905, American Society of Testing and Materials, Philadelphia, PA, 1984, pp. 741–774.

28. Joyce, J.A. and Hacket, E.M., "An Advanced Procedure for *J-R* Curve Testing Using a Drop Tower." ASTM STP 995, American Society for Testing and Materials, Philadelphia, PA, 1989, pp. 298–317.

29. Robertson, T.S., "Brittle Fracture of Mild Steel." *Engineering,* Vol. 172, 1951, pp. 445–448.

30. Naus, D.J., Nanstad, R.K., Bass, B.R., Merkle, J.G., Pugh, C.E., Corwin, W.R., and Robinson, G.C., "Crack-Arrest Behavior in SEN Wide Plates of Quenched and Tempered A 533 Grade B Steel Tested under Nonisothermal Conditions." NUREG/CR-4930, U.S. Nuclear Regulatory Commission, Washington, DC, 1987.

31. E 1221-96, "Standard Method for Determining Plane-Strain Crack-Arrest Toughness, K_{Ia}, of Ferritic Steels." American Society for Testing and Materials, Philadelphia, PA, 1996 (Reapproved 2002).

32. Crosley, P.B., Fourney, W.L., Hahn, G.T., Hoagland, R.G., Irwin, G.R., and Ripling, E.J., "Final Report on Cooperative Test Program on Crack Arrest Toughness Measurements." NUREG/CR-3261, U.S. Nuclear Regulatory Commission, Washington, DC, 1983.

33. Barker, D.B., Chona, R., Fourney, W.L., and Irwin, G.R., "A Report on the Round Robin Program Conducted to Evaluate the Proposed ASTM Test Method of Determining the Crack Arrest Fracture Toughness, K_{Ia}, of Ferritic Materials." NUREG/CR-4996, 1988.

34. BS 7448: Part 2, "Fracture Mechanics Toughness Tests, Part 1, Method for Determination of K_{IC}, critical CTOD and Critical *J* Values of Welds in Metallic Materials." British Standards Institution, London, 1997.

35. Dawes, M.G., Pisarski, H.G., and Squirrell, H.G., "Fracture Mechanics Tests on Welded Joints." ASTM STP 995, American Society for Testing and Materials, Philadelphia, PA, 1989, pp. II-191–II-213.

36. Satok, K. and Toyoda, M., "Guidelines for Fracture Mechanics Testing of WM/HAZ." Working Group on Fracture Mechanics Testing of Weld Metal/HAZ, International Institute of Welding, Commission X, IIW Document X-1113-86.

37. RP 2Z, "Recommended Practice for Preproduction Qualification of Steel Plates for Offshore Structures." American Petroleum Institute, Washington, DC, 1987.

38. Towers, O.L. and Dawes, M.G., "Welding Institute Research on the Fatigue Precracking of Fracture Toughness Specimens." ASTM STP 856, American Society for Testing and Materials, Philadelphia, PA, 1985, pp. 23–46.

39. Wallin, K., "Fracture Toughness Transition Curve Shape for Ferritic Structural Steels." In: *Proceedings of the Joint FEFG/ICF International Conference on Fracture of Engineering Materials,* Singapore, 1991, pp. 83–88.

40. Stienstra, D.I.A., "Stochastic Micromechanical Modeling of Cleavage Fracture in the Ductile-Brittle Transition Region." Ph.D. Dissertation, Texas A&M University, College Station, TX, 1990.

41. Merkle, J.G., Wallin, K., and McCabe, D.E., "Technical Basis for an ASTM Standard on Determining the Reference Temperature, T_o, for Ferritic Steels in the Transition Range." NUREG/CR-5504, U.S. Nuclear Regulatory Commission, Washington, DC, 1998.

42. E 23-02a, "Standard Test Methods for Notched Bar Impact Testing of Metallic Materials." American Society for Testing and Materials, Philadelphia, PA, 2002.

43. E 208-95a, "Standard Test Method for Conducting Drop-Weight Test to Determine Nil-Ductility Transition Temperature of Ferritic Steels." American Society for Testing and Materials, Philadelphia, PA, 1995 (Reapproved 2000).

44. E 436-03, "Standard Method for Drop-Weight Tear Tests of Ferritic Steels." American Society for Testing and Materials, Philadelphia, PA, 2003.

45. E 604-83, "Standard Test Method for Dynamic Tear Testing of Metallic Materials." American Society for Testing and Materials, Philadelphia, PA, 1983 (Reapproved 2002).

46. Marandet, B. and Sanz, G., "Evaluation of the Toughness of Thick Medium Strength Steels by LEFM and Correlations Between K_{Ic} and CVN." ASTM STP 631, American Society for Testing and Materials, Philadelphia, PA, 1977, pp. 72–95.

47. Rolfe, S.T. and Novak, S.T., "Slow Bend K_{Ic} Testing of Medium Strength High Toughness Steels." ASTM STP 463, American Society for Testing and Materials, Philadelphia, PA, 1970, pp. 124–159.

48. Barsom, J.M. and Rolfe, S.T., "Correlation Between K_{Ic} and Charpy V Notch Test Results in the Transition Temperature Range." ASTM STP 466, American Society for Testing and Materials, Philadelphia, PA, 1970, pp. 281–301.

49. Sailors, R.H. and Corten, H.T., "Relationship between Material Fracture Toughness Using Fracture Mechanics and Transition Temperature Tests." ASTM STP 514, American Society for Testing and Materials, Philadelphia, PA, 1973, pp. 164–191.

50. Begley, J.A. and Logsdon, W.A., "Correlation of Fracture Toughness and Charpy Properties for Rotor Steels." Westinghouse Report, Scientific Paper 71-1E7, MSLRF-P1-1971.

51. Ito, T., Tanaka, K., and Sato, M., "Study of Brittle Fracture Initiation from Surface Notch in Welded Fusion Line." IIW Document X-704-730, 1973.

52. Wallin, K., "A Simple Theoretical Charpy-V-K-Correlation for Irradiation Embrittlement." *Innovative Approaches to Irradiation Damage and Fracture Analysis,* ASME PVP Volume 170, American Society of Mechanical Engineers, New York, 1989, pp. 93–100.

53. Wullaert, R.A., "Applications of the Instrumented Charpy Impact Test." ASTM STP 466, American Society for Testing and Materials, Philadelphia, PA, 1970, pp. 148–164.

54. Turner, C.E., "Measurement of Fracture Toughness by Instrumented Impact Test." ASTM STP 466, American Society for Testing and Materials, Philadelphia, PA, 1970, pp. 93–114.

55. Towers, O.L., "Stress Intensity Factors, Compliances, and Elastic η Factors for Six Test Geometries." Report 136/1981, The Welding Institute, Abington, UK, 1981.

56. Kumar, V., German, M.D., and Shih, C.F., "An Engineering Approach for Elastic-Plastic Fracture Analysis." EPRI Report NP-1931, Electric Power Research Institute, Palo Alto, CA, 1981.

57. Kumar, V., German, M.D., Wilkening, W.W., Andrews, W.R., deLorenzi, H.G., and Mowbray, D.F., "Advances in Elastic-Plastic Fracture Analysis." EPRI Report NP-3607, Electric Power Research Institute, Palo Alto, CA, 1984.

8 Fracture Testing of Nonmetals

The procedures for fracture toughness testing of metals, which are described in Chapter 7, are fairly well established. Fracture testing of plastics, composites, and ceramics is relatively new, however, and there are a number of unresolved issues.

Although many aspects of fracture toughness testing are similar for metals and nonmetals, there are several important differences. In some cases, metals fracture testing technology is inadequate on theoretical grounds. For example, the mechanical behavior of plastics can be highly rate dependent, and composites often violate continuum assumptions (see Chapter 6). There are also more pragmatic differences between fracture testing of metals and nonmetals. Ceramics, for instance, are typically very hard and brittle, which makes specimen fabrication and testing more difficult.

This chapter briefly summarizes the current procedures for measuring the fracture toughness in plastics, fiber-reinforced composites, and ceramics. The reader should be familiar with the material in Chapter 7, since much of the same methodology (e.g., specimen design, instrumentation, and fracture parameters) is currently being applied to nonmetals.

8.1 FRACTURE TOUGHNESS MEASUREMENTS IN ENGINEERING PLASTICS

Engineers and researchers who have attempted to measure the fracture toughness of plastics have relied almost exclusively on metals testing technology. Existing experimental approaches implicitly recognize the potential for time-dependent deformation, but do not specifically address viscoelastic behavior in most instances. Schapery's viscoelastic J integral [1,2], which was introduced in Chapter 4, has not seen widespread application to laboratory testing.

The Mode I stress-intensity factor K_I and the conventional J integral were originally developed for time-independent materials, but may also be suitable for viscoelastic materials in certain cases. The restrictions on these parameters are explored below, followed by a summary of procedures for K and J testing on plastics. Section 8.1.5 briefly outlines possible approaches for taking account of viscoelastic behavior and time-dependent yielding in fracture toughness measurements.

8.1.1 THE SUITABILITY OF K AND J FOR POLYMERS

A number of investigators [3–7] have reported K_{Ic}, J_{Ic}, and J-R curve data for plastics. They applied testing and data analysis procedures that are virtually identical to metals approaches (See Chapter 7). The validity of K and J is not guaranteed, however, when a material exhibits rate-dependent mechanical properties. For example, neither J nor K are suitable for characterizing creep crack growth in metals (Section 4.2);[1] an alternate parameter C^* is required to account for the time-dependent material behavior. Schapery [1,2] has proposed an analogous parameter J_v to characterize viscoelastic materials (Section 4.3).

---

[1] The stress-intensity factor is suitable for high-temperature behavior in limited situations. At short times, when the creep zone is confined to a small region surrounding the crack tip, K uniquely characterizes crack-tip conditions, while C^* is appropriate for large-scale creep.

Let us examine the basis for applying K and J to viscoelastic materials, as well as the limitations on these parameters.

8.1.1.1 *K*-Controlled Fracture

In linear viscoelastic materials, remote loads and local stresses obey the same relationships as in the linear elastic case. Consequently, the stresses near the crack tip exhibit a $1/\sqrt{r}$ singularity:

$$\sigma_{ij} = \frac{K_I}{\sqrt{2\pi r}} f_{ij}(\theta) \tag{8.1}$$

and K_I is related to remote loads and geometry through the conventional linear elastic fracture mechanics (LEFM) equations introduced in Chapter 2. The strains and displacements depend on the viscoelastic properties, however. Therefore, the critical stress-intensity factor for a viscoelastic material can be rate dependent; a K_{Ic} value from a laboratory specimen is transferable to a structure only if the local crack-tip strain histories of the two configurations are similar. Equation (8.1) applies only when yielding and nonlinear viscoelasticity are confined to a small region surrounding the crack tip.

Under plane strain linear viscoelastic conditions, K_I is related to the viscoelastic J integral J_v, as follows [1]:

$$J_v = \frac{K_I^2(1-\nu^2)}{E_R} \tag{8.2}$$

where E_R is a reference modulus, which is sometimes defined as the short-time relaxation modulus.

Figure 8.1 illustrates a growing crack at times t_o and $t_o + t_\rho$.[2] Linear viscoelastic material surrounds a Dugdale strip-yield zone, which is small compared to specimen dimensions. Consider a point A, which is at the leading edge of the yield zone at t_o and is at the trailing edge at $t_o + t_\rho$. The size of the yield zone and the crack-tip-opening displacement (CTOD) can be approximated as follows (see Chapter 2 and Chapter 3):

$$\rho_c = \frac{\pi}{8}\left(\frac{K_{Ic}}{\sigma_{cr}}\right)^2 \tag{8.3}$$

and

$$\delta_c \approx \frac{K_{Ic}^2}{\sigma_{cr}E(t_\rho)} \tag{8.4}$$

where σ_{cr} is the crazing stress. Assume that crack extension occurs at a constant CTOD. The time interval t_ρ is given by

$$t_\rho = \frac{\rho_c}{\dot{a}} \tag{8.5}$$

---

[2] This derivation, which was adapted from Marshall et al. [8], is only heuristic and approximate. Schapery [9] performed a more rigorous analysis that led to a result that differs slightly from Equation (8.9).

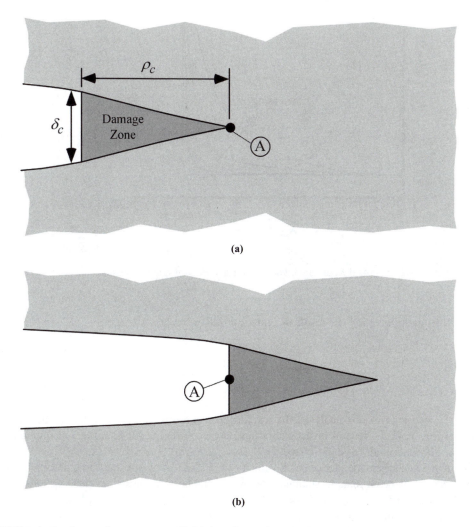

FIGURE 8.1 Crack growth at a constant CTOD in a linear viscoelastic material: (a) crack-tip position at time t_o, and (b) crack-tip position at time $t_o + t_\rho$.

where $\dot{a}$ is the crack velocity. For many polymers, the time dependence of the relaxation modulus can be represented by a simple power law:

$$E(t) = E_1 t^{-n} \tag{8.6}$$

where E_1 and n are material constants that depend on temperature. If crazing is assumed to occur at a critical strain that is time independent, the crazing stress is given by

$$\sigma_{cr} = E(t)\varepsilon_{cr} \tag{8.7}$$

Substituting Equation (8.5) to Equation (8.7) into Equation (8.4) leads to

$$K_{Ic}^2 = \delta_c \varepsilon_{cr} E_1^2 t^{-2n}$$

$$= \delta_c \varepsilon_{cr} E_1^2 \left(\frac{\rho_c}{\dot{a}}\right)^{-2n} \tag{8.8}$$

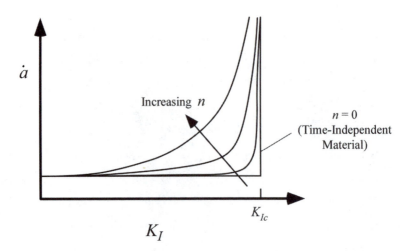

FIGURE 8.2 Effect of applied K_I on crack velocity for a variety of material responses.

Solving for ρ_c and inserting the result in Equation (8.8) gives

$$K_{Ic} = \sqrt{\delta_c \varepsilon_{cr}} \left(\frac{8\varepsilon_{cr}}{\pi \delta_c} \right)^n E_1 \dot{a}^n \qquad (8.9)$$

Therefore, according to this analysis, the fracture toughness is proportional to $\dot{a}^n$, and the crack velocity varies as $K_I^{1/n}$. Several investigators have derived relationships similar to Equation (8.9), including Marshall et al. [8] and Schapery [9].

Figure 8.2 is a schematic plot of crack velocity vs. K_I for various n values. In a time-independent material, $n = 0$; the crack remains stationary below K_{Ic} and becomes unstable when $K_I = K_{Ic}$. In such materials, K_{Ic} is a unique material property. Most metals and ceramics are nearly time independent at ambient temperature. When $n > 0$, crack propagation can occur over a range of K_I values. If, however, n is small, the crack velocity is highly sensitive to stress intensity, and the $\dot{a}$-K_I curve exhibits a sharp knee. For example, if $n = 0.1$, the crack velocity is proportional to K_I^{10}. In typical polymers below T_g, $n < 0.1$.

Consider a short-time K_{Ic} test on a material with $n \leq 0.1$, where K_I increases monotonically until the specimen fails. At low K_I values (i.e., in the early portion of the test), the crack growth would be negligible. The crack velocity would accelerate rapidly when the specimen reached the knee in the $\dot{a}$-K_I curve. The specimen would then fail at a critical K_{Ic} that would be relatively insensitive to rate. Thus, if the knee in the crack velocity–stress intensity curve is sufficiently sharp, a short-time K_{Ic} test can provide a meaningful material property.

One must be careful in applying a K_{Ic} value to a polymer structure, however. While a statically loaded structure made from a time-independent material will not fail as long as $K_I < K_{Ic}$, slow crack growth below K_{Ic} does occur in viscoelastic materials. Recall from Chapter 1 the example of the polyethylene pipe that failed by time-dependent crack growth over a period of several years. The power-law form of Equation (8.9) enables long-time behavior to be inferred from short-time tests, as Example 8.1 illustrates.

Equation (8.9) assumes that the critical CTOD for crack extension is rate independent, which is a reasonable assumption for materials that are well below T_g. For materials near T_g, where E is highly sensitive to temperature and rate, the critical CTOD often exhibits a rate dependence [3].

EXAMPLE 8.1

Short-time fracture toughness tests on a polymer specimen indicate a crack velocity of 10 mm/sec at $K_{Ic} = 5$ MPa$\sqrt{\text{m}}$. If a pipe made from this material contains a flaw such that $K_I = 2.5$ MPa$\sqrt{\text{m}}$, estimate the crack velocity, assuming $n = 0.08$.

Solution: Since the crack velocity is proportional to $K_I^{12.5}$, the growth rate at 2.5 MPa $\sqrt{\text{m}}$ is given by

$$\dot{a} = 10 \text{ mm/sec} \left(\frac{2.5 \text{ MPa} \sqrt{\text{m}}}{5 \text{ MPa} \sqrt{\text{m}}} \right)^{12.5} = 0.0017 \text{ mm/sec} = 6.2 \text{ mm/h}$$

8.1.1.2 *J*-Controlled Fracture

Schapery [1,2] has introduced a viscoelastic *J* integral J_v that takes into account various types of linear and nonlinear viscoelastic behavior. For any material that obeys the assumed constitutive law, Schapery showed that J_v uniquely defines the crack-tip conditions (Section 4.3.2). Thus, J_v is a suitable fracture criterion for a wide range of time-dependent materials. Most practical applications of fracture mechanics to polymers, however, have considered only the conventional *J* integral, which does not account for time-dependent deformation.

Conventional *J* tests on polymers can provide useful information, but is important to recognize the limitations of such an approach. One way to assess the significance of critical *J* data for polymers is by evaluating the relationship between *J* and J_v. The following exercise considers a constant rate fracture test on a viscoelastic material.

Recall from Chapter 4 that strains and displacements in viscoelastic materials can be related to pseudo-elastic quantities through hereditary integrals. For example, the pseudo-elastic displacement Δ^e is given by

$$\Delta^e = E_R^{-1} \int_0^t E(t - \tau) \frac{\partial \Delta}{\partial \tau} d\tau \tag{8.10}$$

where Δ is the actual load-line displacement and τ is an integration variable. Equation (8.10) stems from the correspondence principle, and applies to linear viscoelastic materials for which Poisson's ratio is constant. This approach also applies to a wide range of *nonlinear* viscoelastic material behavior, although $E(t)$ and E_R have somewhat different interpretations in the latter case.

For a constant displacement rate fracture test, Equation (8.10) simplifies to

$$\Delta^e = \dot{\Delta} E_R^{-1} \int_0^t E(t - \tau) d\tau$$

$$= \Delta \frac{\overline{E}(t)}{E_R} \tag{8.11}$$

where $\dot{\Delta}$ is the displacement rate and $\overline{E}(t)$ is a time-average modulus, defined by

$$\overline{E}(t) = \frac{1}{t} \int_0^t E(t - \tau) d\tau \tag{8.12}$$

Figure 8.3 schematically illustrates load-displacement and load pseudo-displacement curves for constant rate tests on viscoelastic materials. For a linear viscoelastic material (Figure 8.3(a)), the P-Δ^e curve is linear, while the P-Δ curve is nonlinear due to time dependence. Evaluation of pseudo strains and displacements effectively removes the time dependence. When Δ^e is evaluated

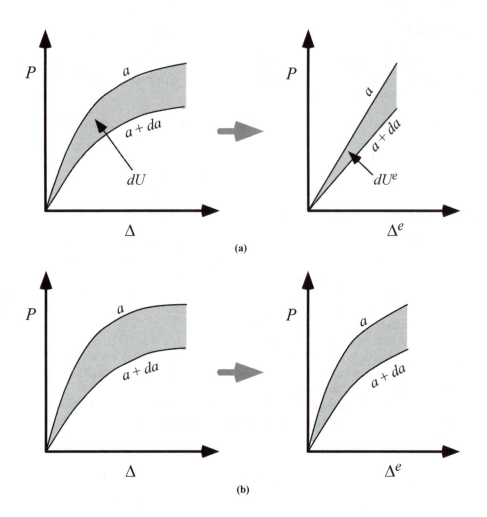

FIGURE 8.3 Load-displacement and load-pseudo-displacement curves for viscoelastic materials: (a) linear viscoelastic material, and (b) nonlinear viscoelastic material.

for a nonlinear viscoelastic material (Figure 8.3(b)), the material nonlinearity can be decoupled from the time-dependent nonlinearity.

The viscoelastic J integral can be defined from the load-pseudo-displacement curve:

$$J_v = -\frac{\partial}{\partial a}\left[\int_0^{\Delta^e} P\,d\Delta^e\right]_{\Delta^e} \tag{8.13}$$

where P is the applied load in a specimen of unit thickness. Assume that the P-Δ^e curve obeys a power law:

$$P = M(\Delta^e)^N \tag{8.14}$$

where M and N are time-independent parameters; N is a material property, while M depends on both the material and geometry. For a linear viscoelastic material, $N = 1$, and M is the elastic stiffness. Inserting Equation (8.14) into Equation (8.13) leads to

$$J_v = -\frac{(\Delta^e)^{N+1}}{N+1}\left(\frac{\partial M}{\partial a}\right)_{\Delta^e} \tag{8.15}$$

Solving for J_v in terms of physical displacement (Equation (8.11)) gives

$$J_v = -\frac{\Delta^{N+1}}{N+1}\left(\frac{\bar{E}(t)}{E_R}\right)^{N+1}\left(\frac{\partial M}{\partial a}\right)_\Delta \tag{8.16}$$

Let us now evaluate J from the same constant rate test:

$$J = -\frac{\partial}{\partial a}\left[\int_0^\Delta P\,d\Delta\right]_\Delta \tag{8.17}$$

The load can be expressed as a function of physical displacement by combining Equation (8.11) and Equation (8.14):

$$P = M\Delta^N\left(\frac{\bar{E}(t)}{E_R}\right)^N \tag{8.18}$$

Substituting Equation (8.18) into Equation (8.17) leads to

$$J = -\Delta^{N+1}\left(\frac{\partial M}{\partial a}\right)_\Delta \frac{1}{t^{N+1}}\int_0^t\left(\frac{\bar{E}(\tau)}{E_R}\right)^N \tau^N d\tau \tag{8.19}$$

since $\Delta = \dot{\Delta}t$. Therefore

$$J_v = J\phi(t) \tag{8.20}$$

where

$$\phi(t) = \frac{[t\bar{E}(t)]^{N+1}}{(N+1)E_R}\left\{\int_0^t [\bar{E}(\tau)]^N \tau^N d\tau\right\}^{-1} \tag{8.21}$$

Thus J and J_v are related through a dimensionless function of time in the case of a constant rate test. For a linear viscoelastic material in plane strain, the relationship between J and K_I is given by

$$J = \frac{K_I^2(1-v^2)}{E_R\phi(t)} \tag{8.22}$$

The conventional J integral uniquely characterizes the crack-tip conditions in a viscoelastic material for a *given time*. A critical J value from a laboratory test is transferable to a structure, provided the failure times in the two configurations are the same.

A constant rate J test apparently provides a rational measure of fracture toughness in polymers, but applying such data to structural components may be problematic. Many structures are statically loaded at either a fixed load or remote displacement. Thus a constant load creep test or a load relaxation test on a cracked specimen might be more indicative of structural conditions than a constant displacement rate test. It is unlikely that the J integral would uniquely characterize viscoelastic crack-growth behavior under all loading conditions. For example, in the case of viscous creep in metals, plots of J vs. da/dt fail to exhibit a single trend, but C^* (which is a special case of J_v) correlates crack-growth data under different loading conditions (see Chapter 4).

The application of fracture mechanics to polymers presents additional problems for which both J and J_v may be inadequate. At sufficiently high stresses, polymeric materials typically experience irreversible deformation, such as yielding, microcracking, and microcrazing. This nonlinear material behavior exhibits a different time dependence than viscoelastic deformation; computing pseudo strains and displacements may not account for rate effects in such cases.

In certain instances, the J integral may be approximately applicable to polymers that exhibit large-scale yielding. Suppose that there exists a quantity J_y that accounts for time-dependent yielding in polymers. A conventional J test will reflect the material fracture behavior if J and J_y are related through a separable function of time [10]:

$$J_y = J\phi_y(t) \tag{8.23}$$

Section 8.1.5 outlines a procedure for determining J_y experimentally.

In metals, the J integral ceases to provide a single-parameter description of crack-tip conditions when the yielding is excessive. Critical J values become geometry dependent when the single-parameter assumption is no longer valid (see Chapter 3). A similar situation undoubtedly exists in polymers: the single-parameter assumption becomes invalid after sufficient irreversible deformation. Neither J nor J_y will give geometry-independent measures of fracture toughness in such cases. Specimen size requirements for a single-parameter description of fracture behavior in polymers have yet to be established, although there has been some research in this area (see Section 8.1.3 and Section 8.1.4).

Crack growth presents further complications when the plastic zone is large. Material near the crack tip experiences nonproportional loading and unloading when the crack grows, and the J integral is no longer path independent. The appropriate definition of J for a growing crack is unclear in metals (Section 3.4.2), and the problem is complicated further when the material is rate sensitive. The rate dependence of unloading in polymers is often different from that of loading.

In summary, the J integral can provide a rational measure of toughness for viscoelastic materials, but the applicability of J data to structural components is suspect. When the specimen experiences significant time-dependent yielding prior to fracture, J may give a reasonable characterization of fracture initiation from a stationary crack, as long as the extent of yielding does not invalidate the single-parameter assumption. Crack growth in conjunction with time-dependent yielding is a formidable problem that requires further study.

8.1.2 Precracking and Other Practical Matters

As with metals, fracture toughness tests on polymers require that the initial crack be sharp. Precracks in plastic specimens can be introduced by a number of methods, including fatigue and razor notching.

Fatigue precracking in polymers can be very time consuming. The loading frequency must be kept low in order to minimize hysteresis heating, which can introduce residual stresses at the crack tip.

Because polymers are soft relative to metals, plastic fracture toughness specimens can be precracked by pressing a razor blade into a machined notch. Razor notching can produce a sharp crack in a fraction of the time required to grow a fatigue crack, and the measured toughness is not adversely affected if the notching is done properly [4].

Two types of razor notching are common: razor-notch guillotine and razor sawing. In the former case, the razor blade is simply pressed into the material by a compressive force, while razor sawing entails a lateral slicing motion in conjunction with the compressive force. Figure 8.4(a) and Figure 8.4(b) are photographs of fixtures for the razor-notch guillotine and razor-sawing procedures, respectively.

(a)

(b)

FIGURE 8.4 Razor notching of polymer specimens: (a) razor-notched guillotine, and (b) razor sawing. Photographs provided by M. Cayard.

In order to minimize material damage and residual stresses that result from razor notching, Cayard [4] recommends a three-step procedure: (1) fabrication of a conventional machined notch, (2) extension of the notch with a narrow slitting saw, and (3) final sharpening with a razor blade by either of the techniques described above. Cayard found that such an approach produced very sharp cracks with minimal residual stresses. The notch-tip radius is typically much smaller than the radius of the razor blade, apparently because a small pop-in propagates ahead of the razor notch.

While the relative softness of plastics aids the precracking process, it can cause problems during testing. The crack-opening force that a clip gage applies to a specimen (Figure 7.8) is negligible for metal specimens, but this load can be significant in plastic specimens. The conventional cantilever clip gage design may be too stiff for soft polymer specimens; a ring-shaped gage may be more suitable.

One may choose to infer the specimen displacement from the crosshead displacement. In such cases it is necessary to correct for extraneous displacements due to the indentation of the specimen by the test fixture. A displacement calibration can be inferred from a load-displacement curve for an unnotched specimen. If the calibration curve is linear, the correction to displacement is relatively simple:

$$\Delta = \Delta_{tot} - C_i P \qquad (8.24)$$

where Δ_{tot} is the measured displacement and C_i is the compliance due to indentation. Since the deformation of the specimen is time dependent, the crosshead rate in the calibration experiment should match that in the actual fracture toughness tests.

8.1.3 K_{Ic} Testing

The American Society for Testing and Materials (ASTM) has published a number of standards for fracture testing of metals, which Chapter 7 describes. Committee D20 within ASTM developed a standard method for K_{Ic} testing of plastics [10]. This standard was based on a protocol developed by the European Structural Integrity Society (ESIS), which in turn was based on the ASTM K_{Ic} standard for metals, E 399 [12].

The ASTM K_{Ic} standard for plastics is very similar to E 399. Both test methods define an apparent crack initiation load P_Q by a 5% secant construction (Figure 7.13). This load must be greater than 1.1 times the maximum load in the test for the result to be valid. The provisional fracture toughness K_Q must meet the following specimen size requirements:

$$B, a \geq 2.5 \left(\frac{K_Q}{\sigma_{YS}} \right)^2 \qquad (8.25a)$$

$$0.45 \leq \frac{a}{W} \leq 0.55 \qquad (8.25b)$$

where
    B = specimen thickness
    a = crack length
    W = specimen width, as defined in Figure 7.1

The yield strength σ_{YS} is defined in a somewhat different manner for plastics. Figure 8.5 schematically illustrates a typical stress-strain curve for engineering plastics. When a polymer

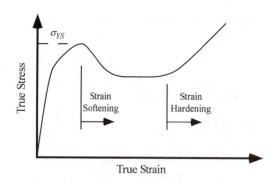

FIGURE 8.5 Typical stress-strain response of engineering plastics.

yields, it often experiences strain softening followed by strain hardening. The yield strength is defined at the peak stress prior to strain softening, as Figure 8.5 illustrates. Because the flow properties are rate dependent, the ASTM K_{Ic} standard for plastics requires that the time to reach σ_{YS} in a tensile test coincide with the time to failure in the fracture test to within ±20%.

The size requirements for metals (Equation (8.25)) have been incorporated into the ASTM K_{Ic} standard for plastics, apparently without assessing the suitability of these criteria for polymers. Recall from Chapter 2 and Chapter 7 the purported reasons for the K_{Ic} size requirements:

- The plastic zone should be small compared to in-plane dimensions to ensure the presence of an elastic singularity zone ahead of the crack tip.
- The plastic zone (supposedly) should be small compared to the thickness to ensure predominantly plane strain conditions at the crack tip.

In recent years, the second requirement has been called into question. As discussed in Section 2.10, the apparent thickness dependence of fracture toughness in metals is a result of the mixture of two fracture morphologies: flat fracture and shear fracture. Moreover, plane strain conditions can exist near the crack tip even in the fully plastic regime; *the plastic zone need not be small relative to thickness to ensure high triaxiality near the crack tip.*

Section 7.2.2 discusses the shortcomings of the ASTM E 399 test procedure and validity requirements when applied to metals. Many of these issues also apply to polymer testing. For example, the 5% secant method often introduces an artificial size dependence in K_Q, as described below.

Figure 8.6 shows the effect of specimen width on K_Q values for a rigid polyvinyl chloride (PVC) and a polycarbonate (PC) [4]. In most cases, the specimens were geometrically similar, with $W = 2B$ and $a/W = 0.5$. For specimen widths greater that 50 mm in the PC, the thickness was fixed at 25 mm, which corresponds to the plate thickness. The size dependence in K_Q is a direct result of inferring P_Q from a 5% secant construction (Figure 7.13).

As discussed in Section 7.2, nonlinearity in the load-displacement curve can come from two sources: stable crack growth and crack-tip yielding (or crazing). In the former case, a 5% deviation from linearity corresponds to the crack growth through approximately 2% of the specimen ligament. For materials that exhibit a rising crack-growth resistance curve, the 2% crack growth criterion results in a size-dependent K_Q, as Figure 7.17 illustrates. For materials that fracture without prior stable crack growth, nonlinearity in the load-displacement curve is largely due to yielding or crazing at the crack tip. In such cases, 5% nonlinearity corresponds to the point at which the crack-tip plastic zone (or damage zone) is on the order of a few percent of the specimen ligament. The measured K_Q is proportional to $\sqrt{W-a}$ until the specimen size is sufficient for fracture to occur before a 5% nonlinearity is achieved.

In the case of PVC and PC (Figure 8.6), the nonlinearity in the load-displacement curve is likely due to crack-tip yielding or crazing. Note that K_Q is more size dependent in the PC than in the PVC. The ASTM E 399 size requirements for in-plane dimensions appear to be adequate for the latter but not for the former when K_Q is defined by a 5% secant construction. The different behavior for the two polymer systems can be partially attributed to strain-softening effects. Figure 8.7 shows the stress-strain curves for these two materials. Note that the PC exhibits significant strain softening, while the rigid PVC stress-strain curve is relatively flat after yielding. Strain softening probably increases the size of the yielded zone. If one defines σ_{YS} as the lower flow stress plateau, the size requirements are more restrictive for materials that strain soften. Figure 8.6(b) shows the E 399 in-plane requirements corresponding to the lower yield strength; in the polycarbonate. Even with this adjustment, however, the E 399 methodology is not sufficient to ensure a size-independent fracture toughness estimate in the PC. That is not to say that the fracture toughness is actually more sensitive to the specimen size in the PC. Rather, the real problem is that the 5% secant construction introduces an artificial size dependence in the toughness estimate K_Q.

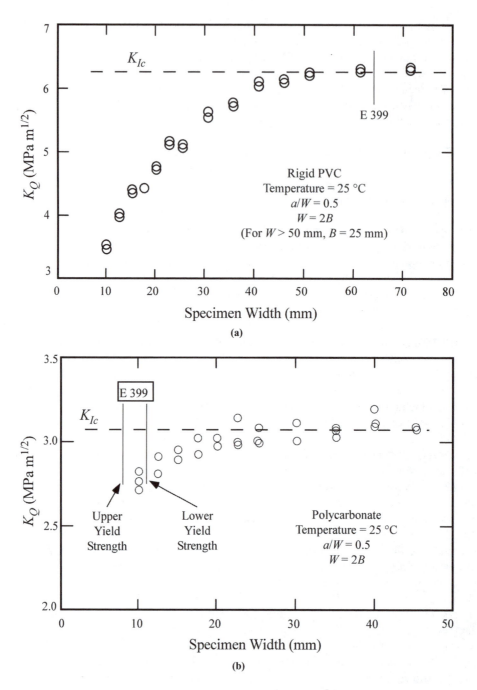

FIGURE 8.6 Effect of specimen width on K_Q in two engineering plastics: (a) rigid PVC and (b) polycarbonate. Taken from Cayard, M., "Fracture Toughness Testing of Polymeric Materials." Ph.D. Dissertation, Texas A&M University, College Station, TX, 1990.

Figures 8.8(a) and 8.8(b) are plots of fracture toughness vs. thickness for the PVC and the PC, respectively [4]. Although all of the experimental data for the PVC are below the required thickness (according to Equation (8.25)), these data do not exhibit a thickness dependence; Figure 8.8(a) indicates that the E 399 thickness requirement is too severe for this material. In the case of the PC, most of the data are above the E399 thickness requirement. These data also do not exhibit a thickness

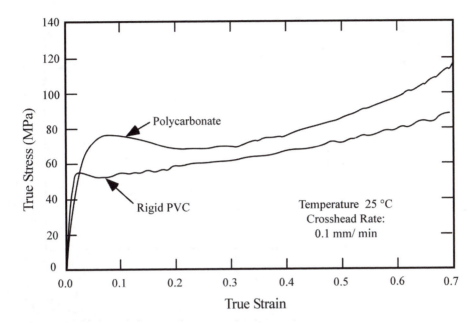

FIGURE 8.7 Stress-strain curves for the rigid PVC and polycarbonate. Taken from Cayard, M., "Fracture Toughness Testing of Polymeric Materials." Ph.D. Dissertation, Texas A&M University, College Station, TX, 1990.

dependence, which implies that the E 399 requirement is at least adequate for this material. Further testing of thinner sections would be required to determine if the E 399 thickness requirement is overly conservative for the PC.

The apparent lack of a significant thickness effect on the fracture toughness in PVC and PC is broadly consistent with the observed behavior in metals. Recall from Section 2.10 and Section 7.2 that the apparent thickness effect on fracture toughness is due primarily to shear lip formation in materials that fail by ductile crack growth. When a metal fails by cleavage, the fracture surface is usually flat (no shear lips), and the fracture toughness is much less sensitive to the specimen thickness. The PVC and PC specimens failed in a relatively brittle manner, and shear lips were not evident on the fracture surface. Consequently, one would not expect a significant thickness effect.

One final observation regarding the ASTM K_{Ic} standard for plastics is that the procedure for estimating P_Q ignores time effects. Recall from the earlier discussion that nonlinearity in the load-displacement curve in K_{Ic} tests can come from two sources: yielding and crack growth. In the case of polymers, viscoelasticity can also contribute to nonlinearity in the load-displacement curve. Consequently, at least a portion of the 5% deviation from linearity at P_Q could result from a decrease in the modulus during the test.

For most practical situations, however, viscoelastic effects are probably negligible during K_{Ic} tests. In order to obtain a valid K_{Ic} result in most polymers, the test temperature must be well below T_g, where rate effects are minimal at short times. The duration of a typical K_{Ic} test is on the order of several minutes, and the elastic properties *probably* will not change significantly prior to fracture. The rate sensitivity should be quantified, however, to evaluate the assumption that E does not change during the test.

8.1.4 *J* TESTING

A number of researchers have applied *J* integral test methods to polymers [3–6] over the last couple of decades. More recently, ASTM has published a standard for *J-R* curve testing of plastics [13]. In addition, the European Structural Integrity Society (ESIS) has developed a protocol for *J* testing of plastics [7]. Both *J* test methods are based on standards that were developed for metals such as ASTM E 1820 [14].

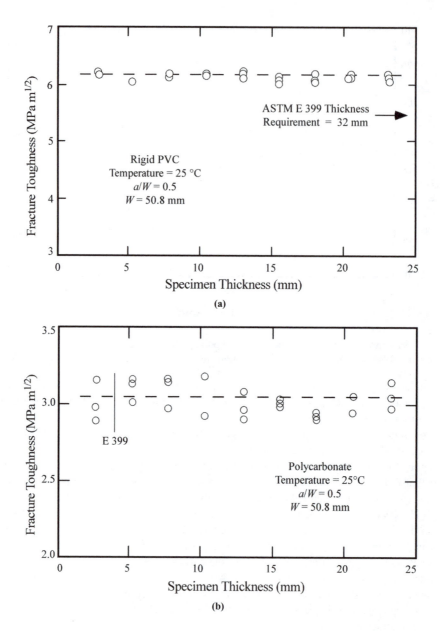

FIGURE 8.8 Effect of specimen thickness on fracture toughness of plastics: (a) rigid PVC and (b) polycarbonate. Taken from Cayard, M., "Fracture Toughness Testing of Polymeric Materials." Ph.D. Dissertation, Texas A&M University, College Station, TX, 1990.

For common test specimens such as the compact and bend geometries, J is typically inferred from the area under the load vs. load-line displacement curve:

$$J = \frac{\eta}{B_N b_o} \int_0^{\Delta} P \, d\Delta \qquad (8.26)$$

where

B_N = net thickness (for side-grooved specimens)

b_o = initial ligament length

η = dimensionless parameter that depends on geometry

Note that the J estimate in Equation (8.26) is not corrected for crack growth, nor is J separated into elastic and plastic components as it is in ASTM E 1820.

The most common approach for inferring the J-R curve for a polymer is the multiple specimen method. A set of nominally identical specimens are loaded to various displacements, and then unloaded. The initial crack length and stable crack growth are measured optically from each specimen, resulting in a series of data points on a J-Δa plot. Determining the initial crack length a_o is relatively easy, but inferring the precise position of the final crack front in a given specimen is difficult. Several techniques have been employed, including:

- Cooling the specimen and fracturing at either normal loading rates or at high impact rates
- Fracturing the specimen by impact loading at ambient temperature
- Fatigue cycling after the test
- Injecting ink onto the notch to mark the crack front
- Slicing the specimen into several sections and optically measuring the crack depth in each section while it is under load

Each technique has advantages and disadvantages. When breaking open the specimen, whether at normal or high loading rates, it is often difficult to discern the difference between crack propagation during the test and posttest fracture. In some polymers, cooling in liquid nitrogen or fracturing at high loading rates produces thumbnail-shaped features on the fracture surface, which lead to significant overestimates of stable crack growth. Fatigue postcracking can be an effective means to identify stable crack growth during the test, but it can also be time consuming. Sectioning is also time consuming.

Single specimen techniques, such as unloading compliance, may also be applied to the measurement of J_{Ic} and the J-R curve. Time-dependent material behavior can complicate unloading compliance measurements, however. Figure 8.9 schematically illustrates the unload-reload behavior of a viscoelastic material. If rate effects are significant during the time frame of the unload-reload, the resulting curve can exhibit a hysteresis effect. One possible approach to account for viscoelasticity in such cases is to relate instantaneous crack length to *pseudo* elastic displacements (see Section 8.1.5).

Critical J values for polymers exhibit less size dependence than K_Q values. Figure 8.10 compares K_Q values for the polycarbonate (Figure 8.6) with K_{Jc} values, which were obtained by converting critical J values at fracture to an equivalent critical K through the following relationship:

$$K_{Jc} = \sqrt{\frac{J_{crit}E}{1-v^2}} \tag{8.27}$$

As discussed in Section 8.1.3 above, the 5% secant construction method leads to K_Q values that scale with $\sqrt{W-a}$ when nonlinearity in the load-displacement curve is due to yielding or crazing at the crack tip. However, when the specimen size is sufficient that fracture occurs with negligible

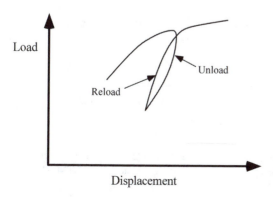

FIGURE 8.9 Schematic unloading behavior in a polymer. Hysteresis in the unload-reload curve complicates unloading compliance measurements.

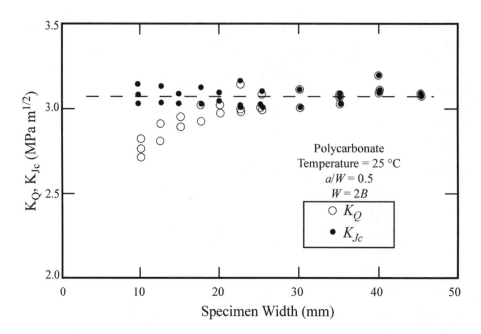

FIGURE 8.10 Size dependence of K_Q and J-based fracture toughness for PC. Taken from Cayard, M., "Fracture Toughness Testing of Polymeric Materials." Ph.D. Dissertation, Texas A&M University, College Station, TX, 1990.

nonlinearity in the load-displacement curve, $K_{Jc} = K_Q = K_{Ic}$. Note that the K_{Jc} values are independent of specimen size over the range of available data, despite material nonlinearity in the smaller specimens.

Crack-growth resistance curves can be highly rate dependent. Figure 8.11 shows J-R curves for a polyethylene pipe material that was tested at three crosshead rates [3]. Increasing the crosshead rate from 0.254 to 1.27 mm/min (0.01 and 0.05 in./min, respectively) results in nearly a threefold increase in J_{Ic} in this case.

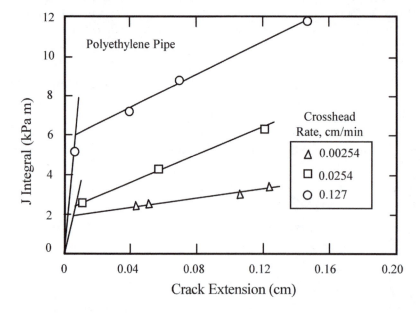

FIGURE 8.11 Crack-growth resistance curves for polyethylene pipe at three crosshead rates. Taken from Jones, R.E. and Bradley, W.L., "Fracture Toughness Testing of Polyethylene Pipe Materials." ASTM STP 995, Vol. 1, American Society for Testing and Materials, Philadelphia, PA, 1989, pp. 447–456.

8.1.5 EXPERIMENTAL ESTIMATES OF TIME-DEPENDENT FRACTURE PARAMETERS

While J_{Ic} values and J-R curves may be indicative of a polymer's relative toughness, the existence of a unique correlation between J and the crack-growth rate is unlikely. Parameters such as J_v may be more suitable for some viscoelastic materials. For polymers that experience large-scale yielding, neither J nor J_v may characterize crack growth.

This section outlines a few suggestions for inferring crack-tip parameters that take into account the time-dependent deformation of engineering plastics. Since most of these approaches have yet to be validated experimentally, much of what follows contains an element of conjecture. These proposed methods, however, are certainly no worse than conventional J integral approaches, and may be considerably better for many engineering plastics.

The viscoelastic J integral J_v can be inferred by converting physical displacements to pseudo displacements. For a constant rate test, Equation (8.11) gives the relationship between Δ and Δ^e. The viscoelastic J integral is given by Equation (8.13); J_v can also be evaluated directly from the area under the P-Δ^e curve:

$$J_v = \frac{\eta}{b} \int_0^{\Delta^e} P \, d\Delta^e \tag{8.28}$$

for a specimen with unit thickness. If the load-pseudo displacement is a power law (Equation (8.18)), Equation (8.28) becomes

$$J_v = \frac{\eta M (\Delta^e)^{N+1}}{b(N+1)} \tag{8.29}$$

Comparing Equation (8.29) and Equation (8.15) leads to

$$\eta = -\frac{b}{M} \left(\frac{\partial M}{\partial a} \right)_{\Delta^e} \tag{8.30}$$

Since M does not depend on time, the dimensionless η factor is the same for both J and J_v.

Computing pseudo-elastic displacements might also remove hysteresis effects in unloading compliance tests. If the unload-reload behavior is linear viscoelastic, the P-Δ^e unloading curves would be linear, and the crack length could be correlated to the *pseudo-elastic compliance*.

Determining pseudo displacements from Equation (8.11) or the more general expression (Equation (8.10)) requires a knowledge of $E(t)$. A separate experiment to infer $E(t)$ would not be particularly difficult, but such data would not be relevant if the material experienced large-scale yielding in a fracture test. An alternative approach to inferring crack-tip parameters that takes time effects into account is outlined below.

Schapery [10] has suggested evaluating a J-like parameter from isochronous (fixed time) load-displacement curves. Consider a series of fracture tests that are performed over a range of crosshead rates (Figure 8.12(a)). If one selects a fixed time and determines the various combinations of load and displacement that correspond to this time, the resulting locus of points forms an isochronous load-displacement curve (Figure 8.12(b)). Since the viscoelastic and yield properties are time dependent, the isochronous curve represents the load-displacement behavior for fixed material properties, as if time stood still while the test was performed. A fixed-time J integral can be defined as follows:

$$J_t = \frac{\eta}{b} \left(\int_0^{\Delta} P \, d\Delta \right)_{t=\text{constant}} \tag{8.31}$$

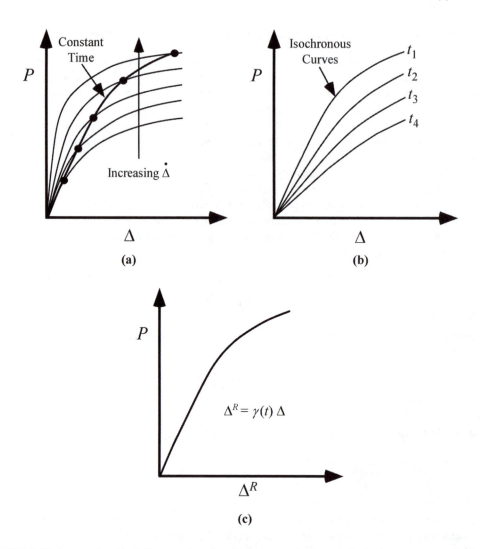

FIGURE 8.12 Proposed method for removing time dependence from load-displacement curves. (a) first, a set of tests are performed over a range of displacement rates (b) next, isochronous load-displacement curves are inferred, and (c) finally, the displacement axis of each curve is multiplied by a function $\gamma(t)$, resulting in a single curve.

Suppose that the displacements at a given load are related by a separable function of time, such that it is possible to relate all the displacements (at that particular load) to a reference displacement:

$$\Delta^R = \Delta\gamma(t) \qquad (8.32)$$

The isochronous load-displacement curves could then be collapsed onto a single trend by multiplying each curve by $\gamma(t)$, as Figure 8.12(c) illustrates. It would also be possible to define a reference J:

$$J^R = J_t\gamma(t) \qquad (8.33)$$

Note the similarity between Equation (8.23) and Equation (8.33).

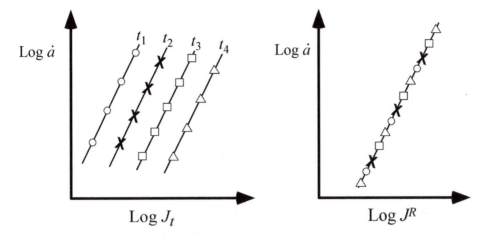

FIGURE 8.13 Postulated crack-growth behavior in terms of J_t and J^R.

The viscoelastic J is a special case of J^R. For a constant rate test, comparing Equation (8.11) and Equation (8.32) gives

$$J_v = J_t \frac{\overline{E}(t)}{E_R} \tag{8.34}$$

Thus, for a linear viscoelastic material in plane strain,

$$J_t = \frac{K_I^2(1-v^2)}{\overline{E}(t)} \tag{8.35}$$

Isochronous load-displacement curves would be linear for a linear viscoelastic material, since the modulus is constant at a fixed time.

The parameter J^R is more general than J_v; the former may account for time dependence in cases where extensive yielding occurs in the specimen. The reference J should characterize crack initiation and growth in materials where Equation (8.33) removes the time dependence of displacement. Figure 8.13 schematically illustrates the postulated relationship between J_t, J^R, and crack velocity. The $J_t - \dot{a}$ curves should be parallel on a log-log plot, while a $J^R - \dot{a}$ plot should yield a unique curve. Even if it is not possible to produce a single $J^R - \dot{a}$ curve for a material, the J_t parameter should still characterize fracture at a fixed time.

Although J^R may characterize fracture initiation and the early stages of crack growth in a material that exhibits significant time-dependent yielding, this parameter would probably not be capable of characterizing extensive crack growth, since unloading and nonproportional loading occur near the growing crack tip. (See Section 8.1.1.)

8.1.6 QUALITATIVE FRACTURE TESTS ON PLASTICS

The ASTM standard D 256 [15] describes impact testing of notched polymer specimens. This test method is currently the most common technique for characterizing the toughness of engineering plastics. Over the years, impact testing has been performed on both Charpy and Izod specimens (Figure 7.36), but D 256 covers only the Izod specimen.

The procedure for impact testing of plastics is very similar to the metals approach, which is outlined in ASTM E 23 [16] (see Section 7.9). A pendulum strikes a notched specimen, and the

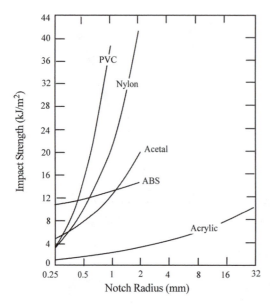

FIGURE 8.14 Effect of notch radius on the Izod impact strength of several engineering plastics. Taken from D 256-03, "Standard Test Methods for Determining the Izod Pendulum Impact Resistance of Plastics." American Society for Testing and Materials, Philadelphia, PA, 1988.

energy required to fracture the specimen is inferred from the initial and final heights of the pendulum (Figure 7.37). In the case of the Izod test, the specimen is a simple cantilever beam that is restrained at one end and struck by the pendulum at the other. One difference between the metals and plastics test methods is that the absorbed energy is normalized by the net ligament area in plastics tests, while tests according to ASTM E 23 report only the total energy. The normalized fracture energy in plastics is known as the *impact strength*.

The impact test for plastics is pervasive throughout the plastics industry because it is a simple and inexpensive measurement. Its most common application is as a material-screening criterion. The value of impact strength measurements is questionable, however.

One problem with this test method is that the specimens contain blunt notches. Figure 8.14 [17] shows the Izod impact strength values for several polymers as a function of notch radius. As one might expect, the fracture energy decreases as the notch becomes sharper. The slope of the lines in Figure 8.14 is a measure of the *notch sensitivity* of the material. Some materials are highly notch sensitive, while others are relatively insensitive to the radius of the notch. Note that the relative ordering of the materials' impact strengths in Figure 8.14 changes with notch acuity. Thus a fracture energy for a particular notch radius may not be an appropriate criterion for ranking material toughness. Moreover, the notch strength is often not a reliable indicator of how the material will behave when it contains a sharp crack.

Since Izod and Charpy tests are performed under impact loading, the resulting fracture energy values are governed by the short-time material response. Many polymer structures, however, are loaded quasistatically and must be resistant to slow, stable crack growth. The ability of a material to resist crack growth at long times is not necessarily related to the fracture energy of a blunt-notched specimen in impact loading.

The British Standards Institution (BSI) specification for unplasticized polyvinyl chloride (PVC-U) pipe, BS 3505 [18], contains a procedure for fracture toughness testing. Although the toughness test in BS 3506 is primarily a qualitative screening criterion, it is much more relevant to structural performance than the Izod impact test.

Appendices C and D of BS 3506 outline a procedure for inferring the toughness of a PVC-U pipe after exposure to an aggressive environment. A C-shaped section is removed from the pipe of interest and is submerged in dichloromethane liquid. After 15 min of exposure, the specimen is removed from the liquid and the surface is inspected for bleaching or whitening. A sharp notch is placed on the inner surface of the specimen, which is then dead-loaded for 15 min or until cracking

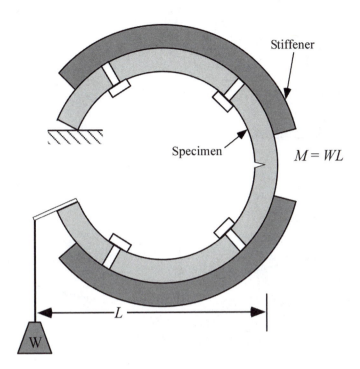

Stiffener

Specimen

$M = WL$

L

W

FIGURE 8.15 Loading apparatus for evaluating the toughness of a PVC-U pipe according to BS 3506. Taken from BS 3505:1986, "British Standard Specification for Unplasticized Polyvinyl Chloride (PVC-U) Pressure Pipes for Cold Potable Water." British Standards Institution, London, 1986.

or total fracture is observed. Figure 8.15 is a schematic drawing of the testing apparatus. The loading is such that the notch region is subject to a bending moment. If the specimen cracks or fails completely during the test, the fracture toughness of the material can be computed from the applied load and notch depth by means of standard K_I formulas. If no cracking is observed during the 15-min test, the toughness can be quantified by testing additional specimens at higher loads. The BS 3506 standard includes a semiempirical size correction for small pipes and high toughness materials that do not behave in an elastic manner.

8.2 INTERLAMINAR TOUGHNESS OF COMPOSITES

Chapter 6 outlined some of the difficulties in applying fracture mechanics to fiber-reinforced composites. The continuum assumption is often inappropriate, and cracks may not grow in a self-similar manner. The lack of a rigorous framework to describe fracture in composites has led to a number of qualitative approaches to characterize toughness.

Interlaminar fracture is one of the few instances where fracture mechanics formalism is applicable to fiber-reinforced composites on a global scale. A zone of delamination can be treated as a crack; the resistance of the material to the propagation of this crack is the fracture toughness. Since the crack typically is confined to the matrix material between plies, the continuum theory is applicable, and the crack growth is self similar.

A number of researchers have performed delamination experiments on fiber-reinforced composites over the past several decades [19–22]. Several standardized test methods for measuring interlaminar fracture toughness have been published recently. For example, ASTM D 5528 addresses Mode I delamination testing [23].

Figure 8.16 illustrates three common specimen configurations for interlaminar fracture toughness measurements. The double cantilever beam (DCB) specimen is probably the most common configuration

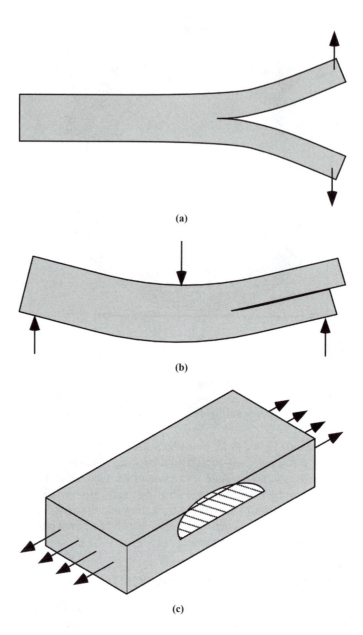

FIGURE 8.16 Common configurations for evaluating interlaminar fracture toughness: (a) double cantilever beam specimen, (b) end-notched flexure specimen, and (c) edge-delamination specimen.

for this type of test. One advantage of this specimen geometry is that it permits measurements of Mode I, Mode II, or mixed mode fracture toughness. The end-notched flexure (ENF) specimen has essentially the same geometry as the DCB specimen, but the latter is loaded in three-point bending, which imposes Mode II displacements of the crack faces. The edge delamination specimen simulates the conditions in an actual structure. Recall from Chapter 6 that tensile stresses normal to the ply are highest at the free edge (Figure 6.16); thus delamination zones often initiate at the edges of a panel.

The initial flaw in a DCB specimen is normally introduced by placing a thin film (e.g., aluminum foil) between plies prior to molding. The film should be coated with a release agent so that it can be removed prior to testing.

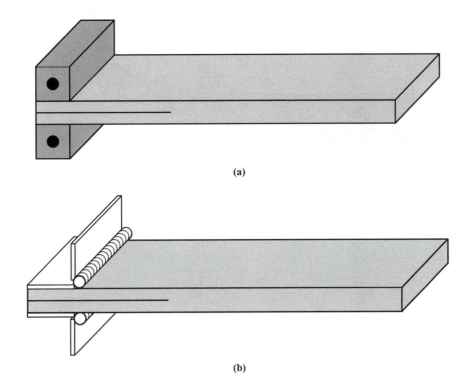

FIGURE 8.17 Loading fixtures for DCB specimens: (a) end blocks and (b) piano hinges.

Figure 8.17 illustrates two common fixtures that facilitate loading the DCB specimen. The blocks or hinges are normally adhesively bonded to the specimen. These fixtures must allow the free rotation of the specimen ends with a minimum of stiffening.

The DCB specimen can be tested in Mode I, Mode II, or mixed-mode conditions, as Figure 8.18 illustrates. Recall from Chapter 2 that the energy release rate of this specimen configuration can be inferred from the beam theory.

For pure Mode I loading (Figure 8.18(a)), elastic beam theory leads to the following expression for energy release rate (see Example 2.2):

$$\mathcal{G}_I = \frac{P_I^2 a^2}{BEI} \tag{8.36}$$

where

$$EI = \frac{2P_I a^3}{3\Delta_I} \tag{8.37}$$

The corresponding relationship for Mode II (Figure 8.18(b)) is given by

$$\mathcal{G}_{II} = \frac{3P_{II}^2 a^2}{4BEI} \tag{8.38}$$

assuming linear beam theory. Mixed-loading conditions can be achieved by unequal tensile loading of the upper and lower portions of the specimens, as Figure 8.18(c) illustrates. The applied loads

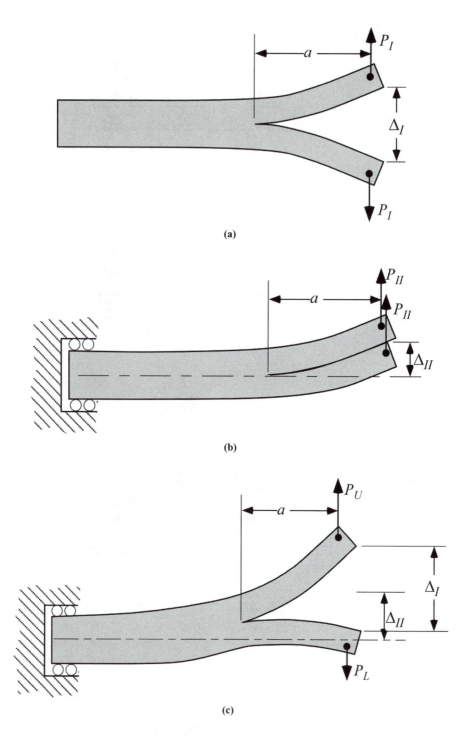

FIGURE 8.18 Mode I, II, and mixed mode loading of DCB specimens: (a) Mode I, (b) Mode II, and (c) Mixed mode.

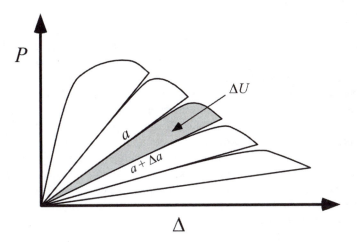

FIGURE 8.19 Schematic load-displacement curve for a delamination toughness measurement.

can be resolved into Mode I and Mode II components as follows:

$$P_I = |P_L| \tag{8.39a}$$

$$P_{II} = \frac{|P_U| - |P_L|}{2} \tag{8.39b}$$

where P_U and P_L are the upper and lower loads, respectively. The components of $\mathcal{G}$ can be computed by inserting P_I and P_{II} into Equation (8.36) and Equation (8.38). Recall from Chapter 2 that Mode I and Mode II components of energy release rate are additive.

Linear beam theory may result in erroneous estimates of energy release rate, particularly when the specimen displacements are large. The area method [21, 22] provides an alternative measure of energy release rate. Figure 8.19 schematically illustrates a typical load-displacement curve, where the specimen is periodically unloaded. The loading portion of the curve is typically nonlinear, but the unloading curve is usually linear and passes through the origin. The energy release rate can be estimated from the incremental area inside the load-displacement curve, divided by the change in crack area:

$$\mathcal{G} = \frac{\Delta U}{B \Delta a} \tag{8.40}$$

The Mode I and Mode II components of $\mathcal{G}$ can be inferred from the P_I-Δ_I and P_{II}-Δ_{II} curves, respectively. The corresponding loads and displacements for Mode I and Mode II are defined in Figure 8.18 and Equation (8.39).

Figure 8.20 illustrates a typical delamination resistance curve for Mode I. After initiation and a small amount of growth, delamination occurs at a steady-state $\mathcal{G}_{Ic}$ value, provided the global behavior of the specimen is elastic.

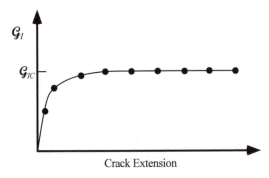

Crack Extension

FIGURE 8.20 Schematic R curve inferred from a delamination experiment.

8.3 CERAMICS

Fracture toughness is usually the limiting property in ceramic materials. Ceramics tend to have excellent creep properties and wear resistance, but are excluded from many load-bearing applications because they are relatively brittle. The latest generation of ceramics have enhanced toughness (Section 6.2), but brittle fracture is still a primary area of concern in these materials.

Because toughness is a crucial property for ceramic materials, rational fracture toughness measurements are absolutely essential. Unfortunately, fracture toughness tests on ceramics can be very difficult and expensive. Specimen fabrication, for example, requires special grinding tools, since ordinary machining tools are inadequate. Precracking by fatigue is extremely time consuming; some investigators have reported precracking times in excess of one week per specimen [24]. During testing, it is difficult to achieve stable crack growth with most specimen configurations and testing machines.

Several test methods have been developed to overcome some of the difficulties associated with fracture toughness measurements in ceramics. The chevron-notched specimen [25–28] eliminates the need for precracking, while the bridge indentation approach [24, 29–33] is a novel method for introducing a crack without resorting to a lengthy fatigue-precracking process.

8.3.1 CHEVRON-NOTCHED SPECIMENS

A chevron notch has a V-shaped ligament, such that the notch depth varies through the thickness, with the minimum notch depth at the center. Figure 8.21 shows two common configurations of chevron-notched specimens: the short bar and the short rod. In addition, single-edge-notched bend (SENB) and compact specimens (Figure 7.1) are sometimes fabricated with chevron notches. The chevron notch is often utilized in conventional fracture toughness tests on metals because this shape facilitates the initiation of the fatigue precrack. For fracture toughness tests on brittle materials, the unique properties of the chevron notch can eliminate the need for precracking altogether.

Figure 8.22 schematically compares the stress-intensity factor vs. the crack length for chevron and straight notch configurations. When the crack length $= a_o$, the stress-intensity factor in the chevron-notched specimen is very high, because a finite load is applied over a very small net thickness. When $a \geq a_1$, the K_I the values for the two notch configurations are identical, since the chevron notch no longer has an effect. The K_I for the chevron-notched specimen exhibits a minimum at a particular crack length a_m, which is between a_o and a_1.

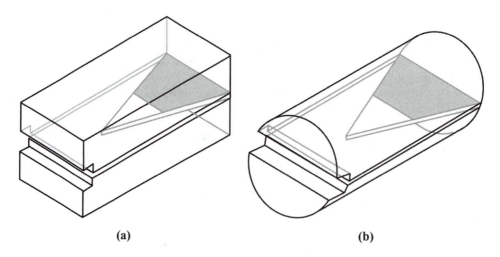

(a)   **(b)**

FIGURE 8.21 Two common designs of chevron-notched specimens: (a) short bar and (b) short rod. Taken from E 1304-97, "Standard Test Method for Plane-Strain (Chevron Notch) Fracture Toughness of Metallic Materials." American Society for Testing and Materials, Philadelphia, PA, 1997 (Reapproved 2002).

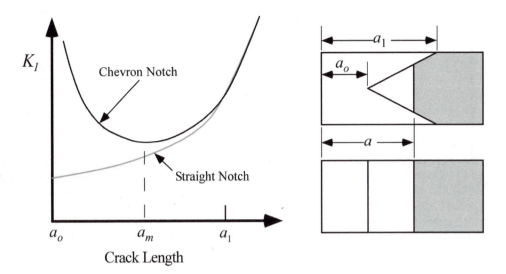

FIGURE 8.22 Comparison of stress-intensity factors in specimens with chevron and straight notches. Note that K_I exhibits a minimum in the chevron-notched specimens.

The K_I vs. crack length behavior of the chevron-notched specimen makes this specimen particularly suitable for measuring the toughness in brittle materials. Consider a material in which the R curve reaches a steady-state plateau soon after the crack initiates (Figure 8.23). The crack should initiate at the tip of the chevron upon application of a small load, since the local K_I is high. The crack is stable at this point, because the driving force decreases rapidly with crack advance; thus additional load is required to grow the crack further. The maximum load in the test, P_M, is achieved when the crack grows to a_m, the crack length corresponding to the minimum in the K_I-a curve. At this point, the specimen will be unstable if the test is conducted in load control, but stable crack growth may be possible beyond a_m if the specimen is subject to crosshead control. The point of instability in the latter case depends on the compliance of the testing machine, as discussed in Section 2.5.

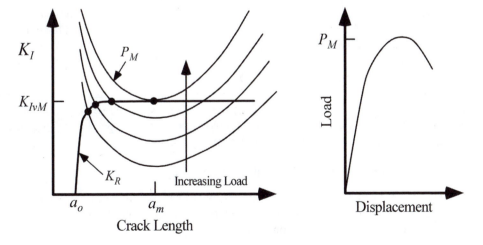

FIGURE 8.23 Fracture toughness testing of a material with a flat R curve. The maximum load in the test occurs when $a = a_m$.

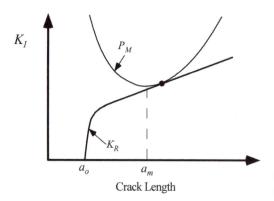

FIGURE 8.24 Application of the chevron-notched specimen to a material with a rising R curve.

Since final instability occurs at a_m, and a_m is known *a priori* (from the K_I vs. crack length relationship), it is necessary only to measure the maximum load in this test. The fracture toughness is given by

$$K_{IvM} = \frac{P_M}{B\sqrt{W}} f(a_m/W) \qquad (8.41)$$

where K_{IvM} is the chevron-notched toughness defined at maximum load, and $f(a/W)$ is the geometry correction factor. Early researchers developed simple models to estimate $f(a/W)$ for chevron-notched specimens, but more recent (and more accurate) estimates are based on three-dimensional finite element and boundary element analysis of this configuration [27].

The maximum load technique for inferring toughness does not work as well when the material exhibits a rising R curve, as Figure 8.24 schematically illustrates. The point of tangency between the driving force and the R curve may not occur at a_m in this case, resulting in an error in the stress-intensity calculation. Moreover, the value of K_R at the point of tangency is geometry dependent when the R curve is rising.

If both load and crack length are measured throughout the test, it is possible to construct the R curve for the material under consideration. Optical observation of the growing crack is not usually feasible for a chevron-notched specimen, but the crack length can be inferred through an unloading compliance technique [22], in which the specimen is periodically unloaded and the crack length is computed from the elastic compliance.

Two ASTM standards for chevron-notched specimens are currently available. The first such standard developed for this specimen geometry, E 1304 [25], applies to brittle metals such as high strength aluminum alloys. A more recent test method, ASTM C 1421 [25], addresses fracture toughness measurement in advanced ceramics.[3]

The chevron-notched specimen has proved to be very useful in characterizing the toughness of brittle materials. The advantages of this test specimen include its compact geometry, the simple instrumentation requirements (in the case of the K_{IvM} measurement), and the fact that no precracking is required. One of the disadvantages of this specimen is its complicated design, which leads to higher machining costs. Also, this specimen is poorly suited to high-temperature testing, and the K_{IvM} measurement is inappropriate for material with rising R curves.

8.3.2 BEND SPECIMENS PRECRACKED BY BRIDGE INDENTATION

A novel technique for precracking ceramic bend specimens has recently been developed in Japan [29]. A number of researchers [24, 29–33] have adopted this method, which has been incorporated into an ASTM standard for fracture toughness testing of ceramics, C 1421 [25]. Warren et al. [30],

---

[3] ASTM C 1421 covers both the chevron-notched test specimen and the edge-cracked bend specimen. The latter is precracked using the bridge indentation method described in Section 8.3.2.

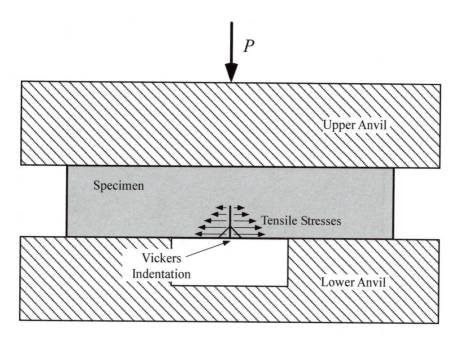

FIGURE 8.25 The bridge indentation method for precracking. Taken from Nose, T. and Fujii, T., "Evaluation of Fracture Toughness for Ceramic Materials by a Single-Edge-Precracked-Beam Method." *Journal of the American Ceramic Society,* Vol. 71, 1988, pp. 328–333.

who were among the first to apply this precracking technique, have termed it the "bridge indentation" method.

Figure 8.25 is a schematic drawing of the loading fixtures for the bridge indentation method of precracking. A starter notch is introduced into a bend specimen by means of a Vickers hardness indentation. The specimen is compressed between two anvils, as Figure 8.25 illustrates. The top anvil is flat, while the bottom anvil has a gap in the center. This arrangement induces a local tensile stress in the specimen, which leads to a pop-in fracture. The fracture arrests because the propagating crack experiences a falling K field.

The bridge indentation technique is capable of producing highly uniform crack fronts in bend specimens. After precracking, these specimens can be tested in three- or four-point bending with conventional fixtures. Nose and Fujii [24] showed that fracture toughness values obtained from bridge-precracked specimens compared favorably with data from conventional fatigue-precracked specimens.

Bar-On et al. [31] investigated the effect of precracking variables on the size of the crack that is produced by this technique. Figure 8.26 shows that the length of the pop-in in alumina decreases with increasing Vickers indentation load. Also note that the pop-in load decreases with increasing indentation load. Large Vickers indentation loads produce significant initial flaws and tensile residual stresses, which enable the pop-in to initiate at a lower load; the crack arrests sooner at lower loads because there is less elastic energy available for crack propagation. Thus, it is possible to control the length of the precrack though the Vickers indentation load.

The bridge indentation technique is obviously much more economical than the fatigue precracking of ceramic specimens. The edge-cracked bend configuration is simple, and therefore less expensive to fabricate. Also, three- and four-point bend fixtures are suitable for high-temperature testing. One problem with the bend specimen is that it consumes more material than the chevron-notched specimens illustrated in Figure 8.21. This is a major shortcoming when evaluating new materials, where only small samples are available. Another disadvantage of the beam configuration is that it tends to be unstable; most test machines are too compliant to achieve stable crack growth in brittle bend specimens [32, 33].

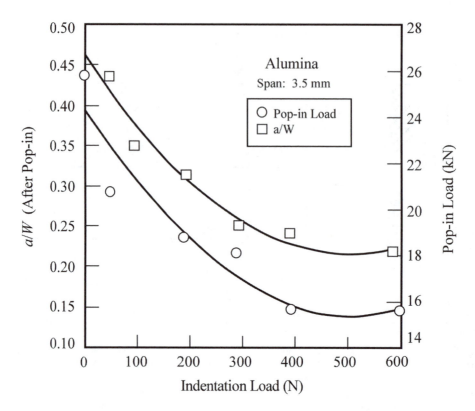

FIGURE 8.26 Effect of bridge indentation load on the crack length after pop-in. Taken from Bar-On, I., Beals, J.T., Leatherman, G.L., and Murray, C.M., "Fracture Toughness of Ceramic Precracked Bend Bars." *Journal of the American Ceramic Society,* Vol. 73, 1990, pp. 2519–2522.

REFERENCES

1. Schapery, R.A., "Correspondence Principles and a Generalized *J* Integral for Large Deformation and Fracture Analysis of Viscoelastic Media." *International Journal of Fracture,* Vol. 25, 1984, pp. 195–223.
2. Schapery, R.A., "Time-Dependent Fracture: Continuum Aspects of Crack Growth." *Encyclopedia of Materials Science and Engineering,* Pergamon Press, Oxford, 1986, pp. 5043–5054.
3. Jones, R.E. and Bradley, W.L., "Fracture Toughness Testing of Polyethylene Pipe Materials." ASTM STP 995, Vol. 1, American Society for Testing and Materials, Philadelphia, PA, 1989, pp. 447–456.
4. Cayard, M., "Fracture Toughness Testing of Polymeric Materials." Ph.D. Dissertation, Texas A&M University, College Station, TX, 1990.
5. Williams, J.G., *Fracture Mechanics of Polymers,* Halsted Press, New York, 1984.
6. Lu, M.L. and Chang, F-C., "Fracture Tougham of PC/PBT Blend Based on J-Integral Methods." *Journal of Applied Polymer Science*, Vol. 56, pp. 1065–1075.
7. Moore, D.R., Pavan, A., and Williams, J.G., *Fracture Mechanics Test Methods for Polymers, Adhesives and Composites,* ESIS Publication 28, Elsevier, London, 2001.
8. Marshall, G.P., Coutts, L.H., and Williams, J.G., "Temperature Effects in the Fracture of PMMA." *Journal of Materials Science,* Vol. 9, 1974, pp. 1409–1419.
9. Schapery, R.A., "A Theory of Crack Initiation and Growth in Viscoelastic Media—I. Theoretical Development." *International Journal of Fracture,* Vol. 11, 1975, pp. 141–159.
10. Schapery, R.A., Private communication, 1990.
11. D 5045-99 "Standard Test Methods for Plane Strain Fracture Toughness and Strain Energy Release Rate of Plastic Materials." American Society for Testing and Materials, Philadelphia, PA, 1999.

12. E 399-90, "Standard Test Method for Plane Strain Fracture Toughness of Metallic Materials." American Society for Testing and Materials, Philadelphia, PA, 1990 (Reapproved 1997).

13. D 6068-96, "Standard Test Method for Determining J-R Curves of Plastic Materials." American Society for Testing and Materials, Philadelphia, PA, 1996 (Reapproved 2002).

14. E 1820-01, "Standard Test Method for Measurement of Fracture Toughness." American Society for Testing and Materials, Philadelphia, PA, 2001.

15. D 256-03, "Standard Test Methods for Determining the Izod Pendulum Impact Resistance of Plastics." American Society for Testing and Materials, Philadelphia, PA, 1988.

16. E 23-02a, "Standard Test Methods for Notched Bar Impact Testing of Metallic Materials." American Society for Testing and Materials, Philadelphia, PA, 2002.

17. *Engineered Materials Handbook, Volume 2: Engineering Plastics.* ASM International, Metals Park, OH, 1988.

18. BS 3505:1986, "British Standard Specification for Unplasticized Polyvinyl Chloride (PVC-U) Pressure Pipes for Cold Potable Water." British Standards Institution, London, 1986.

19. Whitney, J.M., Browning, C.E., and Hoogsteden, W., "A Double Cantilever Beam Test for Characterizing Mode I Delamination of Composite Materials." *Journal of Reinforced Plastics and Composites,* Vol. 1, 1982, pp. 297–313.

20. Prel, Y.J., Davies, P., Benzeggah, M.L., and de Charentenay, F.-X., "Mode I and Mode II Delamination of Thermosetting and Thermoplastic Composites." ASTM STP 1012, American Society for Testing and Materials, Philadelphia, PA, 1989, pp. 251–269.

21. Corleto, C.R. and Bradley, W.L., "Mode II Delamination Fracture Toughness of Unidirectional Graphite/Epoxy Composites." ASTM STP 1012, American Society for Testing and Materials, Philadelphia, PA, 1989, pp. 201–221.

22. Hibbs, M.F., Tse, M.K., and Bradley, W.L., "Interlaminar Fracture Toughness and Real-Time Fracture Mechanism of Some Toughened Graphite/Epoxy Composites." ASTM STP 937, American Society for Testing and Materials, Philadelphia, PA, 1987, pp. 115–130.

23. D 5528-01, "Standard Test Method for Mode I Interlaminar Fracture Toughness of Unidirectional Fiber-Reinforced Polymer Matrix Composites." American Society for Testing and Materials, Philadelphia, PA, 2001.

24. Nose, T. and Fujii, T., "Evaluation of Fracture Toughness for Ceramic Materials by a Single-Edge-Precracked-Beam Method." *Journal of the American Ceramic Society,* Vol. 71, 1988, pp. 328–333.

25. E 1304-97, "Standard Test Method for Plane-Strain (Chevron Notch) Fracture Toughness of Metallic Materials." American Society for Testing and Materials, Philadelphia, PA, 1997 (Reapproved 2002).

26. C 1421-01b, "Standard Test Methods for Determination of Fracture Toughness of Advanced Ceramics at Ambient Temperatures." American Society for Testing and Materials, Philadelphia, PA, 2001.

27. Newman, J.C., "A Review of Chevron-Notched Fracture Specimens." ASTM STP 855, American Society for Testing and Materials, Philadelphia, PA, 1984, pp. 5–31.

28. Shannon, J.L., Jr. and Munz, D.G., "Specimen Size Effects on Fracture Toughness of Aluminum Oxide Measured with Short-Rod and Short Bar Chevron-Notched Specimens." ASTM STP 855, American Society for Testing and Materials, Philadelphia, PA, 1984, pp. 270–280.

29. Nunomura, S. and Jitsukawa, S., "Fracture Toughness for Bearing Steels by Indentation Cracking under Multiaxial Stress." *Tetsu to Hagane,* Vol. 64, 1978 (in Japanese).

30. Warren, R. and Johannsen, B., "Creation of Stable Cracks in Hard Metals Using 'Bridge' Indentation." *Powder Metallurgy,* Vol. 27, 1984, pp. 25–29.

31. Bar-On, I., Beals, J.T., Leatherman, G.L., and Murray, C.M., "Fracture Toughness of Ceramic Precracked Bend Bars." *Journal of the American Ceramic Society,* Vol. 73, 1990, pp. 2519–2522.

32. Barratta, F.I. and Dunlay, W.A., "Crack Stability in Simply Supported Four-Point and Three-Point Loaded Beams of Brittle Materials." *Proceedings of the Army Symposium on Solid Mechanics, 1989—Mechanics of Engineered Materials and Applications,* U.S. Army Materials Technology Laboratory, Watertown, MA, 1989, pp. 1–11.

33. Underwood, J.H., Barratta, F.I., and Zalinka, J.J., "Fracture Toughness Tests and Displacement and Crack Stability Analyses of Round Bar Bend Specimens of Liquid-Phase Sintered Tungsten." *Proceedings of the 1990 SEM Spring Conference on Experimental Mechanics,* Albuquerque, NM, 1990, pp. 535–542.

9 Application to Structures

Recall Figure 1.7 in Chapter 1, which illustrates the so-called fracture mechanics triangle. When designing a structure against fracture, there are three critical variables that must be considered: stress, flaw size, and toughness. Fracture mechanics provides mathematical relationships between these quantities. Figure 9.1 is an alternative representation of the three key variables of fracture mechanics. The stress and flaw size provide the *driving force* for fracture, and the fracture toughness is a measure of the material's *resistance* to crack propagation. Fracture occurs when the driving force reaches or exceeds the material resistance. Chapter 7 and Chapter 8 describe experimental methods to measure fracture toughness in metals and nonmetals, respectively. This chapter focuses on approaches for computing the fracture driving force in structural components that contain cracks.

Several parameters are available for characterizing the fracture driving force. In Chapter 2, the stress-intensity factor K and the energy release rate $\mathcal{G}$ were introduced. These parameters are suitable when the material is predominately elastic. The J integral and crack-tip-opening displacement (CTOD) are appropriate driving force parameters in the elastic-plastic regime. Recall from Chapter 3 that the J integral is a generalized formulation of the energy release rate; in the limit of linear elastic material behavior, $J = \mathcal{G}$.

Techniques for computing fracture driving force range from very simple to complex. The most appropriate methodology for a given situation depends on geometry, loading, and material properties. For example, a three-dimensional finite element analysis (Chapter 12) may be necessary when both the geometry and loading are sufficiently complicated that a simple hand calculation will not suffice. When significant yielding precedes fracture, an analysis based on linear elastic fracture mechanics (LEFM) may not be suitable.[1]

This chapter focuses on fracture initiation and instability in structures made from linear elastic and elastic-plastic materials. A number of engineering approaches are discussed; the basis of these approaches and their limitations are explored. This chapter covers only quasistatic methodologies, but such approaches can be applied to rapid loading and crack arrest in certain circumstances (see Chapter 4). The analyses presented in this chapter do not address time-dependent crack growth. Chapter 10 and Chapter 11 consider fatigue crack growth and environmental cracking, respectively.

9.1 LINEAR ELASTIC FRACTURE MECHANICS

Analyses based on LEFM apply to structures where crack-tip plasticity is small. Chapter 2 introduced many of the fundamental concepts of LEFM. The fracture behavior of a linear elastic structure can be inferred by comparing the applied K (the driving force) to a critical K or a K-R curve (the fracture toughness). The elastic energy release rate $\mathcal{G}$ is an alternative measure of driving force, and a critical value of $\mathcal{G}$ quantifies the material toughness.

For Mode I loading (Figure 2.14), the stress-intensity factor can be expressed in the following form:

$$K_I = Y\sigma\sqrt{\pi a} \tag{9.1}$$

---

[1] There are a number of simplified elastic-plastic analysis methods that use LEFM parameters such as K, combined with an appropriate adjustment for plasticity. Two such methodologies are described in the present chapter: the reference stress approach and the failure assessment diagram (FAD).

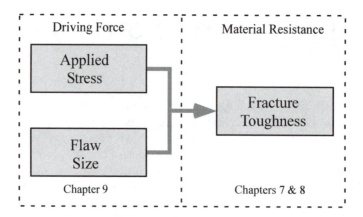

FIGURE 9.1 Relationship between the three critical variables in fracture mechanics. The stress and flaw size contribute to the driving force, and the fracture toughness is a measure of the material resistance.

where

    Y = dimensionless geometry correction factor

    σ = characteristic stress

    a = characteristic crack dimension

If the geometry factor is known, the applied K_I can be computed for any combination of σ and a. The applied stress intensity can then be compared to the appropriate material property, which may be a K_{Ic} value, a K-R curve, environment-assisted cracking data, or, in the case of cyclic loading, fatigue crack growth data (see Chapter 10).

Fracture analysis of a linear elastic structure becomes relatively straightforward once a K solution is obtained for the geometry of interest. Stress-intensity solutions can come from a number of sources, including handbooks, the published literature, experiments, and numerical analysis.

A large number of stress-intensity solutions have been published over the past 50 years. Several handbooks [1–3] contain compilations of solutions for a wide variety of configurations. The published literature contains many more solutions. It is often possible to find a K solution for a geometry that is similar to the structure of interest.

When a published K solution is not available, or the accuracy of such a solution is in doubt, one can obtain the solution experimentally or numerically. Deriving a closed-form solution is probably not a viable alternative, since this is possible only with simple geometries and loading, and nearly all such solutions have already been published. Experimental measurement of K is possible through optical techniques, such as photoelasticity [4, 5] and the method of caustics [6], or by determining $\mathcal{G}$ from the rate of change in compliance with crack length (Equation (2.30)) and computing K from $\mathcal{G}$ (Equation (2.58)). However, these experimental methods for determining fracture driving force parameters have been rendered virtually obsolete by advances in computer technology. Today, nearly all new K solutions are obtained numerically. Chapter 12 describes a number of computational techniques for deriving stress intensity and energy release rate.

An alternative to finite element analysis and other computational techniques is to utilize the principle of elastic superposition, which enables new K solutions to be constructed from known cases. Section 2.6.4 outlined this approach, and demonstrated that the effect of a far-field stress on K can be represented by an appropriate crack-face pressure. Influence coefficients [7], described below, are an application of the superposition principle. Section 2.6.5 introduced the concept of weight functions [8, 9], from which K solutions can be obtained for arbitrary loading. Examples of the application of the weight function approach are presented in Section 9.1.3.

9.1.1 K_I for Part-Through Cracks

Laboratory fracture toughness specimens usually contain idealized cracks, but naturally occurring flaws in structures are under no obligation to live up to these ideals. Structural flaws are typically irregular in shape and are part-way through the section thickness. Moreover, severe stress gradients often arise in practical situations, while laboratory specimens experience relatively simple loading.

Newman and Raju [10] have published a series of K_I solutions for part-through cracks. Figure 9.2 illustrates the assumed geometries. Newman and Raju approximate buried cracks, surface cracks, and corner cracks as ellipses, half ellipses, and quarter ellipses, respectively. These solutions apply to linear stress distributions, where the stress normal to the flaw can be resolved into bending and membrane components, respectively (Figure 9.3). If the stress distribution is not perfectly linear, the equivalent membrane stress can be inferred from the integrated average stress through the thickness, while the equivalent bending stress can be inferred by computing a resultant moment (per unit width) and dividing by $6t^2$.

The Newman and Raju solutions for part-through flaws subject to membrane and bending stresses are expressed in the following form:

$$K_I = (\sigma_m + H\sigma_b)F\sqrt{\frac{\pi a}{Q}} \qquad (9.2)$$

where F and H are geometry factors, which Newman and Raju obtained from finite element analysis. The parameters F and H depend on a/c, a/t, ϕ (Figure 9.2), and plate width. The appendix at the end of this chapter lists polynomial fits for F and H that correspond to each of the crack shapes in Figure 9.2. Q is the flaw-shape parameter, which is based on the solution of an elliptical integral of the second kind (see Section A2.4 in the appendix following Chapter 2). The following expression gives a good approximation of Q:

$$Q = 1 + 1.464\left(\frac{a}{c}\right)^{1.65} \qquad \text{for } a \leq c \qquad (9.3)$$

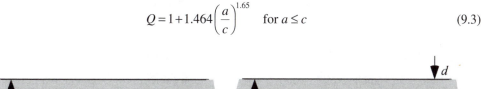

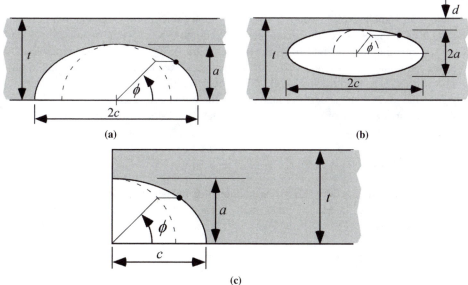

FIGURE 9.2 Idealized part-through crack geometries: (a) semielliptical surface crack, (b) elliptical buried flaw, and (c) quarter-elliptical corner crack.

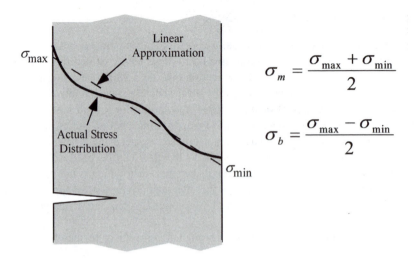

$$\sigma_m = \frac{\sigma_{max} + \sigma_{min}}{2}$$

$$\sigma_b = \frac{\sigma_{max} - \sigma_{min}}{2}$$

FIGURE 9.3 Approximating a nonuniform stress distribution as linear, and resolving the stresses into membrane and bending components.

The Newman and Raju solutions apply to flat plates, but provide a reasonable approximation for cracks in curved shells as long as the radius of curvature R is large relative to the shell thickness t. Recently, Anderson et al. [11] published a comprehensive set of K solutions for surface cracks in cylindrical and spherical shells with a wide range of R/t values.

Equation (9.2) is reasonably flexible, since it can account for a range of stress gradients, and includes pure membrane loading and pure bending as special cases. This equation, however, is actually a special case of the influence coefficient approach, which is described below.

9.1.2 INFLUENCE COEFFICIENTS FOR POLYNOMIAL STRESS DISTRIBUTIONS

Recall Figure 2.25 in Chapter 2, where a remote boundary traction $P(x)$ results in a normal stress distribution $p(x)$ on Plane A-B of this uncracked configuration. Next, we introduce a crack on Plane A-B while maintaining the far-field traction (Figure 2.26(a)). By invoking the principle of superposition, we can replace the boundary traction with a crack-face pressure $p(x)$ and obtain the same K_I. In other words, a far-field traction $P(x)$ and a crack-face pressure of $p(x)$ result in the same K_I, where $p(x)$ is the normal stress across Plane A-B in the absence of a crack.

Consider a surface crack of depth a with power-law crack-face pressure (Figure 9.4):

$$p(x) = p_n \left(\frac{x}{a} \right)^n \tag{9.4}$$

where p_n is the pressure at $x = a$ and n is a nonnegative integer. For the special case of uniform crack-face pressure, $n = 0$. The Mode I stress intensity for this loading can be written in the following form:

$$K_I = G_n p_n \sqrt{\frac{\pi a}{Q}} \tag{9.5}$$

where G_n is an *influence coefficient*, and Q is given by Equation (9.3). The value of the influence coefficient is a function of geometry, crack dimensions, and the power-law exponent n.

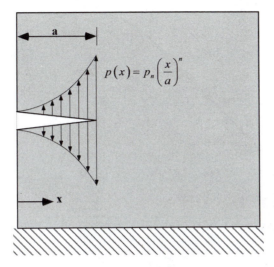

FIGURE 9.4 Power-law stress distribution applied to the crack face.

Now consider a nonuniform normal stress distribution (in the absence of a crack) that can be represented by a polynomial expression:

$$\sigma(x) = \sigma_o + \sigma_1\left(\frac{x}{t}\right) + \sigma_2\left(\frac{x}{t}\right)^2 + \sigma_3\left(\frac{x}{t}\right)^3 + \sigma_4\left(\frac{x}{t}\right)^4 \qquad (9.6)$$

where t is the section thickness. Figure 9.5 illustrates the nonuniform stress distribution and defines the x coordinate. If we introduce a surface crack at the location where the above stress distribution applies, the application of the principle of superposition leads to the following expression for K_I:

$$K_I = \left[\sigma_o G_o + \sigma_1 G_1\left(\frac{a}{t}\right) + \sigma_2 G_2\left(\frac{a}{t}\right)^2 + \sigma_3 G_3\left(\frac{a}{t}\right)^3 + \sigma_4 G_4\left(\frac{a}{t}\right)^4\right]\sqrt{\frac{\pi a}{Q}} \qquad (9.7)$$

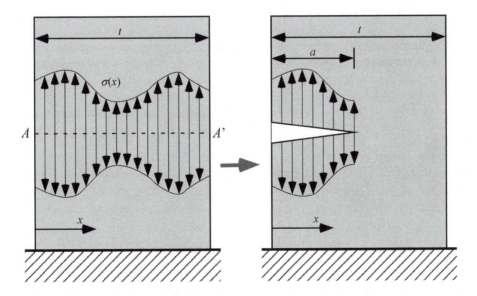

FIGURE 9.5 Nonuniform stress distribution that can be fit to a four-term polynomial (Equation (9.6)).

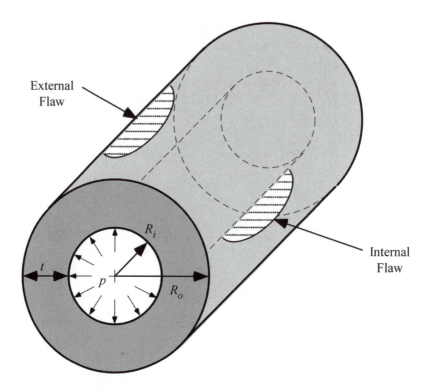

FIGURE 9.6 Internal and external axial surface flaws in a pressurized cylinder.

That is, we have superimposed K_I solutions for power-law loading with $n = 0, 1, 2, 3, 4$ to obtain the solution for the polynomial distribution. By comparing each term in the above expression with Equation (9.6), we see that

$$P_n = \sigma_n \left(\frac{a}{t} \right)^n$$

Consider the example of a pressurized cylinder with an internal axial surface flaw, as illustrated in Figure 9.6. In the absence of the crack, the hoop stress in a thick wall pressure vessel is as follows:

$$\sigma_{\theta\theta} = \frac{pR_i^2}{R_o^2 - R_i^2} \left[1 + \left(\frac{R_o}{r} \right)^2 \right] \tag{9.8}$$

where p is the internal pressure and the other terms are defined in Figure 9.6. If we define the origin at the inner wall ($x = r - R_i$) and perform a Taylor series expansion about $x = 0$, Equation (9.8) becomes

$$\sigma_{\theta\theta} = \frac{pR_o^2}{R_o^2 - R_i^2} \left[1 + \left(\frac{R_i}{R_o} \right)^2 - 2 \left(\frac{x}{R_i} \right) + 3 \left(\frac{x}{R_i} \right)^2 - 4 \left(\frac{x}{R_i} \right)^3 + 5 \left(\frac{x}{R_i} \right)^4 + \cdots \right] \quad (0 \le x/R_i \le 1) \tag{9.9}$$

where x is in the radial direction with the origin at R_i. The first five terms of this expansion give the desired fourth-order polynomial. An alternate approach would be to curve-fit a polynomial to the stress field. This latter method is necessary when the stress distribution does not have a closed-form solution, such as when nodal stresses from a finite element analysis are used.

When computing K_I for the internal surface flaw, we must also take account of the pressure loading on the crack faces. Superimposing p on Equation (9.9), and applying Equation (9.7) to each term in the series leads to the following expression for K_I [7,12]:

$$K_I = \frac{pR_o^2}{R_o^2 - R_i^2}\left[2G_o - 2\left(\frac{a}{R_i}\right)G_1 + 3\left(\frac{a}{R_i}\right)^2 G_2 - 4\left(\frac{a}{R_i}\right)^3 G_3 + 5\left(\frac{a}{R_i}\right)^4 G_4\right]\sqrt{\frac{\pi a}{Q}} \qquad (9.10)$$

Applying a similar approach to an external surface flaw leads to

$$K_I = \frac{pR_i^2}{R_o^2 - R_i^2}\left[2G_o + 2\left(\frac{a}{R_o}\right)G_1 + 3\left(\frac{a}{R_o}\right)^2 G_2 + 4\left(\frac{a}{R_o}\right)^3 G_3 + 5\left(\frac{a}{R_o}\right)^4 G_4\right]\sqrt{\frac{\pi a}{Q}} \qquad (9.11)$$

The origin in this case was defined at the outer surface of the cylinder, and a series expansion was performed as before. Thus K_I for a surface flaw in a pressurized cylinder can be obtained by substituting the appropriate influence coefficients into Equation (9.10) or Equation (9.11).

The influence coefficient approach is useful for estimating K_I values for cracks that emanate from stress concentrations. Figure 9.7 schematically illustrates a surface crack at the toe of a fillet weld. This geometry produces local stress gradients that affect the K_I for the crack. Performing a finite element analysis of this structural detail with a crack is generally preferable, but the influence coefficient method can give a reasonable approximation. If the stress distribution at the weld toe is known for the uncracked case, these stresses can be fit to a polynomial (Equation (9.6)), and K_I can be estimated by substituting the influence coefficients and polynomial coefficients into Equation (9.7).

The methodology in the previous example is only approximate, however. If the influence coefficients were obtained from an analysis of a flat plate, there may be slight errors if these G_n values are applied to the fillet weld geometry. The actual weld geometry has a relatively modest effect on the G_n values. As long as the stress gradient emanating from the weld toe is taken into account, computed K_I values will usually be within 10% of values obtained from a more rigorous analysis.

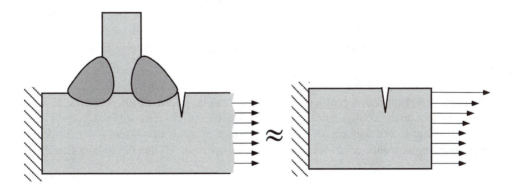

FIGURE 9.7 Application of the influence coefficient approach to a complex structural detail such as a fillet weld.

Since the flaw in Figure 9.7 is near a weld, there is a possibility that weld residual stresses will be present. These stresses must be taken into account in order to obtain an accurate estimate of K_I. Weld residual stresses are discussed further in Section 9.1.4.

9.1.3 WEIGHT FUNCTIONS FOR ARBITRARY LOADING

While the influence coefficient approach is useful, it has limitations. It requires that the stress distribution be fitted to a polynomial. There are many instances where a polynomial expression (fourth order or lower) does not provide a good representation of the actual stress field. For example, the five-term Taylor series expansion of the hoop stress in a thick-wall pressure vessel (Equation (9.8) and Equation (9.9)) is not accurate for $R_i/t < 4$.

The weight function method, which was introduced in Chapter 2, provides a means to compute stress-intensity solutions for arbitrary loading. Consider a surface crack of depth a, subject to a normal stress that is an arbitrary function of x, where x is oriented in the crack depth direction and is measured from the free surface.[2] The Mode I stress-intensity factor is given by

$$K_I = \int_0^a h(x, a)\sigma(x)\,dx \tag{9.12}$$

where $h(x, a)$ is the weight function.

For the deepest point of a semielliptical crack ($\phi = 90°$), the weight function can be represented by an equation of the following form [13,14]:

$$h_{90} = \frac{2}{\sqrt{2\pi(a-x)}}\left[1 + M_1\left(1 - \frac{x}{a}\right)^{1/2} + M_2\left(1 - \frac{x}{a}\right) + M_3\left(1 - \frac{x}{a}\right)^{3/2}\right] \tag{9.13}$$

where the coefficients M_1 to M_3 depend on the geometry and crack dimensions. The corresponding expression for the free surface ($\phi = 0°$) is given by [13,14]

$$h_0 = \frac{2}{\sqrt{\pi x}}\left[1 + N_1\left(\frac{x}{a}\right)^{1/2} + N_2\left(\frac{x}{a}\right) + N_3\left(\frac{x}{a}\right)^{3/2}\right] \tag{9.14}$$

Each of these expressions contains three unspecified coefficients. However, one unknown can be eliminated by invoking boundary conditions on the weight function [13,14]. For the deepest point of the crack, imposing the condition that the second derivative of the weight function is zero at $x = 0$ implies that $M_2 = 3$. Setting $h_0 = 0$ at $x = a$ results in the following relationship:

$$N_1 + N_2 + N_3 + 1 = 0 \tag{9.15}$$

Therefore, the weight function coefficients M_i and N_i can be inferred from reference K_I solutions for two load cases on the configuration of interest. The choice of reference load cases is arbitrary, but it is convenient to use uniform and linear crack-face pressure ($n = 0$ and $n = 1$, respectively). The corresponding influence coefficients for these load cases are G_0 and G_1. Setting Equation (9.5)

---

[2] Note that the present discussion is restricted to a one-dimensional normal stress distribution. That is, the pressure normal to the crack face may vary over the depth of the flaw, but it is constant along the flaw length at a given x value. Equation (2.53) describes the general case where the traction varies arbitrarily over the crack surface.

equal to Equation (9.12) for the two reference cases and assuming a unit value for p_n results in simultaneous integral equations:

$$G_0 \sqrt{\frac{\pi a}{Q}} = \int_0^a h(x)\, dx \tag{9.16a}$$

$$G_1 \sqrt{\frac{\pi a}{Q}} = \int_0^a h(x)\left(\frac{x}{a}\right) dx \tag{9.16b}$$

Substituting Equation (9.13) and Equation (9.14) into the above expressions and applying the aforementioned boundary conditions leads to expressions for M_i and N_i. At the deepest point of the crack, the weight function coefficients are given by

$$M_1 = \frac{2\pi}{\sqrt{2Q}}(3G_1 - G_0) - \frac{24}{5} \tag{9.17a}$$

$$M_2 = 3 \tag{9.17b}$$

$$M_3 = \frac{6\pi}{\sqrt{2Q}}(G_0 - 2G_1) + \frac{8}{5} \tag{9.17c}$$

where the influence coefficients G_0 and G_1 are evaluated at $\phi = 90°$. The weight function coefficients at the free surface are as follows:

$$N_1 = \frac{3\pi}{\sqrt{Q}}(2G_0 - 5G_1) - 8 \tag{9.18a}$$

$$N_2 = \frac{15\pi}{\sqrt{Q}}(3G_1 - G_0) + 15 \tag{9.18b}$$

$$N_3 = \frac{3\pi}{\sqrt{Q}}(3G_0 - 10G_1) - 8 \tag{9.18c}$$

where G_0 and G_1 are evaluated at $\phi = 0°$.

Once M_i and N_i have been determined for a given geometry and crack size, the stress distribution for the problem of interest must be substituted into Equation (9.12). Numerical integration is normally required, especially if $\sigma(x)$ is characterized by nodal stress results from a finite element analysis. For some stress distributions, a closed-form integration of Equation (9.12) is possible. For example, closed-form solutions exist for power-law crack-face pressure (Equation (9.4)). Consequently, it is possible to solve for higher-order influence coefficients using the weight function method.

At the deepest point of the crack ($\phi = 90°$), the influence coefficients for $n = 2$ to 4 are given by

$$G_2 = \frac{\sqrt{2Q}}{\pi}\left(\frac{16}{15} + \frac{1}{3}M_1 + \frac{16}{105}M_2 + \frac{1}{12}M_3\right) \tag{9.19a}$$

$$G_3 = \frac{\sqrt{2Q}}{\pi}\left(\frac{32}{35} + \frac{1}{4}M_1 + \frac{32}{315}M_2 + \frac{1}{20}M_3\right) \tag{9.19b}$$

$$G_4 = \frac{\sqrt{2Q}}{\pi}\left(\frac{256}{315} + \frac{1}{5}M_1 + \frac{256}{3465}M_2 + \frac{1}{30}M_3\right) \tag{9.19c}$$

The corresponding influence coefficients for the free surface ($\phi = 0°$) are as follows:

$$G_2 = \frac{\sqrt{Q}}{\pi}\left(\frac{4}{5} + \frac{2}{3}N_1 + \frac{4}{7}N_2 + \frac{1}{2}N_3\right) \tag{9.20a}$$

$$G_3 = \frac{\sqrt{Q}}{\pi}\left(\frac{4}{7} + \frac{1}{2}N_1 + \frac{4}{9}N_2 + \frac{2}{5}N_3\right) \tag{9.20b}$$

$$G_4 = \frac{\sqrt{Q}}{\pi}\left(\frac{4}{9} + \frac{2}{5}N_1 + \frac{4}{11}N_2 + \frac{1}{3}N_3\right) \tag{9.20c}$$

Therefore, if one wishes to apply the influence coefficient approach, it is not necessary to compute the higher-order G_n solutions with finite element analysis. If solutions for uniform and linear crack-face loading are available, the weight function coefficients can be computed from Equation (9.17) and Equation (9.18). For more complex load cases, K_I can be inferred by integrating Equation (9.12). Alternatively, if the stress distribution can be represented by a polynomial expression, the higher-order influence coefficients can be computed from Equation (9.19) and Equation (9.20) and K_I can be inferred from Equation (9.7). The advantage of the latter approach is that numerical integration is not required. Consequently, the influence coefficient approach is more conducive to spreadsheet calculations than the weight function method.

Equations (9.13)–(9.20) apply only to two locations on the crack front: $\phi = 0°$ and $\phi = 90°$. Wang and Lambert [15] have developed a weight function expression that applies to the full range $0° \le \phi \le 90°$ for semielliptical surface cracks. Anderson et al. [11] have integrated this expression to solve for the influence coefficients (G_i) at arbitrary crack front angles. The resulting equations are rather lengthy, and consequently are omitted for the sake of brevity.

9.1.4 PRIMARY, SECONDARY, AND RESIDUAL STRESSES

Section 2.4 introduced the concept of load control and displacement control, where a structure or specimen is subject to applied forces or imposed displacements, respectively. The applied energy release rate, and therefore the stress intensity factor, is the same, irrespective of whether a given stress distribution is the result of applied loads or imposed displacements. Section 2.5 and Section 3.4.1 discuss the relative stability of cracked components in load control and displacement control. Crack growth tends to be unstable in load control but can be stable in displacement control.

There are very few practical situations in which a cracked body is subject to pure displacement control. Figure 2.12, which is a simple representation of the more typical case, shows a cracked plate subject to a fixed remote displacement Δ_t. The spring in series represents the system compliance C_m. For example, a long cracked plate in which the ends are fixed would have a large system compliance. If C_m is large, there is essentially no difference between (remote) displacement control and load control as far as the crack is concerned. See Section 9.3.3 for further discussion of the effect of system compliance on crack stability.

Some design codes for structures such as pressure vessels and piping refer to load-controlled stresses as *primary* and displacement-controlled stresses as *secondary*. Hoop stress due to internal pressure in a pipe or pressure vessel is an example of a primary stress because the pressure constitutes applied forces on the boundary of the structure. Thermal expansion (or contraction) leads to imposed displacements, so thermal stresses are usually considered secondary. As long as the total stress is well below the yield strength, the classification of stresses as primary and secondary is not important. When plastic deformation occurs, however, secondary stresses redistribute and may relax from their initial values. For this reason, design codes that classify stresses as primary and secondary usually permit higher allowable values of the latter.

A special case of a displacement-controlled stress is residual stress due to weld shrinkage. Weld residual stresses are usually not taken into account in design because they do not affect the failure stress of a ductile material in the absence of significant crack-like flaws. When a crack is present, however, residual stresses do contribute to the fracture driving force and must be included in the analysis.

In linear elastic analyses, primary, secondary, and residual stresses are treated in an identical fashion. The total stress intensity is simply the sum of the primary and secondary components:

$$K_I^{total} = K_I^P + K_I^S + K_I^R \tag{9.21}$$

where the superscripts P, S, and R denote primary, secondary, and residual stress quantities, respectively.

The distinction between primary and secondary stresses is important only in elastic-plastic and fully plastic analyses. Section 9.2 and Section 9.4 describe the treatment of primary, secondary, and residual stresses in such cases.

9.1.5 A WARNING ABOUT LEFM

Performing a purely linear elastic fracture analysis and *assuming* that LEFM is valid is potentially dangerous, because the analysis gives no warning when it becomes invalid. The user must rely on experience to know whether or not plasticity effects need to be considered. A general rule of thumb is that plasticity becomes important at around 50% of the yield, but this is by no means a universal rule.

The safest approach is to adopt an analysis that spans the entire range from linear elastic to fully plastic behavior. Such an analysis accounts for the two extremes of brittle fracture and plastic collapse. At low stresses, the analysis reduces to LEFM, but predicts collapse if the stresses are sufficiently high. At intermediate stresses, the analysis automatically applies a plasticity correction when necessary; the user does not have to decide whether or not such a correction is needed. The failure assessment diagram (FAD) approach, described in Section 9.4, is an example of a general methodology that spans the range from linear elastic to fully plastic material behavior.

9.2 THE CTOD DESIGN CURVE

The CTOD concept was applied to structural steels beginning in the late 1960s. The British Welding Research Association, now known as The Welding Institute (TWI), and other laboratories performed CTOD tests on structural steels and welds. At that time there was no way to apply these results to welded structures because CTOD driving force equations did not exist. Burdekin and Stone [16] developed the CTOD equivalent of the strip-yield model in 1966. Although their model provides a basis for a CTOD driving force relationship, they were unable to modify the strip-yield model to account for residual stresses and stress concentrations. (These difficulties were later overcome when a strip-yield approach became the basis of the R6 fracture analysis method, as discussed in Section 9.4).

In 1971, Burdekin and Dawes [17] developed the CTOD design curve, a semiempirical driving force relationship, which was based on an idea that Wells [18] originally proposed. For linear elastic conditions, fracture mechanics theory was reasonably well developed, but the theoretical framework required to estimate the driving force under elastic-plastic and fully plastic conditions did not exist until the late 1970s. Wells, however, suggested that global strain should scale linearly with CTOD under large-scale yielding conditions. Burdekin and Dawes based their elastic-plastic driving force relationship on Wells' suggestion and an empirical correlation between small-scale CTOD tests and wide double-edge-notched tension panels made from the same material. The wide plate specimens were loaded to failure, and the failure strain and crack size of a given large-scale specimen were correlated with the critical CTOD in the corresponding small-scale test.

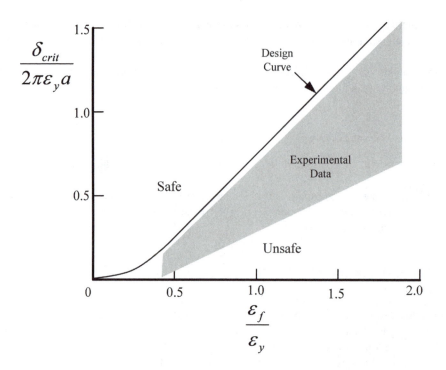

FIGURE 9.8 The CTOD design curve.

The correlation that resulted in the CTOD design curve is illustrated schematically in Figure 9.8. The critical CTOD is nondimensionalized by the half crack length a of the wide plate and is shown on the ordinate of the graph. The nondimensional CTOD is plotted against the failure strain in the wide plate, normalized by the elastic yield strain ε_y. Based on a plot similar to Figure 9.8, Burdekin and Dawes [17,19] proposed the following two-part relationship:

$$\frac{\delta_{crit}}{2\pi\varepsilon_y a} = \left(\frac{\varepsilon_f}{\varepsilon_y}\right)^2 \quad \text{for} \frac{\varepsilon_f}{\varepsilon_y} \leq 0.5 \tag{9.22a}$$

and

$$\frac{\delta_{crit}}{2\pi\varepsilon_y a} = \frac{\varepsilon_f}{\varepsilon_y} - 0.25 \quad \text{for} \frac{\varepsilon_f}{\varepsilon_y} \leq 0.5 \tag{9.22b}$$

Equation (9.22a), which was derived from LEFM theory, includes a safety factor of 2.0 on crack size. Equation (9.22b) represents an upper envelope of the experimental data.

The applied strain and flaw size in a structure, along with the critical CTOD for the material, can be plotted on Figure 9.8. If the point lies above the design curve, the structure is considered safe because all observed failures are below the design line. Equation (9.22a) and Equation (9.22b) conform to the classical view of a fracture mechanics analysis, in relating stress (or strain in this case) to fracture toughness (δ_{crit}) and flaw size (a). The CTOD design curve is conservative, however, and does not relate *critical* combinations of these variables.

In 1980, the CTOD design curve approach was incorporated into the British Standards document PD 6493 [20]. This document addresses flaws of various shapes by relating them back to an equivalent through-thickness dimension $\bar{a}$. For example, if a structure contains a surface flaw of length $2c$ and depth a, the equivalent through-thickness flaw produces the same stress intensity when loaded to the same stress as the structure with the surface flaw. Thus $\bar{a}$ is a generalized measure of a flaw's severity. The CTOD design curve can be applied to any flaw by replacing a with $\bar{a}$ in Equation (9.22).

The original CTOD design curve was based on correlations with flat plates loaded in tension. Real structures, however, often include complex shapes that result in stress concentrations. In addition, the structure may be subject to bending and residual stresses, as well as tensile (membrane) stresses. The PD 6493:1980 approach accounts for complex stress distributions simply and conservatively by estimating the maximum total strain in the cross section and assuming that this strain acts through the entire cross section. The maximum strain can be estimated from the following equation:

$$\varepsilon_1 = \frac{1}{E}[k_t(P_m + P_b) + (S + R)] \tag{9.23}$$

where

k_t = elastic stress concentration factor

P_m = primary membrane stress

P_b = primary bending stress

S = secondary stress

R = residual stresses

Since the precise distribution of residual stresses was usually unknown, R was typically assumed to equal the yield strength in an as-welded weldment.

When Burdekin and Dawes developed the CTOD design curve, the CTOD and wide plate data were limited; and the curve they constructed laid above all available data. In 1979, Kamath [21] reassessed the design curve approach with additional wide plate and CTOD data generated between 1971 and 1979. In most cases, there were three CTOD tests for a given condition. Kamath used the lowest measured CTOD value to predict failure in the corresponding wide plate specimen. When he plotted the results in the form of Figure 9.8, a few data points fell above the design curve, indicating Equation (9.22) was nonconservative in these instances. The CTOD design curve, however, was conservative in most cases. Kamath estimated the average safety factor on crack size to be 1.9, although individual safety factors ranged from less than 1 to greater than 10. With this much scatter, the concept of a safety factor is of little value. A much more meaningful quantity is the confidence level. Kamath estimated that the CTOD design curve method corresponds to a 97.5% confidence of survival. That is, the method in PD 6493:1980 is conservative, approximately 97.5% of the time.

The CTOD design curve and PD 6493:1980 have been superseded by procedures that are based more on fundamental principles than empirical correlations. Two such approaches for elastic-plastic fracture mechanics analysis are direct evaluation of the J integral and the FAD approach. These methodologies are described in Section 9.3 and Section 9.4, respectively.

9.3 ELASTIC-PLASTIC J-INTEGRAL ANALYSIS

The most rigorous method to compute J is to perform an elastic-plastic finite element analysis on the structural component that contains a crack. Chapter 12 outlines techniques for incorporating a crack into a finite element mesh, and it describes the algorithms for computing the J integral from finite element results. Some commercial finite element codes have built-in J-integral analysis capabilities.

There are a number of simplified methods for estimating J in lieu of elastic-plastic finite element analysis. Two such methods, the Electric Power Research Institute(EPRI) J estimation scheme and the reference stress approach, are described in Section 9.3.1 and Section 9.3.2, respectively.

For materials that fail by an unstable mechanism such as cleavage (Chapter 5), fracture analysis is simply a matter of comparing the applied J to the critical J. In ductile materials, however, fracture toughness is typically characterized by a resistance curve rather than a single J value. Section 9.3.3 describes ductile instability analysis, where the driving force is compared with a J resistance curve.

9.3.1 The EPRI J-Estimation Procedure

The J integral was first used as a fracture toughness parameter in the early 1970s. At that time, there was no convenient way to compute the applied J in a structural component. Stress-intensity-factor handbooks were available, but a corresponding handbook for elastic-plastic analysis did not exist. Finite element analysis was a relatively new tool that was available only to a few specialists.

In the late 1970s, the EPRI decided to develop a series of J handbooks that would enable users to estimate the fracture driving force under elastic-plastic conditions. Shih and Hutchinson [22] developed the basic methodology for cataloging J solutions in handbook form. A series of finite element analyses were performed at General Electric Corporation in Schenectady, New York, and the first J handbook was an engineering handbook by EPRI in 1981 [23].

A few additional J handbook volumes were published subsequent to the original [24–26], but this collection of J-integral solutions covers a very limited range of cases. Most of the solutions are for simple two-dimensional geometries such as flat plates with through cracks and edge cracks. Because of these limitations, the EPRI J handbooks are of little value for most real-world problems.

As a practical engineering tool, the ERPI J handbooks have not lived up to original expectations. However, the research funded by EPRI in the late 70s and early 80s did contribute to our understanding of elastic-plastic fracture mechanics. The EPRI research had an influence on current methodologies. For example, the development of a J-based failure assessment diagram (Section 9.4.2) was largely inspired by the EPRI J estimation approach. Given its historical significance, it is appropriate to include a description of the EPRI approach in this chapter.

The EPRI procedure provides a means for estimating the applied J integral under elastic-plastic and fully plastic conditions. The elastic and plastic components of J are computed separately and added to obtain the total J:

$$J_{tot} = J_{el} + J_{pl} \tag{9.24}$$

The elastic J is actually the elastic energy release rate $\mathcal{G}$, which can be computed from K_I using Equation (2.54). Fully plastic J solutions were inferred from finite element analysis and were tabulated in a dimensionless form, as described later in this chapter.

9.3.1.1 Theoretical Background

Consider a cracked structure with a fully plastic ligament, where elastic strains are negligible. Assume that the material follows a power-law stress-strain curve:

$$\frac{\varepsilon_{pl}}{\varepsilon_o} = \alpha \left(\frac{\sigma}{\sigma_o} \right)^n \tag{9.25}$$

which is the second term in the Ramberg-Osgood model (Equation (3.22)). The parameters α, n, ε_o, and σ_o are defined in Section 3.2.3. Close to the crack tip, under J-controlled conditions, the stresses are given by the HRR singularity:

$$\sigma_{ij} = \sigma_o \left(\frac{J}{\alpha \varepsilon_o \sigma_o I_n r} \right)^{\frac{1}{n+1}} \tilde{\sigma}_{ij}(n, \theta) \tag{9.26}$$

which is a restatement of Equation (3.24a). Solving for J in the HRR equation gives

$$J = \alpha \varepsilon_o \sigma_o I_n r \left(\frac{\sigma_{ij}}{\sigma_o} \right)^{n+1} \tilde{\sigma}_{ij}^{\ n+1} \tag{9.27}$$

For J-controlled conditions, the loading must be proportional. That is, the local stresses must increase in proportion to the remote load P. Therefore, Equation (9.27) can be written in terms of P:

$$J = \alpha \varepsilon_o \sigma_o h L \left(\frac{P}{P_o} \right)^{n+1} \tag{9.28}$$

where

h = dimensionless function of geometry and n
L = characteristic length dimension for the structure
P_o = reference load

Both L and P_o can be defined arbitrarily, and h can be determined by numerical analysis of the configuration of interest.

It turns out that the assumptions of J dominance at the crack tip and proportional loading are not necessary to show that J scales with P^{n+1} for a power-law material, but these assumptions were useful for deriving the correct form of the J-P relationship.

Equation (9.28) is an estimate of the *fully plastic J*. Under linear elastic conditions, J must scale with P^2. The EPRI J estimation procedure assumes that the total J is equal to the sum of the elastic and plastic components (Equation (9.24)).

9.3.1.2   Estimation Equations

The fully plastic equations for J, crack-mouth-opening displacement V_p, and load line displacement Δ_p have the following form for most geometries:

$$J_{pl} = \alpha \varepsilon_o \sigma_o b h_1 (a/W, n) \left(\frac{P}{P_o} \right)^{n+1} \tag{9.29}$$

$$V_p = \alpha \varepsilon_o a h_2 (a/W, n) \left(\frac{P}{P_o} \right)^{n} \tag{9.30}$$

$$\Delta_p = \alpha \varepsilon_o a h_3 (a/W, n) \left(\frac{P}{P_o} \right)^{n} \tag{9.31}$$

where

b = uncracked ligament length
a = crack length
h_1, h_2, and h_3 = dimensionless parameters that depend on geometry and hardening exponent

The h factors for various geometries and n values, for both plane stress and plane strain, are tabulated in several EPRI reports [23–26]. The appendix at the end of this chapter includes tables of h factors for several simple configurations.

The reference load P_o is usually defined by a limit load solution for the geometry of interest; P_o normally corresponds to the load at which the net cross section yields.

Several configurations have J expressions that are slightly different from Equation (9.29). For example, the fully plastic J integral for a center-cracked panel and a single-edge-notched tension panel is given by

$$J_{pl} = \alpha \varepsilon_o \sigma_o \frac{ba}{W} h_1(a/W, n) \left(\frac{P}{P_o} \right)^{n+1} \tag{9.32}$$

where, in the case of the center-cracked panel, a is the half crack length and W is the half width. This modification was made in order to reduce the sensitivity of h_1 to the crack length/width ratio.

The elastic J is equal to $\mathcal{G}(a_{eff})$, the energy release rate for an effective crack length, which is based on a modified Irwin plastic zone correction:

$$a_{eff} = a + \frac{1}{1+(P/P_o)^2} \frac{1}{\beta\pi} \left(\frac{n-1}{n+1} \right) \left(\frac{K_I}{\sigma_o} \right)^2 \tag{9.33}$$

where $\beta = 2$ for plane stress and $\beta = 6$ for plane strain conditions. Equation (9.33) is a first-order correction, in which a_{eff} is computed from the elastic K_I, rather than K_{eff}; thus iteration is not necessary. This is an empirical adjustment that was applied in order to match the J values from the estimation procedure with corresponding values from rigorous elastic-plastic finite element analysis. This correction has a relatively small effect on computed J values.

The CTOD can be estimated from a computed J value as follows:

$$\delta = d_n \frac{J}{\sigma_o} \tag{9.34}$$

where d_n is a dimensionless constant that depends on flow properties [27]. Figure 3.18 shows plots of d_n for both plane stress and plane strain. Equation (9.34) must be regarded as approximate in the elastic-plastic and fully plastic regimes, because the J-CTOD relationship is geometry dependent in large-scale yielding.

EXAMPLE 9.1

Consider a single-edge-notched tensile panel with $W = 1$ m, $B = 25$ mm, and $a = 125$ mm. Calculate J vs. applied load assuming plane stress conditions. Neglect the plastic zone correction.
Given: $\sigma_o = 414$ MPa, $n = 10$, $\alpha = 1.0$, $E = 207,000$ MPa, $\varepsilon_o = \sigma_o/E = 0.002$

Solution: From Table A9.13, the reference load for this configuration is given by`

$$P_o = 1.072\eta\sigma_o bB$$

where

$$\eta = \sqrt{1+\left(\frac{a}{b}\right)^2} - \frac{a}{b} = 0.867 \text{ for } a/b = 125/875 = 0.143$$

Solving for P_o gives

$$P_o = 8.42 \text{ MN}$$

For $a/W = 0.125$ and $n = 10$, $h_1 = 4.14$ (from Table A9.13). Thus the fully plastic J is given by

$$J_{pl} = (1.0)(0.002)(414,000 \text{kPa}) \frac{(0.875 \text{ m})(1.125 \text{ m})}{1.0 \text{ m}} (4.14) \left(\frac{P}{8.42 \text{ MN}} \right)^{11}$$

$$= 2.486 \times 10^{-8} P^{11}$$

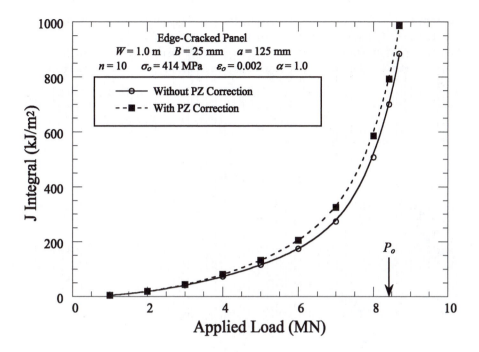

FIGURE 9.9 Applied J vs. applied load in an edge-cracked panel (Figure 9.1).

where P is in MN and J_{pl} is in kJ/m². The elastic J is given by

$$J_{el} = \frac{K_I^2}{E} = \frac{P^2 f^2(a/W)}{B^2\, W\, E}$$

From the polynomial expression in Table 2.4, $f(a/W) = 0.770$ for $a/W = 0.125$. Thus

$$J_{el} = \frac{1000\, P^2 (0.770)^2}{(0.025\,\mathrm{m})^2 (1.0\,\mathrm{m})(207{,}000\,\mathrm{MPa})} = 4.584\, P^2$$

where P is in MN and J_{el} is in kJ/m². The total J is the sum of J_{el} and J_{pl}:

$$J = 4.584 P^2 + 2.486 \times 10^{-8} P^{11}$$

Figure 9.9 shows a plot of this equation. An analysis that includes the plastic zone correction (Equation (9.33)) is also plotted for comparison.

9.3.1.3 Comparison with Experimental *J* Estimates

Typical equations for estimating J from a laboratory specimen have the following form:

$$J = \frac{K^2}{E'} + \frac{\eta_p}{b} \int_0^{\Delta_p} P d\Delta_p \qquad (9.35)$$

assuming unit thickness and a stationary crack. Equation (9.35) is convenient for experimental measurements because it relates J to the area under the load vs. the load-line displacement curve.

Since Equation (9.31) gives an expression for the P-Δ_p curve for a stationary crack, it is possible to compare J_{pl} estimates from Equation (9.29) and Equation (9.32) with Equation (9.35). According to Equation (9.31), the P-Δ_p curve follows a power law, where the exponent is the same as in the material's true stress–true strain curve. The plastic energy absorbed by the specimen is as follows:

$$\int_0^{\Delta_p} P d\Delta_p = \frac{n}{n+1} P \Delta_p$$

$$= \frac{n}{n+1} P_o \alpha \varepsilon_o a h_3 \left(\frac{P}{P_o} \right)^{n+1} \tag{9.36}$$

Thus the plastic J is given by

$$J_{pl} = \frac{n}{n+1} \eta_p P_o \alpha \varepsilon_o \frac{a}{b} h_3 \left(\frac{P}{P_o} \right)^{n+1} \tag{9.37}$$

Comparing Equation (9.29) and Equation (9.37) and solving for η_p gives

$$\eta_p = \frac{n+1}{n} \frac{\sigma_o b^2 h_1}{P_o W h_3} \tag{9.38a}$$

Alternatively, if J_{pl} is given by Equation (9.34)

$$\eta_p = \frac{n+1}{n} \frac{\sigma_o b^2 h_1}{P_o W h_3} \tag{9.38b}$$

Consider an edge-cracked bend specimen in plane strain. The EPRI fully plastic J solution for this configuration is tabulated in Table A9.8. The reference load, assuming unit thickness and the standard span of $4W$, is given by

$$P_o = \frac{0.364 \sigma_o b^2}{W} \tag{9.39}$$

Substituting Equation (9.39) into Equation (9.38a) gives

$$\eta_p = \frac{1}{0.364} \frac{n+1}{n} \frac{W}{a} \frac{h_1}{h_3} \tag{9.40}$$

Equation (9.40) is plotted in Figure 9.10 for $n = 5$ and $n = 10$. According to the equation that was derived in Section 3.2.5, $\eta_p = 2$. This derivation, however, is valid only for deep cracks, since it assumes that the ligament length b is the only relevant length dimension. Figure 9.10 indicates that Equation (9.40) approaches the deep crack limit with increasing a/W. For $n = 10$, the deep crack formula appears to be reasonably accurate beyond $a/W \sim 0.3$. Note that the η_p values computed from Equation (9.40) for deep cracks fluctuate about an average of ~ 1.9, rather than the theoretical value of 2.0. The deviations from the theoretical value for deep cracks may be indicative of numerical errors in the h_1 and h_3 values.

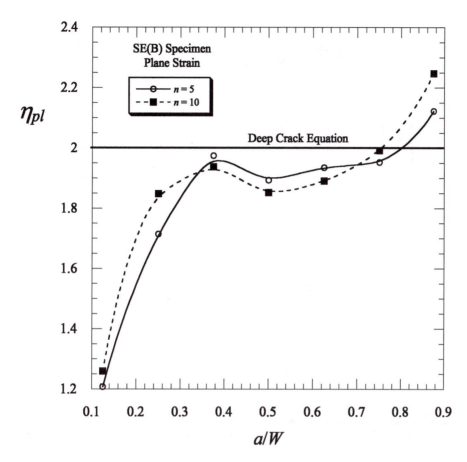

FIGURE 9.10 Comparison of the plastic η factor inferred from the EPRI handbook with the deep crack value of 2.0 derived in Chapter 3.

Equation (3.32) was derived for a double-edge-notched tension panel, but also applies to a deeply notched center-cracked panel. A comparison of Equation (9.32) with the second term of Equation (3.32) leads to the following relationship for a center-cracked panel in plane stress:

$$\frac{J_{EPRI}}{J_{DC}} = \frac{n+1}{n-1}\frac{b}{W}\frac{h_1}{h_3} \qquad (9.41)$$

where J_{EPRI} is the plastic J computed from Equation (9.32) and J_{DC} is the plastic J from the deep crack formula. Figure 9.11 is a plot of Equation (9.41). The deep crack formula underestimates J at small a/W ratios, but coincides with J_{EPRI} when a/W is sufficiently large. Note that the deep crack formula applies to a wider range of a/W for $n = 10$. The deep crack formula assumes that all plasticity is confined to the ligament, a condition that is easier to achieve in low-hardening materials.

9.3.2 THE REFERENCE STRESS APPROACH

Prior to the publication of the EPRI J solutions, Ainsworth [28][3] developed a methodology to estimate the C^* parameter for creep crack growth, which is a J-like parameter that accounts for

---

[3] The actual publication date of Ainsworth's work on creep crack growth was 1982, which was after the first EPRI handbook had been published. However, Ainsworth actually began his work in the late 1970s, before tabulated fully plastic J solutions were available.

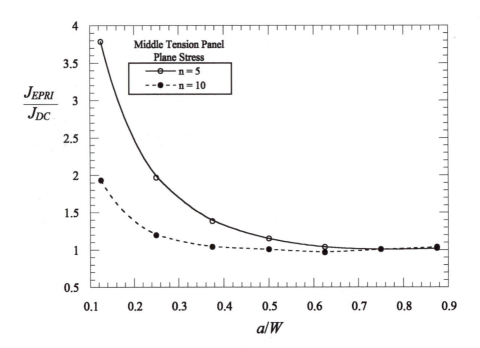

FIGURE 9.11 Comparison of J estimates from the EPRI handbook with the deep crack formula for a center-cracked panel.

time-dependent creep deformation (Chapter 4). He expressed the creep crack driving force C^* as a function of a parameter that he called *reference stress*. Several years later, Ainsworth [29] combined the reference stress concept with the EPRI J estimation procedure to introduce a new fracture assessment methodology that is more versatile than the original EPRI approach.

The EPRI equations for fully plastic J, Equation (9.29) and Equation (9.32), assume that the material's stress-plastic strain curve follows a simple power law. Many materials, however, have flow behavior that deviates considerably from a power law. For example, most low carbon steels exhibit a plateau in the flow curve immediately after yielding. Applying Equation (9.29) or Equation (9.32) to such a material, results in significant errors. Ainsworth [29] modified the EPRI relationships to reflect more closely the flow behavior of real materials. He defined a reference stress as follows:

$$\sigma_{ref} = (P/P_o)\sigma_o \tag{9.42}$$

He further defined the reference strain as the total axial strain when the material is loaded to a uniaxial stress of σ_{ref}. Substituting these definitions into Equation (9.29) gives

$$J_{pl} = \sigma_{ref}bh_1\left(\varepsilon_{ref} - \frac{\sigma_{ref}\varepsilon_o}{\sigma_o}\right) \tag{9.43}$$

For materials that obey a power law, Equation (9.43) agrees precisely with Equation (9.29), but the former is more general, in that it is applicable to all types of stress-strain behavior.

Equation (9.43) still contains h_1, the geometry factor that depends on the power-law-hardening exponent n. Ainsworth proposed redefining P_o for a given configuration to produce another constant h_1' that is insensitive to n. He noticed, however, that even without the modification of P_o, h_1 was relatively insensitive to n except at high n values (low-hardening materials). Ainsworth believed that accurate estimates of h_1 were less crucial for high n values because at that time, the strip-yield failure assessment diagram (Section 9.4.1) was considered adequate for low-hardening materials.

He proposed the following approximation.

$$h_1(n) \approx h_1(1) \tag{9.44}$$

where $h_1(n)$ is the geometry constant for a material with a strain hardening exponent of n and $h_1(1)$ is the corresponding constant for a linear material. By substituting $h_1(1)$ into Equation (9.29) or Equation (9.32), Ainsworth was able to relate the plastic J to the linear elastic stress-intensity factor:

$$J_{pl} = \frac{\mu K_I^2}{E}\left(\frac{E\varepsilon_{ref}}{\sigma_{ref}} - 1\right) \tag{9.45}$$

where $\mu = 0.75$ for plane strain and $\mu = 1.0$ for plane stress.

Ainsworth's work has important ramifications. When applying the EPRI approach, one must obtain a stress-intensity factor solution to compute the elastic J, and a separate solution for h_1 in order to compute the plastic term. The h_1 constant is a plastic geometry correction factor. However, Equation (9.45) makes it possible to estimate J_{pl} from an elastic geometry correction factor. The original EPRI handbook [23] and subsequent additions [24–26] contain h_1 solutions for a relatively small number of configurations, but there are thousands of stress-intensity factor solutions in handbooks and the literature. Thus Equation (9.45) is not only simpler than Equation (9.29), but also more widely applicable.

Ainsworth made additional simplifications and modifications to the reference stress model in order to express it in terms of a FAD. The expression for the FAD based on the reference stress concept is given in Section 9.4.3.

9.3.3 DUCTILE INSTABILITY ANALYSIS

Section 3.4.1 outlined the theory of stability of J-controlled crack growth. Crack growth is stable as long as the rate of change in the driving force (J) is less than or equal to the rate of change of the material resistance (J_R). Equation (3.49) and Equation (3.50) defined the tearing modulus, which is a nondimensional representation of the derivatives of both the driving force and the resistance:

$$T_{app} = \frac{E}{\sigma_o^2}\left(\frac{dj}{da}\right)_{\Delta_T} \qquad \text{and} \qquad T_R = \frac{E}{\sigma_o^2}\frac{dj_R}{da} \tag{9.46}$$

where Δ_T is the remote displacement:

$$\Delta_T = \Delta + C_M P \tag{9.47}$$

where C_M is the system compliance (Figure 2.12). Recall that the value of C_M influences the relative stability of the structure; $C_M = \infty$ corresponds to dead loading, which tends to be unstable, while $C_M = 0$ represents the other extreme of displacement control, which tends to be more stable. Crack growth is unstable when

$$T_{app} > T_R \tag{9.48}$$

The rate of change in driving force at a fixed remote displacement is given by[4]

$$\left(\frac{dJ}{da}\right)_{\Delta_T} = \left(\frac{\partial J}{\partial a}\right)_P - \left(\frac{\partial J}{\partial P}\right)_a\left(\frac{\partial \Delta}{\partial a}\right)_P\left[C_m + \left(\frac{\partial \Delta}{\partial P}\right)_a\right]^{-1} \tag{9.49}$$

---

[4] The distinction between local and remote displacements (Δ and $C_M P$, respectively) is arbitrary as long as all displacements due to the crack are included in Δ. The local displacement Δ can contain any portion of the "no crack" elastic displacements without affecting the term in square brackets in Equation (9.49).

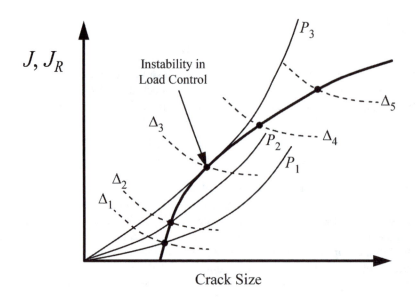

FIGURE 9.12 Schematic driving force diagram for load control and displacement control.

Methods for assessing structural stability include crack driving force diagrams and stability assessment diagrams. The former is a plot of J and J_R vs. crack length, while a stability assessment diagram is a plot of tearing modulus vs. J. These diagrams are merely alternative methods for plotting the same information.

Figure 9.12 shows a schematic driving force diagram for both load control and displacement control. In this example, the structure is unstable at P_3 and Δ_3 in load control, but the structure is stable in displacement control. Figure 9.13 illustrates driving force curves for this same structure, but with fixed remote displacement Δ_T and finite system compliance C_M. The structure is unstable at $\Delta_{T(4)}$ in this case.

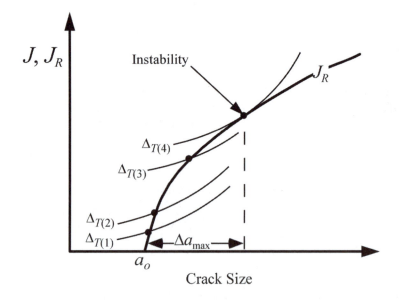

FIGURE 9.13 Schematic driving force diagram for a fixed remote displacement.

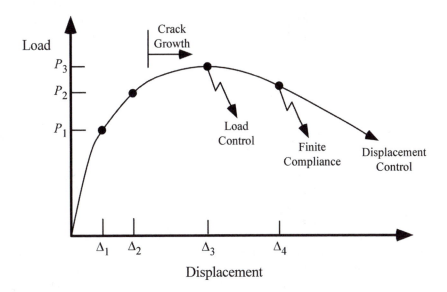

FIGURE 9.14 Schematic load-displacement curve for the material in Figure 9.12 and Figure 9.13.

Figure 9.14 illustrates the load-displacement curve for this hypothetical structure. A maximum load plateau occurs at P_3 and Δ_3, and the load decreases with further displacement. In load control, the structure is unstable at P_3, because the load cannot increase further. The structure is always stable in pure displacement control ($C_M = 0$), but is unstable at Δ_4 (and $\Delta_{T(4)} = \Delta_4 + C_M P_4$) for the finite compliance case.

Figure 9.15 is a schematic stability assessment diagram. The applied and material tearing moduli are plotted against J and J_R, respectively. Instability occurs when the T_{app}-J curve crosses the T_R-J_R curve. The latter curve is relatively easy to obtain, since J_R depends only on the amount of crack growth:

$$J_R = J_R(a - a_o) \tag{9.50}$$

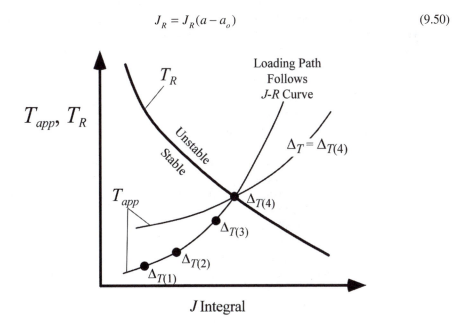

FIGURE 9.15 Schematic stability assessment diagram for the material in the three previous figures.

Thus, there is a unique relationship between T_R and J_R, and the T_R-J_R curve can be defined unambiguously. Suppose, for example, that the J-R curve is fit to a power law:

$$J_R = C_1(a - a_o)^{C_2} \qquad (9.51)$$

The material tearing modulus is given by

$$T_R = \frac{E}{\sigma_o^2} \frac{C_2 J_R}{(a - a_o)} = \frac{E}{\sigma_o^2} C_2 C_1^{1/C_2} J_R^{(C_2-1)/C_2} \qquad (9.52)$$

The applied tearing modulus curve is less clearly defined, however. There are a number of approaches for defining the T_{app}-J curve, depending on the application. Figure 9.15 illustrates two possible approaches, which are discussed next.

Suppose that the initial crack size a_o is known, and one wishes to determine the loading conditions (P, Δ, and Δ_T) at failure. In this case, the T_{app} should be computed at various points on the R curve. Since $J = J_R$ during stable crack growth, the applied J at a given crack size can be inferred from the J-R curve (Equation (9.50)). The remote displacement Δ_T increases as the loading progresses up the J-R curve (Figure 9.13); instability occurs at $\Delta_{T(4)}$. The final load, local displacement, crack size, and stable crack extension can be readily computed, once the critical point on the J-R curve has been identified.

The T_{app}-J curve can also be constructed by fixing one of the loading conditions (P, Δ, or Δ_T), and determining the critical crack size at failure, as well as a_o. For example, if we fix Δ_T at $\Delta_{T(4)}$ in the structure, we would predict the same failure point as the previous analysis but the T_{app}-J curve would follow a different path (Figure 9.15). If, however, we fix the remote displacement at a different value, we would predict failure at another point on the T_R-J_R curve; the critical crack size, stable crack extension, and a_o would be different from the previous example.

9.3.4 SOME PRACTICAL CONSIDERATIONS

If the material is sufficiently tough or if crack-like flaws in the structure are small, the structure will not fail unless it is loaded into the fully plastic regime. When performing fracture analyses in this regime, there are a number of important considerations that many practitioners overlook.

In the fully plastic regime, the J integral varies with P^{n+1} for a power-law material; a slight increase in load leads to a large increase in the applied J. The J vs. the crack-length driving force curves are also very steep in this regime. Consequently, the failure stress and critical crack size are insensitive to toughness in the fully plastic regime; rather, failure is governed primarily by the flow properties of the material. The problem is reduced to a limit load situation, where the main effect of the crack is to reduce the net cross section of the structure.

Predicting the failure stress or critical crack size under fully plastic conditions need not be complicated. A detailed tearing instability analysis and a simple limit load analysis should lead to similar estimates of failure conditions.

Problems arise, however, when one tries to compute the applied J at a given load and crack size. Since J is very sensitive to load in the fully plastic regime, a slight error in P produces a significant error in the estimated J. For example, a 10% overestimate in the yield strength σ_o will produce a corresponding error in P_o, which will lead to an underestimate of J by a factor of 3.2 for $n = 10$. Since flow properties typically vary by several percent in different regions of a plate, and heat-to-heat variations can be much larger, accurate estimates of the applied J at a fixed load are extremely difficult.

If estimates of the applied J are required in the fully plastic regime, the *displacement*, not the load, should characterize conditions in the structure. While the plastic J is proportional to P^{n+1}, J_{pl} scales with $\Delta_p^{(n+1)/n}$ according to Equation (9.29) and Equation (9.31). Thus a J-Δ plot is nearly linear in the fully plastic regime, and displacement is a much more sensitive indicator of the applied

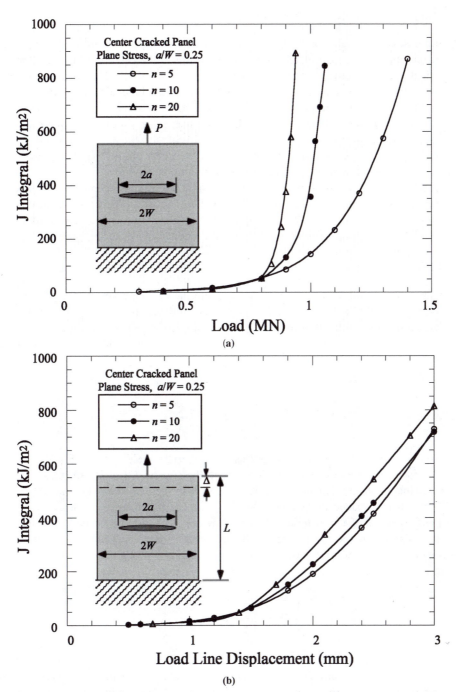

FIGURE 9.16 Comparison of *J*-load and *J*-displacement curves for a center-cracked panel ($W = 50$ mm, $B = 25$ mm, $L = 400$ mm, $\sigma_o = 420$ MPa, $\varepsilon_o = 0.002$, $\alpha = 1.0$): (a) *J* vs. load and (b) *J* vs. load-line displacement.

J in a structure. Figure 9.16 compares *J-P* and *J-Δ* plots for a center-cracked panel with three strain hardening exponents.

Recall Section 9.2, where the empirical correlation of CTOD and wide plate data that resulted in the CTOD design curve was plotted in terms of strain (i.e., displacement over a fixed gage length) rather than stress [17,19]. A correlation based on stress would not have worked, because the failure stresses in the wide plate specimens were clustered around the flow stress of the material.

9.4 FAILURE ASSESSMENT DIAGRAMS

The Failure Assessment Diagrams (FAD) is probably the most widely used methodology for elastic-plastic fracture mechanics analysis of structural components. The original FAD was derived from the strip-yield plastic zone correction, as described in Section 9.4.1. The strip-yield model has limitations, however. For example, it does not account for strain hardening. A more accurate FAD can be derived from an elastic-plastic J-integral solution (Section 9.4.2). In addition, there are simplified versions of the FAD that account for strain hardening but do not require a rigorous J integral solution (Section 9.4.3).

The FAD approach is easy to implement. A highly nonlinear problem (elastic-plastic fracture) is solved in terms of two parameters that vary linearly with applied load.

The FAD approach is also very versatile. It spans a wide range of material behavior, from brittle fracture under linear elastic conditions to ductile overload in the fully plastic regime. The FAD method is appropriate for welded components because it can account for residual stresses (Section 9.4.5). The FAD can also be used for ductile tearing analysis (Section 9.4.7).

9.4.1 Original Concept

Dowling and Townley [30] and Harrison et al. [31] introduced the concept of a two-criterion FAD to describe the interaction between brittle fracture and fully ductile rupture. The first FAD was derived from a modified version of the strip-yield model, as described next.

The effective stress intensity factor for a through crack in an infinite plate, according to the strip-yield model, is given by

$$K_{eff} = \sigma_{YS} \sqrt{\pi a} \left[\frac{8}{\pi^2} \ln \sec \left(\frac{\pi \sigma}{2 \sigma_{YS}} \right) \right]^{1/2} \tag{9.53}$$

As discussed in Chapter 2, this relationship is asymptotic to the yield strength. Equation (9.53) can be modified for real structures by replacing σ_{YS} with the collapse stress σ_c for the structure. This would ensure that the strip-yield model predicts failure as the applied stress approaches the collapse stress. For a structure loaded in tension, collapse occurs when the stress on the net cross-section reaches the flow stress of the material. Thus σ_c depends on the tensile properties of the material and the flaw size relative to the total cross section of the structure. The next step in deriving a failure assessment diagram from the strip-yield model entails dividing the effective stress intensity by the linear elastic K:

$$\frac{K_{eff}}{K_I} = \frac{\sigma_c}{\sigma} \left[\frac{8}{\pi^2} \ln \sec \left(\frac{\pi}{2} \frac{\sigma}{\sigma_c} \right) \right]^{1/2} \tag{9.54}$$

This modification not only expresses the driving force in a dimensionless form but also eliminates the square root term that contains the half-length of the through crack. Thus Equation (9.54) removes the geometry dependence of the strip-yield model.[5] This is analogous to the PD 6493:1980 approach, where the driving force relationship was generalized by defining an equivalent through thickness flaw $\bar{a}$. As a final step, we can define the stress ratio S_r and the K ratio K_r as follows:

$$K_r = \frac{K_I}{K_{eff}} \tag{9.55}$$

---

[5] This generalization of the strip-yield model is not rigorously correct for all configurations, but it is a good approximation.

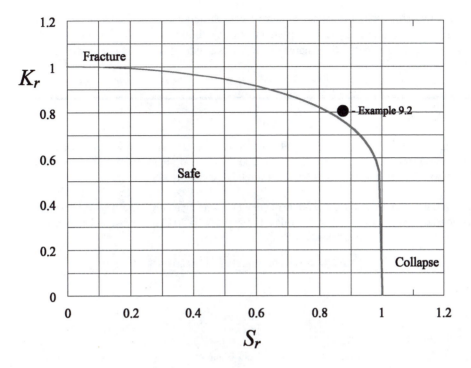

FIGURE 9.17 The strip-yield failure assessment diagram. Taken from Dowling, A.R. and Townley, C.H.A., "The Effects of Defects on Structural Failure: A Two-Criteria Approach." *International Journal of Pressure Vessels and Piping,* Vol. 3, 1975, pp. 77–137; Harrison, R.P., Loosemore, K., and Milne, I., "Assessment of the Integrity of Structures Containing Defects." CEGB Report R/H/R6, Central Electricity Generating Board, UK, 1976.

and

$$S_r = \frac{\sigma}{\sigma_c} \tag{9.56}$$

The failure assessment diagram is then obtained by inserting the above definitions into Equation (9.54) and taking the reciprocal:

$$K_r = S_r \left[\frac{8}{\pi^2} \ln \sec \left(\frac{\pi}{2} S_r \right) \right]^{1/2} \tag{9.57}$$

Equation (9.57) is plotted in Figure 9.17. The curve represents the locus of predicted failure points. Fracture is predicted when $K_{eff} = K_{mat}$, where K_{mat} is the fracture toughness in terms of stress-intensity units.[6] If the toughness is very large, the structure fails by collapse when $S_r = 1.0$. A brittle material will fail when $K_r = 1.0$. In intermediate cases, collapse and fracture interact, and both K_r and S_r are less than 1.0 at failure. All points inside of the FAD are considered safe; points outside of the diagram are unsafe.

In order to assess the significance of a particular flaw in a structure, one must determine the toughness ratio as follows:

$$K_r = \frac{K_I}{K_{mat}} \tag{9.58}$$

[6] K_{mat} is not necessarily a linear elastic toughness such as K_{Ic}. Toughness can be measured in terms of the J integral or CTOD and converted to the equivalent K_{mat} through relationships provided in Section 9.4.2.

The stress ratio for the component of interest can be defined as the ratio of the applied stress to the collapse stress. Alternatively, the applied S_r can be defined in terms of axial forces or moments. If the assessment point with coordinates (S_r, K_r) falls inside of the FAD curve, the analysis predicts that the component is safe.

EXAMPLE 9.2

A middle tension (MT) panel (Figure 7.1(e)) 1 m wide and 25 mm thick with a 200 mm crack must carry a 7.00 MN load. For the material $K_{mat} = 200$ MPa,$\sqrt{\text{m}}$ $\sigma_{YS} = 350$ MPa, and $\sigma_{TS} = 450$ MPa. Use the strip-yield FAD to determine whether or not this panel will fail.

Solution: We can take account of work hardening by assuming a flow stress that is the average of yield and tensile strength. Thus $\sigma_{flow} = 400$ MPa. The collapse load is then defined when the stress on the remaining cross section reaches 400 MPa:

$$P_c = (400 \text{ MPa})(0.025 \text{ m})(1 \text{ m} - 0.200 \text{ m}) = 8.00 \text{ MN}$$

Therefore

$$S_r = \frac{7.00 \text{ MN}}{8.00 \text{ MN}} = 0.875$$

The applied stress intensity can be estimated from Equation (2.46) (without the polynomial term):

$$K_I = \frac{7.00 \text{ MN}}{(0.025 \text{ m})(1.0 \text{ m})} \sqrt{\pi(0.100 \text{ m})\sec\left(\frac{\pi(0.100 \text{ m})}{1.00 \text{ m}}\right)} = 161 \text{ MPa}\sqrt{\text{m}}$$

Thus

$$K_r = \frac{161}{200} = 0.805$$

The point (0.875, 0.805) is plotted in Figure 9.17. Since this point falls outside of the failure assessment diagram, the panel will fail before reaching 7 MN. Note that a collapse analysis or brittle fracture analysis alone would have predicted a "safe" condition. The interaction of fracture and plastic collapse causes failure in this case.

In 1976, the Central Electricity Generating Board (CEGB) in Great Britain incorporated the strip-yield failure assessment into a fracture analysis methodology, which became known as the R6 approach [31]. There have been several revisions of the R6 procedure over the years. The R6 approach is still based on the FAD methodology, but Equation (9.57) is no longer used. Section 9.4.2 and Section 9.4.3 describe the modern formulation of the FAD approach.

9.4.2 *J*-Based FAD

Bloom [32] and Shih et al. [33] showed that a *J*-integral solution from the EPRI handbook could be plotted in terms of a FAD. Of course, a more rigorous *J* solution based on elastic-plastic finite element analysis can also be plotted as a FAD.

The failure assessment diagram is nothing more than an alternative method for plotting the fracture driving force. The shape of the FAD curve is a function of plasticity effects. As described above, the first FAD was derived from the strip-yield plastic zone correction, which assumes a nonhardening material. A *J* solution merely provides a more accurate description of the FAD curve.

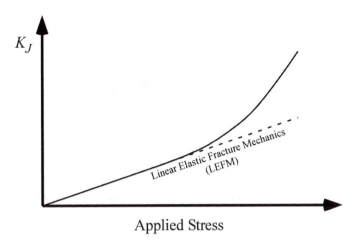

FIGURE 9.18 Schematic plot of crack driving force vs. applied stress.

Figure 9.18 is a schematic plot of the fracture driving force vs. the applied stress. The applied J can be converted to an equivalent K through the following relationship:

$$K_J = \sqrt{\frac{JE}{1 - v^2}} \qquad (9.59)$$

In the linear elastic range, $K_J = K_I$ and stresses near the crack tip are characterized by a $1/\sqrt{r}$ singularity (Chapter 2). In the elastic-plastic range, the plot of K_J vs. stress deviates from linearity and a $1/\sqrt{r}$ stress singularity no longer exists. Note that K_J is a special case of K_{eff}, the stress-intensity factor adjusted for plasticity effects (Section 2.8). However, K_J has a sounder theoretical basis than K_{eff} values estimated from the Irwin plastic zone correction or the strip-yield model.

Following the approach of the original derivation of the FAD from the strip-yield model, let us define a K ratio as follows:

$$K_r = \frac{K_I}{K_J} \qquad (9.60)$$

Figure 9.19 is a schematic plot of K_r vs. applied stress. In the limit of small applied stresses, linear elastic conditions prevail and $K_r = 1$. This ratio decreases as the applied stress increases.

Next, we normalize the horizontal axis by introducing a load ratio:

$$L_r = \frac{\sigma_{ref}}{\sigma_{YS}} \qquad (9.61)$$

where σ_{ref} is the reference stress, which was introduced in Section 9.3.2. Traditionally, the reference stress has been based on yield load or limit load solutions for the configuration of interest (Equation (9.42)). However, this approach introduces geometry dependence into the FAD curve. Section 9.4.4 presents an alternative definition of σ_{ref}.

The final step in creating the FAD is introducing a cut-off value $L_{r(max)}$ on the horizontal axis. This cutoff represents a limit load criterion.

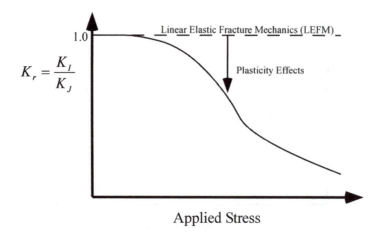

FIGURE 9.19 Driving force curve from Figure 9.18, replotted in terms of K_r vs. applied stress.

The FAD curve is a representation of the driving force. In order to assess the likelihood of failure, we need to incorporate the fracture toughness into the analysis. This is accomplished by plotting an assessment point on the FAD. The y coordinate of this point is defined as follows:

$$K_r = \frac{K_I}{K_{mat}} \tag{9.62}$$

where K_{mat} is the material's fracture toughness in stress-intensity units. The x coordinate of the assessment point is computed from Equation (9.61).

In most low- and medium-strength structural alloys, it is not practical to obtain a K_{Ic} value that is valid according to the ASTM 399 procedure (Section 7.2). Consequently, fracture toughness is usually characterized by either J or CTOD. The conversions to K_{mat} for these two parameters are as follows:

$$K_{mat} = \sqrt{\frac{J_{crit}E}{1-v^2}} \tag{9.63a}$$

or

$$K_{mat} = \sqrt{\frac{\chi\sigma_{YS}\delta_{crit}E}{1-v^2}} \tag{9.63b}$$

where χ is a constraint factor, which typically ranges from 1.5 to 2 for most geometries and materials.

Figure 9.20 illustrates a hypothetical assessment point plotted on the FAD. If the assessment point falls inside the FAD, the structure is considered safe. Failure is predicted when the point falls outside of the FAD. The nature of the failure is a function of where the point falls. When both the toughness and applied stress are low (small L_r and large K_r), the failure occurs in the linear elastic range and usually is brittle. At the other extreme (large L_r and small K_r), the failure mechanism is ductile overload. For cases that fall between these extremes, fracture is preceded by plastic deformation.

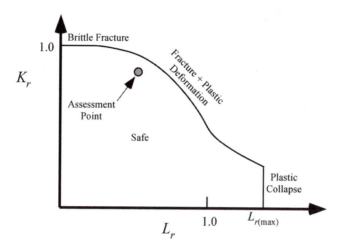

FIGURE 9.20 Failure assessment diagram (FAD), which spans the range of fully brittle to fully ductile behavior.

Provided $L_r < L_{r(max)}$, the failure criterion in the FAD method can be inferred from a comparison of Equation (9.60) and Equation (9.62):

$$\frac{K_I}{K_{mat}} \geq \frac{K_I}{K_J} \tag{9.64}$$

which is equivalent to

$$K_J \geq K_{mat} \tag{9.65}$$

Therefore, there is no substantive difference between the FAD method and a conventional J analysis. The only difference, which is purely cosmetic, is the way in which the driving force and material resistance are presented graphically.

In the original formulation of the J-based FAD, the y axis was actually defined as the square root of the ratio of the elastic J to the total J. However, by applying Equation (9.59) to the numerator and denominator of this ratio, it can easily be shown that such a formulation is identical to Equation (9.60):

$$\sqrt{J_r} = \sqrt{\frac{J_{el}}{J_{tot}}} = \sqrt{\frac{K_I^2(1-v^2)}{E} \times \frac{E}{K_J^2(1-v^2)}} = \frac{K_I}{K_J} = K_r \tag{9.66}$$

The y axis of the FAD can also be expressed in terms of a CTOD ratio $\sqrt{\delta_r}$. This formulation is also identical to Equation (9.62), provided the same constraint factor χ in the CTOD–K conversion (Equation (9.63b)) is applied to both the numerator and denominator.

9.4.3 APPROXIMATIONS OF THE FAD CURVE

The most rigorous method to determine a FAD curve for a particular application is to perform an elastic-plastic J integral analysis and define K_r by Equation (9.60). Such an analysis can be complicated and time consuming, however. Simplified approximations of the FAD curve are available. Two such approaches are outlined below.

In general, the shape of the FAD curve depends on material properties and geometry. However, the geometry dependence can be virtually eliminated through a proper definition of L_r, as described in Section 9.4.4. The following expression, which is based on the reference stress approach described in Section 9.3.3, accounts for the material dependence in the FAD curve but assumes that it is geometry independent [34]:

$$K_r = \left(\frac{E\varepsilon_{ref}}{L_r\sigma_{YS}} + \frac{L_r^3\sigma_{YS}}{2E\varepsilon_{ref}} \right)^{-1/2} \quad \text{for } L_r \leq L_{r(max)} \tag{9.67}$$

where L_r is given by Equation (9.60) and the reference strain ε_{ref} is inferred from the true stress – true strain curve at various σ_{ref} values. The FAD curve obtained from Equation (9.67) reflects the shape of the stress-strain curve. Therefore, Equation (9.67) predicts a FAD that is unique for each material.

When stress-strain data are not available for the material of interest, one of the following generic FAD expressions may be used [34,35]:

$$K_r = [1 - 0.14(L_r)^2]\{0.3 + 0.7\exp[-0.65(L_r)^6]\} \quad \text{for } L_r \leq L_{r(max)} \tag{9.68a}$$

$$K_r = [1 + 0.5(L_r)^2]^{-1/2}\{0.3 + 0.7\exp[-0.6(L_r)^6]\} \quad \text{for } L_r \leq L_{r(max)} \tag{9.68b}$$

Note that these expressions assume that the FAD is independent of both geometry and material properties. Equation (9.68a) was obtained from an empirical fit of FAD curves generated with Equation (9.67) with a range of stress-strain curves [34]. This fit was biased toward the lower bound of the family of FAD curves obtained with Equation (9.67). Equation (9.68b), which gives a FAD curve that is within 3% of Equation (9.68a), is a recent modification that is intended to provide a better fit of Equation (9.67) at intermediate L_r values [35].

Figure 9.21 is a plot that compares Equation (9.67) and Equation (9.68). In the case of the material-specific FAD (Equation (9.67)), the Ramberg-Osgood stress-strain curve was assumed with three different hardening exponents. As strain-hardening increases (i.e., as n decreases), there is a more gradual "tail" in the FAD curve. The FAD curves produced by Equation (9.68a) and Equation (9.68b) are very close to one another, although the new expression does indeed provide better agreement with Equation (9.67) than the original expression. Note that the material dependence in the FAD curve manifests itself primarily in the fully plastic regime ($L_r > 1$). For $L_r < 1$, the difference between the various FAD curves plotted in Figure 9.21 is minimal. Equation (9.68a) or Equation (9.68b) should be adequate for most practical applications because design stresses are usually below yield. When performing a fracture analysis in the fully plastic regime, the material-specific FAD (Equation (9.67)) or an elastic-plastic J analysis should be used.

9.4.4 ESTIMATING THE REFERENCE STRESS

Currently, most FAD approaches normalize the x axis by the limit load or yield load solution. Unfortunately, this practice can lead to apparent geometry dependence in the FAD curve. Figure 9.22 shows an example of this phenomenon. The EPRI J handbook procedure, described in Section 9.3.1, was used to generate FAD curves for various normalized crack lengths in a middle

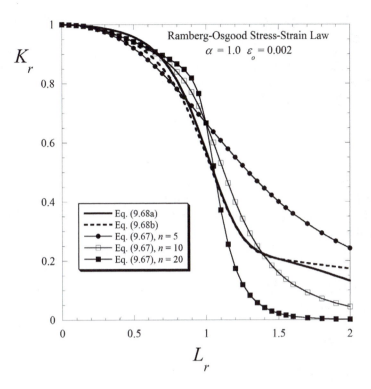

FIGURE 9.21 Comparison of simplified FAD expressions (Equation (9.67)) and Equation (9.68)).

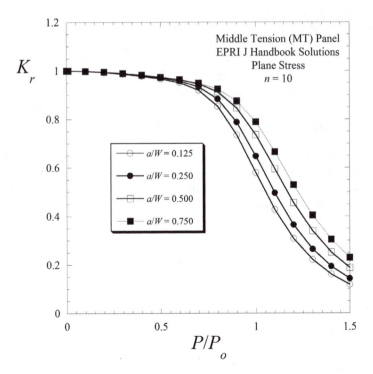

FIGURE 9.22 EPRI handbook *J*-integral solutions for middle tension panels, plotted in terms of FAD curves. The *x* axis is normalized by the yield load P_o.

tension (MT) specimen (Figure 7.1d). When the applied load is normalized by the yield load P_o on the x axis, the resulting FAD curves depend on the relative crack length.

Recently, an alternative approach for normalizing the x axis of the FAD was proposed [36]. Consider the material-specific FAD expression (Equation (9.67)). Setting $L_r = 1$ in this equation and solving for the ratio of the total J to the elastic component leads to

$$\left.\frac{J}{J_{elastic}}\right|_{L_r=1} = 1 + \frac{0.002E}{\sigma_{YS}} + \frac{1}{2}\left(1 + \frac{0.002E}{\sigma_{YS}}\right)^{-1} \tag{9.69}$$

where σ_{YS} is the 0.2% offset yield strength. An elastic-plastic J solution, obtained from finite element analysis or the EPRI J handbook, can be normalized such that the above expression is satisfied at $L_r = 1$. The reference stress, which is used to compute L_r (Equation (9.61)), is proportional to the nominal applied stress:

$$\sigma_{ref} = \sigma_{nominal} F \tag{9.70}$$

where F is a geometry factor. The value of $\sigma_{nominal}$ at which Equation (9.69) is satisfied can be inferred from the elastic-plastic J solution. The geometry factor is given by

$$F = \frac{\sigma_{YS}}{\sigma_{nominal}\big|_{L_r=1}} \tag{9.71}$$

Figure 9.23 shows the FAD curves from Figure 9.22 normalized by the reference stress, as defined by the above procedure. The geometry dependence disappears when the x axis is defined by this procedure. This method forces all curves to pass through the same point at $L_r = 1$. Since the curves have a nearly identical shape, they are in close agreement at other L_r values.

Figure 9.24 compares J-based FAD curves for two geometries with Equation (9.67), the material-specific and geometry-independent FAD expression. The three curves are in precise agreement at $L_r = 1$ because Equation (9.69) was derived from Equation (9.67). At other L_r values, there is good agreement. Therefore, the shape of the FAD curve is relatively insensitive to geometry, and the material-specific FAD expression agrees reasonably well with a rigorous J solution, provided the reference stress is defined by Equation (9.69) to Equation (9.71).

Figure 9.25 is a plot of the reference stress geometry factor F for a surface crack as a function of crack front position and strain hardening. The reference stress solution is relatively insensitive to the location on the crack front angle ϕ (Figure 9.2), but F is a strong function of the hardening exponent. Fortunately, the hardening dependence of F follows a predictable trend. The following empirical expression relates F values at two hardening exponents:

$$F(n_2) = F(n_1)\frac{\left(1 + 2n_2^{1.15}\right)}{\left(1 + 2n_1^{1.15}\right)} \tag{9.72}$$

The above expression was developed specifically for this chapter, and has not been published elsewhere as of this writing. Figure 9.26 is a plot of F vs. n for a surface crack in a plate subject to membrane loading. The predicted curve was computed from Equation (9.72) using a reference-hardening exponent n_1 of 10. Note that F is insensitive to the ratio σ_o/E for $n > 3$. Equation (9.72) is less accurate for very high hardening materials, but most engineering alloys have hardening exponents in the range $5 \le n \le 15$.

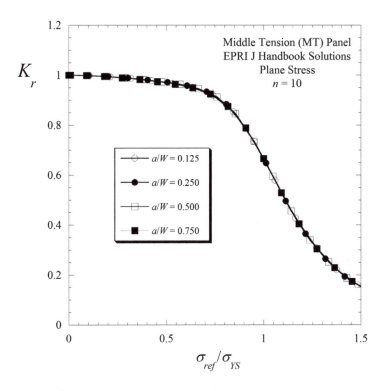

FIGURE 9.23 FAD curves from Figure 9.22, but with reference stress defined according to the procedure in Equation (9.69) to Equation (9.71).

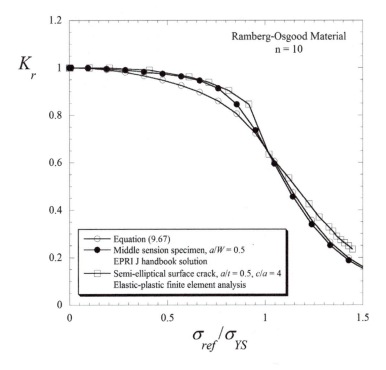

FIGURE 9.24 Comparison of the simplified material-dependent FAD (Equation (9.67)) with *J*-based FAD curves for two geometries.

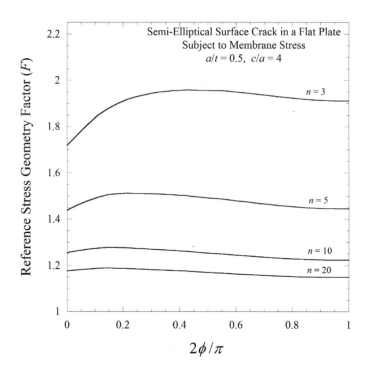

FIGURE 9.25 Reference stress geometry factor as a function of crack front position and hardening exponent for a semielliptical surface crack in a flat plate (see Figure 9.2 for a definition of dimensions and the angle ϕ).

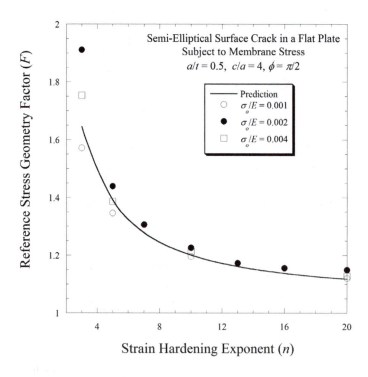

FIGURE 9.26 Correlation between the reference stress geometry factor and the hardening exponent for a semielliptical surface crack in a plate subject to a membrane stress.

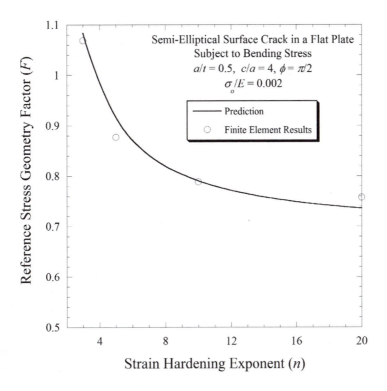

FIGURE 9.27 Correlation between the reference stress geometry factor and the hardening exponent for a semielliptical surface crack in a plate subject to a bending stress.

Figure 9.27 shows that Equation (9.72) also works for a plate in bending. Therefore, given a reference stress solution for a specific geometry and hardening exponent, Equation (9.72) can be used to estimate F for that same geometry with a different hardening exponent.

The above approach for defining the reference stress also applies to cases where the loading is more complex. For example, if the component is subject to both membrane and bending stress, Equation (9.70) can be written in terms of the membrane stress, and F would depend on the σ_b/σ_m ratio. Alternatively, separate estimates of reference stress can be made for the membrane and bending components and the effects can be combined to give the total reference stress. Unfortunately, linear superposition does not apply to reference stress because it is derived from elastic-plastic analysis. A method for inferring a reference stress for combined loading is outlined next.

Suppose that reference stress geometry factors for a given configuration are available for both pure membrane loading and pure bending:

$$\sigma_{ref}^{m} = \sigma_{m}F_{m} \tag{9.73a}$$

$$\sigma_{ref}^{b} = \sigma_{b}F_{b} \tag{9.73b}$$

If linear superposition were applicable, the reference stress for combined membrane and bending loads could be inferred simply by summing the contributions from each. In reality, such an approach will usually lead to an overestimate of the true reference stress. Figure 9.28 illustrates a more accurate approach, in which the two contributions to reference stress are treated as vectors oriented

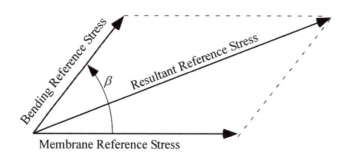

FIGURE 9.28 Vector summation method for combining reference stress solutions for two load cases. The magnitude of the resultant reference stress is related to the two components through a phase angle β.

at an angle β from one another. The total reference stress for combined loading is equal to the sum of these vectors. That is

$$\sigma_{ref}^{total} = \sqrt{\left(\sigma_{ref}^{m}\right)^2 + \left(\sigma_{ref}^{b}\right)^2 + 2\sigma_{ref}^{m}\sigma_{ref}^{b}\cos(\beta)} \qquad (9.74)$$

In order to determine the phase angle β, it is necessary to perform at least one elastic-plastic analysis for combined loading at a fixed σ_b/σ_m ratio. Once the phase angle is established for the geometry of interest, Equation (9.74) can be used to compute reference stress for other σ_b/σ_m ratios.

The above methodology can be validated by performing a series of elastic-plastic finite element analyses with a range of σ_b/σ_m ratios and inferring the angle β from Equation (9.74). Figure 9.29

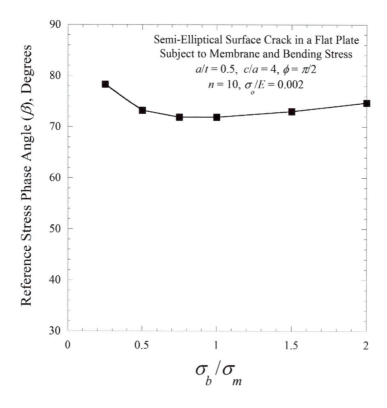

FIGURE 9.29 Phase angle inferred from a series of elastic-plastic finite element analyses on plates with surface cracks with a range of bending/membrane stress ratios.

shows the results of such an exercise. The inferred phase angle is insensitive to the σ_b/σ_m ratio, indicating that the vector summation method is appropriate for superimposing reference stress solutions for two load cases.

When combined loading is present, such as membrane and bending stresses, the loading must be proportional in order to apply the methodology outlined above. That is, the σ_b/σ_m ratio must be constant as load is applied. In the plastic range, the sequence of loading is important. If, for example, the membrane stress is applied first, followed by the bending, a different J might result than if the two stress components were applied at a fixed σ_b/σ_m ratio. The FAD method and the reference stress concept assume proportional loading, and may not be suitable for nonproportional elastic-plastic loading. The only accurate way to model nonproportional loading in the plastic range is through finite element analysis.

Welded components that include residual stresses represent a special case of nonproportional loading where the FAD method is routinely applied. Residual stresses occur at the time of welding and applied stresses are imposed in service, so the loading sequence is obviously not proportional. Procedures for applying the FAD method to welded components are described next.

9.4.5 APPLICATION TO WELDED STRUCTURES

Welds introduce a number of complexities into a fracture analysis. The welding process invariably creates residual stresses in and around the weld. Geometric anomalies such as weld misalignment create additional local stresses. The weld metal and heat-affected zone (HAZ) typically have different material properties than the base metal. The toughness properties of the weld must, of course, be taken into account in the material resistance. In addition, the different stress-strain responses of the weld metal and base metal can have a significant effect on the crack driving force. The FAD method can be modified to consider each of these complexities, as described next.

9.4.5.1 Incorporating Weld Residual Stresses

Conventional welding processes entail fusing two parts together with molten metal. Welding is similar to soldering and brazing in that respect. Unlike soldering and brazing, however, the weld metal has a chemical composition that is close to that of the base metal. When weld metal cools to the ambient temperature, the resulting thermal contraction is restrained by the surrounding base metal. This restraint leads to residual stresses. In multi-pass welds the residual stress pattern reflects the complex thermal history of the weld region. In recent years, finite element simulation of welding has advanced to the point where realistic estimates of residual stress are possible [37].

Section 9.1.4 introduced the concept of primary, secondary, and residual stresses. Weld residual stress is usually not considered in most design codes because it does not have a significant effect on the tensile strength of the welded joint, provided the material is ductile. When a crack is present, however, residual stresses must be included in the crack driving force. Under linear elastic conditions, residual stresses are treated the same as any other stress, as Equation (9.21) indicates. When there is local or global plastic deformation, residual stresses may relax or redistribute. The FAD method can be modified to account for residual stresses, as outlined next.

Figure 9.30 is a schematic plot of crack driving force, in terms of K_J, vs. the applied primary stress. The two curves compare the driving force for primary stresses only with a case where both primary and tensile residual stresses are present. In the latter case, $K_J > 0$ when the applied primary stress is zero because the residual stresses contribute to the crack driving force. At intermediate applied stresses, the K_J vs. stress curve is nonlinear because the combination of primary and residual stresses result in crack-tip plasticity. At higher applied stresses, global plasticity results in relaxation of residual stresses. This phenomenon is known as *mechanical stress relief*. The two driving force curves in Figure 9.30 coincide in the fully plastic range because residual stresses have relaxed completely.

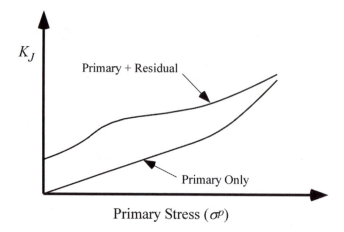

FIGURE 9.30 Schematic plot of crack driving force vs. applied primary stress, with and without an imposed residual stress.

The crack driving force can be plotted in FAD coordinates, as described in Section 9.4.2. The FAD curves for the two cases described above are illustrated in Figure 9.31. For primary stresses alone, the FAD curve is defined as before:

$$K_r = \frac{K_I^P}{K_J} = f_1(L_r) \tag{9.75}$$

When residual stresses are present, the shape of the FAD curve is a function of the magnitude of the residual stresses:

$$K_r^* = \frac{K_I^P + K_I^R}{K_J} = f_2\left(L_r, K_I^R\right) \tag{9.76}$$

Note that only primary stresses are included in L_r. The unusual shape of the FAD curve for the weld with residual stress is due to crack-tip plasticity at intermediate L_r values and mechanical

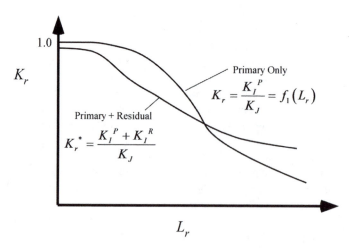

FIGURE 9.31 Driving force curves from Figure 9.30, plotted in FAD coordinates.

stress relief at high L_r values. The FAD curve where residual stresses are present crosses the FAD curve for primary stresses only, because K_r^* is computed based on the *original elastic* values of K_I^P and K_I^R. When residual stresses have relaxed, $K_r^* > K_r$ for a given L_r.

It is not particularly convenient to apply a FAD curve whose shape is a function of residual stress. An alternative formulation, where the residual stress effects are decoupled from the FAD curve, is preferable. Let us define the FAD curve as $f_1(L_r)$, which corresponds to the FAD in the absence of residual stress (Equation (9.75)). When residual stresses are present, the K ratio can be redefined such that the resulting FAD curve exhibits the desired shape. This can be accomplished by multiplying K_I^R by an adjustment factor:

$$K_r = f_1(L_r) = \frac{K_I^P + \Phi K_I^R}{K_J} \tag{9.77}$$

Solving for Φ gives

$$\Phi = \frac{f_1(L_r)K_J - K_I^P}{K_I^R} \tag{9.78}$$

The Φ factor can be viewed as a plasticity adjustment. Figure 9.32 schematically illustrates the relationship between Φ and applied stress. As L_r increases, the crack-tip plasticity magnifies the total driving force, so $\Phi > 1$. Eventually Φ reaches a peak and then decreases due to mechanical stress relief. Figure 9.33 compares FAD curves defined by Equation (9.76) and Equation (9.77). The two curves cross when $\Phi = 1$.

The Φ factor can be derived from elastic-plastic finite element analysis. Various initial residual stress distributions are imposed on a finite element model that contains a crack, and then primary loads are applied. The J-integral results are plotted in terms of FAD curves (Figure 9.31) and Φ is inferred from Equation (9.78). Ainsworth et al. [38] have developed expressions for Φ based on earlier work by Hooton and Budden [39].

The y coordinate of the assessment point on the FAD is given by

$$K_r = \frac{K_I^P + \Phi K_I^R}{K_{mat}} \tag{9.79}$$

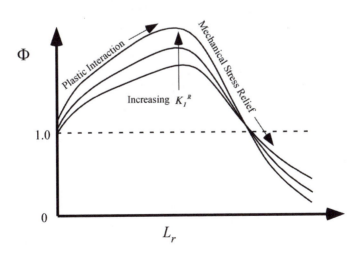

FIGURE 9.32 Schematic plot of the plasticity adjustment factor on residual stress, Φ, vs. applied primary stress.

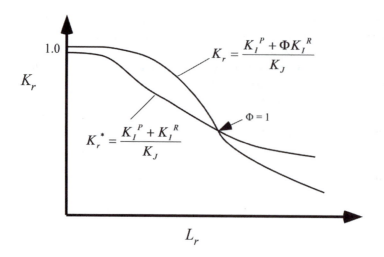

FIGURE 9.33 Effect of the plasticity adjustment factor Φ on the FAD curve for a weldment with residual stress.

and L_r is computed from Equation (9.61). Comparing Equation (9.77) and Equation (9.79), we see that the failure criterion is $K_J \geq K_{mat}$, as is the case for the J-based FAD for primary stresses only (Section 9.4.2).

9.4.5.2 Weld Misalignment

When plates or shells are welded, there is invariably some degree of misalignment. Figure 9.34 illustrates two common types of misalignment: centerline offset and angular misalignment. In both instances, the misalignment creates a local bending stress. This local stress usually does not make a significant contribution to static overload failure, provided the material is ductile. Misalignment stresses can, however, increase the risk of brittle fracture and shorten the fatigue life of a welded joint.

Equations for misalignment stresses have been developed for various weld geometries and types of misalignment [12, 40, 41]. For example, consider a weld between two plates of equal thickness with a centerline offset e (Figure 9.34(a)). Applying standard beam equations leads to the following relationship between the local bending stress and the remotely applied membrane stress:

$$\sigma_b^{local} = \sigma_m^{remote} \left(\frac{6e}{t} \right) \tag{9.80}$$

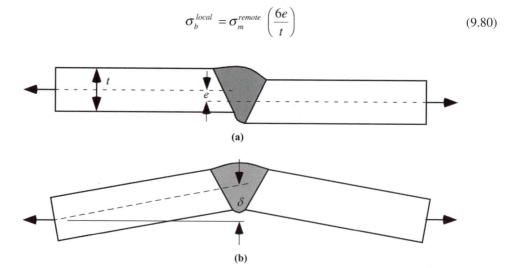

FIGURE 9.34 Examples of weld misalignment: (a) centerline offset and (b) angular misalignment.

When applying the FAD method, it is customary to treat misalignment stresses in the same way as weld residual stresses. That is, they are not included in the calculation of L_r, and the applied stress-intensity factor due to misalignment stresses is multiplied by Φ.

In some respects, misalignment stresses are similar to residual stresses in the way they influence structural behavior. They are normally not considered in design but should be included in crack driving force calculations. As is the case with residual stress, misalignment stresses may relax with plastic deformation. With angular misalignment in welded plates, for example, applied stresses at or above the yield strength tend to straighten the weldment, thereby reducing the bending moment.

There is, however, a key difference between misalignment stress and weld residual stress. The former is directly proportional to the applied stress, but residual stresses can exist in the absence of applied loads. Therefore, misalignment stresses do not behave in exactly the same manner as residual stresses, so the plasticity correction factor Φ in Equation (9.79) may not be rigorously correct when applied to misalignment stresses.

9.4.5.3 Weld Strength Mismatch

In most welded structures, the base metal and weld metal have different tensile properties. The weld metal is typically stronger than the base metal, but there are instances where the weld metal has lower strength. A weldment is said to be *overmatched* when the weld metal has higher strength than the base metal. The reverse situation is known as an *undermatched* weldment.

The mismatch in strength properties affects the crack driving force in the elastic-plastic and fully plastic regimes. Mismatch in properties is normally not a significant issue in the elastic range because the weld metal and base metal typically have similar elastic constants. Figure 9.35 is a schematic plot of the crack driving force for a crack in a base metal as well as for a crack of the same size in an overmatched weld. For the purpose of this illustration, the effect of residual stress on crack driving force is ignored, and the base metal and weld metal are assumed to have the same

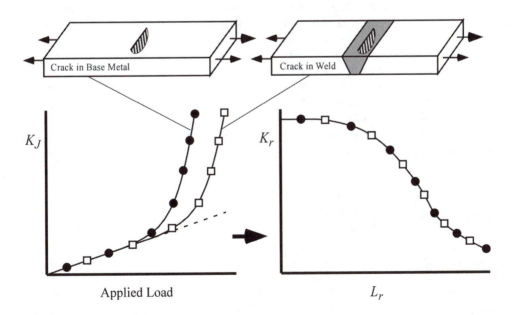

FIGURE 9.35 Effect of weld strength mismatch on crack driving force. Strength mismatch effects can be taken into account in the FAD method through the reference stress solution for the weldment. In this schematic, weld residual stress is neglected, and the weld and base metal are assumed to have similar hardening characteristics.

strain-hardening characteristics. Because the weld metal has higher yield strength than the base metal, the upswing in the driving force curve occurs at a higher load in the weldment. At a fixed load in the elastic-plastic regime, the driving force in the cracked weldment is significantly lower than in the cracked base plate.

The effect of weld strength mismatch can be taken into account in the FAD method through an appropriate definition of L_r [42], as Figure 9.35 illustrates. The reference stress for a weldment should be defined from the elastic-plastic J solution using the approach in Section 9.4.4 or another self-consistent approach.

9.4.6 PRIMARY VS. SECONDARY STRESSES IN THE FAD METHOD

The term *secondary stress* was introduced in Section 9.1.4. Stresses that are typically classified as secondary may include displacement-controlled loads such as thermal expansion as well as weld misalignment stresses. In the FAD method, it is customary to treat secondary stresses the same as residual stresses. Equation (9.79) can be modified to include secondary stresses:

$$K_r = \frac{K_I^P + \Phi K_I^{SR}}{K_{mat}} \qquad (9.81)$$

where K_I^{SR} is the stress-intensity factor due to the combined effects of secondary and residual stresses. Only primary stresses are used in the calculation of L_r.

The classification of stresses as primary and secondary is somewhat arbitrary, and it is not always clear what the "correct" classification should be. In some cases, loads that are categorized as secondary by design codes are indistinguishable from primary loads as far as a crack is concerned. Thermal expansion stresses in piping, for example, normally should be treated as primary in a FAD analysis despite being categorized as secondary in the governing design code. Recall Section 9.3.3, which considered the stability of a cracked structure subject to a fixed remote displacement. This configuration was represented by a spring in series with a compliance of C_M. If C_M is large, there is a significant amount of stored elastic energy in the system, and the behavior approaches that of load control. In a large piping system, the system compliance and stored elastic energy are typically very large, so thermal expansion behaves as a load-controlled stress for all practical purposes. Load-controlled stresses should always be treated as primary.

One of the traditional criteria for determining whether a stress should be treated as primary or secondary in a FAD analysis is the degree to which the load might contribute to ductile overload (plastic collapse). The aforementioned piping example is a case in point. An improperly designed piping system could experience buckling or ductile rupture due to thermal loads. Weld residual stress and misalignment stress, however, do not contribute significantly to plastic collapse, so they have traditionally not been treated as primary in a FAD analysis.

The above criterion for stress classification in the FAD method originally stems from the view that K_r is the "fracture axis" and L_r is the "collapse axis." All stresses potentially contribute to brittle fracture, but only primary loads need be included in a plastic collapse or limit load analysis. The traditional definition of L_r in terms of the limit load solution for the cracked body has contributed to the perception that the x and y coordinates of the FAD entail independent assessments of collapse and fracture, respectively. This two-criterion viewpoint may have been appropriate for the original FAD based on the strip-yield model (Section 9.4.1), but is less valid for a J-based FAD.

As Figure 9.19 illustrates, the shape of the FAD curve is a measure of the deviation from linear elasticity. Therefore, the downward trend of K_r with increasing L_r represents a plasticity correction on the crack driving force. When reference stress is defined according to the procedure in Section 9.4.4 rather than from a limit load solution, it is not appropriate to view L_r as a collapse parameter. Rather, L_r is an indication of the degree of crack-tip plasticity.

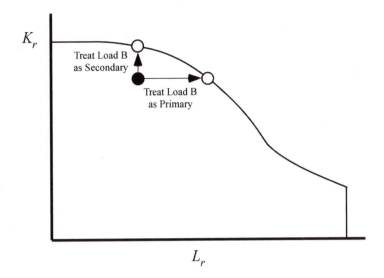

FIGURE 9.36 Schematic illustration of the effect of stress classification on the location of the assessment point. Load *A* is treated as primary but the correct classification of load *B* is unclear. In this hypothetical example, the net outcome of the analysis is the same irrespective of the classification of load *B*.

The Φ factor that is applied to secondary and residual stress in Equation (9.79) is also a plasticity correction. Unlike the FAD curve, however, there is not a simple monotonic relationship between Φ and L_r, as Figure 9.32 illustrates. At low to intermediate values of L_r, where $\Phi > 1$, it may make little difference whether a given stress is included in K_I^P or K_I^{SR}, as the example given below illustrates.

Consider a case where a cracked structural component is subject to two types of load. Assume that load *A* is definitely primary and is treated as such, but that it is unclear whether to treat load *B* as primary or secondary. The FAD in Figure 9.36 illustrates the effect of the stress classification on the assessment point. The solid point represents a baseline case, where both loads *A* and *B* are included in K_r but no plasticity correction has applied to load *B* yet. If load *B* is treated as secondary, the corresponding stress intensity must be multiplied by Φ, which results in an upward shift in K_r. On the other hand, if load *B* is treated as primary, it must be included in L_r, resulting in a shift of the assessment point to the right. In this hypothetical example, the end result of both assumptions was the same. Namely, the assessment point falls on the FAD in either case. In general, the choice of stress classification will make some difference on the outcome of the assessment, but the effect may be small in many instances.

One key difference between the Φ factor and the plasticity correction inherent in the FAD curve is that the magnitude of the former *decreases* at large L_r values due to mechanical stress relief. Perhaps then the key question should be: *Does the stress in question relax with plastic deformation in a matter similar to residual stress?* If the answer is *no*, then the stress should be treated as primary for the purpose of a FAD analysis. By this criterion, many instances of thermal stress would be considered primary. In some instances, it may be appropriate to include weld misalignment stress in K_I^{SR}, but it is not clear that such a classification is universally correct for all cases of weld misalignment. Even if a stress does relax with plastic flow, the rate at which this relaxation occurs is a consideration. If large plastic strains are required to achieve significant stress relaxation, then it might be appropriate to classify the stress as primary.

Probably the best advice is: *When in doubt, treat stresses as primary in a FAD analysis.* This is generally the most conservative approach.

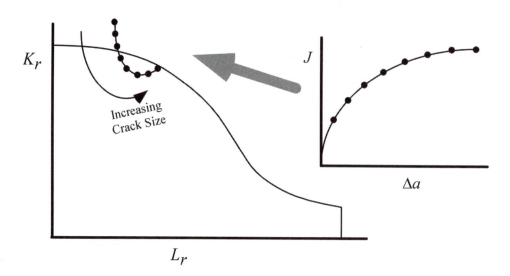

FIGURE 9.37 Ductile-tearing analysis with the FAD. A locus of assessment points is generated from the resistance curve for the material of interest. Taken from Chell, G.G. and Milne, I., "Ductile Tearing Instability Analysis: A Comparison of Available Techniques." ASTM STP 803, American Society for Testing and Materials and Testing, Philadelphia, PA, 1983, pp. II-179–II-205.

9.4.7 Ductile-Tearing Analysis with the FAD

A ductile-tearing analysis can be performed with the FAD method. The approach is essentially identical to that described in Section 9.3.3, except that the driving force and material resistance are plotted in terms of K_r and L_r. Recall from Section 9.4.2 that the FAD curve is a dimensionless representation of the driving force and that the assessment point represents the fracture toughness. When toughness is given in terms of a resistance curve, such data are plotted as a locus of assessment points on the FAD.

Figure 9.37 illustrates the construction of the assessment points from the resistance curve [43]. Note that both the crack size and fracture toughness are updated with crack growth. The assessment points form a curve that moves downward and to the right initially. The curve may reach a minimum and then decrease, depending on the amount of crack growth.

Figure 9.38 illustrates three possible outcomes of a ductile-tearing analysis. If all assessment points lie inside the FAD, no crack growth occurs. The case where the first few assessment points fall outside of the FAD, but others lie inside the FAD, corresponds to a finite amount of stable crack growth. If all assessment points fall outside of the FAD, the structure is unstable. The onset of unstable crack growth occurs where the locus of assessment points is tangent to the FAD [43].

One limitation of the FAD method is that it implicitly assumes load control for all analyses. Consequently, a ductile instability analysis would be conservative if the loading conditions were actually closer to displacement control.

9.4.8 Standardized FAD-Based Procedures

In 1976, the original FAD derived from the strip-yield model (Section 9.4.1) was incorporated into a flaw assessment procedure for the British electric power industry [31]. This procedure became known as the R6 method, which has subsequently undergone several revisions. The latest version of R6, as of this writing, was published in 2001 and is the 4th revision [40]. The most recent version of the R6 method no longer uses the strip-yield FAD. Rather, it reflects various improvements in the FAD approach that have taken place over the past few decades. The R6 method offers a choice between simple, closed-form FAD expressions (Section 9.4.3) and more rigorous FAD descriptions based on elastic-plastic J-integral results. It addresses both fracture instability with a single toughness value (Figure 9.20) and ductile-tearing analysis with a resistance curve (Figure 9.38).

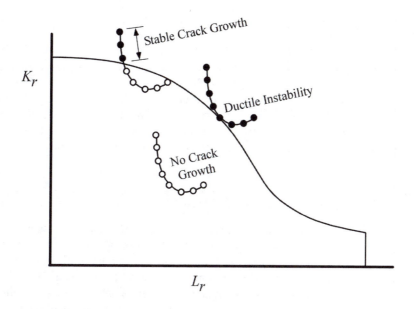

FIGURE 9.38 Examples of possible outcomes of a ductile-tearing assessment. Instability occurs when the locus of assessment points is tangent to the FAD curve. Taken from Chell, G.G. and Milne, I., "Ductile Tearing Instability Analysis: A Comparison of Available Techniques." ASTM STP 803, American Society for Testing and Materials and Testing, Philadelphia, PA, 1983, pp. II-179–II-205.

There are a number of other internationally recognized codified flaw assessment procedures that implement the FAD method. For example, the British Standards Institute (BSI) has published BS 7910:1999 [41],[7] and the European Union conducted a cooperative research program that culminated in the publication of the SINTAP document [35], which is an abbreviation for "structural integrity assessment procedures for European industry." The R6, BS 7910, and SINTAP methods are very similar to one another, probably because many of the same individuals were involved in creating all the three documents. A fourth European FAD-based approach was developed in Sweden [44].

In the U.S., the American Petroleum Institute has published API 579 [12], which is a comprehensive fitness-for-service guide that addresses various types of flaws and damage, including cracks, general corrosion, local corrosion, pitting, bulging, and weld misalignment. The API 579 assessment of cracks implements the FAD method, and is similar in many respects to R6, BS 7910, and SINTAP. The API document, however, has a more extensive library of K_I solutions than the other documents.

Reference stress solutions in the current standardized FAD methods are based on limit load and yield load solutions. Section 9.4.4 discusses the disadvantages of such an approach and introduces a more rational definition of σ_{ref}. This alternative procedure for defining σ_{ref} from an elastic-plastic J solution (Equation (9.69) to Equation (9.71)) is described in Appendix B of API 579. The API Fitness-for-Service Task Group has recently commissioned a finite element study in which new reference stress solutions are being generated using the approach outlined in Section 9.4.4. These new solutions will incorporate the effect of weld strength mismatch.

Another area of active research among the various committees charged with developing and maintaining the FAD-based standards is the estimation of residual stress for common welded joints. In early flaw assessment methods for welded structures, the weld residual stress was typically assumed to be uniform through the cross section with a magnitude equal to the yield strength of the material. This assumption was very conservative and has resulted in many unnecessary weld repairs. In recent years, finite element methods have been used to simulate the welding process and compute weld residual stress. Most of the FAD-based standards include compendia of residual

---

[7] Amendment No. 1 to BS 7910:1999 was released in 2000.

stress distributions that were inferred from finite element analysis. These compendia are continually being updated as new residual stress solutions become available.

The latest version of the R6 document [40] includes a procedure to modify the FAD based on a loss of crack-tip triaxiality. This constraint adjustment is based on some of the models and theories presented in Chapter 3 (e.g., the Q parameter and the T stress). However, most of the research into constraint effects on fracture toughness has focused on base metals and relatively simple configurations. There does not yet appear to be a sufficient technical basis to apply these models to cracks in complex welded components. The traditional approach, where similitude between standard laboratory toughness specimens and structural components is assumed, is conservative.

9.5 PROBABILISTIC FRACTURE MECHANICS

Most fracture mechanics analyses are deterministic, i.e., a single value of fracture toughness is used to estimate the failure stress or critical crack size. Much of what happens in the real world, however, is not predictable. Since fracture toughness data in the ductile-brittle transition region are widely scattered, it is not appropriate to view fracture toughness as a single-valued material constant. Other factors also introduce uncertainty into fracture analyses. A structure may contain a number of flaws of various sizes, orientations, and locations. Extraordinary events such as hurricanes, tidal waves, and accidents can result in stresses significantly above the intended design level. Advances in finite element modeling of welding has led to improved estimates of residual stress, but there are uncertainties associated with such estimates. Because of these complexities, fracture should be viewed probabilistically rather than deterministically.

Figure 9.39 is a schematic probabilistic fracture analysis. The curve on the left represents the distribution of driving force in the structure K_J, while the curve on the right is the toughness distribution K_{mat}. The former distribution depends on the uncertainties in stress and flaw size. When the distributions of the applied K_J and K_{mat} overlap, there is a finite probability of failure, indicated by the shaded area. For example, suppose the cumulative distribution of the driving

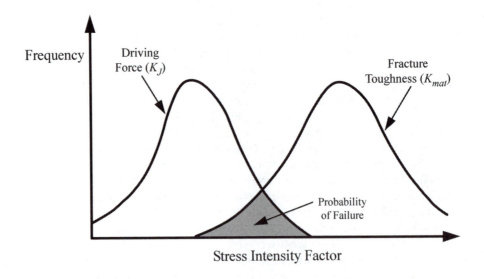

FIGURE 9.39 Schematic probabilistic fracture analysis.

force is $F_1(K_J)$ and the cumulative toughness distribution is $F_2(K_{mat})$. The failure probability P_f is given by

$$P_f = \int [1 - F_1(K_J)] dF_2(K_{mat}) \qquad (9.82)$$

Time-dependent crack growth, such as fatigue and stress corrosion cracking, can be taken into account by applying the appropriate growth law to the flaw distribution. Flaw growth would cause the applied K_J distribution to shift to the right with time, thereby increasing failure probability.

The overlap of two probability distributions (Figure 9.39) represents a fairly simple case. In most practical situations, there is randomness or uncertainty associated with several variables, and a simple numerical integration to solve for P_f (Equation (9.82)) is not possible. Monte Carlo simulation can estimate failure probability when there are multiple random variables. Such an analysis is relatively easy to perform, since it merely involves incorporating a random number generator into a deterministic model. Monte Carlo analysis is very inefficient, however, as numerous "trials" are required for convergence. There are other more efficient numerical algorithms for probabilistic analysis, but they are more complex and more difficult to implement than Monte Carlo analysis. The relative inefficiency of Monte Carlo analysis is no longer a significant hindrance in most instances because of the speed of modern-day computers.

The Monte Carlo and FAD methods are well suited to one another. Figure 9.40 shows the results of a Monte Carlo FAD analysis. Each Monte Carlo trial results in a single assessment point. Uncertainties in the input parameters (e.g., toughness, flaw size, applied stress, residual stress, and tensile properties) lead to uncertainties in K_r and L_r. The FAD in Figure 9.40 presents a compelling visual presentation of this uncertainty. The failure probability is defined as the number of points that fall outside of the FAD curve divided by the total number of trials:

$$P_f = \frac{N_{failures}}{N_{total}} \qquad (9.83)$$

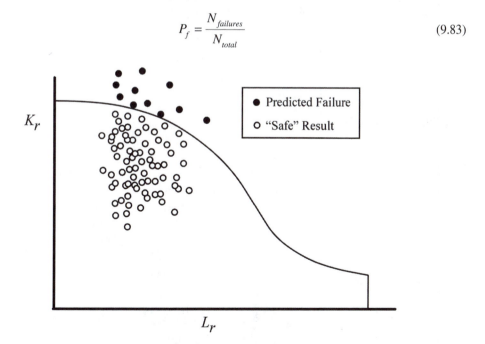

FIGURE 9.40 Results of a Monte Carlo probabilistic analysis with the FAD method. Uncertainty in input parameters is reflected in scatter in the assessment points.

APPENDIX 9:   STRESS INTENSITY AND FULLY PLASTIC J SOLUTIONS FOR SELECTED CONFIGURATIONS

Tables A9.1 to A9.5 list stress-intensity solutions for part-wall cracks in flat plates [10]. Surface, buried, and corner cracks are idealized as semielliptical, elliptical, and quarter-elliptical, respectively. These solutions have the following form:

$$K_I = (\sigma_m + H\sigma_b)\sqrt{\frac{\pi a}{Q}}\,F \tag{A9.1}$$

where F, H, and Q are geometry factors.

TABLE A9.1
Stress-Intensity Solution for a Semielliptical Surface Flaw in a Flat Plate for $a \leq c$ [10].

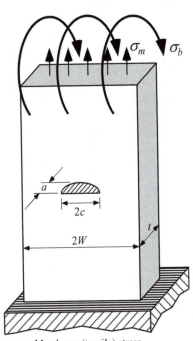

$$K_I = (\sigma_m + H\sigma_b)\sqrt{\frac{\pi a}{Q}}\,F\left(\frac{a}{t}, \frac{a}{c}, \frac{c}{W}, \phi\right)$$

where

$$Q = 1 + 1.464\left(\frac{a}{c}\right)^{1.65}$$

$$F = \left[M_1 + M_2\left(\frac{a}{t}\right)^2 + M_3\left(\frac{a}{t}\right)^4\right]f_\phi f_w g$$

$$M_1 = 1.13 - 0.09\left(\frac{a}{c}\right)$$

$$M_2 = -0.54 + \frac{0.89}{0.2 + \frac{a}{c}}$$

$$M_3 = 0.5 - \frac{1.0}{0.65 + \frac{a}{c}} + 14\left(1.0 - \frac{a}{c}\right)^{24}$$

$$f_\phi = \left[\left(\frac{a}{c}\right)^2 \cos^2\phi + \sin^2\phi\right]^{1/4}$$

$$f_w = \left[\sec\left(\frac{\pi c}{2W}\sqrt{\frac{a}{t}}\right)\right]^{1/2}$$

$$g = 1 + \left[0.1 + 0.35\left(\frac{a}{t}\right)^2\right](1 - \sin\phi)^2$$

$$H = H_1 + (H_2 - H_1)(\sin\phi)^p$$

$$p = 0.2 + \frac{a}{c} + 0.6\left(\frac{a}{t}\right)$$

$$H_1 = 1 - 0.34\frac{a}{t} - 0.11\frac{a}{c}\left(\frac{a}{t}\right)$$

$$H_2 = 1 + G_1\left(\frac{a}{t}\right) + G_2\left(\frac{a}{t}\right)^2$$

$$G_1 = -1.22 - 0.12\left(\frac{a}{c}\right)$$

$$G_2 = 0.55 - 1.05\left(\frac{a}{c}\right)^{0.75} + 0.47\left(\frac{a}{c}\right)^{1.5}$$

σ_m - Membrane (tensile) stress
σ_b - Bending stress

$$\sigma_b = \frac{Mt}{2I}$$

where

$$I = \frac{Wt^3}{6}$$

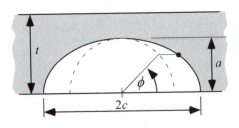

TABLE A9.2
Stress-Intensity Solution for a Semielliptical Surface Flaw in a Flat Plate for $a/c > 1$ [10].

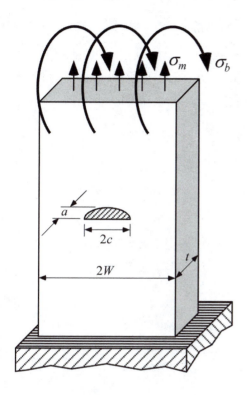

σ_m - Membrane (tensile) stress
σ_b - Bending stress

$$\sigma_b = \frac{Mt}{2I}$$

where

$$I = \frac{Wt^3}{6}$$

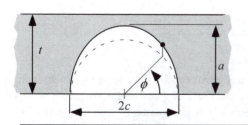

$$K_I = (\sigma_m + H\sigma_b)\sqrt{\frac{\pi a}{Q}} F\left(\frac{a}{t}, \frac{a}{c}, \frac{c}{W}, \phi\right)$$

where

$$Q = 1 + 1.464\left(\frac{c}{a}\right)^{1.65}$$

$$F = \left[M_1 + M_2\left(\frac{a}{t}\right)^2 + M_3\left(\frac{a}{t}\right)^4\right] f_\phi f_w g$$

$$M_1 = \sqrt{\frac{c}{a}}\left[1 + 0.04\left(\frac{c}{a}\right)\right]$$

$$M_2 = 0.2\left(\frac{c}{a}\right)^4$$

$$M_3 = -0.11\left(\frac{c}{a}\right)^4$$

$$g = 1 + \left[0.1 + 0.35\left(\frac{c}{a}\right)\left(\frac{a}{t}\right)^2\right](1 - \sin\phi)^2$$

$$f_\phi = \left[\left(\frac{c}{a}\right)^2 \sin^2\phi + \cos^2\phi\right]^{1/4}$$

$$f_w = \left[\sec\left(\frac{\pi c}{2W}\sqrt{\frac{a}{t}}\right)\right]^{1/2}$$

$$p = 0.2 + \frac{c}{a} + 0.6\left(\frac{a}{t}\right)$$

$$H_1 = 1 + G_{11}\left(\frac{a}{t}\right) + G_{12}\left(\frac{a}{t}\right)^2$$

$$H_2 = 1 + G_{21}\left(\frac{a}{t}\right) + G_{22}\left(\frac{a}{t}\right)^2$$

$$G_{11} = -0.04 - 0.41\left(\frac{c}{a}\right)$$

$$G_{12} = 0.55 - 1.93\left(\frac{c}{a}\right)^{0.75} + 1.38\left(\frac{c}{a}\right)^{1.5}$$

$$G_{21} = -2.11 - 0.77\left(\frac{c}{a}\right)$$

$$G_{22} = 0.55 - 0.72\left(\frac{c}{a}\right)^{0.75} + 0.14\left(\frac{c}{a}\right)^{1.5}$$

TABLE A9.3
Stress-Intensity Solution for an Elliptical Buried Flaw in a Flat Plate [10].

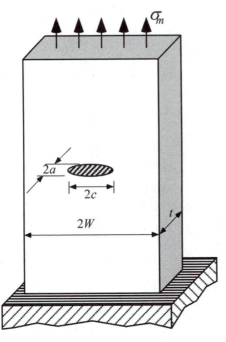

σ_m - Membrane (tensile) stress

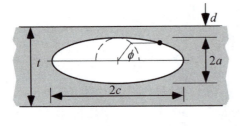

$$K_I = \sigma_m F \sqrt{\frac{\pi a}{Q}}$$

where

$$F = (M_1 + M_2\lambda^2 + M_3\lambda^4)gf_\phi f_w$$

$$\lambda = \frac{a}{a+d}$$

$$f_w = \left[\sec\left(\frac{\pi c}{2W}\sqrt{\frac{2a}{t}}\right)\right]^{1/2}$$

$$M_2 = \frac{0.05}{0.11+\left(\frac{a}{c}\right)^{1.5}}$$

$$M_3 = \frac{0.29}{0.23+\left(\frac{a}{c}\right)^{1.5}}$$

$$g = 1 - \frac{\lambda^4\sqrt{2.6-2\lambda}}{1+4\lambda}|\cos\phi|$$

For $a/c \leq 1$:

$$Q = 1 + 1.464\left(\frac{a}{c}\right)^{1.65}$$

$$f_\phi = \left[\left(\frac{a}{c}\right)^2\cos^2\phi + \sin^2\phi\right]^{1/4}$$

$$M_1 = 1$$

For $a/c > 1$:

$$Q = 1 + 1.464\left(\frac{c}{a}\right)^{1.65}$$

$$f_\phi = \left[\left(\frac{c}{a}\right)^2\sin^2\phi + \cos^2\phi\right]^{1/4}$$

$$M_1 = \sqrt{\frac{c}{a}}$$

Tables A9.6 to A9.15 list fully plastic J and displacement solutions for selected geometries from the original EPRI plastic fracture handbook [23]. Note that the total J and displacement are obtained by including the elastic contribution. Refer to Section 9.3.1 for the complete estimation procedure.

Recall from Chapter 3 that load-line displacement can be partitioned into "crack" and "no crack" components:

$$\Delta_{total} = \Delta_{nc} + \Delta_c \tag{A9.2}$$

TABLE A9.4
Stress-Intensity Solution for a Quarter-Elliptical Corner Crack in a Flat Plate for $a \leq c$ [10].

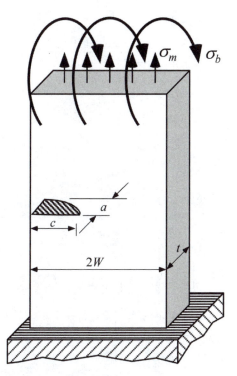

σ_m - Membrane (tensile) stress
σ_b - Bending stress

$$\sigma_b = \frac{Mt}{2I}$$

where

$$I = \frac{Wt^3}{12}$$

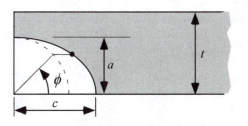

$$K_I = (\sigma_m + H\sigma_b)F\sqrt{\frac{\pi a}{Q}}$$

$$F = \left[M_1 + M_2\left(\frac{a}{t}\right)^2 + M_3\left(\frac{a}{t}\right)^4\right]g_1 g_2 f_\phi f_w$$

$$H = H_1 + (H_2 - H_1)(\sin\phi)^p$$

$$Q = 1 + 1.464\left(\frac{a}{c}\right)^{1.65}$$

$$M_1 = 1.08 - 0.03\left(\frac{a}{c}\right)$$

$$M_2 = -0.44 + \frac{1.06}{0.3 + \frac{a}{c}}$$

$$M_3 = -0.5 - 0.25\left(\frac{a}{c}\right) + 14.8\left(1 - \frac{a}{c}\right)^{15}$$

$$g_1 = 1 + \left[0.08 + 0.4\left(\frac{a}{t}\right)^2\right](1 - \sin\phi)^3$$

$$g_2 = 1 + \left[0.08 + 0.15\left(\frac{a}{t}\right)^2\right](1 - \cos\phi)^3$$

$$f_\phi = \left[\left(\frac{a}{c}\right)^2 \cos^2\phi + \sin^2\phi\right]^{1/4}$$

$$f_w = \left[\sec\left(\frac{\pi c}{2W}\sqrt{\frac{a}{t}}\right)\right]^{1/2}$$

$$g = 1 + \left[0.1 + 0.35\left(\frac{a}{t}\right)^2\right](1 - \sin\phi)^2$$

$$H = H_1 + (H_2 - H_1)(\sin\phi)^p$$

$$p = 0.2 + \frac{a}{c} + 0.6\left(\frac{a}{t}\right)$$

$$H_1 = 1 - 0.34\frac{a}{t} - 0.11\frac{a}{c}\left(\frac{a}{t}\right)$$

$$H_2 = 1 + G_1\left(\frac{a}{t}\right) + G_2\left(\frac{a}{t}\right)^2$$

$$G_1 = -1.22 - 0.12\left(\frac{a}{c}\right)$$

$$G_2 = 0.64 - 1.05\left(\frac{a}{c}\right)^{0.75} + 0.47\left(\frac{a}{c}\right)^{1.5}$$

TABLE A9.5
Stress-Intensity Solution for a Quarter-Elliptical Corner Crack in a Flat Plate for $a > c$ [10].

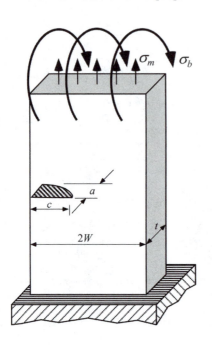

σ_m - Membrane (tensile) stress
σ_b - Bending stress

$$\sigma_b = \frac{Mt}{2I}$$

where

$$I = \frac{Wt^3}{12}$$

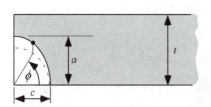

$$K_I = (\sigma_m + H\sigma_b)F\sqrt{\frac{\pi a}{Q}}$$

$$F = \left[M_1 + M_2 \left(\frac{a}{t}\right)^2 + M_3 \left(\frac{a}{t}\right)^4 \right] g_1 g_2 f_\phi f_w$$

$$H = H_1 + (H_2 - H_1)(\sin\phi)^P$$

$$Q = 1 + 1.464 \left(\frac{c}{a}\right)^{1.65}$$

$$M_1 = \sqrt{\frac{c}{a}} \left[1.08 + 0.03 \left(\frac{c}{a}\right) \right]$$

$$M_2 = 0.375 \left(\frac{c}{a}\right)^2$$

$$M_3 = -0.25 \left(\frac{c}{a}\right)^2$$

$$g_1 = 1 + \left[0.08 + 0.4 \left(\frac{c}{t}\right)^2 \right](1 - \sin\phi)^3$$

$$g_2 = 1 + \left[0.08 + 0.15 \left(\frac{c}{t}\right)^2 \right](1 - \cos\phi)^3$$

$$f_\phi = \left[\left(\frac{c}{a}\right)^2 \sin^2\phi + \cos^2\phi \right]^{1/4}$$

$$f_w = \left[\sec\left(\frac{\pi c}{2W}\sqrt{\frac{a}{t}}\right) \right]^{1/2}$$

$$p = 0.2 + \frac{c}{a} + 0.6 \left(\frac{a}{t}\right)$$

$$H_1 = 1 + G_{11} \left(\frac{a}{t}\right) + G_{12} \left(\frac{a}{t}\right)^2$$

$$H_2 = 1 + G_{21} \left(\frac{a}{t}\right) + G_{22} \left(\frac{a}{t}\right)^2$$

$$G_{11} = -0.04 - 0.41 \left(\frac{c}{a}\right)$$

$$G_{12} = 0.55 - 1.93 \left(\frac{c}{a}\right)^{0.75} + 1.38 \left(\frac{c}{a}\right)^{1.5}$$

$$G_{21} = -2.11 - 0.77 \left(\frac{c}{a}\right)$$

$$G_{22} = 0.64 - 0.72 \left(\frac{c}{a}\right)^{0.75} + 0.14 \left(\frac{c}{a}\right)^{1.5}$$

TABLE A9.6
Fully Plastic _J_ and Displacement for a Compact Specimen in Plane Strain [23].

a/W:		n = 1	n = 2	n = 3	n = 5	n = 7	n = 10	n = 13	n = 16	n = 20
0.250	h_1	2.23	2.05	1.78	1.48	1.33	1.26	1.25	1.32	1.57
	h_2	17.9	12.5	11.7	10.8	10.5	10.7	11.5	12.6	14.6
	h_3	9.85	8.51	8.17	7.77	7.71	7.92	8.52	9.31	10.9
0.375	h_1	2.15	1.72	1.39	0.970	0.693	0.443	0.276	0.176	0.098
	h_2	12.6	8.18	6.52	4.32	2.97	1.79	1.10	0.686	0.370
	h_3	7.94	5.76	4.64	3.10	2.14	1.29	0.793	0.494	0.266
0.500	h_1	1.94	1.51	1.24	0.919	0.685	0.461	0.314	0.216	0.132
	h_2	9.33	5.85	4.30	2.75	1.91	1.20	0.788	0.530	0.317
	h_3	6.41	4.27	3.16	2.02	1.41	0.888	0.585	0.393	0.236
0.625	h_1	1.76	1.45	1.24	0.974	0.752	0.602	0.459	0.347	0.248
	h_2	7.61	4.57	3.42	2.36	1.81	1.32	0.983	0.749	0.485
	h_3	5.52	3.43	2.58	1.79	1.37	1.00	0.746	0.568	0.368
0.750	h_1	1.71	1.42	1.26	1.033	0.864	0.717	0.575	0.448	0.345
	h_2	6.37	3.95	3.18	2.34	1.88	1.44	1.12	0.887	0.665
	h_3	4.86	3.05	2.46	1.81	1.45	1.11	0.869	0.686	0.514
→ 1	h_1	1.57	1.45	1.35	1.18	1.08	0.950	0.850	0.730	0.630
	h_2	5.39	3.74	3.09	2.43	2.12	1.80	1.57	1.33	1.14
	h_3	4.31	2.99	2.47	1.95	1.79	1.44	1.26	1.07	0.909

$$J_{pl} = \alpha\varepsilon_o\sigma_o bh_1(a/W,n)\left(\frac{P}{P_o}\right)^{n+1}$$

$$V_p = \alpha\varepsilon_o ah_2(a/W,n)\left(\frac{P}{P_o}\right)^{n}$$

$$\Delta_p = \alpha\varepsilon_o ah_3(a/W,n)\left(\frac{P}{P_o}\right)^{n}$$

$$P_o = 1.455\eta Bb\sigma_o$$

where

$$\eta = \sqrt{\left(\frac{2a}{b}\right)^2 + \frac{4a}{b} + 2} - \left(\frac{2a}{b}+1\right)$$

Thickness = B

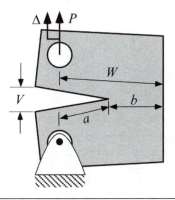

where Δ_{nc} is the displacement in the absence of the crack and Δ_c is the additional displacement due to the crack. Where appropriate, the solutions listed in Tables A9.6 to A9.15 distinguish between Δ_c and Δ_{nc}.

TABLE A9.7
Fully Plastic J and Displacement for a Compact Specimen in Plane Stress [23].

a/W:		n = 1	n = 2	n = 3	n = 5	n = 7	n = 10	n = 13	n = 16	n = 20
0.250	h_1	1.61	1.46	1.28	1.06	0.903	0.729	0.601	0.511	0.395
	h_2	17.6	12.0	10.7	8.74	7.32	5.74	4.63	3.75	2.92
	h_3	9.67	8.00	7.21	5.94	5.00	3.95	3.19	2.59	2.023
0.375	h_1	1.55	1.25	1.05	0.801	0.647	0.484	0.377	0.284	0.220
	h_2	12.4	8.20	6.54	4.56	3.45	2.44	1.83	1.36	1.02
	h_3	7.80	5.73	4.62	3.25	2.48	1.77	1.33	0.990	0.746
0.500	h_1	1.40	1.08	0.901	0.686	0.558	0.436	0.356	0.298	0.238
	h_2	9.16	5.67	4.21	2.80	2.12	1.57	1.25	1.03	0.814
	h_3	6.29	4.15	3.11	2.09	1.59	1.18	0.938	0.774	0.614
0.625	h_1	1.27	1.03	0.875	0.695	0.593	0.494	0.423	0.370	0.310
	h_2	7.47	4.48	3.35	2.37	1.92	1.54	1.29	1.12	0.928
	h_3	5.42	3.38	2.54	1.80	1.47	1.18	0.988	0.853	0.710
0.750	h_1	1.23	0.977	0.833	0.683	0.598	0.506	0.431	0.373	0.314
	h_2	6.25	3.78	2.89	2.14	1.78	1.44	1.20	1.03	0.857
	h_3	4.77	2.92	2.24	1.66	1.38	1.12	0.936	0.800	0.666
→ 1	h_1	1.13	1.01	0.775	0.680	0.650	0.620	0.490	0.470	0.420
	h_2	5.29	3.54	2.41	1.91	1.73	1.59	1.23	1.17	1.03
	h_3	4.23	2.83	1.93	1.52	1.39	1.27	0.985	0.933	0.824

$$J_{pl} = \alpha \varepsilon_o \sigma_o bh_1 (a/W, n) \left(\frac{P}{P_o} \right)^{n+1}$$

$$V_p = \alpha \varepsilon_o ah_2 (a/W, n) \left(\frac{P}{P_o} \right)^{n}$$

$$\Delta_p = \alpha \varepsilon_o ah_3 (a/W, n) \left(\frac{P}{P_o} \right)^{n}$$

$$P_o = 1.072 \eta Bb\sigma_o$$

where

$$\eta = \sqrt{ \left(\frac{2a}{b} \right)^2 + \frac{4a}{b} + 2 } - \left(\frac{2a}{b} + 1 \right)$$

Thickness = B

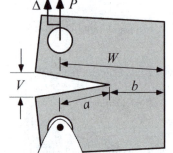

TABLE A9.8
Fully Plastic J and Displacement for a Single-Edge-Notched Bend (Se(B)) Specimen in Plane Strain Subject to Three-point Bending [23].

a/W:		$n=1$	$n=2$	$n=3$	$n=5$	$n=7$	$n=10$	$n=13$	$n=16$	$n=20$
0.125	h_1	0.936	0.869	0.805	0.687	0.580	0.437	0.329	0.245	0.165
	h_2	6.97	6.77	6.29	5.29	4.38	3.24	2.40	1.78	1.19
	h_3	3.00	22.1	20.0	15.0	11.7	8.39	6.14	4.54	3.01
0.250	h_1	1.20	1.034	0.930	0.762	0.633	0.523	0.396	0.303	0.215
	h_2	5.80	4.67	4.01	3.08	2.45	1.93	1.45	1.09	0.758
	h_3	4.08	9.72	8.36	5.86	4.47	3.42	2.54	1.90	1.32
0.375	h_1	1.33	1.15	1.02	0.084	0.695	0.556	0.442	0.360	0.265
	h_2	5.18	3.93	3.20	2.38	1.93	1.47	1.15	0.928	0.684
	h_3	4.51	6.01	5.03	3.74	3.02	2.30	1.80	1.45	1.07
0.500	h_1	1.41	1.09	0.922	0.675	0.495	0.331	0.211	0.135	0.0741
	h_2	4.87	3.28	2.53	1.69	1.19	0.773	0.480	0.304	0.165
	h_3	4.69	4.33	3.49	2.35	1.66	1.08	0.669	0.424	0.230
0.625	h_1	1.46	1.07	0.896	0.631	0.436	0.255	0.142	0.084	0.0411
	h_2	4.64	2.86	2.16	1.37	0.907	0.518	0.287	0.166	0.0806
	h_3	4.71	3.49	2.70	1.72	1.14	0.652	0.361	0.209	0.102
0.750	h_1	1.48	1.15	0.974	0.693	0.500	0.348	0.223	0.140	0.0745
	h_2	4.47	2.75	2.10	1.36	0.936	0.618	0.388	0.239	0.127
	h_3	4.49	3.14	2.40	1.56	1.07	0.704	0.441	0.272	0.144
0.875	h_1	1.50	1.35	1.20	1.02	0.855	0.690	0.551	0.440	0.321
	h_2	4.36	2.90	2.31	1.70	1.33	1.00	0.782	0.613	0.459
	h_3	4.15	3.08	2.45	1.81	1.41	1.06	0.828	0.646	0.486

$$J_{pl} = \alpha \varepsilon_o \sigma_o b h_1 (a/W, n)\left(\frac{P}{P_o}\right)^{n+1}$$

$$V_p = \alpha \varepsilon_o a h_2 (a/W, n)\left(\frac{P}{P_o}\right)^{n}$$

$$\Delta_p = \alpha \varepsilon_o a h_3 (a/W, n)\left(\frac{P}{P_o}\right)^{n}$$

$$P_o = \frac{1.455 B b^2 \sigma_o}{S}$$

Thickness $= B$

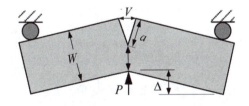

TABLE A9.9

Fully Plastic _J_ and Displacement for a Single-Edge-Notched Bend (Se(B)) Specimen in Plane Stress Subject to Three-point Bending [23].

a/W:		n = 1	n = 2	n = 3	n = 5	n = 7	n = 10	n = 13	n = 16	n = 20
0.125	h_1	0.676	0.600	0.548	0.459	0.383	0.297	0.238	0.192	0.148
	h_2	6.84	6.30	5.66	4.53	3.64	2.72	2.12	1.67	1.26
	h_3	2.95	20.1	14.6	12.2	9.12	6.75	5.20	4.09	3.07
0.250	h_1	0.869	0.731	0.629	0.479	0.370	0.246	0.174	0.117	0.0593
	h_2	5.69	4.50	3.68	2.61	1.95	1.29	0.897	0.603	0.307
	h_3	4.01	8.81	7.19	4.73	3.39	2.20	1.52	1.01	0.508
0.375	h_1	0.963	0.797	0.680	0.527	0.418	0.307	0.232	0.174	0.105
	h_2	5.09	3.73	2.93	2.07	1.58	1.13	0.841	0.626	0.381
	h_3	4.42	5.53	4.48	3.17	2.41	1.73	1.28	0.948	0.575
0.500	h_1	1.02	0.767	0.621	0.453	0.324	0.202	0.128	0.0813	0.0298
	h_2	4.77	3.12	2.32	1.55	1.08	0.655	0.410	0.259	0.0974
	h_3	4.60	4.09	3.09	2.08	1.44	0.874	0.545	0.344	0.129
0.625	h_1	1.05	0.786	0.649	0.494	0.357	0.235	0.173	0.105	0.0471
	h_2	4.55	2.83	2.12	1.46	1.02	0.656	0.472	0.286	0.130
	h_3	4.62	3.43	2.60	1.79	1.26	0.803	0.577	0.349	0.158
0.750	h_1	1.07	0.786	0.643	0.474	0.343	0.230	0.167	0.110	0.0442
	h_2	4.39	2.66	1.97	1.33	0.928	0.601	0.427	0.280	0.114
	h_3	4.39	3.01	2.24	1.51	1.05	0.680	0.483	0.316	0.129
0.875	h_1	1.086	0.928	0.810	6.46	0.538	0.423	0.332	0.242	0.205
	h_2	4.28	2.76	2.16	1.56	1.23	0.922	0.702	0.561	0.428
	h_3	4.07	2.93	2.29	1.65	1.30	0.975	0.742	0.592	0.452

$$J_{pl} = \alpha \varepsilon_o \sigma_o b h_1 (a/W, n) \left(\frac{P}{P_o} \right)^{n+1}$$

$$V_p = \alpha \varepsilon_o a h_2 (a/W, n) \left(\frac{P}{P_o} \right)^n$$

$$\Delta_p = \alpha \varepsilon_o a h_3 (a/W, n) \left(\frac{P}{P_o} \right)^n$$

$$P_o = \frac{1.072 B b^2 \sigma_o}{S}$$

Thickness = B

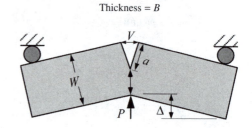

TABLE A9.10
Fully Plastic J and Displacement for a Middle Tension (MT) Specimen in Plane Strain [23].

a/W:		$n = 1$	$n = 2$	$n = 3$	$n = 5$	$n = 7$	$n = 10$	$n = 13$	$n = 16$	$n = 20$
	h_1	2.80	3.61	4.06	4.35	4.33	4.02	3.56	3.06	2.46
0.125	h_2	3.05	3.62	3.91	4.06	3.93	3.54	3.07	2.60	2.06
	h_3	0.303	0.574	0.840	1.30	1.63	1.95	2.03	1.96	1.77
	h_1	2.54	3.01	3.21	3.29	3.18	2.92	2.63	2.34	2.03
0.250	h_2	2.68	2.99	3.01	2.85	2.61	2.30	1.97	1.71	1.45
	h_3	0.536	0.911	1.22	1.64	1.84	1.85	1.80	1.64	1.43
	h_1	2.34	2.62	2.65	2.51	2.28	1.97	1.71	1.46	1.19
0.375	h_2	2.35	2.39	2.23	1.88	1.58	1.28	1.07	0.890	0.715
	h_3	0.699	1.06	1.28	1.44	1.40	1.23	1.05	0.888	0.719
	h_1	2.21	2.29	2.20	1.97	1.76	1.52	1.32	1.16	0.978
0.500	h_2	2.03	1.86	1.60	1.23	1.00	0.799	0.664	0.564	0.466
	h_3	0.803	1.07	1.16	1.10	0.968	0.796	0.665	0.565	0.469
	h_1	2.12	1.96	1.76	1.43	1.17	0.863	0.628	0.458	0.300
0.625	h_2	1.71	1.32	1.04	0.707	0.524	0.358	0.250	0.178	0.114
	h_3	0.844	0.937	0.879	0.701	0.522	0.361	0.251	0.178	0.115
	h_1	2.07	1.73	1.47	1.11	0.895	0.642	0.461	0.337	0.216
0.750	h_2	1.35	0.857	0.596	0.361	0.254	0.167	0.114	0.0810	0.0511
	h_3	0.805	0.700	0.555	0.359	0.254	0.168	0.114	0.0813	0.0516
	h_1	2.08	1.64	1.40	1.14	0.987	0.814	0.688	0.573	0.461
0.875	h_2	0.889	0.428	0.287	0.181	0.139	0.105	0.0837	0.0682	0.0533
	h_3	0.632	0.400	0.291	0.182	0.140	0.106	0.0839	0.0683	0.0535

$$J_{pl} = \alpha \varepsilon_o \sigma_o \frac{ba}{W} h_1(a/W, n)\left(\frac{P}{P_o}\right)^{n+1}$$

$$V_p = \alpha \varepsilon_o a h_2(a/W, n)\left(\frac{P}{P_o}\right)^{n}$$

$$\Delta_{p(c)} = \alpha \varepsilon_o a h_3(a/W, n)\left(\frac{P}{P_o}\right)^{n}$$

$$P_o = \frac{4}{\sqrt{3}} Bb\sigma_o$$

$$\Delta_{p(nc)} = \frac{\sqrt{3}}{2}\alpha \varepsilon_o L \left(\frac{\sqrt{3}P}{4BW\sigma_o}\right)^{n}$$

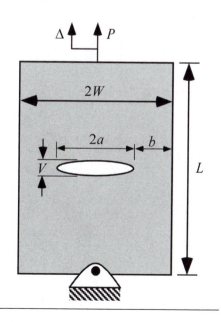

TABLE A9.11
Fully Plastic J and Displacement for a Middle Tension (MT) Specimen in Plane Stress [23].

a/W:		n = 1	n = 2	n = 3	n = 5	n = 7	n = 10	n = 13	n = 16	n = 20
0.125	h_1	2.80	3.57	4.01	4.47	4.65	4.62	4.41	4.13	3.72
	h_2	3.53	4.09	4.43	4.74	4.79	4.63	4.33	4.00	3.55
	h_3	0.350	0.661	0.997	1.55	2.05	2.56	2.83	2.95	2.92
0.250	h_1	2.54	2.97	3.14	3.20	3.11	2.86	2.65	2.47	2.20
	h_2	3.10	3.29	3.30	3.15	2.93	2.56	2.29	2.08	1.81
	h_3	0.619	1.01	1.35	1.83	2.08	2.19	2.12	2.01	1.79
0.375	h_1	2.34	2.53	2.52	2.35	2.17	1.95	1.77	1.61	1.43
	h_2	2.71	2.62	2.41	2.03	1.75	1.47	1.28	1.13	0.988
	h_3	0.807	1.20	1.43	1.59	1.57	1.43	1.27	1.13	0.994
0.500	h_1	2.21	2.20	2.06	1.81	1.63	1.43	1.30	1.17	1.00
	h_2	2.34	2.01	1.70	1.30	1.07	0.871	0.757	0.666	0.557
	h_3	0.927	1.19	1.26	1.18	1.04	0.867	0.758	0.668	0.560
0.625	h_1	2.12	1.91	1.69	1.41	1.22	1.01	0.853	0.712	0.573
	h_2	1.97	1.46	1.13	0.785	0.617	0.474	0.383	0.313	0.256
	h_3	0.975	1.05	0.970	0.763	0.620	0.478	0.386	0.318	0.273
0.750	h_1	2.07	1.71	1.46	1.21	1.08	0.867	0.745	0.646	0.532
	h_2	1.55	0.970	0.685	0.452	0.361	0.262	0.216	0.183	0.148
	h_3	0.929	0.802	0.642	0.450	0.361	0.263	0.216	0.183	0.149
0.875	h_1	2.08	1.57	1.31	1.08	0.972	0.862	0.778	0.715	0.630
	h_2	1.03	0.485	0.310	0.196	0.157	0.127	0.109	0.0971	0.0842
	h_3	0.730	0.452	0.313	0.198	0.157	0.127	0.109	0.0973	0.0842

$$J_{pl} = \alpha \varepsilon_o \sigma_o \frac{ba}{W} h_1(a/W, n) \left(\frac{P}{P_o} \right)^{n+1}$$

$$V_p = \alpha \varepsilon_o a h_2(a/W, n) \left(\frac{P}{P_o} \right)^{n}$$

$$\Delta_{p(c)} = \alpha \varepsilon_o a h_3(a/W, n) \left(\frac{P}{P_o} \right)^{n}$$

$$P_o = 2Bb\sigma_o$$

$$\Delta_{p(nc)} = \alpha \varepsilon_o L \left(\frac{P}{2BW\sigma_o} \right)^{n}$$

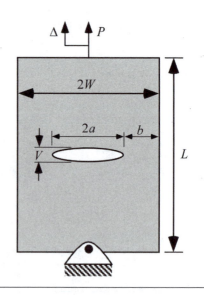

TABLE A9.12
Fully Plastic J and Displacement for an Edge-Cracked Tension Specimen in Plane Strain [23].

a/W:		n = 1	n = 2	n = 3	n = 5	n = 7	n = 10	n = 13	n = 16	n = 20
0.125	h_1	4.95	6.93	8.57	11.5	13.5	16.1	18.1	19.9	21.2
	h_2	5.25	6.47	7.56	9.46	11.1	12.9	14.4	15.7	16.8
	h_3	26.6	25.8	25.2	24.2	23.6	23.2	23.2	23.5	23.7
0.250	h_1	4.34	4.77	4.64	3.82	3.06	2.17	1.55	1.11	0.712
	h_2	4.76	4.56	4.28	3.39	2.64	1.81	1.25	0.875	0.552
	h_3	10.3	7.64	5.87	3.70	2.48	1.50	0.970	0.654	0.404
0.375	h_1	3.88	3.25	2.63	1.68	1.06	0.539	0.276	0.142	0.0595
	h_2	4.54	3.49	2.67	1.57	0.946	0.458	0.229	0.116	0.048
	h_3	5.14	2.99	1.90	0.923	0.515	0.240	0.119	0.060	0.0246
0.500	h_1	3.40	2.30	1.69	0.928	0.514	0.213	0.0902	0.0385	0.0119
	h_2	4.45	2.77	1.89	0.954	0.507	0.204	0.0854	0.0356	0.0110
	h_3	3.15	1.54	0.912	0.417	0.215	0.085	0.0358	0.0147	0.0045
0.625	h_1	2.86	1.80	1.30	0.697	0.378	0.153	0.0625	0.0256	0.0078
	h_2	4.37	2.44	1.62	0.081	0.423	0.167	0.0671	0.0272	0.0082
	h_3	2.31	1.08	0.681	0.329	0.171	0.067	0.0268	0.0108	0.0033
0.750	h_1	2.34	1.61	1.25	0.769	0.477	0.233	0.116	0.059	0.0215
	h_2	4.32	2.52	1.79	1.03	0.619	0.296	0.146	0.0735	0.0267
	h_3	2.02	1.10	0.765	0.435	0.262	0.125	0.0617	0.0312	0.0113
0.875	h_1	1.91	1.57	1.37	1.10	0.925	0.702			
	h_2	4.29	2.75	2.14	1.55	1.23	0.921			
	h_3	2.01	1.27	0.988	0.713	0.564	0.424			

$$J_{pl} = \alpha \varepsilon_o \sigma_o \frac{ba}{W} h_1(a/W, n) \left(\frac{P}{P_o} \right)^{n+1}$$

$$V_p = \alpha \varepsilon_o a h_2(a/W, n) \left(\frac{P}{P_o} \right)^{n}$$

$$\Delta_{p(c)} = \alpha \varepsilon_o a h_3(a/W, n) \left(\frac{P}{P_o} \right)^{n}$$

$$P_o = 1.455 \eta B b \sigma_o$$

where

$$\eta = \sqrt{1 + \left(\frac{a}{b} \right)^2} - \frac{a}{b}$$

$$\Delta_{p(nc)} = \frac{\sqrt{3}}{2} \alpha \varepsilon_o L \left(\frac{\sqrt{3}P}{4BW\sigma_o} \right)^{n}$$

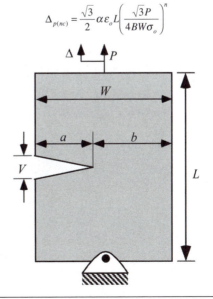

TABLE A9.13
Fully Plastic J and Displacement for an Edge-Cracked Tension Specimen in Plane Stress [23].

a/W:		$n = 1$	$n = 2$	$n = 3$	$n = 5$	$n = 7$	$n = 10$	$n = 13$	$n = 16$	$n = 20$
	h_1	3.58	4.55	5.06	5.30	4.96	4.14	3.29	2.60	1.92
0.125	h_2	5.15	5.43	6.05	6.01	5.47	4.46	3.48	2.74	2.02
	h_3	26.1	21.6	18.0	12.7	9.24	5.98	3.94	2.72	2.0
	h_1	3.14	3.26	2.92	2.12	1.53	0.960	0.615	0.400	0.230
0.250	h_2	4.67	4.30	3.70	2.53	1.76	1.05	0.656	0.419	0.237
	h_3	10.1	6.49	4.36	2.19	1.24	0.630	0.362	0.224	0.123
	h_1	2.88	2.37	1.94	1.37	1.01	0.677	0.474	0.342	0.226
0.375	h_2	4.47	3.43	2.63	1.69	1.18	0.762	0.524	0.372	0.244
	h_3	5.05	2.65	1.60	0.812	0.525	0.328	0.223	0.157	0.102
	h_1	2.46	1.67	1.25	0.776	0.510	0.286	0.164	0.0956	0.0469
0.500	h_2	4.37	2.73	1.91	1.09	0.694	0.380	0.216	0.124	0.0607
	h_3	3.10	1.43	0.871	0.461	0.286	0.155	0.088	0.0506	0.0247
	h_1	2.07	1.41	1.105	0.755	0.551	0.363	0.248	0.172	0.107
0.625	h_2	4.30	2.55	1.84	1.16	0.816	0.523	0.353	2.42	0.150
	h_3	2.27	1.13	0.771	0.478	0.336	0.215	0.146	0.100	0.0616
	h_1	1.70	1.14	0.910	0.624	0.447	0.280	0.181	0.118	0.0670
0.750	h_2	4.24	2.47	1.81	1.15	0.798	0.490	0.314	0.203	0.115
	h_3	1.98	1.09	0.784	0.494	0.344	0.211	0.136	0.0581	0.0496
	h_1	1.38	1.11	0.962	0.792	0.677	0.574			
0.875	h_2	4.22	2.68	2.08	1.54	1.27	1.04			
	h_3	1.97	1.25	0.969	0.716	0.591	0.483			

$$J_{pl} = \alpha \varepsilon_o \sigma_o \frac{ba}{W} h_1(a/W,n)\left(\frac{P}{P_o}\right)^{n+1}$$

$$\Delta_{p(nc)} = \alpha \varepsilon_o L \left(\frac{P}{2BW\sigma_o}\right)^{n}$$

$$V_p = \alpha \varepsilon_o a h_2(a/W,n)\left(\frac{P}{P_o}\right)^{n}$$

$$\Delta_{p(c)} = \alpha \varepsilon_o a h_3(a/W,n)\left(\frac{P}{P_o}\right)^{n}$$

$$P_o = 1.072\,\eta Bb\sigma_o$$

where

$$\eta = \sqrt{1+\left(\frac{a}{b}\right)^2} - \frac{a}{b}$$

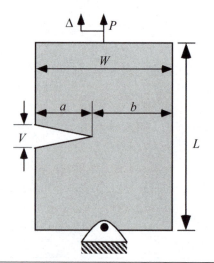

TABLE A9.14
Fully Plastic J and Displacement for a Double-Edge-Notched Tension (DENT) Specimen in Plane Strain [23].

a/W:		n = 1	n = 2	n = 3	n = 5	n = 7	n = 10	n = 13	n = 16	n = 20
0.125	h_1	0.572	0.772	0.922	1.13	1.35	1.61	1.86	2.08	2.44
	h_2	0.732	0.852	0.961	1.14	1.29	1.50	1.70	1.94	2.17
	h_3	0.063	0.126	0.200	0.372	0.571	0.911	1.30	1.74	2.29
0.250	h_1	1.10	1.32	1.38	1.65	1.75	1.82	1.86	1.89	1.92
	h_2	1.56	1.63	1.70	1.78	1.80	1.81	1.79	1.78	1.76
	h_3	0.267	0.479	0.698	1.11	1.47	1.92	2.25	2.49	2.73
0.375	h_1	1.61	1.83	1.92	1.92	1.84	1.68	1.49	1.32	1.12
	h_2	2.51	2.41	2.35	2.15	1.94	1.68	1.44	1.25	1.05
	h_3	0.637	1.05	1.40	1.87	2.11	2.20	2.09	1.92	1.67
0.500	h_1	2.22	2.43	2.48	2.43	2.32	2.12	1.91	1.60	1.51
	h_2	3.73	3.40	3.15	2.71	2.37	2.01	1.72	1.40	1.38
	h_3	1.26	1.92	2.37	2.79	2.85	2.68	2.40	1.99	1.94
0.625	h_1	3.16	3.38	3.45	3.42	3.28	3.00	2.54	2.36	2.27
	h_2	5.57	4.76	4.23	3.46	2.97	2.48	2.02	1.82	1.66
	h_3	2.36	3.29	3.74	3.90	3.68	3.23	2.66	2.40	2.19
0.750	h_1	5.24	6.29	7.17	8.44	9.46	10.9	11.9	11.3	17.4
	h_2	9.10	7.76	7.14	6.64	6.83	7.48	7.79	7.14	11.1
	h_3	4.73	6.26	7.03	7.63	8.14	9.04	9.40	8.58	13.5
0.875	h_1	14.2	24.8	39.0	78.4	140.0	341.0	777.0	1570.0	3820.0
	h_2	20.1	19.4	22.7	36.1	58.9	133.0	294.0	585.0	1400.0
	h_3	12.7	18.2	24.1	40.4	65.8	149.0	327.0	650.0	1560.0

$$J_{pl} = \alpha \varepsilon_o \sigma_o \frac{ba}{W} h_1(a/W, n) \left(\frac{P}{P_o} \right)^{n+1}$$

$$V_p = \alpha \varepsilon_o a h_2(a/W, n) \left(\frac{P}{P_o} \right)^{n}$$

$$\Delta_{p(c)} = \alpha \varepsilon_o a h_3(a/W, n) \left(\frac{P}{P_o} \right)^{n}$$

$$P_o = \left(0.72 + 1.82 \frac{b}{W} \right) Bb\sigma_o$$

$$\Delta_{p(nc)} = \frac{\sqrt{3}}{2} \alpha \varepsilon_o L \left(\frac{\sqrt{3}P}{4BW\sigma_o} \right)^{n}$$

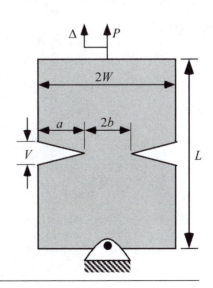

TABLE A9.15
Fully Plastic J and Displacement for a Double-Edge-Notched Tension (DENT) Specimen in Plane Stress [23].

a/W:		$n=1$	$n=2$	$n=3$	$n=5$	$n=7$	$n=10$	$n=13$	$n=16$	$n=20$
0.125	h_1	0.583	0.825	1.02	1.37	1.71	2.24	2.84	3.54	4.62
	h_2	0.853	1.05	1.23	1.55	1.87	2.38	2.96	3.65	4.70
	h_3	0.0729	0.159	0.26	0.504	0.821	1.41	2.18	3.16	4.73
0.250	h_1	1.01	1.23	1.36	1.48	1.54	1.58	1.59	1.59	1.59
	h_2	1.73	1.82	1.89	1.92	1.91	1.85	1.80	1.75	1.70
	h_3	0.296	0.537	0.770	1.17	1.49	1.82	2.02	2.12	2.20
0.375	h_1	1.29	1.42	1.43	1.34	1.24	1.09	0.970	0.873	0.674
	h_2	2.59	2.39	2.22	1.86	1.59	1.28	1.07	0.922	0.709
	h_3	0.658	1.04	1.30	1.52	1.55	1.41	1.23	1.07	0.830
0.500	h_1	1.48	1.47	1.38	1.17	1.01	0.845	0.732	0.625	0.208
	h_2	3.51	2.82	2.34	1.67	1.28	0.944	0.762	0.630	0.232
	h_3	1.18	1.58	1.69	1.56	1.32	1.01	0.809	0.662	0.266
0.625	h_1	1.59	1.45	1.29	1.04	0.882	0.737	0.649	0.466	0.0202
	h_2	4.56	3.15	2.32	1.45	1.06	0.790	0.657	0.473	0.0277
	h_3	1.93	2.14	1.95	1.44	1.09	0.809	0.665	0.487	0.0317
0.750	h_1	1.65	1.43	1.22	0.979	0.834	0.701	0.630	0.297	
	h_2	5.90	3.37	2.22	1.30	0.966	0.741	0.636	0.312	
	h_3	3.06	2.67	2.06	1.31	0.978	0.747	0.638	0.318	
0.875	h_1	1.69	1.43	1.22	0.979	0.845	0.738	0.664	0.614	0.562
	h_2	8.02	3.51	2.14	1.27	0.971	0.775	0.663	0.596	0.535
	h_3	5.07	3.18	2.16	1.30	0.980	0.779	0.665	0.597	0.538

$$J_{pl} = \alpha \varepsilon_o \sigma_o \frac{ba}{W} h_1(a/W, n) \left(\frac{P}{P_o} \right)^{n+1}$$

$$V_p = \alpha \varepsilon_o a h_2(a/W, n) \left(\frac{P}{P_o} \right)^n$$

$$\Delta_{p(c)} = \alpha \varepsilon_o a h_3(a/W, n) \left(\frac{P}{P_o} \right)^n$$

$$P_o = \frac{4}{\sqrt{3}} Bb\sigma_o$$

$$\Delta_{p(nc)} = \alpha \varepsilon_o L \left(\frac{P}{2BW\sigma_o} \right)^n$$

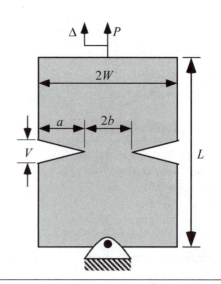

REFERENCES

1. Rooke, D.P. and Cartwright, D.J., *Compendium of Stress Intensity Factors.* Her Majesty's Stationery Office, London, 1976.
2. Tada, H., Paris, P.C., and Irwin, G.R., *The Stress Analysis of Cracks Handbook.* 3rd Ed., American Society of Mechanical Engineers, New York, 2000.
3. Murakami, Y., *Stress Intensity Factors Handbook.* Pergamon Press, New York, 1987.
4. Sanford, R.J. and Dally, J.W., "A General Method for Determining Mixed-Mode Stress Intensity Factors from Isochromatic Fringe Patterns." *Engineering Fracture Mechanics,* Vol. 11, 1979, pp. 621–633.
5. Chona, R., Irwin, G.R., and Shukla, A., "Two and Three Parameter Representation of Crack Tip Stress Fields." *Journal of Strain Analysis,* Vol. 17, 1982, pp. 79–86.
6. Kalthoff, J.F., Beinart, J., Winkler, S., and Klemm, W., "Experimental Analysis of Dynamic Effects in Different Crack Arrest Test Specimens." ASTM STP 711, American Society for Testing and Materials, Philadelphia, PA, 1980, pp. 109–127.
7. Raju, I.S. and Newman, J.C., Jr., "Stress-Intensity Factors for Internal and External Surface Cracks in Cylindrical Vessels." *Journal of Pressure Vessel Technology,* Vol. 104, 1982, pp. 293–298.
8. Rice, J.R., "Some Remarks on Elastic Crack-Tip Stress Fields." *International Journal of Solids and Structures,* Vol. 8, 1972, pp. 751–758.
9. Rice, J.R., "Weight Function Theory for Three-Dimensional Elastic Crack Analysis." ASTM STP 1020, American Society for Testing and Materials, Philadelphia, PA, 1989, pp. 29–57.
10. Newman, J.C. and Raju, I.S., "Stress-Intensity Factor Equations for Cracks in Three-Dimensional Finite Bodies Subjected to Tension and Bending Loads." NASA Technical Memorandum 85793, NASA Langley Research Center, Hampton, VA, April 1984.
11. Anderson, T.L., Thorwald, G., Revelle, D.J., Osage, D.A., Janelle, J.L., and Fuhry, M.E., "Development of Stress Intensity Factor Solutions for Surface and Embedded Cracks in API 579." WRC Bulletin 471, Welding Research Council, New York, 2002.
12. API 579, "Fitness for Service." American Petroleum Institute, Washington, DC, 2000.
13. Shen, G. and Glinka, G., "Determination of Weight Functions from Reference Stress Intensity Solutions." *Theoretical and Applied Fracture Mechanics,* Vol. 15, 1991, pp. 237–245.
14. Zheng, X.J., Kiciak, A., and Glinka, G., "Weight Functions and Stress Intensity Factors for Internal Surface Semi-Elliptical Crack in Thick-Walled Cylinder." *Engineering Fracture Mechanics,* Vol. 58, 1997, pp. 207–221.
15. Wang, X. and Lambert, S.B., "Local Weight Functions and Stress Intensity Factors Surface Semi-Elliptical Surface Cracks in Finite Width Plates." *Theoretical and Applied Fracture Mechanics,* Vol. 15, 1991, pp. 237–245.
16. Burdekin, F.M. and Stone, D.E.W., "The Crack Opening Displacement Approach to Fracture Mechanics in Yielding Materials." *Journal of Strain Analysis,* Vol. 1, 1966, pp. 144–153.
17. Burdekin, F.M. and Dawes, M.G., "Practical Use of Linear Elastic and Yielding Fracture Mechanics with Particular Reference to Pressure Vessels." *Proceedings of the Institute of Mechanical Engineers Conference,* London, May 1971, pp. 28–37.
18. Wells, A.A., "Application of Fracture Mechanics at and Beyond General Yielding." *British Welding Journal,* Vol. 10, 1963, pp. 563–570.
19. Dawes, M.G., "Fracture Control in High Yield Strength Weldments." *Welding Journal,* Vol. 53, 1974, pp. 369–380.
20. PD 6493:1980, "Guidance on Some Methods for the Derivation of Acceptance Levels for Defects in Fusion Welded Joints." British Standards Institution, London, March 1980.
21. Kamath, M.S., "The COD Design Curve: An Assessment of Validity Using Wide Plate Tests." The Welding Institute Report 71/1978/E, September 1978.
22. Shih, C.F. and Hutchinson, J.W., "Fully Plastic Solutions and Large-Scale Yielding Estimates for Plane Stress Crack Problems." *Journal of Engineering Materials and Technology,* Vol. 98, 1976, pp. 289–295.
23. Kumar, V., German, M.D., and Shih, C.F., "An Engineering Approach for Elastic-Plastic Fracture Analysis." EPRI Report NP-1931, Electric Power Research Institute, Palo Alto, CA, 1981.

24. Kumar, V., German, M.D., Wilkening, W.W., Andrews, W.R., deLorenzi, H.G., and Mowbray, D.F., "Advances in Elastic-Plastic Fracture Analysis." EPRI Report NP-3607, Electric Power Research Institute, Palo Alto, CA, 1984.

25. Kumar, V. and German, M.D., "Elastic-Plastic Fracture Analysis of Through-Wall and Surface Flaws in Cylinders." EPRI Report NP-5596, Electric Power Research Institute, Palo Alto, CA, 1988.

26. Zahoor, A., "Ductile Fracture Handbook, Volume 1: Circumferential Throughwall Cracks." EPRI Report NP-6301-D, Electric Power Research Institute, Palo Alto, CA, 1989.

27. Shih, C.F., "Relationship between the J-Integral and the Crack Opening Displacement for Stationary and Extending Cracks." *Journal of the Mechanics and Physics of Solids,* Vol. 29, 1981, pp. 305–326.

28. Ainsworth, R.A., "Some Observations on Creep Crack Growth." *International Journal of Fracture,* Vol. 20, 1982, pp. 417–159.

29. Ainsworth, R.A., "The Assessment of Defects in Structures of Strain Hardening Materials." *Engineering Fracture Mechanics,* Vol. 19, 1984, p. 633.

30. Dowling, A.R. and Townley, C.H.A., "The Effects of Defects on Structural Failure: A Two-Criteria Approach." *International Journal of Pressure Vessels and Piping,* Vol. 3, 1975, pp. 77–137.

31. Harrison, R.P., Loosemore, K., and Milne, I., "Assessment of the Integrity of Structures Containing Defects." CEGB Report R/H/R6, Central Electricity Generating Board, UK, 1976.

32. Bloom, J.M., "Prediction of Ductile Tearing Using a Proposed Strain Hardening Failure Assessment Diagram." *International Journal of Fracture,* Vol. 6, 1980, pp. R73–R77.

33. Shih, C.F., German, M.D., and Kumar, V., "An Engineering Approach for Examining Crack Growth and Stability in Flawed Structures." *International Journal of Pressure Vessels and Piping,* Vol. 9, 1981, pp. 159–196.

34. Milne, I., Ainsworth, R.A., Dowling, A.R., and Stewart, A.T., "Background to and Validation of CEGB Report R/H/R6—Revision 3." *International Journal of Pressure Vessels and Piping,* Vol. 32, 1988, pp. 105–196.

35. SINTAP, "Structural Integrity Assessment Procedures for European Industry—Final Procedure." European Union Project No. BE95-1426, 1999.

36. Anderson, T.L. and Osage, D.A., "API 579: A Comprehensive Fitness-for-Service Guide." *International Journal of Pressure Vessels and Piping,* Vol. 77, 2000, pp. 953–963.

37. Dong, P. and Hong, J.K., "Recommendations for Determining Residual Stresses in Fitness-for-Service Assessment." *WRC Bulletin 476,* Welding Research Council, New York, November 2002.

38. Ainsworth, R.A., Sharples, J.K., and Smith, S.D., "Effects of Residual Stress on Fracture Behavior—Experimental Results and Assessment Methods." *Journal of Strain Analysis,* Vol. 35, 2000, pp. 307–316.

39. Hooton, D.G. and Budden, P.J., "R6 Developments in The Treatment of Secondary Stresses." *ASME PVP,* Vol. 304, 1995, pp. 503–509.

40. R6, Revision 4, "Assessment of the Integrity of Structures Containing Defects." British Energy Generation, Gloucester, 2001.

41. BS 7970:1999, "Guidance on Methods for Assessing the Acceptability of Flaws in Metallic Structures." Amendment No. 1, British Standards Institution, London, 2000.

42. Ainsworth, R.A. and Lei, Y., "Strength Mis-Match in Estimation Schemes." In: Schwalbe, K.-H. and Kocak, M. (eds.), *Mismatching of Interfaces and Welds,* GKSS, Germany, 1997, pp. 35–54.

43. Chell, G.G. and Milne, I., "Ductile Tearing Instability Analysis: A Comparison of Available Techniques." ASTM STP 803, American Society for Testing and Materials and Testing, Philadelphia, PA, 1983, pp. II-179–II-205.

44. Andersson, P., Bergman, M., Brickstad, B., Dahlberg, L., Nilsson, F., and Sattari-Far, I., *A Procedure for Safety Assessment of Components with Cracks—Handbook.* 3rd Ed., SAQ/FoU—Report 96/08, SAQ Kontroll AB, Stockholm, 1996.

10 Fatigue Crack Propagation

Most of the material in the preceding chapters dealt with static or monotonic loading of cracked bodies. This chapter considers crack growth in the presence of cyclic stresses. The focus is on fatigue of metals, but many of the concepts presented in this chapter apply to other materials as well.

In the early 1960s, Paris et al. [1,2] demonstrated that fracture mechanics is a useful tool for characterizing crack growth by fatigue. Since that time, the application of fracture mechanics to fatigue problems has become fairly routine. There are, however, a number of controversial issues and unanswered questions in this field.

The procedures for analyzing constant amplitude fatigue[1] under small-scale yielding conditions are fairly well established, although a number of uncertainties remain. Variable amplitude loading, large-scale plasticity, and short cracks introduce additional complications that are not fully understood.

This chapter summarizes the fundamental concepts and practical applications of the fracture mechanics approach to fatigue crack propagation. Section 10.1 outlines the similitude concept, which provides the theoretical justification for applying fracture mechanics to fatigue problems. This is followed by a summary of the more common empirical and semiempirical equations for characterizing fatigue crack growth. Subsequent sections discuss crack closure, the fatigue threshold, variable amplitude loading, retardation, and growth of short cracks. The micromechanisms of fatigue are also discussed briefly. The final two sections are geared to practical applications: Section 10.8 outlines procedures for experimental measurements of fatigue crack growth and Section 10.9 summarizes the damage tolerance approach to fatigue safe design. Appendix 10 at the end of this chapter addresses the applicability of the J integral to cyclic loading.

10.1 SIMILITUDE IN FATIGUE

The concept of similitude, when it applies, provides the theoretical basis for fracture mechanics. Similitude implies that the crack-tip conditions are uniquely defined by a single loading parameter such as the stress-intensity factor. In the case of a stationary crack, two configurations will fail at the same critical K value, provided an elastic singularity zone exists at the crack tip (Section 2.9). Under certain conditions, fatigue crack growth can also be characterized by the stress-intensity factor, as discussed next.

Consider a growing crack in the presence of a constant amplitude cyclic stress intensity (Figure 10.1). A cyclic plastic zone forms at the crack tip, and the growing crack leaves behind a plastic wake. If the plastic zone is sufficiently small that it is embedded within an elastic singularity zone, the conditions at the crack tip are uniquely defined by the current K value,[2] and the crack growth rate

---

[1] In this chapter, constant amplitude loading is defined as a constant *stress-intensity* amplitude rather than constant stress amplitude.

[2] The justification for the similitude assumption in fatigue is essentially identical to the dimensional argument for steady-state crack growth (Section 3.5.2 and Appendix 3.5.2). If the tip of the growing crack is sufficiently far from its initial position, and external boundaries are remote, the plastic zone size and width of the plastic wake will reach steady-state values.

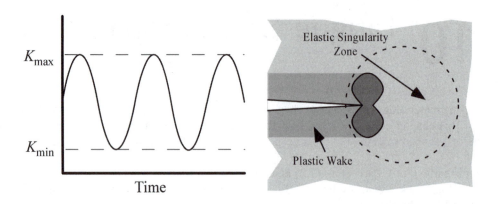

FIGURE 10.1 Constant amplitude fatigue crack growth under small-scale yielding conditions.

is characterized by K_{min} and K_{max}. It is convenient to express the functional relationship for crack growth in the following form:

$$\frac{da}{dN} = f_1(\Delta K, R) \tag{10.1}$$

where

$\Delta K = (K_{max} - K_{min})$
$R = K_{min}/K_{max}$
$da/dN = $ crack growth per cycle

The influence of the plastic zone and plastic wake on crack growth is implicit in Equation (10.1), since the size of the plastic zone depends only on K_{min} and K_{max}.

A number of expressions for f_1 have been proposed, most of which are empirical. Section 10.2 outlines some of the more common fatigue crack growth relationships. Equation (10.1) can be integrated to estimate fatigue life. The number of cycles required to propagate a crack from an initial length a_o to a final length a_f is given by

$$N = \int_{a_o}^{a_f} \frac{da}{f_1(\Delta K, R)} \tag{10.2}$$

If K_{max} or K_{min} varies during cyclic loading, the crack growth in a given cycle may depend on the loading history as well as the current values of K_{min} and K_{max}:

$$\frac{da}{dN} = f_2(\Delta K, R, \mathcal{H}) \tag{10.3}$$

where $\mathcal{H}$ indicates the history dependence, which results from prior plastic deformation. Equation (10.3) violates the similitude assumption; two configurations cyclically loaded at the same ΔK and R will not exhibit the same crack growth rate unless both configurations are subject to the same prior history.

The similitude assumption may not be valid in certain instances of variable amplitude loading, particularly when there are occasional overloads and underloads during the loading history. Section 10.5 discusses the reasons for history-dependent fatigue. Fatigue crack growth analyses become considerably more complicated when prior loading history is taken into account. Consequently, equations of the form of Equation (10.1) are applied whenever possible. It must be recognized, however, that such analyses are potentially subject to error in the case of variable amplitude loading.

Excessive plasticity during fatigue can violate similitude, since K no longer characterizes the crack-tip conditions in such cases. A number of researchers [3, 4] have applied the J integral to fatigue accompanied by large-scale yielding; they have assumed a growth law of the form

$$\frac{da}{dN} = f_3(\Delta J, R) \tag{10.4}$$

where ΔJ is a contour integral for cyclic loading, analogous to the J integral for monotonic loading (see Appendix 10). Equation (10.4) is valid in the case of constant amplitude fatigue in small-scale yielding because of the relationship between J and K under linear elastic conditions.[3] The validity of Equation (10.4) in the presence of significant plasticity is less clear, however.

Recall from Chapter 3 that deformation plasticity (i.e., nonlinear elasticity) is an essential component of J integral theory. When unloading occurs in an elastic-plastic material, deformation plasticity theory no longer models the actual material response (see Figure 3.7). Consequently, the ability of the J integral to characterize fatigue crack growth in the presence of large-scale cyclic plasticity is questionable, to say the least.

There is, however, some theoretical and experimental evidence in support of Equation (10.4). If certain assumptions are made with respect to the loading and unloading branches of a cyclic stress-strain curve, it can be shown that ΔJ is path independent, and it uniquely characterizes the change in stresses and strains in a given cycle [5, 6]. Appendix 10 summarizes this analysis. Experimental data [3, 4] indicate that ΔJ correlates crack growth data reasonably well in certain cases. Several researchers have found that crack-tip-opening displacement (CTOD) may also be a suitable parameter for fatigue under elastic-plastic conditions [7].

The validity of Equation (10.4) has not been proven conclusively, but this approach appears to be useful for many engineering problems. Of course, Equation (10.4) is subject to the same restrictions on prior history as Equation (10.1). The crack growth rate may exhibit a history effect if ΔJ or R varies during cyclic loading.

10.2 EMPIRICAL FATIGUE CRACK GROWTH EQUATIONS

Figure 10.2 is a schematic log-log plot of da/dN vs. ΔK, which illustrates typical fatigue crack growth behavior in metals. The sigmoidal curve contains three distinct regions. At intermediate ΔK values, the curve is linear, but the crack growth rate deviates from the linear trend at high and low ΔK levels. At the low end, da/dN approaches zero at a threshold ΔK, below which the crack will not grow. Section 10.4 explores the causes of this threshold. In some materials, the observed growth rate increases rapidly at high ΔK values. There are two possible explanations for the Region III behavior. Some researchers have hypothesized that the crack growth rate accelerates as K_{max} approaches K_c, the fracture toughness of the material. According to this hypothesis, microscopic fracture events (e.g., pop-ins) contribute to crack growth, resulting in a higher overall growth rate. An alternative hypothesis is that the apparent acceleration in da/dN is not real but is due to the influence of crack-tip plasticity on the true driving force for fatigue. At high K_{max} values, linear elastic fracture mechanics is no longer valid, and a parameter like ΔJ might be more appropriate to characterize fatigue.

The linear region of the log-log plot in Figure 10.3 can be described by a power law:

$$\frac{da}{dN} = C\Delta K^m \tag{10.5}$$

---

[3] $\Delta J = \Delta K^2/E'$ in the case of small-scale yielding. Thus ΔJ cannot be interpreted as the range of applied J values. That is, $\Delta J \neq J_{max} - J_{min}$ in general. See Appendix 10 for additional background on the definition of ΔJ.

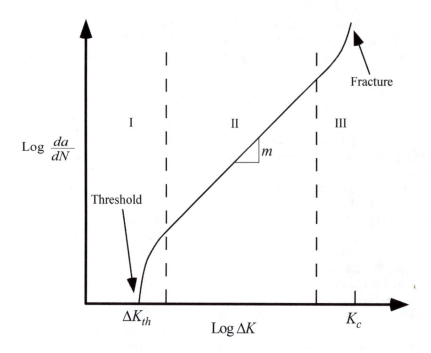

FIGURE 10.2 Typical fatigue crack growth behavior in metals.

where C and m are material constants that are determined experimentally. According to Equation (10.5), the fatigue crack growth rate depends only on ΔK; da/dN is insensitive to the R ratio in Region II.

Paris and Erdogan [2] were apparently the first to discover the power-law relationship for fatigue crack growth in Region II. They proposed an exponent of 4, which was in line with their experimental data. Subsequent studies over the past three decades, however, have shown that m can range from 2 to 4 for most metals in the absence of a corrosive environment. Equation (10.5) has become widely known as the *Paris Law*.

A number of researchers have developed equations that model all or part of the sigmoidal da/dN and ΔK relationship [8–14]. Most of these equations are empirical, although some are based on physical considerations. Forman [8] proposed the following relationship for Region II and Region III:

$$\frac{da}{dN} = \frac{C\Delta K^m}{(1-R)K_c - \Delta K} \tag{10.6}$$

This equation can be rewritten in the following form:

$$\frac{da}{dN} = \frac{C\Delta K^{m-1}}{\frac{K_c}{K_{max}} - 1} \tag{10.7}$$

Thus the crack growth rate becomes infinite as K_{max} approaches K_c. Of course, this equation is based on the assumption that Region III behavior is caused by a superposition of fracture and fatigue rather than plastic zone effects. Note that the above relationship contains an R ratio dependence, while Equation (10.5) assumes that da/dN depends only on ΔK. Another important

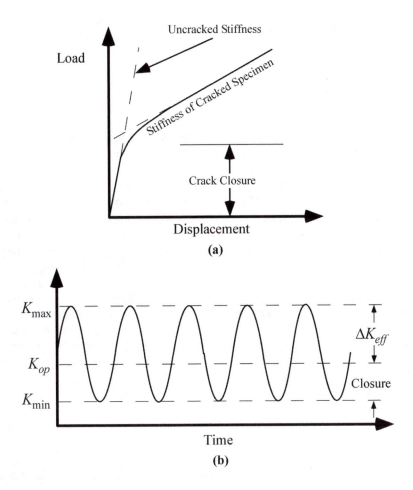

FIGURE 10.3 Crack closure during fatigue crack growth. The crack faces contact at a positive load (a), resulting in a reduced driving force for fatigue, ΔK_{eff} (b): (a) load-displacement behavior and (b) definition of effective stress intensity range.

point is that the material constants C and m in the Forman equation do not have the same numerical values or units as in the Paris-Erdogan equation (Equation (10.5)).

Klesnil and Lukas [10] modified Equation (10.5) to account for the threshold:

$$\frac{da}{dN} = C\left(\Delta K^m - \Delta K_{th}^m\right) \tag{10.8}$$

Donahue [11] suggested a similar equation, but with the exponent m applied to the quantity $(\Delta K - \Delta K_{th})$. In both cases, the threshold is a fitting parameter to be determined experimentally. One problem with these equations is that ΔK_{th} often depends on the R ratio (see Section 10.4).

A number of equations attempt to describe the entire crack growth curve, taking account of both the threshold and Stage III behavior. Most such relationships assume that Stage III occurs when K_{max} approaches K_c, although this assumption does not appear to have been based on solid experimental evidence.

The most common expression to describe fatigue crack growth in all three regions was developed at NASA and was first published by Forman and Mettu [14]. This equation, in a slightly simplified form,[4] is given by

$$\frac{da}{dN} = C\Delta K^m \frac{\left(1 - \frac{\Delta K_{th}}{\Delta K}\right)^p}{\left(1 - \frac{K_{max}}{K_c}\right)^q} \tag{10.9}$$

where C, m, p, and q are material constants. At intermediate ΔK values were $\Delta K \gg \Delta K_{th}$ and $K_{max} \ll K_c$, Equation (10.9) reduces to Equation (10.5). Therefore, the C and m values for Equation (10.5) and Equation (10.9) are equivalent.

Equation (10.5) to Equation (10.9) all have the form of Equation (10.1). Each of these equations can be integrated to infer fatigue life (Equation (10.2)). The most general of these expressions contains six material constants:[5] C, m, p, q, K_{crit}, and ΔK_{th}. For a given material, the fatigue crack growth rate depends only on the loading parameters ΔK and R, at least according to Equation (10.5) to Equation (10.9). Therefore, all of the preceding expressions assume elastic similitude of the growing crack; none of these equations incorporate a history dependence, and, thus, are strictly valid only for constant (stress intensity) amplitude loading. Many of these formulas, however, were developed with variable amplitude loading in mind. Although there are many situations where similitude is a good assumption for variable amplitude loading, one must always bear in mind the potential for history effects. See Section 10.5 for additional discussion of this issue.

Dowling and Begley [3] applied the J integral to fatigue crack growth under large-scale yielding conditions where K is no longer valid. They fit the growth rate data to a power-law expression in ΔJ:

$$\frac{da}{dN} = C\Delta J^m \tag{10.10}$$

Appendix 10 outlines the theoretical justification and limitations of J-based approaches for fatigue.

EXAMPLE 10.1

Derive an expression for the number of stress cycles required to grow a semicircular surface crack from an initial radius a_o to a final size a_f, assuming the Paris-Erdogan equation describes the growth rate. Assume that a_f is small compared to plate dimensions, the crack maintains its semicircular shape, and that the stress amplitude $\Delta\sigma$ is constant.

Solution: The stress-intensity amplitude for a semicircular surface crack in an infinite plate (Figure 2.19) can be approximated by

$$\Delta K \approx \frac{1.04}{\sqrt{2.464}} \Delta\sigma \sqrt{\pi a} = 0.663\Delta\sigma\sqrt{\pi a}$$

if we neglect the ϕ dependence of the surface correction factor λ_s. Substituting this expression into Equation (10.5) gives

$$\frac{da}{dN} = C(0.663\Delta\sigma)^m (\pi a)^{m/2}$$

---

[4] The general version of the Forman-Mettu equation contains a multiplying factor on ΔK to account for crack closure effects (Section 10.3).
[5] The threshold stress-intensity range ΔK_{th} is not a true material constant since it usually depends on the R ratio (Section 10.4).

which can be integrated to determine fatigue life:

$$N = \frac{1}{C(0.663\sqrt{\pi\Delta\sigma})^m} \int_{a_o}^{a_f} a^{-m/2} da$$

$$= \frac{a_o^{1-m/2} - a_f^{1-m/2}}{C\left(\frac{m}{2}-1\right)(0.663\sqrt{\pi\Delta\sigma})^m} \quad \text{(for } m \neq 2\text{)}$$

Closed-form integration is possible in this case because the K expression is relatively simple. In most instances, numerical integration is required.

10.3 CRACK CLOSURE

An accidental discovery by Elber [15] in 1970 resulted in several decades of research into a phenomenon known as *crack closure*. He noticed an anomaly in the elastic compliance of several fatigue specimens, which Figure 10.3(a) schematically illustrates. At high loads, the compliance ($d\Delta/dP$) agreed with standard formulas for fracture mechanics specimens (Appendix 7), but at low loads, the compliance was close to that of an uncracked specimen. Elber believed that this change in compliance was due to the contact between crack surfaces (i.e., crack closure) at loads that were low but greater than zero.

Elber postulated that crack closure decreased the fatigue crack growth rate by reducing the effective stress-intensity range. Figure 10.3(b) illustrates the closure concept. When a specimen is cyclically loaded at K_{max} and K_{min}, the crack faces are in contact below K_{op}, the stress intensity at which the crack opens. Elber assumed that the portion of the cycle that is below K_{op} does not contribute to fatigue crack growth because there is no change in crack-tip strain during cyclic loading of a closed crack. He defined an effective stress-intensity range as follows:

$$\Delta K_{eff} \equiv K_{max} - K_{op} \tag{10.11}$$

He also introduced an effective stress-intensity ratio:

$$U \equiv \frac{\Delta K_{eff}}{\Delta K}$$

$$= \frac{K_{max} - K_{op}}{K_{max} - K_{min}} \tag{10.12}$$

Elber then proposed a modified Paris-Erdogan equation:

$$\frac{da}{dN} = C\Delta K_{eff}^m \tag{10.13}$$

Equation (10.13) has been reasonably successful in correlating fatigue crack growth data at various R ratios. The numerical value of the material constant C is different in Equation (10.5) and Equation (10.13) if closure occurs in Region II, such that $\Delta K_{eff} < \Delta K$.

Since Elber's original study, numerous researchers have confirmed that crack closure does in fact occur during fatigue crack propagation. Suresh and Ritchie [16] identified five mechanisms for fatigue crack closure, which are illustrated in Figure 10.4.

Plasticity-induced closure, Figure 10.4(a), results from residual stresses in the plastic wake. Budiansky and Hutchinson [17] applied the Dugdale-Barenblatt strip-yield model to this problem and showed that residual stretch in the plastic wake causes the crack faces to close at a positive remote stress. Although quantitative predictions from the Budiansky and Hutchinson model do not agree with experimental data [18], this model is useful for demonstrating qualitatively the effect

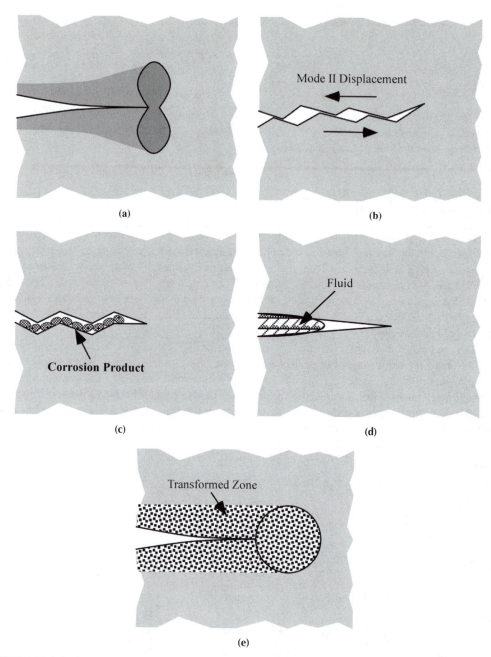

FIGURE 10.4 Fatigue crack closure mechanisms in metals: (a) plasticity-induced closure (b) roughness-induced closure, (c) oxide-induced closure, (d) closure induced by a viscous fluid, and (e) transformation-induced closure. Taken from Suresh, S. and Ritchie, R.O., "Propagation of Short Fatigue Cracks." *International Metallurgical Reviews*, Vol. 29, 1984, pp. 445–476.

of plasticity on crack closure. A number of investigators [19–24] have studied plasticity-induced closure with finite element analysis. James and Knott [25] performed an experiment that provided compelling evidence for plasticity-induced closure. They grew a fatigue crack in a standard specimen and then used electron discharge machining to remove a small amount of material from the crack faces up to 0.5 mm from the crack tip. When they placed the sample back into the test machine, the crack growth rate increased by an order of magnitude compared to prior measurements.

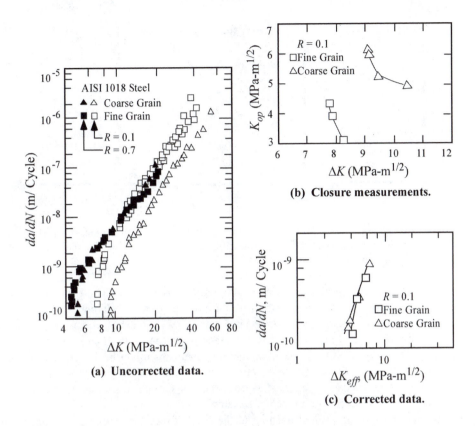

FIGURE 10.5 Effect of grain size on fatigue crack growth in mild steel. Taken from Gray, G.T., Williams, J.C., and Thompson, A.W., "Roughness Induced Crack Closure: An Explanation for Microstructurally Sensitive Fatigue Crack Growth." *Metallurgical Transactions*, Vol. 14A, 1983, pp. 421–433.

They concluded that the plastic wake behind the crack was causing closure and a reduced crack growth rate, and that crack growth rate accelerated because the zone of compressive residual stresses in the plastic wake was removed by machining.

Roughness-induced closure, which is illustrated in Figure 10.4(b), is influenced by the microstructure. Although fatigue cracks propagate in pure Mode I conditions on a global scale, crack deflections due to microstructural heterogeneity can lead to mixed mode conditions on the microscopic level. When the crack path deviates from the Mode I symmetry plane, the crack is subject to Mode II displacements, as Figure 10.4(b) illustrates. These displacements cause mismatch between upper and lower crack faces, which in turn results in contact of crack faces at a positive load. Coarse-grained materials usually produce a higher degree of surface roughness in fatigue, and correspondingly higher closure loads [26]. Figure 10.5 illustrates the effect of grain size on fatigue crack propagation in 1018 steel. At the lower R ratio, where closure effects are most pronounced, the coarse-grained material has a higher ΔK_{th}, due to a higher closure load that is caused by greater surface roughness (Figure 10.5(b)). Note that grain size effects disappear when the data are characterized by ΔK_{eff} (Figure (10.5(c)).

Oxide-induced closure, Figure 10.4(c), is usually associated with an aggressive environment. Oxide debris or other corrosion products become wedged between crack faces. Crack closure can also be introduced by a viscous fluid, as Figure 10.4(d) illustrates. The fluid acts as a wedge between crack faces, somewhat like the oxide mechanism. A stress-induced martensitic transformation at the tip of the growing crack can result in a process zone wake,[6] as Figure 10.4(e) illustrates.

---

[6] See Section 6.2.2 for a discussion of stress-induced martensitic transformations in ceramics.

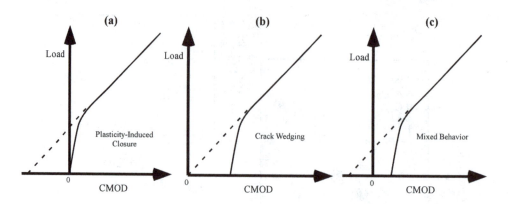

FIGURE 10.6 Expected load-displacement behavior for: (a) plasticity-induced closure, (b) wedging mechanisms, and (c) mixed behavior. In order to observe these trends, however, the clip gage must be zeroed prior to the start of fatigue cracking.

Residual stresses in the transformed zone can lead to crack closure. The relative importance of the various closure mechanisms depends on microstructure, yield strength, and environment.

Three of the closure mechanisms illustrated in Figure 10.4 (roughness, corrosion product, and viscous fluid) involve crack wedging. That is, the crack is prevented from closing completely by an obstruction of some type. Strictly speaking, these mechanisms might be described more appropriately by the term *residual crack opening* rather than crack closure. The plasticity and phase transformation mechanisms can be regarded as true crack closure because residual stress applies a closure force to the crack faces. Both types of mechanisms can result in a diminished ΔK_{eff}, either through premature closure or residual opening of the crack. Since the wedging mechanisms have traditionally been referred to as "crack closure," that convention will be adopted here despite its shortcomings. Wedging mechanisms are discussed in greater detail in Section 10.3.1.

In principle, it should be possible to distinguish between plasticity-induced closure and wedging mechanisms by observing the load-displacement curve. (Transformation-induced closure is rare and will be excluded from the present discussion.) Figure 10.6 schematically compares three scenarios: plasticity-induced closure, crack wedging, and mixed behavior. Load is plotted vs. crack-mouth-opening displacement (CMOD). Because the crack is actually closing as a result of the plasticity mechanism, the load-CMOD curve should pass through the origin. In the case of a wedging mechanism, there will be a residual displacement at zero load, but the load-CMOD curve at high loads (when the crack is fully open) should extrapolate to the origin. When both mechanisms are present, there will be a positive residual displacement and the load-CMOD curve at high loads will extrapolate to a negative displacement, as the third schematic in Figure 10.6 illustrates.

In an actual experiment, it would be possible to distinguish between the various behaviors illustrated in Figure 10.6 *only if the clip gage is zeroed before the start of cracking.* In many experiments, the specimen is precracked without a clip gage, and the clip gage is inserted and zeroed later, after closure has had a chance to develop in the initial fatigue crack. In that case, it would not be possible to determine a true zero CMOD reference from the load-clip gage record.

10.3.1 A CLOSER LOOK AT CRACK-WEDGING MECHANISMS

Consider an idealized scenario where a rigid wedge is inserted into an open crack, as Figure 10.7 illustrates. Suppose that the shape of this wedge is such that the crack perfectly conforms to it when the load is removed. That is, assume that the crack faces contact the wedge at all points simultaneously when the load is removed. As the applied load decreases to zero, the crack is held open at a fixed displacement. An applied stress intensity of K_{wedge} exists at the crack tip despite the lack of an externally

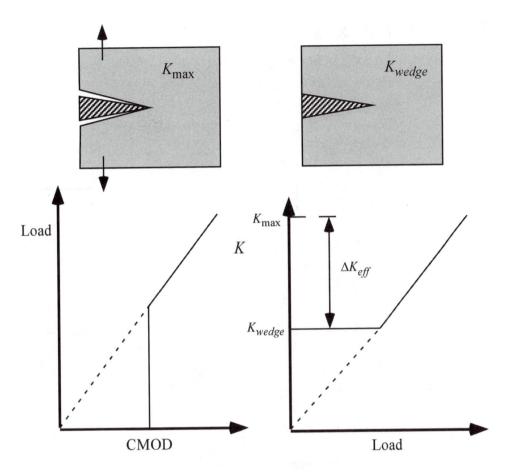

FIGURE 10.7 Load-displacement behavior and ΔK_{eff} for an ideal wedge, which is rigid and conforms perfectly to the crack-opening profile.

applied load. When load is reapplied, K does not change until load is a sufficient cause for a crack-opening displacement greater than the width of the wedge. If this configuration were loaded cyclically, such that the crack faces were in contact with the wedge during the portion of each cycle, ΔK_{eff} would equal the difference between K_{max} and K_{wedge}, as Figure 10.7 illustrates.

Now assume that the crack contains a single rigid particle, as Figure 10.8 illustrates. As the load is removed and the crack faces close, they eventually contact the particle. When this initial contact occurs, the slope of the load-CMOD curve becomes steeper. As load is reduced further, portions of the crack face not in contact with the particle continue to close, resulting in a continual decrease in the applied K. There is a small residual K when the load is removed completely. Unlike the ideal wedge in Figure 10.7, however, ΔK_{eff} for cyclic loading is greater than that would be inferred by defining K_{op} at the point where the compliance changes. That is, one would overestimate closure effects (and underestimate fatigue driving force) if he or she defined the opening load where the slope changes.

Figure 10.7 and Figure 10.8 depict two extremes of crack-wedging behavior. Figure 10.9 illustrates a more realistic case, where the crack is filled with particles of various sizes. As the load is relaxed, the slope of the load-CMOD curve gradually changes as more particles make contact with the crack. Eventually, no further contact occurs and the CMOD attains a constant value, assuming the particles are rigid. In the real world, the final slope of the load-CMOD curve would

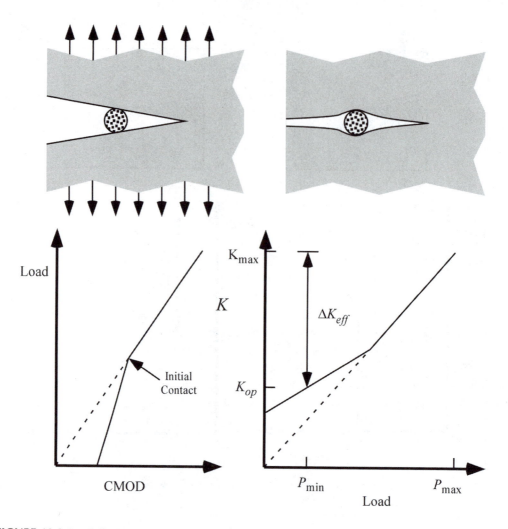

FIGURE 10.8 Load-displacement behavior and ΔK_{eff} for a single particle or asperity.

reflect the elastic properties of the objects that fill the crack.[7] As was the case for the single-particle scenario, defining K_{op} at the point of initial slope change could lead to errors.

The important factor to keep in mind is how the wedging action (or plasticity-induced closure, for that matter) affects the true driving force for fatigue. Defining ΔK_{eff} for an ideal wedge is unambiguous and straightforward. More realistic cases are not quite so simple. Traditionally, the measurement of K_{op} has been highly subjective, and different individuals and laboratories have obtained widely varying results for a given material under the same test conditions. Section 10.8.3 proposes a K_{op} definition that removes the subjectivity and provides a more accurate indication of the effective ΔK. This new definition is based on the relationship between the applied load and the *true* applied K. Figure 10.7 to Figure 10.9 show simplified examples of the proposed definition of ΔK_{eff}.

---

[7] For laboratory specimens, there is an additional contribution from compliance of the machined notch. Even if the fatigue crack is fully closed or wedged open, the faces of the notch are not in contact. The slope of the load vs. clip gage displacement curve reflect the compliance of a notched but uncracked specimen, as Figure 10.3 illustrates.

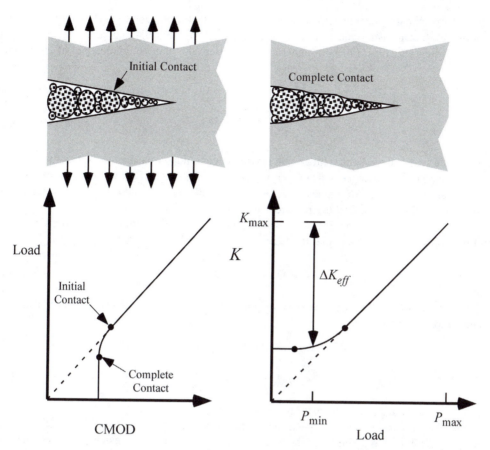

FIGURE 10.9 Load-displacement behavior and ΔK_{eff} for a crack that is filled with multiple particles or asperities of various sizes.

10.3.2 EFFECTS OF LOADING VARIABLES ON CLOSURE

Most experts believe that the stress intensity for crack closure is not actually a material constant, but depends on a number of factors. Over the years, a number of investigators have attempted to correlate K_{op} to loading variables such as R ratio, but with limited success. Part of the problem lies in the subjective nature of closure measurements, as discussed above. Another problem is that different closure behavior occurs in different alloys and in different loading regimes for a given alloy. No single empirical correlation works for all situations. What follows is an overview of some of the attempts to correlate closure to loading variables.

Elber [15] measured the closure stress intensity in 2023-T3 aluminum at various load levels and R ratios, and obtained the following empirical relationship:

$$U = 0.5 + 0.4R \quad (-0.1 \leq R \leq 0.7) \tag{10.14}$$

Subsequent researchers [27–29] inferred similar empirical expressions for other alloys.

According to Equation (10.14), U depends only on R. Shih and Wei [30, 31], however, argued that the Elber expression, as well as many of the subsequent equations, are oversimplified. Shih and Wei observed a dependence on K_{max} for crack closure in a Ti-6Al-4V titanium alloy. They also showed that experimental data of earlier researchers, when replotted, exhibit a definite K_{max} dependence.

There has been a great deal of confusion and controversy about the K_{max} dependence of U. Hudak and Davidson [18] cited contradictory examples from the literature where various researchers reported U to increase, decrease, or remain constant with increasing K_{max}. Hudak and Davidson performed closure measurements on a 7091 aluminum alloy and 304 stainless steel over a wide range of loading variables. For both materials, they inferred a closure relationship of the form:

$$U = 1 - \frac{K_o}{K_{max}}$$

$$= 1 - \frac{K_o(1-R)}{\Delta K} \tag{10.15}$$

where K_o is a material constant. Hudak and Davidson concluded that the inconsistent results from the literature could be attributed to a number of factors: (1) the range of ΔK values in the experiments was too narrow; (2) the ΔK values were not sufficiently close to the threshold; and (3) the measurement techniques lacked the required sensitivity.

The effect of loading variables on K_{op} can be inferred by substituting the relationships for U into Equation (10.12). For example, the Elber implies the following relationship for K_{op}:

$$K_{op} = \Delta K \left(\frac{1}{1-R} - 0.5 - 0.4R \right) \tag{10.16}$$

Note that the ratio $K_{op}/\Delta K$ depends only on the R ratio. Equation (10.15), however, leads to a different expression for K_{op}:

$$K_{op} = K_o(1-R) + K_{max}R$$

$$= K_o(1-R) + \frac{\Delta KR}{1-R} \tag{10.17}$$

Thus K_o is the opening stress intensity for $R = 0$.

Hudak and Davidson [18] attributed the confusion and controversy over the effect of loading variables on closure to experimental factors. McClung [32] conducted an extensive review of experimental and analytical closure results and concluded that there are three distinct regimes of crack closure. Near the threshold, closure levels decrease with increasing stress intensity, while U is independent of K_{max} at intermediate ΔK levels. At high ΔK values, the specimen experiences a loss in constraint, and U decreases with increasing stress intensity. McClung found that no single equation could describe closure in all three regimes. According to McClung, most of the seemingly contradictory data in the literature can be reconciled by considering the regimes in which the data were collected.

Microstructural effects can also lead to differences in observed closure behavior in various materials. Figure 10.5, for example, shows the effect of grain size on crack closure in 1018 steel. Moreover, various closure mechanisms can occur in different materials and environments, as Figure 10.4 illustrates. A correlation developed for, say, plasticity-induced closure cannot be expected to apply to a situation where roughness-induced closure dominates.

10.4 THE FATIGUE THRESHOLD

The fatigue threshold ΔK_{th} is the point below which a fatigue crack will not grow. Experimental measurements of the threshold are usually inferred from a load-shedding procedure, where ΔK is gradually reduced until the crack growth rate reaches a very small value. In most experiments in the threshold range, either K_{max} or the R ratio is held constant, while ΔK is reduced. The way in which the test is conducted can affect the measured ΔK_{th} for reasons described later. Section 10.8 describes the experimental procedures for measuring da/dN and ΔK_{th}.

Most experts believe that the threshold consists of two components: an intrinsic threshold that is a material property, and an extrinsic component that is a function of loading variables such as the R ratio. The precise mechanism for the intrinsic threshold has not been established, but several researchers have developed models based on dislocation emission from the crack tip [33] or blockage of slip bands by grain boundaries [34] (see Section 10.7.2). Most experts believe that the R ratio effects on the threshold are due to crack closure. This viewpoint is discussed in Section 10.4.1.

There is a minority opinion in the fatigue community that believes that closure plays a minor role in fatigue behavior near the threshold. Instead, they believe that there are two intrinsic thresholds: a ΔK threshold and a K_{max} threshold. Section 10.4.2 describes this model.

Both points of view find support in the published data. While the empirical evidence supports a K_{max} threshold, such a threshold is also consistent with the crack closure argument. The potential proof of who is correct lies in closure measurements. That is, can closure measurements reconcile the R ratio effect on ΔK_{th}?

Unfortunately, this issue is clouded by the fact that traditional closure measurements are notoriously subjective and inconsistent. There are examples in the literature where investigators purport to correct for R ratio effects in ΔK_{th} with closure measurements. Such measurements, however, can potentially be biased by knowledge of the desired outcome. That is, knowing the intrinsic threshold after extrinsic effects are removed could influence the measurement of the apparent closure loads. Unless correlation comes from closure measurements that are made blindly without knowledge of the desired outcome, or if a completely unambiguous definition of closure is adopted, there will continue to be debates about the role of crack closure in the fatigue threshold.

10.4.1 The Closure Model for the Threshold

Figure 10.10 schematically illustrates the effect of crack closure on the apparent fatigue threshold. Let us assume that a given material has an intrinsic threshold ΔK_{th}^* and that K_{op} is also a material constant that is independent of the R ratio. The relationship between the apparent threshold ΔK_{th} and the intrinsic threshold is given by

$$U\Delta K_{th} = \Delta K_{th}^* \qquad (10.18)$$

where U is defined in Equation (10.12). Rewriting U in terms of ΔK_{th} and R gives

$$U = \min\left[\left(\frac{1}{1-R} - \frac{K_{op}}{\Delta K_{th}}\right), 1\right] \qquad (10.19)$$

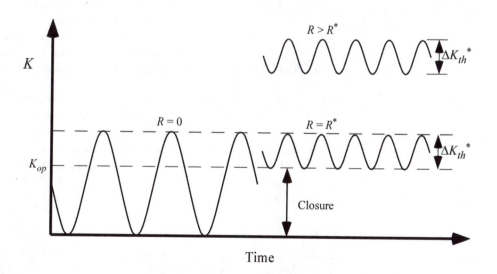

FIGURE 10.10 Schematic illustration of the relationship between closure behavior and the R ratio, assuming K_{op} is constant.

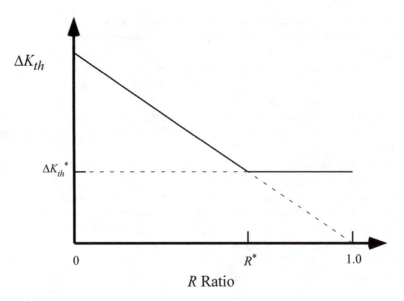

FIGURE 10.11 Schematic illustration of the effect of R ratio on ΔK_{th}, as implied by Figure 10.10.

Substituting Equation (10.19) into Equation (10.18) and solving for ΔK_{th} leads to

$$\Delta K_{th} = \begin{cases} \left(K_{op} + \Delta K_{th}^*\right)(1-R), & R \le R^* \\ \Delta K_{th}^*, & R > R^* \end{cases} \qquad (10.20)$$

where R^* is the R ratio above which closure no longer exerts an influence, as Figure 10.10 illustrates; R^* is given by

$$R^* = 1 - \frac{\Delta K_{th}^*}{K_{op} + \Delta K_{th}^*} \qquad (10.21)$$

Figure 10.11 is a schematic plot of Equation (10.20). This expression predicts that the threshold stress intensity range varies linearly with R below R^* and is constant at higher R ratios. Figure 10.12 is a plot of actual threshold data for a variety of steels. Equation (10.20) appears to provide a very good description of the data. For most of the steels on this plot, R^* appears to be around 0.8 and ΔK_{th}^* is between 2 and 3 MPa $\sqrt{\text{m}}$. The exception to the trend in Figure 10.12 is a 1590 MPa (230 ksi) yield strength steel that does not exhibit an R ratio dependence. It turns out that Equation (10.20) accurately describes the ratio dependence of ΔK_{th} in a wide range of alloys.

Figure 10.13 shows crack growth data for a mild steel in the threshold range at various R ratios [7]. Crack growth is plotted against both ΔK and ΔK_{eff}, where the latter quantity corresponds to data corrected for closure by Equation (10.11). For the highest R ratios, closure was not observed, so $\Delta K = \Delta K_{eff}$. When data at lower R ratios are corrected for closure, the R ratio effect disappears and all data exhibit the same threshold, which corresponds to ΔK_{th}^* for the material. Provided the closure measurements in Ref. [7] are accurate and unbiased, Figure 10.13 presents strong evidence in favor of the closure mechanism in the threshold range.

10.4.2 A TWO-CRITERION MODEL

Vasudevan et al. [35–40] have proposed an alternative way of looking at threshold behavior. They argue that there are two intrinsic thresholds, one on ΔK and one on K_{max}. They believe that the latter parameter is not associated with closure. Rather, they contend that the K_{max} threshold reflects the minimum stress-intensity level that must be achieved at the crack tip for fatigue damage to occur,

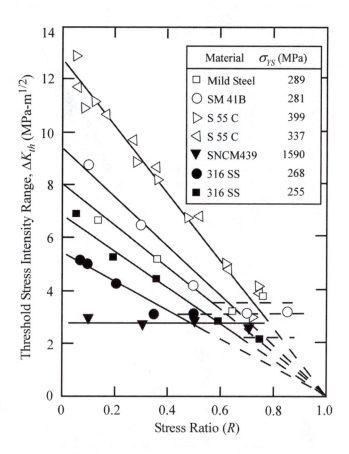

FIGURE 10.12 Effect of R ratio on the threshold stress-intensity range for various steels. Taken from Tanaka, K., "Mechanics and Micromechanics of Fatigue Crack Propagation." ASTM STP 1020, American Society for Testing and Materials, Philadelphia, PA, 1989, pp. 151–183.

and it is an intrinsic material property. They point out that their approach is analogous to the Goodman diagram for S-N curves, where fatigue life is a function of both stress amplitude and mean stress.

Figure 10.14 is a schematic illustration of the two-criterion threshold concept. The two thresholds on cyclic stress intensity ΔK_{th}^* and maximum stress intensity K_{max}^* form an L-shaped curve. Both thresholds must be exceeded for crack growth to occur, according to this model. Figure 10.15 shows an L-shaped threshold plot for Ti – 6Al – V [35].

Referring to the right-hand side of Figure 10.14, we see that when an L-shaped threshold curve is plotted as ΔK_{th} vs. R, the resulting trend is identical to Figure 10.11. This trend can be expressed mathematically as follows:

$$\Delta K_{th} = \begin{cases} K_{max}^*(1-R), & R \le R^* \\ \Delta K_{th}^* & R > R^* \end{cases}$$

Therefore, the existence of thresholds for both K_{max} and ΔK is wholly consistent with the closure argument. If the R ratio effects on the measured threshold are, indeed, a result of crack closure, then the K_{max} threshold is simply the sum of the opening K and the intrinsic ΔK threshold:

$$K_{max}^* = K_{op} + \Delta K_{th}^* \tag{10.22}$$

However, Vasudevan et al. [35–40] contend that K_{max}^* is an intrinsic material property, not the result of crack closure. Their main arguments are threefold:

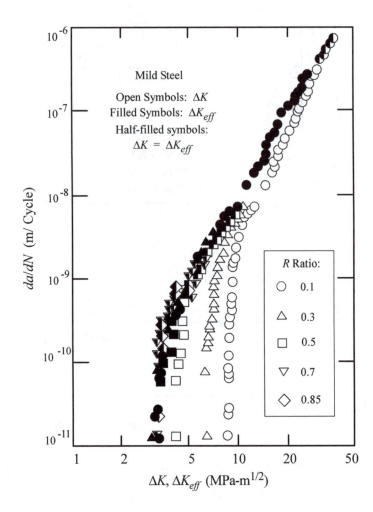

FIGURE 10.13 Fatigue crack growth data near the threshold for mild steel at various R ratios. Taken from Tanaka, K., "Mechanics and Micromechanics of Fatigue Crack Propagation." ASTM STP 1020, American Society for Testing and Materials, Philadelphia, PA, 1989, pp. 151–183.

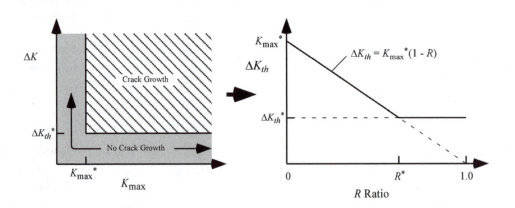

FIGURE 10.14 Schematic illustration of the dual threshold model. According to this model, thresholds on both ΔK and K_{max} must be exceeded for crack growth to occur. Note that this model predicts the same R ratio dependence on ΔK_{th} as the closure model (Figure 10.11).

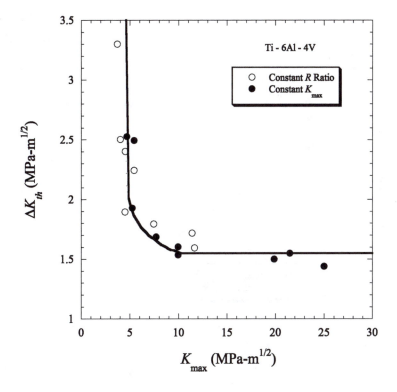

FIGURE 10.15 Threshold ΔK vs. K_{max} for Ti – 6Al – 4V. These threshold data were generated using the load-shedding method, where ΔK decreases at a constant K_{max} or R. Taken from Vasudevan, A.K., Sadananda, K., and Loutat, N., "A Review of Crack Closure, Fatigue Crack Threshold, and Related Phenomena." *Materials Science and Engineering*, Vol. A188, 1994, pp. 1–22.

1. A comparison of threshold data for air and vacuum environments indicates that plasticity-induced closure is not a significant factor in the threshold region.
2. While roughness-induced closure can occur to some degree, traditional methods that define closure from changes in the slope of the load-displacement curve grossly overestimate K_{op}.
3. Closure measurements tend to be highly scattered, but the R ratio dependence of ΔK_{th} is highly consistent and reproducible. If closure were responsible for the R ratio effects, they argue, ΔK_{th} should exhibit scatter comparable to that of K_{op}.

The first point appears to have some merit, as discussed later, but the other two arguments are less convincing. The contention that closure measurements grossly overestimate the true closure behavior is based largely on an analysis similar to the single-particle model illustrated in Figure 10.8. If crack wedging, either through roughness or a corrosion product, were indeed controlled by a single obstruction or asperity, Vasudevan and his colleagues would be correct that defining K_{op} at the point at which the slope changes would lead to a gross overestimate of the effect of closure on the fatigue driving force. Figure 10.9 illustrates a more realistic case, where there are multiple obstructions or asperities along the entire crack face. Although there is a finite load range where the compliance is continually changing due to an increasing number of obstructions coming into contact with the crack face, the true K_{op}, as defined in Figure 10.9, generally occurs relatively close to the initial contact point. The region of nonlinearity in the load-displacement curve has led to somewhat subjective definitions of closure. The ambiguity in closure measurements is undoubtedly partially responsible for the observed scatter. If a self-consistent and physically based definition of K_{op} is adopted (Section 10.8.3), the scatter in closure measurements may well reduce to a level of scatter comparable to that observed in ΔK_{th} measurements.

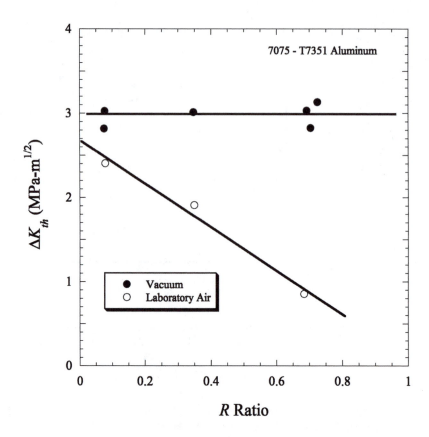

FIGURE 10.16 Comparison of ΔK_{th} for an aluminum alloy in air and vacuum environments. Taken from Vasudevan, A.K., Sadananda, K., and Glinka, G., "Critical Parameters for Fatigue Damage." *International Journal of Fatigue*, Vol. 23, 2001, pp. S39–S53.

10.4.3 THRESHOLD BEHAVIOR IN INERT ENVIRONMENTS

When fatigue crack growth experiments are performed in a vacuum environment, the data can be significantly different from that in air. One does not normally think of air as a corrosive environment, but there is definitely an environmental interaction when air is present at the tip of a fatigue crack. These effects can be particularly pronounced in the threshold region. The primary corrosive agents in air are apparently moisture combined with oxygen, because the environmental interaction increases with relative humidity. By comparing the threshold behavior for a given material in air and vacuum environments, it is possible to infer information about the role, if any, of plasticity-induced closure on ΔK_{th}.

Figure 10.16 compares air and vacuum threshold values for a 7075 aluminum alloy [40]. The threshold in air is significantly less than in a purely inert environment, especially at high R ratios. Moreover, the vacuum data do not exhibit an R ratio dependence for positive values of R. It appears that significant crack closure is definitely not occurring in the vacuum. Most fatigue experts view the existence of an R ratio dependence on ΔK_{th} as an indication of closure effects. The reverse argument is that the *absence* of an R ratio dependence on ΔK_{th} in a specific data set would imply the *lack* of significant closure.[8] The formation of a plastic zone in front of the crack and the plastic

---

[8] The lack of an R ratio effect on the threshold is not absolute proof of the absence of closure. It is theoretically possible for ΔK_{th} to be independent of the R ratio in the presence of closure if K_{op} exhibits an R ratio dependence that exactly cancels the effect indicated by Equation (10.20). Such a coincidence appears unlikely, however. If K_{op} did have such an R ratio dependence in a vacuum environment, one would expect the same dependence to occur in air if K_{op} was driven by plasticity-induced closure.

wake behind the crack tip should not be affected by the presence or absence of air in the crack. If significant closure is not occurring in the vacuum data, the R ratio dependence of ΔK_{th} for the air data cannot be due to plasticity-induced closure.

Even if plasticity-induced closure does not play a significant role near the threshold, there is compelling evidence that other closure mechanisms are important in this regime. The published literature contains multiple data sets like Figure 10.5 and Figure 10.13, where the R ratio dependence of ΔK_{th} has been correlated with closure measurements. There are many sets of corrosion fatigue data where the measured ΔK_{th} is elevated (relative to the value in air) due to corrosion product wedging in the crack.

The most likely explanation for the different closure behavior in air and vacuum is associated with surface morphology. Fatigue crack growth in moist air is typically transgranular in the threshold regime, as discussed in Section 10.7. This cleavage-like crack path is apparently due to environmental interactions, and it apparently results in roughness-induced closure. Fatigue crack growth in an inert environment is relatively planar, so the roughness-induced closure is not as significant.

The differing behavior between air and inert environments in Figure 10.16 is fairly typical for a range of alloys. Vasudevan et al. [35] have compiled published threshold data for a variety of alloy systems and plotted ΔK_{th} vs. K_{max} to obtain L-shaped threshold curves like Figure 10.14 and Figure 10.15. The dual thresholds inferred from such plots, ΔK_{th}^{*} and K_{max}^{*}, were then correlated to each other for both air and vacuum environments. Figure 10.17 shows the results of this exercise. There is approximately a 1:1 correlation between the two thresholds in a vacuum environment. Referring to Equation (10.22), this implies that $K_{op} \approx 0$ in a vacuum environment near the threshold. The trend for the air data has a slope of 0.4, which implies that $K_{op} \approx 0.6 K_{max}^{*}$, assuming closure is responsible for the observed R ratio effects.

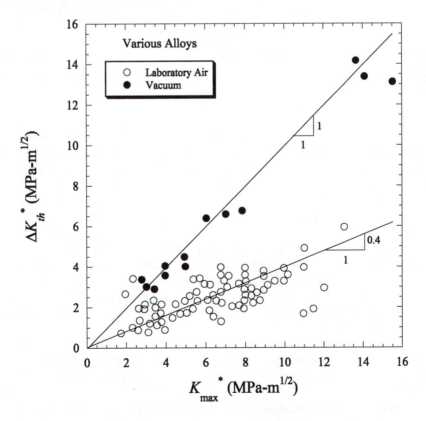

FIGURE 10.17 Correlation between ΔK_{th}^{*} and K_{max}^{*} in air and vacuum environments. Taken from Vasudevan, A.K., Sadananda, K., and Loutat, N., "A Review of Crack Closure, Fatigue Crack Threshold, and Related Phenomena." *Materials Science and Engineering*, Vol. A188, 1994, pp. 1–22.

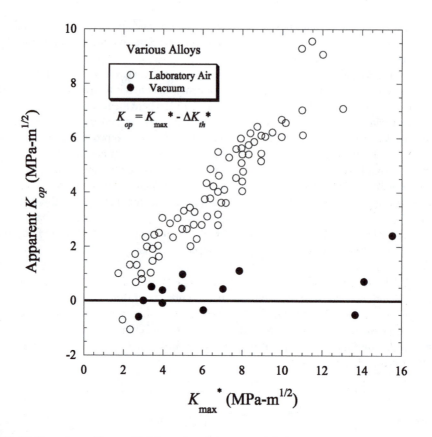

FIGURE 10.18 Data from Figure 10.17, replotted in terms of the apparent K_{op}, as estimated from Equation (10.22).

Figure 10.18 is a replot of the data from Figure 10.17 in terms of the apparent K_{op}, which was inferred from Equation (10.22). The authors who originally created the $\Delta K_{th}^* - K_{max}^*$ plot would probably object to having their data re-plotted in this manner because they contend that the K_{max} threshold is not driven by closure. Nevertheless, Figure 10.18 supports their opinion that plasticity-induced closure does not play a significant role near the threshold. The apparent K_{op} for the vacuum data is approximately zero, which would not be the case if plasticity-induced closure was occurring to any significant degree.

The fatigue threshold in a vacuum environment exhibits an R ratio dependence at negative R values. Figure 10.19 schematically illustrates the typical threshold behavior at positive and negative R ratios for both air and vacuum environments. It is intuitively obvious that closure would play a role when the R ratio is negative, because compressive loads would force the crack faces together. For the special case where $R < 0$ and $K_{op} = 0$, $\Delta K_{eff} = K_{max}$, and only the tensile part of the cycle contributes to fatigue damage.

Finally, it should be noted that the role of plasticity-induced closure in the threshold regime is a controversial topic, and is the subject of ongoing studies. Proponents of plasticity-induced closure cite more recent data (unpublished as of this writing) where tests in vacuum do exhibit an R ratio dependence.

There are certain instances where plasticity does appear to influence the threshold measurement, whether in air or vacuum environments. In the load-shedding test procedure, for example, where ΔK is reduced at a constant K_{max} or constant R ratio, the measured threshold can be artificially elevated if ΔK decreases too abruptly. Such results imply that the plastic wake produced at higher ΔK levels

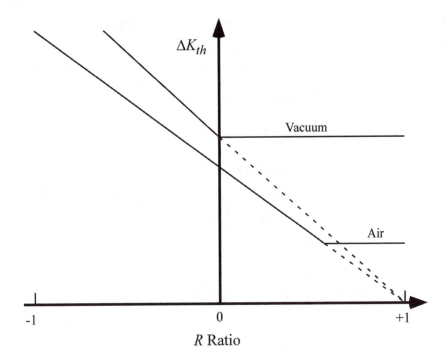

FIGURE 10.19 Schematic illustration of the threshold behavior in air and vacuum environments, at both positive and negative R ratios.

affects the threshold measurement unless the decrease in ΔK is sufficiently gradual that the crack tip can grow beyond the influence of the earlier plasticity. New test procedures are being developed for the threshold regime to avoid the potential pitfalls of the load-shedding method. One such technique involves testing at constant load amplitude following precracking in compression [41].

10.5 VARIABLE AMPLITUDE LOADING AND RETARDATION

Similitude of crack-tip conditions, which implies a unique relationship between da/dN, ΔK, and R, is rigorously valid only for constant amplitude loading (i.e., $dK/da = 0$). Real structures, however, seldom conform to this ideal. A typical structure experiences a spectrum of stresses over its lifetime. In such cases, the crack growth rate at any moment in time may depend on the prior history as well as current loading conditions. Equation (10.3) is a general mathematical representation of the dependence on past and present conditions.

Variable amplitude fatigue analyses that account for prior loading history are considerably more cumbersome than analyses that assume similitude. Therefore, the latter type of analysis is desirable if the similitude assumption is justified. There are many practical situations where such an assumption is reasonable. Such cases normally involve cyclic loading at high R ratios, where crack closure effects are negligible. Steel bridges, for example, have high dead loads due to their own weight, which translates into high R ratios. Fatigue of welds that have not been stress relieved often obey similitude because tensile residual stresses, which are static, increase the effective R ratio.

When similitude applies, at least approximately, the *linear damage model* is suitable for variable amplitude loading. There are a variety of other variable amplitude fatigue models that account for history effects. The linear damage model is described in the next section. This is followed by a discussion of the mechanisms for history dependence, as well as an overview of existing models that attempt to include history effects.

10.5.1   LINEAR DAMAGE MODEL FOR VARIABLE AMPLITUDE FATIGUE

Consider a case where a cracked component experiences N_1 cycles at a cyclic stress of $\Delta\sigma_1$, N_2 cycles at $\Delta\sigma_2$, and so on. The total crack growth during this load history is Δa. If we assume that $\Delta a \ll a$, such that da/dN for a given cyclic stress does not change significantly during the loading history,[9] the crack growth can be estimated as follows:

$$\Delta a \approx \left(\frac{da}{dN}\right)_1 N_1 + \left(\frac{da}{dN}\right)_2 N_2 + \cdots \tag{10.23}$$

If the crack growth is described by the Paris equation (Equation (10.5)), the above expression becomes:

$$\Delta a \approx C[(\Delta K_1)^m N_1 + (\Delta K_2)^m N_2 + \cdots]$$
$$= CY^m (\pi a)^{m/2} \left(\Delta\sigma_1^m N_1 + \Delta\sigma_2^m N_2 + \cdots \right) \tag{10.24}$$

where Y is a geometry factor in the stress-intensity solution. It is possible to define an equivalent constant amplitude stress range from Equation (10.24):

$$\Delta\bar{\sigma} = \left(\frac{\sum_{i=1}^{n} \Delta\sigma_i^m N_i}{N_{tot}} \right)^{1/m} \tag{10.25}$$

This weighted average cyclic stress corresponds to an average growth rate for the loading spectrum at a given crack size:

$$\frac{d\bar{a}}{dN} = C(Y\Delta\bar{\sigma}\sqrt{\pi a})^m \tag{10.26}$$

Life prediction is performed as if the loading were constant amplitude, with $\Delta\bar{\sigma}$ outside of the integral:

$$N = \frac{1}{C(\Delta\bar{\sigma}\sqrt{\pi})^m} \int_{a_o}^{a_f} \frac{da}{Ya^{m/2}} \tag{10.27}$$

The above procedure for handling variable amplitude loading must be modified when accounting for a threshold. Consider the following simplified growth law:

$$\frac{da}{dN} = \begin{cases} C\Delta K^m, & \Delta K > \Delta K_{th} \\ 0, & \Delta K \le \Delta K_{th} \end{cases} \tag{10.28}$$

In this case, an equivalent constant amplitude stress can still be computed from Equation (10.25). However, only cyclic stresses that contribute to fatigue should be included in the summation. This concept is illustrated in Figure 10.20. The threshold cyclic stress is given by

$$\Delta\sigma_{th} = \frac{\Delta K_{th}}{Y\sqrt{\pi a}} \tag{10.29}$$

Stresses below $\Delta\sigma_{th}$ do not contribute to fatigue damage, and thus should not be included in the numerator in Equation (10.25). However, N_{tot} in the denominator should include *all* cycles, even

---

[9] The assumption of small crack growth during the load history is not actually necessary, but such an assumption makes the derivation easier to follow. A more rigorous derivation would consider the integrated average growth rate over the interval a to $a + \Delta a$. The final result would not change, however.

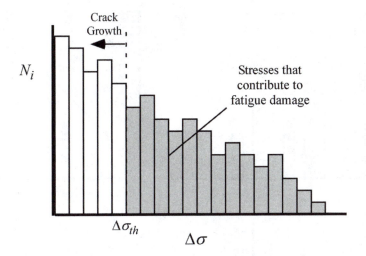

FIGURE 10.20 Schematic cyclic stress histogram. Only cycles below $\Delta\sigma_{th}$ contribute to fatigue damage. This threshold cyclic stress decreases with crack growth in accordance with Equation (10.29).

those that do not contribute to fatigue damage. One complication of introducing a threshold is that $\Delta\sigma_{th}$ is a function of crack size. As the crack grows, $\Delta\sigma_{th}$ decreases and a larger fraction of the loading spectrum contributes to fatigue. Consequently, $\Delta\bar{\sigma}$ must remain inside of the integral (Equation (10.27)) because it is a function of crack size.

We can generalize the linear damage model for an arbitrary growth law by computing an average crack growth rate for the loading spectrum:

$$\frac{d\bar{a}}{dN} = \frac{1}{N_{tot}} \sum_{i=1}^{n} \left(\frac{da}{dN}\right)_i N_i \qquad (10.30)$$

Figure 10.21 illustrates the construction of a da/dN histogram from a cyclic stress histogram and the crack growth law. Note that the unshaded histogram bar on the left-hand side of the diagram represents cyclic stress levels where $da/dN = 0$. The total number of cycles in the above expression, N_{tot}, must include *all* cycles, including those where $da/dN = 0$.[10] Life prediction is achieved through integration of the average growth rate:

$$N = \int_{a_o}^{a_f} \left(\frac{d\bar{a}}{dN}\right)^{-1} da \qquad (10.31)$$

Of course, the da/dN histogram and the average crack growth rate are a function of crack size, so they must be continually recomputed during the life calculation.

10.5.2 Reverse Plasticity at the Crack Tip

History effects in fatigue are a direct result of the history dependence of plastic deformation. Figure 10.22 schematically illustrates the cyclic stress-strain response of an elastic-plastic material which is loaded beyond yield in both tension and compression. If we wish to know the stress at a particular strain ε^*, it is not sufficient merely to specify the strain. For the loading path in

---

[10] Consider the following analogy. A professor gives a final exam to a class of 30 students. However, three out of the 30 students do not show up for the exam, so they each receive a score of zero. When computing the average score for the exam, the professor adds up the total scores of all students and divides by 30. The three scores of zero bring down the average score for the class in the same way that fatigue cycles below the threshold reduce the average da/dN for the loading spectrum.

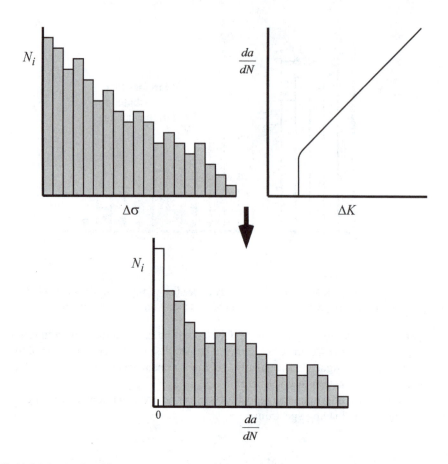

FIGURE 10.21 Derivation of a *da/dN* histogram from a cyclic stress histogram and the growth law.

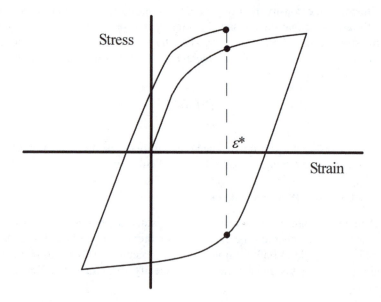

FIGURE 10.22 Schematic stress-strain response of a material that is yielded in both tension and compression. The stress at a given strain ε^*, depends on prior loading history.

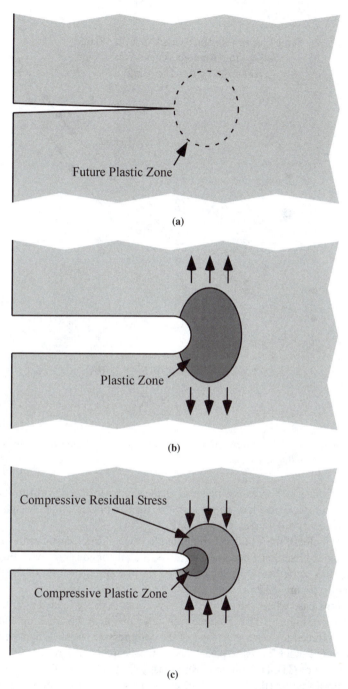

FIGURE 10.23 Deformation at the tip of a crack subject to a single load-unload cycle: (a) crack in a virgin material, (b) crack blunting and plastic zone formation resulting from an applied tensile load, and (c) compressive residual stresses at the crack tip after unloading.

Figure 10.22, there are three different stresses that correspond to ε^*; we must specify not only ε^*, but also the deformation history that preceded this strain.

Figure 10.23 illustrates the crack-tip deformation resulting from a single load-unload cycle. As the load is applied, a plastic zone forms and the crack-tip blunts. When the load is removed, the

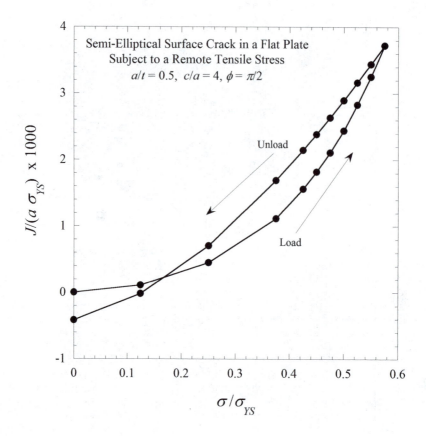

FIGURE 10.24 *J*-integral results obtained from an elastic-plastic finite element analysis of a load-unload cycle in a plate with a semielliptical surface crack.

surrounding elastic material forces the material at the crack-tip to conform to its original shape. This results in compressive residual stresses in the former plastic zone. A smaller compressive plastic zone forms close to the crack tip [42, 43]

Figure 10.24 is a plot of the *J* integral computed from an elastic-plastic finite element analysis[11] of a load-unload cycle. Note that *J* becomes negative when the load is removed. Figure 10.25 shows the CTOD during this same load-unload cycle. The normal stress in front of the crack tip for this analysis is plotted in Figure 10.26.

Now consider the case where the cracked plate is compressively loaded after having previously been loaded in tension. Figure 10.27 illustrates the deformation at the crack tip under this scenario. The crack faces are forced together as a result of the compressive stress. When the load is removed, the previously blunted crack is resharpened and a tensile residual stress forms at the crack tip. Figure 10.28 is a plot of CTOD from an elastic-plastic finite element analysis of the compressive load-unload scenario. Figure 10.29 is a plot of crack-tip stresses from this same analysis.

10.5.3 THE EFFECT OF OVERLOADS AND UNDERLOADS

Consider the fatigue loading history illustrated in Figure 10.30. Constant amplitude loading is interrupted by a single overload, after which the *K* amplitude resumes its previous value. Prior to

---

[11] The elastic-plastic finite element results presented in Figure 10.24, Figure 10.25, Figure 10.26, Figure 10.28, and Figure 10.29 were generated by the author especially for this book.

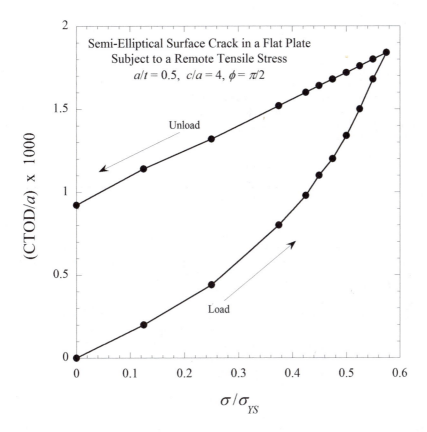

FIGURE 10.25 Crack-tip-opening displacement (CTOD) results obtained from an elastic-plastic finite element analysis of a load-unload cycle in a plate with a semielliptical surface crack.

the overload, the plastic zone would have reached a steady-state size, but the overload cycle produces a significantly larger plastic zone. When the load drops to the original $K_{\min}$ and $K_{\max}$, the residual stresses that result from the overload plastic zone (Figure 10.26) are likely to influence subsequent fatigue behavior.

Figure 10.31 illustrates the typical behavior following an overload. *Retardation*, where the crack growth rate decelerates by an order of magnitude or more, generally results from an overload. In some instances, there is a brief period of acceleration in *da/dN* following an overload, and retardation occurs later. As the crack grows following the overload, the rate eventually approaches that observed for constant amplitude loading immediately prior to the overload.

Three possible mechanisms have been proposed to explain retardation following an overload:

1. The crack blunts following an overload, and crack growth is delayed while the crack tip resharpens.
2. The compressive residual stresses in front of the crack tip (Figure 10.23 and Figure 10.26) retard the crack growth rate.
3. As the crack grows into the overload zone, residual stresses *behind* the crack tip result in plasticity-induced closure.

The crack blunting theory was popular at one time, but it has since been rejected by most experts. The second explanation is favored by proponents of a dual $\Delta K - K_{\max}$ threshold (Section 10.4.2).

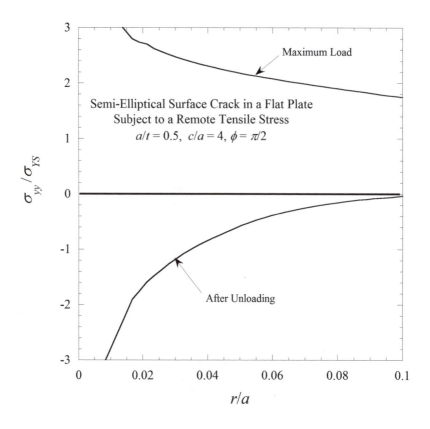

FIGURE 10.26 Normal stress vs. distance from the crack tip computed from an elastic-plastic finite element analysis of a load-unload cycle in a plate with a semielliptical surface crack.

The compressive residual stresses in front of the crack, they argue, result in a reduction in K_{max}, and thus a reduction in crack growth rate. However, the majority opinion among fatigue specialists is that plasticity-induced closure is responsible for retardation effects.

Perhaps the most compelling evidence in favor of the closure argument is the phenomenon of *delayed retardation*. That is, retardation generally does not occur immediately following the application of an overload. In some cases, the crack growth rate actually accelerates for a brief period after the overload, as Figure 10.31 illustrates. If retardation were driven by a reduced K_{max} due to compressive stresses in front of the crack tip, one would expect the effect to be immediate. The elastic-plastic finite element results presented in Figure 10.24 to Figure 10.26 clearly show that the crack-tip stresses are compressive following the application and removal of a tensile stress. Moreover, the J integral is actually negative when the applied load is removed. Therefore, the reduction in K_{max} is immediate following an overload, but the reduction in crack growth rate is delayed.

The closure mechanism provides a plausible explanation for the momentary acceleration of crack growth rate following an overload. If closure is occurring during constant amplitude loading, $\Delta K_{eff} < \Delta K$ and the crack growth rate is less than it would be in the absence of closure. When an overload is applied, the resulting crack blunting causes the crack faces to move apart. Closure does not occur in the cycles immediately following the overload, so the crack growth rate is momentarily higher than it was prior to the overload. Once the crack grows a short distance into the overload zone, compressive residual stresses result in plasticity-induced closure, which in turn results in retardation.

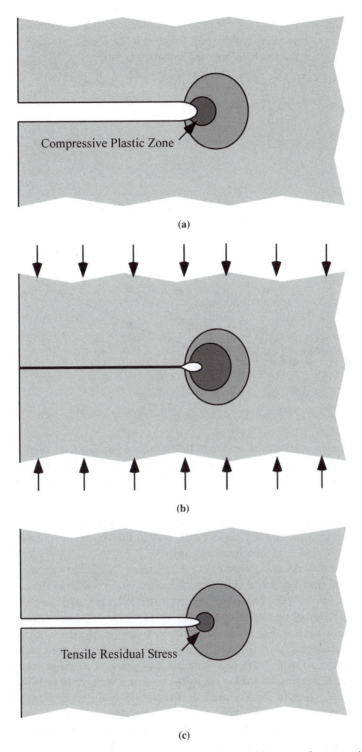

FIGURE 10.27 Compressive underload of a blunted crack with an initial zone of compressive residual stress at the crack tip: (a) initial state prior to application of compressive load, (b) application of compressive load, which results in crack face contact, and removal of compressive load. The crack tip resharpens and a tensile residual stress forms at the crack tip.

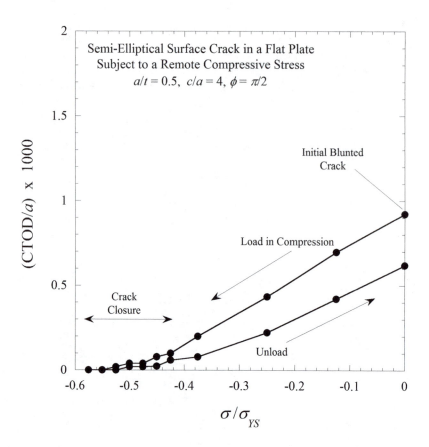

FIGURE 10.28 Crack-tip-opening displacement (CTOD) obtained from an elastic-plastic finite element analysis of the compressive underload scenario illustrated in Figure 10.27. The compressive load was applied following the tensile load-unload cycle corresponding to the finite element results plotted in Figure 10.24 to Figure 10.26.

Figure 10.32 illustrates the closure mechanism for retardation. The blunting following the overload may result in a momentary acceleration due to the absence of closure, but retardation occurs once the crack grows into the overload zone.

Proponents of the dual $\Delta K - K_{max}$ threshold concept reject the closure mechanism for retardation. They explain delayed retardation and momentary acceleration by invoking the concept of "internal stresses" which influence the effective fatigue driving force [38]. They argue that tensile stresses occur immediately behind the crack tip following an overload. However, this argument seems to run counter to their assertion that stresses *in front of* the crack tip drive fatigue. Moreover, the elastic-plastic finite element analysis results in Figure 10.24 to Figure 10.26 indicate that the crack driving force, as quantified by the *J* integral, is negative following the application of a tensile overload.

Underloads (i.e., a compressive load or a tensile load that is well below previous minimum loads) can result in an acceleration of crack growth or a reduction in the level of retardation. In certain circumstances, an underload can produce tensile residual stresses at the crack tip. Figure 10.27 illustrates this phenomenon. A blunted crack is forced closed by a compressive underload. The crack resharpens and a tensile residual stress forms ahead of the crack tip, as the finite element results in Figure 10.28 and Figure 10.29 illustrate. In order for an underload to produce a tensile residual stress zone, the crack must be blunted and open at the time of the underload. If the crack faces are in contact at the time an underload is applied, a tensile residual stress zone will not form at the crack tip, but compressive forces on the crack faces may flatten asperities. This may have the effect of reducing the magnitude of roughness-induced closure in subsequent crack growth.

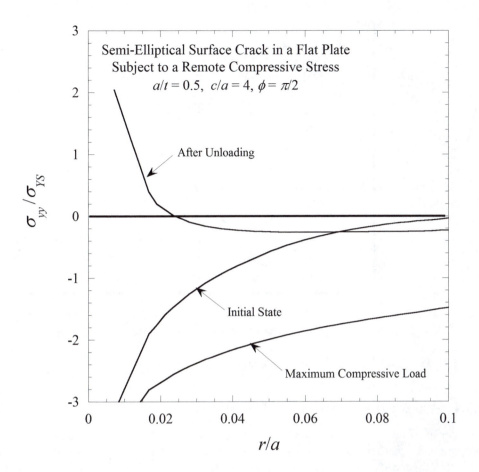

FIGURE 10.29 Normal stress vs. distance from the crack tip, obtained from an elastic-plastic finite element analysis of the compressive underload scenario illustrated in Figure 10.27. The compressive load was applied following the tensile load-unload cycle corresponding to the finite element results plotted in Figure 10.24 to Figure 10.26.

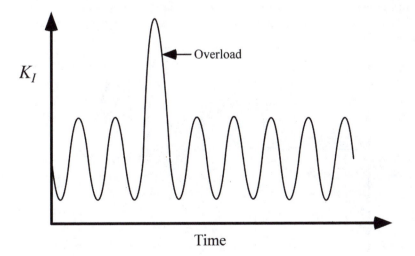

FIGURE 10.30 A single overload during cyclic loading.

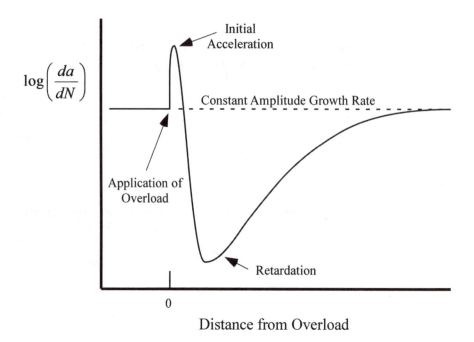

FIGURE 10.31 Typical crack growth behavior following the application of a single overload.

10.5.4 Models for Retardation and Variable Amplitude Fatigue

A number of models have been proposed to account for load interaction effects in fatigue crack propagation. The earliest models assume that history effects stem from residual stresses in front of the crack tip due to overloads and underloads. More recent models are based on the assumption that plasticity-induced closure is responsible for load interaction effects.

Retardation models developed by Wheeler [44] and Willenborg et al. [45] are based on the premise that residual stresses in front of the crack influence the growth rate. Figure 10.33 illustrates fatigue crack growth following an overload. The overload produced a plastic zone of size $r_{y(o)}$ and the crack has grown a distance Δa following the overload. The current plastic zone size is $r_{y(o)}$. Both the Wheeler and Willenborg models relate the current crack growth rate to the distance the crack has grown into the overload zone.

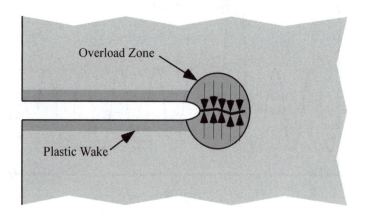

FIGURE 10.32 Retardation caused by plasticity-induced closure.

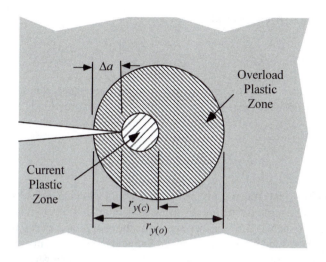

FIGURE 10.33 Growth of a fatigue crack following an overload. The Wheeler and Willenborg retardation models relate the current crack growth rate to the distance the crack has grown into the overload zone. Taken from Wheeler, O.E., "Spectrum Loading and Crack Growth." *Journal of Basic Engineering*, Vol. 94, 1972, pp. 181–186. Willenborg, J., Engle, R.M., Jr., and Wood, R.A., "A Crack Growth Retardation Model Using an Effective Stress Concept." Air Force Flight Dynamics Laboratory Report AFFDL-TM-71-1-FBR, January 1971.

The Wheeler model contains a retardation factor ϕ_R, which is a function of the relative amount of crack growth into the overload zone.

$$\phi_R = \left(\frac{\Delta a + r_{y(c)}}{r_{y(o)}} \right)^{\gamma} \tag{10.32}$$

where γ is a fitting parameter. The crack growth rate is reduced from a baseline value by ϕ_R:

$$\left(\frac{da}{dN} \right)_R = \phi_R \frac{da}{dN} \tag{10.33}$$

The overload plastic zone size can be estimated from the following expression:

$$r_{y(o)} = \frac{1}{\beta \pi} \left(\frac{K_o}{\sigma_{YS}} \right)^2 \tag{10.34}$$

where K_o is the stress intensity at the peak overload, and $\beta = 2$ for plane stress and $\beta = 6$ for plane strain. The plastic zone size that corresponds to the current $K_{\max}$ is given by

$$r_{y(c)} = \frac{1}{\beta \pi} \left(\frac{K_{\max}}{\sigma_{YS}} \right)^2 \tag{10.35}$$

Wheeler assumed that retardation effects persist as long as the current plastic zone is contained within the overload zone, but the effects disappear when the current plastic zone touches the outer boundary of the overload zone.

Unlike the Wheeler model, the Willenborg model does not contain a fitting parameter. Willenborg et al. defined a residual stress-intensity factor as follows:

$$K_R = \begin{cases} K_o \left(1 - \dfrac{\Delta a}{r_{y(o)}} \right)^{1/2} - K_{\max}, & \Delta a \leq r_{y(o)} \\[3mm] 0, & \Delta a > r_{y(o)} \end{cases} \tag{10.36}$$

They then introduced an effective R ratio:

$$R_{eff} = \frac{K_{min} - K_R}{K_{max} - K_R} \qquad (10.37)$$

In order for the Willenborg model to have any effect on the fatigue calculation, the growth equation must include an R ratio dependence. For example, the Walker expression [13] is a modification of the Paris power law:

$$\frac{da}{dN} = C\left[\frac{\Delta K}{(1-R)^n}\right]^m \qquad (10.38)$$

where n is a material constant. When applying the Willenborg model, R_{eff} is substituted into Equation (10.38) in place of R. The same is true for any growth equation that includes an R ratio dependence.

Retardation models such as the Wheeler and Willenborg approaches can be applied to variable amplitude loading. These models are implemented in much the same manner as for overload cases, except that K_o and the current K_{max} must be evaluated for each cycle in a variable amplitude problem. Figure 10.34 illustrates the basic concept. The most recent overload is used to compute the current crack growth rate. In the example depicted in Figure 10.34, the second overload is used to compute $r_{y(o)}$ because this overload zone extends further ahead of the current crack tip.

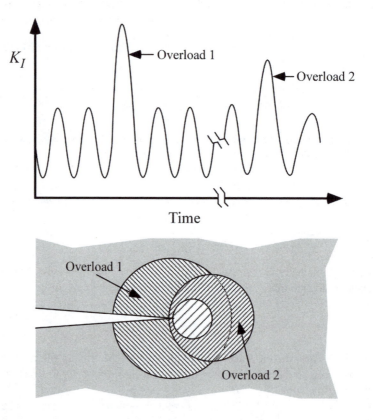

FIGURE 10.34. Simple example of variable amplitude loading. The Wheeler and Willenborg retardation models would compute the current crack growth rate based on Overload 2 because the overload zone extends further ahead of the crack tip. Taken from Wheeler, O.E., "Spectrum Loading and Crack Growth." *Journal of Basic Engineering*, Vol. 94, 1972, pp. 181–186; Willenborg, J., Engle, R.M., Jr., and Wood, R.A., "A Crack Growth Retardation Model Using an Effective Stress Concept." Air Force Flight Dynamics Laboratory Report AFFDL-TM-71-1-FBR, January 1971.

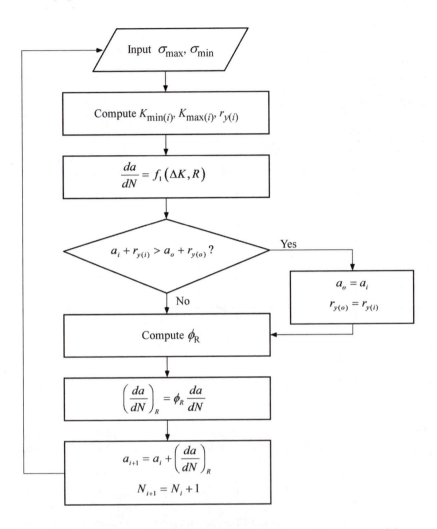

FIGURE 10.35 Flow chart for variable amplitude fatigue analysis with the Wheeler model.

Figure 10.35 shows a flow chart for computing fatigue crack growth with the Wheeler model. Although the algorithm is relatively simple, the analysis can be very time consuming, since a cycle-by-cycle summation is required. The stress input consists of two components: the spectrum and the sequence. The former is a statistical distribution of stress amplitudes, which quantifies the relative frequency of low, medium, and high stress cycles. The sequence, which defines the order of the various stress amplitudes, can be either random or a regular pattern.

An alternative to retardation models that focus on residual stress in front of the crack tip is a closure-based model. A number of investigators have extended the Dugdale-Barenblatt strip-yield model to a growing crack. Budiansky and Hutchinson [17] developed an analytical model for plasticity-induced closure based modified strip-yield model, which incorporates residual stresses in the wake behind the crack tip. Newman [46] has developed a numerical crack closure model that is also a modification of the Dugdale-Barenblatt strip-yield concept. The Newman numerical model can be applied to variable amplitude fatigue analysis, but the Budiansky and Hutchinson analytical model is not well suited to engineering problems.

Figure 10.36 illustrates the Newman closure model. The plastic zone is divided into discrete segments. As the crack advances, segments are broken and become part of the plastic wake. The residual stress in each broken segment is computed from the maximum stretch the segment was

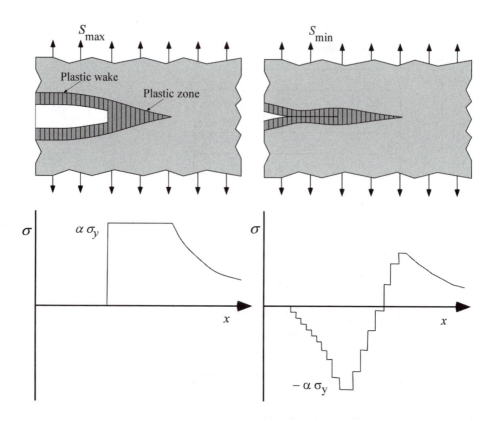

FIGURE 10.36 The Newman closure model. Taken from Newman, J.C., "Prediction of Fatigue Crack Growth under Variable Amplitude and Spectrum Loading Using a Closure Model." ASTM STP 761, American Society for Testing and Materials, Philadelphia, PA, 1982, pp. 255–277.

subjected to when it was intact. At the maximum far-field stress S_{max}, the crack is fully open, and plastic zone is stressed to $\alpha\sigma_y$, where $\alpha = 1$ for plane stress and $\alpha = 3$ for plane strain. At the minimum stress S_{min}, the crack is closed. The residual stress distribution in the plastic wake determines the far-field opening stress S_o. The effective stress intensity K_{eff} is then computed from the effective stress amplitude $S_{max} - S_o$. Load interaction and history effects are accounted for through their influence on the calculated opening stress.

10.6 GROWTH OF SHORT CRACKS

The concepts of fracture mechanics similitude and a ΔK threshold break down near the point of crack initiation. Were this not the case, fatigue cracks would never nucleate from a smooth surface because $\Delta K = 0$ when crack size = 0. Figure 10.37 illustrates the effect of crack size on the threshold stress for fatigue $\Delta\sigma_{th}$. Equation (10.29) describes the relationship between $\Delta\sigma_{th}$ and crack size when fracture mechanics similitude applies. According to Equation (10.29), $\Delta\sigma_{th}$ is proportional to $a^{-1/2}$, provided the crack is sufficiently small that the geometry factor Y is constant. This relationship implies that the stress required to nucleate a crack is infinite. In reality, $\Delta\sigma_{th}$ reaches a plateau for very small cracks. This plateau represents the minimum cyclic stress required to nucleate a crack from a smooth surface. The *short crack regime* corresponds to the transition from crack nucleation regime to the long crack regime where fracture mechanics similitude applies.

The fatigue behavior of short cracks is often very different from that of longer cracks. There is not a precise definition of what constitutes a "short" crack, but most experts consider cracks less than 1 mm deep to be small. Because most fatigue cracks spend the vast majority of their lives as short cracks, the behavior of such flaws is of significant practical importance.

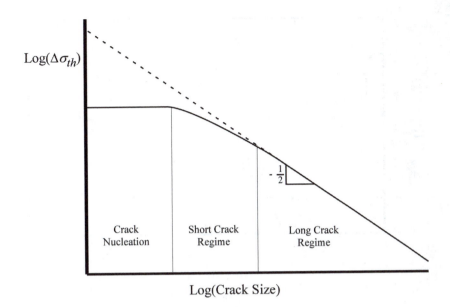

FIGURE 10.37 Relationship between threshold cyclic stress $\Delta\sigma_{th}$ and crack size. Fracture mechanics similitude does not apply in the crack nucleation regime.

Figure 10.38 compares short crack data with long crack data near the threshold [47]. In this case, the short cracks were initiated from a blunt notch. Note that the short cracks exhibit finite growth rates well below ΔK_{th} for long cracks. Also, the trend in da/dN is inconsistent with expected behavior; the crack growth rate actually decreases with ΔK when the stress range is 60 MPa, and the $da/dN - \Delta K$ curve exhibits a minimum at the other stress level.

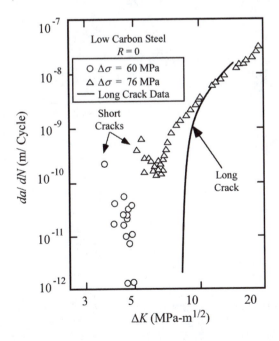

FIGURE 10.38 Growth of short cracks in a low-carbon steel. Taken from Tanaka, K. and Nakai, Y., "Propagation and Non-Propagation of Short Fatigue Cracks at a Sharp Notch." *Fatigue of Engineering Materials and Structures*, Vol. 6, 1983, pp. 315–327.

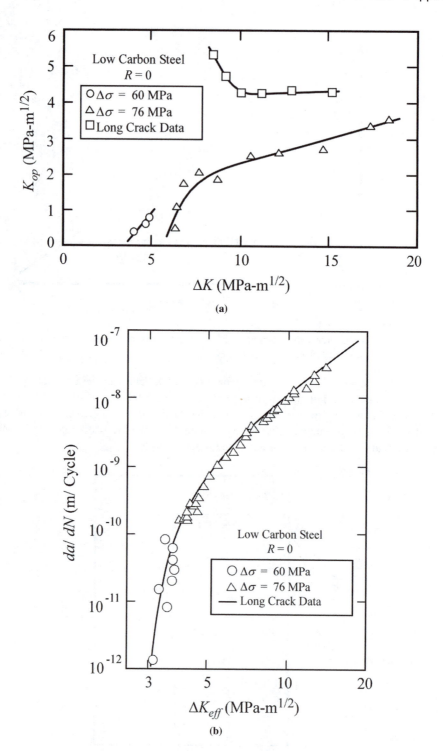

FIGURE 10.39 Short crack fatigue crack growth data from Figure 10.38, corrected for closure: (a) crack closure data for short and long cracks and (b) closure-corrected data. Taken from Tanaka, K. and Nakai, Y., "Propagation and Non-Propagation of Short Fatigue Cracks at a Sharp Notch." *Fatigue of Engineering Materials and Structures*, Vol. 6, 1983, pp. 315–327.

A number of factors can contribute to the anomalous behavior of small fatigue cracks. The fatigue mechanisms depend on whether the crack is microstructurally short or mechanically short, as described in the next section.

10.6.1 MICROSTRUCTURALLY SHORT CRACKS

A microstructurally short crack has dimensions that are on the order of the grain size. Cracks less than 100 μm long are generally considered microstructurally short. The material no longer behaves as a homogeneous isotropic continuum at such length scales; the growth is strongly influenced by microstructural features in such cases. The growth of microstructurally short cracks is often very sporadic; the crack may grow rapidly at certain intervals, and then virtually arrest when it encounters barriers such as grain boundaries and second-phase particles [7]. Potirniche et al. [48] have simulated small fatigue crack growth near a grain boundary using finite element analysis and constitutive material relations derived from crystal plasticity theory, which considers the plastically anisotropic behavior of individual grains.

10.6.2 MECHANICALLY SHORT CRACKS

A crack that is between 100 μm and 1 mm in depth is mechanically short. The size is sufficient to apply continuum theory, but the mechanical behavior is not the same as in longer cracks. Mechanically short cracks typically grow much faster than long cracks at the same ΔK level, particularly near the threshold (Figure 10.38).

Two factors have been identified as contributing to faster growth of short cracks: plastic zone size and crack closure.

When the plastic zone size is significant compared to the crack length, an elastic singularity does not exist at the crack tip, and K is invalid. The effective driving force can be estimated by adding an Irwin plastic zone correction. El Haddad et al. [49] introduced an "intrinsic crack length" which, when added to the physical crack size, brings short crack data in line with the corresponding long crack results. The intrinsic crack length is merely a fitting parameter, however, and does not correspond to a physical length scale in the material. Tanaka [7], among others, proposed adjusting the data for crack-tip plasticity by characterizing da/dN with ΔJ rather than ΔK.

According to the closure argument, short cracks exhibit different crack closure behavior than long cracks, and data for different crack sizes can be rationalized through ΔK_{eff}. Figure 10.39(a) [47] shows K_{op} measurements for the short and long crack data in Figure 10.38. The closure loads are significantly higher in the long cracks, particularly at low ΔK levels. Figure 10.39(b) shows the small and large crack data lie on a common curve when da/dN is plotted against ΔK_{eff}, thereby lending credibility to the closure theory of short crack behavior.

10.7 MICROMECHANISMS OF FATIGUE

Figure 10.40 summarizes the failure mechanisms for metals in the three regions of the fatigue crack growth curve. In Region II, where da/dN follows a power law, the crack growth rate is relatively insensitive to microstructure and tensile properties, while da/dN at either extreme of the curve is highly sensitive to these variables. The fatigue mechanisms in each region are described in more detail in the following sections.

10.7.1 FATIGUE IN REGION II

In Region II, the fatigue crack growth rate is not a strong function of microstructure or monotonic flow properties. Two aluminum alloys with vastly different mechanical properties, for instance, are

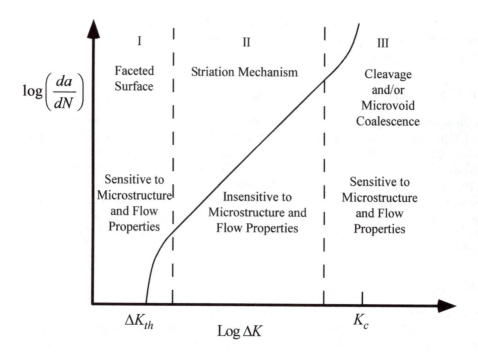

FIGURE 10.40 Micromechanisms of fatigue in metals.

likely to have very similar fatigue crack growth characteristics. Steel and aluminum, however, exhibit significantly different fatigue behavior. Thus, *da/dN* is not sensitive to microstructure and tensile properties *within a given material system.*

One explanation for the lack of sensitivity to metallurgical variables is that *cyclic* flow properties, rather than monotonic tensile properties, control fatigue crack propagation. Figure 10.41 schematically compares monotonic and cyclic stress-strain behavior for two alloys of a given

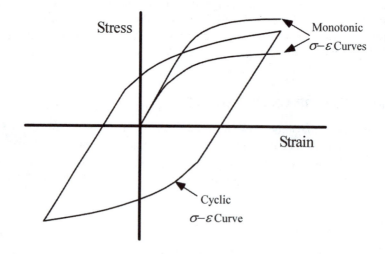

FIGURE 10.41 Schematic comparison of monotonic and cyclic stress-strain curves.

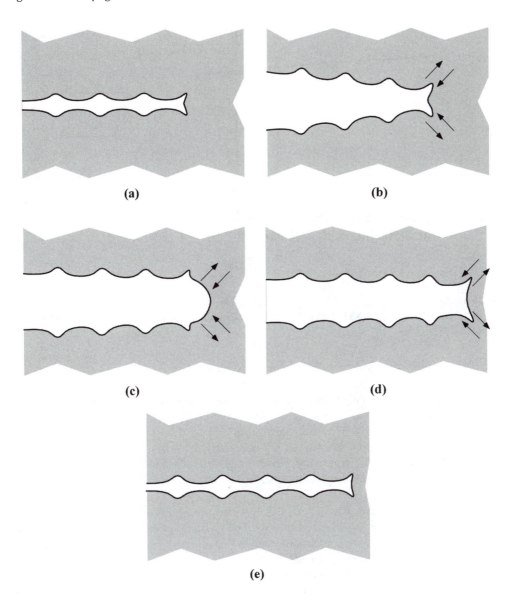

FIGURE 10.42 The crack-blunting mechanism for striation formation during fatigue crack growth. Taken from Laird, C., "Mechanisms and Theories of Fatigue." *Fatigue and Microstructure*, American Society for Metals, Metals Park, OH, 1979, pp. 149–203.

material. The low strength alloy tends to strain harden, while the strong alloy tends to strain *soften* with cyclic loading. In both cases, the cyclic stress-strain curve tends toward a steady-state hysteresis loop, which is relatively insensitive to the initial strength level.

Propagating fatigue cracks often produce *striations* on the fracture surface. Striations are small ridges that are perpendicular to the direction of crack propagation. Figure 10.42 illustrates one proposed mechanism for striation formation during fatigue crack growth [50]. The crack tip blunts as the load increases, and an increment of growth occurs as a result of the formation of a stretch zone. Local slip is concentrated at ±45° from the crack plane. When the load decreases, the direction of slip reverses, and the crack tip folds inward. The process is repeated with subsequent cycles, and each cycle produces a striation on the upper and lower crack

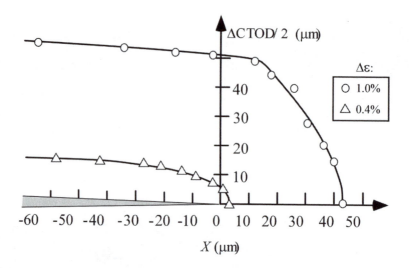

FIGURE 10.43 Crack-opening profile in copper. Taken from Tanaka, K., "Mechanics and Micromechanics of Fatigue Crack Propagation." ASTM STP 1020, American Society for Testing and Materials, Philadelphia, PA, 1989, pp. 151–183.

faces. The striation spacing, according to this mechanism, is equal to the crack growth per cycle (*da/dN*).

An alternative view of fatigue is the damage accumulation mechanism, which states that a number of cycles are required to produce a critical amount of damage, at which time the crack grows a small increment [51]. This mechanism was supported by Lankford and Davidson [52], who observed that the striation spacing did not necessarily correspond to crack growth after one cycle. Several cycles may be required to produce one striation, depending on the ΔK level; the number of cycles per striation apparently decreases with increasing ΔK, and striation spacing = *da/dN* at high ΔK values.

A number of researchers have attempted to relate the observed crack growth rate to the micromechanism of fatigue, with limited success. The blunting mechanism, where crack advance occurs through the formation of a stretch zone, implies that the crack growth per cycle is proportional to ΔCTOD. This, in turn, implies that *da/dN* should be proportional to ΔK^2. Actual Paris law exponents, however, are typically between three and four for metals. One possible explanation for this discrepancy is that the blunting mechanism is incorrect. An alternate explanation for exponents greater than 2 is that the shape of the blunted crack is not geometrically similar at high and low *K* values [7]. Figure 10.43 [7] shows the crack-opening profile for copper at two load levels. Note that the shapes of the blunted cracks are different.

10.7.2 MICROMECHANISMS NEAR THE THRESHOLD

The fracture surface that results from fatigue near the threshold has a flat, faceted appearance that resembles cleavage [53]. The crack apparently follows specific crystallographic planes, and changes directions when it encounters a barrier such as a grain boundary.

The fatigue crack growth rate in this region is sensitive to grain size, in part because coarse-grained microstructures produce rough surfaces and roughness-induced closure (Figure 10.5). Grain size can also affect the *intrinsic* threshold in certain cases. One model for ΔK_{th}^* [34] states that the threshold occurs when grain boundaries block slip bands and prevent them from propagating into the adjoining grain. This apparently happens when the plastic zone size is approximately equal to the average grain diameter, which suggests the following relationship between ΔK_{th}^* and grain size:

$$\sqrt{d} = A \frac{\Delta K_{th}^*}{\sigma_{YS}} \tag{10.39}$$

where d is the average grain diameter and A is a constant. Thus, the intrinsic threshold increases with grain size, assuming σ_{YS} is constant. The Hall-Petch relationship, however, predicts that yield strength decreases with grain coarsening:

$$\sigma_{YS} = \sigma_i + k_y d^{-1/2} \tag{10.40}$$

Consequently, the grain size dependence of yield strength offsets the tendency for the intrinsic threshold to increase with grain coarsening.

10.7.3 Fatigue at High ΔK Values

In Region III, there is an apparent acceleration in da/dN. In some instances, this may be due to an interaction between fatigue and fracture mechanisms. Fracture surfaces in this region may include a mixture of fatigue striations, microvoid coalescence, and (depending on the material and the temperature) cleavage facets. The overall growth rate can be estimated by summing the effects of the various mechanisms:

$$\left.\frac{da}{dN}\right|_{total} = \left.\frac{da}{dN}\right|_{fatigue} + \left.\frac{da}{dN}\right|_{MVC} + \left.\frac{da}{dN}\right|_{cleavage} \tag{10.41}$$

where the subscript MVC refers to microvoid coalescence (Chapter 5). The relative contribution of fatigue decreases with increasing K_{max}. At K_c, crack growth is completely dominated by microvoid coalescence, cleavage, or both.

An alternative explanation for the apparent upswing in the crack growth rate at high ΔK values involves plasticity effects. If the plastic zone produced at K_{max} comprises a significant fraction of the remaining ligament, the stress-intensity factor (computed from linear elastic relationships) is an underestimate of the true crack driving force. If an appropriate plasticity correction were applied to ΔK, the Paris power law would extend to higher load levels.

The dominant mechanism in Region III probably depends on material properties. One would expect the superimposed fracture/fatigue phenomenon to occur in low-toughness materials. The plasticity mechanism would dominate in tough materials, where significant plastic flow precedes fracture.

In any event, the crack growth behavior in Region III is of little practical importance in most instances. Typical fatigue failures involve the growth of a small initial flaw to a critical size. In such cases, the growing fatigue crack probably spends less than 1% of its life in Region III.

10.8 FATIGUE CRACK GROWTH EXPERIMENTS

The American Society for Testing and Materials (ASTM) has published Standard E 647 [54], which outlines a test method for fatigue crack growth measurements. This standard, described later, includes a recommended practice for estimating the crack-opening load P_{op}. However, the standard admits that its recommended P_{op} measurement procedure does not necessarily give an accurate indication of the true ΔK_{eff} in the presence of crack closure. A new proposed procedure for computing ΔK_{eff} is described in Section 10.8.3.

10.8.1 Crack Growth Rate and Threshold Measurement

The *Standard Test Method for Measurement of Fatigue Crack Growth Rates*, ASTM E 647 [54], describes how to determine da/dN as a function of ΔK from an experiment. The crack is grown by cyclic loading, and K_{min}, K_{max}, and crack length are monitored throughout the test.

The test fixtures and specimen design are essentially identical to those required for fracture toughness testing, which are described in Chapter 7. The E 647 document allows tests on compact specimens and middle tension panels (Figure 7.1).

The ASTM standard for fatigue crack growth measurements requires that the behavior of the specimen be predominantly elastic during the tests. This standard specifies the following requirement for the uncracked ligament of a compact specimen:

$$ W - a \geq \frac{4}{\pi} \left(\frac{K_{max}}{\sigma_{YS}} \right)^2 \tag{10.42} $$

There are no specific requirements on specimen thickness. This standard is often applied to thin sheet alloys for aerospace applications. The fatigue properties, however, can depend on thickness, so the thickness of the test specimen should match the section thickness of the structure of interest.

All specimens must be fatigue precracked prior to the actual test. The K_{max} at the end of fatigue precracking should not exceed the initial K_{max} in the fatigue test. Otherwise, retardation effects may influence the growth rate.

During the test, the crack length must be measured periodically. Crack length measurement techniques include optical, compliance, and potential drop methods (see Chapter 7). Accurate optical crack length measurements require a traveling microscope. One disadvantage of this method is that it can only detect growth on the surface. In thick specimens, the crack length measurements must be corrected for tunneling, which cannot be done until the specimen is broken open after the test. Another disadvantage of the optical technique is that the crack length measurements are usually recorded manually,[12] while the other techniques can be automated. The unloading compliance technique may involve interrupting the test for each crack length measurement. If the specimen is statically loaded for a finite length of time, material in the plastic zone may creep. In an aggressive environment, long hold times may result in environmentally assisted cracking or the deposition of an oxide film on the crack faces. Consequently, the compliance measurements should be made as quickly as possible. The ASTM standard requires that hold times be limited to 10 min; it should be possible to perform an unloading compliance measurement in less than 1 min. Depending on the instrumentation, it may be possible to obtain accurate compliance measurements during cyclic loading, thereby eliminating the need to interrupt the test.

The ASTM Standard E 647 outlines two types of fatigue tests: (1) constant load amplitude tests where K increases and (2) K-decreasing tests. In the latter case, the load amplitude decreases during the test to achieve a negative K gradient. The K-increasing test is suitable for crack growth rates greater that 10^{-8} m/cycle, but is difficult to apply at lower rates because of fatigue precracking considerations. The K-decreasing procedure is preferable when near-threshold data are required. In a typical K-decreasing test, either K_{max} or the R ratio is held constant while ΔK decreases. These two approaches usually result in different behavior in the threshold region due to the R ratio effect on ΔK_{th} (Section 10.4).

---

[12] It is possible to automate optical crack length measurements with image analysis hardware and software, but most mechanical testing laboratories do not have this capability.

Because of the potential for history effects when the K amplitude varies, ASTM E 647 requires that the normalized K gradient be computed and reported:

$$G \equiv \frac{1}{K}\frac{dK}{da} = \frac{1}{\Delta K}\frac{d\Delta K}{da} = \frac{1}{K_{min}}\frac{dK_{min}}{da} = \frac{1}{K_{max}}\frac{dK_{max}}{da} \qquad (10.43)$$

The K-decreasing test is more likely to produce history effects, because prior cycles produce larger plastic zones, which can retard crack growth. Retardation in a rising K test is not a significant problem, since the plastic zone produced by a given cycle is slightly larger than that in the previous cycle. A K-increasing test is not immune to history effects, however; the width of the plastic wake increases with crack growth, which may result in different closure behavior than in a constant K amplitude test.

The ASTM standard recommends that the algebraic value of G be greater than -0.08 mm^{-1} in a K-decreasing test. If the test is computer controlled, the load can be programmed to decrease continuously to give the desired K gradient. Otherwise, the load amplitude can be decreased in steps, provided the step size is less than 10% of the current ΔP. In either case, the load should be decreased until the desired crack growth rate is achieved. It is usually not practical to collect data below $da/dN = 10^{-10}$ m/cycle.

The E 647 standard outlines a procedure for assessing whether or not history effects have occurred in a K-decreasing test. First, the test is performed at a negative G value until the crack growth rate reaches the intended value. Then the K gradient is reversed, and the crack is grown until the growth rate is well out of the threshold region. The K-decreasing and K-increasing portions of the test should yield the same $da/dN - \Delta K$ curve. This two-step procedure is time consuming, but it need only be performed once for a given material and R ratio to ensure that the true threshold behavior is achieved by subsequent K-decreasing tests.

Figure 10.44 schematically illustrates typical crack length vs. N curves. These curves must be differentiated to infer da/dN. The ASTM Standard E 647 suggests two alternative numerical methods to compute the derivatives. A linear differentiation approach is the simplest, but it is subject to scatter. The derivative at a given point on the curve can also be obtained by fitting several neighboring points to a quadratic polynomial (i.e., a parabola).

The linear method computes the slope from two neighboring data points: (a_i, N_i) and (a_{i+1}, N_{i+1}). The crack growth rate for $a = \bar{a}$ is given by

$$\left(\frac{da}{dN}\right)_{\bar{a}} = \frac{a_{i+1} - a_i}{N_{i+1} - N_i} \qquad (10.44)$$

where $\bar{a} = (a_{i-1} + a_i)/2$.

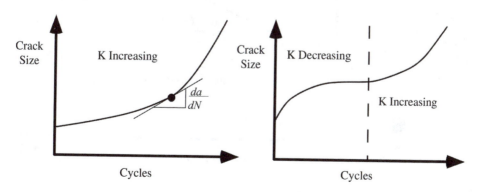

FIGURE 10.44 Schematic fatigue crack growth curves. da/dN is inferred from numerical differentiation of these curves.

The incremental polynomial approach involves fitting a quadratic equation to a local region of the crack length vs. N curve, and solving for the derivative mathematically. A group of $(2n + 1)$ neighboring points are selected, where n is typically 1, 2, 3, or 4, and (a_i, N_i) is the middle value in the $(2n + 1)$ points. The following equation is fitted to the range $a_{i-1} \le a \le a_{i+1}$:

$$\hat{a}_j = b_o + b_1\left(\frac{N_j - C_1}{C_2}\right) + b_2\left(\frac{N_j - C_1}{C_2}\right)^2 \quad (i - n \le j \le i + n) \tag{10.45}$$

where b_o, b_1, and b_2 are the curve-fitting coefficients, and $\hat{a}_j$ is the fitted value of crack length at N_j. The coefficients $C_1 = (N_{i-n} + N_{i+1})/2$ and $C_2 = (N_{i-n} - N_{i+1})/2$ scale the data in order to avoid numerical difficulties. (N_j is often a large number.) The crack growth rate at $\hat{a}_i$ is determined by differentiating Equation (10.45):

$$\left(\frac{da}{dN}\right)_{\hat{a}_i} = \frac{b_1}{C_2} + \frac{2B_2(N_i - C_1)}{C_2^2} \tag{10.46}$$

10.8.2 Closure Measurements

A number of experimental techniques for measurement of closure loads in fatigue have been applied. Allison [55] has reviewed the existing procedures. A brief summary of the more common techniques is given below, including the recommended practice in ASTM E 647 [54].

Most measurements of closure conditions are inferred from compliance. Figure 10.45 schematically illustrates the load-displacement behavior of a specimen that exhibits crack closure. The precise opening load is ill-defined because there is often a significant range of loads where the crack is partially closed. The crack load can be defined by a deviation in linearity in either the fully closed or fully open case (P_1 and P_3, respectively), or by extrapolating the fully closed and fully open load-displacement curve to the point of intersection (P_2). See Section 10.8.3 for a definition of P_{op} that is intended to give a true indication of ΔK_{eff}.

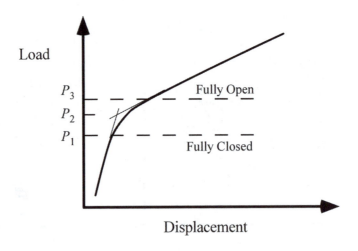

FIGURE 10.45 Alternative definitions of the closure load.

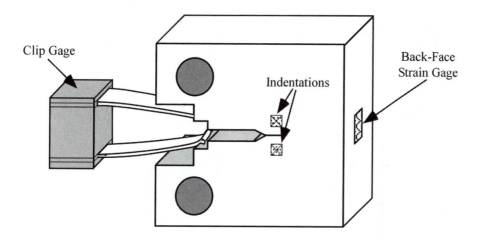

FIGURE 10.46 Instrumentation for the three most common closure measurement techniques.

Figure 10.46 illustrates the required instrumentation for the three most common compliance techniques for closure measurements. The closure load can be inferred from clip gage displacement at the crack mouth, back-face strain measurements, or laser interferometry applied to surface indentations. Specimen alignment is critical when inferring closure loads from compliance measurements.

Displacement measurements remote from the crack tip often lack sensitivity, which can result in scatter in compliance estimates. A signal-processing technique called *compliance offset* can enhance sensitivity of global displacement measurements. This is the technique that ASTM E 647 recommends for the measurement of P_{op}. A baseline compliance is inferred from the fully open portion of the load-displacement curve, and the measured compliance is compared with this baseline. The compliance offset is defined as follows:

$$\text{Compliance offset} = \frac{C_o - C}{C_o} \times 100\% \tag{10.47}$$

where C_o is the open-crack compliance. Figure 10.47 schematically illustrates a plot of load vs. compliance offset. The ASTM standard recommends estimating opening loads corresponding to 1%, 2%, and 4% compliance offset.

Interferometric techniques provide a local measurement of crack closure [56]. Monochromatic light from a laser is scattered off of two indentations on either side of the crack. The two scattered beams interfere constructively and destructively, resulting in fringe patterns. The fringes change as the indentations move apart.

Crack closure is a three-dimensional phenomenon. The interior of a specimen exhibits different closure behavior than the surface. The clip gage and back-face strain gage methods provide a thickness-average measure of closure, while laser interferometry is strictly a surface measurement.

More elaborate experimental techniques are available to study three-dimensional effects. For example, optical interferometry [57] has been applied to transparent polymers to infer closure behavior through the thickness. Fleck [58] has developed a special gage to measure crack-opening displacements at the interior of a specimen.

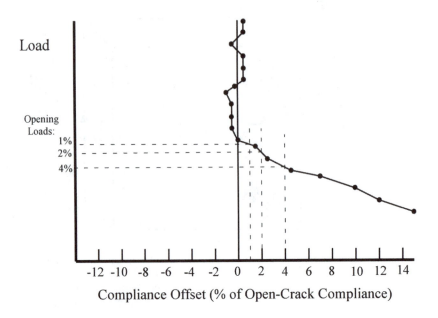

FIGURE 10.47 The compliance offset method. This technique enhances the sensitivity of closure measurements from clip gage displacement.

10.8.3 A PROPOSED EXPERIMENTAL DEFINITION OF ΔK_{eff}

As was previously discussed, crack closure measurements tend to be highly subjective because closure is a gradual process that occurs over a finite load range (Figure 10.45). The important point to keep in mind is that closure is of practical concern because of its effect on fatigue driving force. Therefore, the goal of any closure measurement should be the determination of the true ΔK_{eff}. A definition of P_{op} that is consistent with this goal is outlined below.

The first part of this proposed measurement entails adjusting the clip gage displacement to correct for the compliance of the notch. A typical laboratory specimen contains a machined notch of finite width. When the fatigue crack faces are in contact, the notch faces are separated. In such cases, the slope of the load vs. clip gage displacement curve reflects compliance of a notched but uncracked specimen (Figure 10.3). Removing this contribution to compliance should have a beneficial effect on the sensitivity of closure measurements. Therefore, let us define an adjusted clip gage displacement as follows:

$$V^* \equiv V - C(a_N)P$$

where $C(a_N)$ is the compliance of an uncracked specimen with notch depth a_N.

The left-hand side of Figure 10.48 is a schematic plot of load vs. adjusted clip gage displacement. The corresponding plot of K_I vs. load is on the right-hand side of Figure 10.48. The part of the load-displacement curve labeled Region 1 corresponds to the regime where the crack is fully open and the applied K_I is driven by the remotely applied load. In Region 2, the crack is partially or fully closed, and K_I is driven by the displacement of the crack faces. Even when the fatigue crack is completely closed, the P vs. V^* curve may have a finite slope due to the compliance of asperities or corrosion products in the crack. The *true* applied K_I at P_{min} can be inferred through the construction that is illustrated in Figure 10.48. The steps in this procedure are as follows:

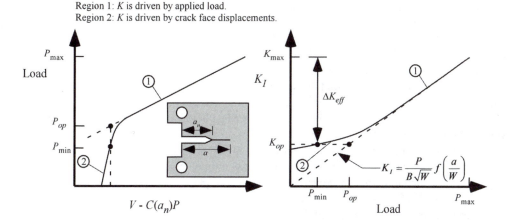

Region 1: K is driven by applied load.
Region 2: K is driven by crack face displacements.

FIGURE 10.48 Proposed definition of P_{op}, designed to give a true indication of ΔK_{eff}.

1. Determine the adjusted clip gage displacement at P_{min}.
2. Construct a vertical line at the V^* value from Step 1.
3. Extrapolate the load-displacement curve for the fully open condition (Region 1) down to lower load levels. The opening load P_{op} is defined at the point where the extrapolated load-displacement curve for Region 1 intersects the vertical line constructed in Step 2.
4. Compute K_{op} from P_{op} using the standard relationship for the test specimen.

As the right-hand side of Figure 10.48 illustrates, the K_{op} value inferred from the above procedure corresponds to the actual applied K_I at P_{min}. That is, K_{op} is defined in such a way that $(K_{max} - K_{op}) = \Delta K_{eff}$ reflects the true driving force for fatigue.

To understand the physical basis of this method, consider the case where closure is due to a pure wedging mechanism, such as a corrosion product in the crack. Figure 10.6(b) illustrates this scenario, in which the load-displacement curve in the fully open regime extrapolates to the origin. The above procedure should apply equally well to plasticity-induced closure or mixed conditions, where there is an offset in the load-displacement curve (Figure 10.6(a) and Figure 10.6(c)), but the simplified case in Figure 10.6(b) makes the following explanation more straightforward. Suppose that the initial applied load is sufficient that the crack faces are not in contact with the wedge. In this case, the applied K_I can be computed from the load using standard formulas. As the load is gradually removed and the crack faces begin to contact the wedge, the slope of the load-displacement curve gradually changes. In this case, the true applied K_I is a function of the applied crack opening rather than the applied load. This would be the case whether the crack is in full contact with the wedge or in partial contact. At P_{min}, there is a residual crack opening that results in $K_I = K_{op}$. If we now remove the wedge, the applied $K_I = K_{min}$, which is the stress intensity corresponding to the applied load P_{min}. In order to achieve the condition $K_I = K_{op}$ *in the absence of the wedge*, the applied load must be increased to P_{op}. This higher load is necessary to achieve the same crack opening that was previously imposed by the wedge.

10.9 DAMAGE TOLERANCE METHODOLOGY

This section describes how to apply fatigue data and growth models to structures, as part of a damage tolerance design scheme. The term *damage tolerance* has a variety of meanings, but normally refers to a design methodology in which fracture mechanics analyses predict remaining life, and quantify inspection intervals. This approach is usually applied to structures that are susceptible to time-dependent flaw growth (e.g., fatigue, environmental-assisted cracking, creep crack growth). As its name suggests, the damage tolerance philosophy allows flaws to remain in the structure, provided they are well below the critical size.

Fracture control procedures vary considerably among various industries; a detailed description of each available approach is beyond the scope of this book. This section outlines a generic damage tolerance methodology and discusses some of the practical considerations. Although fatigue is the primary subject of this chapter, the approaches described below can, in principle, be applied to all types of time-dependent crack growth.

One of the first tasks of a damage tolerance analysis is the estimation of the critical flaw size a_c. Chapter 9 describes approaches for computing critical crack size. Depending on material properties, ultimate failure may be governed by fracture or plastic collapse. Consequently, an elastic-plastic fracture mechanics analysis that includes the extremes of brittle fracture and collapse as special cases is preferable. The possibility of geometric instabilities, such as buckling, should also be considered.

Once the critical crack size has been estimated, a safety factor is normally applied to determine the *tolerable* flaw size a_t. The safety factor is often chosen arbitrarily, but a more rational definition should be based on uncertainties in the input parameters (e.g., stress and toughness) in the fracture analysis. Another consideration in specifying the tolerable flaw size is the crack growth rate; a_t should be chosen such that da/dt at this flaw size is relatively small, and a reasonable length of time is required to grow the flaw from a_t to a_c.

Fracture mechanics analysis is closely tied to nondestructive evaluation (NDE) in fracture control procedures. The NDE provides input to the fracture analysis, which in turn helps to define inspection intervals. A structure is inspected at the beginning of its life to determine the size of initial flaws. If no significant flaws are detected, the initial flaw size is set at an assumed value a_o, which corresponds to the largest flaw *that might be missed* by NDE. This flaw size should not be confused with the NDE detectability limit, which is the smallest flaw *that can be detected* by the NDE technique (on a good day). In most cases, a_o is significantly larger than the detectability limit, due to the variability in operating conditions and the skill of the operator.

Figure 10.49(a) illustrates the procedure for determining the first inspection interval in the structure. The lower curve defines the "true" behavior of the worst flaw in the structure, while the predicted curve assumes the initial flaw size is a_o. The time required to grow the flaw from a_o to a_t (the tolerable flaw size) is computed. The first inspection interval I_1 should be less than this time, in order to preclude flaw growth beyond a_t before the next inspection. If no flaws greater than a_o are detected, the second inspection interval I_2 is equal to I_1, as Figure 10.49(b) illustrates. Suppose that the next inspection reveals a flaw of length a_1, which is larger than a_o. In this instance, a flaw growth analysis must be performed to estimate the time required to grow from a_1 to a_t. The next inspection interval I_3 might be shorter than I_2, as Figure 10.49(c) illustrates. Inspection intervals would then become progressively shorter as the structure approaches the end of its life. The structure is repaired or taken out of service when the flaw size reaches the maximum tolerable size, or when required inspections become too frequent to justify continued operation.

In many applications, a variable inspection interval is not practical; inspections must be often carried out at regular times that can be scheduled well in advance. In such instances a variation of the above approach is required. The main purpose of any damage tolerance assessment is to ensure that flaws will not grow to failure between inspections. The precise methods for achieving this goal depend on practical circumstances.

The schematic in Figure 10.49(c) illustrates a flaw growth analysis that is conservative. If retardation effects are not taken into account, the analysis will be considerably simpler and will tend to overestimate growth rates. If a more detailed analysis is applied, a comparison of actual and predicted flaw sizes after each inspection interval can be used to calibrate the analysis.

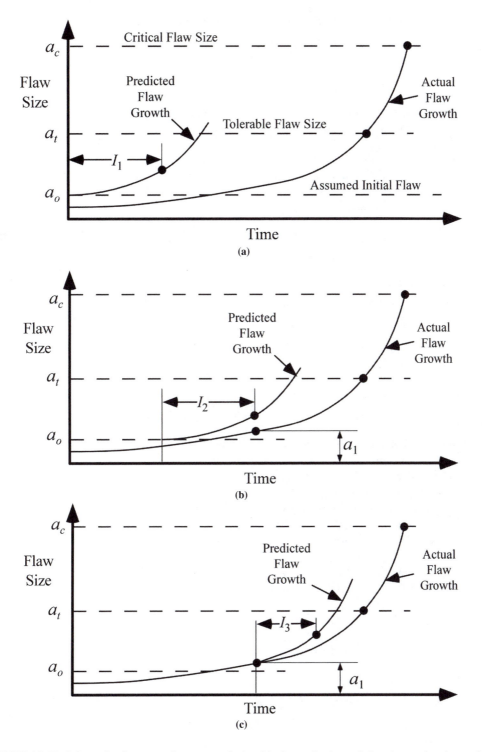

FIGURE 10.49 Schematic damage tolerance analysis: (a) determination of first inspection interval I_1, (b) determination of second inspection interval I_2, and (c) determination of third inspection interval I_3.

APPENDIX 10: APPLICATION OF THE *J* CONTOUR INTEGRAL
TO CYCLIC LOADING

A10.1 DEFINITION OF Δ*J*

Material ahead of a growing fatigue crack experiences cyclic elastic-plastic loading, as Figure A10.1 illustrates. The material deformation can be characterized by the stress range $\Delta\sigma_{ij}$ and the strain range $\Delta\varepsilon_{ij}$ in a given cycle.

Consider the *loading* branch of the stress-strain curve, where the stresses and strains have initial values $\sigma_{ij}^{(1)}$ and $\varepsilon_{ij}^{(1)}$, and increase to $\sigma_{ij}^{(2)}$ and $\varepsilon_{ij}^{(2)}$. It is possible to define a *J*-like integral as follows [3–6]:

$$\Delta J = \int_{\Gamma}\left(\psi(\Delta\varepsilon_{ij})dy - \Delta T_i\frac{\partial\Delta u_i}{\partial x}ds\right) \tag{A10.1}$$

where Γ defines the integration path around the crack tip, and ΔT_i and Δu_i are the changes in traction and displacement between points (1) and (2). The quantity ψ is analogous to the strain energy density:

$$\psi(\Delta\varepsilon_{kl}) = \int_0^{\Delta\varepsilon_{kl}}\Delta\sigma_{ij}d(\Delta\varepsilon_{ij})$$

$$= \int_{\varepsilon_{kl}^{(1)}}^{\varepsilon_{kl}^{(2)}}\left(\sigma_{ij} - \sigma_{ij}^{(1)}\right)d\varepsilon_{ij} \tag{A10.2}$$

Note that ψ represents the stress work per unit volume performed during *loading*, rather than the stress work in a complete cycle. The latter corresponds to the area inside the hysteresis loop (Figure A10.1). For the special case where $\sigma_{ij}^{(1)} = \varepsilon_{ij}^{(1)} = 0$, $\Delta J = J$. Thus ΔJ is merely a generalization of the *J* integral, in which the origin is not necessarily at zero stress and strain.

Although ΔJ is normally defined from the loading branch of the cyclic stress-strain curve, it is also possible to define a ΔJ from the unloading branch. The two definitions coincide if the cyclic stress-strain curve forms a closed loop, and the loading and unloading branches are symmetric.

Just as it is possible to estimate *J* experimentally from a load-displacement curve (Chapter 3 and Chapter 7), ΔJ can be inferred from the cyclic load-displacement behavior. Consider a specimen

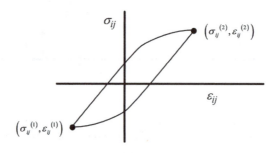

FIGURE A10.1 Schematic cyclic stress-strain behavior ahead of a growing fatigue crack.

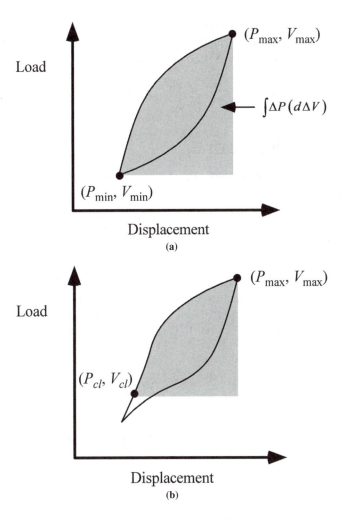

FIGURE A10.2 Cyclic load displacement behavior for fatigue under large-scale yielding conditions: (a) no closure and (b) with crack closure.

with thickness B and uncracked ligament length b that is cyclically loaded between the loads P_{min} and P_{max} and the load-line displacements V_{min} and V_{max}, as Figure A10.2(a) illustrates.[13] The ΔJ can be computed from an equation of the form:

$$\Delta J = \frac{\eta}{Bb} \int_0^{\Delta V} \Delta P d(\Delta V)$$

$$= \frac{\eta}{Bb} \int_{V_{min}}^{V_{max}} (P - P_{min}) dV \tag{A10.3}$$

where the dimensionless constant η has the same value as for monotonic loading. For example, $\eta = 2.0$ for a deeply notched bend specimen.

---

[13] The convention of previous chapters, where Δ represents the load-line displacement, is suspended here to avoid confusion with the present use of this symbol.

Because the ΔJ parameter is often applied to crack growth under large-scale yielding conditions, plasticity induced closure often has a significant effect on the results. If the crack is closed below P_{cl} and V_{cl} (Figure 10.2(b)), Equation (A10.3) can be modified as follows:[14]

$$\Delta J_{eff} = \frac{\eta}{Bb} \int_{V_{cl}}^{V_{max}} (P - P_{cl})dV \qquad (A10.4)$$

A10.2 PATH INDEPENDENCE OF ΔJ

If ψ exhibits the properties of a potential, the stresses can be derived by differentiating ψ with respect to the strains:

$$\Delta\sigma_{ij} = \frac{\partial\psi}{\partial(\Delta\varepsilon_{ij})} \qquad (A10.5)$$

The validity of Equation (A10.5) is both necessary and sufficient for path independence of ΔJ. The proof of path independence is essentially identical to the analysis in Appendix 3.2, except that stresses, strains, and displacements are replaced by the *changes* in these quantities from states (1) to (2). Evaluating ΔJ along a closed contour Γ^* (Figure A3.2) and invoking Green's theorem gives

$$\Delta J^* = \int_{A*} \left[\frac{\partial\psi}{\partial x} - \frac{\partial}{\partial x_j} \left(\Delta\sigma_{ij} \frac{\partial\Delta u_j}{\partial x} \right) \right] dx\, dy \qquad (A10.6)$$

where A^* is the area enclosed by Γ^*. By assuming ψ displays the properties of a potential (Equation (A10.5)), the first term in the integrand can be written as

$$\frac{\partial\psi}{\partial x} = \frac{\partial\psi}{\partial(\Delta\varepsilon_{ij})} \frac{\partial(\Delta\varepsilon_{ij})}{\partial x} = \Delta\sigma_{ij} \frac{\partial(\Delta\varepsilon_{ij})}{\partial x} \qquad (A10.7)$$

By invoking the strain-displacement relationships for small strains, it can be shown that Equation (A10.7) is equal to the absolute value of the second term in the integrand in Equation (A10.6). (See Equation (A3.16) to Equation (A3.18) for the mathematical details.) Thus $\Delta J^* = 0$ for any closed contour. Path independence of ΔJ evaluated along a crack-tip integral can thus be readily demonstrated by considering the contour illustrated in Figure A3.3, and noting that $J_1 = -J_2$.

The validity of Equation (A10.5) is crucial in demonstrating path independence of ΔJ. This relationship is automatically satisfied when there is proportional loading on each branch of the cyclic stress-strain curves. That is, $\Delta\sigma_{ij}$ must increase (or decrease) in proportion to $\Delta\sigma_{kl}$, and the shapes of the $\Delta\sigma_{ij} - \Delta\varepsilon_{ij}$ hysteresis loops must be similar to one another.

Proportional loading also implies a single-parameter characterization of crack-tip conditions. Consequently, ΔJ uniquely defines the changes in stress and strain near the crack tip when there is proportional loading in this region.

In the case of monotonic loading, the J integral ceases to provide a single-parameter description of crack-tip conditions when there is excessive plastic flow or crack growth (Section 3.6). Similarly, one would not expect ΔJ to characterize fatigue crack growth beyond a certain level of plastic deformation. The limitations of ΔJ have yet to be established.

---

[14] The global displacement at closure V_{cl}, is not necessarily zero. The crack-tip region may be closed, while the crack mouth is open. Thus, V_{cl} is often positive.

A10.3 SMALL-SCALE YIELDING LIMIT

When the cyclic plastic zone is small compared to specimen dimensions, ΔJ should characterize fatigue crack growth, since it is related to ΔK. The precise relationship between ΔK and ΔJ under small-scale yielding conditions can be inferred by evaluating Equation (A10.1) along a contour in the elastic singularity dominated zone. For a given ΔK_I, the changes in the stresses, strains, and displacements are given by

$$\Delta \sigma_{ij} = \frac{\Delta K_I}{\sqrt{2\pi r}} f_{ij}(\theta) \tag{A10.8a}$$

$$\Delta \varepsilon_{ij} = \frac{\Delta K_I}{\sqrt{2\pi r}} g_{ij}(\theta) \tag{A10.8a}$$

$$\Delta u_i = \frac{\Delta K_I}{2\mu} \sqrt{\frac{r}{2\pi}} h_{ij}(\theta) \tag{A10.8c}$$

where f_{ij} and h_{ij} are given in Table 2.1 and Table 2.2, and g_{ij} can be inferred from Hooke's law or the strain-displacement relationships.

Inserting Equation (A10.8a) to Equation (A10.8c) into Equation (A10.1) and evaluating J along a circular contour of radius r leads to

$$\Delta J = \frac{\Delta K_I^2}{E'} \tag{A10.9}$$

where $E' = E$ for plane stress conditions and $E' = E/(1 - v^2)$ for plane strain. Note that although $\Delta K = (K_{max} - K_{min})$, $\Delta J \neq (J_{max} - J_{min})$, since

$$\Delta K^2 = (K_{max})^2 - 2K_{max}K_{min} + (K_{min})^2$$

REFERENCES

1. Paris, P.C., Gomez, M.P., and Anderson, W.P., "A Rational Analytic Theory of Fatigue." *The Trend in Engineering*, Vol. 13, 1961, pp. 9–14.
2. Paris, P.C. and Erdogan, F., "A Critical Analysis of Crack Propagation Laws." *Journal of Basic Engineering*, Vol. 85, 1960, pp. 528–534.
3. Dowling, N.E. and Begley, J.A., "Fatigue Crack Growth During Gross Plasticity and the *J*-Integral." ASTM STP 590, American Society for Testing and Materials, Philadelphia, PA, 1976, pp. 82–103.
4. Lambert, Y., Saillard, P., and Bathias, C., "Application of the *J* Concept to Fatigue Crack Growth in Large-Scale Yielding." ASTM STP 969, American Society for Testing and Materials, Philadelphia, PA, 1988, pp. 318–329.
5. Lamba, H.S., "The *J*-Integral Applied to Cyclic Loading." *Engineering Fracture Mechanics*, Vol. 7, 1975, pp. 693–703.
6. Wüthrich, C., "The Extension of the *J*-Integral Concept to Fatigue Cracks." *International Journal of Fracture*, Vol. 20, 1982, pp. R35–R37.
7. Tanaka, K., "Mechanics and Micromechanics of Fatigue Crack Propagation." ASTM STP 1020, American Society for Testing and Materials, Philadelphia, PA, 1989, pp. 151–183.
8. Foreman, R.G., Keary, V.E., and Engle, R.M., "Numerical Analysis of Crack Propagation in Cyclic-Loaded Structures." *Journal of Basic Engineering*, Vol. 89, 1967, pp. 459–464.
9. Weertman, J., "Rate of Growth of Fatigue Cracks Calculated from the Theory of Infinitesimal Dislocations Distributed on a Plane." *International Journal of Fracture Mechanics*, Vol. 2, 1966, pp. 460–467.

10. Klesnil, M. and Lukas, P., "Influence of Strength and Stress History on Growth and Stabilisation of Fatigue Cracks." *Engineering Fracture Mechanics*, Vol. 4, 1972, pp. 77–92.

11. Donahue, R.J., Clark, H.M., Atanmo, P., Kumble, R., and McEvily, A.J., "Crack Opening Displacement and the Rate of Fatigue Crack Growth." *International Journal of Fracture Mechanics*, Vol. 8, 1972, pp. 209–219.

12. McEvily, A.J., "On Closure in Fatigue Crack Growth." ASTM STP 982, American Society for Testing and Materials, Philadelphia, PA, 1988, pp. 35–43.

13. Walker, K., "The Effect of Stress Ratio During Crack Propagation and Fatigue for 2024-T3 and 7075-T6 Aluminum." ASTM STP 462, American Society for Testing and Materials, Philadelphia, PA, 1970, pp. 1–14.

14. Forman, R.G. and Mettu, S.R., "Behavior of Surface and Corner Cracks Subjected to Tensile and Bending Loads in Ti – 6Al – 4V Alloy." ASTM STP 1131, American Society for Testing and Materials, Philadelphia, PA, 1992, pp. 519–546.

15. Elber, W., "Fatigue Crack Closure under Cyclic Tension." *Engineering Fracture Mechanics*, Vol. 2, 1970, pp. 37–45.

16. Suresh, S. and Ritchie, R.O., "Propagation of Short Fatigue Cracks." *International Metallurgical Reviews*, Vol. 29, 1984, pp. 445–476.

17. Budiansky, B. and Hutchinson, J.W., "Analysis of Closure in Fatigue Crack Growth." *Journal of Applied Mechanics*, Vol. 45, 1978, pp. 267–276.

18. Hudak, S.J., Jr. and Davidson, D.L., "The Dependence of Crack Closure on Fatigue Loading Variables." ASTM STP 982, American Society for Testing and Materials, Philadelphia, PA, 1988, pp. 121–138.

19. Newman, J.C., "A Finite Element Analysis of Fatigue Crack Closure." ASTM STP 590, American Society for Testing and Materials, Philadelphia, PA, 1976, pp. 281–301.

20. McClung, R.C. and Raveendra, S.T., "On the Finite Element Analysis of Fatigue Crack Closure—1. Basic Modeling Issues." *Engineering Fracture Mechanics*, Vol. 33, 1989, pp. 237–252.

21. RoyChowdhury, S. and Dodds, R.H., Jr., "Three Dimensional Effects on Fatigue Crack Closure in the Small-Scale Yielding Regime." *Fatigue & Fracture of Engineering Materials and Structures*, Vol. 26, 2003, pp. 663–673.

22. RoyChowdhury, S. and Dodds, R.H., Jr., "A Numerical Investigation of 3-D Small-Scale Yielding Fatigue Crack Growth." *Engineering Fracture Mechanics,* Vol. 70, 2003, pp. 2363–2383.

23. RoyChowdhury, S. and Dodds, R.H., Jr., "Effect of *T*-Stress on Fatigue Crack Closure in 3-D Small-Scale Yielding." *International Journal for Solids and Structures*, Vol. 41, 2004, pp. 2581–2606.

24. Solanki, K., Daniewicz, S.R., Newman, J.C., "Finite Element Analysis of Plasticity-Induced Fatigue Crack Closure: An Overview." *Engineering Fracture Mechanics*, Vol. 71, 2004, pp. 149–171.

25. James, M.N. and Knott, J.F., "An Assessment of Crack Closure and the Extent of the Short Crack Regime in Q1N (HY 80) Steel." *Fatigue of Engineering Materials and Structures*, Vol. 8, 1985, pp. 177–191.

26. Gray, G.T., Williams, J.C., and Thompson, A.W., "Roughness Induced Crack Closure: An Explanation for Microstructurally Sensitive Fatigue Crack Growth." *Metallurgical Transactions*, Vol. 14A, 1983, pp. 421–433.

27. Schijve, J., "Some Formulas for the Crack Opening Stress Level." *Engineering Fracture Mechanics*, Vol. 14, 1981, pp. 461–465.

28. Gomez, M.P., Ernst, H., and Vazquez, J., "On the Validity of Elber's Results on Fatigue Crack Closure for 2024-T3 Aluminum." *International Journal of Fracture*, Vol. 12, 1976, pp. 178–180.

29. Clerivet, A. and Bathias, C., "Study of Crack Tip Opening under Cyclic Loading Taking into Account the Environment and *R* Ratio." *Engineering Fracture Mechanics*, Vol. 12, 1979, pp. 599–611.

30. Shih, T.T. and Wei, R.P., "A Study of Crack Closure in Fatigue." *Engineering Fracture Mechanics*, Vol. 6, 1974, pp. 19–32.

31. Shih, T.T. and Wei, R.P., "Discussion." *International Journal of Fracture*, Vol. 13, 1977, pp. 105–106.

32. McClung, R.C., "The Influence of Applied Stress, Crack Length, and Stress Intensity Factor on Crack Closure." *Metallurgical Transactions*, Vol. 22A, 1991, pp. 1559–1571.

33. Yokobori, T., Yokobori, A.T., Jr., and Kamei, A., "Dislocation Dynamic Theory for Fatigue Crack Growth." *International Journal of Fracture*, Vol. 11, 1975, pp. 781–788.

34. Tanaka, K., Akiniwa, Y., and Yamashita, M., "Fatigue Growth Threshold of Small Cracks." *International Journal of Fracture*, Vol. 17, 1981, pp. 519–533.

35. Vasudevan, A.K., Sadananda, K., and Loutat, N., "A Review of Crack Closure, Fatigue Crack Threshold, and Related Phenomena." *Materials Science and Engineering*, Vol. A188, 1994, pp. 1–22.

36. Vasudevan, A.K. and Sadananda, K., "Classification of Fatigue Crack Growth Behavior." *Metallurgical and Materials Transactions*, Vol. 26A, 1995, pp. 1221–1234.

37. Vasudevan, A.K., Sadananda, K., and Rajan, K., "Role of Microstructures on the Growth of Long Fatigue Cracks." *International Journal of Fatigue*, Vol. 19, 1997, pp. S151–S159.

38. Sadananda, K., Vasudevan, A.K., Holtz, R.L., and Lee, E.U., "Analysis of Overload Effects and Related Phenomena." *International Journal of Fatigue*, Vol. 21, 1999, pp. S233–S246.

39. Vasudevan, A.K., Sadananda, K., and Glinka, G., "Critical Parameters for Fatigue Damage." *International Journal of Fatigue*, Vol. 23, 2001, pp. S39–S53.

40. Sadananda, K. and Vasudevan, A.K., "Crack Tip Driving Forces and Crack Growth Representation under Fatigue." *International Journal of Fatigue*, Vol. 26, 2004, pp. S39–S47.

41. Forth, S.C., Newman, J.C., and Forman, R.G., "On Generating Fatigue Crack Growth Thresholds." *International Journal of Fatigue*, Vol. 25, 2003, pp. 9–15.

42. Rice, J.R., "Mechanics of Crack-Tip Deformation and Extension by Fatigue." ASTM STP 415, American Society for Testing and Materials, Philadelphia, PA, 1967, pp. 247–309.

43. McClung, R.C., "Crack Closure and Plastic Zone Sizes in Fatigue." *Fatigue and Fracture of Engineering Materials and Structures*, Vol. 14, 1991, pp. 455–468.

44. Wheeler, O.E., "Spectrum Loading and Crack Growth." *Journal of Basic Engineering*, Vol. 94, 1972, pp. 181–186.

45. Willenborg, J., Engle, R.M., Jr., and Wood, R.A., "A Crack Growth Retardation Model Using an Effective Stress Concept." Air Force Flight Dynamics Laboratory Report AFFDL-TM-71-1-FBR, January 1971.

46. Newman, J.C., "Prediction of Fatigue Crack Growth under Variable Amplitude and Spectrum Loading Using a Closure Model." ASTM STP 761, American Society for Testing and Materials, Philadelphia, PA, 1982, pp. 255–277.

47. Tanaka, K. and Nakai, Y., "Propagation and Non-Propagation of Short Fatigue Cracks at a Sharp Notch." *Fatigue of Engineering Materials and Structures*, Vol. 6, 1983, pp. 315–327.

48. Potirniche, G.P., Daniewicz, S.R., and Newman, J.C., "Simulating Small Crack Growth Behavior Using Crystal Plasticity Theory and Finite Element Analysis." *Fatigue and Fracture of Engineering Materials and Structures*, Vol. 27, 2004, pp. 59–71.

49. El Haddad, M.H., Topper, T.H., and Smith, K.N., "Prediction of Non-Propagating Cracks." *Engineering Fracture Mechanics*, Vol. 11, 1979, pp. 573–584.

50. Laird, C., "Mechanisms and Theories of Fatigue." *Fatigue and Microstructure*, American Society for Metals, Metals Park, OH, 1979, pp. 149–203.

51. Starke, E.A. and Williams, J.C., "Microstructure and the Fracture Mechanics of Fatigue Crack Propagation." ASTM STP 1020, American Society for Testing and Materials, Philadelphia, PA, 1989, pp. 184–205.

52. Lankford, J. and Davidson, D.L., "Fatigue Crack Micromechanisms in Ingot and Powder Metallurgy 7XXX Aluminum Alloys in Air and Vacuum." *Acta Metallurgica*, Vol. 31, 1983, pp. 1273–1284.

53. Hertzberg, R.W., *Deformation and Fracture of Engineering Materials*, John Wiley & Sons, New York, 1989.

54. E 647-93, "Standard Method for Measurement of Fatigue Crack Growth Rates." American Society for Testing and Materials, Philadelphia, PA, 1993.

55. Allison, J.E., "The Measurement of Crack Closure during Fatigue Crack Growth." ASTM STP 945, American Society for Testing and Materials, Philadelphia, PA, 1988, pp. 913–933.

56. Sharpe, W.N. and Grandt, A.F., "A preliminary study of Fatigue Crack Retardation using Laser Egterferometry to Measure Crack Surface Displacement." ASTM STP 590. American Society for Testing and Materials, Philadelphia, PA, 1976, pp. 302–320.

57. Pitoniak, F.J., Grandt, A.F., Jr., Montulli, L.T., and Packman, P.F., "Fatigue Crack Retardation and Closure in Polymethylmethacrylate." *Engineering Fracture Mechanics*, Vol. 6, 1974, pp. 663–670.

58. Fleck, N.A. and Smith, R.A., "Crack Closure—Is it Just a Surface Phenomenon?" *International Journal of Fatigue*, Vol. 4, 1982, pp. 157–160.

11 Environmentally Assisted Cracking in Metals

Environmentally assisted cracking (EAC) is a common problem in a variety of industries. In the petroleum industry, for example, EAC is pervasive. Offshore platforms are susceptible to corrosion-assisted fatigue. Equipment in refineries and petrochemical plants are exposed to a myriad of aggressive environments that lead to stress corrosion cracking and hydrogen embrittlement. Similar problems exist in other settings, including fossil and nuclear power plants, pulp and paper plants, ships, bridges, and aircrafts. Environmental cracking can occur even when there are no visible signs of corrosion.

The published literature is full of experimental data, as well as theoretical models that attempt to explain environmental cracking phenomena. There is no shortage of controversy in the environmental cracking literature. It is not unusual to find two articles that present data or models that directly contradict one another.

This chapter presents an overview of a highly complex subject. The focus is on aspects of environmental cracking that are relevant to a fracture mechanics specialist. Basic principles of corrosion and electrochemistry are summarized, and the various mechanisms for environmental cracking are introduced, but detailed discussions on theoretical models are avoided. For readers who desire a more in-depth understanding of this subject, a number of excellent books and reviews have been published in the past two decades [1–5].

11.1 CORROSION PRINCIPLES

11.1.1 ELECTROCHEMICAL REACTIONS

All corrosion processes involve electrochemical reactions. Figure 11.1 illustrates a simple electrochemical cell. The *anode* and *cathode* are physically connected to one another and are immersed in a conductive medium called an *electrolyte*.[1] Atoms from the anode material give up electrons, resulting in ions being released into the electrolyte and electrons flowing to the cathode. Note that the corrosion cell forms an electrical circuit. There is a voltage drop, ΔE, between the anode and cathode. Over time, the anode is consumed (i.e., corrodes), as it releases ions into the electrolyte.

In cases where the two electrodes in an electrochemical cell are different metals, the anode is the metal that has a higher propensity to give up electrons (oxidize). For example, in an electrochemical cell with gold and iron electrodes, iron would be the anode because it oxidizes more readily than gold. Such a configuration is an example of *galvanic coupling*.

If an external power source is applied to an electrochemical cell, the flow of electric current can be reversed, such that the anode becomes the cathode and vice versa. This approach is used in electroplating. External power sources can also be used as a form of *cathodic protection*, as discussed in Section 11.1.4.

An electrochemical cell need not include a bond between dissimilar metals. A single metal in contact with an electrolyte may be sufficient to form a corrosion cell, depending on the respective

---

[1] Corrosion and cracking in a gas environment involve chemical reactions where an electrolyte is not present.

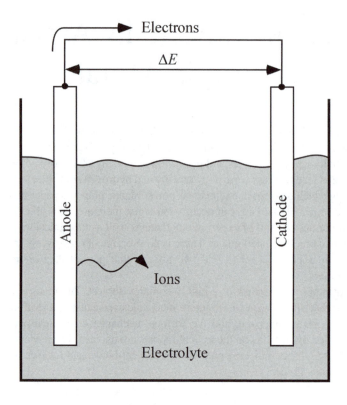

FIGURE 11.1 Schematic illustration of a simple electrochemical cell.

chemical compositions of the metal and electrolyte. For example, consider a coupon of iron immersed in hydrochloric acid (HCl). The chemical reaction is

$$Fe + 2HCl \rightarrow FeCl_2 + H_2 \tag{11.1}$$

The iron is consumed by this reaction and hydrogen gas (H_2) is generated. If we consider only the interaction between iron and hydrogen, the above reaction can be written in the following form:

$$Fe + 2H^+ \rightarrow Fe^{2+} + 2H \rightarrow Fe^{2+} + H_2 \tag{11.2}$$

Therefore, iron reacts with hydrogen ions to form iron ions, atomic hydrogen, and hydrogen gas. This reaction can be divided into two parts:

$$Fe \rightarrow Fe^{+2} + 2e \tag{11.3a}$$

$$2H^+ + 2e \rightarrow 2H \rightarrow H_2 \tag{11.3b}$$

Iron is *oxidized* to iron ions and hydrogen ions are *reduced* to H atoms that can either be absorbed by the electrode or recombine and evolve into the electrolyte as hydrogen gas.[2] The former is an anodic reaction and the latter is a cathodic reaction. An oxidizing or anodic reaction involves the production of electrons, while the consumption of electrons indicates a reducing or cathodic reaction.

---

[2] The former scenario, where atomic hydrogen is absorbed into the electrode, can lead to hydrogen embrittlement, as discussed in Section 11.4.

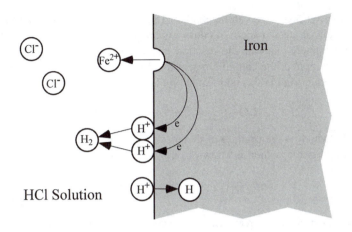

FIGURE 11.2 Anodic and cathodic reactions for iron exposed to hydrochloric acid. Hydrogen gas is released into the electrolyte and atomic hydrogen is absorbed by the iron electrode.

Figure 11.2 schematically illustrates the anodic and cathodic reactions involved in the corrosion of iron by hydrochloric acid. In this particular case, both the anodic and cathodic reactions occur at the same physical location. In other instances, such as a galvanic cell or cathodic protection, the anode and cathode reactions can occur at different locations. However, irrespective of the precise configuration of the electrochemical cell and relative locations of the anodic and cathodic reactions, *both reactions must occur simultaneously and at the same rate*.

Every corrosion process consists of an anodic and cathodic reaction. The anodic reaction normally involves the oxidation of a metal to its ion. The general form for the anodic reaction is given by

$$M \rightarrow M^{+n} + ne \qquad (11.4)$$

where n is the number of electrons produced, which equals the valence of the ion. Most metallic corrosion processes involve one or more of the cathodic (reduction) reactions listed below [1].

- Hydrogen evolution:

$$2H^+ + 2e \rightarrow H_2 \qquad (11.5)$$

- Oxygen reduction (acid solutions):

$$O_2 + 4H^+ + 4e \rightarrow 2H_2O \qquad (11.6)$$

- Oxygen reduction (neutral or basic solutions):

$$O_2 + 2H_2O + 4e \rightarrow 4OH^- \qquad (11.7)$$

- Metal ion reduction:

$$M^{3+} + e \rightarrow M^{2+} \qquad (11.8)$$

- Metal deposition:

$$M^{n+} + ne \rightarrow M \qquad (11.9)$$

Nearly all corrosion problems can be explained in terms of the above reactions. For example, consider the corrosion of steel (which, of course, is predominately iron) immersed in water that is exposed to the atmosphere. The anodic reaction in this case is Equation (11.3a). Since water is nearly neutral and is aerated, the cathodic reaction is Equation (11.7). The overall reaction can be inferred by adding Equation (11.3a) and Equation (11.7):

$$2Fe + 2H_2O + O_2 \rightarrow 2Fe^{2+} + 4OH^- \rightarrow 2Fe(OH)_2 \tag{11.10}$$

Ferrous hydroxide, which is the product of the above reaction, is unstable in oxygenated water. It oxidizes to ferric hydroxide, which is known to the layperson as rust:

$$2Fe(OH)_2 + H_2O + \frac{1}{2}O_2 \rightarrow 2Fe(OH)_3 \tag{11.11}$$

Note that both water and oxygen are required to corrode steel. Steel that is completely submerged in water normally corrodes very slowly because the cathodic reaction is starved for oxygen. Steel corrodes most quickly when there is an ample supply of both moisture and oxygen, such as in a climate with high relative humidity and frequent rain showers. The corrosion rate is also accelerated if steel is coupled galvanically to a more noble metal.

Consider a steel structure in a seawater environment, such as an offshore platform. The most aggressive environments occur just above and below the water surface. In the splash zone above the surface, both oxygen and water are plentiful. Within the first few feet below the surface, the water is oxygen rich because wave motion traps air bubbles. This relatively simple situation is complicated by tight crevice geometries (Section 11.2.2), the presence of additional dissolved ions in the electrolyte, and the imposed cathodic protection (Section 11.1.4).

11.1.2 Corrosion Current and Polarization

Since corrosion is an electrochemical process, the magnitude of the electric current in the corrosion cell is a fundamental measure of the corrosion rate. As stated earlier, both the anodic and cathodic reactions occur simultaneously and at the same rate. The corrosion current can be reduced by inhibiting either reaction, or by reducing the conductivity of the electrolyte.

When an electrochemical reaction is retarded by one or more environmental factors, it is said to be *polarized*. There are three types of polarization: *activation polarization, concentration polarization*, and *resistance polarization*. Activation polarization refers to processes that are controlled by the rate of the reaction at the metal-electrolyte interface. Concentration polarization occurs when the rate-limiting step is diffusion of ions in the electrolyte. Resistance polarization is a consequence of the resistivity of the electrolyte. A reaction can also be polarized by an externally applied current (galvanostatic polarization) or potential (potentiostatic polarization).

Resistance polarization is a major factor in the corrosiveness of seawater compared to tap water and de-ionized water. Seawater is very conductive because there is an ample supply of sodium and chloride ions, while de-ionized water has relatively low electrical conductivity. Normal tap water falls somewhere between these extremes.

11.1.3 Electrode Potential and Passivity

A key factor that controls the corrosion current is the *electrode potential*. Recall the simple corrosion cell in Figure 11.1, which showed an electric potential drop (ΔE) between the anode and cathode. The electrode potential refers to the half-cell potential of the electrode. It is defined as the potential difference between the electrode of interest and a reference electrode, such as a standard hydrogen electrode (SHE). The magnitude of the electrode potential is a function of the chemical composition of the electrode and the oxidizing power of the electrolyte. The oxidizing power is a function of the reagents that are present as well as their concentration.

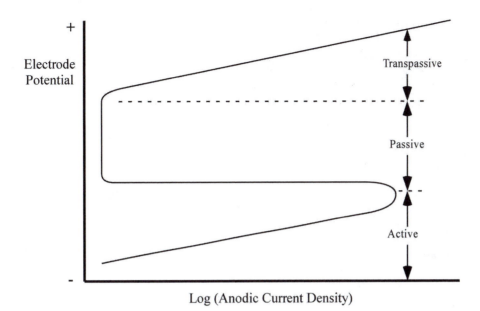

FIGURE 11.3 Polarization diagram of a metal that exhibits passivity effects.

Normally, the corrosion current increases exponentially with increasing electrode potential. However, many technologically important materials (e.g., steel, aluminum, and titanium alloys) exhibit a more complex behavior called *passivity*.

Figure 11.3 illustrates the typical behavior of a metal that exhibits passivity effects. There are three distinct regimes: active, passive, and transpassive. In the active region, a small increase in electrode potential causes a large increase in corrosion rate. A plot of electrode potential vs. the logarithm of current density is a straight line in the active region. As electrode potential is increased further by any of the polarization processes described in Section 11.1.2, the current density exhibits a sudden decrease at the beginning of the passive region. The corrosion rate in the passive region is typically 3 to 6 orders of magnitude slower than one would predict by extrapolating the trend in the active region. In the passive region, a surface film that acts as a protective barrier forms on the surface. This surface film remains stable over a wide range of electrode potential. The surface film breaks down in the transpassive region due to the presence of very powerful oxidizers. The highly protective surface films are very thin, perhaps tens of nanometers. Such films are easily damaged by mechanical means, but quickly reform to protect the metal from corrosion.

11.1.4 CATHODIC PROTECTION

Cathodic protection entails suppressing corrosion by reversing the direction of electric current. In the absence of such protection, corrosion occurs when current passes from the metal surface to the electrolyte and the rate of the anodic (dissolution) reaction is high. Therefore, the corrosion reaction can be suppressed if current enters the metal from the electrolyte. One can view cathodic protection as supplying electrons to the metal surface that is to be protected. As a result, one or more of the cathodic reactions are stimulated on the metal surface in place of the anodic reaction.

Cathodic protection can be achieved either through an external power source or appropriate galvanic coupling. In the former case, the external power source imposes an electric current in the opposite direction of the naturally occurring electrochemical reaction. The galvanic coupling method entails the use of a sacrificial anode, which corrodes instead of the protected material.

The sacrificial anode must oxidize more readily than the protected material, and it must be electrically connected to the protected material. Typical materials for sacrificial anodes include magnesium, zinc, and aluminum-tin alloys.

It is important to understand the active environmental degradation mechanism in a given application before applying a remediation technique such as cathodic protection. This is particularly true for environmental cracking. Cathodic protection can be an effective means to prevent or minimize general corrosion or stress corrosion cracking, but it can actually exacerbate cracking due to hydrogen embrittlement.

11.1.5 Types of Corrosion

Corrosion can manifest itself in a number of forms, and there are generally accepted categories of corrosion based on the appearance and electrochemical processes. The types of corrosion include (but are not limited to):

- Uniform attack
- Galvanic (two-metal) corrosion
- Crevice corrosion
- Pitting
- Intergranular corrosion
- Erosion corrosion
- Environmental cracking

The last item is, of course, the subject of this chapter and is considered in detail. Within the context of the present chapter, the other forms of corrosion are relevant to the extent that they influence environmental cracking. For example, corrosion pits and crevices often act as crack nucleation sites. Not only do such surface features concentrate stress, they also provide a chemical concentration, as discussed in Section 11.2.2. Uniform corrosion is of less concern from the standpoint of crack initiation, but can influence crack propagation. For example, if the cracking is driven by a cathodic reaction at the crack tip (i.e., hydrogen embrittlement) a corresponding anodic reaction on the surface outside of the crack can play a role.

11.2 ENVIRONMENTAL CRACKING OVERVIEW

Section 11.3 to Section 11.5 go into detail about specific types of environmental cracking. There are, however, certain observations one can make about environmental cracking that are not specific to one particular mechanism. Some of these more general phenomena, such as occluded chemistry, threshold stress intensity, fluctuating vs. static loads, and crack morphology, are discussed in this section. These discussions are preceded by brief descriptions of the four types of environmental cracking.

11.2.1 Terminology and Classification of Cracking Mechanisms

The terminology of environmental cracking varies in the published literature. This chapter adopts the most common nomenclature. The term *environmentally assisted cracking* (EAC) is meant to be generic, as it refers to all cracking in metals that is aided by a chemical environment. There are four recognized types of EAC.

- Stress corrosion cracking (SCC)
- Hydrogen embrittlement (HE)
- Corrosion fatigue (CF)
- Liquid metal embrittlement (LME)

SCC refers to crack propagation that is driven by an anodic corrosion reaction at the crack tip. In the past, the term *stress corrosion cracking* was used to refer to a broad range of environmental cracking mechanisms, but this chapter adopts the more recent and more restrictive definition of SCC. Section 11.3 describes SCC in more detail.

HE involves the loss of a metal's bond strength due to the presence of atomic hydrogen at grain boundaries and interstitial sites in the crystal lattice. Of particular interest in the present context are situations where the presence of atomic hydrogen leads to crack propagation. In many of these instances, an electrochemical corrosion reaction acts as a hydrogen source at the crack tip. What distinguishes hydrogen embrittlement cracking from stress corrosion cracking is that the *cathodic* reaction drives crack propagation in the former. Section 11.4 covers HE in greater detail.

This chapter separates environmental cracking mechanisms into anodic (SCC) and cathodic (HE) processes for the sake of simplicity and convenience. In reality, the distinction between anodic- and cathodic-driven crack growth is not always clear. For example, there are cases where environmental cracking initiates through an anodic process but propagates as a result of cathodic hydrogen production. Alternatively, both anodic and cathodic reactions can occur near the crack tip, with both contributing to propagation.

CF is defined as the acceleration of fatigue failure in a chemical environment compared to the fatigue life in an inert environment. Damage results from a synergistic interaction between plastic deformation and electrochemical reactions at the crack tip. See Section 11.5 for a more detailed discussion of CF.

LME or *liquid metal cracking* normally does not involve an electrochemical corrosion reaction. Cracking occurs when liquid metal penetrates grain boundaries of another metal that (obviously) is in a solid state. Several alloy systems are susceptible to LME, including aluminum, titanium, stainless steel, and nickel-based alloys. Metals with low melting points, such as lead, mercury, and zinc, tend to be the materials that cause cracking in certain situations. One of the more common occurrences of liquid metal cracking involves contamination of austenitic stainless steel by zinc. Such contamination normally occurs as a result of contact between galvanized parts and the stainless steel at elevated temperatures.

11.2.2 OCCLUDED CHEMISTRY OF CRACKS, PITS, AND CREVICES

Shielded areas such as cracks, corrosion pits, and crevices often experience different corrosion behavior than the surrounding material because of restricted mass transport between the bulk and local environments. The phenomenon of *crevice corrosion* refers to preferential attack associated with small volumes of stagnant solutions at holes, lap joints, gasket surfaces, and under bolt and rivet heads. The different corrosion behavior at shielded or *occluded* areas is a very important factor in nearly all forms of environmental cracking.

Figure 11.4 schematically illustrates a corrosion pit and a crack exposed to an electrolyte. Electrochemical reactions in these occluded volumes result in a different local chemistry, which influences subsequent reactions. At corrosion pits, the occluded chemistry at the bottom of the pit can lead to initiation of environmental cracks. The occluded chemistry at the tip of an environmental crack often has a significant effect on the propagation rate.

To illustrate the electrochemical processes that lead to occluded chemistry, consider a steel component with a surface crack exposed to aerated seawater. Assuming the seawater is neutral (pH 7), the overall reaction consists of the dissolution of iron (Equation (11.3(a))) and the reduction of oxygen to hydroxide ions (Equation (11.7)). Charge conservation is maintained in both the steel and the water. Each electron that is produced during the oxidation of iron is immediately consumed by the oxygen reduction reaction. Over time, oxygen in the crack is depleted, and the oxygen reduction reaction ceases. However, the oxidation of iron continues, which results in an excess of positive iron ions. In order to maintain charge balance, chloride ions migrate into the crack. The iron and chloride ions react with water to form ferric hydroxide hydrogen ions in the form of

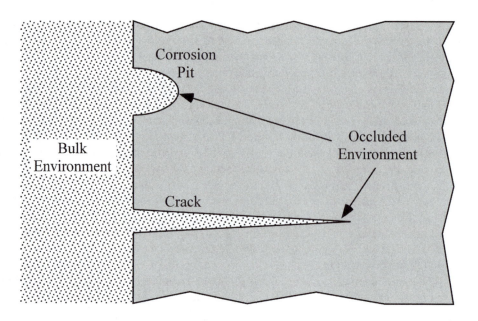

FIGURE 11.4 Occluded environment at the bottom of a pit and the tip of a crack differs from the bulk environment.

hydrochloric acid. A crack or crevice exposed to a neutral dilute sodium chloride solution may contain as much as 10 times the bulk concentration of chlorine and have a pH in the range of 2 to 3.

The occluded crack-tip chemistry can accelerate both anodic stress-corrosion cracking and hydrogen-embrittlement cracking. In the former case, the local acidic and chloride-rich solution tends to be more aggressive than the bulk environment. In the case of hydrogen embrittlement, occluded environments aid the production of hydrogen, which is then absorbed into the metal at the crack tip.

11.2.3 Crack Growth Rate vs. Applied Stress Intensity

As is the case with fatigue (Chapter 10), environmental cracking is subcritical and time dependent. Moreover, fracture mechanics similitude can be applied to the crack growth rate. That is, equal crack growth rates are observed for equal applied stress-intensity values. In the case of corrosion fatigue, the stress-intensity range ΔK is the main characterizing parameter as it is with fatigue in inert environments (Chapter 10). For static loading, the crack growth rate da/dt can be correlated to the applied K.

Figure 11.5 illustrates typical environmental crack growth behavior for static loading. In Stage I, the crack growth rate da/dt is highly sensitive to the applied K. One can define a threshold stress intensity K_{IEAC}, below which the anticipated crack growth is negligible. The threshold concept is discussed further in Section 11.2.4. In Stage II, there are two types of behavior that have been observed. With Type A behavior (Figure 11.5(a)), the crack growth rate is insensitive to the applied K in Stage II. With Type B behavior (Figure 11.5(b)), the crack growth increases with the applied K in Stage II, but the slope of the log(da/dt) vs. K curve is much lower than in Stage I. Stage III corresponds to final fracture when the applied K reaches the fracture toughness of the material.

Figure 11.6 and Figure 11.7 show examples of Type A and Type B environmental cracking behavior, respectively. Figure 11.6 is a plot of environmental cracking data for an aluminum alloy in a saturated salt water solution [6]. Figure 11.7 shows data for hydrogen-embrittlement cracking in a high strength steel [7]. Type A behavior, where there is a plateau in Stage II, is more common than Type B behavior.

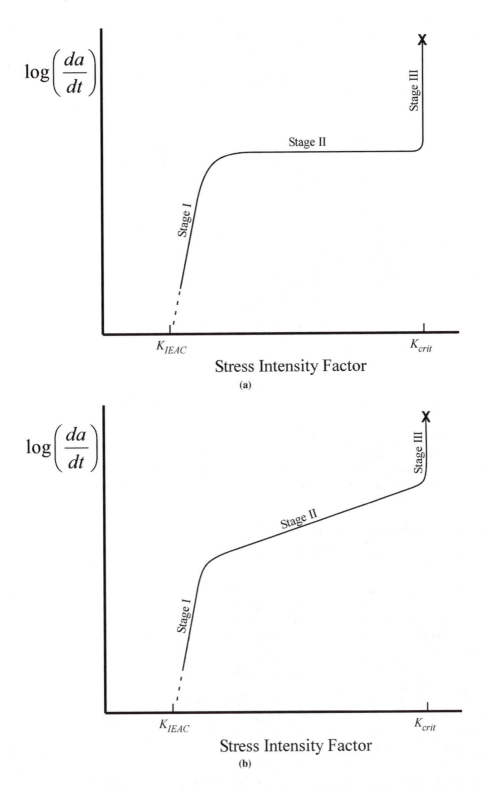

FIGURE 11.5 Typical environmental crack growth behavior for static loading: (a) Type A: crack growth rate independent of K in Stage II and (b) Type B: crack growth rate sensitive to K in Stage II.

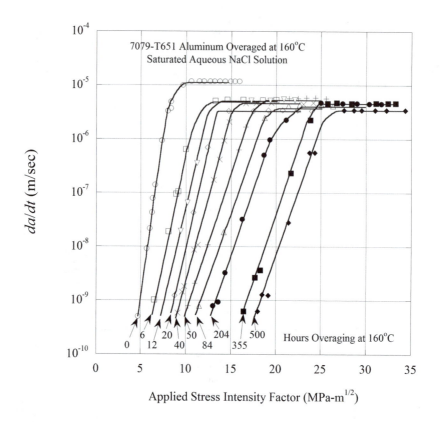

FIGURE 11.6 Effect of an overaging heat treatment on the stress corrosion cracking behavior of 7079-T651 aluminum exposed to a saturated salt water solution. This material–environment combination exhibits Type A behavior in Stage II. Taken from Speidel, M.O., "Current Understanding of Stress Corrosion Crack Growth in Aluminum Alloys." *The Theory of Stress Corrosion Cracking in Alloys.* NATO Scientific Affairs Division, Brussels, Belgium, 1971, pp. 289–344.

11.2.4 THE THRESHOLD FOR EAC

The concept of a threshold stress intensity for environmental cracking K_{IEAC} has been a subject of considerable debate over the years. When fracture mechanics was first applied to environmental cracking problems, many viewed K_{IEAC} as a unique property that applied to a particular combination of material and environment. More recently, environmental cracking experts have become skeptical about the existence of a true threshold stress intensity. When conducting environmental cracking experiments, the apparent threshold tends to decrease as the duration of the test increases. Some experts have gone so far as to argue that the *true* threshold is zero. They believe that the *apparent* threshold, as measured in an experiment, will decrease indefinitely as the time of exposure increases.

Since it is not possible to run an experiment of infinite duration, we may never obtain scientific proof of the existence or absence of a true threshold for environmental cracking. While this may be an interesting issue for theoreticians, it is of little practical significance to engineers because they usually need not be concerned with environmental cracking that occurs over geologic time scales.

It is important to remember that the crack growth rate is plotted on a logarithmic scale, and that the Stage I curve tends to be very steep. In fact, the Stage I crack growth curve for hydrogen cracking of 4130 steel (Figure 11.7) is almost vertical. In Stage I, the crack growth rate can decrease by several orders of magnitude with a modest decrease in the applied K. Consequently, it is possible to establish a *practical* threshold for environmental cracking.

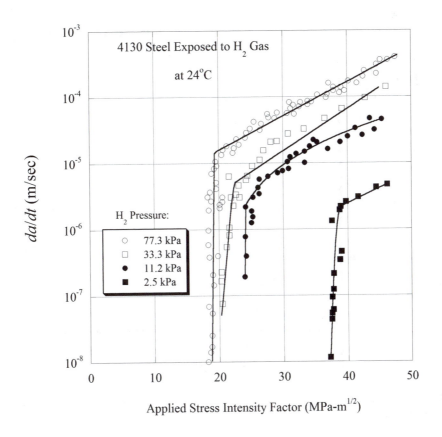

FIGURE 11.7 Cracking of 4130 steel exposed to gaseous hydrogen. This material–environment combination exhibits Type B behavior. Taken from Nelson, H.G. and Williams, D.P., *Stress Corrosion Cracking and Hydrogen Embrittlement of Iron-Base Alloys*, NACE International, Houston, TX, 1977, pp. 390–404.

Figure 11.8 illustrates a procedure to estimate a practical threshold. Crack growth rate data in Stage I can be used to establish the slope of the log(da/dt) vs. K line. This line can then be extrapolated down to an acceptably low target crack growth rate. For example, suppose that 1 mm of crack growth in 10 years constitutes insignificant propagation over the life of the structure. This corresponds to a target growth rate of 3.2×10^{-12} m/sec. Of course, extrapolation introduces errors and uncertainties, so it is best to obtain data at the lowest growth rate that is practical, given experimental time constraints and instrument sensitivity.

11.2.5 SMALL CRACK EFFECTS

As with fatigue crack propagation (Chapter 10), one must be extremely cautious when applying threshold concepts to design. This is especially true for small cracks, where applied K values may be very low but crack propagation occurs nonetheless.

Figure 11.9 is a schematic plot of the threshold stress for environmental cracking vs. crack size. A constant K_{IEAC} for a given material and environment implies that the threshold stress should vary as $1/\sqrt{a}$ (if we neglect the crack size dependence on the geometry factor Y). Such a model predicts that threshold stress approaches infinity as $a \rightarrow 0$. This model obviously cannot be correct at small crack sizes. Otherwise, environmental cracking would never occur on initially smooth surfaces. The threshold K model cannot explain crack initiation from smooth surfaces, nor can it explain propagation of small cracks. In the small crack regime, there are both mechanical and electrochemical factors that enable crack propagation to occur below the apparent threshold.

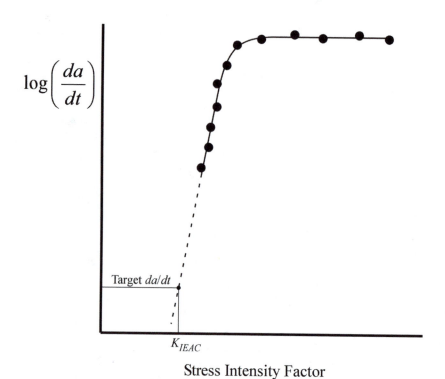

FIGURE 11.8 Obtaining a practical estimate of the threshold by extrapolating data in Stage I to a sufficiently low target growth rate.

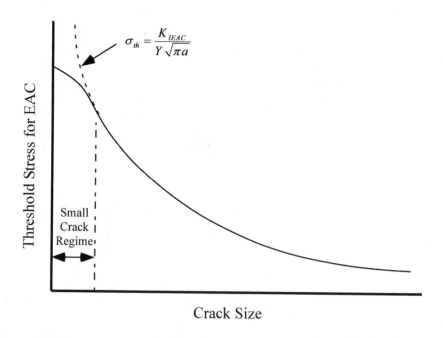

FIGURE 11.9 Effect of crack size on the threshold stress for environmental cracking. The concept of a threshold stress intensity breaks down in the small crack regime.

One mechanical factor that influences small crack behavior is that the plastic zone size may be of the same order as the crack size, which invalidates linear elastic fracture mechanics. Even if plastic zone effects are taken into account in the crack driving force, fracture mechanics similitude may not apply to crack nucleation and the early stages of growth.

The crack size can influence the nature and rate of electrochemical reactions. For example, there are instances where the anode reaction occurs outside of the crack and the cathode reaction occurs at the crack tip (or vice versa). Since an electrochemical reaction requires transport of ions and electrons between the anode and the cathode (Figure 11.1) and this ion transport can be the rate-controlling step in some instances. This reaction may proceed at a faster rate when there is less distance for ions to travel, such as when the crack is shallow. Moreover, the occluded chemistry at the tip of a short crack may be different from that of a deep crack.

11.2.6 Static, Cyclic, and Fluctuating Loads

Environmental cracking data, such as plotted in Figure 11.6 and Figure 11.7, are typically obtained from static tests. As discussed in Section 11.6, most such tests are usually performed in either constant load or constant displacement conditions. In many applications, however, the applied loads fluctuate. Corrosion fatigue experiments can characterize a material in the presence of both cyclic loads and a corrosive environment. In this case, factors such as the frequency of loading and the waveform can influence the crack growth rate. Refer to Section 11.5 for a more detailed discussion of corrosion fatigue.

There are applications in which loads fluctuate mildly or infrequently. Such situations are not normally considered corrosion fatigue, because fatigue damage would be negligible in an inert environment. However, minor or infrequent load fluctuations can have an effect on the environmental cracking rate. Both stress corrosion cracking and hydrogen embrittlement can be sensitive to the crack-tip strain rate. During periods where the K is increasing with time ($dK/dt = \dot{K} > 0$) the material may be more susceptible to crack propagation. As a result, data from static tests may underestimate the crack growth rate and overestimate the threshold when $\dot{K} > 0$. Examples of this behavior for hydrogen cracking are discussed in Section 11.4.

11.2.7 Cracking Morphology

Figure 11.10 illustrates three typical cracking morphologies that can occur through environmental means. Cracking can be either transgranular or intergranular. There may be a single planar crack or a branched network. Branched cracking can be either transgranular or intergranular. The cracking morphology is a result of the interaction between the material and the environment.

Branched cracking is problematic from an analytical standpoint. There is no practical way to compute rigorous stress-intensity solutions for severely branched cracks. Even if one decided to model such a crack with finite element analysis or another analytical technique, existing nondestructive examination (NDE) technology is not capable of providing detailed information about the configuration of a branched network. The good news is that treating branched cracking as a single planar crack with the equivalent depth is conservative. Multiple crack tips in a branched network tend to shield one another, resulting in a decrease in the effective crack driving force.

11.2.8 Life Prediction

In principle, life prediction for EAC can be performed in much the same way as it is for fatigue crack propagation. For example, suppose that the crack growth rate is a known function of stress intensity:

$$\frac{da}{dt} = f(K) \tag{11.12}$$

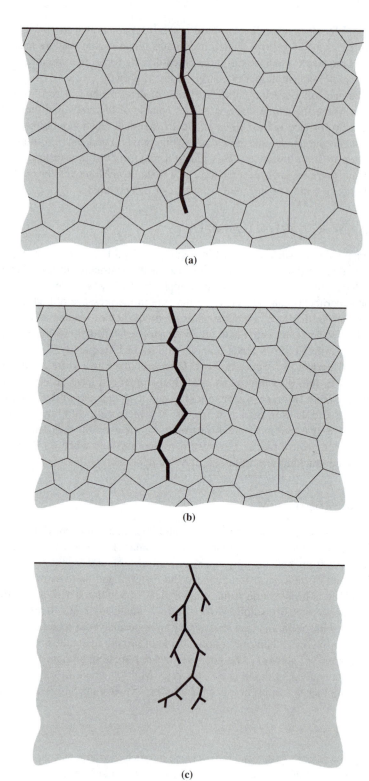

FIGURE 11.10 Examples of cracking morphology: (a) transgranular cracking, (b) intergranular cracking, and (c) branched cracking, which can be transgranular or intergranular.

The time required for a crack to propagate from an initial size to a final size can be computed by integrating the growth law:

$$t_f = \int_{a_o}^{a_f} \frac{da}{f(K)} \tag{11.13}$$

In many practical situations, however, accurate life predictions are virtually impossible. Cracking velocities, as well as the threshold, tend to be highly sensitive to environmental and metallurgical variables. Published environmental cracking data are available only for a limited number of environment–material combinations. Even when experimental data are available for the material and environment of interest, life prediction may still be difficult. Laboratory EAC data are typically obtained under controlled environmental conditions, while the service environment may fluctuate and be ill-defined.

11.3 STRESS CORROSION CRACKING

The term *stress corrosion cracking* (SCC), as it is used in this chapter, refers to crack propagation due to an anodic reaction at the crack tip. The crack propagates because the material at the crack tip is consumed by the corrosion reaction. In many cases, SCC occurs when there is little visible evidence of general corrosion on the metal surface, and is commonly associated with metals that exhibit substantial passivity.

Figure 11.11 is a simple illustration of SCC. In order for the crack to propagate by this mechanism, the corrosion rate at the crack tip must be much greater than the corrosion rate at the walls of the crack. If the crack faces and crack tip corrode at similar rates, the crack will blunt. Under conditions that are favorable to SCC, a passive film (usually an oxide) forms on the crack walls. This protective layer suppresses the corrosion reaction on the crack faces. High stresses at the crack tip cause the protective film to rupture locally, which exposes the metal surface to the electrolyte, resulting in crack propagation due to anodic dissolution. See Section 11.3.1 for further discussion of the film rupture model.

Because of the need for a passive layer to form on the crack faces, conditions that favor SCC often do not favor general corrosion. Figure 11.12 illustrates the typical regions of SCC susceptibility on a polarization diagram. Stress corrosion cracking tends to occur in the transition between active and passive behavior, as well as the transition between passive and transpassive behavior. In the latter case, the region of SCC susceptibility coincides with susceptibility to corrosion pitting. Corrosion pits often act as nucleation sites for SCC due to local stress concentration and occluded chemistry effects.

Plots of electrode potential vs. pH, which are called Pourbaix diagrams, can be used to assess the relative susceptibility to SCC as a function of environmental variables. Figure 11.13 shows the

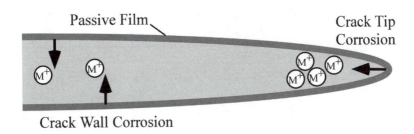

FIGURE 11.11 Simple illustration of anodic SCC. The crack-tip corrosion rate must be much greater than the corrosion rate at the crack walls. Such a condition requires that a passive film form on the crack walls.

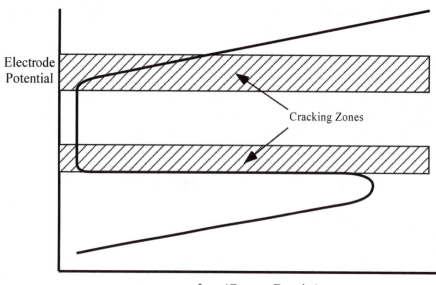

FIGURE 11.12 Polarization diagram, which illustrates the zones that tend to favor SCC. Taken from Jones, R.H. (ed.), *Stress-Corrosion Cracking: Materials Performance and Evaluation*. ASM International, Metals Park, OH, 1992.

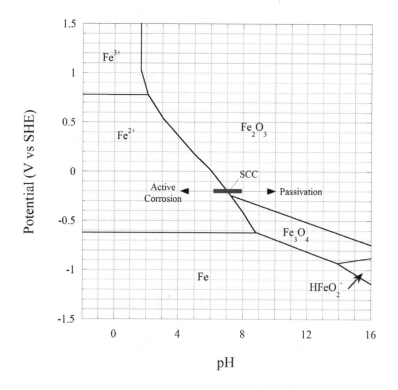

FIGURE 11.13 Pourbaix diagram for iron in water. At a potential of 0.2V vs. standard hydrogen electrode (SHE), the optimal pH for stress corrosion cracking is 7. Taken from Jones, R.H. (ed.), *Stress-Corrosion Cracking: Materials Performance and Evaluation*. ASM International, Metals Park, OH, 1992.

Pourbaix diagram for iron in water. Low pH levels (i.e., acidic conditions) combined with potentials above −0.6 V vs. standard hydrogen electrode (SHE) correspond to regions of active corrosion. In this active region, iron will oxidize to Fe^{2+} or Fe^{3+}. Consider a potential of −0.2 V and a pH ranging from 4 to 10, as indicated in Figure 11.13. At the upper end of this range, the oxide is stable and the iron surface passivates. The low end of this region is in an area of active corrosion. The optimal pH for SCC in this case is 7. At higher pH levels, the oxide film becomes stable, and the corrosion rate is very slow due to passivation. At lower pH levels, a passive film does not form, and general corrosion occurs instead of SCC.

While Pourbaix diagrams may be useful in determining the susceptibility to SCC initiation, they are often of less value when applied to crack propagation. The occluded environment at the crack tip normally results in a different potential and pH compared to the bulk environment. Measuring the potential and pH at the crack-tip is not practical in most instances. Ongoing research is aimed at developing predictive models that enable the crack-tip environment to be inferred from the bulk environment, and significant successes have been recorded over the past two decades.

11.3.1 THE FILM RUPTURE MODEL

As Figure 11.11 illustrates, the crack-tip corrosion rate must be much greater than the corrosion rate on the crack faces in order for SCC to occur. One way to establish such a condition is for the crack tip to be exposed to the electrolyte while the crack faces are protected by a passive film. According to the *film rupture model*, plastic strain at the crack tip causes the passive film to fracture, thereby exposing fresh metal to the environment. This exposed metal corrodes, resulting in SCC growth.

A number of versions of the film rupture model have been proposed [3]. In one version of the model, the crack tip does not repassivate once the film is ruptured, so crack propagation is continuous. Figure 11.14 illustrates the continuous cracking model. Figure 11.15 depicts an alternate version of the film rupture model, which states that cracking is discontinuous. Local plastic deformation at the crack tip exposes fresh metal, which then corrodes to produce an increment of crack growth. After a brief period, the crack tip repassivates as the film thickens and metal dissolution is mitigated. At a critical combination of film thickness and accumulated strain, the process repeats itself. Crack arrest marks may be seen on the fracture surface in the case of discontinuous propagation. There is experimental evidence for both the continuous and discontinuous cracking mechanisms.

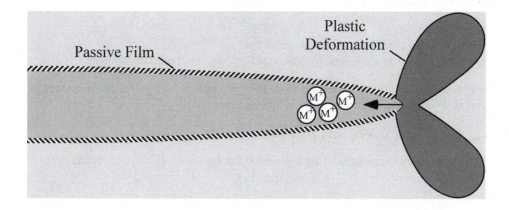

FIGURE 11.14 Film rupture model, where plastic strain at the crack tip fractures the passive film, resulting in preferential corrosion at the crack tip. Taken from Jones, R.H. (ed.), *Stress-Corrosion Cracking: Materials Performance and Evaluation.* ASM International, Metals Park, OH, 1992.

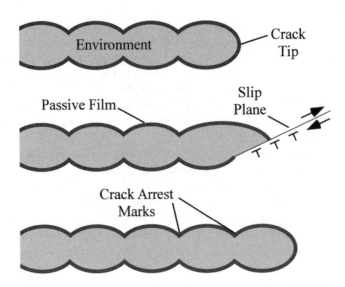

FIGURE 11.15 Alternate version of the film rupture model, where crack growth is discontinuous. The crack tip re-passivates after a small crack advance. Taken from Jones, R.H. (ed.), *Stress-Corrosion Cracking: Materials Performance and Evaluation.* ASM International, Metals Park, OH, 1992.

The film rupture model provides an explanation for Stage I cracking behavior (Figure 11.5), where crack velocity is highly sensitive to the applied K. In Stage I, the crack growth rate is controlled by deformation processes at the tip. Below the threshold, there is insufficient plastic strain at the crack tip to rupture the protective film, and crack velocity is negligible. At the onset of Stage II, there is ample plastic strain for film rupture; so chemical factors control the cracking rate, as discussed in the following section.

11.3.2 CRACK GROWTH RATE IN STAGE II

Most SCC processes exhibit Type A behavior in Stage II (Figure 11.5(a)), where crack velocity is independent of the applied K. In this regime, the crack growth rate is controlled by chemical factors. The overall speed of the electrochemical reaction is governed by the rate-controlling step. Several processes must occur simultaneously, and it is the slowest process that governs the corrosion rate and consequently the crack growth rate. The processes that may occur during SCC include:

- Surface reactions
- Reactions in the solution near the crack tip
- Mass transport in the electrolyte, the crack tip, and the bulk environment.

Unlike hydrogen embrittlement (Section 11.4) there are no time-dependent reaction steps or mass transport that occurs ahead of the crack tip in the plastic zone.

11.3.3 METALLURGICAL VARIABLES THAT INFLUENCE SCC

There are three main characteristics of a material that influence its cracking behavior in a given environment:

- Alloy chemistry
- Microstructure
- Strength

Each of these factors is discussed briefly here.

The various elements and compounds that are present in an alloy dictate which electrochemical reactions can occur. The relative amounts of alloying elements and impurities influence the rate of the electrochemical reactions and the stability of various compounds. Variations in alloy chemistry within a given material class can modify the polarization and Pourbaix diagrams, which can have a pronounced effect on the environmental cracking behavior at a given potential and pH. Alloy content can also affect slip morphology, which plays a significant role in the film rupture process.

Microstructural heterogeneity can result in preferred paths for SCC. For example, intergranular SCC is normally associated with either grain boundary precipitation or grain boundary segregation. An example of grain boundary precipitation is carbide formation on grain boundaries in austenitic stainless steels, which results in chromium depletion adjacent to the grain boundaries. Grain boundary segregation of impurities such as phosphorous and sulfur can occur as a result of temper embrittlement of alloy steels. Such impurity segregation can produce very thin grain boundary films that make the material susceptible to intergranular SCC.

The conventional wisdom states that high-strength alloys are more susceptible to SCC than low-strength alloys. This is by no means a universal rule, however. Much of the experimental data that have been cited as evidence of a strength effect on environmental cracking actually pertain to a hydrogen embrittlement mechanism rather than anodic SCC. Figure 11.6, for example, shows a dramatic increase in K_{IEAC} with overaging of an initially high strength aluminum alloy. In this instance, cracking is driven by hydrogen production at the crack tip rather than anodic dissolution. Moreover, it is not clear whether the K_{IEAC} increase is due to a strength increase or to the microstructural changes that accompany overaging.

11.3.4 CORROSION PRODUCT WEDGING

When anodic dissolution occurs inside a crack, the crack can become filled with a corrosion product, as has been observed in aluminum alloys in moist environments. A similar precipitation can also occur in the crack electrolyte under predominately cathodic conditions, such as has been observed in cathodically protected steel in seawater. The volume of this corrosion product, which is usually an oxide, can be several times greater than the metal that was consumed in the anodic reaction. The expansion of corrosion product inside a crack can result in applied loading on the crack faces.

Several researchers have proposed corrosion product wedging as a mechanism to drive SCC [8–11]. While this mechanism has been demonstrated experimentally in a couple of instances, crack propagation driven by corrosion product wedging does not appear to be widespread. Theoretically, the applied loading due to expanding corrosion product in a crack can lead to significant applied K_I values. On the other hand, the electrochemical conditions under which significant corrosion product is generated in a crack may not be favorable to SCC. That is, the generation of significant corrosion product may be associated with the general corrosion regime, where the crack faces corrode and the crack tip blunts. Another potential mitigating factor is that the corrosion product may obstruct the electrolyte's access to the crack tip.

Corrosion product wedging does play a significant role in corrosion fatigue, particularly in the threshold regime. See Section 11.5.4 for a detailed discussion of this mechanism.

11.4 HYDROGEN EMBRITTLEMENT

When atomic hydrogen is introduced into an alloy, the toughness and ductility can be reduced dramatically, and subcritical crack growth can occur. Many engineers and metallurgists think of hydrogen embrittlement as a mechanism that affects only steel. This perception is due in part to the fact that steel is a much more common structural material than other alloy systems. Hydrogen embrittlement actually affects most, if not all, important alloy systems, including titanium, aluminum, nickel-based alloys and, of course, both ferritic and austenitic steel.

Hydrogen embrittlement is responsible for much of what has traditionally been referred to as "stress corrosion cracking." For example, environmental cracking of high strength steel, aluminum, and titanium alloys in aqueous solutions is usually driven by hydrogen production at the crack tip (i.e., the cathodic reaction) rather than anodic SCC.

11.4.1 Cracking Mechanisms

Hydrogen atoms are small compared to most metallic atoms such as iron, aluminum, and titanium. As a result, hydrogen atoms can fit within interstitial sites in a metallic crystal, as well as at grain boundaries. Moreover, atomic hydrogen readily diffuses through metals, even at room temperature.

The atomistic mechanism for hydrogen embrittlement has been the subject of much debate over the years, and several competing models have been proposed. Currently, the majority opinion is that atomic hydrogen reduces the bond strength between metal atoms, thereby making fracture easier. The reduction in cohesive strength due to the presence of atomic hydrogen can be explained on theoretical grounds, but there is no direct experimental evidence for this mechanism.

Hydrogen embrittlement can lead to a reduction in fracture toughness as well as subcritical crack growth. It is the latter that is of interest in this chapter. Hydrogen embrittlement is a very common mechanism for EAC.

Gangloff [12] suggests that crack propagation due to hydrogen embrittlement should be divided into two categories: *hydrogen-environment-assisted cracking* (HEAC) and *internal-hydrogen-assisted cracking* (IHAC). Figure 11.16 illustrates both types of cracking. The key difference between these two designations is the source of the hydrogen.

In both HEAC and IHAC, hydrogen is concentrated at the fracture process zone near the crack tip. The high degree of stress triaxiality near the crack tip causes the crystal lattice to expand, which increases the hydrogen solubility locally. The high local concentration of hydrogen causes the process zone to be embrittled. This embrittlement, along with the high local stresses, results in microcracking in the process zone. The microcracks that form in the process zone link up with the main crack, resulting in crack extension. The main crack propagates over time, as the local crack-tip processes of hydrogen uptake and microcracking occur continuously. The precise location of the microcrack is controversial, but it is likely somewhere between the crack-tip surface and several

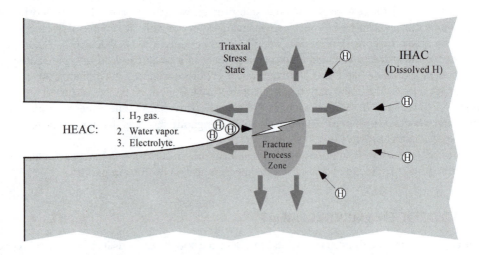

FIGURE 11.16 Environmentally assisted cracking driven by hydrogen embrittlement. Hydrogen-environment-assisted cracking (HEAC) provides atomic hydrogen to the fracture process zone by means of a surface reaction at the crack tip. With internal-hydrogen-assisted cracking (IHAC), the dissolved hydrogen in the bulk material diffuses to the process zone. Taken from Gangloff, R.P., "Hydrogen-Assisted Cracking in High-Strength Alloys." *Comprehensive Structural Integrity, Volume 6: Environmentally-Assisted Fracture.* Elsevier, Oxford, 2003.

crack-tip-opening displacements inward. Because of the different sources of hydrogen, microcracking probably occurs closer to the crack tip for HEAC than it does for IHAC.

HEAC involves hydrogen entering the material at the crack tip. An example of HEAC is when a material is exposed to H_2 gas. Atomic hydrogen is produced at the crack tip when H_2 molecules disassociate. In the absence of a crack under stress, the amount of atomic hydrogen absorbed into the material is negligible at ambient temperature. However, the triaxial stress at the tip of a crack under an applied load affects the equilibrium between H_2 and atomic hydrogen. When coupled with a phenomenon called hydrogen trapping, the local stress field results in very high hydrogen concentrations near the crack tip. Figure 11.16 lists two other sources of hydrogen that can drive HEAC: water vapor and an electrolyte. In the latter case, a corrosion reaction occurs inside the crack as in anodic SCC, but it is the *cathodic* reaction that drives HEAC. Substantial levels of hydrogen are absorbed on the surface through such electrochemical processes.

IHAC occurs when there is dissolved hydrogen in the material. The solubility of atomic hydrogen in most materials is very low at ambient temperature but is significant at elevated temperatures. A material can become hydrogen charged at an elevated temperature when it is exposed to H_2 gas or other compounds that contain hydrogen, such as H_2S. Upon cooling to ambient temperature, atomic hydrogen diffuses out of the material because it is supersaturated. Hydrogen outgassing takes time, however, particularly in thick sections. Consequently, dissolved hydrogen can remain in the material for a significant period of time. When a hydrogen-charged material contains a crack under stress, dissolved hydrogen diffuses into the fracture process zone, as Figure 11.16 illustrates. The concentration of hydrogen in the fracture process zone can be one or more orders of magnitude greater than the bulk concentration.

11.4.2 VARIABLES THAT AFFECT CRACKING BEHAVIOR

Both the threshold K and the crack growth rate for hydrogen-driven cracking are controlled by a complex interaction between the environment, the material properties, and the applied loading. Some of the key factors that influence cracking behavior are described below.

11.4.2.1 Loading Rate and Load History

Most environmental cracking experiments are performed under constant load or constant displacement conditions, but loads typically fluctuate in service environments. The effect of load variations on hydrogen-driven cracking is complex. On the one hand, hydrogen embrittlement is enhanced by an actively straining crack tip, which implies that a rising load should produce more cracking than static conditions. On the other hand, a finite time is required for the various steps involved in the embrittlement process illustrated in Figure 11.16. As a result of these competing factors, load fluctuations can either increase or decrease the susceptibility to EAC, depending on the material, environment, and the rate of load fluctuation.

Following the principles of fracture mechanics similitude, the rate of variation of load or displacement is best characterized by the rate of change in the applied stress intensity dK/dt or $\dot{K}$. For a constant load test, $dK/dt > 0$ if the crack is growing. In the case of a fixed crack-mouth-opening displacement (CMOD), K decreases with crack growth in most specimen geometries. Near the threshold, where cracking velocities are very small, $dK/dt \approx 0$ for both constant load and constant CMOD experiments. This issue is explored further in Section 11.6.2. A variety of dK/dt conditions may be encountered in practice.

Figure 11.17 is a plot of the apparent threshold K vs. dK/dt for a high strength steel subject to a rising load and a gaseous hydrogen environment [13]. The threshold, defined as the onset of measurable crack growth, increases monotonically with loading rate. Cracking, or lack thereof, in this instance is controlled by the surface reactions at the crack tip and diffusion of atomic hydrogen to the fracture process zone. At a sufficiently low dK/dt, the measured threshold from rising load tests agrees with that measured from fixed load or fixed CMOD tests.

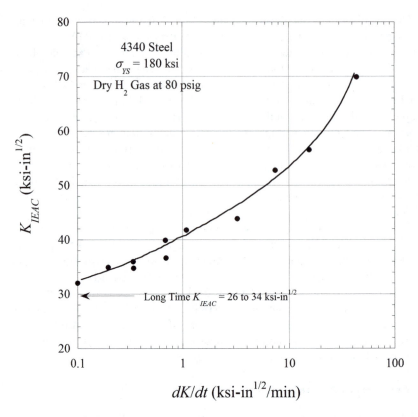

FIGURE 11.17 Effect of loading rate on the apparent threshold for environmental cracking in hydrogen gas. 1 ksi√in = 1.099 MPa√m. Taken from Clark, W.G., Jr. and Landes, J.D., "An Evaluation of Rising Load K_{ISCC} Testing of 4340 Steel." ASTM STP 601, American Society for Testing and Materials, Philadelphia, PA, 1976, pp. 108–127.

For material/environment combinations that exhibit the behavior in Figure 11.17, the dK/dt required to obtain a lower-bound threshold measurement from a rising load test is directly related to the crack growth rate in Stage II [12]. In other words, if a material exhibits rapid crack growth in Stage II, the measured threshold is insensitive to dK/dt up to relatively high loading rates.

Figure 11.18 compares measured threshold values for rising load and fixed CMOD tests in a hydrogen-charged Cr-Mo steel [12]. At low strength levels, the rising load threshold is significantly lower than the corresponding fixed CMOD value. At high strength levels, slowly rising load and fixed CMOD tests give similar threshold estimates. The trend in Figure 11.18 may indicate that the threshold is sensitive to the crack-tip strain rate at lower strength levels. An alternate explanation is that the fixed CMOD specimens, which were tested over a much longer time than the rising load specimens, lost hydrogen during the experiment.

Above the threshold, crack growth rates can increase with increasing dK/dt for low to moderate loading rates. At high loading rates, the crack velocity is limited by electrochemical processes or the rate of diffusion of hydrogen to the process zone.

In certain alloy/environment systems, HEAC can be exacerbated by small amplitude cyclic loading (small ΔK and high R ratio). This so-called *ripple loading* can significantly reduce the threshold K and significantly increase the Stage II growth rate relative to static loading. Acceleration of cracking due to ripple loading has been observed at very low amplitudes ($\Delta K < 1$ MPa√m), where fatigue damage would be negligible in an inert environment. However, some have referred to this phenomenon as a form of corrosion fatigue. Whether or not one chooses to attach such a label to HEAC under ripple loading depends on one's definition of corrosion fatigue.

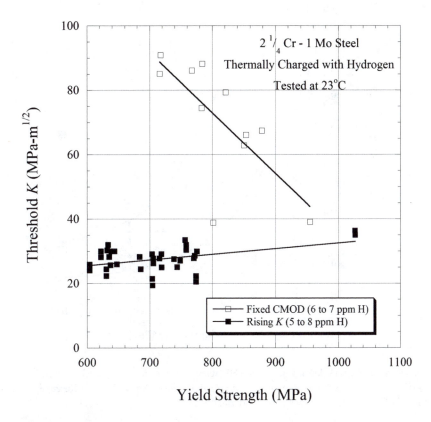

FIGURE 11.18 Comparison of the apparent IHAC threshold for fixed crack-mouth-opening displacement (CMOD) and rising K. Taken from Gangloff, R.P., "Hydrogen-Assisted Cracking in High-Strength Alloys." *Comprehensive Structural Integrity, Volume 6: Environmentally-Assisted Fracture.* Elsevier, Oxford, 2003.

The cause of ripple loading effects is the subject of ongoing research [5,14]. These efforts are focused on explaining the impact of small K fluctuations on hydrogen production, uptake, and damage at the crack tip. For low-strength alloys, Parkins [5] has stated that ripple loading effect can be explained through passive film rupture considerations.[3]

11.4.2.2 Strength

There is a large body of experimental evidence that indicates that the susceptibility to cracking by a hydrogen embrittlement mechanism increases dramatically with increasing strength. Typically, the threshold K decreases and the Stage II growth rate increases with strength.

Figure 11.19 shows a significant drop in the threshold stress intensity with increasing yield strength in martensitic low-alloy steels stressed during immersion in a sodium chloride solution [15]. The effect is most pronounced at yield strength values below about 1350 MPa, which corresponds roughly to 200 ksi.

Figure 11.20 is a plot of the Stage II crack growth rate vs. yield strength for turbine rotor steels [20]. Note that da/dt increases exponentially with yield strength.

One possible explanation for the strength effect on hydrogen-driven EAC is that the stresses inside the fracture process zone are proportional to yield strength. Higher stress in the process zone

---

[3] Film rupture mechanisms do not pertain only to anodic SCC. In the case of HEAC, passive films inhibit hydrogen uptake at the crack tip. Therefore, exposing a fresh metal surface to the environment by rupturing the passive film promotes HEAC.

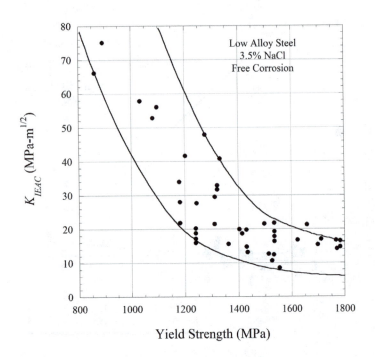

FIGURE 11.19 Yield strength dependence on the HEAC threshold of low-alloy steel in an NaCl solution. Taken from Gangloff, R.P., *Corrosion Prevention and Control*, 33rd Sagamore Army Materials Research Conference, U.S. Army Laboratory Command, Watertown, PA, 1986, pp. 64–111.

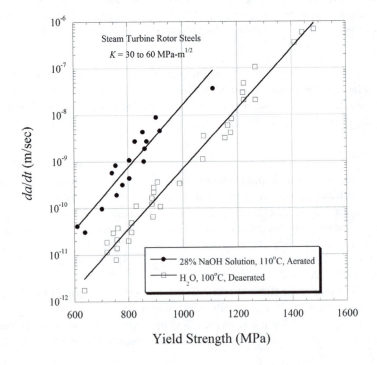

FIGURE 11.20 Effect of yield strength on the Stage II crack growth rate in steam turbine rotor steels. Taken from Speidel, M.O. and Atrens, A., *Corrosion of Power Generating Equipment*. Plenum Press, New York, 1984.

results in more distortion of the crystal lattice, which provides a more welcoming environment for atomic hydrogen to concentrate.

It is often difficult to separate the influence of strength from other metallurgical factors, because variations in strength in a given alloy are usually achieved by modifying the microstructure. Figure 11.6 shows the effect of overaging in 7079 aluminum. Overaging softens the material by precipitate particle coarsening. It is not clear how much of the improvement in cracking resistance is due purely to yield strength effects vs. the change in microstructure. In some alloy systems, nanoscale features that contribute to strength also provide effective sites for H segregation [12].

11.4.2.3 Amount of Available Hydrogen

As one would expect, the relative amount of hydrogen available to the crack tip is an important factor in both HEAC and IHAC. In general, the threshold decreases and the Stage II crack growth rate increases when more hydrogen is available from either the crack-tip environment or the material surrounding the fracture process zone. Figure 11.21 illustrates the effect of the diffusible hydrogen content[4] on the threshold for IHAC in AerMet® 100 [17]. When the concentration falls below 1 part per million in terms of weight (wppm), the threshold K increases dramatically. Figure 11.22 is a plot of the threshold for HEAC in high-strength steel vs. H_2 gas pressure [15]. Given that H_2 pressure is plotted on a logarithmic scale, the effect on the threshold is more gradual than it is for dissolved hydrogen. For HEAC due to an electrolyte in the crack, the threshold K and the Stage II growth rate are a function of the amount of hydrogen that is absorbed on the crack surfaces near the tip.

11.4.2.4 Temperature

The temperature dependence of hydrogen-driven EAC is a reflection of the kinetics of the various processes that control cracking. For example, the rate of IHAC is often controlled by the rate of hydrogen diffusion to the fracture process zone. Diffusivity obeys an Arrhenius law:

$$D = D_o e^{-Q/RT} \tag{11.4}$$

where
  Q = activation energy
  R = gas constant
  T = absolute temperature

Most of the processes that occur in HEAC also exhibit an Arrhenius (exponential) temperature dependence. Consequently, crack growth rate also follows an Arrhenius relationship, at least over a limited temperature range.

Figure 11.23 illustrates the typical behavior for both HEAC and IHAC. The slope of the curve is a reflection of the activation energy of the rate-controlling step in crack growth [12]. The crack growth rate reaches a maximum and then drops abruptly. This peak growth rate typically occurs within 50–100°C of ambient temperature.

The threshold K for most steels increases monotonically with temperature, as Figure 11.24 schematically illustrates. A typical K_{IEAC} vs. temperature curve has a similar shape as a toughness-temperature curve for steels. Just as toughness in steel undergoes a brittle-to-ductile transition with

---

[4] A portion of dissolved hydrogen is typically trapped and is unable to diffuse. Only the diffusible portion of the dissolved hydrogen is available to the fracture process zone.

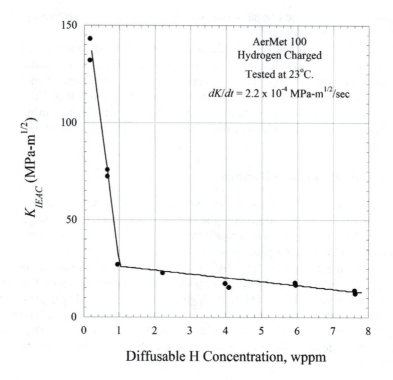

FIGURE 11.21 Effect of dissolved hydrogen concentration on the IHAC threshold in AerMet 100. Taken from Thomas, R.L.S., Scully, J.R., and Gangloff, R.P., "Internal Hydrogen Embrittlement of Ultrahigh-Strength AERMET 100 Steel." *Metallurgical and Materials Transactions*, Vol. 34A, 2003, pp. 327–344.

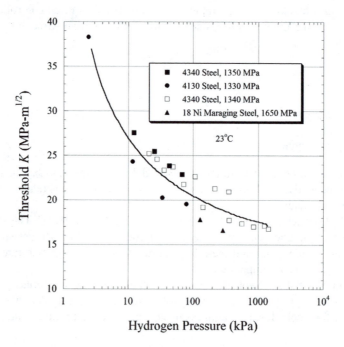

FIGURE 11.22 Effect of H_2 pressure on the HEAC threshold for high-strength steel. Taken from Gangloff, R.P., *Corrosion Prevention and Control*, 33rd Sagamore Army Materials Research Conference, U.S. Army Laboratory Command, Watertown, PA, 1986, pp. 64–111.

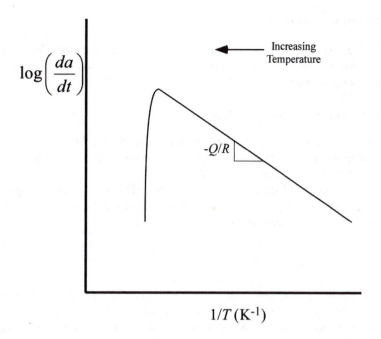

$$\log\left(\frac{da}{dt}\right)$$

Increasing Temperature

$-Q/R$

$$1/T\ (\text{K}^{-1})$$

FIGURE 11.23 Effect of temperature on Stage II cracking velocity for HEAC and IHAC. Taken from Gangloff, R.P., "Hydrogen-Assisted Cracking in High-Strength Alloys." *Comprehensive Structural Integrity, Volume 6: Environmentally-Assisted Fracture.* Elsevier, Oxford, 2003.

increasing temperature (Chapter 5), the susceptibility of a material to hydrogen embrittlement apparently disappears at warmer temperatures.

The abrupt drop in crack growth rate at a critical temperature (Figure 11.23) is consistent with the temperature dependence of the threshold (Figure 11.24). As temperature increases, a point is eventually reached where $K \le K_{\text{IEAC}}$, at which time $da/dt \to 0$.

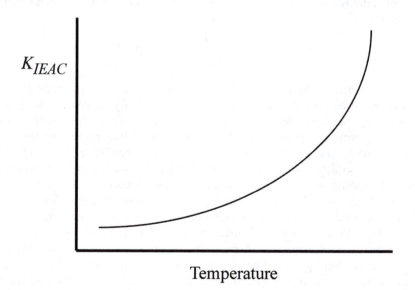

K_{IEAC}

Temperature

FIGURE 11.24 Effect of temperature on the threshold for HEAC and IHAC.

11.5   CORROSION FATIGUE

Corrosion fatigue can be defined as the acceleration of fatigue crack growth due to interaction with the environment. Such acceleration of fatigue cracking can occur even in situations where EAC is negligible under static loading (e.g., cyclic loads where $K_{max} < K_{IEAC}$).

11.5.1   TIME-DEPENDENT AND CYCLE-DEPENDENT BEHAVIOR

Figure 11.25 illustrates three types of corrosion fatigue [18]. Corrosion fatigue can be cycle dependent (Figure 11.25(a)), time dependent (Figure 11.25(b)), or a combination of both (Figure 11.25(c)). Each type of behavior is discussed in more detail in this section.

Cycle-dependent corrosion fatigue, which is illustrated in Figure 11.25(a), corresponds to a simple acceleration of the fatigue crack growth that is insensitive to the loading frequency. The crack growth rate can be represented by an acceleration factor Φ, multiplied by the inert growth rate:

$$\left(\frac{da}{dN} \right)_{aggressive} = \Phi \left(\frac{da}{dN} \right)_{inert} \tag{11.15}$$

This expression can be applied above the fatigue threshold ΔK_{th} for the inert environment. The acceleration factor Φ may be a constant or it may vary with ΔK. Cycle-dependent corrosion fatigue normally occurs in environments that do not result in significant EAC under static loading and where mass transport and electrochemical reactions that contribute to fatigue acceleration are very rapid.

Time-dependent corrosion fatigue (Figure 11.25(b)) can be modeled by a simple superposition of the inert fatigue crack growth rate with the environmental cracking rate.

$$\left(\frac{da}{dN} \right)_{aggressive} = \left(\frac{da}{dN} \right)_{inert} + \frac{1}{f} \left(\frac{d\bar{a}}{dt} \right)_{EAC} \tag{11.16}$$

where $(d\bar{a}/dt)_{EAC}$ is the average environmental crack growth rate over a loading cycle, and f is the loading frequency. Figure 11.26 illustrates the superposition of the two cracking mechanisms, and Figure 11.27 defines the average EAC growth rate.

Most material–environment combinations exhibit both cycle-dependent and time-dependent behavior. Combining Equation (11.15) and Equation (11.16) gives a more general expression for corrosion fatigue:

$$\left(\frac{da}{dN} \right)_{aggressive} = \Phi \left(\frac{da}{dN} \right)_{inert} + \frac{1}{f} \left(\frac{d\bar{a}}{dt} \right)_{EAC} \tag{11.17}$$

Figure 11.28 illustrates the predicted effect of loading frequency on crack growth rate for corrosion fatigue, based on the simple superposition model. By definition, the growth rate is independent of frequency for cycle-dependent corrosion fatigue. Time-dependent corrosion fatigue is sensitive to frequency, as Equation (11.16) indicates. At high frequencies, the crack growth rate approaches the inert rate because the EAC growth per cycle is negligible. At low frequencies, the environmental crack growth per cycle dominates over fatigue, and the rate is proportional to $1/f$. When there is a combination of time-dependent and cycle-dependent acceleration of crack growth rate, the former dominates at low frequencies and the latter dominates at high frequencies.

Another variable that can affect the corrosion fatigue crack growth rate is the waveform of the cyclic loading. This effect can be partially understood by studying Figure 11.27, which defines the average environmental crack growth rate over a given fatigue cycle. If, for example, the cyclic loading followed a square waveform instead of a sine wave, one might expect the average EAC growth rate to be greater because the maximum load is held for a sustained period with a square waveform. A saw-tooth waveform might be expected to produce less environmental cracking per cycle, all else being equal. A mitigating factor is that environmental cracking is often faster during periods when K

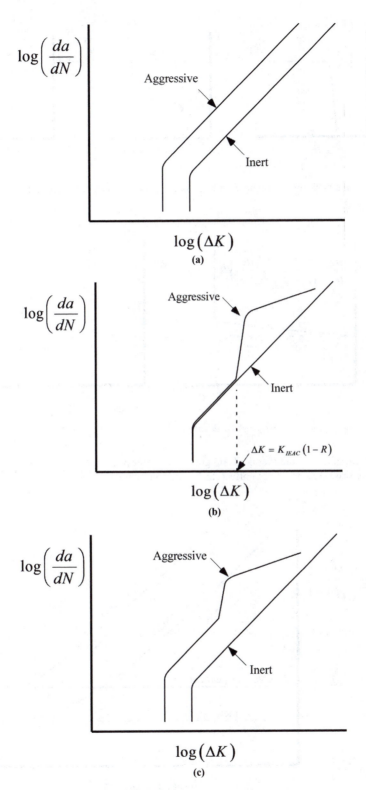

FIGURE 11.25 Three types of corrosion fatigue behavior: (a) cycle-dependent corrosion fatigue, (b) time-dependent corrosion fatigue, and, (c) cycle- and time-dependent corrosion fatigue. Taken from McEvily, A.J. and Wei, R.P., "Fracture Mechanics and Corrosion Fatigue." *Corrosion Fatigue: Chemistry, Mechanics and Microstructures*, NACE International, Houston, TX, 1972, pp. 25–30.

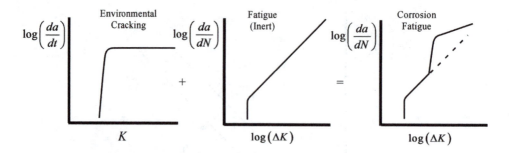

FIGURE 11.26 Superposition of fatigue with EAC in the case of purely time-dependent corrosion fatigue.

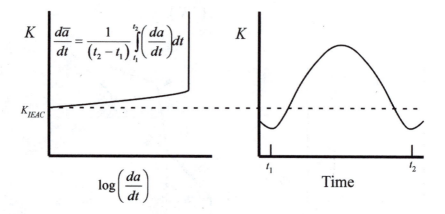

FIGURE 11.27 Definition of the average EAC growth rate over a single fatigue cycle. The waveform (e.g., sine wave, square wave, saw-tooth) can affect this average growth rate.

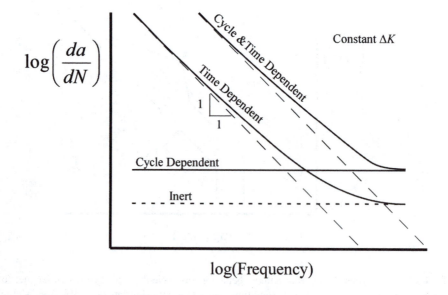

FIGURE 11.28 Effect of frequency on the various types of corrosion fatigue.

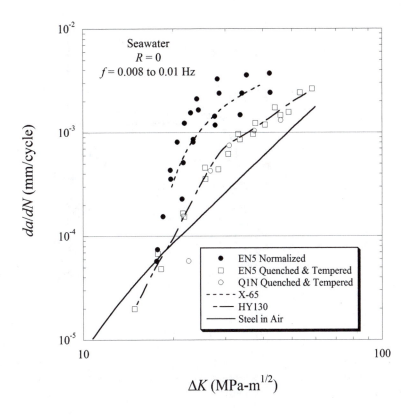

FIGURE 11.29 Corrosion fatigue behavior of several C–Mn and alloy steels in seawater. Taken from Gangloff, R.P., "Corrosion Fatigue Crack Propagation in Metals." *Environment-Induced Cracking of Metals*, NACE International, Houston, TX, 1990, pp. 55–109.

is increasing. In those instances, a square waveform might actually result in less environmental cracking per cycle because a sustained load is less damaging than a continually rising load.

11.5.2 TYPICAL DATA

Figure 11.29 shows corrosion fatigue data for C–Mn and alloy steels in seawater [19]. When compared to the corresponding data for crack propagation in air, these data exhibit the trends one would expect for time-dependent corrosion fatigue (Figure 11.26(b)).

Figure 11.30 is a plot of corrosion fatigue data for a high strength alloy steel in an H_2 gas environment as well as a hydrogen-oxygen mixture and air [19]. There is a slight effect of frequency in the H_2 gas environment, which is indicative of time-dependent behavior. When H_2 gas is combined with oxygen, the resulting behavior is identical to that for air, and there is no frequency dependence.

Figure 11.31 shows the effect of frequency on the corrosion fatigue behavior of two aluminum alloys in a sodium chloride solution [19]. The 7079-T651 alloy exhibits a frequency dependence that corresponds closely to the expected 1:1 trend (on a logarithmic plot) for time-dependent corrosion fatigue, as schematically illustrated in Figure 11.28. The 2219-T87 alloy does not exhibit a frequency dependence, but has an elevated growth rate relative to a purely inert environment.

Figure 11.32 is a plot of *da/dN* vs. frequency for Inconel 600 in a sodium hydroxide solution [19]. The crack growth rate exhibits a frequency dependence that asymptotically approaches the 1:1 line at low frequencies. As frequency increases, *da/dN* asymptotically approaches the rate for this material in air. Note, however, that the crack growth rate in air is elevated relative to the vacuum environment, which indicates that air is not a purely inert environment with respect to corrosion

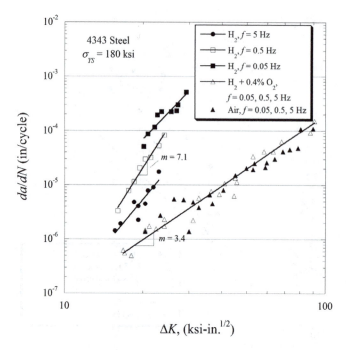

FIGURE 11.30 Corrosion fatigue behavior of 4340 steel in gaseous hydrogen and $H_2 + O_2$ mixtures. Taken from Gangloff, R.P., "Corrosion Fatigue Crack Propagation in Metals." *Environment-Induced Cracking of Metals*, NACE International, Houston, TX, 1990, pp. 55–109.

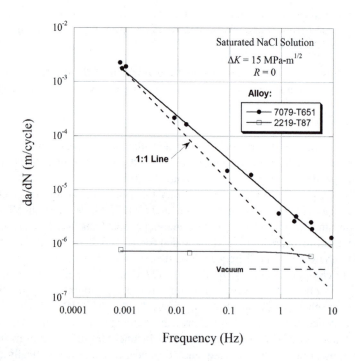

FIGURE 11.31 Frequency dependence of corrosion fatigue of two aluminum alloys in a saturated sodium chloride solution. Taken from Gangloff, R.P., "Corrosion Fatigue Crack Propagation in Metals." *Environment-Induced Cracking of Metals*, NACE International, Houston, TX, 1990, pp. 55–109.

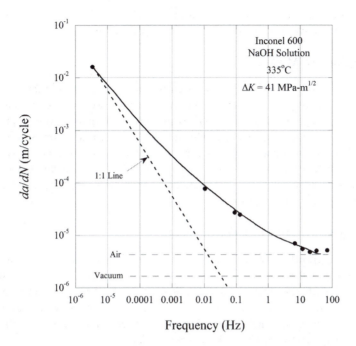

FIGURE 11.32 Frequency dependence of corrosion fatigue of Inconel 600 in a sodium hydroxide solution. Taken from Gangloff, R.P., "Corrosion Fatigue Crack Propagation in Metals." *Environment-Induced Cracking of Metals*, NACE International, Houston, TX, 1990, pp. 55–109.

fatigue. The air environment apparently produces cycle-dependent (time-independent) corrosion fatigue. The relative shift between air and vacuum environments (or between air and an inert gas environment such as argon) has been observed in a variety of alloy systems. The possible reasons for this behavior are explored in the section on mechanisms.

Note that the frequency dependence of da/dN will approach the 1:1 slope on a logarithmic scale ($1/f$ dependence) only if the following conditions are satisfied:

1. The simple superposition model of Equation (11.16) and Equation (11.17) provides a reasonably accurate description of the actual corrosion fatigue behavior.
2. The frequency is sufficiently low or the time-dependent cracking rate is sufficiently high that the second term on the right-hand side of Equation (11.16) or Equation (11.17) dominates.

The data for 7079-T651 aluminum in Figure 11.31 exhibits the 1:1 trend at lower frequencies because this alloy is highly susceptible to time-dependent environmental cracking in the NaCl environment. Recently, however, Gasem and Gangloff [20] presented data for other 7000-series aluminum alloys in a variety of environments that shows much less sensitivity to frequency. The rate of time-dependent environmental cracking in the Gasem-Gangloff experiments was much less than in the earlier 7079-T651 data. Consequently, logarithmic plots of da/dN vs. frequency never approach the theoretical 1:1 slope. Another potential contributing factor to the lower observed sensitivity to frequency in some data sets may be that the simple superposition model is inadequate for these alloy/environment systems. More sophisticated mechanism-based corrosion fatigue models are being developed [20,21].

11.5.3 MECHANISMS

Efforts to explain corrosion fatigue behavior have focused primarily on the same mechanisms that are believed to control environmental cracking under static loading, namely, film rupture/anodic dissolution mechanisms and hydrogen embrittlement. Additional considerations in corrosion fatigue are the interaction between cyclic plastic deformation and local chemical or electrochemical reactions, as well as convective mixing effects on mass transport kinetics.

11.5.3.1 Film Rupture Models

Film rupture/anodic dissolution models have been proposed for corrosion fatigue in a number of alloy/environment systems. Such models are an extension of those proposed for anodic SCC under static loading (Figure 11.14 and Figure 11.15). In this view, cyclic plastic strain ruptures a protective film at the crack tip, resulting in transient anodic dissolution, followed by repassivation. The amount of environmental crack propagation per fatigue cycle depends on the kinetics of the reaction on the clean metal surface as well as the time between film ruptures. The latter depends on the crack-tip strain rate and film ductility, and models that incorporate such effects have been used to explain the frequency effect on crack growth rate.

11.5.3.2 Hydrogen Environment Embrittlement

Because hydrogen embrittlement appears to be the predominate mechanism for environmental cracking under static and monotonic loading in a variety of alloy/environmental systems, it is reasonable to invoke this mechanism for cyclic loading in those same systems. As Figure 11.16 illustrates, crack propagation occurs as a result of hydrogen absorption by the fracture process zone near the crack tip, which leads to local embrittlement and microcracking. As with film rupture/anodic dissolution mechanisms, hydrogen embrittlement is time dependent, which leads to a frequency effect.

Figure 11.30 shows an example of acceleration of fatigue crack propagation in 4340 steel due to a gaseous hydrogen environment. When O_2 is mixed with H_2, crack-tip embrittlement does not occur because an oxygen reduction reaction (Equation 11.6) creates an oxide barrier on the surface that blocks H uptake.

Under static loading, 7079-T651 aluminum is susceptible to HEAC, as the data in Figure 11.6 indicate. Figure 11.31 shows that this alloy is also susceptible to time-dependent corrosion fatigue. A reasonable inference is that hydrogen uptake from the electrolyte is responsible for the acceleration in crack growth rate, given that this mechanism operates under static loading. Note that 2219-T87 aluminum does not exhibit time-dependent corrosion fatigue behavior, which would seem to be an indication that this alloy is not susceptible to hydrogen embrittlement.

11.5.3.3 Surface Films

Early studies of corrosion fatigue initiation from smooth surfaces focused on the effect of environmentally produced thin films on local slip processes that govern cyclic plastic deformation. More recently, the possible impact of surface films on crack-tip behavior has been considered, but much of the discussion in the published literature is speculative in nature, and experimental results that provide even circumstantial evidence for this theory are extremely limited. The view is that surface films may interfere with reversible slip during cycle loading, and thus have an impact on crack growth.

Surface film effects have been proposed to explain the cycle-dependent acceleration of fatigue of some alloy systems in air relative to an inert environment. In aluminum alloys, for example, it has been hypothesized that the faster growth rate in air is due to irreversible slip and crack blunting, which produces striations. Flat, striation-free surfaces have been observed for fatigue of aluminum in a vacuum environment, while well-defined striations occur when fatigue cracks propagate in air.

11.5.4 The Effect of Corrosion Product Wedging on Fatigue

While a corrosive environment often results in an acceleration of fatigue crack growth rates, there are instances where the environment has the opposite effect. Anodic dissolution inside a crack can result in the crack being filled with an oxide or other corrosion product. The volume of the corrosion product

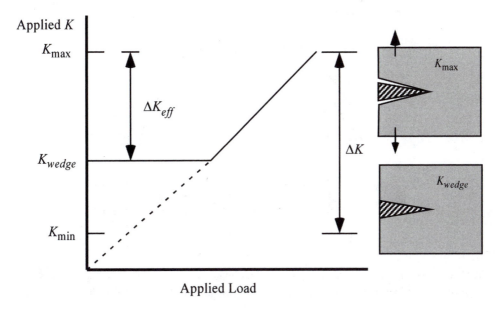

FIGURE 11.33 Effect of corrosion product wedging on the effective cyclic stress intensity.

may be two or three times that of the metal that was consumed in the anodic reaction. An expanding corrosion product inside the crack tip can lead to wedge loading, as discussed in Section 11.3.4.

While the propensity for corrosion product wedging to influence anodic SCC under static loading is a matter of debate, the effect on fatigue crack propagation is well established. Figure 11.33 illustrates the influence of a corrosion product on the effective ΔK. At low applied load levels, K_I is independent of load because it is controlled by the wedging action of the corrosion product. In order for the applied stress intensity to increase above K_{wedge}, the load must be sufficient to open the crack to a displacement that is greater than the corrosion product thickness. The result is a diminished ΔK_{eff}, which in turn results in a reduced crack growth rate. This effect is most pronounced at low R ratios and near the threshold ΔK.

Corrosion product wedging results in an increase in the threshold and enhanced sensitivity to the R ratio. See Chapter 10 for a more extensive discussion of the effect of crack wedging (from corrosion products and other sources) on fatigue.

11.6 EXPERIMENTAL METHODS

Traditional experimental methods to evaluate the behavior of materials subjected to aggressive environments while under stress involved tests on smooth specimens. Such tests reflected both initiation and propagation stages of environmental cracking. Beginning in the 1960s, researchers performed environmental cracking experiments on pre-cracked specimens and applied the principles of linear elastic fracture mechanics. Today, both smooth-specimen tests and fracture-mechanics-based tests are routinely performed. The most important aspects of either type of test are: (1) strict control of the environment to which the specimen is exposed and (2) accurate measurements of crack initiation and propagation.

Smooth specimen tests are largely qualitative, and they can provide an overly optimistic assessment of the material performance in the presence of a corrosive environment. In many instances, materials that exhibit little or no susceptibility to environmental cracking in smooth-specimen tests show a high degree of susceptibility when a precracked specimen of that material is tested in the same environment. Titanium alloys are particularly susceptible to this behavior. Occluded chemistry

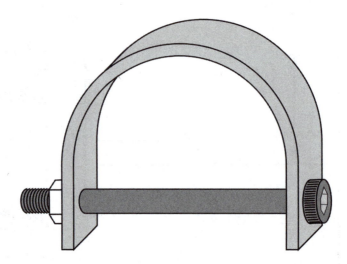

FIGURE 11.34 The U-bend test. Taken from G 30-97, "Standard Practice for Making and Using U-Bend Stress-Corrosion Test Specimens." American Society for Testing and Materials, Philadelphia, PA, 2003.

effects, discussed in Section 11.2.2, are partially responsible for the different behavior at a smooth surface vs. the tip of a pit, crevice, or crack. Local stress concentration effects are another factor.

Common test methods for both smooth specimens and cracked specimens are described in the following section.

11.6.1 TESTS ON SMOOTH SPECIMENS

Two of the oldest test methods for assessing environmental cracking susceptibility are the U-bend test and the bent-beam test. The test configurations are illustrated in Figure 11.34 and Figure 11.35, respectively. The American Society for Testing and Materials (ASTM) has published standard procedures for both test methods [22,23]. For each test method, the specimen and loading apparatus are self-contained, so there is no need for a test machine. The entire assembly can be placed in the environment of interest.

The bent-beam test is suitable for testing in the elastic range. The applied stress in the specimen can be computed with standard beam deflection equations. A U-bend specimen is typically loaded into the plastic range, so the precise stress and strain state are usually not known.

The slow strain rate test, as the name suggests, involves slowly loading a tensile specimen (Figure 11.36) exposed to a potentially corrosive environment. The advantage of this test over the U-bend and bent-beam tests is that it is normally faster because the active straining accelerates crack initiation and propagation. A disadvantage of this test is that a special test machine is required. An ASTM Standard [24] provides a detailed procedure for performing slow strain rate tests.

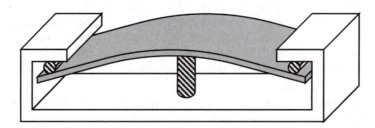

FIGURE 11.35 The bent-beam test. Taken from G 39-99, "Standard Practice for Preparation and Use of Bent-Beam Stress-Corrosion Test Specimens." American Society for Testing and Materials, Philadelphia, PA, 1999.

FIGURE 11.36 Tensile specimen for the slow strain rate test. Taken from G 129-00, "Standard Practice for Slow Strain Rate Testing to Evaluate the Susceptibility of Metallic Materials to Environmentally Assisted Cracking." American Society for Testing and Materials, Philadelphia, PA, 2000.

11.6.2 FRACTURE MECHANICS TEST METHODS

Most tests of cracked specimens in aggressive environments fall into one of the following categories:

1. Constant load
2. Constant displacement
3. Cyclic loading (corrosion fatigue)
4. Controlled K history

The first two types of tests are the most common and are covered by the ASTM Standard E 1681 [25]. Constant load and displacement testing are discussed in more detail below. The procedures for performing corrosion fatigue experiments on precracked specimens are essentially the same as for tests in air, but with the added complexity associated with exposing the specimen to the appropriate environment. An example of the fourth type of experiment is the rising K test, where dK/dt is controlled. The data in Figure 11.17 were obtained from a series of rising K tests. Both cyclic and K-controlled testing require a closed loop servo-hydraulic or servo-electric test machine. Constant load and constant displacement testing can each be performed with a simpler and less costly apparatus. All such specimens require a fatigue precrack.

Figure 11.37 illustrates a cantilever bend test. This is the most common type of constant load test. An edge-cracked bend specimen is fixed at one end and a constant weight is applied to the opposite end. The central portion of the specimen, which contains a fatigue precrack, is exposed to the environment of interest.

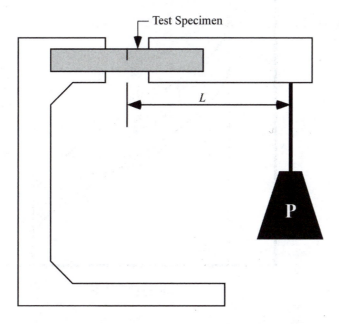

FIGURE 11.37 Constant load cantilever bend test. Taken from E 1681-03, "Standard Test Method for Determining a Threshold Stress Intensity Factor for Environment-Assisted Cracking of Metallic Materials." American Society for Testing and Materials, Philadelphia, PA, 2003.

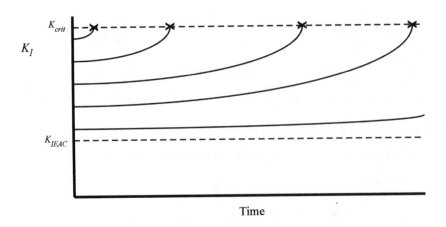

FIGURE 11.38 Environmental crack growth under constant applied load.

Figure 11.38 is a schematic illustration of the crack growth behavior in a constant load test. The specimen is loaded to an initial K_I, which increases with crack growth. Failure occurs when the applied K_I reaches the fracture toughness of the material. When the initial K_I is close to the threshold for environmental cracking, the duration of the test may be very long. It is easy to overestimate K_{IEAC} in a constant load test because there may be no detectable cracking for a significant period of time despite the fact that the applied K_I is above the true threshold.

Figure 11.39 illustrates the inherent difficulty of estimating K_{IEAC} from a constant load test [26]. When a precracked specimen is first exposed to the corrosive environment, there is an incubation time that precedes crack propagation. The incubation time often increases with decreasing K_I. The incubation time can be very long when the initial applied K_I is close to K_{IEAC}. Therefore, when

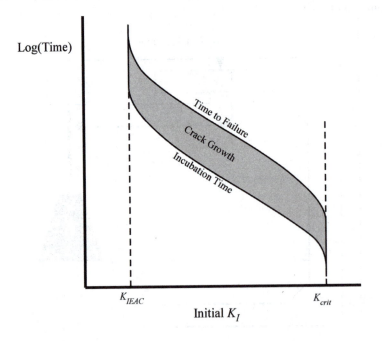

FIGURE 11.39 Incubation time and time to failure as a function of the initial applied K_I in a constant load test. Taken from Wei, R.P., Novak, S.R., and Williams, D.P., "Some Important Considerations in the Development of Stress Corrosion Cracking Test Methods." *Materials Research and Standards*, Vol. 12, 1972, pp. 25–30.

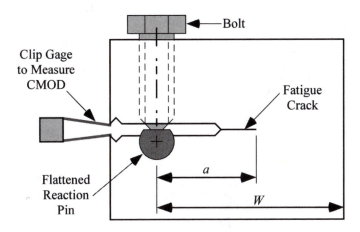

FIGURE 11.40 Constant CMOD bolt-loaded compact specimen. Taken from E 1681-03, "Standard Test Method for Determining a Threshold Stress Intensity Factor for Environment-Assisted Cracking of Metallic Materials." American Society for Testing and Materials, Philadelphia, PA, 2003.

there is no discernable cracking over a prolonged period of time, there is no way to know whether the applied K_I is above or below K_{IEAC}.

A constant CMOD test is a much more efficient means for estimating the threshold for cracking in a given material and environment. A standard specimen configuration for such a test is illustrated in Figure 11.40. The bolt-loaded specimen is a modified version of the standard compact specimen (Figure 7.1). The bolt is tightened to produce the desired CMOD. The applied K_I is related to CMOD through the following expression [25]:

$$K_I = \frac{VE}{\sqrt{W}} f\left(\frac{a}{W}\right) \tag{11.18}$$

where

$V =$ CMOD
$E =$ Young's modulus
$f(a/W) =$ dimensionless geometry factor

$$f\left(\frac{a}{W}\right) = \sqrt{1-\frac{a}{W}}\left[0.654 - 1.88\left(\frac{a}{W}\right) + 2.66\left(\frac{a}{W}\right)^2 - 1.233\left(\frac{a}{W}\right)^3\right] \tag{11.19}$$

Figure 11.41 is a plot of Equation (11.19). Note that K_I decreases with crack size when CMOD is fixed in the bolt-loaded compact specimen. A decreasing K_I is an important feature of this specimen configuration. A high initial K_I can be imposed at the beginning of the test in order to minimize incubation time (Figure 11.39). The applied K_I decreases with crack growth. Once K_I reaches the threshold, the crack arrests. Therefore, K_{IEAC} can be inferred by exposing the specimen for a sufficiently long time, measuring the final crack length (after fracturing the specimen), and then substituting the final crack size and applied CMOD into Equation (11.18) and Equation (11.19).

If practical, it is desirable to measure the crack length at various points during a test on a bolt-loaded compact specimen. Such measurements can confirm whether or not the crack has, in fact, arrested and thus the threshold has been reached. Moreover, continuous crack length measurements can be used to infer crack growth rate prior to reaching the threshold. Real-time crack length measurements can be accomplished in one of two ways. The compliance method can be applied if a special bolt equipped with a load cell is used. Alternatively, electric potential methods can be used to infer crack size. See Chapter 7 for more information on compliance and electric potential methods.

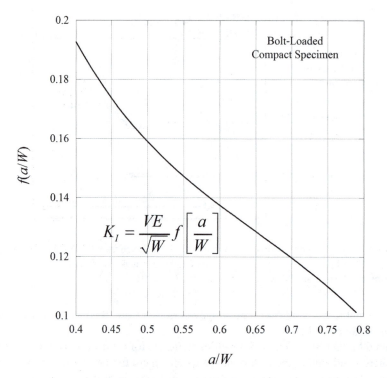

FIGURE 11.41 Nondimensional stress-intensity solution for the bolt-loaded compact specimen. Taken from E 1681-03, "Standard Test Method for Determining a Threshold Stress Intensity Factor for Environment-Assisted Cracking of Metallic Materials." American Society for Testing and Materials, Philadelphia, PA, 2003.

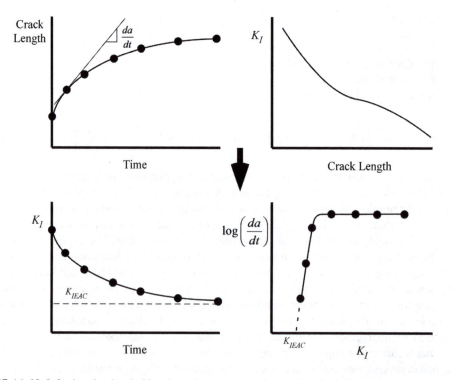

FIGURE 11.42 Inferring the threshold and crack growth rate from a bolt-loaded compact specimen test.

Figure 11.42 illustrates the procedure for determining crack growth rate and threshold from a test on a bolt-loaded compact specimen. Periodic crack length measurements are used to compute crack growth rate and K_I as a function of time. The threshold can be inferred by extrapolating the growth rate down to a suitably low value, as illustrated in Figure 11.8.

Data obtained from bolt-loaded compact specimens should be used with caution when service conditions include loads that vary with time. As Figure 11.17 and Figure 11.18 illustrate, the threshold can be affected by loading rate. A rising load can either increase or decrease K_{IEAC}, depending on strength and other factors. The crack growth rate is often accelerated by a rising load. Certain material–environment combinations are susceptible to ripple load effects, as discussed in Section 11.4.2.

Setting aside the practical difficulties associated with threshold measurements in constant load tests, there should be no significant difference between K_{IEAC} values obtained from constant load vs. constant CMOD tests, provided the former is of sufficient duration to overcome incubation effects. Consider a conventional fracture mechanics specimen in which K_I is related to the applied load and specimen dimensions. Referring to Chapter 7, K_I expressions for such specimens typically have the following form:

$$K_I = \frac{P}{B\sqrt{W}}\, f\left(\frac{a}{W}\right) \tag{11.20}$$

Differentiating the above expression with respect to time gives

$$\dot{K}_I = \frac{\dot{P}}{B\sqrt{W}}\, f\left(\frac{a}{W}\right) + \frac{P}{BW^{3/2}}\, f'\left(\frac{a}{W}\right)\frac{da}{dt} \tag{11.21}$$

If the applied load is constant, the first term in Equation (11.21) vanishes:

$$\dot{K}_I = \frac{P}{BW^{3/2}}\, f'\left(\frac{a}{W}\right)\frac{da}{dt} \tag{11.22}$$

Similarly, the rate of change in K_I in a constant CMOD test can be determined by differentiating Equation (11.18):

$$\dot{K}_I = \frac{VE}{W^{3/2}}\, f'\left(\frac{a}{W}\right)\frac{da}{dt} \tag{11.23}$$

If the crack is growing, $\dot{K}_I > 0$ for a constant load test and $\dot{K}_I < 0$ for a constant CMOD test due to the sign on $f'(a/W)$ in the above expressions. Near the threshold, however, da/dt is very small, so $\dot{K}_I \approx 0$ for both cases. Therefore, since there are no significant differences in the crack-tip conditions near the threshold for constant load and constant CMOD tests, the latter test method is usually preferable because it is considerably simpler and less time consuming. However, the constant CMOD specimen is prone to corrosion product wedging in some alloy/environment systems, which can obscure the threshold measurement.

The rising-K test has recently gained popularity as a means to estimate the threshold for environmental cracking [12]. The advantage of this test method is that its duration is typically much shorter than either the constant load or constant CMOD tests. A disadvantage is that it requires a computer-controlled test machine that is equipped with a suitable environmental chamber. Moreover, as the data in Figure 11.17 illustrate, a rising-K test can overestimate the threshold in high strength alloys if $\dot{K}_I$ is too high.

REFERENCES

1. Fontana, M.G., *Corrosion Engineering*, 3rd ed., McGraw Hill, New York, 1986.
2. Uhlig, H.H., *Corrosion and Corrosion Control*, 3rd ed., John Wiley & Sons, New York, 1985.
3. Jones, R.H. (ed.), *Stress-Corrosion Cracking: Materials Performance and Evaluation*. ASM International, Metals Park, OH, 1992.
4. Petit, J. and Scott, P. (eds.), *Comprehensive Structural Integrity, Volume 6: Environmentally-Assisted Fracture*. Elsevier, Oxford, 2003.
5. Parkins, R.N. (ed.), *Life Prediction of Corrodible Structures*, Vol. I. NACE International, Houston, TX, 1994.
6. Speidel, M.O., "Current Understanding of Stress Corrosion Crack Growth in Aluminum Alloys." *The Theory of Stress Corrosion Cracking in Alloys*. NATO Scientific Affairs Division, Brussels, Belgium, 1971, pp. 289–344.
7. Nelson, H.G. and Williams, D.P., *Stress Corrosion Cracking and Hydrogen Embrittlement of Iron-Base Alloys*, NACE International, Houston, TX, 1977, pp. 390–404.
8. Pickering, H.W., Beck, F.H., and Fontana, M.G., "Wedging Action of Solid Corrosion Product during Stress Corrosion of Austenitic Stainless Steels." *Corrosion*, Vol. 18, 1962.
9. Heald, P.T., "The Oxide Wedging of Surface Cracks." *Materials Science and Engineering,* Vol. 35, 1978, pp. 165–169.
10. Hudak, S.J. and Page, R.A., "Analysis of Oxide Wedging during Environment Assisted Crack Growth." *Corrosion*, Vol. 39, 1983, pp. 285–290.
11. Robinson, M.J., "The Role of Wedging Stresses in the Exfoliation Corrosion of High Strength Aluminum Alloys." *Corrosion Science*, Vol. 23, 1983, pp. 887–899.
12. Gangloff, R.P., "Hydrogen-Assisted Cracking in High-Strength Alloys." *Comprehensive Structural Integrity, Volume 6: Environmentally-Assisted Fracture*. Elsevier, Oxford, 2003.
13. Clark, W.G., Jr. and Landes, J.D., "An Evaluation of Rising Load $K_{\rm ISCC}$ Testing of 4340 Steel." ASTM STP 601, American Society for Testing and Materials, Philadelphia, PA, 1976, pp. 108–127.
14. Somerday, B.P., Young, L.M., and Gangloff, R.P., "Crack Tip Mechanics Effects on Environment-Assisted Cracking of Beta-Titanium Alloys in Aqueous NaCl." *Fatigue and Fracture of Engineering Materials and Structures*, Vol. 23, 2000, pp. 39–58.
15. Gangloff, R.P., *Corrosion Prevention and Control*, 33rd Sagamore Army Materials Research Conference, U.S. Army Laboratory Command, Watertown, PA, 1986, pp. 64–111.
16. Speidel, M.O. and Atrens, A., *Corrosion of Power Generating Equipment*. Plenum Press, New York, 1984.
17. Thomas, R.L.S., Scully, J.R., and Gangloff, R.P., "Internal Hydrogen Embrittlement of Ultrahigh-Strength AERMET 100 Steel." *Metallurgical and Materials Transactions*, Vol. 34A, 2003, pp. 327–344.
18. McEvily, A.J. and Wei, R.P., "Fracture Mechanics and Corrosion Fatigue." *Corrosion Fatigue: Chemistry, Mechanics and Microstructures*, NACE International, Houston, TX, 1972, pp. 25–30.
19. Gangloff, R.P., "Corrosion Fatigue Crack Propagation in Metals." *Environment-Induced Cracking of Metals*, NACE International, Houston, TX, 1990, pp. 55–109.
20. Gasem, Z. and Gangloff, R.P., "Rate-Limiting Processes in Environmental Fatigue Crack Propagation in 7000-Series Aluminum Alloys." *Chemistry and Electrochemistry of Corrosion and Stress Corrosion Cracking*, TMS, Warrendale, PA, 2001, pp. 501–521.
21. Gangloff, R.P., "Environment Sensitive Fatigue Crack Tip Processes and Propagation in Aerospace Aluminum Alloys." *Fatigue 2002*, Materials Advisory Services, West Midlands, UK, 2002, pp. 3401–3433.
22. G 30-97, "Standard Practice for Making and Using U-Bend Stress-Corrosion Test Specimens." American Society for Testing and Materials, Philadelphia, PA, 2003.
23. G 39-99, "Standard Practice for Preparation and Use of Bent-Beam Stress-Corrosion Test Specimens." American Society for Testing and Materials, Philadelphia, PA, 1999.
24. G 129-00, "Standard Practice for Slow Strain Rate Testing to Evaluate the Susceptibility of Metallic Materials to Environmentally Assisted Cracking." American Society for Testing and Materials, Philadelphia, PA, 2000.
25. E 1681-03, "Standard Test Method for Determining a Threshold Stress Intensity Factor for Environment-Assisted Cracking of Metallic Materials." American Society for Testing and Materials, Philadelphia, PA, 2003.
26. Wei, R.P., Novak, S.R., and Williams, D.P., "Some Important Considerations in the Development of Stress Corrosion Cracking Test Methods." *Materials Research and Standards*, Vol. 12, 1972, pp. 25–30.

12 Computational Fracture Mechanics

Computers have had an enormous influence in virtually all branches of engineering, and fracture mechanics is no exception. Numerical modeling has become an indispensable tool in fracture analysis, since relatively few practical problems have closed-form analytical solutions.

Stress-intensity solutions for literally hundreds of configurations have been published, the majority of which were inferred from numerical models. Elastic-plastic analyses to compute the *J* integral and crack-tip-opening displacement (CTOD) are also becoming relatively common. In addition, researchers are applying advanced numerical techniques to special problems, such as fracture at interfaces, dynamic fracture, and ductile crack growth.

Rapid advances in computer technology are primarily responsible for the exponential growth in applications of computational fracture mechanics. The personal computers that most engineers have on their desks are more powerful than mainframe computers of 10–20 years ago.

Hardware does not deserve all of the credit for the success of computational fracture mechanics, however. More efficient numerical algorithms have greatly reduced solution times in fracture problems. For example, the domain integral approach (Section 12.3) enables one to generate *K* and *J* solutions from finite element models with surprisingly coarse meshes. Commercial numerical analysis codes have become relatively user friendly, and many codes have incorporated fracture mechanics routines.

This chapter will not turn the reader into an expert on computational fracture mechanics, but it should serve as an introduction to the subject. The sections that follow describe some of the traditional approaches in numerical analysis of fracture problems, as well as some recent innovations.

The format of this chapter differs from earlier chapters, in that the main body of this chapter contains several relatively complicated mathematical derivations; previous chapters confined such material to appendices. This information is unavoidable when explaining the basis of the common numerical techniques. Readers who are intimidated by the mathematical details should at least skim this material and attempt to understand its significance.

12.1 OVERVIEW OF NUMERICAL METHODS

It is often necessary to determine the distribution of stresses and strains in a body that is subject to external loads or displacements. In limited cases, it is possible to obtain a closed-form analytical solution for the stresses and strains. If, for example, the body is subject to either plane stress or plane strain loading and is composed of an isotropic linear elastic material, it may be possible to find a stress function that leads to the desired solution. Westergaard [1] and Williams [2] used such an approach to derive solutions for the stresses and strains near the tip of a sharp crack in an elastic material (see Appendix 2). In most instances, however, closed-form solutions are not possible, and the stresses in the body must be estimated numerically.[1]

A variety of numerical techniques have been applied to problems in solid mechanics, including finite difference [3], finite element [4], and boundary integral equation methods [5]. In recent years, the latter two numerical methods have been applied almost exclusively. The vast majority of analyses

---

[1] Experimental stress analysis methods, such as photoelasticity, Moire' interferometry, and caustics are available, even these techniques often require numerical analysis to interpret experimental observations.

of cracked bodies use finite elements, although the boundary integral method may be useful in limited circumstances.

12.1.1 THE FINITE ELEMENT METHOD

In the finite element method, the structure of interest is subdivided into discrete shapes called *elements*. The most common element types include one-dimensional beams, two-dimensional plane stress or plane strain elements, and three-dimensional bricks or tetrahedrons. The elements are connected at *node* points where continuity of the displacement fields is enforced. The dimensionality of the structure need not correspond to the element dimension. For example, a three-dimensional truss can be constructed from beam elements.

The *stiffness finite element method* [4] is usually applied to stress analysis problems. This approach is outlined below for the two-dimensional case.

Figure 12.1 shows an *isoparametric* continuum element for two-dimensional plane stress or plane strain problems, together with local and global coordinate axes. The local coordinates, which are also called *parametric* coordinates, vary from −1 to +1 over the element area; the node at the lower left-hand corner has parametric coordinates $(-1, -1)$ while upper right-hand corner is at $(+1, +1)$ in the local system. Note that the parametric coordinate system is not necessarily orthogonal. Consider a point on the element at (ξ, η). The global coordinates of this point are given by

$$x = \sum_{i=1}^{n} N_i(\xi, \eta) x_i$$

$$y = \sum_{i=1}^{n} N_i(\xi, \eta) y_i$$

(12.1)

where n is the number of nodes in the element and N_i are the shape functions corresponding to the node i, whose coordinates are (x_i, y_i) in the global system and (ξ_i, η_i) in the parametric system.

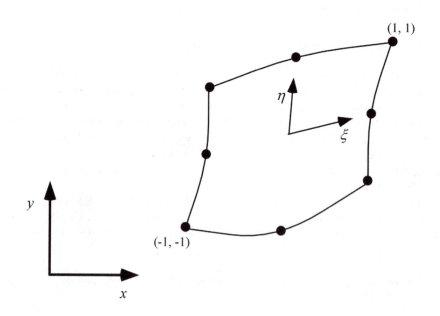

FIGURE 12.1 Local and global coordinates for a two-dimensional element.

The shape functions are polynomials that interpolate field quantities within the element. The degree of the polynomial depends on the number of nodes in the element. If, for example, the element contains nodes only at the corners, N_i are linear. Figure 12.1 illustrates a four-sided, eight-node element, which requires a quadratic interpolation. Appendix 12 gives the shape functions for the latter case.

The displacements within an element are interpolated as follows:

$$u = \sum_{i=1}^{n} N_i(\xi, \eta) u_i$$

$$v = \sum_{i=1}^{n} N_i(\xi, \eta) v_i$$

(12.2)

where (u_i, v_i) are the nodal displacements in the x and y directions, respectively. The strain matrix at (x, y) is given by

$$\left\{ \begin{array}{c} \varepsilon_x \\ \varepsilon_y \\ \gamma_{xy} \end{array} \right\} = [\mathbf{B}] \left\{ \begin{array}{c} u_i \\ v_i \end{array} \right\}$$

(12.3)

where

$$[\mathbf{B}] = \begin{bmatrix} \dfrac{\partial N_i}{\partial x} & 0 \\ 0 & \dfrac{\partial N_i}{\partial y} \\ \dfrac{\partial N_i}{\partial y} & \dfrac{\partial N_i}{\partial x} \end{bmatrix}$$

(12.4)

and

$$\left\{ \begin{array}{c} \dfrac{\partial N_i}{\partial x} \\ \dfrac{\partial N_i}{\partial y} \end{array} \right\} = [\mathbf{J}]^{-1} \left\{ \begin{array}{c} \dfrac{\partial N_i}{\partial \xi} \\ \dfrac{\partial N_i}{\partial \eta} \end{array} \right\}$$

(12.5)

where $[\mathbf{J}]$ is the Jacobian matrix, which is given by

$$[\mathbf{J}] = \begin{bmatrix} \dfrac{\partial x}{\partial \xi} & \dfrac{\partial y}{\partial \xi} \\ \dfrac{\partial x}{\partial \eta} & \dfrac{\partial y}{\partial \eta} \end{bmatrix} = \begin{bmatrix} \cdots \dfrac{\partial N_i}{\partial x} \cdots \\ \cdots \dfrac{\partial N_i}{\partial y} \cdots \end{bmatrix} \begin{bmatrix} \vdots \\ x_i y_i \\ \vdots \end{bmatrix}$$

(12.6)

The stress matrix is computed as follows:

$$\{\sigma\} = [\mathbf{D}]\{\varepsilon\}$$

(12.7a)

where $[\mathbf{D}]$ is the stress-strain constitutive matrix. For problems that incorporate incremental plasticity, stress and strain are computed incrementally and $[\mathbf{D}]$ is updated at each load step:

$$\{\Delta\sigma\} = [\mathbf{D}(\varepsilon, \sigma)]\{\Delta\varepsilon\}$$

(12.7b)

Thus the stress and strain distribution throughout the body can be inferred from the nodal displacements and the constitutive law. The stresses and strains are usually evaluated at several Gauss points or integration points within each element. For two-dimensional elements, 2×2 Gaussian integration is typical, where there are four integration points on each element.

The displacements at the nodes depend on the element stiffness and the nodal forces. The elemental stiffness matrix is given by

$$[k] = \int_{-1}^{1} \int_{-1}^{1} [\mathbf{B}]^T [\mathbf{D}][\mathbf{B}] \det | \mathbf{J} | d\xi d\eta \tag{12.8}$$

where the superscript T denotes the transpose of the matrix. Equation (12.8) can be derived from the principle of minimum potential energy [4].

The elemental stiffness matrices are assembled to give the global stiffness matrix $[\mathcal{K}]$. The global force, displacement, and stiffness matrices are related as follows:

$$[\mathcal{K}][\mathbf{u}] = [\mathbf{F}] \tag{12.9}$$

12.1.2 THE BOUNDARY INTEGRAL EQUATION METHOD

Most problems in nature cannot be solved mathematically without specifying appropriate boundary conditions. In solid mechanics, for example, a well posed problem is one in which either the tractions or the displacements (but not both) are specified over the entire surface. In the general case, the surface of a body can be divided into two regions: S_u, where displacements are specified, and S_T, where tractions are specified. One cannot specify both traction and displacement on the same area, since one quantity depends on the other. Given these boundary conditions, it is theoretically possible to solve for the tractions on S_u and the displacements on S_T, as well as the stresses, strains, and displacements within the body.

The boundary integral equation (BIE) method [5–9] is a very powerful technique for solving for unknown tractions and displacements on the surface. This approach can also provide solutions for internal field quantities, but finite element analysis is more efficient for this purpose.

The BIE method stems from Betti's reciprocal theorem, which relates work done by two different loadings on the same body. In the absence of body forces, Betti's theorem can be stated as follows:

$$\int_S T_i^{(1)} u_i^{(2)} dS = \int_S T_i^{(2)} u_i^{(1)} dS \tag{12.10}$$

where T_i and u_i are components of the traction and displacement vectors, respectively, and the superscripts denote loadings (1) and (2). The standard convention is followed in this chapter, where repeated indices imply summation. Equation (12.10) can be derived from the principle of virtual work, together with the fact that $\sigma_{ij}^{(1)}\varepsilon_{ij}^{(2)} = \sigma_{ij}^{(2)}\varepsilon_{ij}^{(1)}$ for a linear elastic material.

Let us assume that (1) is the loading of interest and (2) is a reference loading with a known solution. Figure 12.2 illustrates the conventional reference boundary conditions for BIE problems. A unit force is applied at an interior point p in each of the three coordinate directions x_i, resulting in displacements and tractions at surface point Q in the x_j direction. For example, a unit force in the x_1 direction may produce displacements and tractions at Q in all three coordinate directions. Consequently, the resulting displacements and tractions at Q, $\mathbf{u}_{ij}$ and $\mathbf{T}_{ij}$, are second-order tensors. The quantities $\mathbf{u}_{ij}(p,Q)$ and $\mathbf{T}_{ij}(p,Q)$ have closed-form solutions for several cases, including a point force on the surface of a semi-infinite elastic body [5].

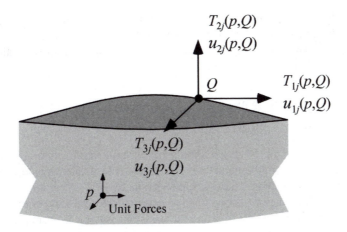

FIGURE 12.2 Reference boundary conditions for a boundary integral element problem. Unit forces are applied in each of the coordinate directions at point p, resulting in tractions and displacements on the surface point Q.

Applying the Betti reciprocal theorem to the boundary conditions described above [5] leads to

$$u_i(p) = -\int_S \mathbf{T}_{ij}(p,Q)u_j(Q)dS + \int_S \mathbf{u}_{ij}(p,Q)T_j(Q)dS \qquad (12.11)$$

where $u_i(p)$ is the displacement vector at the interior point p, $u_j(Q)$ and $T_j(Q)$ are the reference displacement and traction vectors at the boundary point Q. Note that $u_i(p)$, $u_j(Q)$, and $T_j(Q)$ correspond to the loading of interest, i.e., loading (1). At a given point Q on the boundary, either traction or displacement is known *a priori*, and it is necessary to solve for the other quantity. If we let $p \rightarrow P$, where P is a point on the surface, Equation (12.11) becomes [5]

$$\frac{1}{2}u_i(P) + \int_S \mathbf{T}_{ij}(P,Q)u_j(Q)dS = \int_S \mathbf{u}_{ij}(P,Q)T_j(Q)dS \qquad (12.12)$$

assuming the surface is smooth. (This relationship is modified slightly when P is near a corner or other discontinuity.) Equation (12.12) represents a set of integral constraint equations that relate surface displacements to surface tractions. In order to solve for the unknown boundary data, the surface must be subdivided into segments (i.e., elements), and Equation (12.12) approximated by a system of algebraic equations. If it is assumed that u_i and T_i vary linearly between discrete nodal points on the surface, Equation (12.12) can be written as

$$\left(\left[\frac{1}{2}\delta_{ij} \right] + [\Delta T_{ij}] \right)\{u_j\} = [\Delta u_{ij}]\{T_j\} \qquad (12.13)$$

where δ_{ij} is the Kronecker delta. Equation (12.13) represents a set of $3n$ equations for a three-dimensional problem, where n is the number of nodes. Once all of the boundary quantities are known, displacements at internal points can be inferred from Equation (12.11).

The boundary elements have one less dimension than the body being analyzed. That is, the boundary of a two-dimensional problem is surrounded by one-dimensional elements, while the surface of a three-dimensional solid is paved with two-dimensional elements. Consequently, boundary element analysis can be very efficient, particularly when the boundary quantities are of primary interest. This method tends to be inefficient, however, when solving for internal field quantities.

The boundary integral equation method is usually applied to linear elastic problems, but this technique can also be utilized for elastic-plastic analysis [6,9]. As with the finite element technique, nonlinear BIE analyses are typically performed incrementally, and the stress-strain relationship is assumed to be linear within each increment.

12.2 TRADITIONAL METHODS IN COMPUTATIONAL FRACTURE MECHANICS

This section describes several of the earlier approaches for inferring fracture mechanics parameters from numerical analysis. Most of these methods have been made obsolete by more recent techniques that are significantly more accurate and efficient (Section 12.3).

The approaches outlined below can be divided into two categories: point matching and energy methods. The former technique entails inferring the stress-intensity factor from the stress or displacement fields in the body, while energy methods compute the energy release rate in the body and relate $\mathcal{G}$ to stress intensity. One advantage of energy methods is that they can be applied to nonlinear material behavior; a disadvantage is that it is more difficult to separate energy release rate into mixed-mode K components.

Most of the techniques described below can be implemented with either finite element or boundary element methods. The stiffness derivative approach (Section 12.2.4), however, was formulated in terms of the finite element stiffness matrix, and thus is not compatible with boundary element analysis.

12.2.1 STRESS AND DISPLACEMENT MATCHING

The stress-intensity factor can be estimated from stresses in front of the crack tip or displacements behind the crack tip, as described below. Figure 12.3 shows the assumed local coordinate system at the crack tip.

Consider a cracked body subject to pure Mode I loading. On the crack plane ($\theta = 0$), K_I is related to the stress normal to the crack plane as follows:

$$K_I = \lim_{r \to 0} [\sigma_{yy} \sqrt{2\pi r}] \quad (\theta = 0) \tag{12.14}$$

The stress-intensity factor can be inferred by plotting the quantity in square brackets against distance from the crack tip, and extrapolating to $r = 0$. Alternatively, K_I can be estimated from a similar

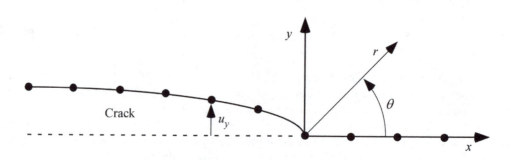

FIGURE 12.3 Local coordinate system for stresses and displacements at the crack tip in a finite element or boundary element model.

extrapolation of crack-opening displacement u_y. For plane strain conditions, K_I is estimated from the following extrapolation:

$$K_I = \lim_{r \to 0} \left[\frac{E u_y}{4(1 - v^2)} \sqrt{\frac{2\pi}{r}} \right] \quad (\theta = \pi) \tag{12.15a}$$

For plane stress loading, the K_I estimate is given by

$$K_I = \lim_{r \to 0} \left[\frac{E u_y}{4} \sqrt{\frac{2\pi}{r}} \right] \quad (\theta = \pi) \tag{12.15b}$$

The above expressions were derived from the Mode I stress and displacement solutions in Table 2.1 and Table 2.2 by setting $\theta = 0$ and $\theta = \pi/2$ for normal stress and normal displacement, respectively. Corresponding stress intensity estimates for Mode II and Mode III can also be inferred from Table 2.1 to Table 2.3.

Equation (12.15a) and Equation (12.15b) tend to give more accurate estimates of K_I than Equation (12.14) because stresses are singular as $r \to 0$ but displacements are proportional to $\sqrt{r}$ near the crack tip. The stress-matching method requires a high degree of mesh refinement to obtain accurate K_I estimates. Section 12.5 presents a series of convergence studies that compare the relative efficiency of the above methods with the modern approach based on the J integral (Section 12.3).

The boundary collocation method [11,12] is an alternative point-matching technique for computing stress-intensity factors. This approach entails finding stress functions that satisfy the boundary conditions at various nodes, and inferring the stress-intensity factor from these functions. For plane stress or plane strain problems, the Airy stress function (Appendix 2) can be expressed in terms of two complex analytic functions, which can be represented as polynomials in the complex variable $z = x + iy$. In a boundary collocation analysis, the coefficients of the polynomials are inferred from nodal quantities. The minimum number of nodes used in the analysis corresponds to the number of unknown coefficients in the polynomials. The results can be improved by analyzing more than the minimum number of nodes and solving for the unknowns by least squares. This approach can be highly cumbersome; displacement matching or energy methods are preferable in most instances.

Early researchers in computational fracture mechanics attempted to reduce the mesh size requirements for point-matching analyses by introducing special elements at the crack tip that exhibit the $1/\sqrt{r}$ singularity [13]. Barsoum [14] later showed that this same effect could be achieved by a slight modification to conventional isoparametric elements (see Section 12.4 and Appendix 12).

12.2.2 ELEMENTAL CRACK ADVANCE

Recall from Chapter 2 that the energy release rate can be inferred from the rate of change in global potential energy with crack growth. If two separate numerical analyses of a given geometry are performed, one with crack length a, and the other with crack length $a + \Delta a$, the energy release rate is given by

$$\mathcal{G} = -\left(\frac{\Delta \Pi}{\Delta a} \right)_{\text{fixed boundary conditions}} \tag{12.16}$$

assuming a two-dimensional body with unit thickness.

This technique requires minimal postprocessing, since total strain energy is output by many commercial analysis codes. This technique is also more efficient than the point-matching methods, since global energy estimates do not require refined meshes.

One disadvantage of the elemental crack advance method is that multiple solutions are required in this case, while other methods infer the desired crack-tip parameter from a single analysis. This may not be a serious shortcoming if the intention is to compute $\mathcal{G}$ (or K) as a function of crack size. The numerical differentiation in Equation (12.16), however, can result in significant errors unless the crack length intervals (Δa) are small.

12.2.3 CONTOUR INTEGRATION

The J integral can be evaluated numerically along a contour surrounding the crack tip. The advantages of this method are that it can be applied both to linear and nonlinear problems, and path independence (in elastic materials) enables the user to evaluate J at a remote contour, where numerical accuracy is greater. For problems that include path-dependent plastic deformation or thermal strains, it is still possible to compute J at a remote contour, provided an appropriate correction term (i.e., an area integral) is applied [15,16].

For three-dimensional problems, however, the contour integral becomes a surface integral, which is extremely difficult to evaluate numerically.

More recent numerical formulations for evaluating J apply an area integration for two-dimensional problems and a volume integration for three-dimensional problems. Area and volume integrals provide much better accuracy than contour and surface integrals, and are much easier to implement numerically. The first such approach was the stiffness derivative formulation of the virtual crack extension method, which is described below. This approach has since been improved and made more general, as Section 12.2.5 and Section 12.3 discuss.

12.2.4 VIRTUAL CRACK EXTENSION: STIFFNESS DERIVATIVE FORMULATION

In 1974, Parks [10] and Hellen [17] independently proposed the following finite element method for inferring energy release rate in elastic bodies. Several years later, Parks [18] extended this method to nonlinear behavior and large deformation at the crack tip. Although the stiffness derivative method is now outdated, it was the precursor to the modern approaches described in Section 12.2.5 and Section 12.3.

Consider a two-dimensional cracked body with unit thickness, subject to Mode I loading. The potential energy of the body, in terms of the finite element solution, is given by

$$\Pi = \frac{1}{2}[\mathbf{u}]^T[\mathbf{K}][\mathbf{u}] - [\mathbf{u}]^T[\mathbf{F}] \tag{12.17}$$

where Π is the potential energy, and the other quantities are as defined in Section 12.1.1. Recall from Chapter 2 that the energy release rate is the derivative of Π with respect to crack area, for both fixed load and fixed displacement conditions. It is convenient in this instance to evaluate $\mathcal{G}$ under fixed load conditions:

$$\mathcal{G} = -\left(\frac{\partial \Pi}{\partial a}\right)_{\text{load}}$$

$$= -\frac{\partial[\mathbf{u}]^T}{\partial a}\{[\mathbf{K}][\mathbf{u}] - [\mathbf{F}]\} - \frac{1}{2}[\mathbf{u}]^T\frac{\partial[\mathbf{K}]}{\partial a}[\mathbf{u}] + [\mathbf{u}]^T\frac{\partial[\mathbf{F}]}{\partial a} \tag{12.18}$$

Comparing Equation (12.9) to the above result, we see that the first term in Equation (12.18) must be zero. In the absence of tractions on the crack face, the third term must also vanish, since

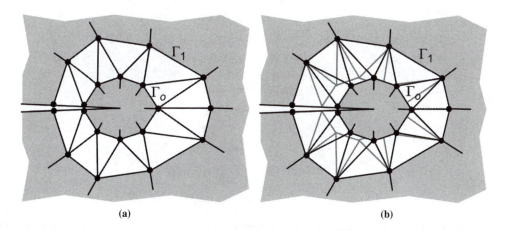

FIGURE 12.4 Virtual crack extension in a finite element model. Elements between Γ_1 and Γ_o are distorted to accommodate a crack advance: (a) initial conditions and (b) after virtual crack advance. Taken from Parks, D.M., "A Stiffness Derivative Finite Element Technique for Determination of Crack Tip Stress Intensity Factors." *International Journal of Fracture*, Vol. 10, 1974, pp. 487–502; Hellen, T.K., "On the Method of Virtual Crack Extensions." *International Journal for Numerical Methods in Engineering*, Vol. 9, 1975, pp. 187–207.

loads are held constant. Thus, the energy release rate is given by

$$\mathcal{G} = \frac{K_I^2}{E'} = -\frac{1}{2}[\mathbf{u}]^T \frac{\partial[\mathbf{K}]}{\partial a}[\mathbf{u}] \tag{12.19}$$

The energy release rate is proportional to the derivative of the stiffness matrix with respect to crack length.

Suppose that we have generated a finite element mesh for a body with crack length a and we wish to extend the crack by Δa. It would not be necessary to change all of the elements in the mesh; we could accommodate the crack growth by moving elements near the crack tip and leaving the rest of the mesh intact. Figure 12.4 illustrates such a process, where elements inside the contour Γ_o are shifted by Δa, and elements outside of the contour Γ_1 are unaffected. Each of the elements between Γ_o and Γ_1 is distorted, such that its stiffness changes. The energy release rate is related to this change in element stiffness:

$$\mathcal{G} = -\frac{1}{2}[\mathbf{u}]^T \left(\sum_{i=1}^{N_c} \frac{\partial[k_i]}{\partial a} \right)[\mathbf{u}] \tag{12.20}$$

where $[k_i]$ are the elemental stiffness matrices and N_c is the number of elements between the contours Γ_o and Γ_1. Parks [10] demonstrated that Equation (12.20) is equivalent to the J integral. The value of $\mathcal{G}$ (or J) is independent of the choice of the inner and outer contours.

It is important to note that in a virtual crack extension analysis, it is not necessary to generate a second mesh with a slightly longer crack. It is sufficient merely to calculate the change in elemental stiffness matrices corresponding to shifts in the nodal coordinates.

One problem with the stiffness derivative approach is that it involves cumbersome numerical differencing. Also, this formulation is poorly suited to problems that include thermal strain. A more recent formulation of the virtual crack extension method overcomes these difficulties, as discussed below.

12.2.5 VIRTUAL CRACK EXTENSION: CONTINUUM APPROACH

Parks [10] and Hellen [17] formulated the virtual crack extension approach in terms of finite element stiffness and displacement matrices. deLorenzi [19,20] improved the virtual crack extension method by considering the energy release rate of a continuum. The main advantages of the continuum

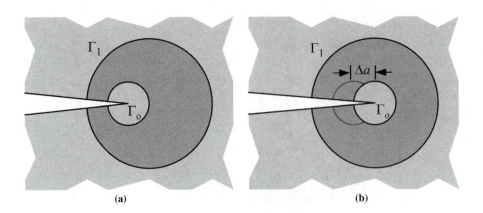

FIGURE 12.5 Virtual crack extension in a two-dimensional elastic continuum: (a) initial conditions and (b) after virtual crack advance.

approach are twofold: first, the methodology is not restricted to the finite element method; and second, deLorenzi's approach does not require numerical differencing.

Figure 12.5 illustrates a virtual crack advance in a two-dimensional continuum. Material points inside Γ_o experience rigid body translation of a distance Δa in the x_1 direction, while points outside of Γ_1 remain fixed. In the region between contours, virtual crack extension causes material points to translate by Δx_1. For an elastic material, or one that obeys deformation plasticity theory, deLorenzi showed that energy release rate is given by

$$\mathcal{G} = \frac{1}{\Delta a} \int_A \left(\sigma_{ij} \frac{\partial u_j}{\partial x_1} - w \delta_{i1} \right) \frac{\partial \Delta x_1}{\partial x_i} dA \tag{12.21}$$

where w is the strain energy density. Equation (12.21) assumes unit thickness, crack growth in the x_1 direction, no body forces within Γ_1, and no tractions on the crack faces. Note that $\partial \Delta x_1/\partial x_i = 0$ outside of Γ_1 and within Γ_o; thus the integration need only be performed over the annular region between Γ_o and Γ_1.

deLorenzi actually derived a more general expression that considers a three-dimensional body, tractions on the crack surface, and body forces:

$$\mathcal{G} = \frac{1}{\Delta A_c} \int_V \left[\left(\sigma_{ij} \frac{\partial u_j}{\partial x_k} - w \delta_{ik} \right) \frac{\partial \Delta x_k}{\partial x_i} - F_i \frac{\partial u_i}{\partial x_j} \Delta x_j \right] dV$$

$$- \frac{1}{\Delta A_c} \int_S T_i \frac{\partial u_j}{\partial x_k} \Delta x_j dS \tag{12.22}$$

where
   ΔA_c = increase in the crack area generated by the virtual crack advance
   V = volume of the body
   F_i = body forces

In this instance, two surfaces enclose the crack front. Material points within the inner surface S_o are displaced by Δa_i, while the material outside of the outer surface S_1 remains fixed. The displacement vector between S_o and S_1 is Δx_i, which ranges from 0 to Δa_i. Equation (12.22) assumes a fixed coordinate system; consequently, the virtual crack advance Δa_i is not necessarily in the x_1 direction when the crack front is curved. The above expression, however, only applies to virtual crack advance normal to the crack front, in the plane of the crack.

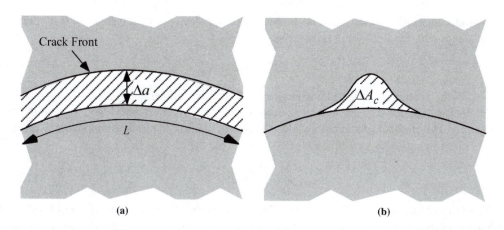

FIGURE 12.6 Virtual crack extension along a three-dimensional crack front: (a) uniform crack advance and (b) advance over an increment of crack front.

In a three-dimensional problem, $\mathcal{G}$ typically varies along the crack front. In computing $\mathcal{G}$, one can consider a uniform virtual crack advance over the entire crack front or a crack advance over a small increment, as Figure 12.6 illustrates. In the former case, $\Delta A_c = \Delta a L$, and the computed energy release rate would be a weighted average. Defining A_c incrementally along the crack front would result in a local measure of $\mathcal{G}$.

For two-dimensional problems, the virtual crack extension formulation of $\mathcal{G}$ requires an area integration, while three-dimensional problems require a volume integration. Such an approach is easier to implement numerically and is more accurate than contour and surface integrations for two- and three-dimensional problems, respectively.

Numerical implementation of the virtual crack extension method entails applying a virtual displacement to nodes within a specified contour. Since the domain integral formulation is very similar to the above method, further discussion on numerical implementation is deferred to Section 12.3.3.

12.3 THE ENERGY DOMAIN INTEGRAL

Shih et al. [21,22] formulated the energy domain integral methodology, which is a general framework for numerical analysis of the J integral. This approach is extremely versatile, as it can be applied to both quasistatic and dynamic problems with elastic, plastic, or viscoplastic material response, as well as thermal loading. Moreover, the domain integral formulation is relatively simple to implement numerically, and it is very efficient. This approach is very similar to the virtual crack extension method.

12.3.1 THEORETICAL BACKGROUND

Appendix 4.2 presents a derivation of a general expression for the J integral that includes the effects of inertia as well as inelastic material behavior. The generalized definition of J requires that the contour surrounding the crack tip be vanishingly small:

$$J = \lim_{\Gamma_o \to 0} \int_{\Gamma_o} \left[(w+T)\delta_{1i} - \sigma_{ij}\frac{\partial u_j}{\partial x_1} \right] n_i d\Gamma \tag{12.23}$$

where T is the kinetic energy density. Various material behavior can be taken into account through the definition of w, the stress work.

Consider an elastic-plastic material loaded under quasistatic conditions ($T = 0$). If thermal strains are present, the total strain is given by

$$\varepsilon_{ij}^{total} = \varepsilon_{ij}^e + \varepsilon_{ij}^p + \alpha\Theta\delta_{ij} = \varepsilon_{ij}^m + \varepsilon_{kk}^t \tag{12.24}$$

where α is the coefficient of thermal expansion and Θ is the temperature relative to a strain-free condition. The superscripts e, p, m, and t denote elastic, plastic, mechanical, and thermal strains, respectively. The mechanical strain is equal to the sum of elastic and plastic components. The stress work is given by

$$w = \int_0^{\varepsilon_{kl}^m} \sigma_{ij} d\varepsilon_{ij}^m \tag{12.25}$$

The form of Equation (12.23) is not suitable for numerical analysis, since it is not feasible to evaluate stresses and strains along a vanishingly small contour. Let us construct a closed contour by connecting inner and outer contours, as Figure 12.7 illustrates. The outer contour Γ_1 is finite, while Γ_o is vanishingly small. For a linear or nonlinear elastic material under quasistatic conditions, J could be evaluated along either Γ_1 or Γ_o, but only the inner contour gives the correct value of J in the general case. For quasistatic conditions, where $T = 0$, Equation (12.23) can be written in terms of the following integral around the closed contour, $\Gamma^* = \Gamma_1 + \Gamma_+ + \Gamma_- - \Gamma_o$ [21,22]:

$$J = \int_{\Gamma^*} \left[\sigma_{ij}\frac{\partial u_j}{\partial x_1} - w\delta_{1i} \right] qm_i d\Gamma - \int_{\Gamma_+ + \Gamma_-} \sigma_{2j}\frac{\partial u_j}{\partial x_1} qd\Gamma \tag{12.26}$$

where
Γ_+ and Γ_- = upper and lower crack faces, respectively
m_i = outward normal on Γ^*
q = arbitrary but smooth function that is equal to unity on Γ_o and zero on Γ_1

Note that $m_i = -n_i$ on Γ_o; also, $m_1 = 0$ and $m_2 = \pm1$ on Γ_+ and Γ_-. In the absence of crack-face tractions, the second integral in Equation (12.26) vanishes.

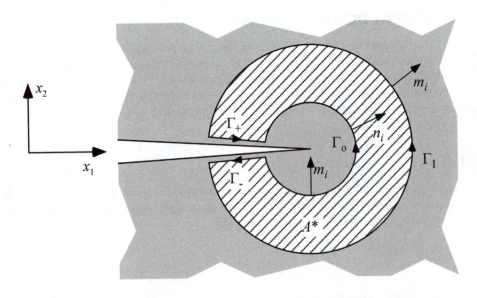

FIGURE 12.7 Inner and outer contours, which form a closed contour around the crack tip when connected by Γ_+ and Γ_-.

For the moment, assume that the crack faces are traction free. Applying the divergence theorem to Equation (12.26) gives

$$J = \int_{A^*} \frac{\partial}{\partial x_i} \left\{ \left[\sigma_{ij} \frac{\partial u_j}{\partial x_1} - w\delta_{1i} \right] q \right\} dA$$

$$= \int_{A^*} \left[\sigma_{ij} \frac{\partial u_j}{\partial x_1} - w\delta_{1i} \right] \frac{\partial q}{\partial x_i} dA + \int_{A^*} \left[\frac{\partial}{\partial x_i} \left(\sigma_{ij} \frac{\partial u_j}{\partial x_1} \right) - \frac{\partial w}{\partial x_1} \right] q dA \tag{12.27}$$

where A^* is the area enclosed by Γ^*. Referring to Appendix 3.2, we see that

$$\frac{\partial}{\partial x_i} \left(\sigma_{ij} \frac{\partial u_j}{\partial x_1} \right) - \frac{\partial w}{\partial x_1} = 0 \tag{12.28}$$

when there are no body forces and w exhibits the properties of an elastic potential:

$$\sigma_{ij} = \frac{\partial w}{\partial \varepsilon_{ij}} \tag{12.29}$$

It is convenient at this point to divide w into elastic and plastic components:

$$w = w^e + w^p = \int_0^{\varepsilon_{kl}^e} \sigma_{ij} d\varepsilon_{ij}^e + \int_0^{\varepsilon_{kl}^p} S_{ij} d\varepsilon_{ij}^p \tag{12.30}$$

where S_{ij} is the deviatoric stress, defined in Equation (A3.62). While the elastic components of w and ε_{ij} satisfy Equation (12.29), plastic deformation does not, in general, exhibit the properties of a potential. (Equation (12.29) may be approximately valid for plastic deformation when there is no unloading.) Moreover, thermal strains would cause the left side of Equation (12.28) to be nonzero. Thus the second integrand in Equation (12.27) vanishes in limited circumstances, but not in general. Taking account of plastic strain, thermal strain, body forces, and crack face-tractions leads to the following general expression for J in two dimensions:

$$J = \int_{A^*} \left\{ \left[\sigma_{ij} \frac{\partial u_j}{\partial x_1} - w\delta_{1i} \right] \frac{\partial q}{\partial x_i} + \left[\sigma_{ij} \frac{\partial \varepsilon_{ij}^p}{\partial x_1} - \frac{\partial w^p}{\partial x_1} + \alpha\sigma_{ii} \frac{\partial \Theta}{\partial x_1} - F_i \frac{\partial u_j}{\partial x_1} \right] q \right\} dA$$

$$- \int_{\Gamma_+ + \Gamma_-} \sigma_{2j} \frac{\partial u_j}{\partial x_1} q d\Gamma \tag{12.31}$$

where the body force contribution is inferred from the equilibrium equations, and the contribution from thermal loading is obtained by substituting Equation (12.24) and Equation (12.30) into Equation (12.27). Inertia can be taken into account by incorporating T, the kinetic energy density, into the group of terms that are multiplied by q. For a linear or nonlinear elastic material under quasistatic conditions, in the absence of body forces, thermal strains, and crack-face tractions, Equation (12.31) reduces to

$$J = \int_{A^*} \left[\sigma_{ij} \frac{\partial u_j}{\partial x_1} - w\delta_{1i} \right] \frac{\partial q}{\partial x_i} dA \tag{12.32}$$

Equation (12.32) is equivalent to Rice's path-independent J integral (Chapter 3). When the sum of the additional terms in the more general expression (Equation 12.31) is nonzero, J is path dependent.

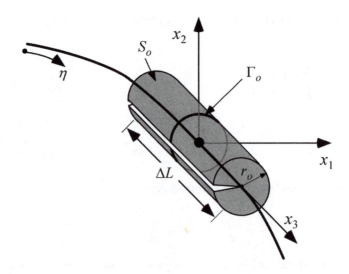

FIGURE 12.8 Surface enclosing an increment of a three-dimensional crack front.

Comparing Equation (12.21) and Equation (12.32) we see that the two expressions are identical if $q = \Delta x_1 / \Delta a$. Thus q can be interpreted as a normalized virtual displacement, although the above derivation does not require such an interpretation. The q function is merely a mathematical device that enables the generation of an area integral, which is better suited to numerical calculations. Section 12.3.3 provides guidelines for defining q.

12.3.2 GENERALIZATION TO THREE DIMENSIONS

Equation (12.23) defines the J integral in both two and three dimensions, but the form of this equation is poorly suited to numerical analysis. In the previous section, J was expressed in terms of an area integral in order to facilitate numerical evaluation. For three-dimensional problems, it is necessary to convert Equation (12.23) into a volume integral.

Figure 12.8 illustrates a planar crack in a three-dimensional body; η corresponds to the position along the crack front. Suppose that we wish to evaluate J at a particular η on the crack front. It is convenient to define a local coordinate system at η, with x_1 normal to the crack front, x_2 normal to the crack plane, and x_3 tangent to the crack front. The J integral at η is defined by Equation (12.23), where the contour Γ_o lies in the x_1–x_2 plane.

Let us now construct a tube of length ΔL and radius r_o that surrounds a segment of the crack front, as Figure 12.8 illustrates. Assuming quasistatic conditions, we can define a *weighted average* J over the crack front segment ΔL as follows:

$$\bar{J}\Delta L = \int_{\Delta L} J(\eta) q \, d\eta$$

$$= \lim_{r_o \to 0} \int_{S_o} \left[w \delta_{1i} - \sigma_{ij} \frac{\partial u_j}{\partial x_1} \right] q n_i \, dS \qquad (12.33)$$

where

$\quad J(\eta)$ = point-wise value of J

$\quad S_o$ = surface area of the tube in Figure 12.8

$\quad q$ = weighting function that was introduced in the previous section

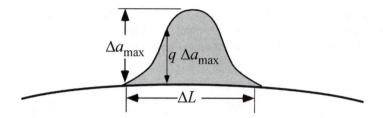

FIGURE 12.9 Interpretation of q in terms of a virtual crack advance along ΔL.

Note that the integrand in Equation (12.33) is evaluated in terms of the local coordinate system, where x_3 is tangent to the crack front at each point along ΔL.

Recall from the previous section that q can be interpreted as a virtual crack advance. For example, Figure 12.9 illustrates an incremental crack advance over ΔL, where q is defined by

$$\Delta a(\eta) = q(\eta)\Delta a_{max} \tag{12.34}$$

and the incremental area of the virtual crack advance is given by

$$\Delta A_c = \Delta a_{max} \int_{\Delta L} q(\eta)d\eta \tag{12.35}$$

The q function need not be defined in terms of a virtual crack extension, but attaching a physical significance to this parameter may aid in understanding.

If we construct a second tube of radius r_1 around the crack front (Figure 12.10), it is possible to define the weighted average J in terms of a closed surface, analogous to the two-dimensional case (Figure 12.7 and Equation 12.26):

$$\bar{J}\Delta L = \int_{S^*}\left[\sigma_{ij}\frac{\partial u_j}{\partial x_1} - w\delta_{1i}\right]qm_i dS - \int_{S_+ + S_-}\sigma_{2j}\frac{\partial u_j}{\partial x_1}q dS \tag{12.36}$$

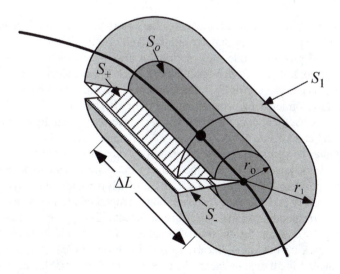

FIGURE 12.10 Inner and outer surfaces, S_o and S_1, which enclose V^*.

where the closed surface $S^* = S_1 + S_+ + S_- - S_o$, and S_+ and S_- are the upper and lower crack faces, respectively, that are enclosed by S_1. From this point, the derivation of the domain integral formulation is essentially identical to the two-dimensional case, except that Equation (12.34) becomes a volume integral:

$$\bar{J}\Delta L = \int_{V^*} \left\{ \left[\sigma_{ij} \frac{\partial u_j}{\partial x_1} - w\delta_{1i} \right] \frac{\partial q}{\partial x_i} + \left[\sigma_{ij} \frac{\partial \varepsilon_{ij}^p}{\partial x_1} - \frac{\partial w^p}{\partial x_1} + \alpha\sigma_{ii} \frac{\partial \Theta}{\partial x_1} - F_i \frac{\partial u_j}{\partial x_1} \right] q \right\} dV$$

$$- \int_{S_+ + S_-} \sigma_{2j} \frac{\partial u_j}{\partial x_1} q d\Gamma$$

(12.37)

Equation (12.37) requires that $q = 0$ at either end of ΔL; otherwise, there may be a contribution to J from the end surfaces of the cylinder. The virtual crack advance interpretation of q (Figure 12.9) fulfills this requirement.

If the point-wise value of the J integral does not vary appreciably over ΔL, to a first approximation, $J(\eta)$ is given by

$$J(\eta) \approx \frac{\bar{J}\Delta L}{\int_{\Delta L} q(\eta, r_o) d\eta}$$

(12.38)

Equation (12.38) is a reasonable approximation if the q gradient along the crack front is steep relative to the variation in $J(\eta)$.

Recall that Equation (12.22) was defined in terms of a fixed coordinate system, while Equation (12.37) assumes a local coordinate system. The domain integral formulation can be expressed in terms of a fixed coordinate system by replacing q with a vector quantity q_i, and evaluating the partial derivatives in the integrand with respect to x_i rather than x_1, where the vectors q_i and x_i are parallel to the direction of crack growth. Several commercial codes that incorporate the domain integral definition of J require that the q function be defined with respect to a fixed origin.

12.3.3 FINITE ELEMENT IMPLEMENTATION

Shih et al. [21] and Dodds and Vargas [23] give detailed instructions for implementing the domain integral approach. Their recommendations are summarized briefly.

In two-dimensional problems, one must define the area over which the integration is to be performed. The inner contour Γ_o is often taken as the crack tip, in which case A^* corresponds to the area inside of Γ_1. The boundary of Γ_1 should coincide with element boundaries. An analogous situation applies in three dimensions, where it is necessary to define the volume of integration. The latter situation is somewhat more complicated, however, since $J(\eta)$ is usually evaluated at a number of locations along the crack front.

The q function must be specified at all nodes within the area or volume of integration. The shape of the q function is arbitrary, as long as q has the correct values on the domain boundaries. In a plane stress or plane strain problem, for example, $q = 1$ at Γ_o, which is usually the crack tip, and $q = 0$ at the outer boundary. Figure 12.11 illustrates two common examples of q functions for two-dimensional problems, with the corresponding virtual nodal displacements. This example shows 4-node square elements and rectangular domains for the sake of simplicity. The pyramid function (Figure 12.11(a)) is equal to 1 at the crack tip but varies linearly to zero in all directions, while the plateau function (Figure 12.11(b)) equals 1 in all regions except the outer ring of elements. Shih et al. [21] have shown that the computed value of J is insensitive to the assumed shape of the q function.

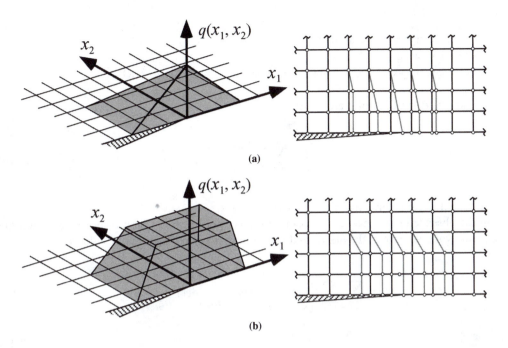

(a)

(b)

FIGURE 12.11 Examples of q functions in two dimensions, with the corresponding virtual nodal displacements: (a) the pyramid function and (b) the plateau function. Taken from Shih, C.F., Moran, B., and Nakamura, T., "Energy Release Rate Along a Three-Dimensional Crack Front in a Thermally Stressed Body." *International Journal of Fracture*, Vol. 30, 1986, pp. 79–102.

Figure 12.12 illustrates the pyramid function along a three-dimensional crack front, where the crack-tip node of interest is displaced a unit amount, and all other nodes are fixed. If desired, $J(\eta)$ can be evaluated at each node along the crack front.

The value of q within an element can be interpolated as follows:

$$q(x_i) = \sum_{I=1}^{n} N_I q_I \tag{12.39}$$

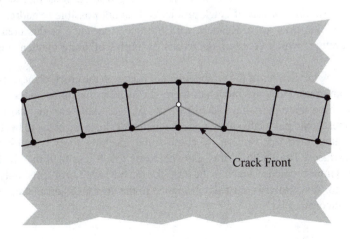

Crack Front

FIGURE 12.12 Definition of q in terms of a virtual nodal displacement along a three-dimensional crack front.

where

   n = number of nodes per element

   q_I = nodal values of q

   N_I = element shape functions, which were introduced in Section 12.1.1

The spatial derivatives of q are given by

$$\frac{\partial q}{\partial x_i} = \sum_{I=1}^{n} \sum_{k=1}^{2 \text{ or } 3} \frac{\partial N_I}{\partial \xi_k} \frac{\partial \xi_k}{\partial x_j} q_I \tag{12.40}$$

where ξ_i are the parametric coordinates for the element.

In the absence of thermal strains, path-dependent plastic strains, and body forces within the integration volume or area, the discretized form of the domain integral is as follows:

$$J = \sum_{A^* \text{ or } V^*} \sum_{p=1}^{m} \left\{ \left[\left(\sigma_{ij} \frac{\partial u_j}{\partial x_1} - w\delta_{1i} \right) \frac{\partial q}{\partial x_i} \right] \det\left(\frac{\partial x_j}{\partial \xi_k} \right) \right\}_p w_p$$

$$- \sum_{\text{crack faces}} \left(\sigma_{2j} \frac{\partial u_j}{\partial x_1} q \right) w \tag{12.41}$$

where m is the number of Gaussian points per element, and w_p and w are weighting factors. The quantities within $\{\ \}p$ are evaluated at the Gaussian points. Note that the integration over crack faces is necessary only when there are nonzero tractions.

12.4 MESH DESIGN

The design of a finite element mesh is as much an art as it is a science. Although many commercial codes have automatic mesh generation capabilities, construction of a properly designed finite element model invariably requires some human intervention. Crack problems, in particular, require a certain amount of judgment on the part of the user.

This section gives a brief overview of some of the considerations that should govern the construction of a mesh for analysis of crack problems. It is not possible to address in a few pages all of the situations that may arise. Readers with limited experience in this area should consult the published literature, which contains numerous examples of finite element meshes for crack problems.

Figure 12.13 shows examples of the arrangement of nodes on the crack plane of two-dimensional models. When both crack faces are modeled, there are usually matching nodes along each crack face. If these matching node pairs have identical coordinates, many commercial mesh generation tools will merge these node pairs into single nodes, which will cause the crack to heal. One technique for avoiding this unwanted intervention by commercial meshing tools is to create a small gap between the crack faces. The width of this gap should be much less then the crack length. When the crack is on a symmetry plane, which is the case for most Mode I problems, only one crack face needs to be modeled. A symmetry constraint is applied to the uncracked cross section, and the crack face is normally unconstrained.[2]

---

[2] For some problems, the crack face may have a traction boundary condition or an imposed displacement field.

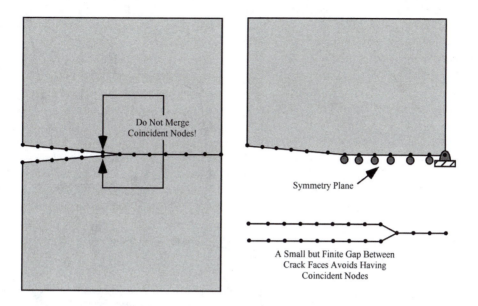

FIGURE 12.13 Examples of nodes on the crack plane in two-dimensional finite element and boundary element models.

Figure 12.14 illustrates the most common continuum element shapes. The elements depicted in Figure 12.14 have mid-side nodes, which imply quadratic shape functions, but corresponding linear elements with nodes only at the element corners are also available. Typical crack analyses use quadrilateral elements for two-dimensional problems and brick elements for three-dimensional problems. Most postprocessing algorithms that evaluate the J integral cannot handle other element types, such as triangular and tetrahedral elements for two- and three-dimensional problems, respectively. Such algorithms typically define the integration domain by searching outward from the crack tip using the element connectivity matrix. In a well-constructed mesh consisting of quadrilateral or brick elements near the crack tip, a search algorithm using the connectively matrix results in regular, concentric integration domains for evaluating J. With triangular or tetrahedral elements, however, an unlimited number of elements can be connected to a single node and the elements are oriented arbitrarily. Consequently defining smooth, regular integration domains from the connectivity matrix is virtually impossible.

As discussed in Section 12.1.1, elements with mid-side nodes have quadratic shape functions. Crack meshes usually use 8-node and 20-node quadratic elements for two- and three-dimensional problems, respectively. Some practitioners prefer 9-node biquadratic Lagrangian elements for two-dimensional problems and 27-node triquadratic Lagrangian elements in three dimensions. Linear elements, which have 4 and 8 nodes for two- and three-dimensional problems, respectively, are also acceptable for crack problems, but additional mesh refinement is required to attain the same level of accuracy as a corresponding mesh consisting of quadratic elements.

At the crack tip, quadrilateral elements (in two-dimensional problems) are usually collapsed down to triangles, as Figure 12.15 illustrates. Note that three nodes occupy the same point in space. Figure 12.16 shows the analogous situation for three dimensions, where a brick element is degenerated to a wedge.

In elastic problems, the nodes at the crack tip are normally tied, and the mid-side nodes moved to the 1/4 points (Figure 12.17(a)). Such a modification results in a $1/\sqrt{r}$ strain singularity in the element, which enhances numerical accuracy. A similar result can be achieved by moving the mid-side nodes to 1/4 points in noncollapsed quadrilateral elements, but the singularity would exist only

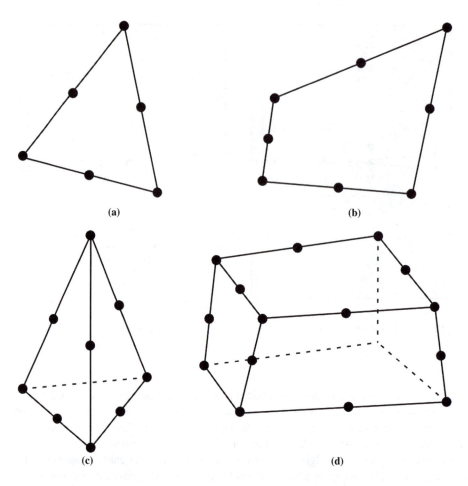

FIGURE 12.14 Common two- and three-dimensional continuum finite elements: (a) triangular element, (b) quadrilateral element, (c) tetrahedral element, and (d) brick element.

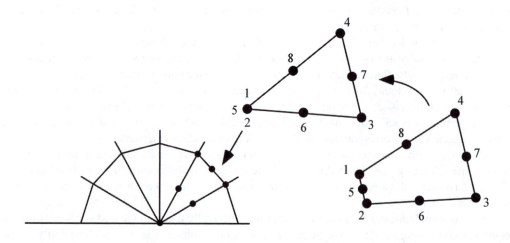

FIGURE 12.15 Degeneration of a quadrilateral element into a triangle at the crack tip.

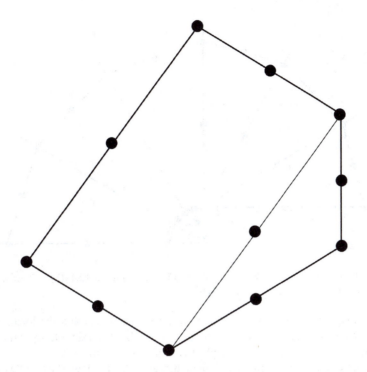

FIGURE 12.16 Degeneration of a brick element into a wedge.

on the element edges [14,24]; collapsed elements are preferable in this case because the singularity exists within the element as well as on the edges. Appendix 12 presents a mathematical derivation that explains why moving the mid-side nodes results in the desired singularity for elastic problems.

When a plastic zone forms, the $1/\sqrt{r}$ singularity no longer exists at the crack tip. Consequently, elastic singular elements are not appropriate for elastic-plastic analyses. Figure 12.15(b) shows an element that exhibits the desired strain singularity under fully plastic conditions. The element is

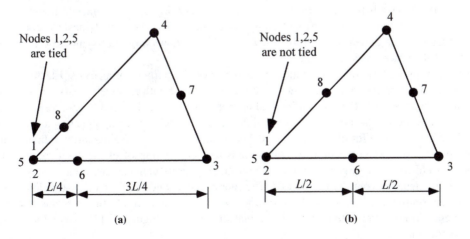

FIGURE 12.17 Crack-tip elements for elastic and elastic-plastic analyses. Element (a) produces a $1/\sqrt{r}$ strain singularity, while (b) exhibits a $1/r$ strain singularity: (a) Elastic singularity element and (b) plastic singularity element.

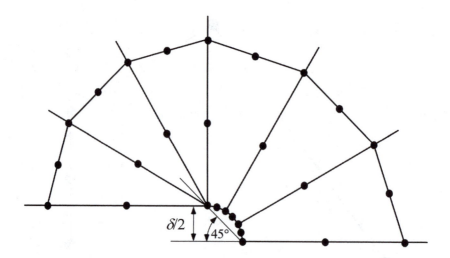

FIGURE 12.18 Deformed shape of plastic singularity elements (Figure 12.17(b)). The crack-tip elements model blunting, and it is possible to measure CTOD.

collapsed to a triangle as before, but the crack-tip nodes are untied and the location of the mid-side nodes is unchanged. This element geometry produces a $1/r$ strain singularity, which corresponds to the actual crack-tip strain field for fully plastic, nonhardening materials.

One side benefit of the plastic singular element design is that it allows the CTOD to be computed from the deformed mesh, as Figure 12.18 illustrates. The untied nodes initially occupy the same point in space, but move apart as the elements deform. The CTOD can be inferred from the deformed crack profile by means of the 90° intercept method (Figure 3.4).

For typical problems, the most efficient mesh design for the crack-tip region has proven to be the "spider-web" configuration, which consists of concentric rings of quadrilateral elements that are focused toward the crack tip. The elements in the innermost ring are degenerated to triangles, as described above. Since the crack tip region contains steep stress and strain gradients, the mesh refinement should be greatest at the crack-tip. The spider-web design facilitates a smooth transition from a fine mesh at the tip to a coarser mesh remote from the tip. In addition, this configuration results in a series of smooth, concentric integration domains (contours) for evaluating the J integral. Figure 12.19 shows a half-symmetric model of a two-dimensional simple cracked body, in which a spider-web mesh transitions to coarse rectangular elements. The spider-web meshing concept can be extended to three-dimensional problems. Figure 12.20 shows a quarter-symmetric model of a semielliptical surface crack in a plate.

The appropriate level of mesh refinement depends on the purpose of analysis. Elastic analyses of stress intensity or energy release rate can be accomplished with relatively coarse meshes since modern methods, such as the domain integral approach, eliminate the need to resolve local crack-tip fields accurately. The area and volume integrations in the newer approaches are relatively insensitive to mesh size for elastic problems. The mesh should include singularity elements at the crack tip, however, when the domain corresponds to the first ring of elements at the tip. If the domain is defined over a larger portion of the mesh, singularity elements are unnecessary because the crack tip elements contribute little to J. The relative contribution of the crack-tip elements can be adjusted through the definition of the q function. For example, in elastic problems, the crack-tip elements do not contribute to J when the plateau function (Figure 12.11(b)) is adopted, since $dq/dx_1 = 0$ at the crack tip.

Elastic-plastic problems require more mesh refinement in regions of the body where yielding occurs. When a body experiences net section yielding, narrow deformation bands often propagate across the specimen (Figure 3.26). The high level of plastic strain in these bands will make a

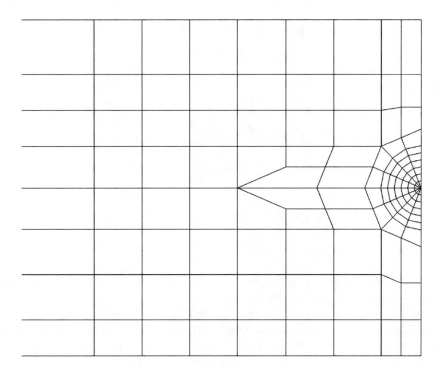

FIGURE 12.19 Half-symmetric two-dimensional model of an edge-cracked plate.

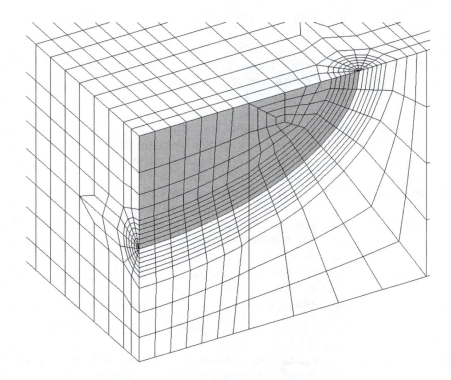

FIGURE 12.20 Quarter-symmetric three-dimensional model of a semielliptical surface crack in a flat plate. The crack face is highlighted in gray.

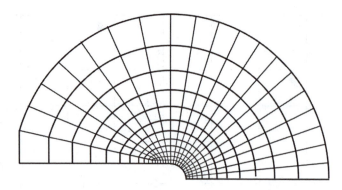

FIGURE 12.21 Crack-tip region of a mesh for large strain analysis. Note that the initial crack-tip radius is finite and the crack-tip elements are not degenerated.

significant contribution to the J integral; the finite element mesh must be sufficiently refined in these regions to capture this deformation accurately.

When the purpose of the analysis is to analyze crack-tip stresses and strains, a very high level of mesh refinement is required [25,26]. As a general rule, it is desirable to have at least 10 elements on a radial line in the region of interest. In addition, if it is necessary to infer crack-tip fields at distances less than twice the CTOD from the crack tip, the analysis code must incorporate large strain theory and a nonlinear geometry kinematic assumption. McMeeking and Parks [26] were among the first to apply such an analysis to the crack-tip region. Figure 3.13 is a plot of some of their results.

Note that the collapsed plastic singular element (Figure 12.17(a)) is not appropriate for analyses that incorporate nonlinear geometry kinematics. In such analyses, the element stiffness and Jacobian matrices are recomputed at every step. The first ring of collapsed elements at the crack tip can become highly distorted as the crack blunts. Severely deformed elements lead to errors and numerical instabilities. For analyses that assume small geometry changes, element deformation is not an issue because the element stiffness and Jacobian matrices are determined from the original configuration.[3]

In a large-strain, nonlinear-geometry analysis, it is customary to begin with a finite radius at the crack tip, as Figure 12.21 illustrates. Note that the crack-tip elements are not collapsed to triangles in this case. Provided the CTOD after deformation is at least five times the initial value, the results should not be affected by the initial blunt notch [26].

This chapter does not address boundary conditions in detail, but it is worth mentioning a common pitfall. Many problems require forces to be applied at the boundaries of the body. For example, a single-edge-notched bend specimen is loaded in the three-point bending, with a load applied at mid-span, and appropriate restraints at each end. In elastic-plastic problems, the manner in which the load is applied can be very important. Figure 12.22 shows both acceptable and unacceptable ways of applying this boundary condition. If the load is applied to a single node (Figure 12.22(a)), a local stress and strain concentration will occur, and the element connected to this node will yield almost immediately. The analysis code will spend an inordinate amount of time solving a punch indentation problem at this node, while the events of interest may be remote from

---

[3] An elastic-plastic analysis assuming small geometry changes predicts stresses that monotonically increase as one approaches the crack tip because the computation is based on the initial sharp-crack configuration. An analysis that accounts for geometry changes at the crack tip predicts a peak in normal stress at approximately twice the CTOD (Figure 3.13). Therefore, while a linear geometry analysis using collapsed singularity elements is numerically stable, the stresses are not accurate close to the crack tip.

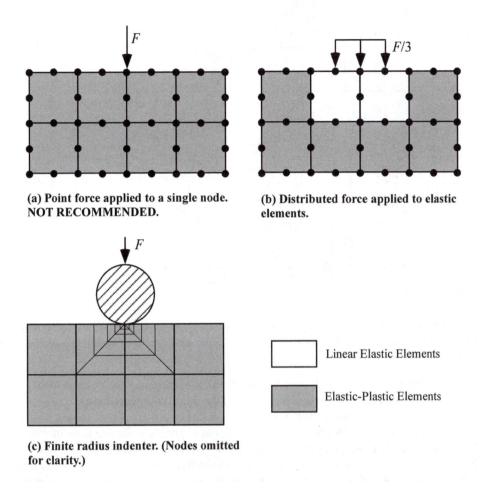

(a) Point force applied to a single node. NOT RECOMMENDED.

(b) Distributed force applied to elastic elements.

(c) Finite radius indenter. (Nodes omitted for clarity.)

Linear Elastic Elements

Elastic-Plastic Elements

FIGURE 12.22 Examples of improper (a) and proper (b and c) methods for applying a force to a boundary.

the boundary. A better way of applying this boundary condition might be to distribute the load over several nodes, and specifying that the elements on which the load acts remain elastic (Figure 12.22(b)); the load will then be transferred to the body without wasting computer time solving a local indentation problem. If, however, the local indentation is of interest, (e.g., if one wants to simulate the effects of the loading fixture) the load can be applied by a rigid or elastic indenter with a finite radius, as Figure 12.22(c) illustrates. Note, however, that greater mesh refinement is required to resolve the plastic deformation at the indenter.

12.5 LINEAR ELASTIC CONVERGENCE STUDY

In order to demonstrate the effect of mesh refinement and solution technique on the accuracy of K_I estimates from finite element analysis, a series of analyses were performed specifically for this chapter. These analyses were performed on a through-thickness crack in a flat plate subject to either a remote membrane stress or a uniform crack face pressure. The plate width was 20 times the crack length, so the model approximated the so-called Griffith plate, where the width is infinite.[4] This configuration was chosen because there is a closed-form theoretical solution for K_I (Equation (2.41))

---

[4] According to Equation (2.46) in Chapter 2, the finite-width correction for a/W = 0.05 is 1.0015, or 0.15% of the infinite body solution.

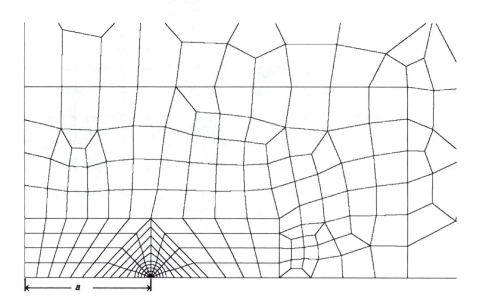

FIGURE 12.23 Close-up of the crack tip region of the baseline quarter-symmetric two-dimensional plane strain model used in the convergence study in Section 12.5. The model represents a through-thickness crack in a flat plate.

with which to compare the finite element results. Both two- and three-dimensional models of the through-thickness crack in a wide plate were created for this convergence study.

Figure 12.23 shows the crack-tip region of the baseline two-dimensional model. This model, which is 1/4-symmetric, includes five concentric rings of elements at the crack tip in the "spider-web" configuration described in the previous section. Two additional models were created in which number of element rings in the crack-tip region was doubled and increased by a factor of 8 relative to the baseline case. All of the two-dimensional models were constructed of 8-node quadrilateral plane strain elements.

Figure 12.24 is a dimensionless plot of stress-intensity values computed from the baseline model using the domain integral approach. The J integral was converted to the Mode I stress-intensity factor through Equation (2.56), and the resulting K_I was normalized by the theoretical solution for the Griffith plate. The first contour corresponds to the first ring of elements, the second contour comprises the first two rings of elements, and so on. Note that the 1/4-point node location in the collapsed elements in the first ring improves the J estimate in the first contour but has little effect on the second and higher contours. Also, note that the stress-intensity factor computed from the J integral is within 0.3% of the theoretical solution despite the fact that the baseline mesh is not particularly refined at the crack tip. Therefore, the J-integral method is an efficient way to compute K_I, in that a high degree of mesh refinement is not required. Also, 1/4-point elastic singularity elements are not necessary, provided the integration domain includes more than just the first ring of elements.

Figure 12.25 is a plot of K_I estimates from the baseline two-dimensional model using the normal stress in front of the crack tip (Equation (12.14)). The stress-matching approach results in a significant underestimate of K_I in this instance. Moreover, a remote stress and a crack-face pressure produce vastly different results, despite the fact that K_I is the same for both load cases, according to the principle of superposition (Section 2.6.4). The 1/4-point elastic singularity elements improve the solution slightly.

The displacement-matching approach (Equation (12.15(a)) produces much better estimates of K_I than the stress-matching method. Figure 12.26 shows a comparison of displacement-based K estimates for crack-face pressure and remote loading. There is virtually no difference between the

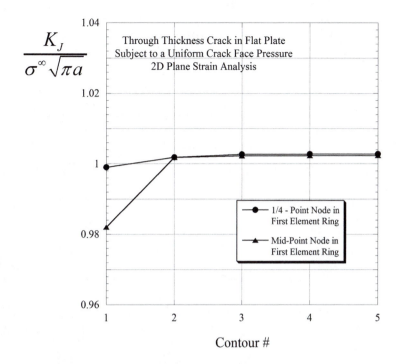

FIGURE 12.24 Dimensionless stress-intensity factor inferred from a *J*-integral analysis of the baseline two-dimensional model (Figure 12.23). The first contour (integration domain) corresponds to the first ring of elements, the second contour encompasses the first two element rings, and so on.

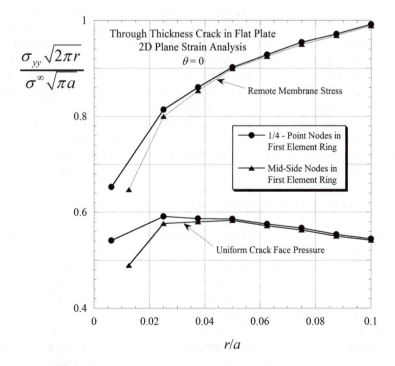

FIGURE 12.25 Dimensionless stress-intensity factor estimated from the normal stresses in front of the crack tip (Equation (12.14)) in the baseline two-dimensional model.

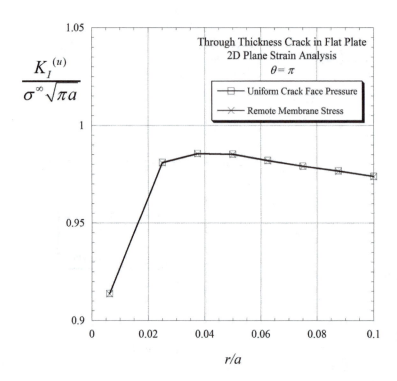

FIGURE 12.26 Dimensionless stress-intensity factor estimated from the displacements behind the crack tip in the baseline two-dimensional model. $K_I^{(u)}$ is the quantity in square brackets in Equation (12.15(a)).

two load cases. If the results for the first two nodes are disregarded, the extrapolation results in a K_I estimate that is within 2% of the theoretical solution. The results in Figure 12.26 were obtained from the baseline mesh with 1/4-point elastic singularity elements. Figure 12.27 compares these results with those from a mesh where the first ring of collapsed elements had mid-side rather than 1/4-point nodes. The elastic singularity elements improve the K_I estimate by 1–2% in the baseline model.

Figure 12.28 illustrates the effect of mesh refinement on K_I estimates from the stress-matching method. Increasing the number of element rings near the crack-tip improves the solution, but the rate of convergence is very slow. Even when the crack tip mesh is refined by a factor of 8 compared to the baseline model, the stress-matching method results in significant errors in K_I estimates. Theoretically, the K_I estimate should approach the correct value with sufficient mesh refinement, but the level of refinement required for acceptable accuracy is much larger than is necessary for both the J integral and displacement methods.

Figure 12.29 shows the results of the mesh refinement convergence study for K_I estimates from crack-opening displacement. As one would expect, mesh refinement results in improved estimates from extrapolation. However, the results from the first few nodes must be excluded from the extrapolation in order to obtain good estimates.

Some commercial finite element programs include an option to compute stress-intensity factors from nodal displacements behind the crack. However, it is risky to place blind trust in such algorithms. As Figure 12.29 illustrates, the K_I estimate can be significantly affected by the choice of nodes that are selected for the extrapolation to $r = 0$. For example, if the computation is based on the displacements of the first two nodes behind the crack tip, significant errors can result. The most reliable approach is to extract and plot the results for all crack-face nodes that are relatively close to the crack tip, and then determine which points should be used in the extrapolation.

Figure 12.30 shows a 1/4-symmetric three-dimensional model of a through-thickness crack in a flat plate. Figure 12.31 is a dimensionless plot of the through-thickness variation in K_J.

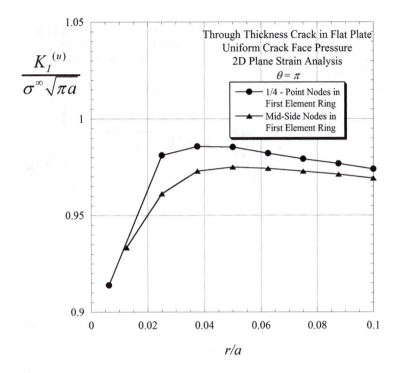

FIGURE 12.27 Effect of elastic singularity elements at the crack tip on K_I estimated from opening displacements behind the crack tip in the baseline model. $K_I^{(u)}$ is the quantity in square brackets in Equation (12.15(a)).

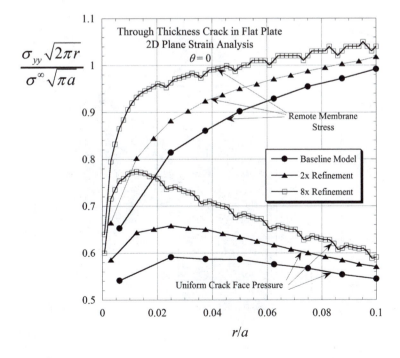

FIGURE 12.28 Effect of mesh refinement at the crack tip on K_I estimated from normal stresses in front of the crack tip.

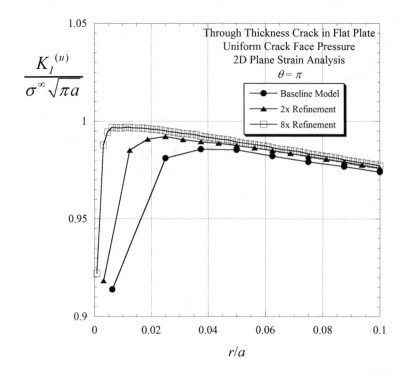

FIGURE 12.29 Effect of mesh refinement at the crack tip on K_I estimated from opening displacements behind crack tip. $K_I^{(u)}$ is the quantity in square brackets in Equation (12.15(a)).

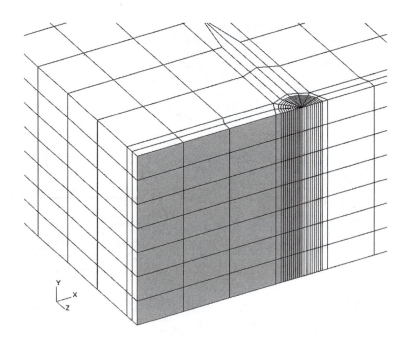

FIGURE 12.30 Closeup of the crack-tip region of a quarter-symmetric two-dimensional finite element model of a through-thickness crack in a wide plate. The crack face is highlighted in gray.

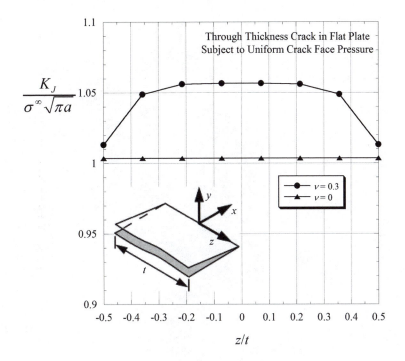

FIGURE 12.31 Through-thickness variation of the Mode I stress-intensity factor inferred from a *J*-integral analysis of the three-dimensional model in Figure 12.30.

For a Poisson's ratio of 0.3, the K_J in the center of the plate is approximately 5% greater than the theoretical solution for the Griffith plate. The classical solution is obtained when Poisson's ratio is set to zero. The results for $\nu = 0.3$ manifest a real three-dimensional effect. The theoretical solution for the Griffith plate is based on a two-dimensional body (plane stress or plane strain), so it cannot account for variations of stress and strain along the crack front. When $\nu = 0$, the three-dimensional model behaves like a two-dimensional plate because there is no out-of-plane deformation.

The effect of mesh refinement along the crack front in the three-dimensional analysis was investigated. Figure 12.32 shows models where the crack front was refined by factors of 3 and 9. The resulting K_J solutions are plotted in Figure 12.33. The K_J value in the center of the plate is insensitive to crack-front refinement, but the value on the free surface continually decreases with refinement along the crack front.

The results in Figure 12.33 further illustrate the difference between two-dimensional and three-dimensional solutions for cracks. Not only does the K_J value at midthickness differ from the classical solution, the computed free surface value is mesh dependent. As discussed in Section 12.3.2, the *J* integral computed at a given point on the crack front in a three-dimensional model actually corresponds to a weighted average over a finite crack-front length. For the node at the free surface, *J* is typically computed over a domain that is one element thick. For reasons described below, *the theoretical value of the J integral on the free surface of a three-dimensional body is zero.* Therefore, the computed *J* value for the node corresponding to where the crack front intersects a free surface will continually decrease with increasing crack-front refinement.[5]

---

[5] The theoretical value of *J* on the free surface is of little practical importance because crack propagation is governed by the crack driving force over a finite distance below the surface. Consider, for example, a semielliptical surface crack of depth *a* and length 2*c* that is growing by fatigue. The fatigue crack growth rate on the free surface d*a*/d*N*, appears to be governed by the nominal Δ*K near* the free surface. Otherwise, we would observe no crack growth on the free surface and the crack would "tunnel" in the *c* direction below the surface.

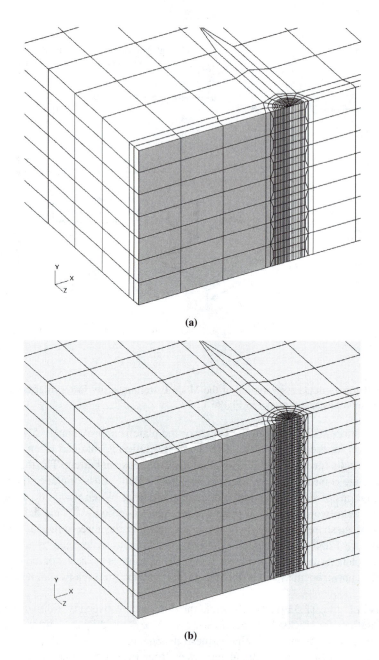

(a)

(b)

FIGURE 12.32 Refinement along the crack front of the three-dimensional model in Figure 12.30: (a) 3x crack-front refinement and (b) 9x crack-front refinement.

A crack intersecting a free surface in a three-dimensional body forms a corner. Such a configuration is decidedly different from a crack in a two-dimensional plane stress model. The corner results in a singularity, but stress varies as power of r that is $\neq -1/2$ for an elastic body. It can be shown that the stress singularity in an elastic material must vary as $r^{-1/2}$ in order for J to be nonzero. Recall the Williams series solution for elastic bodies with cracks (Chapter 2) and the corresponding series solution for fully plastic power-law materials (Chapter 3). It can be shown that only the

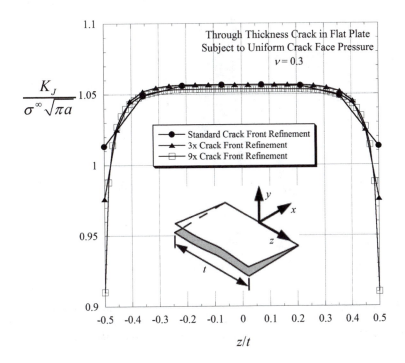

FIGURE 12.33 Effect of crack-front refinement on the through-thickness variation of K_J.

leading singular term in each series contributes to J. In both the elastic and fully plastic series solutions, the product of stress times strain varies as $1/r$ in the leading singular term.

In summary, the domain integral method is the most efficient means to infer stress-intensity factors solutions from finite element analysis. If one's finite element software does not include such capabilities, the displacement-matching technique is an acceptable alternative, provided the mesh refinement is sufficient for convergence. The stress-matching method requires a very high level of mesh refinement, so it is not recommended. Finally, the level of mesh refinement required for convergence is problem-specific, since it depends on the geometry and loading. The convergence results presented here are for purposes of illustration and should not be used as the sole basis for demonstrating convergence for a different problem.

12.6 ANALYSIS OF GROWING CRACKS

Although most computational fracture mechanics analyses are performed on stationary cracks, there are instances where it is desirable to analyze crack growth. Growing a crack in a finite element model often requires a special meshing strategy, and the analysis must include a criterion for crack advance.

The finite element models shown in previous sections all have *focused meshes*, where the highest level of mesh refinement is at the crack tip. The so-called spider-web configuration is an example of a focused mesh. This type of mesh is suitable for the analysis of a stationary crack, but is not appropriate for crack growth unless the focused region moves with the crack tip. Moving the crack tip in a focused mesh normally entails re-meshing. That is, a new focused mesh with a slightly longer crack must be created. Re-meshing is appropriate for elastic problems because stress and strain are not history dependent. For example, fatigue in an elastic body can be modeled by

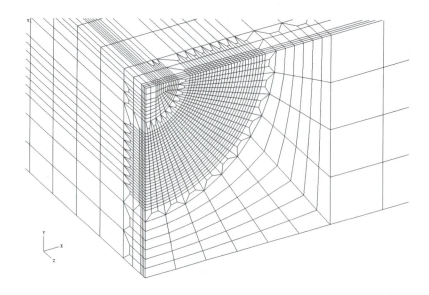

FIGURE 12.34 Cell-type mesh for analysis of crack growth in a semielliptical surface crack in a flat plate. Note the difference between this mesh configuration and the focused mesh in Figure 12.20. The concentric mesh lines on the crack plane correspond to various crack-front positions during the analysis.

computing the stress-intensity range at each time step, and then updating the crack dimensions (with a new mesh) based on the desired growth law.

In elastic-plastic materials, stress and strain are history dependent. Crack growth by re-meshing is possible in principle, provided the prior plastic strain history is properly mapped onto the various models created at each step. However, this approach is highly cumbersome. A better alternative is to create a single mesh that accommodates crack growth. One such mesh configuration is the *cell mesh*. Figure 12.34 shows an example of a cell mesh for a semielliptical surface crack. Rather than a focused refined zone at the crack tip, there are refined bands (or cells) of elements that correspond to the crack-front positions at various stages during the analysis. The disadvantage of this configuration is that the crack path and shape are predetermined by the mesh pattern.

There are three common methods to advance a crack in a cell mesh. Elements along the crack front can be removed from the model once a failure criterion is reached. For example, ductile crack growth can be simulated in a cell mesh using the Gurson–Tvergaard plasticity model described in Chapter 5 [27–29]. Another approach is to release nodes at specific load steps or according to a failure criterion. A third method entails using cohesive elements, which are zero-thickness elements on which a force-displacement law can be specified [30]. In a typical force-displacement curve for cohesive elements, the force reaches a maximum value at an intermediate displacement, and then decreases to zero with further displacement. The crack "unzips" as cohesive elements along the crack front progressively fail.

Irrespective of the numerical crack growth strategy (removing elements, releasing nodes, using cohesive elements) each increment of crack advance corresponds to the element size. For this reason the crack growth response in a finite element simulation is mesh dependent. In real materials, the crack growth response (e.g., the *J* resistance curve) depends on material length scales such as inclusion spacing. A finite element continuum model does not include microstructural features such as inclusions, so element size is the only available length scale to govern crack growth. Crack growth simulations usually need to be tuned to match experimental data. One of the key tuning parameters is the element size in the cell zone on the crack plane.

APPENDIX 12: PROPERTIES OF SINGULARITY ELEMENTS

Certain element/node configurations produce strain singularities. While such behavior is undesirable for most analyses, it is ideal for elastic crack problems. Forcing the elements at the crack tip to exhibit a $1/\sqrt{r}$ strain singularity can improve accuracy and reduces the need for a high degree of mesh refinement at the crack tip. Note that the strain is singular only at the node point at the crack tip. Over the interior of the element, the strain varies as $1/\sqrt{L}$. Consequently, the strain is finite at the gauss points, where it is actually evaluated in the solution.

The derivations that follow show that the desired singularity can be produced in quadratic isoparametric elements by moving the mid-side nodes to the 1/4-points. This behavior was first noted by Barsoum [14] and Henshell and Shaw [24].

From Equation (12.3) and Equation (12.4), the strain matrix for a two-dimensional element can be written in the following form:

$$\{\varepsilon\} = [\mathbf{J}]^{-1}[\mathbf{B}^*]\begin{Bmatrix} u_i \\ v_i \end{Bmatrix} \tag{A12.1}$$

where

$$[\mathbf{B}^*] = \begin{bmatrix} \dfrac{\partial N_i}{\partial \xi} & 0 \\[2mm] 0 & \dfrac{\partial N_i}{\partial \eta} \\[2mm] \dfrac{\partial N_i}{\partial \eta} & \dfrac{\partial N_i}{\partial \xi} \end{bmatrix} \tag{A12.2}$$

where (ξ, η) are the parametric coordinates of a point on the element. Since the nodal displacements $\{u_i, v_i\}$ are bounded, the strain matrix can be singular only if either $[\mathbf{B}^*]$ or $[\mathbf{J}]^{-1}$ is singular.

Consider an 8-noded quadratic isoparametric two-dimensional element (Figure 12.12(a)). The shape functions for this element are as follows [4]:

$$N_i = [(1+\xi\xi_i)(1+\eta\eta_i) - (1-\xi^2)(1+\eta\eta_i) - (1-\eta^2)(1+\xi\xi_i)]\frac{\xi_i^2\eta_i^2}{4}$$

$$+(1-\xi^2)(1+\eta\eta_i)\left(1-\xi_i^2\right)\frac{\eta_i^2}{2} + (1-\eta^2)(1+\xi\xi_i)\left(1-\eta_i^2\right)\frac{\xi_i^2}{2} \tag{A12.3}$$

where (ξ, η) are the parametric coordinates of a point in the element and (ξ_i, η_i) are the coordinates of the ith node.

In general, the shape functions are polynomials. Equation (A12.3), for example, is a quadratic equation. Thus N_i, $\partial N_i/\partial \xi$, or $\partial N_i/\partial \eta$ are all nonsingular, and $[\mathbf{J}]$ must be the cause of the singularity.

A strain singularity can arise if the determinant of the Jacobian matrix vanishes at the crack tip:

$$\det |\,\mathbf{J}\,| = \frac{\partial(x, y)}{\partial(\xi, \eta)} = 0 \tag{A12.4}$$

A12.1 QUADRILATERAL ELEMENT

Consider an 8-noded quadrilateral element with the mid-side nodes at the 1/4 point, as Figure A12.1 illustrates. For convenience, the origin of the x-y global coordinate system is placed at node 1. Let us evaluate the element boundary between nodes 1 and 2. From Equation (A12.3), the shape

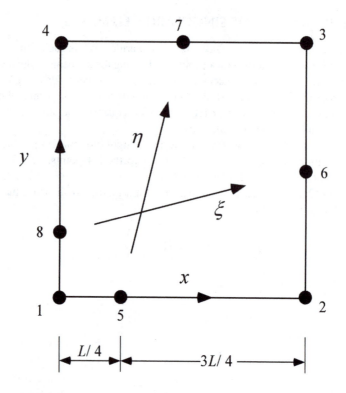

FIGURE A12.1 Quadrilateral isoparametric element with mid-side nodes moved to the quarter points.

functions along this line at nodes 1, 2, and 5 are given by

$$N_1 = -\frac{1}{2}\xi(1-\xi)$$

$$N_2 = \frac{1}{2}\xi(1-\xi) \quad\quad\quad (A12.5)$$

$$N_5 = (1-\xi^2)$$

Inserting these results into Equation (A12.1) gives

$$x = -\frac{1}{2}\xi(1-\xi)x_1 + \frac{1}{2}\xi(1-\xi)x_2 + (1-\xi^2)x_5 \quad\quad (A12.6)$$

Setting $x_1 = 0$, $x_2 = L$, and $x_5 = L/4$ results in

$$x = \frac{1}{2}\xi(1-\xi)L + (1-\xi^2)\frac{L}{4} \quad\quad\quad (A12.7)$$

where L is the length of the element between nodes 1 and 2. Solving for ξ gives

$$\xi = -1 + 2\sqrt{\frac{x}{L}} \qu\quad\quad\quad (A12.8)$$

The relevant term of the Jacobian is given by

$$\frac{\partial x}{\partial \xi} = \frac{L}{2}(1+\xi) = \sqrt{xL} \quad\quad\quad (A12.9)$$

which vanishes at $x = 0$; thus the strain must be singular at this point. Considering only the displacements of points 1, 2, and 5, the displacements along the element edge are as follows:

$$u = -\frac{1}{2}\xi(1-\xi)u_1 + \frac{1}{2}\xi(1-\xi)u_2 + (1-\xi^2)u_5 \qquad (A12.10)$$

Substituting Equation (A12.8) into Equation (A12.10) gives

$$u = -\frac{1}{2}\left(-1+2\sqrt{\frac{x}{L}}\right)\left(2-2\sqrt{\frac{x}{L}}\right)u_1$$

$$+\left(-1+2\sqrt{\frac{x}{L}}\right)\left(2\sqrt{\frac{x}{L}}\right)u_2 + 4\left(\sqrt{\frac{x}{L}}-\frac{x}{L}\right)u_5 \qquad (A12.11)$$

Solving for the strain in the x direction leads to

$$\varepsilon_x = \frac{\partial u}{\partial x} = \frac{\partial \xi}{\partial x}\frac{\partial u}{\partial \xi}$$

$$= -\frac{1}{2}\left(\frac{3}{\sqrt{xL}}-\frac{4}{L}\right)u_1 + \frac{1}{2}\left(-\frac{1}{\sqrt{xL}}+\frac{4}{L}\right)u_2 + \left(\frac{2}{\sqrt{xL}}-\frac{4}{L}\right)u_5 \qquad (A12.12)$$

Therefore, the strain exhibits a $1/\sqrt{r}$ singularity along the element boundary.

A12.2 TRIANGULAR ELEMENT

Let us now construct a triangular element by collapsing nodes 1, 4, and 8 (Figure A12.2). Nodes 5 and 7 are moved to the quarter points in this case. The $1/\sqrt{r}$ strain singularity exists along the 1-5-2 and 4-7-3 edges, as with the quadrilateral element. In this instance, however, the singularity also exists within the element.

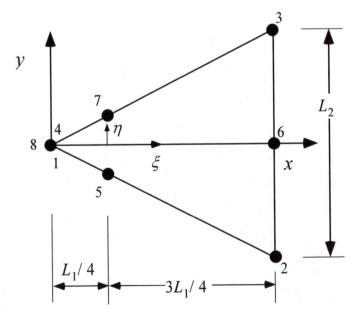

FIGURE A12.2 Degenerated isoparametric element, with mid-side nodes at the quarter points.

Consider the x axis, where $\eta = 0$. The relationship between x and ξ is given by

$$x = (\xi^2 + 2\xi + 1)\frac{L_1}{4} \tag{A12.13}$$

where L_1 is the length of the element in the x direction. Solving for ξ gives

$$\xi = -1 + 2\sqrt{\frac{x}{L_1}} \tag{A12.14}$$

which is identical to Equation (A12.8). Therefore, the strain is singular along the x axis in this element. By solving for strain as before (Equation (A12.10) to Equation (A12.12)) it can easily be shown that the singularity is the desired $1/\sqrt{r}$ type.

REFERENCES

1. Westergaard, H.M., "Bearing Pressures and Cracks." *Journal of Applied Mechanics*, Vol. 6, 1939, pp. 49–53.
2. Williams, M.L., "On the Stress Distribution at the Base of a Stationary Crack." *Journal of Applied Mechanics*, Vol. 24, 1957, pp. 109–114.
3. Lapidus, L. and Pinder, G.F., *Numerical Solution of Partial Differential Equations in Science and Engineering.* John Wiley & Sons, New York, 1982.
4. Zienkiewicz, O.C. and Taylor, R.L., *The Finite Element Method.* (4th ed.,) McGraw-Hill, New York, 1989.
5. Rizzo, F.J., "An Integral Equation Approach to Boundary Value Problems of Classical Elastostatics." *Quarterly of Applied Mathematics*, Vol. 25, 1967, pp. 83–95.
6. Cruse, T.A., *Boundary Element Analysis in Computational Fracture Mechanics.* Kluwer Dordrecht, The Netherlands, 1988.
7. Blandford, G.E. and Ingraffea, A.R., "Two-Dimensional Stress Intensity Factor Computations Using the Boundary Element Method." *International Journal for Numerical Methods in Engineering*, Vol. 17, 1981, pp. 387–404.
8. Cruse, T.A., "An Improved Boundary-Integral Equation for Three Dimensional Elastic Stress Analysis." *Computers and Structures*, Vol. 4, 1974, pp. 741–754.
9. Mendelson, A. and Albers, L.U., "Application of Boundary Integral Equations to Elastoplastic Problems." *Boundary Integral Equation Method: Computational Applications in Applied Mechanics*, AMD-Vol. 11, American Society of Mechanical Engineers, New York, 1975, pp. 47–84.
10. Parks, D.M., "A Stiffness Derivative Finite Element Technique for Determination of Crack Tip Stress Intensity Factors." *International Journal of Fracture*, Vol. 10, 1974, pp. 487–502.
11. Kobayashi, A.S., Cherepy, R.B., and Kinsel, W.C., "A Numerical Procedure for Estimating the Stress Intensity Factor of a Crack in a Finite Plate." *Journal of Basic Engineering*, Vol. 86, 1964, pp. 681–684.
12. Gross, B. and Srawley, J.E., "Stress Intensity Factors of Three Point Bend Specimens by Boundary Collocation." NASA Technical Note D-2603, 1965.
13. Tracey, D.M., "Finite Element Methods for Determination of Crack Tip Elastic Stress Intensity Factors." *Engineering Fracture Mechanics*, Vol. 3, 1971, pp. 255–266.
14. Barsoum, R.S., "On the Use of Isoparametric Finite Elements in Linear Fracture Mechanics." *International Journal for Numerical Methods in Engineering*, Vol. 10, 1976, pp. 25–37.
15. Budiansky, B. and Rice, J.R., "Conservation Laws and Energy Release Rates." *Journal of Applied Mechanics*, Vol. 40, 1973, pp. 201–203.
16. Carpenter, W.C., Read, D.T., and Dodds, R.H., Jr., "Comparison of Several Path Independent Integrals Including Plasticity Effects." *International Journal of Fracture*, Vol. 31, 1986, pp. 303–323.
17. Hellen, T.K., "On the Method of Virtual Crack Extensions." *International Journal for Numerical Methods in Engineering*, Vol. 9, 1975, pp. 187–207.

18. Parks, D.M., "The Virtual Crack Extension Method for Nonlinear Material Behavior." *Computer Methods in Applied Mechanics and Engineering*, Vol. 12, 1977, pp. 353–364.

19. deLorenzi, H.G., "On the Energy Release Rate and the *J*-Integral of 3-D Crack Configurations." *International Journal of Fracture*, Vol. 19, 1982, pp. 183–193.

20. deLorenzi, H.G., "Energy Release Rate Calculations by the Finite Element Method." *Engineering Fracture Mechanics*, Vol. 21, 1985, pp. 129–143.

21. Shih, C.F., Moran, B., and Nakamura, T., "Energy Release Rate Along a Three-Dimensional Crack Front in a Thermally Stressed Body." *International Journal of Fracture*, Vol. 30, 1986, pp. 79–102.

22. Moran, B. and Shih, C.F., "A General Treatment of Crack Tip Contour Integrals." *International Journal of Fracture*, Vol. 35, 1987, pp. 295–310.

23. Dodds, R.H., Jr. and Vargas, P.M., "Numerical Evaluation of Domain and Contour Integrals for Nonlinear Fracture Mechanics." Report UILU-ENG-88-2006, University of Illinois, Urbana, IL, August 1988.

24. Henshell, R.D. and Shaw, K.G., "Crack Tip Finite Elements are Unnecessary." *International Journal for Numerical Methods in Engineering*, Vol. 9, 1975, pp. 495–507.

25. Dodds, R.H., Jr., Anderson, T.L., and Kirk, M.T., "A Framework to Correlate *a/W* Effects on Elastic-Plastic Fracture Toughness (*Jc*)." *International Journal of Fracture*, Vol. 48, 1991, pp. 1–22.

26. McMeeking, R.M. and Parks, D.M., "On Criteria for *J*-Dominance of Crack Tip Fields in Large-Scale Yielding." ASTM STP 668, American Society for Testing and Materials, Philadelphia, PA, 1979, pp. 175–194.

27. Ruggieri, C., Panontin, T.L., and Dodds, R.H., Jr., "Numerical Modeling of Ductile Crack Growth in 3-D Using Computational Cell Elements." *International Journal of Fracture*, Vol. 82, 1996, pp. 67–95.

28. Gao, X., Faleskog, J., Dodds, R.H., Jr., and Shih, C.F., "Ductile Tearing in Part-Through Cracks: Experiments and Cell-Model Predictions." *Engineering Fracture Mechanics*, Vol. 59, 1998, pp. 761–777.

29. Gullerud, A. and Dodds, R.H., Jr., "Simulation of Ductile Crack Growth Using Computational Cells: Numerical Aspects," *Engineering Fracture Mechanics*, Vol. 66, No. 1, 2000, pp. 65–92.

30. Roy, A. and Dodds, R.H., Jr., "Simulation of Ductile Crack Growth in Thin Aluminum Panels Using 3-D Surface Cohesive Elements," *International Journal of Fracture*, Vol. 110, 2001, pp. 21–45.

13 Practice Problems

This chapter contains practice problems that correspond to material in Chapter 1 to Chapter 12. Some of the problems for Chapter 7 to Chapter 11 require a computer program or spreadsheet. This level of complexity was necessary in order to make the application-oriented problems realistic.

All quantitative data are given in SI units, although the corresponding values in English units are also provided in many cases.

13.1 CHAPTER 1

1.1 Compile a list of five mechanical or structural failures that have occurred within the last 20 years. Describe the factors that led to each failure and identify the failures that resulted from misapplication of existing knowledge (Type 1) and those that involved new technology or a significant design modification (Type 2).

1.2 A flat plate with a through-thickness crack (Figure 1.8) is subject to a 100 MPa (14.5 ksi) tensile stress and has a fracture toughness (K_{Ic}) of 50.0 MPa $\sqrt{m}$ (45.5 ksi $\sqrt{in}$.). Determine the critical crack length for this plate, assuming the material is linear elastic.

1.3 Compute the critical energy release rate (G_c) of the material in the previous problem for $E = 207,000$ MPa (30,000 ksi).

1.4 Suppose that you plan to drop a bomb out of an airplane and that you are interested in the time of flight before it hits the ground, but you cannot remember the appropriate equation from your undergraduate physics course. You decide to infer a relationship for time of flight of a falling object by experimentation. You reason that the time of flight, t, must depend on the height above the ground, h, and the weight of the object, mg, where m is the mass and g is the gravitational acceleration. Therefore, neglecting aerodynamic drag, the time of flight is given by the following function:

$$t = f(h, m, g)$$

Apply dimensional analysis to this equation and determine how many experiments would be required to determine the function f to a reasonable approximation, assuming you know the numerical value of g. Does the time of flight depend on the mass of the object?

13.2 CHAPTER 2

2.1 According to Equation (2.25), the energy required to increase the crack area a unit amount is equal to *twice* the fracture work per unit surface area, w_f. Why is the factor of 2 in this equation necessary?

2.2 Derive Equation (2.30) for both load control and displacement control by substituting Equation (2.29) into Equation (2.27) and Equation (2.28), respectively.

2.3 Figure 2.10 illustrates that the driving force is linear for a through-thickness crack in an infinite plate when the stress is fixed. Suppose that a remote displacement (rather than load) were fixed in this configuration. Would the driving force curves be altered? Explain. (*Hint:* see Section 2.5.3).

2.4 A plate $2W$ wide contains a centrally located crack $2a$ long and is subject to a tensile load P. Beginning with Equation (2.24), derive an expression for the elastic compliance $C (= \Delta/P)$ in terms of the plate dimensions and elastic modulus E. The stress in Equation (2.24) is the nominal value, i.e., $\sigma = P/2BW$ in this problem. (*Note:* Equation (2.24) only applies when $a \ll W$; the expression you derive is only approximate for a finite-width plate.)

2.5 A material exhibits the following crack growth resistance behavior:

$$R = 6.95(a - a_o)^{0.5}$$

where a_o is the initial crack size. R has units of kJ/m² and crack size is in millimeters. Alternatively,

$$R = 200(a - a_o)^{0.5}$$

where R has units of in.-lb/in.² and crack size is in inches. The elastic modulus of this material = 207,000 MPa (30,000 ksi). Consider a wide plate with a through crack ($a \ll W$) that is made from this material.

(a) If this plate fractures at 138 MPa (20.0 ksi), compute the following:
   (i) The half crack size at failure (a_c)
   (ii) The amount of stable crack growth (at each crack tip) that precedes failure ($a_c - a_o$)
(b) If this plate has an initial crack length ($2a_o$) of 50.8 mm (2.0 in.) and the plate is loaded to failure, compute the following:
   (i) The stress at failure
   (ii) The half crack size at failure
   (iii) The stable crack growth at each crack tip

2.6 Suppose that a double cantilever beam specimen (Figure 2.9) is fabricated from the same material considered in Problem 2.5. Calculate the load at failure and the amount of stable crack growth. The specimen dimensions are as follows:

$$B = 25.4 \text{ mm (1 in.)}; \quad h = 12.7 \text{ mm (0.5 in.)}; \quad a_o = 152 \text{ mm (6 in.)}$$

2.7 Consider a nominally linear elastic material with a rising R curve (e.g., Problem 2.5 and Problem 2.6). Suppose that one test is performed on wide plate with a through crack (Figure 2.3) and a second test on the same material is performed on a double cantilever beam (DCB) specimen (Figure 2.9). If both tests are conducted in load control, would the $\mathcal{G}_c$ values at instability be the same? If not, which geometry would result in a higher $\mathcal{G}_c$? Explain.

2.8 Example 2.3 showed that the energy release rate $\mathcal{G}$ of the DCB specimen increases with crack growth when the specimen is held at a constant load. Describe (qualitatively) how you could alter the design of the DCB specimen such that a growing crack in load control would experience a constant $\mathcal{G}$.

2.9 Beginning with Equation (2.20), derive an expression for the potential energy of a plate subject to a tensile stress σ with a penny-shaped flaw of radius a. Assume that $a \ll$ plate dimensions.

2.10 Beginning with Equation (2.20), derive expressions for the energy release rate and Mode I stress-intensity factor of a penny-shaped flaw subject to a remote tensile stress. (Your K_I expression should be identical to Equation (2.44)).

2.11 Calculate K_I for a rectangular bar containing an edge crack loaded in three-point bending.

$$P = 35.0 \text{ kN (7870 lb)}; \quad W = 50.8 \text{ mm (2.0 in.)};$$
$$B = 25 \text{ mm (1.0 in.)}; \quad a/W = 0.2; \quad S = 203 \text{ mm (8.0 in.)}.$$

2.12 Consider a material where $K_{IC} = 35 \text{ MPa}\sqrt{m}$ (31.8 ksi$\sqrt{in.}$). Each of the five specimens in Table 2.4 and Figure 2.23 has been fabricated from this material. In each case,

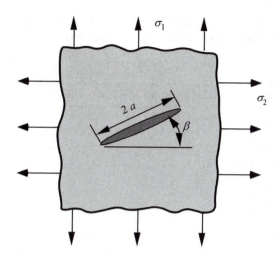

FIGURE 13.1 Through-thickness crack in a biaxially loaded plate (Problem 2.16).

$B = 25.4$ mm (1 in.), $W = 50.8$ mm (2 in.), and $a/W = 0.5$. Estimate the failure load for each specimen. Which specimen has the highest failure load? Which has the lowest?

2.13 A large block of material is loaded to a stress of 345 MPa (50 ksi). If the fracture toughness K_{Ic} is 44 MPa$\sqrt{m}$(40 ksi$\sqrt{in}$.), determine the critical radius of a buried penny-shaped crack.

2.14 A semicircular surface crack in a pressure vessel is 10 mm (0.394 in.) deep. The crack is on the inner wall of the pressure vessel and is oriented such that the hoop stress is perpendicular to the crack plane. Calculate K_I if the local hoop stress = 200 MPa (29.0 ksi) and the internal pressure = 20 MPa (2900 psi). Assume that the wall thickness >> 10 mm.

2.15 Calculate K_I for a semielliptical surface flaw at $\phi = 0°$, 30°, 60°, and 90°. $\sigma = 150$ MPa (21.8 ksi); $a = 8.00$ mm (0.315 in.); $2c = 40$ mm (1.57 in.). Assume the plate width and thickness are large relative to the crack dimensions.

2.16 Consider a plate subject to biaxial tension with a through crack of length $2a$, oriented at an angle β from the σ_2 axis (Figure 13.1). Derive expressions for K_I and K_{II} for this configuration. What happens to each K expression when $\sigma_1 = \sigma_2$?

2.17 A wide flat plate with a through-thickness crack experiences a nonuniform normal stress that can be represented by the following crack-face traction:

$$p(x) = p_o e^{-x/\beta}$$

where $p_o = 300$ MPa and $\beta = 25$ mm. The origin ($x = 0$) is at the left crack tip, as illustrated in Figure 2.27. Using the weight function derived in Example 2.6, calculate K_I at each crack tip for $2a = 25$, 50, and 100 mm. You will need to integrate the weight function *numerically*.

2.18 Calculate K_{eff} (Irwin correction) for a through crack in a plate of width $2W$ (Figure 2.20(b)). Assume plane stress conditions and the following stress-intensity relationship:

$$K_{eff} = \sigma\sqrt{\pi a_{eff}}\left[\sec\left(\frac{\pi a_{eff}}{2W}\right)\right]^{1/2}$$

$\sigma = 250$ MPa (36.3 ksi); $\sigma_{YS} = 350$ MPa (50.8 ksi); $2W = 203$ mm (8.0 in.); $2a = 50.8$ mm (2.0 in.).

2.19 For an infinite plate with a through crack 50.8 mm (2.0 in.) long, compute and tabulate K_{eff}. stress using the three methods indicated below. Assume $\sigma_{YS} = 250$ MPa (36.3 ksi).

Stress	K_{eff} (MPa$\sqrt{m}$ or ksi$\sqrt{in}$.)		
(MPa (ksi))	LEFM	Irwin Correction	Strip-Yield Model
25 (3.63)			
50 (7.25)			
100 (14.5)			
150 (21.8)			
200 (29.0)			
225 (32.6)			
249 (36.1)			
250 (36.3)			

2.20 A material has a yield strength of 345 MPa (50 ksi) and a fracture toughness of 110 MPa$\sqrt{m}$ (100 ksi$\sqrt{in}$.). Determine the minimum specimen dimensions (B, a, W) required to perform a valid K_{Ic} test on this material, according to ASTM E 399. Comment on the feasibility of testing a specimen of this size.

2.21 You have been given a set of fracture mechanics test specimens, all of the same size and geometry. These specimens have been fatigue precracked to various crack lengths. The stress intensity of this specimen configuration can be expressed as follows:

$$K_I = \frac{P}{B\sqrt{W}} f(a/W)$$

where
$\quad\quad P$ = load
$\quad\quad B$ = thickness
$\quad\quad W$ = width
$\quad\quad a$ = crack length
$\quad f(a/W)$ = dimensionless geometry correction factor

Describe a set of experiments you could perform to determine $f(a/W)$ for this specimen configuration. Hint: you may want to take advantage of the relationship between K_I and energy release rate for linear elastic materials.

2.22 Derive the Griffith-Inglis result for the potential energy of a through crack in an infinite plate subject to a remote tensile stress (Equation (2.16)). *Hint:* solve for the work required to close the crack faces; Equation (A2.43b) gives the crack-opening displacement for this configuration.

2.23 Using the Westergaard stress function approach, derive the stress-intensity factor relationship for an infinite array of collinear cracks in a plate subject to biaxial tension (Figure 2.21).

13.3 CHAPTER 3

3.1 Repeat the derivation of Equation (3.1) to Equation (3.3) for the plane strain case.

3.2 A CTOD test is performed on a three-point bend specimen. Figure 13.3 shows the deformed specimen after it has been unloaded. That is, the displacements shown are the *plastic components*.
 (a) Derive an expression for plastic CTOD (δ_p) in terms of Δ_p and specimen dimensions.

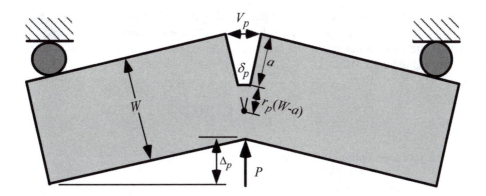

FIGURE 13.2 Three-point bend specimen rotating about a plastic hinge (Problem 3.2).

(b) Suppose that V_p and Δ_p are measured on the same specimen, but that the plastic rotational factor r_p is unknown. Derive an expression for r_p in terms of Δ_p, V_p, and specimen dimensions, assuming the angle of rotation is small.

3.3 Fill in the missing steps between Equation (3.36) and Equation (3.37)

3.4 Derive an expression for the J integral for a deeply notched three-point bend specimen, loaded over a span S, in terms of the area under the load-displacement curve and ligament length b. Figure 13.2 illustrates two displacement measurements on a bend specimen: the load-line displacement (Δ) and the crack-mouth-opening displacement (V). Which of these two displacement measurements is more appropriate for inferring the J integral? Explain.

3.5 Derive an expression for the J integral for an axisymmetrically notched bar in tension (Figure 13.3), where the notch depth is sufficient to confine plastic deformation to the ligament.

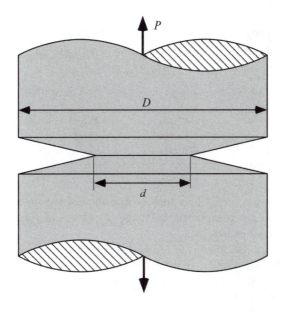

FIGURE 13.3 Axisymmetrically notched bar loaded in tension (Problem 3.5).

3.6 Derive an expression for the J integral in a deeply notched three-point bend specimen in terms of the area under the load-crack-mouth-opening displacement curve. Begin with the corresponding formula for the P-Δ curve, and assume rotation about a plastic hinge (Figure 13.3).

$$J = \frac{K^2}{E'} + \frac{2}{b}\int_0^{\Delta_p} P d\Delta_p$$

for a specimen with unit thickness.

13.4 CHAPTER 4

4.1 A high rate fracture toughness test is to be performed on a high strength steel with $K_{Id} = 110$ MPa$\sqrt{m}$ (100 ksi$\sqrt{in}$.). A three-point bend specimen will be used, with $W = 50.8$ mm (2.0 in.), $a/W = 0.5$, $B = W/2$, and span $= 4W$. Also, $c_1 = 5940$ m/sec (19,500 ft/sec) for steel. Estimate the maximum loading rate at which the quasistatic formula for estimating K_{Id} is approximately valid.

4.2 Unstable fracture initiates in a steel specimen and arrests after the crack propagates 8.0 mm (0.32 in.). The total propagation time was 7.52×10^{-6} sec. The initial ligament length in the specimen was 30.0 mm (1.18 in.) and c_1 for steel $= 5940$ m/sec (19,500 ft/sec). Determine whether or not reflected stress waves influenced the propagating crack.

4.3 Fracture initiates at an edge crack in a 2.0 m (78.7 in.) wide steel plate and rapidly propagates through the material. The stress in the plate is fixed at 300 MPa (43.5 ksi). Plot the crack speed vs. crack size for crack lengths ranging from 10 to 60 mm (or 0.4 to 2.4 in.). The dynamic fracture toughness of the material is given by

$$K_{ID} = \frac{K_{IA}}{1-\left(\frac{V}{V_l}\right)^2}$$

where $K_{IA} = 55$ MPa$\sqrt{m}$ (50 ksi$\sqrt{in}$.) and $V_l = 1500$ m/sec (4920 ft/sec). Use the Rose approximation (Equation (4.17) and Equation (4.18)) for the driving force. The elastic wave speeds for steel are given below.

c_1	5940 m/sec	19,500 ft/sec
c_2	3220 m/sec	10,600 ft/sec
c_r	2980 m/sec	9780 ft/sec

4.4 Derive an expression for C^* in a double-edge-notched tension panel in terms of specimen dimensions, creep exponent, load, and displacement rate. See Section 3.2.5 for the corresponding J expression.

4.5 A three-point bend specimen is tested in displacement control at an elevated temperature. The displacement rate is increased in steps as the test progresses. The load, load-line displacement rate, a/W, and crack velocity are tabulated below. Compute C^*, and construct a log-log plot of crack velocity vs. C^*. The specimen thickness and width are 25 and 50 mm, respectively. The creep exponent $= 5.0$ for the material.

$\dot{\Delta}$ (m/sec)	Load (kN)	a/W	$\dot{a}$(m/sec)
1.0×10^{-7}	10.8	0.52	3.67×10^{-9}
5.0×10^{-7}	13.8	0.54	1.79×10^{-8}
1.0×10^{-6}	14.9	0.56	3.49×10^{-8}
5.0×10^{-6}	19.0	0.58	1.71×10^{-7}
1.0×10^{-5}	20.4	0.60	3.37×10^{-7}
5.0×10^{-5}	24.0	0.65	1.65×10^{-6}

4.6   In a linear viscoelastic material, the pseudo-elastic displacement and the physical displacement are related through a hereditary integral:

$$\Delta^e = \{Ed\Delta\}$$

Simplify this expression for the case of a constant displacement rate.

4.7   Consider a fracture toughness test on a nonlinear viscoelastic material at a constant displacement rate. Assume that the load is related to the pseudo-elastic displacement by a power law:

$$P = M(\Delta^e)^N$$

where M and N are constants that do not vary with time. Show that the viscoelastic J integral and the conventional J integral are related as follows:

$$J_v = J\phi(t)$$

where ϕ is a function of time. Derive an expression for $\phi(t)$. Hint: begin with Equation (3.17) and Equation(4.75). Also, the result from the previous problem may be useful.

4.8   A fracture toughness test on a linear viscoelastic material results in a nonlinear load-displacement curve in a constant rate test. Yielding is restricted to a very small region near the crack tip. Why is the curve nonlinear? Does the stress-intensity factor characterize the crack-tip conditions in this case? Explain. What is the relationship between J and K_I for a linear viscoelastic material? Hint: refer to the second equation in the previous problem.

13.5   CHAPTER 5

5.1   A body-centered cubic (BCC) material contains second phase particles. The size of these particles can be controlled through thermal treatment. Discuss the anticipated effect of particle size on the material's resistance to both cleavage fracture and microvoid coalescence, assuming the volume fraction of the second phase remains constant.

5.2   An aluminum alloy fails by microvoid coalescence when the average void size reaches ten times the initial value. If the voids grow according to Equation (5.11), with σ_{YS} replaced by σ_e, plot the equivalent plastic strain (ε_{eq}) at failure vs. σ_m/σ_e for σ_m/σ_e ranging from 0 to 2.5. Assume the triaxiality ratio remains constant during deformation of a given sample, i.e.,

$$\ln\left(\frac{\overline{R}}{R_o}\right) = 0.283\exp\left(\frac{1.5\sigma_m}{\sigma_e}\right)\int_0^{\varepsilon_{eq}} d\varepsilon_{eq}$$

5.3 The critical microstructural feature for cleavage initiation in a steel sample is a 6.67 μm diameter spherical carbide; failure occurs when this particle forms a microcrack that satisfies the Griffith criterion (Equation (5.18)), where $\gamma_p = 14$ J/m², $E = 207,000$ MPa, and $v = 0.30$ for the material. Assuming Figure 5.14 describes the stress distribution ahead of the macroscopic crack, where $\sigma_o = 350$ MPa, estimate the critical J value of the sample if the particle is located 0.1 mm ahead of the crack tip, on the crack plane. Repeat this calculation for the case where the critical particle is 0.4 mm ahead of the crack tip.

5.4 Cleavage initiates in a ferritic steel at 3.0 μm diameter spherical particles. The fracture energy on a single grain, γ_p, is 14 J/m² and the fracture energy required for propagation across grain boundaries, γ_{gb}, is 50 J/m². At what grain size does propagation across grain boundaries become the controlling step for cleavage fracture?

5.5 Compute the relative size of the 90% confidence band of K_{Ic} data (as in Example 5.1), assuming Equation (5.24) describes the toughness distribution. Compute the confidence bandwidth for $K_o/\Theta_K = 0$, 0.5, 1.0, 2.0, and 5.0. What is the effect of the threshold toughness K_o on the relative scatter? What is the physical significance of Θ_K in this case?

5.6 Compute the relative size of the 90% confidence band of K_{Ic} data (as in Example 5.1), assuming Equation (5.26) describes the toughness distribution. Compute the confidence bandwidth for $K_o/\Theta_K = 0$, 0.5, 1.0, 2.0, and 5.0. What is the effect of the threshold toughness K_o on the relative scatter?

13.6 CHAPTER 6

6.1 For the Maxwell spring and dashpot model (Figure 6.6) derive an expression for the relaxation modulus.

6.2 Fill in the missing steps in the derivation of Equation (6.14).

6.3 At room temperature, tensile specimens of polycarbonate show 60% elongation and no stress whitening, while thick compact specimens used in fracture toughness testing show stress whitening at the crack tip. Explain these observations. Polycarbonate is an amorphous glassy polymer at room temperature.

6.4 A wide and thin specimen of PMMA has a 15 mm (0.59 in.) long through crack with a 1.5 mm (0.059 in.) long craze at each crack tip. If the applied stress is 3.5 MPa (508 psi), calculate the crazing stress in this material.

6.5 When a macroscopic crack grows in a ceramic specimen, a process zone 0.2 mm wide forms. This process zone contains 10,000 penny-shaped microcracks/mm³ with an average radius of 10 μm. Estimate the increase in toughness due to the release of strain energy by these microcracks. The surface energy of the material = 25 J/m².

13.7 CHAPTER 7

7.1 A fracture toughness test is performed on a compact specimen. Calculate K_Q and determine whether or not $K_Q = K_{Ic}$.

$B = 25.4$ mm (1.0 in.); $W = 50.8$ mm (2.0 in.); $a = 27.7$ mm (1.09 in.);

$P_Q = 42.3$ kN (9.52 kip); $P_{max} = 46.3$ kN (10.4 kip); $\sigma_{YS} = 759$ MPa (110 ksi)

7.2 You have been asked to perform a K_{Ic} test on a material with $\sigma_{YS} = 690$ MPa (100 ksi). The toughness of this material is expected to lie between 40 and 60 MPa $\sqrt{m}$ (1 ksi $\sqrt{in}$. = 1.099 MPa$\sqrt{m}$). Design an experiment to measure K_{Ic} (in accordance with ASTM E 399) in this material using a compact specimen. Specify the following quantities:

(a) specimen dimensions, (b) precracking loads, and (c) required load capacity of the test machine.

7.3 A titanium alloy is supplied in 15.9 mm (0.625 in.) thick plate. If $\sigma_{YS} = 807$ MPa (117 ksi), calculate the maximum valid K_{Ic} that can be measured in this material.

7.4 Recall Problem 2.16, where a material with $K_{Ic} = 110$ MPa$\sqrt{\text{m}}$ (100 ksi$\sqrt{\text{in.}}$) required a 254 mm (10.0 in.) thick specimen for a valid K_{Ic} test. Suppose that a compact specimen of the appropriate dimensions has been fabricated. Estimate the required load capacity of the test machine for such a test.

7.5 A 25.4 mm (1 in.) thick steel plate has material properties that are tabulated below. Determine the highest temperature at which it is possible to perform a valid K_{Ic}, according to ASTM E 399.

Temperature (°C)	Yield Strength (MPa)	K_{Ic} (MPa$\sqrt{\text{m}}$)
−10	760	34
−5	725	36
0	690	42
5	655	50
10	620	62
15	586	85
20	550	110
25	515	175

7.6 A fracture toughness test is performed on a compact specimen fabricated from a 5 mm thick sheet aluminum alloy. The specimen width (W) = 50.0 mm and B = 5 mm (the sheet thickness). The initial crack length is 26.0 mm. Young's modulus = 70,000 MPa. Compute the K-R curve from the load-displacement data tabulated in the given table. Assume that all nonlinearity in the P-Δ curve is due to crack growth. (See Appendix 7 for the appropriate compliance and stress-intensity relationships.)

Load (kN)	Load-Line Displacement(mm)	Load (kN)	Load-Line Displacement(mm)
0	0	2.851	0.3698
0.5433	0.0635	2.913	0.3860
1.087	0.1270	2.903	0.3971
1.630	0.1906	2.850	0.4113
2.161	0.2552	2.749	0.4191
2.361	0.2817	2.652	0.4274
2.541	0.3096	2.553	0.4355
2.699	0.3392	2.457	0.4443

1kN = 224.8 lb;   25.4 mm = 1 in.;   1 MPa = 0.145 ksi

7.7 A number of fracture toughness specimens have been loaded to various points and then unloaded. Values of J and crack growth were measured in each specimen and are tabulated in the given table. Using the basic test procedure the J-R curve for this material, determine J_Q and, if possible, J_{Ic}.

$$\sigma_{YS} = 350 \text{ MPa}; \quad \sigma_{TS} = 450 \text{ MPa}; \quad B = 25 \text{ mm}; \quad b_o = 22 \text{ mm}$$

Specimen	$J(kJ/m^2)$	Crack Extension (mm)
1	100	0.30
2	175	0.40
3	185	0.80
4	225	1.20
5	250	1.60
6	300	1.70

25.4 mm = 1 in.;   1 MPa = 0.145 ksi;   1 kJ/m² = 5.71 in-lb/in²

7.8 Recall Problem 2.16, where a material with $K_{Ic} = 110$ MPa$\sqrt{m}$ (100 ksi $\sqrt{in}$.) and $\sigma_{YS} = 345$ MPa (50 ksi) required a specimen 254 mm (10 in.) thick for a valid K_{Ic} test in accordance with ASTM E 399. Estimate the specimen's dimensions required for a valid J_{Ic} test on this material.

$\sigma_{TS} = 483$ MPa (70 ksi); $E = 207,000$ MPa (30,000 ksi); $v = 0.3$

7.9 An unloading compliance test has been performed on a three-point bend specimen. The data obtained at each unloading point are tabulated in the given table.
(a) Compute and plot the J resistance curve according to the procedure outlined in Section 7.4.2.
(b) Determine J_{Ic} according to the procedure illustrated in Figure 7.24.

$B = 25.0$ mm; $W = 50.0$ mm; $a_o = 26.1$ mm; $E = 210,000$ MPa; $v = 0.3$

$\sigma_{YS} = 345$ MPa (50 ksi); $\sigma_{TS} = 483$ MPa (70 ksi)

(c) Compute and plot a J-R curve obtained from the basic procedure (Equation (7.8) and Equation (7.10)). Use the initial crack size for all J calculations. Compare this J-R curve with the one computed in Part (a). At what point does the crack growth correction become significant?

Load (kN)	Plastic Displacement (mm)	Crack Extension (mm)
20.8	0	0.013
31.2	0.0032	0.020
35.4	0.011	0.023
37.4	0.020	0.025
41.6	0.056	0.031
43.7	0.092	0.036
45.7	0.146	0.044
47.6	0.228	0.055
49.9	0.349	0.071
51.6	0.525	0.091
53.5	0.777	0.128
55.3	1.13	0.183
56.6	1.63	0.321
56.7	2.32	0.723
56.5	2.66	0.928

(*Continued*)

Load (kN)	Plastic Displacement (mm)	Crack Extension (mm)
55.8	3.25	1.29
54.7	3.96	1.74
53.7	4.51	2.08
52.5	5.13	2.48
50.1	6.20	3.17
44.4	8.43	4.67
40.0	10.09	5.81
36.6	11.37	6.70
30.9	13.54	8.23
26.8	15.19	9.41

1 kN = 224.8 lb;   25.4 mm = 1 in.;   1 MPa = 0.145 ksi

7.10 A CTOD test was performed on a three-point bend specimen with $B = W = 25.4$ mm (1.0 in.). The crack depth a was 12.3 mm (0.484 in.). Examination of the fracture surface revealed that the specimen failed by cleavage with no prior stable crack growth. Compute the critical CTOD in this test. Be sure to use the appropriate notation (i.e., δ_c, δ_u, δ_i, or δ_m).

$$V_p = 1.05 \text{ mm } (0.0413 \text{ in.}); \quad P_{critical} = 24.6 \text{ kN } (5.53 \text{ ksi});$$
$$E = 207{,}000 \text{ MPa } (30{,}000 \text{ ksi});$$
$$\sigma_{YS} = 400 \text{ MPa } (58.0 \text{ ksi}); \quad v = 0.3.$$

7.11 A crack-arrest test has been performed in accordance with ASTM E 1221. The side-grooved compact crack-arrest specimen has the following dimensions: $W = 100$ mm (3.94 in.), $B = 25.4$ mm (1.0 in.), and $B_N = 19.1$ mm (0.75 in.). The initial crack length = 46.0 mm (1.81 in.) and the crack length at arrest = 63.0 mm (2.48 in.). The corrected clip gage displacements at initiation and arrest are $V_o = 0.582$ mm (0.0229 in.) and $V_a = 0.547$ (0.0215 in.), respectively. $E = 207{,}000$ MPa (30,000 ksi) and $\sigma_{YS(static)} = 483$ MPa (70 ksi). Calculate the stress intensity at initiation K_o and the arrest toughness K_a. Determine whether or not this test satisfies the validity criteria in Equation (7.25). The stress-intensity solution for the compact crack-arrest specimen is as follows:

$$K_I = \frac{EVf(x)\sqrt{B/B_N}}{\sqrt{W}}$$

where

$$x = a/W$$

$$f(x) = \frac{2.24(1.72 - 0.9x + x^2)\sqrt{1-x}}{9.85 - 0.17x + 11x^2}$$

13.8 CHAPTER 8

8.1 A 25.4 mm (1.0 in.) thick plate of PVC has a yield strength of 60 MPa (8.70 ksi). The anticipated fracture toughness (K_{Ic}) of this material is 5 MPa $\sqrt{m}$ (4.5 ksi$\sqrt{in}$.). Design an experiment to measure K_{Ic} of a compact specimen machined from this material. Determine the appropriate specimen dimensions (B, W, a) and estimate the required load capacity of the test machine.

8.2 A 15.9 mm (0.625 in.) thick plastic plate has a yield strength of 50 MPa (7.25 ksi). Determine the largest valid K_{Ic} value that can be measured on this material.

8.3 A K_{Ic} test is to be performed on a polymer with a time-dependent relaxation modulus that has been fit to the following equation:

$$E(t) = (0.417 + 0.0037t^{0.35})^{-1}$$

where E is in GPa and t is in seconds. Assuming P_Q is determined from a 5% secant construction, estimate the test duration (i.e., the time to reach P_Q) at which 90% of the nonlinearity in the load-displacement curve at P_Q is due to viscoelastic effects. Does the 5% secant load give an appropriate indication of material toughness in this case? Explain.

8.4 Derive a relationship between the conventional J integral and the isochronous J integral, J_t, in a constant displacement rate test on a viscoelastic material for which Equation (8.10) and Equation (8.15) describe the load-displacement behavior.

8.5 A 500 mm wide plastic plate contains a through-thickness center crack that is initially 50 mm long. The crack velocity in this material is given by

$$\dot{a} = 10^{-40} K^{10}$$

where K is in kPa$\sqrt{\text{m}}$ and $\dot{a}$ is in mm/sec (1 psi$\sqrt{\text{in.}}$ = 1.1 kPa$\sqrt{\text{m}}$, 1 in. = 24.5 mm). Calculate the time to failure in this plate assuming remote tensile stresses of 5 and 10 MPa (1 ksi = 6.897 MPa). Comment on the sensitivity of the time to failure on the applied stress. (As a first approximation, neglect the finite width correction on K. For an optional exercise, repeat the calculations with this correction to assess its effect on the computed failure times.)

8.6 A composite DCB specimen is loaded to 445 N (100 lb) at which time crack growth begins. Calculate $\mathcal{G}_{Ic}$ for this material assuming linear beam theory.

$$E = 124{,}000 \text{ MPa (18,000 ksi)}; \quad a = 76.2 \text{ mm (3.0 in.)};$$
$$h = 2.54 \text{ mm (0.10 in.)}; \quad B = 25.4 \text{ mm (1.0 in.)}.$$

8.7 One of the problems with testing brittle materials is that crack growth tends to be unstable in conventional test specimens and test machines. Consider, for example, a single-edge-notched bend specimen loaded in a three-point bending. The influence of the test machine can be represented by a spring in series, as Figure 13.4 illustrates. Show that the stress-intensity factor for this specimen can be expressed as a function of crosshead displacement and compliance as follows:

$$K_I = \frac{\Delta_t f(a/W)}{(C + C_m) B \sqrt{W}}$$

where
  Δ_t = crosshead displacement
  C = specimen compliance
  C_m = machine compliance

$f(a/W)$ is defined in Table 12.2. Construct a nondimensional plot of K_I vs. crack size for a fixed crosshead displacement and a/W ranging from 0.25 to 0.75. Develop a family

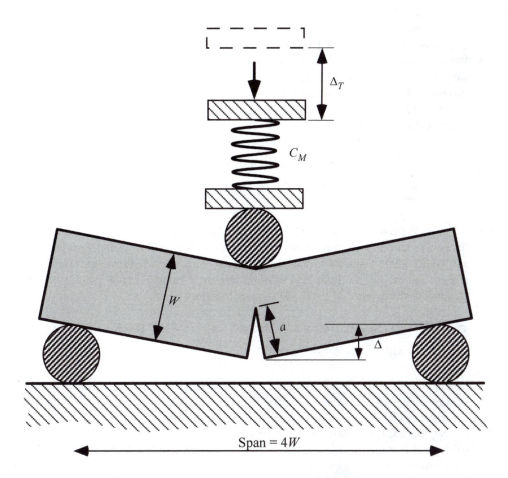

FIGURE 13.4 Single-edge-notched bend specimen loaded in crosshead control (Problem 8.7). The effect of machine compliance is schematically represented by a spring in series.

of these curves for a range of machine compliance. (You will have to express C_m in an appropriate nondimensional form.) What is the effect of machine compliance on the relative stability of the specimen? At what machine compliance would a growing crack experience a relatively constant K_I between $a/W = 0.5$ and $a/W = 0.6$?

13.9 CHAPTER 9

9.1 Develop a computer program or spreadsheet to calculate stress-intensity factors for semielliptical surface cracks in flat plates subject to membrane and bending stress where $a/c \leq 1$ (Table A9.1). Tabulate and plot the geometry factors F and H as a function of a/t and a/c, where $\phi = 90°$ and $c \ll W$.

9.2 Beginning with the Newman-Raju K_I expression for part-through flaws (Equation (9.2)), derive expressions for influence coefficients for uniform and linear loading (G_o and G_1) in terms of the geometry factors F and H.

9.3 For a semielliptical surface flaw in a flat plate where $a/c \leq 1$ and $c \ll W$, tabulate and plot the quadratic influence coefficient G_2 as a function of a/t and a/c, where $\phi = 90°$. *Note: You will need to have worked Problem 9.1 and Problem 9.2 first.*

9.4 A structure contains a through-thickness crack 20 mm long. Strain gages indicate an applied normal strain of 0.0042 when the structure is loaded to its design limit.

The structure is made of a steel with $\varepsilon_y = 0.0020$ and $\delta_{crit} = 0.15$ mm. Is this structure safe, according to the CTOD design curve?

9.5 Write a computer program to compute K_I for a surface crack at $\phi = 90°$ in a flat plate subject to an arbitrary through-thickness normal stress using the weight function method (Section 9.1.3). Assume that $a/c \leq 1$ and $c \ll W$. *Note: You will need to have worked Problem 9.1 and Problem 9.2 first.* Compute and plot K_I vs. crack depth for $a/c = 0.2$, 0.4, 0.6, 0.8, 1, for the following stress distribution:

$$\sigma(x) = 150 \text{ MPa } \exp\left[-5\left(\frac{x}{t}\right)\right]$$

$t = 25.4$ mm (1 in.).

9.6 A flat plate 1.0 m (39.4 in.) wide and 50 mm (2.0 in.) thick which contains a through-thickness crack is loaded in uniaxial tension to 0.75 σ_{YS}. Plot K_r and S_r values on a strip-yield failure assessment diagram for various flaw sizes. Estimate the critical flaw size for failure. For the limit load solution, use the P_o expression in Table A9.11. Set σ_o equal to the average of yield and tensile strength.

$\sigma_{YS} = 345$ MPa (50 ksi); $\quad \sigma_{TS} = 448$ MPa (65 ksi); $\quad E = 207{,}000$ MPa (30,000 ksi);

$K_{mat} = 110$ MPa $\sqrt{\text{m}}$ (100 ksi $\sqrt{\text{in.}}$).

9.7 For the plate in Example 9.1, plot the J results in terms of a failure assessment diagram.
   (a) Compare the FAD curve determined by normalizing the x axis with P/P_o to the FAD curve that is normalized by σ_{ref}/σ_{YS}, where reference stress is defined in Section 9.4.4. Neglect the Irwin plastic zone correction.
   (b) Repeat Part (a), but include the Irwin plastic zone estimate in the first term of the J estimation.

9.8 Suppose the edge-cracked plate in Example 9.1 is subject to a 5 MN tensile load.
   (a) Calculate the applied J integral, both with and without the Irwin plastic zone correction.
   (b) Calculate the load-line displacement over a 5 m gage length.
   (c) Calculate the load-line displacement over a 50 mm gage length.

9.9 For the plate in the previous problem, estimate the following:
   (a) dJ/da for fixed load (5 MN)
   (b) dJ/da for fixed displacement at 5 m gage length. ($P = 5$ MN when $a = 225$ mm.)
   (c) dJ/da for fixed displacement at 50 mm gage length. ($P = 5$ MN when $a = 225$ mm.)

9.10 When a single-edge-notched bend specimen is loaded in the fully plastic range, the deformation can be described by a simple hinge model (Figure 13.3). The plastic rotational factor can be estimated from the load-line displacement and crack-mouth-opening displacement as follows:

$$r_p = \frac{1}{W-a}\left(\frac{WV_p}{\Delta_p} - a\right)$$

assuming a small angle of rotation. Beginning with Equation (9.30) and Equation (9.31) solve for r_p in terms of h_2, h_3, and specimen dimensions. Use the resulting expression to compute r_p for $n = 10$ and $a/W = 0.250$, 0.375, 0.500, 0.625, and 0.750. Repeat for $n = 3$ and the same a/W values. Assume plane strain for all calculations. How do the r_p values estimated from the EPRI handbook compare with the assumed value of 0.44 in ASTM E 1290 and E 1820?

13.10 CHAPTER 10

10.1 Using the Paris equation for fatigue crack propagation, calculate the number of fatigue cycles corresponding to the combinations of initial and final crack radius for a semi-circular surface flaw tabulated in the given table. Assume that the crack radius is small compared to the cross section of the structure.

$$\frac{da}{dN} = 6.87 \times 10^{-12} \, (\Delta K)^3, \text{ where } da/dN \text{ is in m/cycle and}$$

$$\Delta K \text{ is in MPa } \sqrt{m}. \text{ Also, } \Delta\sigma = 200 \text{ MPa.}$$

Initial Crack Radius (mm)	Final Crack Radius (mm)
1	10
1	20
2	10
2	20

1 .1 MPa$\sqrt{in.}$ = 1 ksi$\sqrt{in.}$, 25.4 mm = 1 in.; 1 MPa = 0.145 ksi

Discuss the relative sensitivity of N_{tot} to

- initial crack size
- final crack size

10.2 A structural component made from a high-strength steel is subject to cyclic loading, with $\sigma_{max} = 210$ MPa and $\sigma_{min} = 70$ MPa. This component experiences 100 stress cycles per day. Prior to going into service, the component was inspected by nondestructive evaluation (NDE), and no flaws were found. The material has the following properties: $\sigma_{YS} = 1000$ MPa, $K_{Ic} = 25$ MPa$\sqrt{m}$. The fatigue crack growth rate in this material is the same as in Problem 10.1.
   (a) The NDE technique can find flaws ≥ 2 mm deep. Estimate the maximum safe design life of this component, assuming that subsequent in-service inspections will not be performed. Assume that any flaws that may be present are semicircular surface cracks and that they are small relative to the cross section of the component.
   (b) Repeat part (a), assuming an NDE detectability limit of 10 mm.

10.3 Fatigue tests are performed on two samples of an alloy for aerospace applications. In the first experiment, $R = 0$, while $R = 0.8$ in the second experiment. Sketch the expected trends in the data for the two experiments on a schematic log(da/dN) vs. log(ΔK) plot. Assume that the experiments cover a wide range of ΔK values. Briefly explain the trends in the curves.

10.4 Develop a program or spreadsheet to compute fatigue crack growth behavior in a compact specimen, assuming the fatigue crack growth is governed by the Paris-Erdogan equation.
   Consider a 1T compact specimen (see Section 7.1.1) that is loaded cyclically at a constant load amplitude with $P_{max} = 18$ kN and $P_{min} = 5$ kN. Using the fatigue crack growth data in Problem 10.1, calculate the number of cycles required to grow the crack from $a/W = 0.35$ to $a/W = 0.60$. Plot crack size vs. cumulative cycles for this range of a/W.

10.5 Develop a program or spreadsheet to compute the fatigue crack growth behavior in a flat plate that contains a semielliptical surface flaw with $a/c \leq 1$ that is subject to a cyclic membrane (tensile) stress (Table A9.1). Assume that the flaw remains semielliptical, but take account of the difference in K at $\phi = 0°$ and $\phi = 90°$. That is, compute dc/dN and da/dN at $\phi = 0°$ and $\phi = 90°$, respectively, and advance the crack dimensions based on the relative growth rates. Assume that $c \ll W$, but that a/t is finite. Use the Paris equation to compute the crack growth rate.

Consider a 25.4 mm (1.0 in.) thick plate that is loaded cyclically at a constant stress amplitude of 200 MPa (29 ksi). Given an initial flaw with $a/t = 0.1$ and $a/2c = 0.1$, calculate the number of cycles required to grow the crack to $a/t = 0.8$, using the fatigue crack growth data in Problem 10.1. Construct a contour plot that shows the crack size and shape at $a/t = 0.1, 0.2, 0.4, 0.6,$ and 0.8. What happens to the $a/2c$ ratio as the crack grows?

10.6 You have been asked to perform K-decreasing tests on a material to determine the near-threshold behavior at $R = 0.1$. Your laboratory has a computer-controlled test machine that can be programmed to vary P_{max} and P_{min} on a cycle-by-cycle basis.
(a) Compute and plot P_{max} and P_{min} vs. crack length for the range $0.5 \le a/W \le 0.75$ corresponding to a normalized K gradient of -0.07 mm^{-1} in a 1T compact specimen.
(b) Suppose that the material exhibits the following crack growth behavior near the threshold:

$$\frac{da}{dN} = 4.63 \times 10^{-12}\left(\Delta K^3 - \Delta K_{th}^3\right)$$

where da/dN is in m/cycle and ΔK is in MPa$\sqrt{m}$. For $R = 0.1$, $\Delta K_{th} = 8.50$ MPa$\sqrt{m}$. When the test begins, $a/W = 0.520$ and $da/dN = 1.73 \times 10^{-8}$ m/cycle. As the test continues in accordance with the loading history determined in part (a), the crack growth rate decreases. You stop the test when da/dN reaches 10^{-10} m/cycle. Calculate the following:
(i) The number of cycles required to complete the test
(ii) The final crack length
(iii) The final ΔK

13.11 CHAPTER 11

11.1 You have been asked to review the design of an offshore riser, which is a long vertical pipe that transmits crude oil from below the ocean floor to an offshore platform on the surface. This riser is made up of a series of shorter segments that are bolted together at flange connections. The nuts and bolts are made from a high-strength alloy steel with a yield strength of 1100 MPa (160 ksi). They are to be tightened to a very high torque, such that the stresses in the nuts and bolts are close to yield. The nuts and bolts will not be exposed to the crude oil, but will be immersed in a seawater environment. In order to mitigate corrosion, the riser will be fitted with sacrificial anodes. What concerns, if any, would you have about this design?

11.2 An environmental cracking experiment was performed on a bolt-loaded compact specimen. At the start of the test, the applied K_I was 20 MPa$\sqrt{m}$ and the crack growth rate was 5.00×10^{-6} m/sec and at the end of the test, K_I was 10 MPa$\sqrt{m}$ and da/dt was 1.00×10^{-10} m/sec. A 1T specimen was used, with $W = 50.8$ mm (2.0 in.). At the start of the test, $a/W = 0.45$. The table below gives the computed crack growth rate vs. applied K_I.
(a) Compute the crack-opening displacement, V required to achieve $K_I = 20$ MPa$\sqrt{m}$ at the start of the test.
(b) Assuming V remains constant throughout the test, compute the final a/W at the conclusion of the test.
(c) Given the crack growth data tabulated in the given table, compute and plot crack size vs. time. How long did it take to run this test?

K_I(MPa$\sqrt{m}$)	da/dt(m/sec)	K_I(MPa$\sqrt{m}$)	da/dt(m/sec)
20.00	5.000E-06	13.86	2.669E-06
19.28	5.000E-06	13.65	2.267E-06
18.63	5.000E-06	13.44	1.857E-06
18.04	5.000E-06	13.22	1.439E-06
17.49	5.000E-06	13.00	1.010E-06
16.99	5.000E-06	12.78	5.040E-07
16.52	5.000E-06	12.54	2.466E-07
16.08	5.000E-06	12.30	1.180E-07
15.66	5.000E-06	12.06	5.513E-08
15.25	5.000E-06	11.80	2.510E-08
15.05	5.000E-06	11.53	1.111E-08
14.86	4.614E-06	11.26	4.775E-09
14.66	4.230E-06	10.97	1.988E-09
14.46	3.844E-06	10.68	7.998E-10
14.26	3.456E-06	10.37	3.104E-10
14.06	3.065E-06	10.00	1.000E-10

13.12 CHAPTER 12

12.1 A series of finite element meshes have been generated that model compact specimens with various crack lengths. Plane stress linear elastic analyses have been performed on these models. Nondimensional compliance values as a function of a/W are tabulated in the given table. Estimate the nondimensional stress intensity for the compact specimen from these data and compare your estimates to the polynomial solution in Table 13.2.

$\dfrac{a}{W}$	$\dfrac{\Delta BE}{P}$	$\dfrac{a}{W}$	$\dfrac{\Delta BE}{P}$	$\dfrac{a}{W}$	$\dfrac{\Delta BE}{P}$
0.20	8.61	0.45	29.0	0.70	123
0.25	11.2	0.50	37.0	0.75	186
0.30	14.3	0.55	47.9	0.80	306
0.35	18.1	0.60	63.3	0.85	577
0.40	22.9	0.65	86.3	0.90	1390

12.2 A two-dimensional plane strain finite element analysis is performed on a through crack in a wide plate (Figure 2.3). The remote stress is 100 MPa, and the half-crack length = 25 mm. The stress normal to the crack plane (σ_{yy}) at $\theta = 0$ is determined at node points near the crack tip and is tabulated in the given table. Estimate K_I by means of the stress-matching approach (Equation (12.14)) and compare your estimate to the exact solution for this geometry. Is the mesh refinement sufficient to obtain an accurate solution in this case?

$\dfrac{r}{a}(q=0)$	$\dfrac{\sigma_{yy}}{\sigma^{\infty}}$	$\dfrac{r}{a}(q=0)$	$\dfrac{\sigma_{yy}}{\sigma^{\infty}}$
0.005	11.0	0.080	3.50
0.010	8.07	0.100	3.24
0.020	6.00	0.150	2.83
0.040	4.54	0.200	2.58
0.060	3.89	0.250	2.41

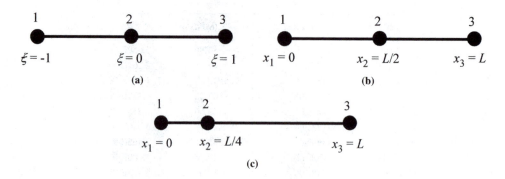

FIGURE 13.5 One-dimensional element with three nodes (Problem 12.4): (a) parametric coordinates, (b) global coordinates, Case (1), and (c) global coordinates, Case (2).

12.3 Displacements at nodes along the upper crack face (u_y at $\theta = \pi$) in the previous problem are tabulated in the given table. The elastic constants are as follows: $E = 208{,}000$ MPa and $v = 0.3$. Estimate K_I by means of the (plane strain) displacement-matching approach (Equation (12.15a)) and compare your estimate to the exact solution for this geometry. Is the mesh refinement sufficient to obtain an accurate solution in this case?

$\dfrac{r}{a}(\theta = \pi)$	$\dfrac{u_y}{a}$	$\dfrac{r}{a}(\theta = \pi)$	$\dfrac{u_y}{a}$
0.005	9.99×10^{-5}	0.080	3.92×10^{-4}
0.010	1.41×10^{-4}	0.100	4.36×10^{-4}
0.020	1.99×10^{-4}	0.150	5.27×10^{-4}
0.040	2.80×10^{-4}	0.200	6.00×10^{-4}
0.060	3.41×10^{-4}	0.250	6.61×10^{-4}

12.4 Figure 13.5 illustrates a one-dimensional element with three nodes. Consider two cases: (1) Node 2 at $x = 0.50L$ and (2) Node 2 at $x = 0.25L$.
   (a) Determine the relationship between the global and parametric coordinates, $x(\xi)$, in each case.
   (b) Compute the axial strain $\varepsilon(\xi)$ for each case in terms of the nodal displacements and parametric coordinate.
   (c) Show that $x_2 = 0.25L$ leads to a $1/\sqrt{x}$ singularity in the axial strain.

Index

A

Advanced ceramics, 291
Airy stress function, 559
American Petroleum Institute, 431
American Society for Testing and Materials (ASTM), 299,
    362, 495, 546
  Committee E08 on fatigue and fracture, 315
  E 399 standard, 309, 312, 329
  E 561 standard, 319
Amorphous polymers, 259
Anode, 511
Application to structures, 385–450
  CTOD design curve, 395–397
  elastic-plastic J-integral analysis, 397–409
    ductile instability analysis, 405–408
    EPRI J-estimation procedure, 398–403
    practical considerations, 408–409
    reference stress approach, 403–405
  failure assessment diagrams, 410–432
    application to welded structures, 423–428
    approximations of FAD curve, 415–416
    ductile-tearing analysis with FAD, 430
    estimating reference stress, 416–423
    J-based FAD, 412–415
    original concept, 410–412
    primary vs. secondary stresses in FAD method,
      428–429
    standardized FAD-based procedures, 430–432
  linear elastic fracture mechanics, 385–395
    influence coefficients for polynomial stress
      distributions, 388–392
    K_I for part-through cracks, 387–388
    primary, secondary, and residual
      stresses, 394–395
    warning about LEFM, 395
    weight functions for arbitrary loading, 392–394
  probabilistic fracture mechanics, 432–433
  stress intensity and fully plastic J solutions for selected
    configurations, 434–448
Applied tearing modulus, 124
ASTM, *see* American Society for Testing
    and Materials
Atomic separation, potential energy and, 26
Atomic spacing, 26
Auger electron spectroscopy, 249

B

BCC materials, *see* Body-centered cubic materials
Bending, edge-cracked plate in, 117, 118
Bend specimens, 380

Betti reciprocal theorem, 557
Biaxiality ratio, 139
Biaxial loading, 84–85
BIE method, *see* Boundary integral equation method
Body-centered cubic (BCC) materials, 234
Bond
  energy, 25
  stiffness, 26
Boundary integral equation (BIE) method, 556
Boundary layer analysis, modified, 138
Bridge indentation, 380, 381, 382
British Standards Institute (BSI), 299, 372, 431
Brittle fracture, 219
BSI, *see* British Standards Institute

C

CAIB, *see* Columbia Accident Investigation Board
Cartesian coordinates, 89
Catastrophic failures, 3
Cathode, 511
Cathodic protection, 511, 515
CEGB, *see* Central Electricity Generating Board
Central Electricity Generating Board
    (CEGB), 412
Ceramics, 282
  advanced, 291
  crucial property for, 378
  ductile phase toughening, 287
  fiber and whisker toughening, 288
  high-toughness, 291
  microcrack toughening, 285
  transformation toughening, 286
Chain scission, 265
Challenger Space Shuttle explosion, 5
Charpy and Izod testing, 341
Charpy-testing machine, 329
Chevron-notched specimens, 378
C^* integral, 191
Cleavage, 234–246
  crack(s)
    arrested, 240
    propagation, 249
  effective driving force for, 149
  events, unsuccessful, 240
  fractography, 234–235
  initiation, probability tree for, 244, 252
  mathematical models of cleavage fracture toughness,
    238–246
  mechanisms of cleavage initiation, 235–238
  propagation, 248, 254
  trigger, 248

Cleavage fracture, 78
  scaling model for
    application of model, 148
    failure criterion, 145
    three-dimensional effects, 147
  statistical modeling of, 250
    conditional probability of propagation, 252–254
    weakest link fracture, 250–252
  toughness data, 246
Clip gage, 333, 460, 499
Closure
  model, threshold, 465
  plasticity-induced, 469, 472
  stress, 64
CMOD, *see* Crack-mouth-opening displacement
Cohesive stress, 26
Cohesive zone model, 291
Collapse axis, 428
Columbia Accident Investigation Board (CAIB), 6
Compliance offset, 499
Composite materials, 270, 373
Compression after impact test, 278
Compressive failure, 275
Computational fracture mechanics, 553–591
  analysis of growing cracks, 585–586
  energy domain integral, 563–570
    finite element implementation, 568–570
    generalization to three dimensions, 566–568
    theoretical background, 563–566
  linear elastic convergence study, 577–585
  mesh design, 570–577
  numerical methods, 553–558
    boundary integral equation method,
      556–558
    finite element method, 554–556
  properties of singularity elements, 587–590
    quadrilateral element, 587–589
    triangular element, 589–590
  traditional methods, 558–563
    contour integration, 560
    elemental crack advance, 559–560
    stress and displacement matching, 558–559
    virtual crack extension (continuum approach),
      561–563
    virtual crack extension (stiffness derivative
      formulation), 560–561
Computer technology, 12
Concrete
  cohesive zone model, 291
  crack growth in, 292
  fictitious crack model, 291, 293
  LEFM, 291
  stress-displacement curve, 291
Conservation integrals, 11
Constant displacement rate fracture test, 357
Continuum theory, 219
Coplanar cracks, 86
Correspondence principle, 200
Corrosion
  crevice, 517
  fatigue

    behavior, 538
    data, 541
    effect of corrosion product wedging, 544
    mechanisms, 543
  principles
    cathodic protection, 515
    corrosion current and polarization, 514
    electrochemical reactions, 511
    electrode potential and passivity, 514
    types of corrosion, 516
  product wedging, 529
Crack(s)
  area, surface area and, 30
  arrest(s), 186, 549
    dynamic fracture and, 173
    rapid crack propagation, 178
    rapid loading of stationary crack, 174
    toughness, 245, 329
  blunting, 103
  closure, 457, 460
    crack wedging mechanisms, 460
    effects of loading variables, 463
    model, 465
  coplanar, 86
  creep, 189
    behavior, 193
    steady-state, 191
    zones, 190
  driving force, 35
  edge, 47
  extension
    CTOD for, 356
    energy criterion and, 12
    force, 35
  fatigue, 494
  formation, 29
  Griffith, 237
  growth, 29
    concrete, 292
    high-triaxiality, 233
    instability analysis, 91
    rate, 496, 518, 218
    stable, 39, 167
    stages of, 132
    steady-state, 132–133, 166, 168
    stick/slip, 268
    strain-energy change and, 9
    time-dependent, 15
    unstable, 39
  –growth resistance curves, 368
    computing J for growing crack, 126
    stable and unstable crack growth, 124
  infinitely sharp, 28
  initiation, 202
  length, intrinsic, 491
  macroscopic, 32
  Mode I, 83
  –mouth-opening displacement (CMOD), 306, 460
  numerical simulation of, 29
  –opening displacement, 60
  parallel, 86

part-through, 387
propagation, rapid, 178
    crack arrest, 186
    crack speed, 180
    dynamic toughness, 184
    elastodynamic crack-tip parameters, 182
radius, 31, 48
rate, 15
short, 488
size, 15
speed, 180, 183
spider-web configuration, 585
stationary, 128, 174
stress analysis of, 42
    effect of finite size, 48
    relationship between K and global behavior, 45
    stress intensity factor, 43
    superposition, 54
    weight functions, 56
through-thickness, 30
−wedging mechanisms, 460
Cracking
behavior
    CMOD, fixed, 531
    variables affecting cracking behavior, 531
branched, 523
environmental, 516
mechanisms, classification of, 516
morphology, 523
stress corrosion, 517, 525
Crack tip
behavior, 267
blunted, 113
constraint, large-scale yielding, 133
    elastic T stress, 137
    J-Q theory, 140
    limitations of two-parameter fracture mechanics, 149
    scaling model for cleavage fracture, 145
−opening angle (CTOA), 164
parameters, elastodynamic, 182
plasticity, 61
    comparison of plastic zone corrections, 66
    effective compliance and, 63
    Irwin approach, 61
    strip-yield model, 64
radius, 304
reverse plasticity at, 475
shielding, 286
singularity, 45, 46, 95
stress(es), 251
    fields, 71
    intensity factor, 60
stress analysis
    in-plane loading, 92
    Westergaard stress function, 95
triaxiality, 73
Crack-tip-opening displacement (CTOD), 11, 103, 231
alternate means for analyzing, 104
approximated, 354
compressive underload scenario, 482
crack extension, 356

design curve, 11, 395
determination of from strip-yield model, 153
elastic-plastic regime and, 385
failure criterion and, 203
fatigue and, 453
linear viscoelastic material, 355
relationships between J and, 120
testing, 326
Creep
behavior, 193
compliance, 204
crack growth, 189
formation, time-dependent, 403–404
primary, 196
steady-state, 191
Crevice corrosion, 517
Critical stress criterion, 31
Cross-linked polymer, 259
Crystalline polymers, 259
CTOA, see Crack-tip-opening angle
CTOD, see Crack-tip-opening displacement
Cup and cone fracture surface, 223, 227
Cyclic loading, application of J contour
            integral to, 504

D

Damage tolerance, 15, 16, 501
DCB specimen, see Double cantilever
            beam specimen
Decohesion stress, 219
Deformation energy, 175
Deformation J, 212
Deformation plasticity
    application of to crack problems, 168
    theory
        incremental theory vs., 169
        validity of, 170
Degree of polymerization, 258
Delamination, 272, 377
Delayed retardation, 480
DENT, see Double-edged notched tension
Difference field, 140
Dimensional analysis
    Buckingham Π-theorem, 18
    fracture mechanics, 19
Dislocation
    model, Goods and Brown, 221
    −particle interactions, 219
Displacement control
    load control vs., 40
    pure, 41
Double cantilever beam (DCB) specimen,
            37, 41, 186, 373
    initial flaw in, 374
    loading fixtures for, 375
    mixed mode loading in, 376
Double-edged notched tension (DENT), 133
Driving force curve, 38
Drop weight test, 342

Ductile-brittle transition, 178, 247–248
  cleavage crack propagation in, 249
  testing of steels in, 338
Ductile crack growth, 231, 232
Ductile failure, void interaction leading to, 231
Ductile instability analysis, 405
Ductile materials, uniaxial tensile deformation of, 220
Ductile phase toughening, 287
Ductile tearing
  crack extension and, 313
  unstable fracture and, 328
Dugdale-Barenblatt strip-yield model, 284, 291
Dynamic contour integrals, 188
Dynamic fracture analysis
  derivation of generalized energy release rate, 209
  ductile crack growth, 231
  elastodynamic crack tip fields, 206
  void growth and coalescence, 222
Dynamic propagation toughness, 184
Dynamic tear tests, 344
Dynamic and time-dependent fracture, 173–215
  creep crack growth, 189–206
    $C*$ integral, 191–193
    linear viscoelasticity, 197–200
    short-time vs. long-time behavior, 193–197
    transition from linear to nonlinear behavior, 204–206
    viscoelastic J integral, 200–203
  dynamic fracture analysis, 206–212
    derivation of generalized energy release rate, 209–212
    elastodynamic crack tip fields, 206–209
  dynamic fracture and crack arrest, 173–189
    dynamic contour integrals, 188–189
    rapid crack propagation and arrest, 178–187
    rapid loading of stationary crack, 174–178

E

EAC, *see* Environmentally assisted cracking
Edge crack, 47, 233
Edge-cracked plates, 20, 117, 118
Edge-cracked specimens, 104
Effective compliance, 63
Effective crack length, definition of, 62
Effective flaw shape parameter, 63
Elastic-plastic fracture mechanics, 103–172
  crack-growth resistance curves, 123–128
    computing J for growing crack, 126–128
    stable and unstable crack growth, 124–126
  crack-tip constraint under large-scale yielding,
      133–153
    elastic T stress, 137–140
    J-Q theory, 140–145
    limitations of two-parameter fracture mechanics,
        149–153
    scaling model for cleavage fracture, 145–149
  crack-tip-opening displacement, 103–107
  J contour integral, 107–119
    J as path-independent line integral, 110–111
    J as stress intensity parameter, 111–112
    laboratory measurement of J, 114–119

    large strain zone, 113–114
    nonlinear energy release rate, 108–110
  J-controlled fracture, 128–133
    J-controlled crack growth, 131–133
    stationary cracks, 128–131
  mathematical foundations of elastic-plastic fracture
      mechanics, 153–170
    analysis of stable crack growth in small-scale yielding,
        162–168
    determining CTOD from strip-yield model, 153–156
    HRR singularity, 159–162
    J contour integral, 156–158
    J as nonlinear elastic energy release rate, 158–159
    notes on applicability of deformation plasticity to
        crack problems, 168–170
  relationship between J and CTOD, 120–123
Elastic-plastic regime, CTOD and, 385
Elastodynamic fracture mechanics, 173
Elastomers, 259
Electric Power Research Institute (EPRI),
      11, 397, 398, 404
Electrolyte, 511
Electron spin resonance (ESR), 265
Elements, 554
Elliptical hole, 28
Elliptical integral, 100
End-notched flexure (ENF) specimen, 374
Energy
  balance
    global, 31
    Griffith, 29
    time integral and, 203
  bond, 25
  criterion, fracture, 12–13
  dissipation, 33
  kinetic, 174, 189
  plastic, 321
  potential, 108
  ratio of kinetic to deformation, 175
  release rate, 13, 34
    definition of, 188
    generalized, 209
    nonlinear, 108, 109
    pseudo-, 212
  strain, increase in, 37
ENF specimen, *see* End-notched flexure specimen
Engineering plastics, 257
  fiber-reinforced plastics, 270
  fracture toughness measurements in, 353
  structure and properties of polymers, 258
  yielding and fracture in polymers, 265
Environmental cracking, 249
Environmentally assisted cracking (EAC), 511, 516
Environmentally assisted cracking in metals, 511–552
  corrosion fatigue, 538–545
    effect of corrosion product wedging on fatigue,
        544–545
    mechanisms, 543–544
    time-dependent and cycle-dependent behavior,
        538–541
    typical data, 541–543

corrosion principles, 511–516
  cathodic protection, 515–516
  corrosion current and polarization, 514
  electrochemical reactions, 511–514
  electrode potential and passivity, 514–515
  types of corrosion, 516
environmental cracking overview, 516–525
  crack growth rate vs. applied stress intensity, 518
  cracking morphology, 523
  life prediction, 523–525
  occluded chemistry of cracks, pits, and crevices,
       517–518
  small crack effects, 521–523
  static, cyclic, and fluctuating loads, 523
  terminology and classification of cracking
       mechanisms, 516–517
  threshold for EAC, 520–521
experimental methods, 545–551
  fracture mechanics test methods, 547–551
  tests on smooth specimens, 546
hydrogen embrittlement, 529–537
  cracking mechanisms, 530–531
  variables affecting cracking behavior, 531–537
stress corrosion cracking, 525–529
  corrosion product wedging, 529
  crack growth rate in Stage II, 528
  film rupture model, 527–528
  metallurgical variables influencing SCC, 528–529
EPRI, *see* Electric Power Research Institute
Equilibrium equations, 91
ESIS, *see* European Structural Integrity Society
ESR, *see* Electron spin resonance
European Structural Integrity Society (ESIS), 362, 365

F

Face-centered cubic (FCC) metals, 234
FAD, *see* Failure assessment diagram
Failure
  compressive, 275
  crack-tip radius at, 304
  ductile, 231
  hazard function for total, 253
  mechanisms, 17, 271
  static overload, 426
  thin sections, 72
Failure assessment diagram (FAD), 66, 410
  analysis, Monte Carlo, 433
  –based procedures, standardized, 430
  creation, 413
  curve approximation, 415
  ductile-tearing analysis with, 430
  J-based, 412
  primary vs. secondary stresses in, 428
Fatigue
  cyclic load displacement behavior, 505
  damage, 280
  micromechanisms of, 491
  precracking, 302, 337
  threshold, 464
    closure model, 465

threshold behavior in inert environments, 470
two-criterion model, 466
Fatigue crack growth
  crack closure during, 455
  experiments, 495
    closure measurements, 498
    crack growth rate and threshold measurement, 496
    proposed experimental definition of ΔK_{eff}, 500
  rate, 494
Fatigue crack propagation, 25, 451–509
  application of *J* contour integral to cyclic loading,
       504–507
    definition of ΔJ, 504–506
    path independence of ΔJ, 506
    small-scale yielding limit, 507
  crack closure, 457–464
    crack-wedging mechanisms, 460–462
    effects of loading variables on closure, 463–464
  damage tolerance methodology, 501–503
  empirical fatigue crack growth equations, 453–457
  fatigue crack growth experiments, 495–501
    closure measurements, 498–499
    crack growth rate and threshold measurement,
         496–498
    proposed experimental definition of ΔK_{eff}, 500–501
  fatigue threshold, 464–473
    closure model for threshold, 465–466
    threshold behavior in inert environments, 470–473
    two-criterion model, 466–469
  growth of short cracks, 488–491
    mechanically short cracks, 491
    microstructurally short cracks, 491
  micromechanisms of fatigue, 491–495
    fatigue at high ΔK values, 495
    fatigue in Region II, 491–494
    micromechanisms near threshold, 494–495
  similitude in fatigue, 451–453
  variable amplitude loading and retardation, 473–488
    effect of overloads and underloads, 478–483
    linear damage model, 474–475
    models for retardation and variable amplitude fatigue,
         484–488
    reverse plasticity at crack tip, 475–478
FCC metals, *see* Face-centered cubic metals
Fiber
  buckling, modeling of, 276
  –reinforced plastics, 270
  toughening, 288
  waviness, 277
Fibrils, 266
Fictitious crack model, 291, 293
Film rupture model, 527, 544
Finite compliance, 41, 42
Finite element method, 554
Finite element stiffness matrix, 558
First law of thermodynamics, 29
Flat plate, elliptical hole in, 28
Flaw(s)
  shape parameter, 101
  size, 12, 15
  stress concentration effect of, 27

Focused meshes, 585
Force-displacement curve, 60
Fractography, 234, 235
Fracture
    anatomic view, 25
    axis, 428
    behavior, 17
    brittle, 219
    cleavage, 78, 145
    design of structure to avoid, 6
    driving force for, 385
    dynamic, 173
    energy criterion, 12–13
    intergranular, 249
    J-controlled, 128, 357
    K-controlled, 69, 354
    material resistance to, 13
    mixed-mode, 80
    nucleation of, 32
    plane strain, 72
    plane stress, 72
    process zone, 77
    research, early, 8
    shear, 223, 344
    strain-controlled, 144
    surface
        cup and cone, 223, 227
        striations on, 493
    testing
        rapid loading in, 329
        weldments, 334
    toughness, 13–14, 35, 69
        cleavage, 238
        distribution, prediction of, 243
        ferritic steels, 247
        plane strain, 75
        specimen size and, 134
        thickness effects on, 75
    –triggering particle, 241
    unstable, 324
    weakest link, 250
Fracture mechanics
    approach to design, 12–15
        energy criterion, 12–14
        stress-intensity approach, 14–15
        time-dependent-crack growth and damage
            tolerance, 15
    dimensional analysis, 18–21
        Buckingham Π-theorem, 18–19
        dimensional analysis in fracture
            mechanics, 19–21
    effect of material properties on fracture, 16–17
    historical perspective, 6–12
        early fracture research, 8–9
        fracture mechanics from 1960 to 1980, 10–12
        fracture mechanics from 1980 to present, 12
        Liberty ships, 9
        post-war fracture mechanics research, 10
    linear elastic, 88
    probabilistic, 432
    research, post-war, 10

    simplified family tree of, 16
    theory, single-parameter, 142
    triangle, 385
    viscoelastic, 196
Free radicals, 258
Fundamental dimensional units, 18

G

Galvanic coupling, 511
Generalized damage integral, 166
Generalized J integral, 201
Geometry correction factor, 62
Goods and Brown dislocation model, 221
Grain size refinement, 238
Griffith crack, 237
Griffith energy balance, 29, 145, 181
Griffith energy criterion, 32
Griffith equation, modified, 32
Griffith plate, 577
GT model, *see* Gurson-Tvergaard model
Gurson model, 223
Gurson-Tvergaard (GT) model, 229

H

HAZ, *see* Heat-affected zone
Hazard function, 252
HCP metals, *see* Hexagonal close-packed metals
HEAC, *see* Hydrogen-environment-assisted cracking
Heat-affected zone (HAZ), 335, 337, 423
Hereditary integrals, 198
Hexagonal close-packed (HCP) metals, 234
High-toughness ceramics, 257
Hillerborg model, 291
Hoff's analogy, 191, 200
Hooke's law, 19
Hoover Dam, 9
HRR singularity, 112, 113, 130, 159
Hydrogen
    amount of available, 535
    embrittlement, 529, 544
        cracking mechanisms, 530
        variables affecting cracking behavior, 531
    –environment-assisted cracking (HEAC), 530
    trapping, 531

I

IHAC, *see* Internal-hydrogen-assisted cracking
Impact strength, 372
Incompressibility condition, 225
Industrial Revolution, choice of building materials
    prior to, 7
Infinitely sharp crack, 28
Influence coefficient, 388
Inglis stress analysis, 31
Initiation toughness, 39
In-plane loading, 92

Intergranular fracture, 249
Internal-hydrogen-assisted cracking (IHAC), 530
International Institute of Standards (ISO), 299
Intrinsic crack length, 491
Iron
   mass production of, 7
   oxidized, 512
Irwin plastic zone adjustment, 62, 63
Isaac Newton, 6
ISO, *see* International Institute of Standards
Isotropic hardening, 168

J

Jacobian matrix, 555
Japan Society of Mechanical Engineers (JSME), 299
J-controlled fracture, 128, 357
   crack growth, 131
   stationary cracks, 128
J integral, 11
   analysis, elastic-plastic, 397
   application of to cyclic loading, 504
   deformation, 127, 212
   elastic-plastic analysis to compute, 553
   energy release rate definition of, 115
   far-field, 167
   fully plastic, 399
   generalized, 201
   laboratory measurement of, 114
   relationships between CTODS and, 120
   as stress intensity parameter, 111
   viscoelastic fracture mechanics, 200
J-Q theory, 140, 142
JSME, *see* Japan Society of Mechanical Engineers

K

K-decreasing test, 497
K-dominated region, 128
Kinetic energy
   deformation energy and, 175
   density, 189
   dissipation of, 174
Kings College Chapel, 7, 8
Kink bands, 276, 277
Kronecker delta, 557

L

Large-scale yielding, crack-tip constraint under, 133
LEFM, *see* Linear elastic fracture mechanics
Leonardo da Vinci, 8, 27
Liberty ships, 3, 9
Linear damage model, 473, 474
Linear elastic fracture mechanics (LEFM), 11, 17, 25–102,
      354
   analyses based on, 385
   atomic view of fracture, 25–27
   concrete and rock, 291

crack-tip plasticity, 61–69
   comparison of plastic zone corrections, 66
   Irwin approach, 61–63
   plastic zone shape, 66–69
   strip-yield model, 64–65
dynamic version of, 173
energy release rate, 34–37
Griffith energy balance, 29–33
   comparison with critical stress criterion,
      31–32
   modified Griffith equation, 32–33
instability and *R* curve, 38–42
   load control vs. displacement control,
      40–41
   reasons for *R* curve shape, 39–40
   structures with finite compliance, 41–42
interaction of multiple cracks, 86–88
   coplanar cracks, 86
   parallel cracks, 86–88
K-controlled fracture, 69–71
mathematical foundations of linear elastic fracture
      mechanics, 88–101
   crack growth instability analysis, 91–92
   crack-tip stress analysis, 92–100
   elliptical integral of second kind, 100–101
   plane elasticity, 88–91
mixed-mode fracture, 80–85
   biaxial loading, 84–85
   equivalent Mode I crack, 83–84
   propagation of angled crack, 81–83
plane strain fracture, 72–80
   crack-tip triaxiality, 73–75
   effect of thickness on apparent fracture toughness,
      75–78
   implications for cracks in structures, 79–80
   plastic zone effects, 78–79
relationship between *K* and *G*, 58–61
stress analysis of cracks, 42–58
   effect of finite size, 48–53
   principle of superposition, 54–55
   relationship between *K* and global behavior,
      45–48
   stress intensity factor, 43–45
   weight functions, 56–58
stress concentration effects of flaws, 27–29
theory, assumption of, 279
warning about, 395
Linear polymers, 259
Linear variable differential transformer (LVDT), 305
Linear viscoelasticity, 197
Load
   control, displacement control vs., 40
   controlled structure, 35
   –displacement curve, 363, 500
   –pseudo-displacement curve, 358
   relaxation test, 359
Loading
   types of, 43
   variables, effects of on closure, 463
   weight functions for arbitrary, 392
LVDT, *see* Linear variable differential transformer

M

Material toughness characterization, 11
Maxwell model, 264
Mechanical stress relief, 423
Mers, 258
Metal(s)
  corrosion processes, 513
  crack propagation through, 33
  creep crack growth in, 353
  FCC, 234
  HCP, 234
  J testing of, 320
  micromechanisms of fracture in, 220
Metals, fracture mechanisms in, 219–256
  cleavage, 234–246
    fractography, 234–235
    mathematical models of cleavage fracture toughness, 238–246
    mechanisms of cleavage initiation, 235–238
  ductile-brittle transition, 247–248
  ductile fracture, 219–233
    ductile crack growth, 231–233
    void growth and coalescence, 222–231
    void nucleation, 219–222
  intergranular fracture, 249–250
  statistical modeling of cleavage fracture, 250–254
    incorporating conditional probability of propagation, 252–254
    weakest link fracture, 250–252
Metals, fracture toughness testing of, 299–352
  CTOD testing, 326–329
  dynamic and crack-arrest toughness, 329–334
    K_{Ia} measurements, 330–334
    rapid loading in fracture testing, 329–330
  fracture testing of weldments, 334–338
    fatigue precracking, 337
    notch location and orientation, 335–337
    posttest analysis, 337–338
    specimen design and fabrication, 334–335
  general considerations, 299–308
    fatigue precracking, 303–305
    instrumentation, 305–306
    side grooving, 307–306
    specimen configurations, 299–301
    specimen orientation, 301–303
  J testing of metals, 320–326
    basic test procedure, 320–322
    critical *J* values for unstable fracture, 324–326
    J-R curve testing, 322–324
  K_{Ic} testing, 308–316
    ASTM 399, 309–312
    shortcomings of E 399 and similar standards, 312–316
  K-R curve testing, 316–319
    experimental measurement of *K-R* curves, 318–319
    specimen design, 317
  qualitative toughness tests, 340–344
    Charpy and Izod impact test, 341–342
    drop weight tear and dynamic tear tests, 344
    drop weight test, 342–343
    stress intensity, compliance, and limit load solutions for laboratory specimens, 344–350
    testing and analysis of steels in ductile-brittle transition region, 338–340
Microcrack(s), 238
  arrest, 245
  toughening, 285
Middle tension (MT) geometry, 317
Mixed-mode fracture, 80–85
  biaxial loading, 84–85
  equivalent Mode I crack, 83
  propagation of angled crack, 81–83
Mode I crack, 83
  loading, 170
  propagation, 182
Mode I stress-intensity factor, 353
Model(s)
  cleavage fracture, 145, 238, 250
  closure, 464
  cohesive zone, 291
  crack closure, 465
  fictitious crack, 293
  film rupture, 527, 544
  Goods and Brown dislocation, 221
  Griffith, 31, 33
  Gurson, 223
  Gurson-Tvergaard, 229
  Hillerborg, 291
  linear damage, 473, 474
  Maxwell, 264
  Ramberg-Osgood, 398
  retardation, 484
  RKR, 239
  strip-yield, 64, 104, 121, 153, 284, 291, 487
  two-criterion, 466
  variable amplitude fatigue, 484
  void coalescence, 230
  Wheeler, 485
  Willenborg, 485
Modified boundary layer analysis, 138
Mohr's circle, 46, 67
Monte Carlo FAD analysis, 433
MSV Kurdistan oil tanker, 4
MT geometry, *see* Middle tension geometry
Multiple cracks, interaction of, 86

N

NASA
  management, criticism of, 6
  testing programs, 72
National Institute of Standards and Technology (NIST), 330
Naval Research Laboratory (NRL), 341
NDE, *see* Nondestructive examination
NIST, *see* National Institute of Standards and Technology
Nondestructive examination (NDE), 15, 502, 523
Nonlinear energy release rate, 108, 109
Nonlinear fracture mechanics, 16
Nonlinear viscoelasticity, 205
Nonmetals, fracture mechanisms in, 257–295

ceramics and ceramic composites, 282–290
  ductile phase toughening, 287–288
  fiber and whisker toughening, 288–290
  microcrack toughening, 285–286
  transformation toughening, 286–287
concrete and rock, 291–293
engineering plastics, 257–281
  fiber-reinforced plastics, 270–281
  structure and properties of polymers, 258–264
  yielding and fracture in polymers, 265–270
Nonmetals, fracture testing of, 353–383
  ceramics, 378–382
    bend specimens precracked by bridge indentation,
      380–382
    chevron-notched specimens, 378–380
  fracture toughness measurements in engineering plastics,
    353–373
    experimental estimates of time-dependent fracture
      parameters, 369–371
    J testing, 365–368
    K_{Ic} testing, 362–365
    precracking and other practical
      matters, 360–362
    qualitative fracture tests on plastics, 371–373
    suitability of K and J for polymers, 353–360
  interlaminar toughness of composites, 373–377
Notch
  location, 335
  sensitivity, 372
  strength, 278
NRL, *see* Naval Research Laboratory
Number average molecular weight, 258
Numerical techniques, 553
  boundary integral equation method, 556
  finite element method, 554
  stiffness finite element method, 554

O

Overmatched weldment, 427

P

Parallel cracks, 86
Parametric coordinates, 554
Paris-Erdogan equation, 455, 457
Paris Law, 454
Particle, fracture-triggering, 241
Part-through cracks, 387
Passivity, 514
Path-independent integral, 110, 111
PE, *see* Polyethylene
Photoelasticity, 386
Pinch clamping, polyethylene, 5, 6
Plane elasticity, 88
  Cartesian coordinates, 89
  polar coordinates, 90
Plane strain
  fracture, 72, 316
    crack-tip triaxiality, 73
    effect of thickness on apparent fracture toughness, 75

implications for cracks in structures, 79
    plastic zone effects, 78
    toughness, 75, 308
  Mode I loading, 138
Plane stress fracture, 72, 73, 308, 316
Plastic deformation, 103
Plastic energy, 321
Plasticity
  reverse, 475
  theory, 169
Plastics
  ASTM standard for, 363, 365
  fiber-reinforced, 270
  qualitative fracture test on, 371
Plastic zone, 61, 304
  corrections, 66, 400
  shape, 66
Plate(s)
  displacement controlled, 36
  edge-cracked, 20
  Griffith, 577
  semi-infinite, 47
Poisson's ratio, 19, 31
Polar coordinates, 90
Polarization, types of, 514
Polydispersity, 259
Polyethylene (PE), 4–5, 368
Polymer(s)
  advantages of over metals, 4–5
  amorphous, 259
  cross-linked, 259
  crystalline, 259
  large-scale yielding, 369
  linear, 259
  molecule, branched, 259
  razor notching of, 361
  semicrystalline, 260
  structures and properties of, 258
  suitability of K and J for, 353
  yielding and fracture in, 265
Polynomial stress distributions, 388
Polyvinyl chloride, unplasticized, 372
Potential energy, 108
Pourbaix diagrams, 525
Power law, 453
  crack-face pressure, 388
  –hardening material, 151
Practice problems, 593–610
Precracking, 303, 337, 360
Primary creep, 196
Primary stress, 394
Probabilistic fracture mechanics, 432
Pseudo-elastic compliance, 369
Pseudo elastic displacements, 367
Pseudo-energy release rate, 212
PVC-U, *see* Unplasticized
                polyvinyl chloride

Q

Quasi-brittle materials, 291

R

Ramberg-Osgood equation, 111, 398
R6 approach, 412
Ratio of kinetic to deformation energy, 175
Razor notching, 360, 361
R curve, see Resistance curve
RDS analysis, see Rice, Drugan,
    and Sham analysis
Reference modulus, 199
Reference stress, 403, 404, 416
Remote tensile stress, 20
  semi-infinite plate, 47
  stress intensity and, 64
  through-thickness crack and, 30
Renaissance Period, construction during, 6
Residual stress, 394, 423
Resistance curve (R curve), 38, 39
Retardation models, 484
Rice, Drugan, and Sham (RDS)
    analysis, 162
Ripple loading, 532
Ritchie, Knott, and Rice (RKR) model, 239
River patterns, formation of, 236
RKR model, see Ritchie, Knott,
    and Rice model
Roman bridge design, 7
Roughness-induced closure, 459
Rubber toughening, 268

S

Scanning electron microscope (SEM) fractographs,
    222, 223
SCC, see Stress corrosion cracking
SCF, see Stress concentration factor
Secondary stress, 394, 428
SEM fractographs, see Scanning electron microscope
    fractographs
Semicrystalline polymers, 260
SENB specimen, 141
  effective thickness for deeply notched, 150
  Q parameter, 141, 142
  thickness, 148
SHE, see Standard hydrogen electrode
Shear
  fracture, 222, 223, 344
  lips, 77, 233
  stress, 54
  yielding, 265
Short crack regime, 488
Side grooving, 307
Similitude, 13–14
Single-parameter fracture mechanics break
    down, 12
Singularity-dominated zone, 45
Small-scale yielding
  limit, 507
  plastic deformation pattern in, 134
  stable crack growth in, 162

Specimen(s)
  bend, 380
  bolt-loaded, 549, 551
  Charpy and Izod, 341
  chevron-notched, 378
  configurations, 299
  DCB, 186, 373
    initial flaw in, 374
    loading fixtures for, 375
    mixed mode loading in, 376
  DENT, 136
  design, 317, 334
  end-notched flexure, 374
  limit load solutions, 344
  load relaxation test on cracked, 359
  nonlinear viscoelasticity, 205
  orientation, 301
  ruptured austenitic stainless steel
    tensile, 226
  SENB, 141, 150
  side-grooved, 77
  size, fracture toughness and, 134
  test on smooth, 546
  thickness, fracture toughness and, 76
  U-bend, 546
Stainless steel
  cup and cone fracture, 227
  ruptured austenitic, 226
Standard hydrogen electrode (SHE), 527
Stationary crack, rapid loading
    of, 174
Steady-state creep, 191
Steel(s)
  fracture toughness of ferritic, 247
  mass production of, 7
  stainless, 226, 227
  testing and analysis of, 338
Stick/slip crack growth, 268
Stiffness finite element method, 554
Strain-displacement relationships
  Cartesian coordinates, 89
  polar coordinates, 90
Strain energy
  complimentary, 108
  density, 110
  increase in, 37
Stress
  closure, 64
  cohesive, 26
  concentration factor (SCF), 27, 278
  corrosion cracking (SCC), 433, 517,
    525, 530
  decohesion, 219
  elastic T, 137
  intensity, 344, 434
    closure stress and, 64
    factor, 10, 14, 43, 60, 327
    parameter, 111
    ratio, 457
  plane, 59
  primary, 394

reference, 403, 404, 416
residual, 394
secondary, 394, 428
shear, 54
singularity, 43
–strain relationships, 89
triaxiality, 231
waves, reflected, 186
Weibull, 242
–whitened region, 266
Strip-yield model, 64, 104, 121, 153, 284, 291, 487
Structure failure, reasons for, 3–6
Superposition, 56, 538
Surface area, crack area and, 30
Surface films, 544
System compliance, finite, 41

T

Tearing modulus, 124
Tensile stress, remote, 20
Tensile tests, yield strength in, 39
Test(s)
  compression after impact, 278
  drop weight, 342
  dynamic tear, 344
  K-decreasing, 497
  load relaxation, 359
  methods, 547
  toughness, 340
Testing
  Charpy and Izod, 341
  CTOD, 326
Threshold
  behavior, inert environments, 470
  closure model, 465
Through-thickness crack, 30
Time-dependent fracture, *see* Dynamic and time-dependent fracture
Total failure, hazard function for, 253
Toughness tests, qualitative, 340
Tower Bridge, 7, 8
Traditional techniques, computational fracture mechanics
  continuum approach, 561
  contour integration, 560
  elemental crack advance, 559
  stiffness derivative formulation, 560
  stress and displacement matching, 558
Transformation toughening, 286
Transition time, 175
The Trend in Engineering, 10
Triaxial stress state, 72
Truncated Weibull distribution, 245
TWI, *see* The Welding Institute
Two-criterion model, 466
Two-parameter fracture mechanics, limitations of, 149
Two-parameter theory, 152
Type 1 failure, 3, 5
Type 2 failure, 4, 5

U

U-bend specimen, 546
Underloads, 482
Undermatched weldment, 427
Unplasticized polyvinyl chloride (PVC-U), 372
Unstable fracture, 324

V

Van der Waals bonds, 260
Variable amplitude loading, 473
  effect of overloads and underloads, 478
  linear damage model, 474
  models for retardation, 484
  reverse plasticity at crack tip, 475
Virtual crack extension analysis, 561
Viscoelastic fracture mechanics, 196
  J integral, 200
  linear viscoelasticity, 197
  transition from linear to nonlinear behavior, 204
  viscoelastic *J* integral, 200
Viscoelasticity, 12
  linear, 197
  nonlinear, 205
Viscoplasticity, 12
Void
  coalescence, modeling of, 230
  growth, 222
  nucleation, 219, 221
  spherical, 228
von Mises equation, 66, 67, 229

W

Weakest link fracture, 250
Weibull distribution, truncated, 245
Weibull-shape parameter, 243
Weibull slope, 243
Weibull stress, 242
Weld
  misalignment, 426
  residual stresses, 423
  strength mismatch, 427
The Welding Institute (TWI), 395
Weldment(s)
  fracture testing of, 334
  overmatched, 427
  undermatched, 427
Westergaard stress function, 95
Wheeler model, 485
Whisker toughening, 288
Willenborg model, 485
WLF relationship, 263

Y

Young's modulus, 13, 26